SOLVED PROBLEMS IN
CLASSICAL ELECTROMAGNETISM

T0177524

Solved Problems in Classical Electromagnetism

Analytical and numerical solutions with comments

J. Pierrus

*School of Chemistry and Physics, University of KwaZulu-Natal,
Pietermaritzburg, South Africa*

OXFORD

UNIVERSITY PRESS

OXFORD
UNIVERSITY PRESS

Great Clarendon Street, Oxford, OX2 6DP,
United Kingdom

Oxford University Press is a department of the University of Oxford.
It furthers the University's objective of excellence in research, scholarship,
and education by publishing worldwide. Oxford is a registered trade mark of
Oxford University Press in the UK and in certain other countries

Published in the United States of America by Oxford University Press
198 Madison Avenue, New York, NY 10016, United States of America

British Library Cataloguing in Publication Data

Data available

Library of Congress Control Number: 2018932794

ISBN 978–0–19–882191–5 (hbk.)
ISBN 978–0–19–882192–2 (pbk.)

DOI: 10.1093/oso/9780198821915.001.0001

Printed and bound by
CPI Group (UK) Ltd, Croydon, CR0 4YY

Links to third party websites are provided by Oxford in good faith and
for information only. Oxford disclaims any responsibility for the materials
contained in any third party website referenced in this work.

Preface

These days there are many excellent textbooks ranging from the introductory to the advanced, and which cover all the core parts of a traditional physics curriculum. The *Solved problems in* ... books (this being the second) were written to fill a gap for those students who prefer self-study. Hopefully, the format is sufficiently appealing to justify entering an already crowded space where there isn't much room for original insight and new points of view.

This book follows its predecessor[1] both in style and approach. It contains nearly 300 questions and solutions on a range of topics in classical electromagnetism that are usually encountered during the first four years of a university physics degree. Most questions end with a series of comments that emphasize important conclusions arising from the problem. Sometimes, possible extensions of the problem and additional aspects of interest are also mentioned. The book is aimed primarily at physics students, although it will be useful to engineering and other physical science majors as well. In addition, lecturers may find that some of the material can be readily adapted for examination purposes.

Wherever possible, an attempt has been made to develop the theme of each chapter from a few fundamental principles. These are outlined either in the introduction or in the first few questions of the chapter. Various applications then follow. Inevitably, the author's personal preferences are reflected in the choice of subject matter, although hopefully not at the expense of providing a balanced overview of the core material. Questions are arranged in a way which leads to a natural flow of the key concepts and ideas, rather than according to their 'degree of difficulty'. Those marked with a ** superscript indicate specialized material and are most likely suitable for postgraduate students. Questions without a superscript will invariably be encountered in middle to senior undergraduate-level courses. A * superscript denotes material which is on the borderline between the two categories mentioned above. In all cases, students are encouraged to attempt the questions on their own before looking at the solutions provided.

It is widely recognized that learning (and teaching!) electromagnetism is one of the most challenging parts of any physics curriculum. In the preface to his book *Modern electrodynamics*, Zangwill explains that 'another stumbling block is the non-algorithmic nature of electromagnetic problem-solving. There are many entry points to a typical electromagnetism problem, but it is rarely obvious which lead to a quick solution and which lead to frustrating complications'. These remarks rather clearly

---

[1]   O. L. de Lange and J. Pierrus, *Solved problems in classical mechanics: Analytical and numerical solutions with comments.* Oxford: Oxford University Press, 2010.

outline the challenge. Certainly, it is my firm belief that students benefit from a high exposure to problem solving. Topics which require the use of a computer are especially valuable because one is forced to ask at each stage in the calculation: 'Is my answer reasonable?' For the most part, the computer cannot assist in this regard. Other considerations play a role. Experience definitely helps. So does that somewhat elusive yet much-prized attribute which we call 'physical intuition'.

All the computational work is carried out using *Mathematica*®, version 10.0. The relevant code (referred to as a notebook) is provided in a shadebox in the text. For easy reference, those questions involving computational work are listed in Appendix J. Readers who use different software for their computer algebra are nevertheless encouraged to read these notebooks and adapt the code—wherever necessary—to suit their own environment. That is to say, students using alternative programming packages should not be 'put off' by our exclusive use of *Mathematica*; this book will certainly be useful to them as well. Also, readers without prior knowledge of *Mathematica* can rapidly learn the basics from the online Help at www.Wolfram.com (or various other places; try a simple internet search). From my experience, students learn enough of the basic concepts to make a reasonable start after only a few hours of training. All graphs of numerical results have been drawn to scale using Gnuplot.

For a book like this there are, of course, certain prerequisites. First, it is assumed that readers have previously encountered the basic phenomena and laws of electricity and magnetism. Second, a working knowledge of standard vector analysis and calculus is required. This includes the ability to solve elementary ordinary differential equations. An acquaintance with some of the special functions of mathematical physics will also be useful. Because readers will have diverse mathematical backgrounds and skills, Chapter 1 is devoted to setting out the important analytical techniques on which the rest of the book depends. As a further aid, nine appendices containing some specialized material have been included. In keeping with the modern trend, SI units are adopted throughout. This has the distinct advantage of producing quantities which are familiar from our daily lives: volts, amps, ohms and watts.

Usually one of the first decisions the author of a physics book must face is the important matter of notation: which symbol to use for which quantity. A cursory look at several standard textbooks immediately reveals notable differences (Φ or V for electric potential, dv or $d\tau$ for a volume element, $\mathbf{S}$ or $\mathbf{N}$ for the Poynting vector, and so on). Because the choice of notation is somewhat subjective, colleagues in the same department often possess divergent opinions on this topic. So my own preferences and prejudices are reflected in the notation used in this book. For easy reference, a comprehensive glossary of symbols is appended.

Chapters 2–4 focus primarily on static electricity and magnetism. Then in Chapters 5 and 6 we begin the transition from quasi-static phenomena to the complete time-dependent Maxwell equations which appear from Chapter 7 onwards. For the most part this is a book that deals with the microscopic theory, except in Chapters 9 and 10, which touch on macroscopic electromagnetism. We end in Chapter 12 with a collection of questions which connect Maxwell's electrodynamics to Einstein's theory of special relativity.

Although the questions and solutions are reasonably self-contained, it may be necessary to consult a standard textbook from time to time. University libraries will usually have a wide selection of these. Some of my favourites, listed by their date of publication, are:

☞ *Classical electrodynamics*, J. D. Jackson, 3rd edition, John Wiley (1998).

☞ *Introduction to electrodynamics*, D. J. Griffiths, 3rd edition, Prentice Hall (1999).

☞ *Electricity and magnetism*, E. M. Purcell and D. J. Morin, Cambridge University Press (2013).

☞ *Modern electrodynamics*, A. Zangwill, Cambridge University Press (2013).

Without the help, guidance and assistance of many people this book would never have reached publication. In particular, I extend my sincere thanks to the following:

☞ Allard Welter for drawing the circuit diagrams of Chapter 6, for his advice on various *Mathematica* queries and for resolving (usually in a good-natured way!) some pedantic issues with LaTeX.

☞ Karl Penzhorn for attending to my other computer-related problems and also for helping with the CorelDRAW software which was used to produce many of the diagrams in this book.

☞ Professor Owen de Lange who conceived the format of these *Solved problems in ...* books, and with whom I co-authored Ref. [1]. Hopefully, at least some of Owen's professionalism and attention to detail has rubbed off onto me since we began collaborating in the early 1990s.

☞ Professor Roger Raab for his encouragement and advice. Roger's research interests have strongly influenced my career, and I still recall our first discussion on the use of Cartesian tensors and the importance of symmetry in problem solving. Indeed, most of Appendix A and several questions at the beginning of Chapter 1 are based on some of his original lecture material.

☞ Former lecturers and colleagues who, in one way or another, helped foster my continuing enjoyment of classical electromagnetic theory. In approximate chronological order they include: Peter Krumm, Dave Walker, Manfred Hellberg, Max Michaelis, Roger Raab, Clive Graham, Paul Jackson, Tony Eagle, Owen de Lange, Frank Nabarro and Assen Ilchev.

☞ Several generations of bright undergraduate and postgraduate students who have provided valuable feedback on lecture notes, tutorial problems and other material from which this book has gradually evolved.

Pietermaritzburg, South Africa   J. Pierrus
December 2017

Contents

1
Some essential mathematics

Nearly all of the questions in this introductory chapter are designed to introduce the essential mathematics required for formulating the theory of electromagnetism. All of the techniques discussed here will be used repeatedly throughout this book, and readers will hopefully find it convenient to have the important mathematical material summarized in a single place. Topics covered include Cartesian tensors, standard vector algebra and calculus, the method of separation of variables, the Dirac delta function, time averaging and the concept of solid angle. Our primary emphasis in this chapter is not on physical content, although certain comments pertaining to electricity and magnetism are made whenever appropriate.

Although the scalar potential Φ, the electric field $\mathbf{E}$ and the magnetic field $\mathbf{B}$ are familiar quantities in electromagnetism, it is not always known that they are examples of a mathematical entity called a tensor. Furthermore, it is sometimes necessary to introduce more complicated tensors than these. This chapter begins with a series of questions involving the use of Cartesian tensors. We will find that the compact nature of tensor notation greatly facilitates the solution of many questions throughout this book. Readers who are unfamiliar with tensors and the associated terminology, or who need to revise the background material, are advised to consult Appendix A before proceeding. At the end of this appendix, we include a 'checklist for detecting errors when using tensor notation'. This guide will be helpful for both the uninitiated and the experienced tensor user.

Question 1.1

Let $\mathbf{r} = x\hat{\mathbf{x}} + y\hat{\mathbf{y}} + z\hat{\mathbf{z}}$ be the position vector of a point in space. Use Cartesian tensors to calculate:

(a) $\nabla_i r_j$,

(b) $\nabla \cdot \mathbf{r}$,

(c) ∇r,

(d) ∇r^k where k is rational,

(e) $\nabla_i (r_j / r^3)$,

(f) $\nabla_i \{(3r_j r_k - r^2 \delta_{jk})/r^5\}$ and,

(g) $\nabla e^{i\mathbf{k} \cdot \mathbf{r}}$ where $\mathbf{k}$ is a constant vector.

Solved Problems in Classical Electromagnetism. J. Pierrus, Oxford University Press (2018).
© J. Pierrus. DOI: 10.1093/oso/9780198821915.001.0001

Solution

(a) The operation $\nabla_i r_j (= \partial r_j / \partial r_i)$ produces a tensor of rank two with nine components. Six of these components have $i \neq j$, and for them $\partial r_j / \partial r_i = 0$. The remaining three components for which $i = j$ all have the value one. Thus

$$\nabla_i r_j = \delta_{ij}, \tag{1}$$

where δ_{ij} is the Kronecker delta defined by (III) of Appendix A.

(b) Expressing $\nabla \cdot \mathbf{r}$ in tensor notation and putting $i = j$ in (1) gives $\nabla \cdot \mathbf{r} = \nabla_i r_i = \delta_{ii}$. Using the Einstein summation convention (see (I) of Appendix A) yields

$$\nabla \cdot \mathbf{r} = \delta_{xx} + \delta_{yy} + \delta_{zz} = 3. \tag{2}$$

(c) Writing $r = \sqrt{\mathbf{r} \cdot \mathbf{r}} = \sqrt{r_j r_j}$ and differentiating give

$$\nabla_i r = \frac{\partial r}{\partial r_i} = \frac{\partial}{\partial r_i} (r_j r_j)^{1/2} = \tfrac{1}{2} (r_j r_j)^{-1/2} \left(\frac{\partial r_j}{\partial r_i} r_j + r_j \frac{\partial r_j}{\partial r_i} \right) = \frac{r_j}{r} \frac{\partial r_j}{\partial r_i} = \frac{r_j}{r} \delta_{ij}$$

because of (1). Using the contraction property of the Kronecker delta gives

$$\nabla_i r = \frac{r_j}{r} \delta_{ij} = \frac{r_i}{r}. \tag{3}$$

But (3) is true for $i = x, y$ and z, and so

$$\nabla r = \frac{\mathbf{r}}{r} = \hat{\mathbf{r}}. \tag{4}$$

(d) Consider the ith component. Then $[\nabla r^k]_i = \nabla_i r^k = \dfrac{\partial r^k}{\partial r_i} = \dfrac{\partial r^k}{\partial r} \dfrac{\partial r}{\partial r_i} = k r^{k-2} r_i$ because of (3). The result is

$$\nabla r^k = k r^{k-2} \mathbf{r} \qquad \text{or} \qquad \nabla r^k = k r^{k-1} \hat{\mathbf{r}}. \tag{5}$$

Putting $k = -1$ gives an important case

$$\nabla \left(\frac{1}{r} \right) = -\frac{\mathbf{r}}{r^3} \qquad \text{or} \qquad \nabla \left(\frac{1}{r} \right) = -\frac{\hat{\mathbf{r}}}{r^2} \tag{6}$$

(see also Question 1.6).

(e) $\nabla_i (r_j / r^3) = \dfrac{\nabla_i r_j}{r^3} + r_j \nabla_i r^{-3} = \dfrac{\nabla_i r_j}{r^3} + r_j \dfrac{\partial r^{-3}}{\partial r_i} = \dfrac{\nabla_i r_j}{r^3} + r_j \dfrac{\partial r^{-3}}{\partial r} \dfrac{\partial r}{\partial r_i}$

$$= \frac{r^2 \delta_{ij} - 3 r_i r_j}{r^5}, \tag{7}$$

where in the last step we use (1) and (3).

(f) Similarly,

$$\nabla_i(3r_j r_k r^{-5}) - \nabla_i(r^{-3}\delta_{jk}) = 3r_k r^{-5}\delta_{ij} + 3r_j r^{-5}\delta_{ik} - 15r_i r_j r_k r^{-7} + 3r_i r^{-5}\delta_{jk}$$

$$= \frac{3r^2(r_i\delta_{jk} + r_j\delta_{ki} + r_k\delta_{ij}) - 15r_i r_j r_k}{r^7}. \tag{8}$$

(g) $\nabla_j e^{i\mathbf{k}\cdot\mathbf{r}} = e^{i\mathbf{k}\cdot\mathbf{r}}\dfrac{\partial(ik_l r_l)}{\partial r_j} = ie^{i\mathbf{k}\cdot\mathbf{r}}k_l\delta_{jl} = ie^{i\mathbf{k}\cdot\mathbf{r}}k_j$, and so $\nabla e^{i\mathbf{k}\cdot\mathbf{r}} = i\mathbf{k}e^{i\mathbf{k}\cdot\mathbf{r}}.$ (9)

Comments

(i) Since $\nabla_i r_j = \nabla_j r_i$ we can write $\delta_{ij} = \delta_{ji}$ (i.e. the Kronecker delta is symmetric in its subscripts). It possesses the following important property:

$$A_i\delta_{ij} = A_x\delta_{xj} + A_y\delta_{yj} + A_z\delta_{zj} = A_j. \tag{10}$$

In the final step leading to (10), j is either x, y or z. Of the three Kronecker deltas (δ_{xj}, δ_{yj} and δ_{zj}) two will always be zero, whilst the third will have the value one. Because of this, δ_{ij} is sometimes also known as the substitution tensor.

(ii) Subscripts that are repeated are said to be contracted. So in (10), i is contracted in $A_i\delta_{ij}$. Equivalently, one can say that $A_i\delta_{ij}$ is contracted with respect to i.

(iii) A tensor is said to be isotropic if its components retain the same values under a proper transformation.[‡] δ_{ij} is an example of an isotropic tensor: any second-rank isotropic tensor T_{ij} can be expressed as a scalar multiple of δ_{ij} (i.e. $T_{ij} = \alpha\,\delta_{ij}$).[1]

Question 1.2

(a) Consider the cross-product $\mathbf{c} = \mathbf{a} \times \mathbf{b}$. Show that

$$c_i = \varepsilon_{ijk}a_j b_k, \tag{1}$$

where ε_{ijk} is the Levi-Civita tensor defined by

$$\varepsilon_{ijk} = \begin{cases} 1 & \text{if } ijk \text{ is taken as any even permutation of } x, y, z \\ -1 & \text{if } ijk \text{ is taken as any odd permutation of } x, y, z \\ 0 & \text{if any two subscripts are equal.} \end{cases} \tag{2}$$

(b) Prove that

$$\nabla \times \mathbf{r} = 0, \tag{3}$$

where $\mathbf{r} = (x, y, z)$.

‡Proper and improper transformations are described in Appendix A.

[1] H. Jeffreys, *Cartesian tensors*, Chap. VII, pp. 66–8. Cambridge: Cambridge University Press, 1952.

Solution

(a) The Cartesian form $\mathbf{c} = \hat{\mathbf{x}}(a_y b_z - a_z b_y) + \hat{\mathbf{y}}(a_z b_x - a_x b_z) + \hat{\mathbf{z}}(a_x b_y - a_y b_x)$ has x-component $c_x = a_y b_z - a_z b_y = \varepsilon_{xyz} a_y b_z + \varepsilon_{xzy} a_z b_y$ as a result of the properties $(2)_1$ and $(2)_2$. Because repeated subscripts imply a summation over Cartesian components, we can write $c_x = \varepsilon_{xjk} a_j b_k$ using $(2)_3$. Similarly, $c_y = \varepsilon_{yjk} a_j b_k$ and $c_z = \varepsilon_{zjk} a_j b_k$. Now the ith component of $\mathbf{c}$ is $(\mathbf{a} \times \mathbf{b})_i$ which is (1).

(b) Following the solution of (a) we write $(\nabla \times \mathbf{r})_i = \varepsilon_{ijk} \nabla_j r_k = \varepsilon_{ijk} \delta_{jk} = \varepsilon_{ijj} = 0$. Here we use the contraction $\varepsilon_{ijk} \delta_{jk} = \varepsilon_{ijj}$ and the property $\varepsilon_{ijj} = 0$ (the same conclusion also follows from (4) of Question 1.5). This result is true for $i = x, y$ and z. Hence (3).

Comments

(i) The Levi-Civita tensor is a third-rank tensor. It is clear from (2) that it is anti-symmetric in any pair of subscripts.

(ii) ε_{ijk} is also known as the alternating tensor or isotropic tensor of rank three: any third-rank *isotropic* tensor T_{ijk} can be expressed as a scalar multiple of ε_{ijk} (i.e. $T_{ijk} = \alpha \varepsilon_{ijk}$).[1]

Question 1.3

(a) Consider the product of two Levi-Civita tensors which have a subscript in common. Show that

$$\varepsilon_{ijk} \varepsilon_{\ell m k} = \delta_{i\ell} \delta_{jm} - \delta_{im} \delta_{j\ell}. \tag{1}$$

Hint: The product $\varepsilon_{ijk} \varepsilon_{\ell m k}$ is an isotropic tensor of rank four. Prove (1) by making a linear combination of products of the Kronecker delta.

(b) Use (1) to prove the identity

$$A_i B_j - A_j B_i = \varepsilon_{ijk} (\mathbf{A} \times \mathbf{B})_k, \tag{2}$$

where $\mathbf{A}$ and $\mathbf{B}$ are arbitrary vectors.

Solution

(a) Because of the hint, $\varepsilon_{ijk} \varepsilon_{\ell m k} = a \delta_{ij} \delta_{\ell m} + b \delta_{i\ell} \delta_{jm} + c \delta_{im} \delta_{j\ell}$ where the constants a, b and c are determined as follows:

$$i = x, \quad j = x, \quad \ell = x, \quad m = x \quad : \quad \varepsilon_{xxk} \varepsilon_{xxk} = 0 = a + b + c.$$
$$i = x, \quad j = y, \quad \ell = x, \quad m = y \quad : \quad \varepsilon_{xyk} \varepsilon_{xyk} = \varepsilon_{xyz} \varepsilon_{xyz} = 1 = b.$$
$$i = x, \quad j = y, \quad \ell = y, \quad m = x \quad : \quad \varepsilon_{xyk} \varepsilon_{yxk} = \varepsilon_{xyz} \varepsilon_{yxz} = -1 = c.$$

Thus $a = 0$ and we obtain (1).

(b) Equations (1) and (2) of Question 1.2 give $(\mathbf{A} \times \mathbf{B})_k = \varepsilon_{k\ell m} A_\ell B_m = \varepsilon_{\ell m k} A_\ell B_m$. Multiplying both sides of this equation by ε_{ijk} and using (1) yield $\varepsilon_{ijk}(\mathbf{A} \times \mathbf{B})_k = \varepsilon_{ijk}\varepsilon_{\ell m k} A_\ell B_m = (\delta_{i\ell}\delta_{jm} - \delta_{im}\delta_{j\ell})A_\ell B_m$. Contracting subscripts gives (2).

Comments

(i) Notice the following contractions that follow from (1):

$$\varepsilon_{ijk}\varepsilon_{ij\ell} = 2\,\delta_{k\ell} \qquad \text{and} \qquad \varepsilon_{ijk}\varepsilon_{ijk} = 6. \tag{3}$$

(ii) Making the replacements $\mathbf{A} \to \nabla$; $\mathbf{B} \to \mathbf{F}$ in (2) gives

$$\nabla_i F_j - \nabla_j F_i = \varepsilon_{ijk}(\nabla \times \mathbf{F})_k, \tag{4}$$

and if $\nabla \times \mathbf{F} = 0$ then

$$\nabla_i F_j = \nabla_j F_i. \tag{5}$$

Question 1.4

Suppose $\mathbf{A}(t)$ and $\mathbf{B}(t)$ are differentiable vector fields which are functions of the parameter t. Prove the following:

(a) $\quad \dfrac{d}{dt}(\mathbf{A} \cdot \mathbf{B}) = \mathbf{B} \cdot \dfrac{d\mathbf{A}}{dt} + \mathbf{A} \cdot \dfrac{d\mathbf{B}}{dt}, \hfill (1)$

(b) $\quad \dfrac{d}{dt}(\mathbf{A} \times \mathbf{B}) = \dfrac{d\mathbf{A}}{dt} \times \mathbf{B} + \mathbf{A} \times \dfrac{d\mathbf{B}}{dt}, \hfill (2)$

(c) $\quad \dfrac{d}{dt}\big[\alpha(t)\mathbf{A}\big] = \mathbf{A}\dfrac{d\alpha}{dt} + \alpha\dfrac{d\mathbf{A}}{dt}. \hfill (3)$

$\big($Here $\alpha(t)$ is a differentiable scalar function of t.$\big)$

Solution

These results are all proved by applying the product rule of differentiation.

(a) $\dfrac{d}{dt}(\mathbf{A} \cdot \mathbf{B}) = \dfrac{d}{dt}(A_i B_i) = B_i\dfrac{dA_i}{dt} + A_i\dfrac{dB_i}{dt}$ which is (1).

(b) From (1) of Question 1.2 it follows that $\dfrac{d}{dt}\big[(\mathbf{A} \times \mathbf{B})\big]_i = \dfrac{d}{dt}\big(\varepsilon_{ijk} A_j B_k\big)$. So

$$\dfrac{d}{dt}\big[(\mathbf{A} \times \mathbf{B})\big]_i = \varepsilon_{ijk}\dfrac{dA_j}{dt}B_k + \varepsilon_{ijk}A_j\dfrac{dB_k}{dt} = \left(\dfrac{d\mathbf{A}}{dt} \times \mathbf{B}\right)_i + \left(\mathbf{A} \times \dfrac{d\mathbf{B}}{dt}\right)_i.$$

Since this is true for $i = x$, y and z, equation (2) follows.

(c) The result is obvious by inspection.

Comment

The parameter t often represents time in physics. Thus $\mathbf{A}(t)$ and $\mathbf{B}(t)$ are time-dependent fields, and accordingly the derivatives (1)–(3) represent their rates of change.

Question 1.5

Suppose s_{ij} and a_{ij} represent second-rank symmetric and antisymmetric tensors respectively. Using the definitions

$$s_{ij} = s_{ji} \quad \text{and} \quad a_{ij} = -a_{ji}, \tag{1}$$

prove that

$$s_{ij}a_{ij} = 0. \tag{2}$$

Solution

The subscript notation is arbitrary, and so

$$s_{ij}a_{ij} = s_{ji}a_{ji}. \tag{3}$$

Substituting (1) in (3) gives $s_{ij}a_{ij} = -s_{ij}a_{ij}$ or $2s_{ij}a_{ij} = 0$, which proves (2).

Comment

Equation (2) is a special case of a general property: the product of a tensor $s_{ijk\ell\dots}$ symmetric in any two of its subscripts with another tensor $a_{mkni\dots}$ that is antisymmetric in the *same* two subscripts is zero. That is,

$$s_{ijk\ell\dots}\,a_{mkni\dots} = 0. \tag{4}$$

Question 1.6

Suppose $\mathbf{r} = (x, y, z)$ and $\mathbf{r}' = (x', y', z')$ represent position vectors[‡] of points P and P' respectively. Prove the following results:

$$\nabla\left(\frac{1}{|\mathbf{r}-\mathbf{r}'|}\right) = -\frac{(\mathbf{r}-\mathbf{r}')}{|\mathbf{r}-\mathbf{r}'|^3} \quad \text{and} \quad \nabla'\left(\frac{1}{|\mathbf{r}-\mathbf{r}'|}\right) = \frac{(\mathbf{r}-\mathbf{r}')}{|\mathbf{r}-\mathbf{r}'|^3}, \tag{1}$$

where $\nabla = \hat{\mathbf{x}}\dfrac{\partial}{\partial x} + \hat{\mathbf{y}}\dfrac{\partial}{\partial y} + \hat{\mathbf{z}}\dfrac{\partial}{\partial z}$ and $\nabla' = \hat{\mathbf{x}}\dfrac{\partial}{\partial x'} + \hat{\mathbf{y}}\dfrac{\partial}{\partial y'} + \hat{\mathbf{z}}\dfrac{\partial}{\partial z'}$ denote differentiation with respect to the unprimed and primed coordinates respectively.

[‡]The common origin O of these vectors is completely arbitrary.

Solution

It is convenient to let $\mathbf{R} = \mathbf{r} - \mathbf{r}'$. Then

$$\nabla_i\left(\frac{1}{R}\right) = \frac{\partial R^{-1}}{\partial r_i} = \frac{\partial R^{-1}}{\partial R}\frac{\partial R}{\partial r_i} = -\frac{1}{R^2}\frac{\partial R}{\partial r_i}. \qquad (2)$$

But $\dfrac{\partial R}{\partial r_i} = \dfrac{\partial}{\partial r_i}(r^2 + r'^2 - 2r_j r'_j)^{1/2} = \dfrac{2r_i - 2r'_j\delta_{ij}}{2R} = \dfrac{r_i - r'_i}{R} = \dfrac{R_i}{R}$ using (1) and (3)

of Question 1.1. Substituting this last result in (2) gives $(1)_1$. Similarly, $(1)_2$ follows, since $\partial R/\partial r_i = -\partial R/\partial r'_i$.

Comment

In electromagnetism, it is important to distinguish between the unprimed coordinates of a field point P and the primed coordinates locating the sources[‡] of the field. As we have seen in the solution above, mathematical operations such as differentiation and integration can be with respect to coordinates of either type.

Question 1.7

Express the Taylor-series expansion of a function $f(x, y, z)$ about an origin O in the form

$$f(x, y, z) = [f(x, y, z)]_0 + [\nabla_i f(x, y, z)]_0\, r_i + \tfrac{1}{2}[\nabla_i \nabla_j f(x, y, z)]_0\, r_i r_j + \cdots. \qquad (1)$$

Solution

The Taylor-series expansion of $f(x, y, z)$ about O is

$$f(x, y, z) = [f(x, y, z)]_0 + \left[\frac{\partial f(x, y, z)}{\partial x}\right]_0 x + \left[\frac{\partial f(x, y, z)}{\partial y}\right]_0 y + \left[\frac{\partial f(x, y, z)}{\partial z}\right]_0 z +$$

$$\frac{1}{2}\left\{\left[\frac{\partial^2 f(x, y, z)}{\partial x^2}\right]_0 x^2 + \left[\frac{\partial^2 f(x, y, z)}{\partial x \partial y}\right]_0 xy + \left[\frac{\partial^2 f(x, y, z)}{\partial x \partial z}\right]_0 xz +\right.$$

$$\left[\frac{\partial^2 f(x, y, z)}{\partial y \partial x}\right]_0 yx + \left[\frac{\partial^2 f(x, y, z)}{\partial y^2}\right]_0 y^2 + \left[\frac{\partial^2 f(x, y, z)}{\partial y \partial z}\right]_0 yz +$$

$$\left.\left[\frac{\partial^2 f(x, y, z)}{\partial z \partial x}\right]_0 zx + \left[\frac{\partial^2 f(x, y, z)}{\partial z \partial y}\right]_0 zy + \left[\frac{\partial^2 f(x, y, z)}{\partial z^2}\right]_0 z^2\right\} + \cdots, \qquad (2)$$

which, in terms of the Einstein summation convention, is (1).

[‡]These being electric charges and currents.

Comments

(i) Note the compact form of the tensor equation (1), and compare this with (2).

(ii) Sometimes the function f is itself a component of a vector (say, the electric field y-component E_y). Then, using tensor notation to express the component of a vector, we have

$$E_i = [E_i]_0 + [\nabla_j E_i]_0 r_j + \tfrac{1}{2}[\nabla_j \nabla_k E_i]_0 r_j r_k + \cdots . \tag{3}$$

Question 1.8

Let **A**, **B**, **C**, f and g represent continuous and differentiable[‡] vector or scalar fields as appropriate. Use tensor notation to prove the following identities:

(a) $\mathbf{A} \cdot (\mathbf{B} \times \mathbf{C}) = (\mathbf{A} \times \mathbf{B}) \cdot \mathbf{C}$ and all other cyclic permutations, $\hfill (1)$

(b) $(\mathbf{A} \times \mathbf{B}) \cdot (\mathbf{A} \times \mathbf{B}) = A^2 B^2 - (\mathbf{A} \cdot \mathbf{B})^2$, $\hfill (2)$

(c) $\mathbf{A} \times (\mathbf{B} \times \mathbf{C}) = \mathbf{B}(\mathbf{A} \cdot \mathbf{C}) - \mathbf{C}(\mathbf{A} \cdot \mathbf{B})$, $\hfill (3)$

(d) $\nabla(fg) = g\nabla f + f\nabla g$, $\hfill (4)$

(e) $\nabla \cdot (f\mathbf{A}) = \mathbf{A} \cdot \nabla f + f(\nabla \cdot \mathbf{A})$, $\hfill (5)$

(f) $\nabla \times (f\mathbf{A}) = \nabla f \times \mathbf{A} + f(\nabla \times \mathbf{A})$, $\hfill (6)$

(g) $\nabla \cdot (\mathbf{A} \times \mathbf{B}) = (\nabla \times \mathbf{A}) \cdot \mathbf{B} - (\nabla \times \mathbf{B}) \cdot \mathbf{A}$, $\hfill (7)$

(h) $\nabla \times (\mathbf{A} \times \mathbf{B}) = (\mathbf{B} \cdot \nabla)\mathbf{A} - (\mathbf{A} \cdot \nabla)\mathbf{B} + \mathbf{A}(\nabla \cdot \mathbf{B}) - \mathbf{B}(\nabla \cdot \mathbf{A})$, $\hfill (8)$

(i) $\nabla \cdot (\nabla \times \mathbf{A}) = 0$, $\hfill (9)$

(j) $\nabla \times \nabla f = 0$, $\hfill (10)$

(k) $\nabla \times (\nabla \times \mathbf{A}) = -\nabla^2 \mathbf{A} + \nabla(\nabla \cdot \mathbf{A})$, $\hfill (11)$

(l) $\nabla \cdot (\nabla f \times \nabla g) = 0$, $\hfill (12)$

(m) $\nabla(\mathbf{A} \cdot \mathbf{B}) = (\mathbf{A} \cdot \nabla)\mathbf{B} + \mathbf{A} \times (\nabla \times \mathbf{B}) + (\mathbf{B} \cdot \nabla)\mathbf{A} + \mathbf{B} \times (\nabla \times \mathbf{A})$. $\hfill (13)$

Solution

(a) The various permutations in (1) may all be proved by invoking the cyclic nature of the subscripts of the Levi-Civita tensor. Consider, for example, $(1)_1$. Using tensor notation for a scalar product and (1) of Question 1.2 gives

$$\mathbf{A} \cdot (\mathbf{B} \times \mathbf{C}) = A_i(\mathbf{B} \times \mathbf{C})_i = A_i \varepsilon_{ijk} B_j C_k = \varepsilon_{ijk} A_i B_j C_k.$$

Now $\varepsilon_{ijk} = \varepsilon_{kij}$, and so $\mathbf{A} \cdot (\mathbf{B} \times \mathbf{C}) = \varepsilon_{kij} A_i B_j C_k = (\mathbf{A} \times \mathbf{B})_k C_k$, which proves the result. The remaining cyclic permutations can be found in a similar way.

[‡]Suppose these fields have continuous second-order derivatives, so $\nabla_i \nabla_j A_k = \nabla_j \nabla_i A_k$, etc.

(b) Clearly, $(\mathbf{A} \times \mathbf{B}) \cdot (\mathbf{A} \times \mathbf{B}) = (\mathbf{A} \times \mathbf{B})_i (\mathbf{A} \times \mathbf{B})_i$

$$= \varepsilon_{ijk} A_j B_k \, \varepsilon_{ilm} A_l B_m$$

$$= (\delta_{jl}\delta_{km} - \delta_{jm}\delta_{kl}) A_j A_l B_k B_m$$

$$= A_i A_i B_j B_j - A_i B_i A_j B_j \quad \text{(subscripts are arbitrary)}$$

$$= (\mathbf{A} \cdot \mathbf{A})(\mathbf{B} \cdot \mathbf{B}) - (\mathbf{A} \cdot \mathbf{B})^2. \quad \text{Hence (2).}$$

(c) It is sufficient to show that $[\mathbf{A} \times (\mathbf{B} \times \mathbf{C})]_i = B_i(\mathbf{A} \cdot \mathbf{C}) - C_i(\mathbf{A} \cdot \mathbf{B})$. From (1) of Question 1.2

$$[\mathbf{A} \times (\mathbf{B} \times \mathbf{C})]_i = \varepsilon_{ijk} A_j (\mathbf{B} \times \mathbf{C})_k$$

$$= \varepsilon_{ijk} A_j \varepsilon_{klm} B_l C_m = \varepsilon_{ijk} \varepsilon_{lmk} A_j B_l C_m = (\delta_{il}\delta_{jm} - \delta_{im}\delta_{jl}) A_j B_l C_m,$$

using the cyclic property of ε_{klm} and (1) of Question 1.3. Contracting the right-hand side gives $A_m B_i C_m - A_l B_l C_i = B_i(\mathbf{A} \cdot \mathbf{C}) - C_i(\mathbf{A} \cdot \mathbf{B})$ as required.

(d) Consider the ith component. Then $\nabla_i(fg) = g\nabla_i f + f\nabla_i g$ by the product rule of differentiation and the result follows.

(e) $\nabla \cdot (f\mathbf{A}) = \nabla_i(f\mathbf{A})_i = \nabla_i(fA_i) = A_i\nabla_i f + f\nabla_i A_i = \mathbf{A} \cdot \nabla f + f(\nabla \cdot \mathbf{A})$.

(f) Consider the ith component. Then

$$[\nabla \times (f\mathbf{A})]_i = \varepsilon_{ijk}\nabla_j(fA_k) = \varepsilon_{ijk}(A_k\nabla_j f + f\nabla_j A_k) = (\nabla f \times \mathbf{A})_i + f(\nabla \times \mathbf{A})_i.$$

(g) $\nabla \cdot (\mathbf{A} \times \mathbf{B}) = \nabla_i(\mathbf{A} \times \mathbf{B})_i = \nabla_i \varepsilon_{ijk} A_j B_k$

$$= \varepsilon_{ijk}(B_k\nabla_i A_j + A_j\nabla_i B_k)$$

$$= (\varepsilon_{kij}\nabla_i A_j)B_k - (\varepsilon_{jik}\nabla_i B_k)A_j \quad \text{(properties of } \varepsilon_{ijk})$$

$$= (\nabla \times \mathbf{A})_k B_k - (\nabla \times \mathbf{B})_j A_j$$

$$= (\nabla \times \mathbf{A}) \cdot \mathbf{B} - (\nabla \times \mathbf{B}) \cdot \mathbf{A}.$$

(h) $[\nabla \times (\mathbf{A} \times \mathbf{B})]_i = \varepsilon_{ijk}\nabla_j \varepsilon_{klm} A_l B_m$

$$= (\delta_{il}\delta_{jm} - \delta_{im}\delta_{jl})\nabla_j(A_l B_m)$$

$$= \nabla_m(A_i B_m) - \nabla_l(A_l B_i) \quad \text{(contract subscripts)}$$

$$= B_m\nabla_m A_i + A_i\nabla_m B_m - B_i\nabla_l A_l - A_l\nabla_l B_i \quad \text{(product rule)}$$

$$= (\mathbf{B} \cdot \nabla)A_i - (\mathbf{A} \cdot \nabla)B_i + A_i(\nabla \cdot \mathbf{B}) - B_i(\nabla \cdot \mathbf{A}),$$

which proves the result.

(i) $\nabla \cdot (\nabla \times \mathbf{A}) = \nabla_i(\nabla \times \mathbf{A})_i = \nabla_i \varepsilon_{ijk}\nabla_j A_k = \varepsilon_{ijk}\nabla_i\nabla_j A_k = 0$,
since $\nabla_i\nabla_j A_k$ is symmetric in i and j, whereas ε_{ijk} is antisymmetric in these subscripts (see Question 1.5). Hence (9).

(j) $[\nabla \times \nabla f]_i = \varepsilon_{ijk}\nabla_j\nabla_k f = 0$ as in (i). Hence (10).

(k) $[\boldsymbol{\nabla} \times (\boldsymbol{\nabla} \times \mathbf{A})]_i = \varepsilon_{ijk}\nabla_j\varepsilon_{klm}\nabla_l A_m$

$\qquad\qquad\quad = \varepsilon_{ijk}\varepsilon_{lmk}\nabla_j\nabla_l A_m$ (cyclic property of ε_{ijk})

$\qquad\qquad\quad = (\delta_{il}\delta_{jm} - \delta_{im}\delta_{jl})\nabla_j\nabla_l A_m$ (contracting subscripts)

$\qquad\qquad\quad = (\nabla_i\nabla_m A_m - \nabla^2 A_i) = \nabla_i(\boldsymbol{\nabla}\cdot\mathbf{A}) - \nabla^2 A_i$

as required.

(l) This result follows immediately from (7) and (10) above.

(m) $\nabla_i(\mathbf{A}\cdot\mathbf{B}) = \nabla_i(A_j B_j)$

$\qquad\qquad = A_j\nabla_i B_j + B_j\nabla_i A_j$

$\qquad\qquad = A_j[\nabla_j B_i + \varepsilon_{ijk}(\boldsymbol{\nabla}\times\mathbf{B})_k] + B_j[\nabla_j A_i + \varepsilon_{ijk}(\boldsymbol{\nabla}\times\mathbf{A})_k],$

where in the last step we use (4) of Question 1.3. This proves the result.

Comments

(i) Equations (1) and (3) are the well-known scalar and vector triple products respectively. We note the following:

☞ In (1) the positions of the dot and cross may be interchanged, provided that the cyclic order of the vectors is maintained.

☞ The identity (3) is used often and is worth remembering. For easy recall, some textbooks call it the '**BAC–CAB** rule'. See, for example, Ref. [2].

(ii) Suppose $\mathbf{A}$, $\mathbf{B}$ and $\mathbf{C}$ are polar vectors.[‡] The transformation $\mathbf{A}\cdot(\mathbf{B}\times\mathbf{C}) \overset{\mathrm{P}}{\rightarrow} -\mathbf{A}\cdot(\mathbf{B}\times\mathbf{C})$ results in the scalar triple product changing sign under inversion, and so it is a pseudoscalar.[♯] If $\mathbf{A}$, $\mathbf{B}$ and $\mathbf{C}$ are the spanning vectors of a crystal lattice, then $\mathbf{A}\cdot(\mathbf{B}\times\mathbf{C})$ is the pseudovolume of the unit cell.[†]

(iii) In electromagnetism (1)–(13) are very useful identities. Although proved here for Cartesian coordinates, the results are valid in all coordinate systems.[♭]

Question 1.9

Consider the scalar functions $f(\mathbf{r})$ and $g(\mathbf{r}(t),t)$. Suppose $\mathbf{r} = \mathbf{r}(t)$ is a time-dependent position vector. Show that

$$\frac{df}{dt} = \left(\frac{d\mathbf{r}}{dt}\cdot\boldsymbol{\nabla}\right)f \quad \text{and} \quad \frac{dg}{dt} = \frac{\partial g}{\partial t} + \left(\frac{d\mathbf{r}}{dt}\cdot\boldsymbol{\nabla}\right)g. \tag{1}$$

[‡] The distinction between polar and axial vectors is described in Appendix A.
[♯] See also Appendix A. In the above, p is the parity operator described on p. 598.
[†] In this example, the *volume* of the unit cell is $|\mathbf{A}\cdot(\mathbf{B}\times\mathbf{C})|$.
[♭] This also applies to other results in this chapter, such as Gauss's theorem and Stokes's theorem.

---

[2]   D. J. Griffiths, *Introduction to electrodynamics*, Chap. 1, p. 8. New York: Prentice Hall, 3 edn, 1999.

Solution

Since both proofs are similar, we consider that for $(1)_2$ only. The total differential of $g(x, y, z, t)$ is

$$dg = \frac{\partial g}{\partial x}dx + \frac{\partial g}{\partial y}dy + \frac{\partial g}{\partial z}dz + \frac{\partial g}{\partial t}dt.$$

Then

$$\frac{dg}{dt} = \frac{\partial g}{\partial t} + \left(\frac{dx}{dt}\frac{\partial}{\partial x} + \frac{dy}{dt}\frac{\partial}{\partial y} + \frac{dz}{dt}\frac{\partial}{\partial z}\right)g,$$

which is $(1)_2$ since $d\mathbf{r}/dt = (dx/dt, dy/dt, dz/dt)$ and $\nabla = (\partial/\partial x, \partial/\partial y, \partial/\partial z)$.

Comments

(i) Equation $(1)_1$ is the chain rule of differentiation. Equation $(1)_2$ is often called the convective derivative. It is composed of two parts: the local or Eulerian derivative $\partial g/\partial t$ and the convective term $(\mathbf{v} \cdot \nabla)g$, where $\mathbf{v} = d\mathbf{r}/dt$ is the velocity of an element of charge or mass as it travels along its trajectory $\mathbf{r}(t)$.

(ii) Suppose $T(\mathbf{r}, t)$ represents a temperature field. The local derivative $\partial T/\partial t$ provides the change in temperature with time at a fixed point in space, whereas the convective term $(\mathbf{v} \cdot \nabla)T$ accounts for the rate at which the temperature changes in a fixed mass of air as it moves, for example, in a convection current.

(iii) For the vector fields $\mathbf{f}(\mathbf{r}(t))$ and $\mathbf{g}(\mathbf{r}(t), t)$, these derivatives are

$$\frac{d\mathbf{f}}{dt} = \left(\frac{d\mathbf{r}}{dt} \cdot \nabla\right)\mathbf{f} \quad \text{and} \quad \frac{d\mathbf{g}}{dt} = \frac{\partial \mathbf{g}}{\partial t} + \left(\frac{d\mathbf{r}}{dt} \cdot \nabla\right)\mathbf{g}. \tag{2}$$

Question 1.10**

The flux ϕ of an arbitrary vector field $\mathbf{F}(\mathbf{r}, t)$ is

$$\phi = \int_s \mathbf{F} \cdot d\mathbf{a},$$

where s is any surface spanning an arbitrary contour c. Suppose the position, size and shape of c (and therefore s) change with time. Show that

$$\frac{d}{dt}\int_s \mathbf{F} \cdot d\mathbf{a} = \int_s \frac{d\mathbf{F}}{dt} \cdot d\mathbf{a}. \tag{1}$$

Hint: Let $\mathbf{r}(u(t), v(t))$ be a parametric representation of s where $u_1 \leq u \leq u_2$ and $v_1 \leq v \leq v_2$ (see Appendix H). Then

$$\phi = \int_{u_1}^{u_2}\int_{v_1}^{v_2} \mathbf{F}\left(\mathbf{r}(u(t), v(t))\right) \cdot \left(\frac{\partial \mathbf{r}}{\partial u} \times \frac{\partial \mathbf{r}}{\partial v}\right) du\, dv. \tag{2}$$

Solution

Differentiating (2) gives

$$\frac{d\phi}{dt} = \frac{d}{dt} \int_{u_1}^{u_2} \int_{v_1}^{v_2} \mathbf{F}\big(\mathbf{r}(u(t), v(t))\big) \cdot \left(\frac{\partial \mathbf{r}}{\partial u} \times \frac{\partial \mathbf{r}}{\partial v}\right) du \, dv$$

$$= \int_{u_1}^{u_2} du \int_{v_1}^{v_2} \left[\frac{d\mathbf{F}}{dt} \cdot \left(\frac{\partial \mathbf{r}}{\partial u} \times \frac{\partial \mathbf{r}}{\partial v}\right) + \mathbf{F} \cdot \frac{d}{dt}\left(\frac{\partial \mathbf{r}}{\partial u} \times \frac{\partial \mathbf{r}}{\partial v}\right)\right] dv. \tag{3}$$

Consider the second term in square brackets in (3). Using (1)$_1$ of Question 1.9 yields

$$\frac{d}{dt}\frac{\partial \mathbf{r}}{\partial u} \times \frac{\partial \mathbf{r}}{\partial v} = \left(\frac{d\mathbf{r}}{dt} \cdot \nabla\right)\left(\frac{\partial \mathbf{r}}{\partial u} \times \frac{\partial \mathbf{r}}{\partial v}\right). \tag{4}$$

Using tensor notation, (4) can be written as

$$\left[\left(\frac{d\mathbf{r}}{dt} \cdot \nabla\right)\left(\frac{\partial \mathbf{r}}{\partial u} \times \frac{\partial \mathbf{r}}{\partial v}\right)\right]_i = \varepsilon_{ijk}\frac{dr_l}{dt}\frac{\partial}{\partial r_l}\frac{\partial r_j}{\partial u}\frac{\partial r_k}{\partial v} \tag{5}$$

$$= \varepsilon_{ikj}\frac{dr_l}{dt}\frac{\partial}{\partial r_l}\frac{\partial r_k}{\partial u}\frac{\partial r_j}{\partial v} \qquad \text{(subscripts are arbitrary)}$$

$$= -\varepsilon_{ijk}\frac{dr_l}{dt}\frac{\partial}{\partial r_l}\frac{\partial r_k}{\partial v}\frac{\partial v}{\partial u}\frac{\partial r_j}{\partial u}\frac{\partial u}{\partial v} \qquad (\varepsilon_{ikj} = -\varepsilon_{ijk})$$

$$= -\varepsilon_{ijk}\frac{dr_l}{dt}\frac{\partial}{\partial r_l}\frac{\partial r_k}{\partial v}\frac{\partial r_j}{\partial u}$$

$$= -\varepsilon_{ijk}\frac{dr_l}{dt}\frac{\partial}{\partial r_l}\frac{\partial r_j}{\partial u}\frac{\partial r_k}{\partial v} \qquad \text{(rearranging terms).} \tag{6}$$

Comparing (5) and (6) shows that $\left[\left(\dfrac{d\mathbf{r}}{dt} \cdot \nabla\right)\dfrac{\partial \mathbf{r}}{\partial u} \times \dfrac{\partial \mathbf{r}}{\partial v}\right]_i = 0$, which is true for all components of this vector. Then (3) becomes

$$\frac{d\phi}{dt} = \int_{u_1}^{u_2} du \int_{v_1}^{v_2} \left[\frac{d\mathbf{F}}{dt} \cdot \left(\frac{\partial \mathbf{r}}{\partial u} \times \frac{\partial \mathbf{r}}{\partial v}\right)\right] dv$$

$$= \int_s \frac{d\mathbf{F}}{dt} \cdot d\mathbf{a},$$

which is (1).

Comment

Equation (1) is a useful result for calculating emfs in non-stationary circuits or media. See Question 5.4.

Question 1.11

Use the relevant definition and the result that $\mathbf{r}$ is a polar vector to determine whether the following vectors are polar or axial: (a) velocity $\mathbf{u}$, (b) linear momentum $\mathbf{p}$, (c) force $\mathbf{F}$, (d) electric field $\mathbf{E}$, (e) magnetic field $\mathbf{B}$ and (f) $\mathbf{E} \times \mathbf{B}$.

(Assume that time, mass and charge are invariant quantities).

Solution

(a) $\mathbf{u} = d\mathbf{r}/dt$ is polar since t is invariant and $\mathbf{r}$ is polar.

(b) $\mathbf{p} = m\mathbf{u}$ is polar since m is invariant and $\mathbf{u}$ is polar.

(c) $\mathbf{F} = d\mathbf{p}/dt$ is polar since t is invariant and $\mathbf{p}$ is polar.

(d) $\mathbf{E} = \mathbf{F}/q$ is polar since q is invariant and $\mathbf{F}$ is polar.

(e) Apply the parity transformation to the force

$$\mathbf{F} \xrightarrow{P} \mathbf{F}' = -\mathbf{F} = q\mathbf{u}' \times \mathbf{B}' = q(-\mathbf{u}) \times \mathbf{B}'.$$

Clearly, $\mathbf{F} = q\mathbf{u} \times \mathbf{B}'$ requires $\mathbf{B}' = \mathbf{B}$ which shows that $\mathbf{B}$ is axial.

(f) $\mathbf{E} \times \mathbf{B} \xrightarrow{P} (-\mathbf{E}) \times \mathbf{B} = -\mathbf{E} \times \mathbf{B}$ which is polar. This vector represents the energy flux per unit time[‡] in the vacuum electromagnetic field (see (7) of Question 7.6).

Comments

(i) The polar (axial) nature of the electric (magnetic) field established in the above solution above can be confirmed by the following intuitive approach. We suppose uniform **E**- and **B**-fields are created by an ideal parallel-plate capacitor and an ideal solenoid respectively, and consider how these fields behave when their sources are inverted. This is illustrated in the figures below; notice that **E** reverses sign, whereas **B** does not.

E-field: cross-section through capacitor perpendicular to the plates

{source q at $\mathbf{r}$} $\xrightarrow{P}$ {source q at $-\mathbf{r}$}

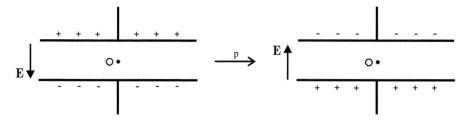

[‡]Apart from a factor μ_0, which is a polar constant of proportionality. See Comment (ii) on p. 14.

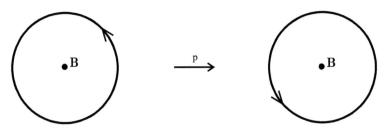

B-field: cross-section through solenoid perpendicular to the symmetry axis

{source Idl at $\mathbf{r}$} $\xrightarrow{P}$ {source $-Idl$ at $-\mathbf{r}$}

(ii) Suppose $\mathbf{c} = \mathbf{a} \times \mathbf{b}$. Clearly, $\mathbf{c}$ is:

☞ polar if either $\mathbf{a}$ or $\mathbf{b}$ is polar and the other is axial (see the $\mathbf{E} \times \mathbf{B}$ example above).

☞ axial if $\mathbf{a}$ and $\mathbf{b}$ are either both polar or both axial.

(iii) Let $\mathbf{a}$ and $\mathbf{b}$ represent arbitrary vectors that satisfy laws of physics which we express algebraically as:

$$\mathbf{b} = \alpha\, \mathbf{a} \quad \text{and} \quad b_i = \beta_{ij} a_j. \tag{1}$$

Here $\mathbf{a}$ is taken to be the 'cause' and $\mathbf{b}$ the 'effect'. The constants of proportionality α and β_{ij} are tensors of rank zero and two respectively.[#] Under rotation of axes, they behave as follows:

☞ α and β_{ij} are polar if $\mathbf{a}$ and $\mathbf{b}$ are either both polar or both axial,

☞ α and β_{ij} are axial if either $\mathbf{a}$ or $\mathbf{b}$ is polar and the other is axial.

So, for example, in the Biot–Savart law $\left(\text{see } (7)_2 \text{ of Question 4.4}\right)$ $d\mathbf{B} = \dfrac{\mu_0}{4\pi}\dfrac{Idl \times \mathbf{r}}{r^3}$, and we conclude that μ_0 is a polar scalar since both $d\mathbf{B}$ and $Idl \times \mathbf{r}$ are axial vectors.

(iv) These results can be generalized to physical tensors and physical property tensors of any rank.[3]

(v) Considerations of symmetry and the spatial nature of tensors can sometimes be exploited to gain useful insight into a physical system. Consider, for example, a sphere of charge which is symmetric about its centre O. Suppose the sphere is spinning about an axis through O. Inversion through O obviously leaves the sphere unchanged as well as all its physical tensors and physical property tensors.

[#]The following terminology is used in the literature (see, for example, Ref. [3]): $\mathbf{a}$, $\mathbf{b}$ are called physical tensors (here they are physical vectors) and α, β_{ij} are physical property tensors (see also Comment (viii) of Question 2.26).

[3]   R. E. Raab and O. L. de Lange, *Multipole theory in electromagnetism*, Chap. 3, pp. 59–72. Oxford: Clarendon Press, 2005.

Because polar vectors change sign under inversion it follows that the electric field at O is necessarily zero, whereas the magnetic field, being an axial vector, may have a finite value at the centre. Symmetry arguments alone cannot reveal the value of B at O; this can be determined either by solving the relevant Maxwell equation or by measurement.

(vi) In addition to characterizing physical tensors and physical property tensors by their spatial properties, it is also possible to consider how such quantities behave under a time-reversal transformation T. In classical physics, time reversal changes the sign of the time coordinate $t \overset{\text{T}}{\to} t' = -t$. For motion in a conservative field the time-reversed trajectory is indistinguishable from the actual trajectory;[3] $\mathbf{r} \overset{\text{T}}{\to} \mathbf{r}' = \mathbf{r}$. With this in mind, we consider the effect of a time-reversal transformation on the following first-rank tensors:

☞ $\mathbf{u} = d\mathbf{r}/dt \overset{\text{T}}{\to} d\mathbf{r}/dt' = -\mathbf{u}$.

☞ $\mathbf{p} = m\mathbf{u} \overset{\text{T}}{\to} -\mathbf{p}$.

☞ $\mathbf{F} = d\mathbf{p}/dt \overset{\text{T}}{\to} \mathbf{F}$.

☞ $\mathbf{E} = \mathbf{F}/q \overset{\text{T}}{\to} \mathbf{E}$.

☞ $q\mathbf{u} \times \mathbf{B} = \mathbf{F} \overset{\text{T}}{\to} q(-\mathbf{u}) \times \mathbf{B}'$ requires $\mathbf{B} \overset{\text{T}}{\to} \mathbf{B}' = -\mathbf{B}$.

Tensors which remain unchanged by time-reversal transformations are called time-even ($\mathbf{F}$ and $\mathbf{E}$ above), whilst those which change sign are time-odd ($\mathbf{u}$, $\mathbf{p}$ and $\mathbf{B}$ above). The space-time symmetry properties of these five vectors are thus:

☞ $\mathbf{u}$ and $\mathbf{p}$ are time-odd polar vectors,

☞ $\mathbf{E}$ and $\mathbf{F}$ are time-even polar vectors, and

☞ $\mathbf{B}$ is a time-odd axial vector.[‡]

(In the bulleted lists above, it has been assumed implicitly that m and q are time-even, polar scalars.[3]) Ref. [3] also provides interesting applications of these symmetry transformations to physical systems. For example, it is shown that the Faraday effect[#] in a fluid (whether optically active or inactive) is not vetoed by a space-time transformation, whereas the electric analogue of this effect, which has never been observed, is vetoed.[3]

(vii) The symmetries referred to in (vi) above are part of a much more general idea based on Neumann's principle which states that every physical property tensor of a system must possess the full space-time symmetry of the system. (This is quite apart from any intrinsic symmetry of the tensor subscripts themselves.)

[‡] An example of a time-even axial vector is torque, $\mathbf{\Gamma} = m\dfrac{d}{dt}(\mathbf{r} \times \mathbf{p})$.

[#] In this effect, a magnetostatic field $\mathbf{B}$ applied parallel to the path of linearly polarized light in a fluid induces a rotation of the plane of polarization through an angle proportional to B.

Question 1.12

Consider a vector field $\mathbf{F}(\mathbf{r})$ with continuous first derivatives in some region of space having volume v bounded by the closed surface s. Use the definition of divergence[‡] to prove that

$$\oint_s \mathbf{F} \cdot d\mathbf{a} = \int_v (\nabla \cdot \mathbf{F}) dv, \tag{1}$$

where $d\mathbf{a}$ is an element of area on s.

Solution

Imagine subdividing the macroscopic volume v into a large number n of infinitesimal elements having volume dv_i, where $i = 1, 2, \ldots, n$ (the elements might, for example, be cuboids with six faces). For the ith element the net outward flux is the sum over six faces. Using the definition of divergence we write

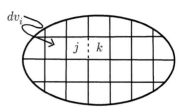

$$\sum_{\text{six faces}} \mathbf{F}_i \cdot d\mathbf{a}_i = (\nabla \cdot \mathbf{F})_i dv_i. \tag{2}$$

The total flux through v is obtained by summing over all volume elements. In this summation the $\mathbf{F}_i \cdot d\mathbf{a}_i$ terms cancel in pairs for all interior surfaces.[♯] The only terms which survive are those on the exterior surfaces for which no cancellation can occur and (2) becomes

$$\sum_{\substack{\text{exterior} \\ \text{faces}}} \mathbf{F}_i \cdot d\mathbf{a}_i = \sum_{\substack{\text{volume} \\ \text{elements}}} (\nabla \cdot \mathbf{F})_i dv_i. \tag{3}$$

In the limit $n \to \infty$, the summation on the left-hand side of (3) becomes an integral over s and that on the right-hand side becomes an integral over v, which is (1).

[‡]The divergence of $\mathbf{F}$ at any point P is defined as follows:

$$\nabla \cdot \mathbf{F} = \lim_{v \to 0} \frac{1}{v} \oint_s \mathbf{F} \cdot d\mathbf{a}.$$

Here P lies within an arbitrary region of space having volume v and bounded by the closed surface s.

[♯]Consider the common face of the volume elements labelled j and k in the above figure (shown, in cross-section, as a dashed boundary line and assumed to be contained entirely within the interior of v). Then the outward flux through this face for element j equals the inward flux through this same face for element k. Since $d\mathbf{a}_j = -d\mathbf{a}_k$, then $\mathbf{F}_j \cdot d\mathbf{a}_j = -\mathbf{F}_k \cdot d\mathbf{a}_k$.

Comments

(i) This important result is known as Gauss's theorem (or sometimes the divergence theorem). It is a mathematical theorem, and should not be confused with Gauss's law which is a law of physics.

(ii) We mention two useful corollaries of the divergence theorem. They are Green's first and second identities:

$$\oint_s (f\nabla g) \cdot da = \int_v (\nabla f \cdot \nabla g + f\nabla^2 g)\, dv, \tag{4}$$

and

$$\oint_s (g\nabla f - f\nabla g) \cdot da = \int_v (g\nabla^2 f - f\nabla^2 g)\, dv, \tag{5}$$

respectively. Here f and g are any two well-behaved scalar fields. Equation (4) is easily proved by substituting $\mathbf{F} = f\nabla g$ in (1) and using (5) of Question 1.8. Equation (5) follows directly from (4).

(iii) Another useful identity, which follows from Gauss's theorem and (7) of Question 1.8, is

$$\oint_s (\mathbf{A} \times \mathbf{B}) \cdot da = \int_v \left[(\nabla \times \mathbf{A}) \cdot \mathbf{B} - (\nabla \times \mathbf{B}) \cdot \mathbf{A} \right] dv. \tag{6}$$

Question 1.13

Consider a vector field $\mathbf{F}(\mathbf{r})$ having continuous first derivatives in a region of space, in which c is an arbitrary closed contour and s any surface spanning c. Prove that

$$\oint_c \mathbf{F} \cdot d\mathbf{l} = \int_s (\nabla \times \mathbf{F}) \cdot d\mathbf{a}, \tag{1}$$

where $d\mathbf{a}$ is an element of area on s.

Solution

To prove (1) we start by evaluating $\oint \mathbf{F} \cdot d\mathbf{l}$ around an infinitesimal rectangular path δc in the xy-plane:

$$(x,\, y,\, z) \to (x + dx,\, y,\, z) \to (x + dx,\, y + dy,\, z) \to (x,\, y + dy,\, z) \to (x,\, y,\, z).$$

If we label the corners of this rectangle 1, 2, 3 and 4, then

$$\oint_{\delta c} \mathbf{F} \cdot d\mathbf{l} = \left(\int_{1\to 2} + \int_{2\to 3} - \left\{ \int_{1\to 4} + \int_{4\to 3} \right\} \right) \mathbf{F} \cdot d\mathbf{l}$$

$$= F_x(x, y, z)\, dx + F_y(x + dx, y, z)\, dy - F_y(x, y, z)\, dy - F_x(x, y + dy, z)\, dx$$

$$= \left(\frac{\partial F_y}{\partial x} - \frac{\partial F_x}{\partial y} \right) dx\, dy$$

$$= (\nabla \times \mathbf{F})_z \, dx\, dy$$

$$= (\nabla \times \mathbf{F}) \cdot \mathbf{n}\, da, \tag{2}$$

where $\mathbf{n}$ is a unit vector perpendicular to the rectangular element of area da. (There is a sign convention (a right-hand rule) implicit in (2), relating the sense in which δc is traversed and the direction of $\mathbf{n}$; see the figure below.) Equation (2) is independent of the choice of coordinates, and applies to an element of any orientation. An arbitrary finite surface s with boundary c can be subdivided into infinitesimal rectangular elements δc_i $(i = 1, 2, \ldots)$. Then

$$\int_c \mathbf{F} \cdot d\mathbf{l} \ = \ \sum_i \oint_{\delta c_i} \mathbf{F} \cdot d\mathbf{l}, \tag{3}$$

because on common segments of adjacent elements the $d\mathbf{l}$ point in opposite directions. So the contributions of $\mathbf{F} \cdot d\mathbf{l}$ to the sum in (3) cancel, whereas no such cancellation occurs on the boundary c. Equations (2) and (3) yield (1). The figure below illustrates the right-hand convention that is assumed here.

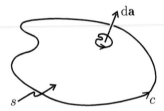

Comment

Equation (1) is known as Stokes's theorem (or sometimes the curl theorem), and it is another very important result.

Question 1.14

(This question and its solution are based on Questions 5.7 and 5.8 of Ref. [4].)

Use Stokes's theorem to prove that a necessary and sufficient condition for a vector field $\mathbf{F}(\mathbf{r})$ to be irrotational[‡] (or conservative) is that $\nabla \times \mathbf{F} = 0$. Split the proof into two parts:

[‡] That is, $\mathbf{F}(\mathbf{r})$ is derivable from a single-valued scalar potential $V(\mathbf{r})$ as $\mathbf{F} = -\nabla V$.

[4]   O. L. de Lange and J. Pierrus, *Solved problems in classical mechanics: Analytical and numerical solutions with comments.* Oxford: Oxford University Press, 2010.

necessary

Assume $\mathbf{F} = -\nabla V$ then show that $\nabla \times \mathbf{F} = 0$. $\hspace{2cm}$ (1)

sufficient

Assume $\nabla \times \mathbf{F} = 0$ then show that $\mathbf{F} = -\nabla V$. $\hspace{2cm}$ (2)

Solution

the necessary condition

If $\mathbf{F} = -\nabla V$, then Stokes's theorem (see (1) of Question 1.13) yields

$$\int_s (\nabla \times \mathbf{F}) \cdot d\mathbf{a} = \oint_c \mathbf{F} \cdot d\mathbf{l} = -\oint_c \nabla V \cdot d\mathbf{l} = -\oint dV(\mathbf{r}) = 0, \hspace{1cm} (3)$$

because $V(\mathbf{r})$ is a single-valued function. The surface s in (3) is arbitrary, and therefore it follows that $\nabla \times \mathbf{F} = 0$ everywhere.

the sufficent condition

This part of the proof is less obvious than the preceding 'necessary' part because one has to prove the existence of the function $V(\mathbf{r})$. If $\nabla \times \mathbf{F} = 0$ *everywhere*, it follows from Stokes's theorem that

$$\oint_c \mathbf{F} \cdot d\mathbf{l} = 0 \hspace{2cm} (4)$$

for *all* closed curves c. According to (4):

$$\int_1 \mathbf{F} \cdot d\mathbf{l} = \int_2 \mathbf{F} \cdot d\mathbf{l}, \hspace{2cm} (5)$$

where 1 and 2 are any two paths from point A to point B. Therefore, the line integral between any two such points is independent of the path followed from A to B: it depends only on the endpoints A and B. Thus, $\mathbf{F} \cdot d\mathbf{l}$ must be the differential of some single-valued scalar function $V(\mathbf{r})$, which we call a perfect differential:

$$\mathbf{F} \cdot d\mathbf{l} = -dV(\mathbf{r}), \hspace{2cm} (6)$$

where a minus sign has been inserted to conform with the standard convention. But

$$dV(\mathbf{r}) = \frac{\partial V}{\partial x} dx + \frac{\partial V}{\partial y} dy + \frac{\partial V}{\partial z} dz = (\nabla V) \cdot d\mathbf{l}. \hspace{1cm} (7)$$

The line element $d\mathbf{l}$ in (6) and (7) is arbitrary, and therefore $\mathbf{F} = -\nabla V$.

Question 1.15

(This question and its solution are based on Questions 5.22 and 5.23 of Ref. [4].)

Use both Stokes's theorem and Gauss's theorem to prove that a necessary and sufficient condition for a vector field $\mathbf{F}(\mathbf{r})$ to be solenoidal[‡] is that $\nabla \cdot \mathbf{F} = 0$. Split the proof into two parts:

necessary

Assume $\mathbf{F} = \nabla \times \mathbf{A}$ then show that $\nabla \cdot \mathbf{F} = 0$.   (1)

sufficient

Assume $\nabla \cdot \mathbf{F} = 0$ then show that $\mathbf{F} = \nabla \times \mathbf{A}$.   (2)

Solution

the necessary condition

Divide the closed surface s into two 'caps', s_1 and s_2, bounded by a common closed curve c, as shown in the figure on p. 21. According to Stokes's theorem

$$\int_{s_1} \mathbf{F} \cdot d\mathbf{a}_1 = \int_{s_1} (\nabla \times \mathbf{A}) \cdot d\mathbf{a}_1 = \oint_c \mathbf{A} \cdot d\mathbf{l} = \int_{s_2} (\nabla \times \mathbf{A}) \cdot d\mathbf{a}_2 = \int_{s_2} \mathbf{F} \cdot d\mathbf{a}_2,$$

where the sense in which c is traversed and the directions of $d\mathbf{a}_1$ and $d\mathbf{a}_2$ are fixed by the right-hand rule. Therefore[♯]

$$\oint_s \mathbf{F} \cdot d\mathbf{a} = \int_{s_1} \mathbf{F} \cdot d\mathbf{a}_1 + \int_{s_2} \mathbf{F} \cdot (-d\mathbf{a}_2) = 0,$$

which by Gauss's theorem means that

$$\int_v (\nabla \cdot \mathbf{F}) \, dv = 0.$$

Because s_1 and s_2 are arbitrary, so is the volume v that they enclose. It therefore follows that $\nabla \cdot \mathbf{F} = 0$.

[‡] That is, $\mathbf{F}(\mathbf{r})$ is derivable from a vector potential $\mathbf{A}(\mathbf{r})$ as $\mathbf{F} = \nabla \times \mathbf{A}$.

[♯] Note that $d\mathbf{a}_2$ is along an inward normal, as shown in the figure on p. 21, and therefore the element to be used in Gauss's theorem is $-d\mathbf{a}_2$.

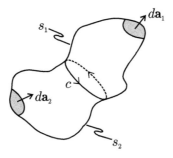

the sufficient condition

The initial part of the proof involves the inverse of the reasoning used in the necessary condition above. If $\nabla \cdot \mathbf{F} = 0$ *everywhere*, then it follows from Gauss's theorem that

$$\oint_s \mathbf{F} \cdot d\mathbf{a} = 0 \tag{3}$$

for *all* closed surfaces s. That is, for 'caps' s_1 and s_2 that share a common bounding curve c, as depicted in the above figure, we have

$$\int_{s_2} \mathbf{F} \cdot d\mathbf{a}_2 = \int_{s_1} \mathbf{F} \cdot d\mathbf{a}_1, \tag{4}$$

meaning that the flux of $\mathbf{F}$ through a cap is unchanged by any deformation of the cap that leaves the bounding curve c unaltered. Therefore, the fluxes in (4) can depend only on the curve c and not on other details of s_1 and s_2. These fluxes can be expressed as the line integral around c of some vector field $\mathbf{A}(\mathbf{r})$:

$$\int_{s_i} \mathbf{F} \cdot d\mathbf{a}_i = \oint_c \mathbf{A} \cdot d\mathbf{l} \qquad (i = 1, 2) \tag{5}$$

$$= \int_{s_i} (\nabla \times \mathbf{A}) \cdot d\mathbf{a}_i \qquad (i = 1, 2), \tag{6}$$

where Stokes's theorem is used in the last step. Since the surface s_i in (6) is arbitrary, we conclude that

$$\mathbf{F} = \nabla \times \mathbf{A}. \tag{7}$$

Question 1.16

(a) Consider the spherically symmetric vector field $\mathbf{F}(\mathbf{r}) = f(r)\,\hat{\mathbf{r}}/r^n$. Use the divergence operator for spherical polar coordinates $\big($see $(\text{XI})_2$ of Appendix C$\big)$ to prove that

$$\nabla \cdot \mathbf{F} = 0 \Rightarrow f(r) = \alpha r^{n-2}, \tag{1}$$

where α is a constant.

(b) Consider the cylindrically symmetric vector field $\mathbf{G}(\mathbf{r}) = g(r)\,\hat{\mathbf{r}}/r^n$. Use the divergence operator for cylindrical polar coordinates (see $(\text{VIII})_2$ of Appendix D) to prove that

$$\nabla \cdot \mathbf{G} = 0 \implies g(r) = \beta r^{n-1}, \tag{2}$$

where β is a constant.

Solution

(a) $\nabla \cdot \mathbf{F} = \dfrac{1}{r^2}\dfrac{\partial}{\partial r}\left(r^2 F_r\right) = \dfrac{1}{r^2}\dfrac{d}{dr}\left(r^{2-n}f\right) = 0$, implying that the term in brackets is a constant. Hence (1).

(b) $\nabla \cdot \mathbf{G} = \dfrac{1}{r}\dfrac{\partial}{\partial r}\left(r\,G_r\right) = \dfrac{1}{r}\dfrac{d}{dr}\left(r^{1-n}g\right) = 0$, implying that the term in brackets is a constant. Hence (2).

Comments

(i) Because $\nabla r^{-(n-1)} = -(n-1)r^{-n}$, it follows that $\dfrac{1}{r^n} = \dfrac{1}{1-n}\nabla\left(\dfrac{1}{r^{n-1}}\right)$, assuming $n \neq 1$. Hence $\nabla \times \mathbf{F} = \nabla \times \mathbf{G} = 0$, since the curl of any gradient is identically zero.

(ii) With $n = 2$ and $\alpha = q/4\pi\epsilon_0$, $\mathbf{F}(\mathbf{r})$ is the electric field $\mathbf{E}$ of a stationary point charge q. Here $\nabla \cdot \mathbf{E} = 0$ (which is one of Maxwell's equations in a source-free vacuum) is valid everywhere except at the location of the charge.

(iii) With $n = 1$ and $\beta = \lambda/2\pi\epsilon_0$, $\mathbf{G}(\mathbf{r})$ is the electric field $\mathbf{E}$ of an infinite electric line charge having uniform density λ. As before, $\nabla \cdot \mathbf{E} = 0$ is valid everywhere except at $r = 0$.

Question 1.17

Below we prove that magnetic fields are always zero. The 'proof' is based on two fundamental equations from electromagnetism (both of which are discussed in later chapters of this book): $\nabla \cdot \mathbf{B} = 0$ and $\mathbf{B} = \nabla \times \mathbf{A}$. Read the 'proof' and then explain where the (fatal) flaw lies.

'proof'

Maxwell's equation: $\nabla \cdot \mathbf{B} = 0 \implies \mathbf{B} = \nabla \times \mathbf{A}$. \hfill (1)

Gauss's theorem and $(1)_1$ give: $\displaystyle\int_v \nabla \cdot \mathbf{B}\, dv = \int_s \mathbf{B} \cdot d\mathbf{a} = 0$. \hfill (2)

Substituting $(1)_2$ in the surface integral (2) and using Stokes's theorem yield

$$\int_s (\nabla \times \mathbf{A}) \cdot d\mathbf{a} = \oint_c \mathbf{A} \cdot d\mathbf{l} = 0. \tag{3}$$

Now (3) implies that $\mathbf{A} = \nabla \phi_m$, $\tag{4}$

where ϕ_m is a scalar potential. Equations $(1)_2$ and (4) then give $\mathbf{B} = \nabla \times \nabla \phi_m$. Since the curl of any gradient is identically zero it necessarily follows that $\mathbf{B} \equiv 0$. Q.E.D.

Solution

In Gauss's theorem s must be a closed surface (see (2) where the theorem is applied incorrectly), but in Stokes's theorem s is open. Therefore, one cannot conclude that the circulation of $\mathbf{A}$ in (3), which follows from (2), is always zero.

Comment

Pay careful attention to all the details in every calculation. Do not assume that nuances in notation are simply a matter of pedantry. This question illustrates how careless execution can lead to incorrect physics (clearly, in this case, spectacularly incorrect).

Question 1.18

Laplace's equation $\nabla^2 \Phi = 0$ is a second-order partial differential equation which arises in many branches of physics. Although there are no general techniques for solving this equation, the 'method of separation of variables' sometimes works. This method is based on a trial solution in which the variables of the problem are separated from one another (see, for example, (1) below). If this trial solution can be made to fit the boundary conditions of the problem (assuming that these have been suitably specified), then its uniqueness is guaranteed.[‡]

(a) Suppose $\Phi = \Phi(x, y, z)$. Attempt solutions to Laplace's equation of the form

$$\Phi(x, y, z) = X(x) Y(y) Z(z), \tag{1}$$

where X, Y and Z are all functions of a single variable. Show that (1) leads to

$$\Phi(x, y, z) = \Phi_0 e^{\pm k_1 x} e^{\pm k_2 y} e^{\pm k_3 z}, \tag{2}$$

where the k_i^2 are real constants.

(b) Find the form of (2) for k_1^2 and k_2^2 both negative.

(c) Find the form of (2) for k_1^2 and k_2^2 both positive.

[‡]See also Question 3.3(d).

Solution

(a) Substituting (1) in $\nabla^2 \Phi = 0$ and dividing by XYZ give

$$\frac{1}{X}\frac{d^2 X}{dx^2} + \frac{1}{Y}\frac{d^2 Y}{dy^2} + \frac{1}{Z}\frac{d^2 Z}{dz^2} = 0. \tag{3}$$

The first term in (3) is independent of y and z, the second term is independent of x and z and the third term is independent of x and y. Now the sum of the terms in (3) is identically zero for *all* x, y and z. This requires that *each* term is independent of x, y and z and is therefore a constant. So

$$\frac{1}{X}\frac{d^2 X}{dx^2} = k_1^2, \qquad \frac{1}{Y}\frac{d^2 Y}{dy^2} = k_2^2 \qquad \text{and} \qquad \frac{1}{Z}\frac{d^2 Z}{dz^2} = k_3^2, \tag{4}$$

where

$$k_1^2 + k_2^2 + k_3^2 = 0. \tag{5}$$

These three ordinary differential equations have the solutions

$$X = X_0 e^{\pm k_1 x}, \qquad Y = Y_0 e^{\pm k_2 y} \qquad \text{and} \qquad Z = Z_0 e^{\pm k_3 z},$$

and together with (1) they yield (2) where $\Phi_0 = X_0 Y_0 Z_0$.

(b) Let $k_1^2 = -\alpha^2$ and $k_2^2 = -\beta^2$ where α and β are real constants. Substituting $k_1 = i\alpha$ and $k_2 = i\beta$ in (5) gives $k_3^2 = -k_1^2 - k_2^2 = \alpha^2 + \beta^2 = \gamma^2$ say. Then $k_3 = \gamma$ is also clearly real, and (2) becomes

$$\Phi(x, y, z) = \Phi_0 e^{\pm i\alpha x} e^{\pm i\beta y} e^{\pm \gamma z}. \tag{6}$$

(c) Now we let $k_1^2 = \alpha^2$ and $k_2^2 = \beta^2$ (again the constants α, β are real). Then $k_1 = \alpha$, $k_2 = \beta$ and $k_3^2 = -k_1^2 - k_2^2 = -\alpha^2 - \beta^2 = -\gamma^2 \Rightarrow k_3 = i\gamma$ (as before γ is real). Substituting these k_i in (2) gives

$$\Phi(x, y, z) = \Phi_0 e^{\pm \alpha x} e^{\pm \beta y} e^{\pm i\gamma z}. \tag{7}$$

Comments

(i) The k_i^2 are known as the separation constants and they may be positive or negative. Their signs are determined by the physics of the problem via the boundary conditions. Choosing k_1^2 and k_2^2 with opposite signs reproduces solutions of the form (6) and (7), but with different permutations of the axes.

(ii) If any one of the various combinations in (6) and (7) is to be 'the' solution to a particular physical problem, then it must be made to satisfy all the boundary conditions of that problem. Furthermore, the boundary conditions can be used to select which (if any) of these possible combinations is a suitable solution. For example, if $\Phi \to 0$ as $z \to \pm\infty$ then a choice involving $e^{\mp \gamma z}$ must be made.

(iii) Linear combinations of these product solutions also satisfy Laplace's equation,[#] and alternative forms of (6) and (7) are therefore

$$\Phi(x,\,y,\,z) \;=\; \Phi_0 \begin{Bmatrix} \cos\alpha x \\ \sin\alpha x \end{Bmatrix} \begin{Bmatrix} \cos\beta y \\ \sin\beta y \end{Bmatrix} \begin{Bmatrix} \cosh\gamma z \\ \sinh\gamma z \end{Bmatrix} \tag{8}$$

and

$$\Phi(x,\,y,\,z) \;=\; \Phi_0 \begin{Bmatrix} \cosh\alpha x \\ \sinh\alpha x \end{Bmatrix} \begin{Bmatrix} \cosh\beta y \\ \sinh\beta y \end{Bmatrix} \begin{Bmatrix} \cos\gamma z \\ \sin\gamma z \end{Bmatrix}, \tag{9}$$

respectively. Because the boundary conditions impose restrictions on the possible values of α, β and γ, they often have a further role in determining the value of the constant Φ_0.

(iv) Separable solutions of Laplace's equation can also be found for other coordinate systems, as for example in spherical polar coordinates. See Question 1.19.

(v) We end these comments with two descriptions of the method of separation of variables. The first, rather colourful, account describes the method 'as one of the most beautiful techniques in all of mathematical physics'.[5] The second description explains that

> the method of separation of variables is perhaps the oldest systematic method for solving partial differential equations. Its essential feature is to transform the partial differential equation by a set of ordinary differential equations. The required solution of the partial differential equation is then exposed as a product $u(x,y) = X(x)Y(y) \neq 0$, or as a sum $u(x,y) = X(x) + Y(y)$, where $X(x)$ and $Y(y)$ are functions of x and y, respectively. Many significant problems in partial differential equations can be solved by the method of separation of variables. This method has been considerably refined and generalized over the last two centuries and is one of the classical techniques of applied mathematics, mathematical physics and engineering science. ... In many cases, the partial differential equation reduces to two ordinary differential equations for X and Y. A similar treatment can be applied to equations in three or more independent variables. However, the question of separability of a partial differential equation into two or more ordinary differential equations is by no means a trivial one. In spite of this question, the method is widely used in finding solutions of a large class of initial boundary-value problems. This method of solution is also known as the Fourier method (or the method of eigenfunction expansion). Thus, the procedure outlined above leads to the important ideas of eigenvalues, eigenfunctions, and orthogonality, all of which are very general and powerful for dealing with linear problems.[6]

[#]This property is proved in Question 3.3(a).

[5]   Source unknown: possibly R. P. Feynman.
[6]   L. Debnath, *Differential equations for scientists and engineers*, Chap. 2, pp. 51–2. Boston: Birkhäuser, 4 edn, 2007.

Question 1.19

Suppose Φ is an axially symmetric potential which satisfies Laplace's equation.

(a) Using ∇^2 for spherical polar coordinates, show that

$$\Phi(r,\theta) = R(r)\Theta(\theta) \tag{1}$$

accomplishes a separation of variables, and leads to the decoupled equations

$$\left.\begin{array}{c} \dfrac{d}{dr}\left(r^2\dfrac{dR}{dr}\right) - kR = 0 \\[3mm] \dfrac{1}{\sin\theta}\dfrac{d}{d\theta}\left(\sin\theta\dfrac{d\Theta}{d\theta}\right) + k\Theta = 0 \end{array}\right\}, \tag{2}$$

where k is a constant.

(b) Hence show that in spherical polar coordinates the general solution of Laplace's equation for boundary-value problems with axial symmetry is

$$\Phi(r,\theta) = \sum_{n=0}^{\infty}\left[A_n r^n + B_n r^{-(n+1)}\right]P_n(\cos\theta), \tag{3}$$

where A_n, B_n are constants and $P_n(\cos\theta)$ is the Legendre polynomial of order n in $\cos\theta$ (see Appendix F).

Hint: Begin with the substitution $R(r) = U(r)/r$ and assume that the separation constant $k = n(n+1)$ where n is a non-negative integer.

Solution

(a) Because of the axial symmetry $\partial\Phi/\partial\phi = 0$ in $\nabla^2\Phi$ (see $(XI)_4$ of Appendix C), and so

$$\frac{1}{r^2}\frac{\partial}{\partial r}\left(r^2\frac{\partial\Phi}{\partial r}\right) + \frac{1}{r^2\sin\theta}\frac{\partial}{\partial\theta}\left(\sin\theta\frac{\partial\Phi}{\partial\theta}\right) = 0. \tag{4}$$

Substituting (1) in (4) and dividing by $R\Theta$ gives

$$\frac{1}{R}\frac{d}{dr}\left(r^2\frac{dR}{dr}\right) = -\frac{1}{\Theta\sin\theta}\frac{d}{d\theta}\left(\sin\theta\frac{d\Theta}{d\theta}\right). \tag{5}$$

Now the left-hand side of (5) is a function of r only, and the right-hand side is a function of θ only. So each must be equal to a constant, k say. Hence (2).

(b) Because of the hint, $(2)_1$ becomes

$$\frac{d^2U}{dr^2} - \frac{n(n+1)}{r^2}U = 0,$$

whose general solution is $U(r) = Ar^{n+1} + Br^{-n}$. Thus

$$R(r) = Ar^n + Br^{-(n+1)}. \tag{6}$$

Next we turn to equation $(2)_2$. Its solutions (as outlined in Appendix F) are of the form

$$\Theta(\theta) = P_n(\cos\theta), \tag{7}$$

apart from an overall constant (which can later be absorbed into other constants). Substituting (6) and (7) in (1), and recalling that the Legendre polynomials form a complete set of functions on the interval $0 \le \theta \le \pi$, it follows that Φ can be expanded as an infinite series

$$\Phi(r,\theta) = \sum_{n=0}^{\infty} \left[A_n r^n P_n(\cos\theta) + B_n r^{-(n+1)} P_{-(n+1)}(\cos\theta) \right].$$

Now

$$P_{-(n+1)}(\cos\theta) = P_n(\cos\theta)$$

as we show below.[‡] Hence (3).

Question 1.20

Let $d\mathbf{a}$ be an infinitesimal area element of some surface s and O any point. The solid angle $d\Omega$ subtended by $d\mathbf{a}$ at O is defined as

$$d\Omega = \frac{d\mathbf{a} \cdot \hat{\mathbf{r}}}{r^2}, \tag{1}$$

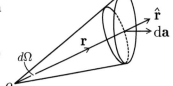

where $\mathbf{r}$ is the vector from O to $d\mathbf{a}$.

(a) Suppose s is the unit sphere centred at O. What is the solid angle Ω subtended by s at O?

(b) Suppose s is a closed surface of arbitrary shape. Show that

$$\Omega = \oint_s d\Omega = \begin{cases} 4\pi & \text{if } O \text{ lies inside } s \\ 0 & \text{if } O \text{ lies outside } s. \end{cases} \tag{2}$$

[‡]If integer n satisfies $k = n(n+1)$, then so does $n' = -(n+1)$, since $n'(n'+1) = -(n+1)(-n) = k$.

Solution

(a) $\Omega = \dfrac{\text{surface area}}{\text{radius squared}} = 4\pi.$ (3)

(b) The rays from O passing through the periphery of $d\mathbf{a}$ generate an infinitesimal cone with apex at O. Similar cones can be generated for all the surface elements of s (say N in total where $N \to \infty$).

O inside s

Fig. (I) shows origin O chosen arbitrarily inside s. Also shown is the unit sphere centred at O. The area element $d\mathbf{a}_A$ of s subtends the same solid angle $d\Omega_1$ at O as the area element $d\mathbf{a}_B$ of the unit sphere. This is true for all the other cones and $\Omega = d\Omega_1 + d\Omega_2 + \cdots + d\Omega_N = 4\pi$ because of (3).

O outside s

The cone shown in Fig. (II) intersects the surface twice. The solid angles subtended at O by the area elements $d\mathbf{a}_1$ and $d\mathbf{a}_2$ are $d\Omega$ and $-d\Omega^{\ddagger}$ respectively, and the sum of these two contributions is zero. This cancellation occurs in pairs for all the other cones and in this case $\Omega = 0$.

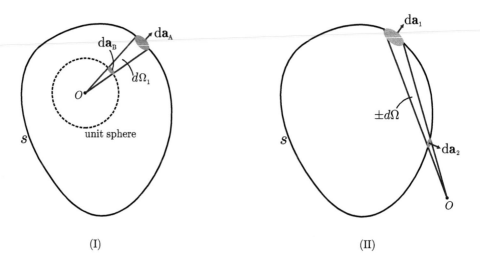

(I)   (II)

Comments

(i) In the SI system the unit of measure of solid angle is called the steradian which is a dimensionless quantity (compare with plane angles which are measured in radians and are also dimensionless).

(ii) Equation (2) can be conveniently expressed as

$\ddagger$By convention the area element $d\mathbf{a}$ is directed along the *outward* normal.

$$\oint_s \frac{d\mathbf{a}\cdot\hat{\mathbf{r}}}{r^2} = 4\pi \int_v \delta(\mathbf{r})\, dv, \tag{4}$$

where $\delta(\mathbf{r})$ is the Dirac delta function. See (X) of Appendix E.

(iii) Consider the vector field $\mathbf{F}(\mathbf{r}) = F(r)\hat{\mathbf{r}}$ where $F(r) = k/r^2$.[#] The flux ψ of $\mathbf{F}$ through any closed surface s follows directly from (1) and (2) and is

$$\psi = \oint_s \mathbf{F}\cdot d\mathbf{a} = k\Omega = \begin{cases} 4\pi k & \text{if the source point } O \text{ lies inside } s, \\ 0 & \text{if the source point } O \text{ lies outside } s. \end{cases} \tag{5}$$

This equation (known as Gauss's law[†]) is a very important law in physics. It relates the flux of $\mathbf{F}$ to its source(s). Familiar examples are:

☞ the gravitational acceleration of a planet having mass M (where $k = GM$), and

☞ the electric field of a point charge q in vacuum (where $k = q/4\pi\epsilon_0$).

(iv) The generality implied by (5) is the reason why Gauss's law is so useful. The flux through the closed surface is *independent* of the location of O. For the source point anywhere inside s we have $\psi = 4\pi k$, and $\psi = 0$ if the source point is anywhere outside s.

(v) The two features of $\mathbf{F}(\mathbf{r})$ upon which Gauss's law critically depends are:

☞ the inverse-square nature of the field, and

☞ the central nature of the field (i.e. $\mathbf{F}$ directed along $\hat{\mathbf{r}}$).

The spherical symmetry present in Newton's law of gravitation and Coulomb's law is not a necessary condition for (5). We show in Question 12.15 that Gauss's law also holds for the non-spherically symmetric, inverse-square, central electric field of a point charge moving relativistically at constant speed.

Question 1.21

Use Gauss's theorem and the definition of solid angle to prove that the Laplacian of r^{-1} (i.e. the divergence of the gradient of r^{-1}) is

$$\nabla^2\left(\frac{1}{r}\right) = -4\pi\delta(\mathbf{r}), \tag{1}$$

where $\delta(\mathbf{r})$ is the Dirac delta function (see Appendix E).

[#]$\mathbf{F}$ is called 'spherically symmetric' because $F(r)$ depends only on the magnitude of $\mathbf{r}$ and not on the direction of $\mathbf{r}$.

[†]As previously mentioned (see Comment (i) of Question 1.12), Gauss's law and Gauss's theorem are separate entities and should not be confused.

Solution

Substituting $\mathbf{F} = -\nabla\left(\dfrac{1}{r}\right) = \dfrac{\mathbf{r}}{r^3}$ in $\oint_s \mathbf{F} \cdot d\mathbf{a} = \int_v (\nabla \cdot \mathbf{F})\, dv$ gives

$$\oint_s \frac{\mathbf{r} \cdot d\mathbf{a}}{r^3} = -\int_v \nabla \cdot \nabla\left(\frac{1}{r}\right) dv = -\int_v \nabla^2\left(\frac{1}{r}\right) dv. \tag{2}$$

Then from (4) of Question 1.20

$$\int_v \nabla^2\left(\frac{1}{r}\right) dv = -4\pi \int_v \delta(\mathbf{r})\, dv, \tag{3}$$

and hence (1) because the volume v is arbitrary.

Comments

(i) Shifting the singularity from $\mathbf{r} = 0$ to $\mathbf{r} = \mathbf{r}'$ gives the more general form of (1):

$$\nabla^2\left(\frac{1}{|\mathbf{r} - \mathbf{r}'|}\right) = -4\pi\delta(\mathbf{r} - \mathbf{r}'). \tag{4}$$

The proof given in the solution above is standard, but see also Ref. [7].

(ii) The form of (1) raises the question: will other derivatives such as $\nabla_i\nabla_j(1/r)$[‡] also contain a delta function? They do.[8] For example,

$$\nabla_i\nabla_j\left(\frac{1}{r}\right) = \frac{3r_ir_j - r^2\delta_{ij}}{r^5} - \frac{4\pi}{3}\delta_{ij}\delta(\mathbf{r}) \tag{5}$$

(a non-rigorous proof of this result is provided in Ref. [9]). The first term on the right-hand side of (5) applies at points *away* from the origin, whilst the second term is zero everywhere except *at* the origin.[♯] For an application involving (5), see Comment (iii) of Question 2.11.

[‡]They should, because the Laplacian of r^{-1} is just a special case of $\nabla_i\nabla_j(1/r)$ when $i = j$.

[♯]Apart from the minus sign, the first term of (5) is (7) of Question 1.1. Evidently, explicit differentiation fails to reveal the δ-function contribution.

---

[7]   V. Hnizdo, 'On the Laplacian of $1/r$', *European Journal of Physics*, vol. 21, pp. L1–L3, 2000.

[8]   See, for example, R. Estrada and R. P. Kanwal, 'The appearance of nonclassical terms in the analysis of point-source fields', *American Journal of Physics*, vol. 63, p. 278, 1995.

[9]   C. P. Frahm, 'Some novel delta-function identities', *American Journal of Physics*, vol. 51, pp. 826–9, 1983.

Question 1.22

Suppose f and $\mathbf{F}$ represent suitably continuous and differentiable scalar and vector fields respectively. Let $\mathbf{b}$ be an arbitrary but constant vector. Using the hint provided alongside each, prove the following integral theorems:

(a) $\quad \oint_s f \, d\mathbf{a} = \int_v (\nabla f) \, dv \qquad$ (in Gauss's theorem let $\mathbf{F} = f\mathbf{b}$), $\qquad\qquad$ (1)

(b) $\quad \oint_s \mathbf{F} \times d\mathbf{a} = -\int_v (\nabla \times \mathbf{F}) \, dv \qquad$ (in Gauss's theorem let $\mathbf{F} \to \mathbf{b} \times \mathbf{F}$), $\qquad$ (2)

$\qquad$ where the region v is bounded by the closed surface s.

(c) $\quad \oint_c f \, d\mathbf{l} = -\int_s \nabla f \times d\mathbf{a} \qquad$ (in Stokes's theorem let $\mathbf{F} = f\mathbf{b}$), $\qquad\qquad$ (3)

(d) $\quad \oint_c \mathbf{F} \times d\mathbf{l} = -\int_s (d\mathbf{a} \times \nabla) \times \mathbf{F} \quad$ (in Stokes's theorem let $\mathbf{F} \to \mathbf{b} \times \mathbf{F}$), $\qquad$ (4)

$\qquad$ where the closed contour c is spanned by the surface s.

Solution

(a) Gauss's theorem becomes $\displaystyle \oint_s f\,\mathbf{b} \cdot d\mathbf{a} = \int_v (\mathbf{b} \cdot \nabla f)\, dv$ which, because $\mathbf{b}$ is a constant vector, can be written as $\displaystyle \mathbf{b} \cdot \left[\oint_s f\, d\mathbf{a} - \int_v (\nabla f)\, dv \right] = 0$. Now $|\mathbf{b}| \neq 0$ and since $\mathbf{b}$ is arbitrary, the cosine of the included angle is not always zero. This equation can only be satisfied if the term in brackets is zero, which proves (1).

(b) Gauss's theorem becomes $\displaystyle \oint_s (\mathbf{b} \times \mathbf{F}) \cdot d\mathbf{a} = -\int_v \mathbf{b} \cdot (\nabla \times \mathbf{F})\, dv$ by (7) of Question 1.8 and $\nabla \times \mathbf{b} = 0$. Using the cyclic property of the scalar triple product (see (1) of Question 1.8), this can be written as $\displaystyle \mathbf{b} \cdot \left[\oint_s \mathbf{F} \times d\mathbf{a} + \int_v (\nabla \times \mathbf{F})\, dv \right] = 0$. As before, this equation can only be satisfied if the term in square brackets is zero, which proves (2).

(c) Stokes's theorem becomes $\displaystyle \oint_c f\,\mathbf{b} \cdot d\mathbf{l} = \int_s (\nabla f \times \mathbf{b}) \cdot d\mathbf{a}$ because $\nabla \times \mathbf{b} = 0$ (see (6) of Question 1.8). Use of the non-commutative property of the cross-product and the cyclic nature of the scalar triple product yields $\displaystyle \oint_c f\,\mathbf{b} \cdot d\mathbf{l} = -\int_s \mathbf{b} \cdot (\nabla f \times d\mathbf{a})$, or $\displaystyle \mathbf{b} \cdot \left[\oint_c f\, d\mathbf{l} + \int_s \nabla f \times d\mathbf{a} \right] = 0$. As in (a) and (b), this equation can only be satisfied if the term in square brackets is zero, which proves (3).

(d) Stokes's theorem becomes $\oint_c (\mathbf{b} \times \mathbf{F}) \cdot d\mathbf{l} = -\int_s [\nabla \times (\mathbf{b} \times \mathbf{F})] \cdot d\mathbf{a}$. Applying (1) and (8) of Question 1.8 to this result, and because $\mathbf{b}$ is a constant vector, we obtain

$$\mathbf{b} \cdot \oint_c \mathbf{F} \times d\mathbf{l} = \int_s [\mathbf{b}(\nabla \cdot \mathbf{F}) - (\mathbf{b} \cdot \nabla)\mathbf{F}] \cdot d\mathbf{a} = \int_s b_i \nabla_k F_k da_i - \int_s (b_i \nabla_i F_j) da_j$$

$$= b_i (\delta_{ij}\delta_{kl} - \delta_{il}\delta_{kj}) \int_s da_j \nabla_l F_k = b_i \varepsilon_{mik}\varepsilon_{mjl} \int_s da_j \nabla_l F_k$$

$$= b_i \varepsilon_{mik} \int_s (d\mathbf{a} \times \nabla)_m F_k = -b_i \varepsilon_{imk} \int_s (d\mathbf{a} \times \nabla)_m F_k$$

$$= b_i \int_s [(d\mathbf{a} \times \nabla) \times \mathbf{F}]_i = \mathbf{b} \cdot \int_s [(d\mathbf{a} \times \nabla) \times \mathbf{F}].$$

Thus $\mathbf{b} \cdot \left[\oint_c \mathbf{F} \times d\mathbf{l} - \int_s (d\mathbf{a} \times \nabla) \times \mathbf{F} \right] = 0$, which leads to (4) in the usual way.

Comment

From (4) we can derive a useful result for the (vector) area $\mathbf{a}$ of a plane surface s:

$$\mathbf{a} = \tfrac{1}{2} \oint_c (\mathbf{r} \times d\mathbf{l}). \tag{5}$$

The proof of (5) is straightforward. Let $\mathbf{F} = \mathbf{r}$ in (4) and use Cartesian tensors to show that $\int_s (d\mathbf{a} \times \nabla) \times \mathbf{r} = -2 \int_s d\mathbf{a}$. The details are left as an exercise for the reader.

Question 1.23

A vector field $\mathbf{F}(\mathbf{r})$ is continuous at all points inside a volume v and on a surface s bounding v, as are its divergence and curl

$$\left. \begin{array}{c} \nabla \cdot \mathbf{F} = S(\mathbf{r}) \\ \nabla \times \mathbf{F} = C(\mathbf{r}) \end{array} \right\}. \tag{1}$$

(Note that $\nabla \cdot \mathbf{C} = 0$ for self-consistency.)

(a) Prove that $\mathbf{F}$ is a unique solution of (1) when it satisfies the boundary condition

$$\mathbf{F} \cdot \hat{\mathbf{n}} \text{ is known everywhere on } s, \tag{2}$$

where $\hat{\mathbf{n}}$ is a unit normal on s.

(b) Repeat (a) for the boundary condition

$$\mathbf{F} \times \hat{\mathbf{n}} \text{ is known everywhere on } s. \tag{3}$$

Hint: Construct a proof by contradiction. Begin by assuming that $\mathbf{F}_1(\mathbf{r})$ and $\mathbf{F}_2(\mathbf{r})$ are two different vector fields having the same divergence and curl and satisfying the same boundary condition. Then let $\mathbf{W} = \mathbf{F}_1 - \mathbf{F}_2$ and use the results of Question 1.12 $\big($use Green's identity (4) for (a); use (6) for (b)$\big)$ to prove that $\mathbf{W} = 0$.

Solution

Since $\mathbf{W} = \mathbf{F}_1 - \mathbf{F}_2$ and because of the hint, it is clear that

$$\left.\begin{array}{c}\boldsymbol{\nabla} \cdot \mathbf{W} = 0 \\ \boldsymbol{\nabla} \times \mathbf{W} = 0\end{array}\right\}. \tag{4}$$

We consider each of the two boundary conditions separately:

(a) Equation $(4)_2$ implies that (apart from a possible minus sign)

$$\mathbf{W} = \boldsymbol{\nabla}V, \tag{5}$$

where $V(\mathbf{r})$ is a scalar potential. It then follows immediately from $(4)_1$ that

$$\nabla^2 V = 0. \tag{6}$$

Substituting $f = g = V$ in Green's first identity $\big($see (4) of Question 1.12$\big)$ gives

$$\oint_s V \boldsymbol{\nabla}V \cdot d\mathbf{a} = \oint_s V \frac{\partial V}{\partial n} \, da = \int_v (\boldsymbol{\nabla}V \cdot \boldsymbol{\nabla}V + V\nabla^2 V) \, dv, \tag{7}$$

since in (7) $d\mathbf{a} = \hat{\mathbf{n}}\,da$ and $\boldsymbol{\nabla}V \cdot \hat{\mathbf{n}} = \partial V/\partial n$ is the normal derivative of V over the boundary surface s. Because of (6), this identity simplifies to

$$\oint_s V \frac{\partial V}{\partial n} \, da = \int_v |\boldsymbol{\nabla}V|^2 \, dv. \tag{8}$$

Now $\mathbf{F}_1$ and $\mathbf{F}_2$ both satisfy the same boundary condition. Hence from (5)

$$(\mathbf{F}_1 - \mathbf{F}_2) \cdot \hat{\mathbf{n}} = \mathbf{W} \cdot \hat{\mathbf{n}} = \boldsymbol{\nabla}V \cdot \hat{\mathbf{n}} = \frac{\partial V}{\partial n} = 0,$$

and (8) becomes

$$\int_v |\boldsymbol{\nabla}V|^2 dv = 0. \tag{9}$$

The integrand in (9) is clearly non-negative, which requires that $\boldsymbol{\nabla}V = 0$ throughout v. Thus $\mathbf{W} = 0$ and $\mathbf{F}_1 = \mathbf{F}_2$ everywhere. A vector field $\mathbf{F}(\mathbf{r})$ which satisfies the boundary condition (2) is therefore a unique solution of (1).

(b) Equation $(4)_1$ implies that

$$\mathbf{W} = \mathbf{\nabla} \times \mathbf{A}, \tag{10}$$

where $\mathbf{A}(\mathbf{r})$ is a vector potential, and so from $(4)_2$

$$\mathbf{\nabla} \times (\mathbf{\nabla} \times \mathbf{A}) = 0. \tag{11}$$

Substituting $\mathbf{B} = \mathbf{\nabla} \times \mathbf{A}$ in the identity (6) of Question 1.12 gives

$$\oint_s [\mathbf{A} \times (\mathbf{\nabla} \times \mathbf{A})] \cdot d\mathbf{a} = \int_v \left\{ (\mathbf{\nabla} \times \mathbf{A}) \cdot (\mathbf{\nabla} \times \mathbf{A}) - \mathbf{A} \cdot [\mathbf{\nabla} \times (\mathbf{\nabla} \times \mathbf{A})] \right\} dv. \tag{12}$$

Now $[\mathbf{A} \times (\mathbf{\nabla} \times \mathbf{A})] \cdot d\mathbf{a} = -[\hat{\mathbf{n}} \times (\mathbf{\nabla} \times \mathbf{A})] \cdot \mathbf{A}\, da$ follows from the properties of the scalar triple product. Then (11) and (12) yield

$$-\oint_s [\hat{\mathbf{n}} \times (\mathbf{\nabla} \times \mathbf{A})] \cdot \mathbf{A}\, da = \int_v (\mathbf{\nabla} \times \mathbf{A})^2\, dv. \tag{13}$$

From (10), and since both $\mathbf{F}_1$ and $\mathbf{F}_2$ satisfy the same boundary condition, we have

$$-\hat{\mathbf{n}} \times (\mathbf{\nabla} \times \mathbf{A}) = \hat{\mathbf{n}} \times (\mathbf{F}_2 - \mathbf{F}_1) = 0.$$

Therefore, (13) becomes

$$\int_v (\mathbf{\nabla} \times \mathbf{A})^2\, dv = 0. \tag{14}$$

Using similar reasoning as in (a), we conclude that $\mathbf{W} = \mathbf{\nabla} \times \mathbf{A} = 0$. Therefore $\mathbf{F}_1 = \mathbf{F}_2$ everywhere, and any vector field $\mathbf{F}(\mathbf{r})$ satisfying the boundary condition (3) is a unique solution of (1).

Comments

(i) The quantities $S(\mathbf{r})$ and $C(\mathbf{r})$ are known as the source and circulation densities respectively. In electromagnetism, $S(\mathbf{r})$ is the electric charge density and $C(\mathbf{r})$ the current density.

(ii) Clearly, the boundary conditions (2) and (3) specify the normal component and the tangential component of $\mathbf{F}$ on s respectively.

Question 1.24*

Suppose $\mathbf{F}(\mathbf{r})$ is any continuous vector field which tends to zero at least as fast as $1/r^2$ as $r \to \infty$.

(a) Prove that $\mathbf{F}(\mathbf{r})$ can be decomposed as follows:

$$\mathbf{F}(\mathbf{r}) = -\nabla V + \nabla \times \mathbf{A}, \tag{1}$$

where $V = V(\mathbf{r})$ is a scalar potential and $\mathbf{A} = \mathbf{A}(\mathbf{r})$ is a vector potential.

Hint: Begin with $(XI)_2$ of Appendix E and make use of the vector identity (11) of Question 1.8: $\nabla \times (\nabla \times \mathbf{w}) = -\nabla^2 \mathbf{w} + \nabla(\nabla \cdot \mathbf{w})$.

(b) Hence show that

$$\left.\begin{aligned}
V(\mathbf{r}) &= \frac{1}{4\pi} \int_v \frac{S(\mathbf{r}')}{|\mathbf{r} - \mathbf{r}'|}\, dv' \\[2mm]
\mathbf{A}(\mathbf{r}) &= \frac{1}{4\pi} \int_v \frac{\mathbf{C}(\mathbf{r}')}{|\mathbf{r} - \mathbf{r}'|}\, dv'
\end{aligned}\right\}, \tag{2}$$

where $S(\mathbf{r}') = \nabla' \cdot \mathbf{F}(\mathbf{r}')$ and $\mathbf{C}(\mathbf{r}') = \nabla' \times \mathbf{F}(\mathbf{r}')$.

(c) Prove that the decomposition (1) is unique.

Solution

(a) Because of the hint,

$$\mathbf{F}(\mathbf{r}) = \int_v \mathbf{F}(\mathbf{r}')\, \delta(\mathbf{r} - \mathbf{r}')\, dv', \tag{3}$$

where v is any region that contains the point $\mathbf{r}$. But $\delta(|\mathbf{r} - \mathbf{r}'|) = -\dfrac{1}{4\pi}\nabla^2(|\mathbf{r} - \mathbf{r}'|)^{-1}$, and so (3) becomes

$$\mathbf{F}(\mathbf{r}) = -\frac{1}{4\pi} \int_v \mathbf{F}(\mathbf{r}')\, \nabla^2(|\mathbf{r} - \mathbf{r}'|)^{-1}\, dv'$$

$$= -\nabla^2 \left[\frac{1}{4\pi} \int_v \frac{\mathbf{F}(\mathbf{r}')}{|\mathbf{r} - \mathbf{r}'|}\, dv' \right], \tag{4}$$

since $\nabla^2 \mathbf{F}(\mathbf{r}')$ is zero.‡ Applying the given identity to (4) yields

$$\mathbf{F}(\mathbf{r}) = -\nabla\left(\nabla \cdot \frac{1}{4\pi} \int_v \frac{\mathbf{F}(\mathbf{r}')}{|\mathbf{r} - \mathbf{r}'|}\, dv' \right) + \nabla \times \left(\nabla \times \frac{1}{4\pi} \int_v \frac{\mathbf{F}(\mathbf{r}')}{|\mathbf{r} - \mathbf{r}'|}\, dv' \right),$$

---

‡Reason: $\mathbf{F}(\mathbf{r}')$ does not depend on the unprimed coordinates of ∇^2.

which is (1) with

$$V(\mathbf{r}) = \nabla \cdot \frac{1}{4\pi} \int_v \frac{\mathbf{F}(\mathbf{r}')}{|\mathbf{r} - \mathbf{r}'|} \, dv'$$

$$A(\mathbf{r}) = \nabla \times \frac{1}{4\pi} \int_v \frac{\mathbf{F}(\mathbf{r}')}{|\mathbf{r} - \mathbf{r}'|} \, dv'$$
(5)

(b) In the proofs below, we will use the following results repeatedly:

1. $\nabla \cdot (f\mathbf{w}) = f\nabla \cdot \mathbf{w} + \mathbf{w} \cdot \nabla f$ or alternatively $\nabla' \cdot (f\mathbf{w}) = f\nabla' \cdot \mathbf{w} + \mathbf{w} \cdot \nabla' f$,

2. $\nabla(|\mathbf{r} - \mathbf{r}'|)^{-1} = -\nabla'(|\mathbf{r} - \mathbf{r}'|)^{-1}$,

3. ∇ acts on a function of $\mathbf{r}$ only; ∇' acts on a function of $\mathbf{r}'$ only.

☞ $V(\mathbf{r})$

From $(5)_1$ it follows that:

$$V(\mathbf{r}) = \frac{1}{4\pi} \int_v \mathbf{F}(\mathbf{r}') \cdot \nabla \left(\frac{1}{|\mathbf{r} - \mathbf{r}'|} \right) dv' = -\frac{1}{4\pi} \int_v \mathbf{F}(\mathbf{r}') \cdot \nabla' \left(\frac{1}{|\mathbf{r} - \mathbf{r}'|} \right) dv'$$

$$= -\frac{1}{4\pi} \int_v \nabla' \cdot \left(\frac{\mathbf{F}(\mathbf{r}')}{|\mathbf{r} - \mathbf{r}'|} \right) dv' + \frac{1}{4\pi} \int_v \frac{\nabla' \cdot \mathbf{F}(\mathbf{r}')}{|\mathbf{r} - \mathbf{r}'|} \, dv'$$

$$= -\frac{1}{4\pi} \oint_s \frac{\mathbf{F}(\mathbf{r}') \cdot d\mathbf{a}'}{|\mathbf{r} - \mathbf{r}'|} + \frac{1}{4\pi} \int_v \frac{\nabla' \cdot \mathbf{F}(\mathbf{r}')}{|\mathbf{r} - \mathbf{r}'|} \, dv',$$

where, in the last step, we use Gauss's theorem. Now if v is chosen over all space, s is a surface at infinity and the surface integral is zero.[#] Hence $(2)_1$.

☞ $A(\mathbf{r})$

From $(5)_2$ it follows that:

$$A(\mathbf{r}) = \frac{1}{4\pi} \int_v \nabla \left(\frac{1}{|\mathbf{r} - \mathbf{r}'|} \right) \times \mathbf{F}(\mathbf{r}') \, dv' = -\frac{1}{4\pi} \int_v \nabla' \left(\frac{1}{|\mathbf{r} - \mathbf{r}'|} \right) \times \mathbf{F}(\mathbf{r}') \, dv'$$

$$= -\frac{1}{4\pi} \int_v \nabla' \times \left(\frac{\mathbf{F}(\mathbf{r}')}{|\mathbf{r} - \mathbf{r}'|} \right) dv' + \frac{1}{4\pi} \int_v \frac{\nabla' \times \mathbf{F}(\mathbf{r}')}{|\mathbf{r} - \mathbf{r}'|} \, dv'$$

$$= \frac{1}{4\pi} \oint_s \frac{\mathbf{F}(\mathbf{r}') \times d\mathbf{a}'}{|\mathbf{r} - \mathbf{r}'|} + \frac{1}{4\pi} \int_v \frac{\nabla' \times \mathbf{F}(\mathbf{r}')}{|\mathbf{r} - \mathbf{r}'|} \, dv',$$

[#]Recall that F scales as $1/r^{2+\epsilon}$ (here the parameter $\epsilon \geq 0$) and da scales as r^2. So, for a distant surface, $\dfrac{\mathbf{F} \cdot d\mathbf{a}'}{|\mathbf{r} - \mathbf{r}'|} \simeq \dfrac{\mathbf{F} \cdot d\mathbf{a}'}{r}$ scales as $1/r^{1+\epsilon}$ which tends to zero as $r \to \infty$.

where, in the last step, we use the version of Gauss's theorem (2) of Question 1.22. The surface integral is zero as before and hence $(2)_2$.

(c) The same reasoning used in (a) of the previous question can be used to prove that the decomposition of $\mathbf{F}(\mathbf{r})$ is unique. We again assume two different solutions $\mathbf{F}_1$ and $\mathbf{F}_2$ and consider their difference $\mathbf{W} = \nabla V = \mathbf{F}_1 - \mathbf{F}_2$, arriving at the result

$$\oint_s V \frac{\partial V}{\partial n}\, da = \int_v |\nabla V|^2\, dv. \tag{6}$$

If the limit is taken in which v becomes infinite, the integral over the surface s at infinity vanishes.[†] Then (6) yields

$$\int_v |\nabla V|^2 dv = 0.$$

Thus $\mathbf{W} = \nabla V = 0$ everywhere so that $\mathbf{F}_1 = \mathbf{F}_2$ and $\mathbf{F}$ is unique.

Comments

(i) Equation (1) shows that any well-behaved vector field can be expressed as the sum of an irrotational field $-\nabla\Phi(\mathbf{r})$ and a solenoidal field $\nabla \times \mathbf{A}(\mathbf{r})$. This is a fundamental result of vector calculus and is known as Helmholtz's theorem. Its relevance to electromagnetism through Maxwell's equations, expressed as they are in terms of divergences and curls, is apparent.

(ii) The quantities $\nabla \cdot \mathbf{F} = S(\mathbf{r})$ and $\nabla \times \mathbf{F} = \mathbf{C}(\mathbf{r})$ serve as source functions which completely determine the field $\mathbf{F}(\mathbf{r})$. Ref. [10] explains that Helmholtz's theorem 'establishes that these serve as complete sources of the field, and that all continuous vector fields can be classified by the two mathematical types, the conservative and the solenoidal'.

Question 1.25

Prove that a uniform vector field $\mathbf{F}_0$ can be expressed as:

(a) an irrotational field

$$\mathbf{F}_0 = -\nabla V, \tag{1}$$

where $V(\mathbf{r}) = -\mathbf{r} \cdot \mathbf{F}_0$.

[†] $\dfrac{\partial V}{\partial n}$, being the normal component of ∇V, scales as $1/r^{2+\epsilon}$ and therefore V scales as $1/r^{1+\epsilon}$. So $V \dfrac{\partial V}{\partial n}\, da$ scales as $1/r^{1+2\epsilon} \to 0$ as $r \to \infty$.

[10] B. P. Miller, 'Interpretations from Helmholtz' theorem in classical electromagnetism', *American Journal of Physics*, vol. 52, pp. 948–50, 1984.

(b) a solenoidal field

$$\mathbf{F}_0 = \nabla \times \mathbf{A}, \tag{2}$$

where $\mathbf{A}(\mathbf{r}) = -\frac{1}{2}(\mathbf{r} \times \mathbf{F}_0)$.

Solution

Considering the ith component of ∇V and $\nabla \times \mathbf{A}$ respectively, and recalling that the F_{0i} are spatially constant, give:

(a) $\nabla_i(\mathbf{r} \cdot \mathbf{F}_0) = \nabla_i(r_j F_{0j}) = F_{0j} \nabla_i r_j = F_{0j} \delta_{ij}$, which contracts to F_{0i} as required.

(b) $(\nabla \times \mathbf{A})_i = -\frac{1}{2}\varepsilon_{ijk}\nabla_j(\mathbf{r} \times \mathbf{F}_0)_k = -\frac{1}{2}\varepsilon_{ijk}\varepsilon_{klm}F_{0m}\nabla_j r_l = -\frac{1}{2}(\delta_{il}\delta_{jm} - \delta_{im}\delta_{jl})F_{0m}\delta_{jl}$.
Contracting subscripts gives F_{0i} as required.

Comments

(i) The results $V(\mathbf{r}) = -\mathbf{r} \cdot \mathbf{F}_0$ and $\mathbf{A}(\mathbf{r}) = -\frac{1}{2}(\mathbf{r} \times \mathbf{F}_0)$ are often convenient potentials for representing uniform electrostatic and magnetostatic fields respectively.

(ii) Because a uniform field does not satisfy the conditions of Helmholtz's theorem (it does not tend to zero at infinity), $\mathbf{F}_0$ has no unique representation. It is easily verified that $\mathbf{F}_0$ can be expressed as a linear combination of (1) and (2) in infinitely many ways.

Question 1.26

Consider a sphere having radius r_0 centred at an arbitrary origin O. Let $\mathbf{r}'$ be the position vector of any point P' inside or on the surface of the sphere; let $\mathbf{r}$ be the position vector of a fixed field point P. Prove the following results:

$$\text{(a)} \ \oint_s \frac{da'}{|\mathbf{r} - \mathbf{r}'|} = \begin{cases} \dfrac{4\pi r_0^2}{r} & r \geq r_0 \\[2mm] 4\pi r_0 & r \leq r_0, \end{cases} \tag{1}$$

$$\text{(b)} \ \int_v \frac{dv'}{|\mathbf{r} - \mathbf{r}'|} = \begin{cases} \dfrac{4\pi}{3}\dfrac{r_0^3}{r} & r \geq r_0 \\[2mm] 2\pi(r_0^2 - \frac{1}{3}r^2) & r \leq r_0, \end{cases} \tag{2}$$

$$\text{(c)} \ \oint_s \frac{da'}{|\mathbf{r} - \mathbf{r}'|} = \begin{cases} \dfrac{4\pi}{3}\left(\dfrac{r_0}{r}\right)^3 \mathbf{r} & r \geq r_0 \\[2mm] \dfrac{4\pi}{3}\mathbf{r} & r \leq r_0. \end{cases} \tag{3}$$

Solution

Orient Cartesian axes so that the point P is located on the z-axis. Then the integrand is axially symmetric about the z-axis and for the point P$'$ it is convenient to use the spherical polar coordinates (r', θ', ϕ').

(a) Here $r' = r_0$ and the element of area $da' = r_0^2 \sin \theta' d\theta' d\phi'$. By the cosine rule, $|\mathbf{r} - \mathbf{r}'| = \sqrt{r^2 + r_0^2 - 2rr_0 \cos \theta'}$, and so

$$\oint_s \frac{da'}{|\mathbf{r} - \mathbf{r}'|} = \int_0^\pi \int_0^{2\pi} \frac{r_0^2 \sin \theta' d\theta' d\phi'}{\sqrt{r^2 + r_0^2 - 2rr_0 \cos \theta'}} = \int_{-1}^1 \frac{2\pi r_0^2 \, d\cos \theta'}{\sqrt{r^2 + r_0^2 - 2rr_0 \cos \theta'}} . \qquad (4)$$

With the substitution $u^2 = r^2 + r_0^2 - 2rr_0 \cos \theta'$, equation (4) becomes

$$2\pi r_0^2 \frac{1}{r_0 r} \int_{\sqrt{(r-r_0)^2}}^{\sqrt{(r+r_0)^2}} du . \qquad (5)$$

For $r > r_0$ the lower limit is $r - r_0$; for $r < r_0$ the lower limit is $r_0 - r$. These limits in (5) yield (1).

(b) We proceed as in (a). Taking the volume element $dv' = r'^2 \sin \theta' dr' d\theta' d\phi'$ and substituting $u^2 = r^2 + r'^2 - 2rr' \cos \theta'$ give

$$\int_v \frac{dv'}{|\mathbf{r} - \mathbf{r}'|} = \int_0^{r_0} \int_0^\pi \int_0^{2\pi} \frac{r'^2 \sin \theta' dr' d\theta' d\phi'}{\sqrt{r^2 + r'^2 - 2rr' \cos \theta'}} = \int_0^{r_0} r'^2 \int_{\sqrt{(r-r')^2}}^{\sqrt{(r+r')^2}} \frac{2\pi}{rr'} \, du \, dr' . \qquad (6)$$

$\boxed{r \geq r_0}$

The lower limit for the integration over u is $r - r'$ (see (a) above) and integration of (6) yields $(2)_1$.

$\boxed{r \leq r_0}$

The point P now lies inside the sphere, which we partition as shown in the figure alongside. The lower limit for the integration over u is either $r - r'$ for $r' < r$ or $r' - r$ for $r' > r$ (see the discussion in (a) above). Equation (6) becomes

$$\int_v \frac{dv'}{|\mathbf{r} - \mathbf{r}'|} = 2\pi \int_0^r r'^2 \int_{(r-r')}^{(r+r')} \frac{1}{rr'} \, du \, dr' + 2\pi \int_r^{r_0} r'^2 \int_{(r'-r)}^{(r+r')} \frac{1}{rr'} \, du \, dr'$$

$$= \frac{4\pi}{r} \int_0^r r'^2 dr' + 4\pi \int_r^{r_0} r' dr'$$

$$= \frac{4\pi}{3} r^2 + 2\pi (r_0^2 - r^2),$$

which is $(2)_2$.

(c) We use (1) of Question 1.22, $\oint_s f\,d\mathbf{a}' = \int_v (\nabla'f)\,dv'$, and put $f = |\mathbf{r} - \mathbf{r}'|^{-1}$. Then

$$\oint_s \frac{d\mathbf{a}'}{|\mathbf{r} - \mathbf{r}'|} = \int_v \nabla'\left(\frac{1}{|\mathbf{r} - \mathbf{r}'|}\right)dv' = -\nabla\int_v \frac{dv'}{|\mathbf{r} - \mathbf{r}'|} \tag{7}$$

since $\nabla'f = -\nabla f$. Substituting (2) in (7) and differentiating give (3).

Comment

Alternative forms of (1)–(3) are sometimes required. For example, if $r \leq r_0$ and the integration is relative to unprimed coordinates, then

$$\oint_s \frac{d\mathbf{a}}{|\mathbf{r} - \mathbf{r}'|} = \begin{cases} \dfrac{4\pi}{3}\left(\dfrac{r_0}{r'}\right)^3 \mathbf{r}' & r' \geq r_0 \\[2ex] \dfrac{4\pi}{3}\mathbf{r}' & r' \leq r_0. \end{cases} \tag{8}$$

Question 1.27

The time average of a dynamical quantity $Q(t)$ is defined as

$$\langle Q(t)\rangle = \langle Q\rangle = \lim_{t\to\infty} \frac{1}{t}\int_0^t Q(t')\,dt'. \tag{1}$$

(a) Suppose $Q(t)$ is periodic having period T. That is, $Q(t + nT) = Q(t)$ where $n = 0, \pm1, \pm2, \ldots.$ Prove that

$$\langle Q\rangle = \frac{1}{T}\int_0^T Q(t')\,dt'. \tag{2}$$

(b) Use (2) to prove the following:

$$\left.\begin{array}{l} \langle\cos\omega t\rangle = \langle\sin\omega t\rangle = 0 \\[1ex] \langle\cos^2\omega t\rangle = \langle\sin^2\omega t\rangle = \frac{1}{2} \\[1ex] \langle\cos^2(kr - \omega t)\rangle = \langle\sin^2(kr - \omega t)\rangle = \frac{1}{2} \end{array}\right\}. \tag{3}$$

Here $\omega = 2\pi/T$, k is a constant and r is the position vector of a point on a plane wavefront.

Solution

(a) From the definition (1) it follows that

$$\langle Q \rangle = \lim_{n \to \infty} \frac{1}{nT} \int_0^{nT} Q(t)\, dt$$

$$= \lim_{n \to \infty} \frac{1}{nT} \left[\int_{(n-1)T}^{nT} Q(t)\, dt + \int_{(n-2)T}^{(n-1)T} Q(t)\, dt + \cdots + \int_0^T Q(t)\, dt \right]$$

$$= \lim_{n \to \infty} \frac{1}{nT} \left[\int_{(n-1)T}^{nT} Q(t - (n-1)T)\, dt + \int_{(n-2)T}^{(n-1)T} Q(t - (n-2)T)\, dt + \right.$$

$$\left. \cdots + \int_0^T Q(t)\, dt \right]. \quad (4)$$

Making the substitutions $t' = t - mT$ where $m = n-1,\, n-2,\, \ldots,\, 0$ in (4) gives

$$\langle Q \rangle = \lim_{n \to \infty} \frac{1}{nT} \left[\int_0^T Q(t')\, dt' + \int_0^T Q(t')\, dt' + \cdots + \int_0^T Q(t')\, dt' \right]$$

$$= \frac{1}{T} \int_0^T Q(t')\, dt',$$

as required.

(b) $\boxed{\langle \cos \omega t \rangle \text{ and } \langle \sin \omega t \rangle}$

These results are obvious by inspection since the definite integrals of $\cos \theta$ and $\sin \theta$ between 0 and 2π are zero.

$\boxed{\langle \cos^2 \omega t \rangle \text{ and } \langle \sin^2 \omega t \rangle}$

Substituting $Q(t) = \cos^2 \omega t$ in (2) and using the identity $\cos^2 \theta = \frac{1}{2}(1 + \cos 2\theta)$ yield

$$\langle Q \rangle = \frac{1}{2\pi/\omega} \int_0^{2\pi/\omega} \cos^2 \omega t\, dt = \frac{\omega}{2\pi} \int_0^{2\pi/\omega} \tfrac{1}{2}(1 + \cos 2\omega t)\, dt.$$

Inserting the limits and cancelling terms give $(3)_2$. Similarly for $\langle \sin^2 \omega t \rangle$.

$\boxed{\langle \cos^2(kr - \omega t) \rangle \text{ and } \langle \sin^2(kr - \omega t) \rangle}$

$\cos(kr - \omega t) = \cos kr \cos \omega t + \sin kr \sin \omega t$, and so

$$\langle \cos^2(kr - \omega t) \rangle = \langle (\cos kr \cos \omega t + \sin kr \sin \omega t)^2 \rangle$$

$$= \cos^2 kr \langle \cos^2 \omega t \rangle + \sin^2 kr \langle \sin^2 \omega t \rangle + \tfrac{1}{2} \sin 2kr \langle \sin 2\omega t \rangle$$

$$= \tfrac{1}{2}(\cos^2 kr + \sin^2 kr) = \tfrac{1}{2},$$

where, in the penultimate step, we use $(3)_1$ and $(3)_2$. Similarly for $\langle \sin^2(kr - \omega t) \rangle$.

Comments

(i) This question provides a formal derivation of (3), although these results are really intuitively obvious if one thinks of a graph (that of $\cos^2 \omega t$ vs t, say).

(ii) Averages like (3) are frequently encountered in electromagnetism, where the time average of a harmonically varying quantity $Q(t)$ is often of more interest than its instantaneous value (e.g. Poynting vectors and dissipated/radiated power).

Question 1.28

Suppose $\mathbf{A} = \mathbf{A}_0 e^{-i\omega t}$ and $\mathbf{B} = \mathbf{B}_0 e^{-i\omega t}$ are time-harmonic fields whose amplitudes $\mathbf{A}_0$ and $\mathbf{B}_0$ are in general complex. Show that

(a) $\langle (\mathrm{Re}\,\mathbf{A})^2 \rangle \;=\; \frac{1}{2}\mathbf{A} \cdot \mathbf{A}^*$, \hfill (1)

(b) $\langle (\mathrm{Re}\,\mathbf{A}) \cdot (\mathrm{Re}\,\mathbf{B}) \rangle \;=\; \frac{1}{2}\mathbf{A} \cdot \mathbf{B}^* \;=\; \frac{1}{2}\mathbf{A}^* \cdot \mathbf{B}$. \hfill (2)

(Here $\mathbf{A}^*$ and $\mathbf{B}^*$ are complex conjugates.)

Solution

(a) Clearly $(\mathrm{Re}\,\mathbf{A}) = \frac{1}{2}(\mathbf{A} + \mathbf{A}^*)$, and so $(\mathrm{Re}\,\mathbf{A})^2 = \frac{1}{4}(A^2 + A^{*2} + 2\mathbf{A} \cdot \mathbf{A}^*)$. Now A^2 and A^{*2} are both time-harmonic functions of frequency 2ω, whereas $\mathbf{A} \cdot \mathbf{A}^*$ is time-independent. From (3) of the previous question we obtain $\langle A^2 \rangle = \langle A^{*2} \rangle = 0$, and hence (1).

(b) Now $(\mathrm{Re}\,\mathbf{A}) \cdot (\mathrm{Re}\,\mathbf{B}) = \frac{1}{2}(\mathbf{A} + \mathbf{A}^*) \cdot \frac{1}{2}(\mathbf{B} + \mathbf{B}^*) = \frac{1}{4}(\mathbf{A} \cdot \mathbf{B} + \mathbf{A} \cdot \mathbf{B}^* + \mathbf{A}^* \cdot \mathbf{B} + \mathbf{A}^* \cdot \mathbf{B}^*)$. Performing a time average of this last result and using the same reasoning as before give

$$\langle (\mathrm{Re}\,\mathbf{A}) \cdot (\mathrm{Re}\,\mathbf{B}) \rangle \;=\; \mathbf{A} \cdot \mathbf{B}^* + \mathbf{A}^* \cdot \mathbf{B} \;=\; \mathbf{A} \cdot \mathbf{B}^* + (\mathbf{A} \cdot \mathbf{B}^*)^*$$
$$= \; 2\mathrm{Re}\,(\mathbf{A} \cdot \mathbf{B}^*) \;=\; 2\mathrm{Re}\,(\mathbf{A}^* \cdot \mathbf{B}).$$

Hence (2).

Comment

It is often convenient to represent time-dependent quantities (e.g. voltages, currents, electric and magnetic fields, etc.) in terms of a complex exponential. This is done mainly for reasons of algebra in a complicated expression, since it is usually easier to manipulate exponential, rather than trigonometric, functions. It is understood that the physically meaningful quantity will always be recovered at the end of a calculation from either the real part (usually) or the imaginary part (less usually).

Question 1.29*

The figure below shows vectors **r** and **r'** both measured relative to an arbitrary origin O.

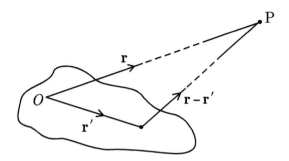

Suppose P is a distant field point and that $r'/r < 1$. Use Cartesian tensors and the binomial theorem‡ to derive the following expansions:

$$\mathbb{F} \ |\mathbf{r}-\mathbf{r'}| = r - \frac{r_i}{r}r'_i - \left[\frac{r_i r_j - r^2 \delta_{ij}}{2r^3}\right]r'_i r'_j - \left[\frac{3r_i r_j r_k - r^2(r_i \delta_{jk} + r_j \delta_{ki} + r_k \delta_{ij})}{6r^5}\right]r'_i r'_j r'_k -$$

$$\left[\frac{10r_i r_j r_k r_\ell - 2(r_i r_j \delta_{k\ell} + r_i r_k \delta_{j\ell} + r_i r_\ell \delta_{jk} + r_j r_k \delta_{i\ell} + r_j r_\ell \delta_{ik} + r_k r_\ell \delta_{ij})r^2 -}{16r^7}\right.$$

$$\left.\frac{(\delta_{ik}\delta_{j\ell} + \delta_{i\ell}\delta_{jk})r^4}{16r^7}\right]r'_i r'_j r'_k r'_\ell - \cdots , \tag{1}$$

$$\mathbb{F} \ |\mathbf{r}-\mathbf{r'}|^{-1} = \frac{1}{r} + \frac{r_i}{r^3}r'_i + \frac{3r_i r_j - r^2 \delta_{ij}}{2r^5}r'_i r'_j + \frac{5r_i r_j r_k - r^2(r_i \delta_{jk} + r_j \delta_{ki} + r_k \delta_{ij})}{2r^7}r'_i r'_j r'_k +$$

$$\left[\frac{35r_i r_j r_k r_\ell - 5(r_i r_j \delta_{k\ell} + r_i r_k \delta_{j\ell} + r_i r_\ell \delta_{jk} + r_j r_k \delta_{i\ell} + r_j r_\ell \delta_{ik} + r_k r_\ell \delta_{ij})r^2 +}{8r^9}\right.$$

$$\left.\frac{(\delta_{ij}\delta_{k\ell} + \delta_{ik}\delta_{j\ell} + \delta_{i\ell}\delta_{jk})r^4}{8r^9}\right]r'_i r'_j r'_k r'_\ell - \cdots . \tag{2}$$

Hint: Exploit, wherever possible, the arbitrary nature of repeated subscripts to express a term in its most symmetric form. So, for example, the symmetric form of $3r^2 r_i \delta_{jk} r'_i r'_j r'_k$ is $r^2(r_i \delta_{jk} + r_j \delta_{ki} + r_k \delta_{ij})r'_i r'_j r'_k$.

---

‡The binomial expansion with $x < 1$ is: $(1+x)^n = 1 + nx + \frac{1}{2!}n(n-1)x^2 + \frac{1}{3!}n(n-1)(n-2)x^3 + \cdots$.

Solution

Applying the cosine rule to the triangle of vectors in the figure on p. 43 gives

$$|\mathbf{r} - \mathbf{r}'| = \sqrt{r^2 + r'^2 - 2\mathbf{r} \cdot \mathbf{r}'} = r\sqrt{1 + \frac{r'^2 - 2\mathbf{r} \cdot \mathbf{r}'}{r^2}}.$$

We now use this result for each of the following expansions:

☞ $|\mathbf{r} - \mathbf{r}'|$

Substituting $x = \dfrac{r'^2 - 2\mathbf{r} \cdot \mathbf{r}'}{r^2}$ in $(1+x)^{1/2} = 1 + \dfrac{x}{2} - \dfrac{x^2}{8} + \dfrac{x^3}{16} - \dfrac{5x^4}{128} + \cdots$ and expanding in powers of r'/r yield:

$$|\mathbf{r} - \mathbf{r}'| = (r^2 + r'^2 - 2\mathbf{r} \cdot \mathbf{r}')^{1/2}$$

$$= r\left[1 - \frac{\mathbf{r} \cdot \mathbf{r}'}{r^2} - \frac{(\mathbf{r} \cdot \mathbf{r}')^2 - r^2 r'^2}{2r^4} - \frac{(\mathbf{r} \cdot \mathbf{r}')^3 - r^2 r'^2(\mathbf{r} \cdot \mathbf{r}')}{2r^6} - \right.$$

$$\left. \frac{5(\mathbf{r} \cdot \mathbf{r}')^4 - 6r^2 r'^2(\mathbf{r} \cdot \mathbf{r}')^2 + r^4 r'^4}{8r^8} + \cdots \right]. \tag{3}$$

Using tensor notation in (3) and remembering the hint give (1).

☞ $|\mathbf{r} - \mathbf{r}'|^{-1}$

Substituting $x = \dfrac{r'^2 - 2\mathbf{r} \cdot \mathbf{r}'}{r^2}$ in $(1 + x)^{-1/2} = 1 - \dfrac{x}{2} + \dfrac{3x^2}{8} - \dfrac{5x^3}{16} + \dfrac{35x^4}{128} - \cdots$ and expanding in powers of r'/r yield:

$$|\mathbf{r} - \mathbf{r}'|^{-1} = (r^2 + r'^2 - 2\mathbf{r} \cdot \mathbf{r}')^{-1/2}$$

$$= \frac{1}{r}\left[1 + \frac{\mathbf{r} \cdot \mathbf{r}'}{r^2} + \frac{3(\mathbf{r} \cdot \mathbf{r}')^2 - r^2 r'^2}{2r^4} + \frac{5(\mathbf{r} \cdot \mathbf{r}')^3 - 3r^2 r'^2(\mathbf{r} \cdot \mathbf{r}')}{2r^6} + \right.$$

$$\left. \frac{35(\mathbf{r} \cdot \mathbf{r}')^4 - 30r^2 r'^2(\mathbf{r} \cdot \mathbf{r}')^2 + 3r^4 r'^4}{8r^8} + \cdots \right]. \tag{4}$$

Using tensor notation in (4) and again applying the hint give (2).

Comment

The results (1) and (2) are required in multipole expansions of the electric scalar and magnetic vector potentials. See Chapters 2, 4, 8 and 11.

Question 1.30*

Consider a bounded distribution of time-dependent charge and current densities $\rho(\mathbf{r}', t)$ and $\mathbf{J}(\mathbf{r}', t)$ in vacuum[‡] which satisfy the equation

$$\nabla \cdot \mathbf{J} + \frac{\partial \rho}{\partial t} = 0. \tag{1}$$

Use (1) and Gauss's theorem to prove the following integral transforms:

$$\text{(a)} \quad \int_v J_i \, dv' = \int_v r'_i \dot{\rho} \, dv', \tag{2}$$

$$\text{(b)} \quad \int_v r'_j J_i \, dv' = -\tfrac{1}{2} \varepsilon_{ijk} \int_v (\mathbf{r}' \times \mathbf{J})_k \, dv' + \tfrac{1}{2} \int_v r'_i r'_j \dot{\rho} \, dv', \tag{3}$$

$$\text{(c)} \quad \int_v r'_j r'_k J_i \, dv' = -\tfrac{1}{3} \varepsilon_{ij\ell} \int_v (\mathbf{r}' \times \mathbf{J})_\ell r'_k \, dv' - \tfrac{1}{3} \varepsilon_{ik\ell} \int_v (\mathbf{r}' \times \mathbf{J})_\ell r'_j \, dv' + \tfrac{1}{3} \int_v r'_i r'_j r'_k \dot{\rho} \, dv'. \tag{4}$$

Hint: The identity $r'_i J_j - r'_j J_i = \varepsilon_{ijk} (\mathbf{r}' \times \mathbf{J})_k$, derived in Question 1.3, is required in the proof of (3) and (4).

Solution

(a) Clearly $\nabla' \cdot (r'_i \mathbf{J}) = \nabla_j (r'_i J_j) = (\delta_{ij} J_j + r'_i \nabla'_j J_j) = (J_i - r'_i \dot{\rho})$. So

$$\int_v \nabla' \cdot (r'_i \mathbf{J}) \, dv' = \int_v (J_i - r'_i \dot{\rho}) \, dv'.$$

Now converting the left-hand side of this equation to a surface integral using Gauss's theorem gives

$$\oint_s r'_i \mathbf{J} \cdot d\mathbf{a}' = \int_v (J_i - r'_i \dot{\rho}) \, dv'.$$

But $\mathbf{J} \cdot d\mathbf{a}' = 0$ everywhere on s and (2) follows immediately because v is arbitrary.

[‡]The term 'bounded distribution' implies that all electric currents are confined to a finite volume v, and that $\mathbf{J} \cdot d\mathbf{a} = 0$ everywhere on the boundary surface s spanning v.

(b) As before, $\nabla' \cdot (r'_i r'_j \mathbf{J}) = \nabla_k (r'_i r'_j J_k) = (\delta_{ik} r'_j J_k + \delta_{jk} r'_i J_k + r'_i r'_j \nabla_k J_k)$. Integrating over v, applying Gauss's theorem and using $\mathbf{J} \cdot d\mathbf{a}' = 0$ yield

$$\int_v (r'_j J_i + r'_i J_j - r'_i r'_j \dot{\rho}) \, dv' = 0,$$

or

$$\int_v r'_j J_i \, dv' = - \int_v r'_i J_j \, dv' + \int_v r'_i r'_j \dot{\rho} \, dv'.$$

Adding $\displaystyle\int_v r'_j J_i \, dv'$ to both sides of this last equation gives

$$2 \int_v r'_j J_i \, dv' = \int_v (r'_j J_i - r'_i J_j) \, dv' + \int_v r'_i r'_j \dot{\rho} \, dv'.$$

Using the identity $r'_j J_i - r'_i J_j = \varepsilon_{jik} (\mathbf{r}' \times \mathbf{J})_k = -\varepsilon_{ijk} (\mathbf{r}' \times \mathbf{J})_k$ (see (2) of Question 1.3) yields (3).

(c) Apart from an additional term, this proof is identical to (b).

Comments

(i) In electromagnetism, (1) is an important result known as the continuity equation for electric charge. It is discussed further in later chapters. See, for example, Question 7.1.

(ii) The identities (2)–(4), and others like them, are used to transform multipole expansions of both the static and dynamic vector potentials. We consider such applications in Chapters 4 and 8.

(iii) The integrals (2)–(4) give the moments of $\mathbf{J}$ about an arbitrary origin. It is clear from the emerging trend that one can write down, by inspection, the moment of $\mathbf{J}$ in any order. So, for example, the next member of the series is:

$$\int_v r'_j r'_k r'_\ell J_i \, dv' = - \tfrac{1}{4} \varepsilon_{ijm} \int_v (\mathbf{r}' \times \mathbf{J})_m r'_k r'_\ell \, dv' - \tfrac{1}{4} \varepsilon_{ikm} \int_v (\mathbf{r}' \times \mathbf{J})_m r'_j r'_\ell \, dv' -$$

$$\tfrac{1}{4} \varepsilon_{i\ell m} \int_v (\mathbf{r}' \times \mathbf{J})_m r'_j r'_k \, dv' + \tfrac{1}{4} \int_v r'_i r'_j r'_k r'_\ell \dot{\rho} \, dv'. \tag{5}$$

2

Static electric fields in vacuum

The first observations of an electrical nature can be traced back to the ancient Greek philosophers,[‡] but it took until the middle of the eighteenth century for the basic facts of electrostatics to be established: the presence in nature of two types of electric charge (which long ago were given the arbitrary labels 'positive' and 'negative'), the conservation of charge[♯] and the existence of conductors and insulators. During the ensuing fifty years, investigators set about the important task of determining the law of force between charges. Through a series of ingenious experiments involving torsion balances and charged spheres, Coulomb generalized the work of Priestley and others. The law of force that today bears Coulomb's name applies to both like and unlike charges. Formally, Coulomb's law can be stated as follows: suppose q_1 and q_2 represent two stationary point charges in vacuum having position vectors $\mathbf{r}_1$ and $\mathbf{r}_2$ respectively (relative to some arbitrary origin O). The force $\mathbf{F}_{12}$ exerted by q_1 on q_2 is given by

$$\mathbf{F}_{12} = \frac{1}{4\pi\epsilon_0} \frac{q_1 q_2}{|\mathbf{r}_1 - \mathbf{r}_2|^3} (\mathbf{r}_1 - \mathbf{r}_2). \tag{I}$$

Notice that (I) depends inversely on the square of the distance between q_1 and q_2, that it satisfies Newton's third law ($\mathbf{F}_{12} = -\mathbf{F}_{21}$) and that if the charges have the same (opposite) sign then the force is repulsive (attractive). This important law is central to the study of electrostatics.

After two preliminary questions, we begin this chapter with the derivation of Maxwell's electrostatic equations (in a vacuum) from Coulomb's law. The integral forms for the electric potential Φ and field $\mathbf{E}$ emerge naturally during this process. These results are then used to determine Φ and $\mathbf{E}$ for various distributions of charge where some inherent symmetry is usually present. Two important methods are used to illustrate this: (1) direct application of Gauss's law and (2) integrating a known charge density over a line, surface or volume. Problems which require computer algebra software (*Mathematica*) to facilitate their solutions are included. A series expansion of $\Phi(\mathbf{r})$ leads to the various electric multipole moments of a static charge distribution, and examples of calculating these moments are presented. Other important topics (like origin independence) are treated along the way.

[‡]For instance, it was discovered that a rubbed amber rod acquired the ability to attract a variety of very light objects like human hair, pieces of straw, etc.

[♯]Experiments which established that charge was also quantized and invariant came much later.

Solved Problems in Classical Electromagnetism. J. Pierrus, Oxford University Press (2018).
© J. Pierrus. DOI: 10.1093/oso/9780198821915.001.0001

Question 2.1

Consider a distribution of n stationary point charges $q_1, q_2, \ldots, q_n$ located in vacuum at positions $\mathbf{r}'_1, \mathbf{r}'_2, \ldots, \mathbf{r}'_n$ relative to an arbitrary origin O. Let P be a point in the field whose position vector (relative to O) is $\mathbf{r}$. Use Coulomb's law and the principle of superposition to show that the electric field[‡] $\mathbf{E}$ at P is given by:

$$\mathbf{E}(\mathbf{r}) = \frac{1}{4\pi\epsilon_0} \sum_{i=1}^{n} q_i \frac{(\mathbf{r} - \mathbf{r}'_i)}{|\mathbf{r} - \mathbf{r}'_i|^3}. \tag{1}$$

Solution

The force exerted on the positive test charge q_0 at P due to any one of these n charges (q_i, say) is given by Coulomb's law:[♯] $\mathbf{F}_{i0} = \dfrac{q_i q_0}{4\pi\epsilon_0} \dfrac{(\mathbf{r} - \mathbf{r}'_i)}{|\mathbf{r} - \mathbf{r}'_i|^3}$. The net force $\mathbf{F}$ on the test charge is the sum of these n two-body forces. Thus

$$\mathbf{F} = \mathbf{F}_{10} + \mathbf{F}_{20} + \cdots + \mathbf{F}_{n0} = \frac{1}{4\pi\epsilon_0} \sum_{i=1}^{n} q_i q_0 \frac{(\mathbf{r} - \mathbf{r}'_i)}{|\mathbf{r} - \mathbf{r}'_i|^3}. \tag{2}$$

The definition $\mathbf{E} = \mathbf{F}/q_0$ yields (1) with $\mathbf{F}$ given by (2).

Comments

(i) It is a remarkable fact that the two-body interaction between the test charge q_0 and any other charge (q_i, say) is unaffected by the presence of the remaining $(n-1)$ charges. This is the crux of the principle of linear superposition which asserts that the net force $\mathbf{F}$ on q_0 is the vector sum of these n two-body forces.

(ii) Crucially, the superposition principle applies to time-dependent electric and magnetic fields as well, and classical electromagnetism, based on Maxwell's equations, is a *linear* theory. Ref. [1] explains that

> at the macroscopic and even at the atomic level, linear superposition is remarkably valid. It is in the subatomic domain that departures from linear superposition can be legitimately sought. As charged particles approach each other very closely, electric field strengths become enormous. ... The final conclusion about linear superposition of fields *in vacuum* is that in the classical domain of sizes and attainable field strengths there is abundant evidence for the validity of linear superposition and no evidence against it. In the atomic and subatomic domain there are small quantum-mechanical nonlinear effects whose origins are in the coupling between charged particles and the electromagnetic field.

[‡]The electric field at P is defined as the force per unit stationary test charge placed at P. That is,
$$\mathbf{E}(\mathbf{r}) = \mathbf{F}(\text{on a test charge } q_0 \text{ at } \mathbf{r}) \div q_0.$$

[♯]In the presence of more than two charges, it is not obvious that Coulomb's law applies. It turns out that it does. See also Comment (i) above.

---

[1] J. D. Jackson, *Classical electrodynamics*, Chap. I, pp. 9–13. New York: Wiley, 3 edn, 1998.

(iii) Suppose n becomes so large that the charge is effectively distributed *continuously* over some region v of space. Replacing q_i in (1) by dq and converting the sum to an integral give

$$\mathbf{E(r)} = \frac{1}{4\pi\epsilon_0} \int \frac{(\mathbf{r} - \mathbf{r'})}{|\mathbf{r} - \mathbf{r'}|^3} \, dq. \tag{3}$$

(iv) Depending on the geometrical nature of the problem (e.g. three-, two- or one-dimensional), the infinitesimal charge dq in (3) may be written as $\rho\, dv'$, $\sigma\, da'$ or $\lambda\, dl'$.[†] The electric field can then be expressed in alternative forms, such as

$$\mathbf{E(r)} = k \int_v \frac{\rho(\mathbf{r'})(\mathbf{r} - \mathbf{r'})}{|\mathbf{r} - \mathbf{r'}|^3} \, dv' \ \text{ or } \ k \int_s \frac{\sigma(\mathbf{r'})(\mathbf{r} - \mathbf{r'})}{|\mathbf{r} - \mathbf{r'}|^3} \, da' \ \text{ or } \ k \int_c \frac{\lambda(\mathbf{r'})(\mathbf{r} - \mathbf{r'})}{|\mathbf{r} - \mathbf{r'}|^3} \, dl', \tag{4}$$

where $k = (4\pi\epsilon_0)^{-1}$. Equation (4) provides a means of determining $\mathbf{E}$ for a known charge distribution, assuming that the relevant integral can be evaluated.

Question 2.2

Express the electric-charge density ρ for the following charge distributions in terms of delta functions (if necessary, review Appendix E now).

(a) Charge q is distributed uniformly along the z-axis of Cartesian coordinates from $-\frac{1}{2}L$ to $\frac{1}{2}L$.

(b) Charge q is distributed uniformly over the surface of a spherical shell of radius a centred on the origin of spherical polar coordinates.

(c) Charge q is distributed uniformly over the surface of a cylinder of length L and radius a aligned along the z-axis of cylindrical polar coordinates.

(d) Charge q is distributed uniformly around the circumference of a circle of radius a centred on the origin of spherical polar coordinates.

(e) Repeat (b) for cylindrical polar coordinates.

(f) Charge q is distributed uniformly over the surface of a circular disc of radius a centred on the origin of cylindrical polar coordinates.

Solution

(a) The charge density is zero everywhere except on the z-axis between $\pm\frac{1}{2}L$. So we let

$$\rho(x', y', z') = \alpha\, \delta(x')\, \delta(y')\, H(\tfrac{1}{2}L - |z'|), \tag{1}$$

[†]Suppose dq is the charge contained in an infinitesimal volume element dv' located at $\mathbf{r'}$. The charge per unit volume or charge density is defined as $\rho(\mathbf{r'}) = dq/dv'$. Analogous definitions for the surface and line densities are $\sigma = dq/da'$ and $\lambda = dq/dl'$, where da' and dl' are elements of area and length respectively.

where α is a constant to be determined and $H(u)$ is the Heaviside function (see (VIII) of Appendix E).

Since $q = \int_v \rho \, dv'$ by definition, it follows that

$$q = \alpha \int_{-\infty}^{\infty} \int_{-\infty}^{\infty} \int_{-\infty}^{\infty} \delta(x')\,\delta(y')\,H(\tfrac{1}{2}L - |z'|)\,dx'\,dy'\,dz'.$$

Now the two integrals involving x' and y' above each have the value one, and so

$$q = \alpha \int_{-\frac{1}{2}L(1-\epsilon)}^{\frac{1}{2}L(1-\epsilon)} dz' \qquad \Rightarrow \qquad \alpha = \frac{q}{L - \epsilon},$$

where ϵ is a parameter very much less than unity. In the limit $\epsilon \to 0$ we obtain $\alpha = q/L$. Substituting this result in (1) gives

$$\rho(x', y', z') = \frac{q}{L}\,\delta(x')\,\delta(y')\,H(\tfrac{1}{2}L - |z'|). \tag{2}$$

(b) Proceeding as in (a) we let $\rho(r') = \alpha\,\delta(r' - a)$. Then

$$q = \alpha \int_0^{2\pi} \int_0^{\pi} \int_0^{\infty} \delta(r' - a)\,r'^2 \sin\theta'\,dr'\,d\theta'\,d\phi' \qquad \Rightarrow \qquad \alpha = \frac{q}{4\pi a^2},$$

and so

$$\rho(r') = \frac{q}{4\pi a^2}\,\delta(r' - a). \tag{3}$$

(c) We let $\rho(r') = \alpha\,\delta(r' - a)$ which gives

$$q = \alpha \int_0^{L} \int_0^{2\pi} \int_0^{\infty} \delta(r' - a)\,r'\,dr'\,d\theta'\,dz' \qquad \Rightarrow \qquad \alpha = \frac{q}{2\pi La},$$

and hence

$$\rho(r') = \frac{q}{2\pi La}\,\delta(r' - a). \tag{4}$$

(d) Now $\rho(r', \theta') = \alpha\,\delta(r' - a)\,\delta(\theta' - \tfrac{1}{2}\pi)$, and so

$$q = \alpha \int_0^{2\pi} \int_0^{\pi} \int_0^{\infty} \delta(r' - a)\,\delta(\theta' - \tfrac{1}{2}\pi)\,r'^2 \sin\theta'\,dr'\,d\theta'\,d\phi'$$

$$= 2\pi\alpha \int_{-1}^{1} \int_0^{\infty} \delta(r' - a)\,\delta(\cos\theta')\,r'^2\,dr'\,d(\cos\theta') \qquad \Rightarrow \qquad \alpha = \frac{q}{2\pi a^2}.$$

Therefore $\rho(r', \theta') = \dfrac{q}{2\pi a^2}\,\delta(r' - a)\,\delta(\theta' - \tfrac{1}{2}\pi)$, or

$$\rho(r', \theta') = \frac{q}{2\pi a^2} \delta(r' - a) \delta(\cos \theta') . \tag{5}$$

(e) As before, $\rho(r', z') = \alpha \delta(r' - a) \delta(z')$ which gives

$$q = \alpha \int_{-\infty}^{\infty} \int_{0}^{2\pi} \int_{0}^{\infty} \delta(r' - a) \delta(z') r' dr' d\theta' dz' \qquad \Rightarrow \qquad \alpha = \frac{q}{2\pi a} ,$$

and hence

$$\rho(\mathbf{r}') = \frac{q}{2\pi a} \delta(r' - a) \delta(z') . \tag{6}$$

(f) Now $\rho(r', z') = \alpha H(a - r') \delta(z')$, and so

$$q = \alpha \int_{-\infty}^{\infty} \int_{0}^{2\pi} \int_{0}^{\infty} H(a - r') \delta(z') r' dr' d\theta' dz'$$

$$= \alpha \int_{-\infty}^{\infty} \int_{0}^{2\pi} \int_{0}^{a} \delta(z') r' dr' d\theta' dz' \qquad \Rightarrow \qquad \alpha = \frac{q}{\pi a^2} .$$

Hence

$$\rho(r', z') = \frac{q}{\pi a^2} H(a - r') \delta(z') . \tag{7}$$

Question 2.3*

Use (3) of Question 2.1 to derive the equations

$$\mathbf{\nabla} \cdot \mathbf{E} = \frac{\rho(\mathbf{r})}{\epsilon_0} \qquad \text{and} \qquad \mathbf{\nabla} \times \mathbf{E} = 0, \tag{1}$$

which apply at a point in vacuum.

Solution

Substituting $\mathbf{\nabla}\left(\dfrac{1}{|\mathbf{r} - \mathbf{r}'|}\right) = -\dfrac{\mathbf{r} - \mathbf{r}'}{|\mathbf{r} - \mathbf{r}'|^3}$ (see (1) of Question 1.6) in (4) of Question 1.1 gives

$$\mathbf{E}(\mathbf{r}) = -\frac{1}{4\pi\epsilon_0} \int_{v} \rho(\mathbf{r}') \mathbf{\nabla}\left(\frac{1}{|\mathbf{r} - \mathbf{r}'|}\right) dv'$$

$$= -\mathbf{\nabla}\left[\frac{1}{4\pi\epsilon_0} \int_{v} \frac{\rho(\mathbf{r}')}{|\mathbf{r} - \mathbf{r}'|} dv'\right]. \tag{2}$$

In the last step we use the fact that the operator $\mathbf{\nabla}$ differentiates the field (unprimed) coordinates only, and so $\mathbf{\nabla}\rho(\mathbf{r}') = 0$. Equation (2) can be written as

$$\mathbf{E}(\mathbf{r}) = -\nabla\Phi(\mathbf{r}), \tag{3}$$

where

$$\Phi(\mathbf{r}) = \frac{1}{4\pi\epsilon_0} \int_v \frac{\rho(\mathbf{r}')}{|\mathbf{r}-\mathbf{r}'|} \, dv'. \tag{4}$$

In (4), $\Phi(\mathbf{r})$ is the electrostatic potential at the field point $\mathbf{r}$ and here it is determined up to an arbitrary additive constant.

☞ Taking the divergence of (2) yields

$$\nabla \cdot \mathbf{E} = -\nabla^2 \left[\frac{1}{4\pi\epsilon_0} \int_v \frac{\rho(\mathbf{r}')}{|\mathbf{r}-\mathbf{r}'|} \, dv' \right]$$

$$= -\frac{1}{4\pi\epsilon_0} \int_v \rho(\mathbf{r}') \nabla^2 \left(\frac{1}{|\mathbf{r}-\mathbf{r}'|} \right) dv' \quad \text{(since } \nabla^2 \rho(\mathbf{r}') = 0)$$

$$= \frac{1}{4\pi\epsilon_0} \int_v 4\pi\rho(\mathbf{r}') \, \delta(\mathbf{r}-\mathbf{r}') \, dv' \quad \text{(using (4) of Question 1.21)}$$

$$= \frac{\rho(\mathbf{r})}{\epsilon_0} \quad \text{(using (XI)}_2 \text{ of Appendix E),}$$

as required.

☞ Equation $(1)_2$ follows immediately from (3) because the curl of a gradient is identically zero $\big($see (10) of Question 1.8$\big)$.

Comments

(i) It follows from (3) that $\mathbf{E}\cdot d\mathbf{l} = -\nabla\Phi\cdot d\mathbf{l} = -\left(\dfrac{\partial\Phi}{\partial x} dx + \dfrac{\partial\Phi}{\partial y} dy + \dfrac{\partial\Phi}{\partial z} dz \right) = -d\Phi,$

or $\displaystyle\int d\Phi = -\int \mathbf{E}\cdot d\mathbf{l}.$ Now if a and b represent two arbitrary points in the field, then

$$\Phi(b) - \Phi(a) = -\int_a^b \mathbf{E}\cdot d\mathbf{l}. \tag{5}$$

The difference in potential $\Phi(b) - \Phi(a)$ is the potential of b relative to a and we write it as Φ_{ab}.

(ii) Equation (1) reveals that electrostatic fields are not, in general, solenoidal but they are always conservative (see also Questions 1.14 and 1.15).

(iii) ☞ Integrating $(1)_1$ over an arbitrary volume gives $\displaystyle\int_v \nabla\cdot\mathbf{E}\,dv' = \frac{1}{\epsilon_0}\int_v \rho(\mathbf{r}')\,dv'.$ Because of Gauss's theorem this becomes

$$\oint_s \mathbf{E} \cdot d\mathbf{a}' = \frac{1}{\epsilon_0} \int_v \rho(\mathbf{r}') \, dv' = \frac{1}{\epsilon_0} \times q_{net}, \tag{6}$$

where $q_{net} = \int_v \rho(\mathbf{r}') \, dv'$ is the net charge enclosed by s. Known as Gauss's law, (6) is a fundamental result.[‡]

☞ It is evident from (5) that around any closed loop

$$\oint_c \mathbf{E} \cdot d\mathbf{l} = 0. \tag{7}$$

(iv) Sometimes (4) is required in the alternative forms

$$\Phi(\mathbf{r}) = \frac{1}{4\pi\epsilon_0} \int \frac{\sigma \, da'}{|\mathbf{r} - \mathbf{r}'|} \quad \text{or} \quad \Phi(\mathbf{r}) = \frac{1}{4\pi\epsilon_0} \int \frac{\lambda \, dl'}{|\mathbf{r} - \mathbf{r}'|}, \tag{8}$$

where σ and λ are surface- and line-charge densities respectively.

(v) The differential equations (1) apply at a point in vacuum, whereas the integral equations (5)–(8) apply over a finite region of space.

Question 2.4

A charge q is distributed uniformly over the surface of a spherical shell of radius a having negligible thickness and centred at the origin O. Calculate the electric field at an arbitrary point P in space using (a) Gauss's law and (b) direct integration.

Solution

(a) Clearly, $\mathbf{E}$ is a spherically symmetric[♯] field. We therefore choose a spherical Gaussian surface G of radius r centred on O and passing through P. The electric flux through G is

$$\oint_s \mathbf{E} \cdot d\mathbf{a} = \oint_s E(r)\hat{\mathbf{r}} \cdot d\mathbf{a} \, \hat{\mathbf{r}} = \oint_s E(r) \, da = 4\pi r^2 E, \tag{1}$$

[‡]Stated in words: the outward electric flux ψ through any closed surface s lying in vacuum equals $\epsilon_0^{-1} \times$ (the net charge enclosed by s).

[♯]Meaning that $\mathbf{E}$ has the following properties. It is:
1. central (i.e. the field is directed towards or away from the origin).
2. dependent, in magnitude, only on the distance from the origin. That is,

$$\mathbf{E}(\mathbf{r}) = E(r)\hat{\mathbf{r}}.$$

where in the last step we use the formula for the surface area of a sphere. Two cases are of interest:

$r \geq a$

Here $q_{net} = q$ and Gauss's law and (1) give $4\pi r^2 E = q/\epsilon_0$, or

$$\mathbf{E}(\mathbf{r}) = \frac{1}{4\pi\epsilon_0}\frac{q}{r^2}\hat{\mathbf{r}}. \tag{2}$$

$r \leq a$

Here $q_{net} = 0$, and so

$$\mathbf{E}(\mathbf{r}) = 0. \tag{3}$$

(b) Two alternative solutions are provided:

Method 1

We begin by calculating the electric potential and then obtain $\mathbf{E}$ by differentiation. From $(8)_1$ of Question 2.3, $\Phi(\mathbf{r}) = \dfrac{1}{4\pi\epsilon_0}\displaystyle\int\dfrac{\sigma da'}{r'}$ where here r' is the distance to P from an infinitesimal band of charge (see the figure) and $\sigma da = \sigma(2\pi a \sin\theta)(a d\theta)$. Thus,

$$\Phi(\mathbf{r}) = \frac{2\pi a^2 \sigma}{4\pi\epsilon_0}\int_0^\pi \frac{\sin\theta}{r'}d\theta. \tag{4}$$

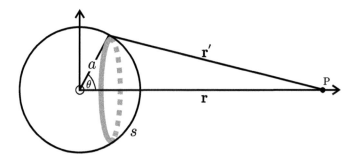

Now by the cosine rule, $r'^2 = r^2 + a^2 - 2ar\cos\theta$. So $2r'dr' = 2ar\sin\theta\,d\theta$ and

$$\int_0^\pi \frac{\sin\theta}{r'}d\theta = \frac{1}{ar}\int_{|r-a|}^{r+a}dr', \tag{5}$$

where the lower limit is either $r - a$ if P lies outside the sphere or $a - r$ if P is inside the sphere. Substituting (5) in (4) gives

$$\Phi(\mathbf{r}) = \frac{a\sigma}{2\epsilon_0 r}\int_{|r-a|}^{r+a}dr' = \frac{1}{4\pi\epsilon_0}\frac{q}{2ar}\int_{|r-a|}^{r+a}dr', \tag{6}$$

because $\sigma = q/4\pi a^2$. As before, we consider the two cases separately:

$r \geq a$

For P outside the sphere, (6) becomes

$$\Phi(\mathbf{r}) = \frac{1}{4\pi\epsilon_0} \frac{q}{2ar} \int_{r-a}^{r+a} dr' = \frac{1}{4\pi\epsilon_0} \frac{q}{r}, \tag{7}$$

and so $\mathbf{E} = -\nabla\Phi = -\frac{\partial\Phi}{\partial r}\hat{\mathbf{r}} = \frac{1}{4\pi\epsilon_0}\frac{q}{r^2}\hat{\mathbf{r}}$, which is (2).

$r \leq a$

For P inside the sphere, (6) becomes

$$\Phi(\mathbf{r}) = \frac{1}{4\pi\epsilon_0} \frac{q}{2ar} \int_{a-r}^{r+a} dr' = \frac{1}{4\pi\epsilon_0} \frac{q}{a}, \tag{8}$$

and so $\mathbf{E} = -\nabla\Phi = 0$ which is (3).

Method 2

Because of the spherical symmetry, we lose no generality by choosing coordinates with P lying on the z-axis. Substituting the charge density $\rho(\mathbf{r}') = \frac{q}{4\pi a^2}\delta(r'-a)$ (see (3) of Question 2.2) in (4) of Question 2.3 gives

$$\Phi_{\mathrm{P}} = \frac{1}{4\pi\epsilon_0} \frac{q}{4\pi a^2} \int_v \frac{\delta(r'-a)\,dv'}{|\mathbf{r}-\mathbf{r}'|},$$

where in spherical coordinates $\mathbf{r} = (z,0,0)$ and $\mathbf{r}' = (r',\theta',\phi')$. Now the volume element $dv' = r'^2 dr' d(\cos\theta')\,d\phi'$ and the distance $|\mathbf{r}-\mathbf{r}'| = \sqrt{z^2 + r'^2 - 2r'z\cos\theta'}$ are given by (VII) and (VIII)$_3$ of Appendix C respectively. Hence

$$\Phi_{\mathrm{P}} = \frac{1}{4\pi\epsilon_0} \frac{q}{4\pi a^2} \int_0^{2\pi} \int_{-1}^{1} \int_0^{\infty} \frac{\delta(r'-a)\,r'^2 dr' d(\cos\theta')\,d\phi'}{\sqrt{z^2 + r'^2 - 2r'z\cos\theta'}}.$$

Now the ϕ'-integration is 2π. Then, because of the sifting property of a delta function, $\displaystyle\int_0^{\infty} \frac{\delta(r'-a)\,r'^2 dr'}{\sqrt{z^2 + r'^2 - 2r'z\cos\theta'}} = \frac{a^2}{\sqrt{z^2 + a^2 - 2az\cos\theta'}}$, and so

$$\Phi_{\mathrm{P}} = \frac{1}{4\pi\epsilon_0} \frac{q}{2} \int_{-1}^{1} \frac{d(\cos\theta')}{\sqrt{z^2 + a^2 - 2az\cos\theta'}}. \tag{9}$$

Making the change of variable $u^2 = z^2 + a^2 - 2az\cos\theta'$ in (9) yields

$$\Phi_{\mathrm{P}} = \frac{1}{4\pi\epsilon_0} \frac{q}{2az} \int_{|z-a|}^{(z+a)} du, \tag{10}$$

which is the same equation as (6). The two cases $z \geq a$ and $z \leq a$ follow as for Method 1, leading to (7) and (8).

Comments

(i) Because of the very high symmetry associated with a sphere, the method involving Gauss's law in (a) provides the simplest solution to the problem. Unfortunately, this approach cannot always be used.[‡] In these cases, the method of direct integration outlined in (b) may be the only possible means of finding an analytical solution. See, for example, Question 2.9.

(ii) The above results find an important practical application for an isolated charged *conducting* sphere (either solid or a thin-walled shell).[♯] Because of this, we make the following summary:

☞ Equations (2) and (7) show that for $r \geq a$ the shell behaves like a point charge located at its centre ($r = 0$).

☞ Equations (3) and (8) show that for $r < a$ the electric field is zero and the potential is constant throughout. The value of this constant equals $\Phi(r = a)$.

The electrostatics of charged conductors (including spheres) is considered further in Chapter 3.

Question 2.5

Consider a uniformly charged circular disc of radius a with negligible thickness carrying a total charge q. Suppose the disc lies in the xy-plane of Cartesian coordinates and is centred on the origin.

(a) Show that the electric potential at an arbitrary point P on the symmetry axis of the disc is given by

$$\Phi(z) = \frac{\sigma}{2\epsilon_0}\left(\sqrt{a^2 + z^2} \mp z\right), \tag{1}$$

where the upper (lower) sign is for $z > 0$ ($z < 0$) and $\sigma = q/\pi a^2$.

(b) Show that the electric potential at an arbitrary point P on the circumference of the disc is given by

$$\Phi_P = \frac{\sigma a}{\pi \epsilon_0}. \tag{2}$$

Hint: We adopt the approach of Ref. [2]. Consider a wedge of charge (shown shaded in the figure on p. 58) with its apex at P. Write down the contribution of the wedge $d\Phi_P$ to the potential, then integrate to obtain Φ_P.

[‡] Of course Gauss's law still holds for problems with low or no symmetry. All we are saying is that the law cannot be deployed *usefully* to determine **E**.

[♯] For a spherical *conductor*, symmetry ensures that the mobile-charge carriers distribute themselves over the surface in such a way that the charge density σ is uniform.

[2]   E. M. Purcell and D. J. Morin, *Electricity and magnetism*, Chap. 2, p. 70. New York: Cambridge University Press, 3 edn, 2013.

Solution

(a) Two alternative solutions are provided:

Method 1

Consider two concentric circles having radii r and $r + dr$ (here $r \le a$) lying in the plane of the disc. The charge dq between these circles is located a distance r' from P where $dq = \sigma(2\pi r)(dr)$, and $r' = \sqrt{r^2 + z^2}$ by Pythagoras. So $2r' dr' = 2r \, dr$, and

$$\Phi_P = \frac{1}{4\pi\epsilon_0} \int \frac{dq}{r'} = \frac{2\pi\sigma}{4\pi\epsilon_0} \int_0^a \frac{r \, dr}{r'} = \frac{\sigma}{2\epsilon_0} \int_{|z|}^{\sqrt{a^2 + z^2}} dr',$$

which is (1) since $|z| = z$ for $z > 0$ and $|z| = -z$ for $z < 0$.

Method 2

The charge density given by (7) of Question 2.2 is $\rho(r', z') = \sigma \, H(a - r') \, \delta(z')$, where the Heaviside function H is defined in (VIII) of Appendix E. Now

$$\Phi_P = \frac{\sigma}{4\pi\epsilon_0} \int_v \frac{H(a - r') \delta(z') \, dv'}{|\mathbf{r} - \mathbf{r'}|},$$

where in these cylindrical coordinates the volume element $dv' = r' dr' d\theta' dz'$ and $|\mathbf{r} - \mathbf{r'}| = \sqrt{z^2 + r'^2}$. So

$$\Phi_P = \frac{\sigma}{4\pi\epsilon_0} \int_{-\infty}^{\infty} \int_0^{2\pi} \int_0^a \frac{\delta(z') \, r' dr' d\theta' dz'}{\sqrt{r'^2 + z^2}} = \frac{2\pi\sigma}{4\pi\epsilon_0} \int_0^a \frac{r' dr'}{\sqrt{r'^2 + z^2}},$$

since $\int_{-\infty}^{\infty} \delta(z') \, dz' = 1$. Making the change of variable $u^2 = r'^2 + z^2$ yields $2u \, du = 2r' dr'$, and then

$$\Phi_P = \frac{\sigma}{2\epsilon_0} \int_{|z|}^{\sqrt{a^2 + z^2}} du,$$

which is (1), as before, since $|z| = z$ for $z > 0$ and $|z| = -z$ for $z < 0$.

(b) Consider a point-like element of charge dq lying inside the shaded wedge at a distance r' from P (see the figure on p. 58). Clearly, $dq = \sigma(r' d\theta)(dr')$, and so $dq/r' = \sigma \, dr' d\theta$. Thus the contribution of the entire wedge to the potential at P is

$$d\Phi_P = \frac{1}{4\pi\epsilon_0} \int \frac{dq}{r'} = \frac{\sigma \, d\theta}{4\pi\epsilon_0} \int_0^R dr' = \frac{\sigma R d\theta}{4\pi\epsilon_0},$$

where $R = 2a \cos \theta$. Hence

$$\Phi_P = \frac{\sigma a}{2\pi\epsilon_0} \int_{-\frac{\pi}{2}}^{\frac{\pi}{2}} \cos \theta \, d\theta = \frac{\sigma a}{2\pi\epsilon_0} \left[\sin \theta \right]_{-\frac{\pi}{2}}^{\frac{\pi}{2}} = \frac{\sigma a}{\pi\epsilon_0},$$

as required.

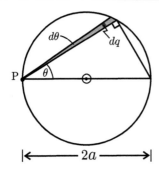

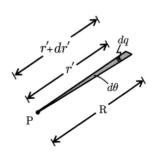

Comments

(i) The electric potential Φ_c at the centre of the disc follows from (1) with $z = 0$ and is $\Phi_c = \dfrac{\sigma a}{2\epsilon_0}$. Clearly, Φ_c is greater than Φ_P by a factor of $\frac{1}{2}\pi$: evidently the disc is not an equipotential surface.

(ii) Along the (positive) z-axis the electric field is given by

$$\mathbf{E} = -\frac{\partial \Phi}{\partial z}\,\hat{\mathbf{z}} = \frac{\sigma}{2\epsilon_0}\left[1 - \frac{z}{\sqrt{a^2 + z^2}}\right]\hat{\mathbf{z}}. \tag{3}$$

Two important results follow from (3):

☞ For an infinite sheet of surface charge $(a \to \infty)$ the magnitude of $\mathbf{E}$ is independent of z and is

$$E = \frac{\sigma}{2\epsilon_0}, \tag{4}$$

which is a well-known elementary result that can readily be confirmed using Gauss's law.

☞ As $z \to 0$ for a finite disc, the electric field is also given by (4). Thus any distribution of surface charge behaves like an infinite sheet for points immediately above (or below) the surface.

(iii) The electric field at a point on the circumference of the disc cannot be determined from (2), since (unlike the z-dependence in (1)) we have no knowledge of the spatial variation of Φ in the immediate vicinity of P.[‡]

(iv) This question and Question 2.4(b) both illustrate a standard technique for determining the electrostatic potential at an arbitrary point P on the symmetry axis of a continuous charge distribution. It is based on the principle of decomposing the distribution into a large number of infinitesimal elements, each of which may be regarded as a 'point' source dq whose field $d\mathbf{E}$ and potential $d\Phi$ at P are given by

---

[‡]In order to calculate the electric field at P from the *derivative* of Φ, we would need to know how the potential changes *spatially* in the immediate *neighbourhood* of P.

$\dfrac{1}{4\pi\epsilon_0}\dfrac{dq}{r^2}\hat{\mathbf{r}}$ and $\dfrac{1}{4\pi\epsilon_0}\dfrac{dq}{r}$ respectively. The net field or potential at P then follows immediately using linear superposition, and the result is an integral over the distribution. This integral may be evaluated either in closed form or numerically. Whether it is better to begin with an **E**-integration or a Φ-integration depends on the problem at hand. There are no fixed rules in this regard. Two obvious advantages for starting with Φ are:

☞ The superposition of elements involves scalar rather than vector addition (i.e. no components are involved).

☞ Differentiation (instead of integration#) then yields **E**.

Notwithstanding the above remarks, keep in mind that a single vector integration may involve less work than a scalar integration followed by a differentiation when the solution to some problem requires only the electric field (i.e. Φ is not needed).

(v) Calculating **E** and Φ at points off the symmetry axis is invariably a far more difficult problem, mainly because the distance from dq to P is not the same for all the differential elements of the charge distribution. In these cases, evaluating the resulting integrals analytically (when this is possible) is usually non-trivial. An example illustrating this point is provided in Question 2.9.

Question 2.6

Consider a spherical charge distribution of radius R that is centred at an origin O. The charge density $\rho(r)$ is defined by

$$\rho(r) = \begin{cases} \rho_0 & \text{for } r \le R \\ 0 & \text{for } r > R, \end{cases} \tag{1}$$

where ρ_0 is a constant.

(a) Use Gauss's law to show that the electric field $\mathbf{E}(r)$ is given by

$$\mathbf{E}(r) = \begin{cases} \dfrac{\rho_0}{3\epsilon_0}\,\mathbf{r} & \text{for } r \le R \\[2mm] \dfrac{\rho_0}{3\epsilon_0}\left(\dfrac{R}{r}\right)^3\mathbf{r} & \text{for } r \ge R. \end{cases} \tag{2}$$

(b) Hence show that the electric potential $\Phi(r)$ is given by

$$\Phi(r) = \begin{cases} \dfrac{\rho_0}{6\epsilon_0}(3R^2 - r^2) & \text{for } r \le R \\[2mm] \dfrac{\rho_0}{3\epsilon_0}\dfrac{R^3}{r} & \text{for } r \ge R. \end{cases} \tag{3}$$

#Differentiation is usually easier to perform than the corresponding integration.

(c) Use $(3)_1$ to show that the average value of Φ inside the charge distribution is

$$\Phi_{\text{av}} = \frac{2\rho_0}{5\epsilon_0} R^2. \tag{4}$$

(d) On the same set of axes, plot graphs of $\Phi(r)/\Phi(0)$ and $E(r)/E(0)$ for $0 \leq r/R \leq 4$.

Solution

(a) The charge density is spherically symmetric.[‡] So are the effects which it produces, like the electric potential and electric field. Because of this symmetry, we choose a spherical Gaussian surface G of radius r centred on O. Everywhere on G the field has a constant magnitude and (for ρ_0 positive) is parallel to the outward normal. Applying (6) of Question 2.3 gives

$$E \times 4\pi r^2 = \frac{1}{\epsilon_0} \times q_{\text{net}} = \begin{cases} \dfrac{\rho_0}{\epsilon_0} \times \frac{4}{3}\pi r^3 & \text{for } r \leq R \\[2mm] \dfrac{\rho_0}{\epsilon_0} \times \frac{4}{3}\pi R^3 & \text{for } r \geq R. \end{cases}$$

But this is (2), because $\mathbf{E}(r) = E(r)\,\hat{\mathbf{r}}$.

(b) Using the result $\Phi_{\text{ab}} = \Phi(b) - \Phi(a) = -\displaystyle\int_a^b \mathbf{E}\cdot d\mathbf{r}$ and putting $\Phi(\infty) = 0$ yield

$$\Phi(r) = -\int_\infty^r \mathbf{E}\cdot d\mathbf{r}. \tag{5}$$

We consider each region separately.

$\boxed{r \leq R}$

Substituting $(2)_1$ into (5) gives:

$$\begin{aligned}
\Phi(r) &= -\int_\infty^R \mathbf{E}\cdot d\mathbf{r} - \int_R^r \mathbf{E}\cdot d\mathbf{r} \\[2mm]
&= -\frac{\rho_0 R^3}{3\epsilon_0}\int_\infty^R \frac{dr}{r^2} - \frac{\rho_0}{3\epsilon_0}\int_R^r r\,dr \\[2mm]
&= \frac{\rho_0 R^3}{3\epsilon_0}\frac{1}{r}\bigg|_\infty^R - \frac{\rho_0}{6\epsilon_0}r^2\bigg|_R^r \\[2mm]
&= \frac{\rho_0 R^3}{3\epsilon_0}\left(\frac{1}{R} - \frac{1}{\infty}\right) - \frac{\rho_0}{6\epsilon_0}(r^2 - R^2),
\end{aligned}$$

which is $(3)_1$.

[‡] That is, $\rho(\mathbf{r}) = \rho(r)$.

$\boxed{r \geq R}$

Similarly, it follows from $(2)_2$ and (5) that

$$\Phi(r) = -\int_{\infty}^{r} \mathbf{E} \cdot d\mathbf{r} = -\frac{\rho_0 R^3}{3\epsilon_0} \int_{\infty}^{r} \frac{dr}{r^2} = \frac{\rho_0}{3\epsilon_0} \frac{R^3}{r},$$

(6)

which is $(3)_2$.

(c) Substituting $(3)_1$ in $\Phi_{av} = \frac{1}{\frac{4}{3}\pi R^3} \int_v \Phi \, dv$ and using $dv = 4\pi r^2 dr$ give

$$\Phi_{av} = \frac{\rho_0}{2\epsilon_0 R^3} \int_0^R (3R^2 - r^2) r^2 \, dr,$$

which is (4).

(d) We obtain the graphs:

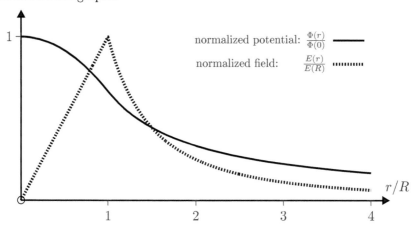

Question 2.7

A non-conducting sphere of radius R centred at O contains a spherical cavity of radius R' centred at O'. Let $\mathbf{d}$ be the displacement of O' relative to O. Throughout the sphere there is a uniform charge density ρ_0 (except inside the cavity, which is uncharged). Assume that the permittivity of the sphere material has the vacuum value ϵ_0.

(a) Use the principle of superposition and the results of Question 2.6 to write down an expression for $\mathbf{E}(\mathbf{r})$ everywhere.

(b) Repeat (a) for the electric potential $\Phi(\mathbf{r})$.

(c) Let O be the origin of Cartesian coordinates. Put $R = 7d$, $R' = 4d$ and $\mathbf{d} = d\hat{\mathbf{x}}$. Introduce the dimensionless coordinates $X = x/d$, $Y = y/d$ and plot graphs of:

☞ $\Phi(X, 0, 0)$ and $E(X, 0, 0)$ for $-3 \leq X \leq 3$;

☞ $\Phi(0, Y, 0)$ and $E(0, Y, 0)$ for $-3 \leq Y \leq 3$.

(d) Take R, R' and $\mathbf{d}$ as given in (c) and let $\rho_0 d^2/6\epsilon_0 = 1\,\mathrm{V}$. Use *Mathematica* to plot the following lines of constant potential in the xy-plane containing O and O': $50\,\mathrm{V}$, $60\,\mathrm{V}$, $80\,\mathrm{V}$, $92\,\mathrm{V}$, $96\,\mathrm{V}$, $100\,\mathrm{V}$, $104\,\mathrm{V}$.

Solution

(a) Imagine the cavity is produced by excising a sphere of radius R' from a larger solid sphere of radius R. This is illustrated below as a 'picture equation', where all three spheres have the same charge density.

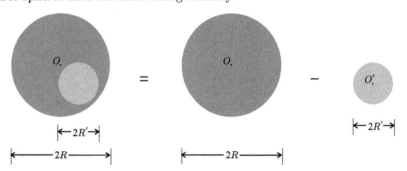

The electric field $\mathbf{E_P}$ (here P is an arbitrary point in space) follows immediately from the principle of superposition:

$$\begin{array}{lll} \mathbf{E_P} \text{ due to a sphere with} & \mathbf{E_P} \text{ due to a solid} & \mathbf{E_P} \text{ due to a smaller} \\ \text{a spherical cavity of} = & \text{sphere of radius} \quad - & \text{solid sphere of radius} \\ \text{radius } R' \text{ centred on } O' & R \text{ centred on } O & R' \text{ centred on } O' \end{array} \Bigg\}, \quad (1)$$

where both terms on the right-hand side of (1) are given by (2) of Question 2.6.[‡] We consider the following regions:

$r \leq R$ with P outside the cavity

$$\mathbf{E(r)} = \frac{\rho_0}{3\epsilon_0}\left[\mathbf{r} - \frac{R'^3}{r'^3}\mathbf{r'}\right] = \frac{\rho_0}{3\epsilon_0}\left[\mathbf{r} - \frac{R'^3}{(r^2 + d^2 - 2\mathbf{r}\cdot\mathbf{d})^{3/2}}(\mathbf{r} - \mathbf{d})\right]. \quad (2)$$

$r \leq R$ with P inside the cavity

$$\mathbf{E(r)} = \frac{\rho_0}{3\epsilon_0}(\mathbf{r} - \mathbf{r'}) = \frac{\rho_0}{3\epsilon_0}\mathbf{d}. \quad (3)$$

$r \geq R$

$$\mathbf{E(r)} = \frac{\rho_0}{3\epsilon_0}\left[\frac{R^3}{r^3}\mathbf{r} - \frac{R'^3}{r'^3}\mathbf{r'}\right]$$

$$= \frac{\rho_0}{3\epsilon_0}\left[\frac{R^3}{r^3}\mathbf{r} - \frac{R'^3}{(r^2 + d^2 - 2\mathbf{r}\cdot\mathbf{d})^{3/2}}(\mathbf{r} - \mathbf{d})\right]. \quad (4)$$

[‡]It is convenient to introduce the following notation. Let $\overline{OP} = \mathbf{r}$ and $\overline{O'P} = \mathbf{r'}$. Then, since $\overline{OO'} = \mathbf{d}$, we have $\mathbf{r'} = \mathbf{r} - \mathbf{d}$ and therefore $r' = \sqrt{r^2 + d^2 - 2\mathbf{r}\cdot\mathbf{d}}$.

(b) Proceeding as in (a):

$$\left. \begin{array}{lll} \Phi_{\rm P} \text{ due to a sphere with} & \Phi_{\rm P} \text{ due to a solid} & \Phi_{\rm P} \text{ due to a smaller} \\ \text{a spherical cavity of} = & \text{sphere of radius} & - \text{solid sphere of radius} \\ \text{radius } R' \text{ centred on } O' & R \text{ centred on } O & R' \text{ centred on } O' \end{array} \right\}.$$

Now (3) of Question 2.6 gives

$$\Phi(\mathbf{r}) = \frac{\rho_0}{6\epsilon_0} \times \begin{cases} \left[(3R^2 - r^2) - \dfrac{2R'^3}{r'} \right] & \text{for } r \le R \text{ with P outside the cavity} \\[2ex] \left[3(R^2 - R'^2) - (r^2 - r'^2) \right] & \text{for } r \le R \text{ with P inside the cavity} \\[2ex] 2\left[\dfrac{R^3}{r} - \dfrac{R'^3}{r'} \right] & \text{for } r \ge R, \end{cases}$$

where $r' = \sqrt{r^2 + d^2 - 2\mathbf{r}\cdot\mathbf{d}}$ as before.

(c) Plotting normalized[#] graphs of the field and potential yields

profile along the x-axis with $y = z = 0$ profile along the y-axis with $x = z = 0$

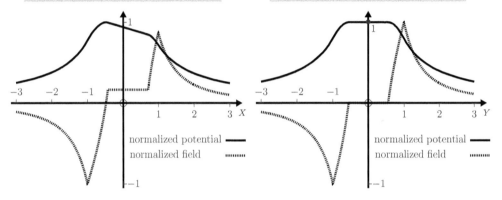

normalized potential ▬
normalized field ▪▪▪▪▪▪

normalized potential ▬
normalized field ▪▪▪▪▪▪

(d) The following equipotentials were calculated using the notebook on p. 64:

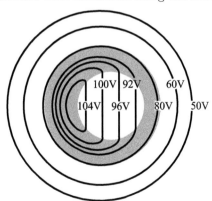

100V 92V   60V

104V 96V   80V   50V

[#] Here normalized means $\Phi/\Phi_{\rm max}$ and $E/|E_{\rm min}|$.

```
In[1]:=  R1 = 7 d;   R2 = 4 d;   d = 1;
```

$$r1[x\_, y\_] := \sqrt{x^2 + y^2} \; ; \quad r2[x\_, y\_] := \sqrt{r1[x, y]^2 + d^2 - 2 \times d} \; ;$$

$$con1[x\_, y\_] := x^2 + y^2 \leq R1^2 \;\&\&\; (x - d)^2 + y^2 \geq R2^2$$

$$con2[x\_, y\_] := x^2 + y^2 \leq R1^2 \;\&\&\; (x - d)^2 + y^2 \leq R2^2$$

$$con3[x\_, y\_] := x^2 + y^2 \geq R1^2$$

```
In[2]:=
```
$$\Phi1[x\_, y\_] := 3\, R1^2 - r1[x, y]^2 - \frac{2\, R2^3}{r2[x, y]}; \qquad \Phi2[x\_, y\_] := 3\,(R1^2 - R2^2) + d^2 - 2 \times d;$$

$$\Phi3[x\_, y\_] := 2 \left(\frac{R1^3}{r1[x, y]} - \frac{R2^3}{r2[x, y]} \right);$$

$$\Phi[x\_, y\_] := \text{Piecewise}[\{\{\Phi1[x, y], con1[x, y]\},$$
$$\{\Phi2[x, y], con2[x, y]\}, \{\Phi3[x, y], con3[x, y]\}\}]$$

$$data = \text{Flatten}\big[\text{Table}[\{x, y, \Phi[x, y]\}, \{x, -12, 12, 0.1\}, \{y, -12, 12, 0.1\}], 1\big];$$

$$\text{ContourList} = \{50, 60, 80, 92, 96, 100, 104\};$$

```
In[3]:=
```
$$gr1 = \text{ListContourPlot}\big[data, \text{ContourShading} \rightarrow \text{None}, \text{Contours} \rightarrow \text{ContourList},$$
$$\text{ContourStyle} \rightarrow \text{Thickness}[.01], \text{PlotRange} \rightarrow \{\{-12, 12\}, \{-12, 12\}\}\big];$$

```
In[4]:=
```
$$gr2 = \text{Graphics}\big[\{\text{Gray}, \text{Disk}[\{0, 0\}, R1]\}\big];$$
$$gr3 = \text{Graphics}\big[\{\text{White}, \text{Disk}[\{d, 0\}, R2]\}\big];$$
$$\text{Show}[gr2, gr3, gr1]$$

```
In[5]:=  Φ0 = 76.6;
```

$$\text{ContourPlot3D}\Bigg[\Bigg\{ \frac{2 \times 7^3}{\sqrt{X^2 + Y^2 + Z^2}} \left(1 - \frac{(4 / 7)^3}{\sqrt{1 + (X^2 + Y^2 + Z^2)^{-1} (1 - 2 X)}} \right) == \Phi0 \Bigg\},$$

$$\{X, -10, 10\}, \{Y, -10, 10\}, \{Z, -10, 10\}\Bigg]$$

Comments

(i) The electric field inside the cavity follows from (3): it has a constant magnitude $\rho_0 d / 3\epsilon_0$ and the same direction as the displacement **d**. The equipotential surfaces are planes perpendicular to this displacement.

(ii) Unlike the solid sphere of Question 2.6, the field of a sphere with a cavity is not spherically symmetric $\big(\text{see } (2)–(4)\big)$. Furthermore, the surface of such a sphere is not an equipotential as is evident in the figure for (c) shown on p. 63.[‡] Interestingly, however, the equipotentials for $r \gtrsim R$ are essentially spherical. Evidently, the effect of the cavity is rapidly smoothed out beyond the surface of the charge distribution.[♯]

[‡]It is interesting to contrast this with objects which conduct electricity (even those containing cavities). The presence of mobile-charge carriers ensures that the entire conductor, including the surface, is an equipotential. See Question 3.1.

[♯]This is easily observed using *Mathematica*'s ContourPlot3D command. See cell 4 in the notebook above. The equipotential surface is drawn for $\Phi_0 = 76.6\,\text{V}$.

(iii) Since Gauss's law also applies to Newton's law of gravitation, we expect the gravitational acceleration **g** inside a planet of constant mass density to increase linearly with distance from the centre. It does.[3] By analogy with this electrostatic example, **g** would be constant within any large spherical cavity that might be present inside a uniformly dense planet.

Question 2.8

(a) The electric field **E** of an infinite uniform line charge λ is an elementary result:†

$$\mathbf{E}(\mathbf{r}) = \frac{1}{2\pi\epsilon_0}\frac{\lambda}{r}\,\hat{\mathbf{r}}, \tag{1}$$

where r is the perpendicular distance from the line charge to an arbitrary field point P. Suppose that the zero of electric potential is chosen at $r = r_0$. Show that

$$\Phi(\mathbf{r}) = \frac{\lambda}{2\pi\epsilon_0}\ln\left[\frac{r_0}{r}\right]. \tag{2}$$

(b) Consider two equal and opposite line charges $\pm\lambda$ lying parallel to the z-axis of Cartesian coordinates and separated by a distance $2a$ as shown. Use (2) to show that

$$\Phi(\mathbf{r}) = \frac{\lambda}{2\pi\epsilon_0}\ln\left[\frac{r_1}{r_2}\right]. \tag{3}$$

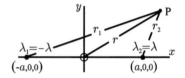

(c) Suppose that the line charges in (b) are now both positive. Prove that

$$\Phi(\mathbf{r}) = \frac{\lambda}{2\pi\epsilon_0}\ln\left[\frac{r_0^2}{r_1 r_2}\right]. \tag{4}$$

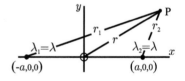

(d) Show that the electric field corresponding to Φ given by (3) and (4) is

$$
\left.
\begin{aligned}
\mathbf{E}(x,y) &= \frac{\lambda}{2\pi\epsilon_0}\frac{2a(x^2 - y^2 - a^2)\,\hat{\mathbf{x}} + 4axy\,\hat{\mathbf{y}}}{a^4 + 2a^2(-x^2 + y^2) + (x^2 + y^2)^2} && (-\lambda_1 = \lambda_2 = \lambda)\\[1em]
\mathbf{E}(x,y) &= \frac{\lambda}{2\pi\epsilon_0}\frac{2x(x^2 + y^2 - a^2)\,\hat{\mathbf{x}} + 2y(x^2 + y^2 + a^2)\,\hat{\mathbf{y}}}{a^4 + 2a^2(-x^2 + y^2) + (x^2 + y^2)^2} && (\lambda_1 = \lambda_2 = \lambda)
\end{aligned}
\right\}. \tag{5}
$$

---

†The result follows immediately by applying Gauss's law to a cylindrical Gaussian surface G of length ℓ and radius r centred on the line charge. Because of the radial symmetry of the field, the outward flux through G is $2\pi r\ell E$ which equals $\lambda\ell/\epsilon_0$. Hence (1).

---

[3]   O. L. de Lange and J. Pierrus, *Solved problems in classical mechanics: Analytical and numerical solutions with comments*, Chap. 11, pp. 367–72. Oxford: Oxford University Press, 2010.

Solution

(a) From (5) of Question 2.3 we have $\Phi(b) - \Phi(a) = -\int_a^b \mathbf{E} \cdot d\mathbf{l}$. So

$$\Phi(r) - \Phi(r_0) = -\int_{r_0}^r \mathbf{E} \cdot d\mathbf{r} = -\frac{\lambda}{2\pi\epsilon_0} \int_{r_0}^r \frac{dr}{r} = -\frac{\lambda}{2\pi\epsilon_0} \ln\left[\frac{r}{r_0}\right],$$

which is (2) since $\Phi(r_0) = 0$.

(b) Because of (2) and the principle of superposition, we have $\Phi(r) = \Phi_{\lambda_1} + \Phi_{\lambda_2}$. Thus

$$\Phi(r) = \frac{(-\lambda)}{2\pi\epsilon_0} \ln\left[\frac{r_0}{r_1}\right] + \frac{\lambda}{2\pi\epsilon_0} \ln\left[\frac{r_0}{r_2}\right] = \frac{\lambda}{2\pi\epsilon_0} \ln\left[\frac{r_1}{r_2}\right],$$

as required.

(c) As before,

$$\Phi(r) = \Phi_{\lambda_1} + \Phi_{\lambda_2} = \frac{\lambda}{2\pi\epsilon_0} \ln\left[\frac{r_0}{r_1}\right] + \frac{\lambda}{2\pi\epsilon_0} \ln\left[\frac{r_0}{r_2}\right],$$

which is (4).

(d) Clearly $r_1 = \sqrt{(x+a)^2 + y^2}$ and $r_2 = \sqrt{(x-a)^2 + y^2}$. These results together with $\mathbf{E} = -\nabla\Phi$ lead, after some algebra, to (5).

Comments

(i) The figures below show the field and some of the equipotentials for the line charges of (b) and (c) above. See Question 2.9 for an example of a *Mathematica* notebook that will produce plots like these.

(ii) The left-hand figure below suggests that for two opposite line charges, the equipotentials and field lines are circular cylinders. This is true and is proved in Question 2.16.

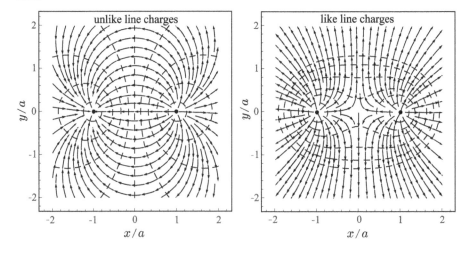

Question 2.9

Positive charge q is distributed uniformly along the z-axis of Cartesian coordinates between $z = \pm\frac{1}{2}d$.

(a) Show that the electric potential at an arbitrary field point P is given by

$$\Phi(x, y, z) = \frac{1}{4\pi\epsilon_0} \frac{q}{d} \ln\left[\frac{d + 2z + \sqrt{4(x^2 + y^2) + (d + 2z)^2}}{-d + 2z + \sqrt{4(x^2 + y^2) + (d - 2z)^2}}\right]. \tag{1}$$

(b) Hence derive the electric field for the following special cases:

$$\begin{aligned} \mathbf{E}(r) &= \frac{q}{4\pi\epsilon_0} \frac{\hat{\mathbf{r}}}{r\sqrt{r^2 + (d/2)^2}} && \text{along a perpendicular bisector} \\[2mm] \mathbf{E}(z) &= \pm\frac{q}{4\pi\epsilon_0} \frac{\hat{\mathbf{z}}}{z^2 - (d/2)^2} && \text{along the } z\text{-axis for } |z| > \frac{1}{2}d \end{aligned} \left.\vphantom{\begin{aligned}A\\B\\C\end{aligned}}\right\}, \tag{2}$$

where $r = \sqrt{x^2 + y^2}$ in $(2)_1$ is a cylindrical polar coordinate and the positive (negative) sign in $(2)_2$ is for $z > 0$ ($z < 0$).

(c) Use (1) to prove that the equipotential surfaces are a family of confocal prolate ellipsoids.[‡]

Hint: Use *Mathematica* to do the integration for (a) and the algebra for (b), (c).

Solution

(a) In this problem, the symmetry of the field is such that Gauss's law cannot be used advantageously. We proceed by direct integration:

$$\Phi(x, y, z) = \frac{1}{4\pi\epsilon_0} \int \frac{dq}{r'} = \frac{\lambda}{4\pi\epsilon_0} \int_{-\frac{d}{2}}^{\frac{d}{2}} \frac{dz'}{r'},$$

where the (uniform) line-charge density $\lambda = q/d$ and $r' = \sqrt{x^2 + y^2 + (z - z')^2}$. Thus

$$\Phi(x, y, z) = \frac{1}{4\pi\epsilon_0} \frac{q}{d} \int_{-\frac{d}{2}}^{\frac{d}{2}} \frac{dz'}{\sqrt{x^2 + y^2 + (z - z')^2}}. \tag{3}$$

[‡]An ellipsoid may be regarded as a three-dimensional analogue of an ellipse. The equation of an ellipsoid, in a Cartesian system of coordinates centred on the origin and aligned with the axes, is

$$\left(\frac{x}{a}\right)^2 + \left(\frac{y}{b}\right)^2 + \left(\frac{z}{c}\right)^2 = 1, \tag{I}$$

where the points $(a, 0, 0)$, $(0, b, 0)$ and $(0, 0, c)$ lie on the surface. Clearly, $a = b = c$ gives the degenerate case of a sphere. The case $a = b < c$ is known as a prolate ellipsoid (prolate spheroid) and the case $a = b > c$ is known as an oblate ellipsoid (oblate spheroid).

Integrating (3) using *Mathematica*'s `Integrate` function (see `cell 1` in the notebook on p. 69) yields (1).

(b) Using $\mathbf{E} = -\nabla\Phi$ and with Φ given by (1), we obtain the desired result (see `cell 2` in the notebook on p. 70).

(c) Consider the equipotential surface corresponding to $\Phi = \Phi_0$. Clearly, $e^{4\pi\epsilon_0\Phi_0 d/q}$ is a positive constant (α, say) and $\alpha > 1$. It follows from (1) that

$$\alpha = \frac{d + 2z + \sqrt{4(x^2 + y^2) + (d + 2z)^2}}{-d + 2z + \sqrt{4(x^2 + y^2) + (d - 2z)^2}}.$$

Rearranging this equation gives

$$d(\alpha + 1) - 2z(\alpha - 1) = \sqrt{4(x^2 + y^2) + (d + 2z)^2} - \alpha\sqrt{4(x^2 + y^2) + (d - 2z)^2}.$$

Squaring twice (to remove the radical) and factorizing[#] yield

$$16(x^2 + y^2)\left[(\alpha^2 - 1)^2 x^2 + (\alpha^2 - 1)^2 y^2 + 4\alpha(\alpha - 1)^2 z^2 - \alpha(\alpha + 1)^2 d^2\right] = 0,$$

which will be satisfied provided the term in square brackets is zero. Thus

$$\frac{(\alpha - 1)^2}{d^2\alpha} x^2 + \frac{(\alpha - 1)^2}{d^2\alpha} y^2 + \frac{4(\alpha - 1)^2}{d^2(\alpha + 1)^2} z^2 = 1. \tag{4}$$

Comparing (4) with (I) in the footnote on p. 67 shows that $a = b = \dfrac{d\sqrt{\alpha}}{(\alpha - 1)}$ and $c = \dfrac{d\,(\alpha + 1)}{2\,(\alpha - 1)}$. Now $\dfrac{a}{c} = \dfrac{2\sqrt{\alpha}}{\alpha + 1} < 1$ since $\alpha > 1$ and the ellipsoid is thus a prolate spheroid. The intersection of this spheroid with the yz-plane passing through the origin is an ellipse with major axis $2c$, minor axis $2a$ and eccentricity $e = \sqrt{1 - c^2/a^2}$. It therefore follows that $e = (\alpha - 1)/(\alpha + 1)$ and the foci of this ellipse are located at $ec = \pm\frac{1}{2}d$, which coincide with the end points of the line charge. The family of prolate spheroids are therefore confocal.

Comments

(i) The figure on p. 69 shows the electric field (in the yz-plane at $x = 0$) and some of the equipotentials for a uniformly charged rod (see `cell 4` in the notebook on p. 70).

(ii) Along a perpendicular bisector of the line charge, (1) reduces with $z = 0$ to

$$\Phi(r) = \frac{1}{4\pi\epsilon_0}\frac{q}{d}\ln\left[\frac{1 + \sqrt{1 + u^2}}{-1 + \sqrt{1 + u^2}}\right], \tag{5}$$

where $u = 2r/d = 2\sqrt{x^2 + y^2}/d$. The two limiting cases of (5) are:

[#]Recall the hint for (b) and use *Mathematica* (see `cell 3` in the notebook on p. 70).

$u \gg 1$

$\Phi(r) \simeq \dfrac{1}{4\pi\epsilon_0} \dfrac{q}{d} \ln\left[\dfrac{1+u}{-1+u}\right] = \dfrac{1}{4\pi\epsilon_0} \dfrac{q}{d} \ln\left[\dfrac{1+1/u}{1-1/u}\right] \simeq \dfrac{1}{4\pi\epsilon_0} \dfrac{2q}{d} \ln(1+1/u)$. Since

$u < 1$, we use the expansion $\ln(1+u) = u - \frac{1}{2}u^2 + \cdots$ to obtain $\Phi(r) \simeq \dfrac{1}{4\pi\epsilon_0} \dfrac{q}{r}$,

which is the expected result (i.e. in this limit, the line behaves like a point charge).

$u \ll 1$

$\Phi(r) \simeq \dfrac{1}{4\pi\epsilon_0} \dfrac{q}{d} \ln\left[\dfrac{2+\frac{1}{2}u^2}{\frac{1}{2}u^2}\right] = \dfrac{1}{4\pi\epsilon_0} \dfrac{q}{d} \ln\left[1 + \left(\dfrac{2}{u}\right)^2\right] \simeq \dfrac{1}{4\pi\epsilon_0} \dfrac{2q}{d} \ln\dfrac{2}{u} = \dfrac{\lambda}{2\pi\epsilon_0} \ln\dfrac{d}{r}$,

which again is the expected result (i.e. in this limit, the finite line behaves like an infinite line charge).

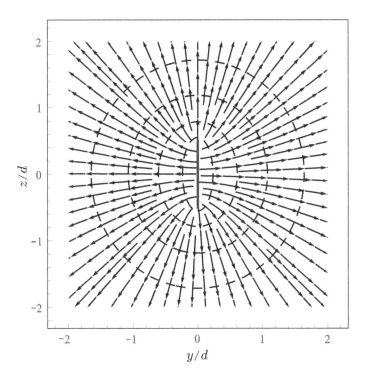

```
In[1]:= f[x_, y_, z_] := 1 / Sqrt[x^2 + y^2 + (z - zp)^2]

assump = {x > 0 && y > 0 && d > 0 && (z > d/2 || z < -d/2)};

Integrate[f[x, y, z], {zp, -d/2, d/2}, Assumptions -> assump]
```

```
In[2]:= Φ[r_, z_] := Log[ (d + 2 z + √(d² + 4 r² + 4 z (z + d)))/(-d + 2 z + √(d² + 4 r² + 4 z (z - d))) ];

     FullSimplify[-D[Φ[r, 0], r]]
     FullSimplify[-D[Φ[0, z], z]]

In[3]:= Simplify[Expand[(d² (α + 1)² + 4 z² (α - 1)² - 4 d z (α² - 1) -
          α² (4 x² + 4 y² + d² + 4 z² - 4 d z) - (4 x² + 4 y² + d² + 4 z² + 4 d z))² -
          4 α² (4 x² + 4 y² + d² + 4 z² - 4 d z) (4 x² + 4 y² + d² + 4 z² + 4 d z)]]

In[4]:= (* We use dimensionless coordinates X=x/d, Y=y/d & Z=z/d *)

     Φ[Y_, Z_] := Log[ (1 + 2 Z + √(1 + 4 Y² + 4 Z (Z + 1)))/(-1 + 2 Z + √(1 + 4 Y² + 4 Z (Z - 1))) ];

     EY[Y_, Z_] = FullSimplify[-D[Φ[Y, Z], Y]];
     EZ[Y_, Z_] = FullSimplify[-D[Φ[Y, Z], Z]];

     gr1 = StreamPlot[{{EY[Y, Z], EZ[Y, Z]}}, {Y, -2, 2}, {Z, -2, 2},
          StreamStyle → {Directive[Thickness[0.004], Black]}];

     gr2 = ContourPlot[{Φ[Y, Z] == 3, Φ[Y, Z] == 1.5, Φ[Y, Z] == 0.9,
          Φ[Y, Z] == .6}, {Y, -2.0, 2.0}, {Z, -2.0, 2.0},
          ContourStyle → {Directive[Thickness[0.004], Black,
          AbsoluteDashing[{10, 10}]]}, PlotPoints → 100];

     gr3 = Graphics[{Thickness[0.008], Line[{{0, 0.5}, {0, -0.5}}]}];

     Show[gr1, gr2, gr3]
```

Question 2.10

Positive charge q is distributed uniformly around the circumference of a circle (having radius r_0), lying in the xy-plane of Cartesian coordinates and centred on the origin.

(a) Show that the electric potential at an arbitrary field point $P(r, \theta, \phi)$ is given by

$$\Phi(r, \theta, \phi) = \frac{1}{4\pi\epsilon_0} \frac{q}{r_0} \frac{2\,K\left[\dfrac{-4(r/r_0)\sin\theta}{1 + r^2/r_0^2 - 2(r/r_0)\sin\theta}\right]}{\pi\sqrt{1 + r^2/r_0^2 - 2(r/r_0)\sin\theta}}, \tag{1}$$

where the function K is a complete elliptic integral of the first kind (in calculations we will use the *Mathematica*-defined EllipticK function).

Hint: Begin by expressing $d\Phi$ in terms of spherical polar coordinates, then use *Mathematica* to perform the integration.

(b) Write a *Mathematica* notebook to calculate the electric field in a plane passing through the origin and perpendicular to the circular line charge (the yz-plane, say). Taking $\dfrac{1}{4\pi\epsilon_0}\dfrac{q}{r_0} = 1\,\mathrm{V}$, plot this field together with some lines of constant potential.

Solution

(a) Consider an infinitesimal charge element dq on the circumference of the circle.

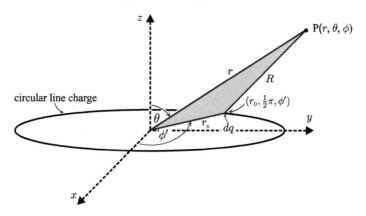

Clearly, $dq = \lambda\,d\ell$ where $\lambda = q/2\pi r_0$ and $d\ell = r_0 d\phi'$. This element makes a contribution to the electric potential at P given by

$$d\Phi \;=\; \frac{1}{4\pi\epsilon_0}\frac{dq}{R} \;=\; \frac{1}{4\pi\epsilon_0}\frac{q}{2\pi}\frac{d\phi'}{R}\,, \tag{2}$$

where the distance from dq to P is given by $R = \sqrt{r_0^2 + r^2 - 2rr_0\sin\theta\cos(\phi - \phi')}$ (see (VII) of Appendix C with $\theta' = \frac{1}{2}\pi$). Substituting R in (2) yields

$$\Phi \;=\; \frac{1}{4\pi\epsilon_0}\frac{q}{2\pi}\int_0^{2\pi}\frac{d\phi'}{\sqrt{r_0^2 + r^2 - 2rr_0\sin\theta\cos(\phi - \phi')}}\,. \tag{3}$$

Evaluating (3) using *Mathematica*'s **Integrate** function (see **cell 1** in the notebook on p. 72) gives (1).

(b) In order to satisfy the requirements of *Mathematica*'s plotting routines, we begin by expressing (1) in Cartesian form. Let $X = x/r_0$, $Y = y/r_0$ and $Z = z/r_0$ be dimensionless coordinates. Then $r^2/r_0^2 = X^2 + Y^2 + Z^2$; $r\sin\theta = \sqrt{x^2 + y^2}$; and $(r/r_0)\sin\theta = \sqrt{X^2 + Y^2}$. Hence

$$\Phi(X,Y,Z) \;=\; \frac{1}{4\pi\epsilon_0}\frac{q}{r_0}\,\frac{2\,\mathrm{K}\!\left[\dfrac{-4\sqrt{X^2+Y^2}}{1+X^2+Y^2+Z^2-2\sqrt{X^2+Y^2}}\right]}{\pi\sqrt{1+X^2+Y^2+Z^2-2\sqrt{X^2+Y^2}}}\,. \tag{4}$$

The field components are calculated from $\mathbf{E} = -\nabla\Phi$, with Φ given by (4). **Cell 2** in the notebook on p. 73 generates the following plot:

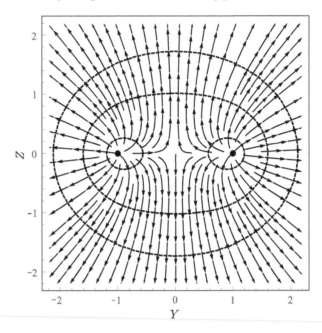

Comments

(i) We see from the plot that the equipotentials near the line charge are toroids which have an approximately circular cross-section. This approximation becomes increasingly more accurate the closer one approaches the line charge. We make use of this observation in Question 3.15.

(ii) A special case of this problem, often treated in many first-year physics textbooks, occurs when P lies on the z-axis. Now the integration of $d\Phi$ is trivial because of the high symmetry, and

$$\Phi(z) = \frac{1}{4\pi\epsilon_0}\frac{q}{\sqrt{r_0^2 + z^2}} \qquad \text{and} \qquad \mathbf{E}(z) = \frac{1}{4\pi\epsilon_0}\frac{qz}{(r_0^2 + z^2)^{3/2}}\hat{\mathbf{z}}. \qquad (5)$$

It is clear from $(5)_2$ that $E(z)$ is zero at the origin, increases in magnitude as $|z|$ increases, reaches a maximum and then decreases to zero. This behaviour is illustrated in the graph on p. 73. Notice the linear variation of the field in the region $z/r_0 \ll 1$.

```
In[1]:= (* We introduce the dimensionless variable u = z/r₀ *)
     list = {u ≥ 0 && 0 ≤ φ ≤ 2 π && 0 ≤ θ ≤ π && (u ≠ 1 || 2 θ ≠ π) && 1 + u² > 2 u Sin[θ]};
      1
     ── Integrate[ ──────────────────────────────────── , {φp, 0, 2 π}, Assumptions → list]
     2 π            √(1 + u² - 2 u Sin[θ] Cos[φp - φ])
```

```
In[2]:= Style1 = {Directive[Thickness[0.004], Black]};
        Style2 = {Directive[Thickness[0.006], Black, AbsoluteDashing[{4, 4}]]};
```

$$\Phi[X\_, Y\_, Z\_] := \frac{2\,\text{EllipticK}\left[-\dfrac{4\sqrt{X^2+Y^2}}{1+X^2+Y^2+Z^2-2\sqrt{X^2+Y^2}}\right]}{\pi\sqrt{1+X^2+Y^2+Z^2-2\sqrt{X^2+Y^2}}};$$

```
        EY[X_, Y_, Z_] = -D[Φ[X, Y, Z], Y];          EZ[X_, Y_, Z_] = -D[Φ[X, Y, Z], Z];
        gr1 = StreamPlot[{EY[0, Y, Z], EZ[0, Y, Z]}, {Y, -2.2, 2.2}, {Z, -2.2, 2.2},
            StreamStyle → Style1];
        gr2 = ContourPlot[{Φ[0, Y, Z] == 1.1, Φ[0, Y, Z] == 0.7, Φ[0, Y, Z] == 0.5},
            {Y, -2.2, 2.2}, {Z, -2.2, 2.2}, ContourStyle → Style2, PlotPoints → 10];
        gr3 = Graphics[{PointSize[0.02], Point[{-1, 0}], Point[{1, 0}]}];
        Show[gr1, gr2, gr3]
```

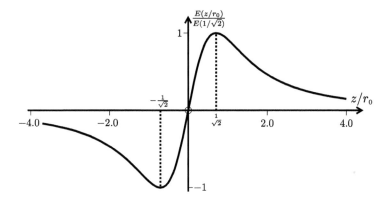

Question 2.11

Two point charges q and $-q$ are a distance $2d$ apart. Cartesian axes are chosen such that the coordinates of these charges are $(0, 0, d)$ and $(0, 0, -d)$, as shown in the figure alongside. Suppose P is an arbitrary point in space far from the origin ($r \gg d$).

(a) Show that the electric potential at P is given by

$$\Phi(\mathbf{r}) = \frac{1}{4\pi\epsilon_0}\frac{\mathbf{p}\cdot\hat{\mathbf{r}}}{r^2}, \tag{1}$$

where $\mathbf{p} = 2qd\hat{\mathbf{z}}$.

(b) Use (1) to derive the field

$$\mathbf{E}(\mathbf{r}) = \frac{1}{4\pi\epsilon_0}\left[\frac{3(\mathbf{p}\cdot\hat{\mathbf{r}})\hat{\mathbf{r}} - \mathbf{p}}{r^3}\right]. \tag{2}$$

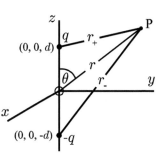

Solution

(a) We begin with the potential of a point charge and the principle of superposition

$$\Phi(\mathbf{r}) = \Phi_+ + \Phi_- = \frac{q}{4\pi\epsilon_0}\left[\frac{1}{r_+} - \frac{1}{r_-}\right],$$

where, by the cosine rule, $r_\pm = \sqrt{r^2 + d^2 \mp 2rd\cos\theta}$. Expanding $r_\pm$ to first order in d/r gives $r_\pm = r(1 \mp \frac{d}{r}\cos\theta)$, since $(1 \mp x)^n \simeq 1 \mp nx$ for $x \ll 1$. Thus

$$\Phi(\mathbf{r}) = \frac{q}{4\pi\epsilon_0}\left[\frac{1}{1 - (d/r)\cos\theta} - \frac{1}{1 + (d/r)\cos\theta}\right] = \frac{1}{4\pi\epsilon_0}\frac{q}{r}\frac{2(d/r)\cos\theta}{(1 - (d/r)^2\cos^2\theta)}.$$

Now the term involving $\cos^2\theta$ in the denominator is $\ll 1$ because $r \gg d$ and

$$\Phi(\mathbf{r}) = \frac{1}{4\pi\epsilon_0}\frac{2qd\cos\theta}{r^2}.$$

This is (1), since $\mathbf{p}\cdot\hat{\mathbf{r}} = p\cos\theta$ and $p = 2qd$ by definition.

(b) From $\mathbf{E} = -\nabla\Phi$ we have $E_i = -\nabla_i\Phi$, and so

$$E_i = -\frac{1}{4\pi\epsilon_0}\nabla_i\left(\frac{p_j r_j}{r^3}\right) = -\frac{1}{4\pi\epsilon_0}p_j\nabla_i(r_j r^{-3}) = \frac{1}{4\pi\epsilon_0}\frac{3(p_j r_j)r_i - r^2 p_i}{r^5}, \qquad (3)$$

because of (7) of Question 1.1. Then (2) follows immediately, since (3) is true for $i = x$, y and z.

Comments

(i) The pair of charges shown in the figure is known as an electric dipole[‡] and the vector $\mathbf{p}$ is the electric dipole moment. Its magnitude is given by $q \times$ the distance between the charges. The direction of $\mathbf{p}$ is from the negative to the positive charge.

(ii) Equations (1) and (2) give the dipole potential and field respectively. Because of the approximation made in deriving Φ, these equations are valid only at points for which $r \gg d$. Now suppose we let $d \to 0$ and $q \to \infty$ in such a way that $\mathbf{p}$ remains finite and constant. Then this physical dipole becomes a 'point' electric dipole[♯] located at the origin. Equations (1) and (2) give the exact potential and field for any $r > 0$.

(iii) Strictly speaking, the dipole field (2) is incomplete because it omits a term that is zero everywhere except at the origin itself. Suppose that in (3) we use (5) of Question 1.21 (instead of (7) of Question 1.1) to differentiate $r_j r^{-3}$. Then

---

[‡]We sometimes refer to this arrangement of charges as a *physical dipole* (as opposed to a *point dipole*). See also Comments (iii) and (iv) above. An example of a physical dipole is the HCl molecule.

[♯]A point dipole arises from the limit of a physical dipole and should be regarded strictly as a mathematical entity.

$$E_i = -\frac{1}{4\pi\epsilon_0}p_j\nabla_i(r_jr^{-3}) = \frac{1}{4\pi\epsilon_0}p_j\nabla_i\nabla_jr^{-1} = \frac{1}{4\pi\epsilon_0}\frac{3p_jr_jr_i - r^2p_i}{r^5} - \frac{1}{3\epsilon_0}p_i\delta(\mathbf{r}),$$

whose vector form is

$$\mathbf{E}(\mathbf{r}) = \frac{1}{4\pi\epsilon_0}\left\{\frac{3(\mathbf{p}\cdot\hat{\mathbf{r}})\hat{\mathbf{r}} - \mathbf{p}}{r^3} - \frac{4\pi}{3}\mathbf{p}\,\delta(\mathbf{r})\right\}. \tag{4}$$

Omitting the delta-function term in (4) usually has no undesirable consequences. However, this is not always the case, as Ref. [4]. explains. See also Question 2.13.

(iv) It is sometimes useful to express (1) and (2) in terms of spherical polar coordinates explicitly. Suppose Cartesian axes are chosen such that $\mathbf{p} = p\hat{\mathbf{z}}$. Then it is easily shown that

$$\Phi(r,\theta) = \frac{1}{4\pi\epsilon_0}\frac{p\cos\theta}{r^2} \quad \text{and} \quad \mathbf{E}(r,\theta) = \frac{1}{4\pi\epsilon_0}\frac{p(2\hat{\mathbf{r}}\cos\theta + \hat{\boldsymbol{\theta}}\sin\theta)}{r^3}. \tag{5}$$

(v) Electric dipoles are an important member of a class of entities called electric multipoles. See Question 2.20 for further discussion of this topic.

Question 2.12*

(a) Consider an arbitrary distribution of charges having density $\rho(\mathbf{r}')$. Let v be a spherical region of space of radius r_0 centred on an origin O. The average electric field $\overline{\mathbf{E}}$ inside v is given by

$$\overline{\mathbf{E}} = \frac{1}{\frac{4}{3}\pi r_0^3}\int_v \mathbf{E}(\mathbf{r})\,dv = \overline{\mathbf{E}}_{\text{int}} + \overline{\mathbf{E}}_{\text{ext}}, \tag{1}$$

where $\overline{\mathbf{E}}_{\text{int}}$ is the average field due to all charges inside v and $\overline{\mathbf{E}}_{\text{ext}}$ is the average field due to all charges outside v. Prove that

$$\left.\begin{aligned}\overline{\mathbf{E}}_{\text{int}} &= -\frac{1}{4\pi\epsilon_0}\frac{\mathbf{p}}{r_0^3} \\[2mm] \overline{\mathbf{E}}_{\text{ext}} &= -\frac{1}{4\pi\epsilon_0}\int_V \frac{\rho(\mathbf{r}')\,\mathbf{r}'}{r'^3}\,dv'\end{aligned}\right\}, \tag{2}$$

where $\mathbf{p}$ is the electric dipole moment (about O) of the internal charges and V is the region of space excluding v where $\rho(\mathbf{r}') \neq 0$.

(b) Two point charges $\pm q$ have the Cartesian coordinates $(0, 0, \pm d)$ respectively. Use (2) to determine the average electric field $\overline{\mathbf{E}}$ inside the spherical region referred to in (a), assuming that ☞ $d < r_0$, and ☞ $d > r_0$.

---

[4] D. J. Griffiths, 'Hyperfine splitting in the ground state of hydrogen', *American Journal of Physics*, vol. 50, pp. 698–703, 1982.

Solution

(a) Substituting $\mathbf{E} = -\nabla\Phi$ in (1) and using (1) of Question 1.22 give

$$\overline{\mathbf{E}} \;=\; \frac{1}{\frac{4}{3}\pi r_0^3}\int_v \mathbf{E}(\mathbf{r})\,dv \;=\; -\frac{1}{\frac{4}{3}\pi r_0^3}\int_v \nabla\Phi(\mathbf{r})\,dv \;=\; -\frac{1}{\frac{4}{3}\pi r_0^3}\oint_s \Phi(\mathbf{r})\,d\mathbf{a}. \tag{3}$$

With $\Phi(\mathbf{r}) \;=\; \dfrac{1}{4\pi\epsilon_0}\displaystyle\int_v \frac{\rho(\mathbf{r}')}{|\mathbf{r}-\mathbf{r}'|}\,dv'$, we obtain from (3):

$$\overline{\mathbf{E}} \;=\; -\frac{1}{\frac{4}{3}\pi r_0^3}\frac{1}{4\pi\epsilon_0}\int_v \rho(\mathbf{r}')\left(\oint_s \frac{d\mathbf{a}}{|\mathbf{r}-\mathbf{r}'|}\right)dv', \tag{4}$$

because the order of integration may be interchanged. But the surface integral in (4) is given by (8) of Question 1.26, and so

$$\left.\begin{aligned}
\overline{\mathbf{E}}_{\text{int}} &= -\frac{1}{4\pi\epsilon_0}\frac{1}{r_0^3}\int_v \rho(\mathbf{r}')\,\mathbf{r}'\,dv' \\[2mm]
\overline{\mathbf{E}}_{\text{ext}} &= -\frac{1}{4\pi\epsilon_0}\int_V \frac{\rho(\mathbf{r}')\,\mathbf{r}'}{r'^3}\,dv'
\end{aligned}\right\}.$$

Now $\mathbf{p} = \displaystyle\int_v \rho(\mathbf{r}')\,\mathbf{r}'dv'$ by definition $\big($see $(6)_2$ of Question 2.20$\big)$. Hence (1).

(b) ☞ $d < r_0$

Since the charges are inside the sphere, $\overline{\mathbf{E}}$ is given by $(2)_1$ with $\mathbf{p} = 2qd\hat{\mathbf{z}}$ and

$$\overline{\mathbf{E}}_{\text{int}} \;=\; -\frac{1}{4\pi\epsilon_0}\frac{2qd}{r_0^3}\,\hat{\mathbf{z}}.$$

☞ $d > r_0$

Substituting $\rho(\mathbf{r}') = q\delta(x')\delta(y')[\delta(z'-d)-\delta(z'+d)]$ in $(2)_2$ gives

$$\overline{\mathbf{E}}_{\text{ext}} = -\frac{q}{4\pi\epsilon_0}\iiint \frac{\delta(x')\delta(y')[\delta(z'-d)-\delta(z'+d)]\,\mathbf{r}'\,dx'dy'dz'}{r'^3} = -\frac{1}{4\pi\epsilon_0}\frac{2q}{d^2}\,\hat{\mathbf{z}}.$$

$\big($The triple integral above follows immediately from $(\text{XI})_2$ of Appendix E.$\big)$

Comments

(i) Equation $(2)_2$ is just Coulomb's law $\big($see $(4)_1$ of Question 2.1 with $\mathbf{r} = 0\big)$. For the dipole of (b), we see that $\overline{\mathbf{E}}_{\text{ext}}$ has the following simple interpretation: the average electric field inside the sphere due to the external charges equals the field which they produce at O.

(ii) A similar analysis may also be carried out for the field of a point magnetic dipole (see Question 4.20).

Question 2.13*

(This question and its solution are based on §II and §III of Ref. [4].)

Consider a point electric dipole located at the origin O of Cartesian coordinates with the z-axis chosen along the direction of $\mathbf{p}$.

(a) Express the field of this dipole in the form

$$\mathbf{E}(\mathbf{r}, \theta, \phi) = \frac{1}{4\pi\epsilon_0} \frac{p[3\sin\theta\cos\theta(\hat{\mathbf{x}}\cos\phi + \hat{\mathbf{y}}\sin\phi) + (3\cos^2\theta - 1)\hat{\mathbf{z}}]}{r^3}, \qquad (1)$$

where r, θ and ϕ are spherical polar coordinates.

(b) The average electric field inside a sphere of radius r_0 centred on O is

$$\overline{\mathbf{E}} = \frac{1}{\frac{4}{3}\pi r_0^3} \int_v \mathbf{E}(\mathbf{r}, \theta, \phi) \, dv. \qquad (2)$$

With $\mathbf{E}(\mathbf{r}, \theta, \phi)$ given by (1) and using dv for spherical polar coordinates, show that—surprisingly—the field $\overline{\mathbf{E}}$ is indeterminate. $\left(\text{Recall that the correct average field was calculated in Question 2.12 and is } -\frac{1}{4\pi\epsilon_0}\frac{\mathbf{p}}{r_0^3}.\right)$

(c) The difficulty exposed in (b) can be resolved as follows: choose an infinitesimal sphere (radius ϵ) centred on O and take

$$\mathbf{E}(\mathbf{r}, \theta, \phi) = \begin{cases} \text{using (1) above} & \text{for } r \geq \epsilon \text{ (region I)}, \\ \dfrac{-\mathbf{p}}{3\epsilon_0}\delta(\mathbf{r}) & \text{for } r < \epsilon \text{ (region II)}. \end{cases} \qquad (3)$$

Show that (3) now gives the correct average field.[‡]

Solution

(a) The result follows immediately when $\hat{\mathbf{r}}$ in (2) of Question 2.11 is expressed in terms of the Cartesian unit vectors using (V) of Appendix C.

(b) Taking $dv = r^2\sin\theta\, dr\, d\theta\, d\phi$ and evaluating the angular integrals give zero. But the integral of $1/r$ between 0 and r_0 is infinite, and so the result is indeterminate.

(c) With $\mathbf{E}(\mathbf{r}, \theta, \phi)$ given by (3), we obtain from (2)

$$\overline{\mathbf{E}} = \frac{1}{\frac{4}{3}\pi r_0^3}\left[\int_I \mathbf{E}\, dv + \int_{II}\left(\frac{-\mathbf{p}\,\delta(\mathbf{r})}{3\epsilon_0}\right) dv\right]. \qquad (4)$$

[‡]See Comment (iii) of Question 2.11 if you are puzzled by the odd-looking equation $(3)_2$.

The first term on the right-hand side of (4) is now unambiguously zero since $\ln r_0/\epsilon$ remains finite (the angular integrals are both zero, as before). The second term gives the expected answer because $\int_{\mathrm{II}} \delta(\mathbf{r})\, dv = 1$ (see Appendix E).

Comments

(i) The calculation in (b) fails because the correct dipole field is given by (4), and not (2), of Question 2.11. This is an example where the delta-function contribution to the field plays a crucial role.

(ii) The entire contribution to the average electric field in a spherical region centred on a point dipole arises from the delta-function term at the origin. Although this term is infinite, its integral over a finite region of space is not.

(iii) A similar analysis may also be carried out for the field of a point magnetic dipole. (See Ref. [4] and Question 4.21.)

Question 2.14

A convenient way of visualizing an electrostatic field $\mathbf{E}$ is through the 'lines of force' and the 'equipotentials'. A line of force is a directed curve at a point in the field and is tangential to $\mathbf{E}$ at that point. Let $d\mathbf{l}$ be an element of this curve, then

$$\mathbf{E} = \lambda\, d\mathbf{l}, \tag{1}$$

where λ is a scale factor. Consider a Cartesian coordinate system and suppose, for example, that we wish to visualize the field in the xy-plane.

(a) Use (1) to show that the lines of force satisfy the equation

$$\frac{dy}{dx} = \frac{E_y}{E_x}. \tag{2}$$

(b) Show also that a line of constant potential (equipotential) satisfies

$$\frac{dy}{dx} = -\frac{E_x}{E_y}. \tag{3}$$

Solution

(a) In the xy-plane $d\mathbf{l} = \hat{\mathbf{x}}\,dx + \hat{\mathbf{y}}\,dy$. So $E_x = \lambda\,dx$ and $E_y = \lambda\,dy$. Hence (2).

(b) Consider $\Phi(x, y)$. Then $d\Phi = \dfrac{\partial \Phi}{\partial x}\,dx + \dfrac{\partial \Phi}{\partial y}\,dy$. By definition $d\Phi = 0$ along a line of constant potential. Putting $\dfrac{\partial \Phi}{\partial x} = -E_x$ and $\dfrac{\partial \Phi}{\partial y} = -E_y$ gives $E_x dx + E_y dy = 0$. Hence (3).

Comments

(i) Notice that $\dfrac{dy}{dx}\Big|_{\text{field lines}} \times \dfrac{dy}{dx}\Big|_{\text{equipot-entials}} = -1$ (i.e. the product of tangents (2) and (3) is negative one). This is not an unexpected result because the field line at any point is always perpendicular to the line of constant potential at that point.

(ii) Since we often represent electrostatic fields in two dimensions, it is useful to express these gradients in terms of polar coordinates. Then $\mathbf{E} = \lambda(\hat{\mathbf{r}}\,dr + r\hat{\boldsymbol{\theta}}\,d\theta)$ leads to

$$\frac{1}{r}\frac{dr}{d\theta} = \frac{E_r}{E_\theta} \quad \text{for a field line} \quad \text{and} \quad \frac{1}{r}\frac{dr}{d\theta} = -\frac{E_\theta}{E_r} \quad \text{for an equipotential.} \quad (4)$$

(iii) The above results generalize to three dimensions: electric field lines are everywhere perpendicular to the equipotential surfaces.

Question 2.15

Consider the model of a point electric dipole $\mathbf{p} = p_0\,\hat{\mathbf{z}}$ located at the origin O of Cartesian coordinates, as the only source of electric field. Suppose $p_0 > 0$.

(a) Derive the polar equations for a dipole line of constant potential and a dipole electric field line in any plane passing through the z-axis.

(b) Use your answers to (a) to draw the field lines in this plane, and on the same plot also show some of the equipotentials.

Note: In the solution, it is convenient to let θ represent the angle between r and the z-axis. In this sense, θ is not the 'usual' plane polar coordinate.

Solution

(a) equipotentials

Putting $\Phi = \Phi_0 =$ a constant in (1) of Question 2.11 and rearranging give $r = k\sqrt{\cos\theta}$ where $k^{-2} = 4\pi\epsilon_0\Phi_0/p_0$ for $\Phi_0 > 0$ and $-\tfrac{1}{2}\pi \le \theta \le \tfrac{1}{2}\pi$. In regions where either Φ_0 or $\cos\theta$ is negative, we take the absolute value. Hence the equation we seek is

$$r = r_0\sqrt{|\cos\theta|}, \quad (1)$$

with $k^{-2} = 4\pi\epsilon_0|\Phi_0|/p_0$.

field lines

For this derivation we use $(4)_1$ of Question 2.14:

$$\frac{1}{r}\frac{dr}{d\theta} = \frac{E_r}{E_\theta} = r\frac{\partial\Phi/\partial r}{\partial\Phi/\partial\theta} = 2\cot\theta, \quad (2)$$

where the last step follows from differentiating the dipole potential $\dfrac{1}{4\pi\epsilon_0}\dfrac{p_0\cos\theta}{r^2}$.

Rearranging (2) yields

$$\int \frac{dr}{r} = 2 \int \cot \theta \, d\theta, \qquad (3)$$

or

$$\ln r = 2 \ln(\sin \theta) + \ln r_0. \qquad (4)$$

Here $\ln r_0$ is a constant of integration, and so

$$r = r_0 \sin^2\theta. \qquad (5)$$

(b) The figure below was drawn using (1) and (5) for various values of r_0 in the interval $[0.4, 1.1]$.

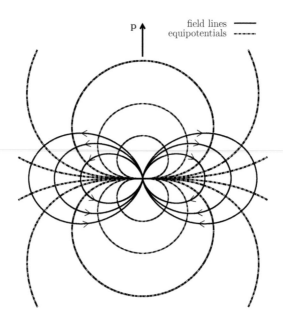

field lines ———
equipotentials ·······

Question 2.16

Two infinitely long opposite line charges $\pm\lambda$ are separated by a distance $2a$ and lie parallel to the z-axis of Cartesian coordinates, as shown in the figure on p. 81. The electric potential at an arbitrary point $P(x, y, z)$, calculated in Question 2.8(b), is given by

$$\Phi = \frac{\lambda}{4\pi\epsilon_0} \ln\left[\frac{r_1^2}{r_2^2}\right] = \frac{\lambda}{4\pi\epsilon_0} \ln\left[\frac{(x+a)^2 + y^2}{(x-a)^2 + y^2}\right]. \qquad (1)$$

Use (1) to prove that all equipotential surfaces (with the exception of the $\Phi = 0$ plane) and all field lines are circular cylinders.

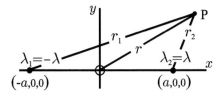

Solution

equipotentials

On an equipotential surface, the term r_1^2/r_2^2 in (1) equals a positive constant, α^2 say. Then

$$\frac{(x+a)^2 + y^2}{(x-a)^2 + y^2} = \alpha^2, \tag{2}$$

which can easily be rearranged to give

$$\left[x - a\left(\frac{\alpha^2+1}{\alpha^2-1}\right)\right]^2 + y^2 = \left(\frac{2\alpha a}{|\alpha^2-1|}\right)^2. \tag{3}$$

In the right half-plane the constant $\alpha^2 > 1$ since $r_1 > r_2$ here, and (3) is the equation of a circle of radius r_0 centred on (x_0, y_0), where

$$r_0 = \frac{2\alpha a}{\alpha^2 - 1} \qquad \text{and} \qquad (x_0, y_0) = \left(\frac{\alpha^2+1}{\alpha^2-1}a, 0\right). \tag{4}$$

In the left half-plane $\alpha^2 < 1$ and (3) is again the equation of a circle, where now

$$r_0 = \frac{2\alpha a}{1 - \alpha^2} \qquad \text{and} \qquad (x_0, y_0) = \left(\frac{\alpha^2+1}{1-\alpha^2}a, 0\right). \tag{5}$$

An entire family of equipotential surfaces that are circular cylinders can be generated with (3)–(5) by regarding α as a parameter greater than zero.

field lines

For this part of the proof we use (2) of Question 2.14. Then

$$\frac{dy}{dx} = \frac{E_y}{E_x} = \frac{\partial \Phi/\partial y}{\partial \Phi/\partial x}. \tag{6}$$

Differentiating (1) with respect to x and y and calculating the ratio E_y/E_x in (6) give

$$\frac{dy}{dx} = \frac{2xy}{x^2 - y^2 - a^2}. \tag{7}$$

In attempting to solve (7) we make the change of variable

$$\left. \begin{array}{l} x = r \cos\theta \\ y = r \sin\theta \end{array} \right\}, \tag{8}$$

hence

$$\frac{dy}{dx} = \frac{2r^2 \sin\theta \cos\theta}{r^2 \cos^2\theta - r^2 \sin^2\theta - a^2}.$$

Now

$$\frac{dy}{dx} = \frac{\dfrac{\partial y}{\partial r} dr + \dfrac{\partial y}{\partial \theta} d\theta}{\dfrac{\partial x}{\partial r} dr + \dfrac{\partial x}{\partial \theta} d\theta} = \frac{\sin\theta \, dr + r \cos\theta \, d\theta}{\cos\theta \, dr - r \sin\theta \, d\theta},$$

and so

$$\frac{\sin\theta \dfrac{dr}{d\theta} + r \cos\theta}{\cos\theta \dfrac{dr}{d\theta} - r \sin\theta} = \frac{2r^2 \sin\theta \cos\theta}{r^2 \cos^2\theta - r^2 \sin^2\theta - a^2}. \tag{9}$$

Solving (9) for $dr/d\theta$ and rearranging yield

$$\int \frac{(r^2 + a^2)}{(r^2 - a^2)} \frac{dr}{r} = \int \cot\theta \, d\theta.$$

Upon integration we obtain

$$\ln\left[\frac{r^2 - a^2}{ra}\right] = \ln(\sin\theta) + \text{a constant}.$$

Choosing the constant of integration equal to $\ln\beta$ (where β is a real number) gives $\ln[(r^2 - a^2)/r\beta a] = \ln(\sin\theta)$, which leads directly to the equation $r^2 - \beta a r \sin\theta = a^2$. Eliminating r and θ in favour of x and y using (8) and rearranging give

$$x^2 + \left(y - \tfrac{1}{2}\beta a\right)^2 = \left(a \sqrt{1 + \tfrac{1}{4}\beta^2}\right)^2. \tag{10}$$

Taking β to be a positive (negative) parameter in (10) generates a family of circular cylinders of radius r_0 centred at (x_0, y_0), where

$$r_0 = a \sqrt{1 + \tfrac{1}{4}\beta^2} \qquad \text{and} \qquad (x_0, y_0) = (0, \tfrac{1}{2}\beta a). \tag{11}$$

Comments

(i) A cross-section of the electrostatic field for these line charges is included in the comments of Question 2.8. See also Question 3.27 for a practical application of the above results.

(ii) The method of conformal transformation considered at the end of Chapter 3 is a more efficient and elegant way of answering this question (at least from a mathematical point of view).

Question 2.17

Consider a charge q_1 located at $\mathbf{r}'_1$ relative to a point O. Let P be a field point at $\mathbf{r}$. Since the potential Φ_1 at P is $\dfrac{1}{4\pi\epsilon_0}\dfrac{q_1}{|\mathbf{r}-\mathbf{r}'_1|}$, the energy required to move a second charge q_2 from infinity to a point $\mathbf{r}'_2$ in the field is[‡]

$$U_{12} \;=\; q_2\left(\Phi_1(\mathbf{r}'_2)-\Phi_1(\infty)\right) \;=\; \frac{1}{4\pi\epsilon_0}\frac{q_1 q_2}{|\mathbf{r}'_2-\mathbf{r}'_1|}\,, \tag{1}$$

because $\Phi_1(\infty)=0$.

(a) Suppose n particles having charges q_1, q_2, $\ldots$, q_n (all initially located infinitely far apart from each other) are moved sequentially to positions $\mathbf{r}'_1$, $\mathbf{r}'_2$, $\ldots$, $\mathbf{r}'_n$ respectively. Carefully explaining your steps, generalize (1) and show that the potential energy U of the fully assembled system is

$$U \;=\; \frac{1}{2}\sum_{i=1}^{n} q_i\,\Phi_i(\mathbf{r}'_i), \tag{2}$$

where

$$\Phi_i(\mathbf{r}'_i) \;=\; \frac{1}{4\pi\epsilon_0}\sum_{j\neq i}^{n}\frac{q_j}{|\mathbf{r}'_j-\mathbf{r}'_i|} \tag{3}$$

is the electrical potential due to all n charges (except i) at the location of q_i.

Hint: Count the contribution from each pair of interacting charges twice (this is done for reasons of convenience). Dividing this quantity by two then yields U.

(b) Now consider a continuous distribution and make the replacements $q_i \to \rho(\mathbf{r}')\,dv'$; $\sum_i \to \int_v$. Transforming (2) in this way gives

$$U \;=\; \frac{1}{2}\int_v \rho(\mathbf{r}')\Phi(\mathbf{r}')\,dv'. \tag{4}$$

The region of integration v in (4) must be large enough to include all the charges. In fact, it does not matter if v is extended to include regions where $\rho(\mathbf{r}') = 0$. Since these regions make no contribution to U anyway, we might as well integrate over all space. So

$$U \;=\; \frac{1}{2}\int_{\substack{\text{all}\\ \text{space}}} \rho(\mathbf{r}')\Phi(\mathbf{r}')\,dv'. \tag{5}$$

Describe the differences between the energies as given by (2) and (5).

(c) Use Gauss's law to eliminate $\rho(\mathbf{r}')$ from (5) using $(1)_1$ of Question 2.3, and hence show that

$$U \;=\; \tfrac{1}{2}\,\epsilon_0\int_{\substack{\text{all}\\ \text{space}}} \mathbf{E}\cdot\mathbf{E}\,dv'. \tag{6}$$

[‡]Electrostatic forces are conservative, so the work done in slowly moving a charge q in the field equals $q \times$ the potential difference between the initial and final states.

Solution

(a) We continue assembling the distribution, particle by particle. Since q_1 and q_2 are already located at positions $\mathbf{r}'_1$ and $\mathbf{r}'_2$ respectively, the work required to move q_3 from infinity to $\mathbf{r}'_3$ is $q_3 \Phi(\mathbf{r}'_3)$, where $\Phi(\mathbf{r}'_3) = \dfrac{1}{4\pi\epsilon_0}\left(\dfrac{q_1}{|\mathbf{r}'_3 - \mathbf{r}'_1|} + \dfrac{q_2}{|\mathbf{r}'_3 - \mathbf{r}'_2|}\right)$.

Now the electrostatic potential energy of this arrangement of three charges is $U_{123} = U_{12} + q_3\,\Phi(\mathbf{r}'_3)$, or

$$U_{123} = U_{12} + U_{13} + U_{23}$$

$$= \frac{1}{4\pi\epsilon_0}\left(\frac{q_1 q_2}{|\mathbf{r}'_2 - \mathbf{r}'_1|} + \frac{q_1 q_3}{|\mathbf{r}'_3 - \mathbf{r}'_1|} + \frac{q_2 q_3}{|\mathbf{r}'_3 - \mathbf{r}'_2|}\right). \qquad (7)$$

By continuing in this way we obtain $U_{123\ldots n}$, or simply U. In view of the hint and from the pattern emerging in (7), we write

$$U = \frac{1}{2} \times \frac{1}{4\pi\epsilon_0} \sum_{i=1}^{n} \sum_{j\neq i}^{n} \frac{q_i q_j}{|\mathbf{r}_j - \mathbf{r}_i|},$$

where the double sum notation reads as follows:

☞ Take $i = 1$ and sum over $j = 2, 3, 4, \ldots, n$, then

☞ Take $i = 2$ and sum over $j = 1, 3, 4, \ldots, n$, and so on through $i = n$.

Thus

$$U = \frac{1}{2}\sum_{i=1}^{n} q_i \sum_{j\neq i} \frac{1}{4\pi\epsilon_0} \frac{q_j}{|\mathbf{r}'_j - \mathbf{r}'_i|} = \frac{1}{2}\sum_{i=1}^{n} q_i \Phi_i,$$

which is (2) with Φ_i given by (3).

(b) Two differences between the energies calculated by (2) and (5) are:

1. The particles described in (a) exist as discrete charges (which are initially infinitely far apart) from each other, and the work done in arranging the distribution is the potential energy (2). However, the distribution described in (b) is continuous and the particles themselves must first be assembled incrementally from infinitesimal amounts of charge $\rho\,dv'$ (starting with $\rho = 0$). Therefore, (5) represents the potential energy of the distribution plus the energy required to create the particles during the accretion process.

2. The potential $\Phi_i(\mathbf{r}'_i)$ in (2) is due to all n charges except q_i, whereas $\Phi(\mathbf{r}')$ in (5) is the *total* potential of all the charge currently assembled: the infinitesimal charge $\rho(\mathbf{r}')\,dv'$ at the macroscopic point at $\mathbf{r}'$ makes no contribution to Φ there.

(c) Substituting $\rho(\mathbf{r}') = \epsilon_0 \boldsymbol{\nabla} \cdot \mathbf{E}(\mathbf{r}')$ in (5) gives $U = \frac{1}{2}\epsilon_0 \displaystyle\int_v \Phi(\boldsymbol{\nabla} \cdot \mathbf{E})\,dv'$, where the region of integration v is a sphere of radius R (shortly we will take the limit

$R \to \infty$). The integrand above can be written as $\Phi(\nabla \cdot \mathbf{E}) = \nabla \cdot (\Phi\,\mathbf{E}) - \mathbf{E} \cdot \nabla\Phi$ (see (5) of Question 1.8), and so

$$U = \tfrac{1}{2}\,\epsilon_0 \int_v \left[\nabla \cdot (\Phi\,\mathbf{E}) + \mathbf{E} \cdot \mathbf{E} \right] dv', \tag{8}$$

because $\mathbf{E} = -\nabla\Phi$ for electrostatic fields. Converting the first integral in (8) to a surface integral using Gauss's theorem gives

$$U = \tfrac{1}{2}\,\epsilon_0 \oint_s \Phi\,\mathbf{E} \cdot d\mathbf{a}' + \tfrac{1}{2}\,\epsilon_0 \int_v \mathbf{E} \cdot \mathbf{E}\, dv'. \tag{9}$$

In the limit $R \to \infty$ this surface integral is zero[‡] and (6) follows.

Comments

(i) A third difference between (2) and (5) now becomes evident. U as given by (5) is positive definite,[#] whereas U given by (2) can be less than zero.

(ii) Clearly, the system of n charges may be assembled in $n!$ different ways. For each of these ways the same amount of work is done because all terms in (2) still occur, albeit in a different order. We therefore conclude that the energy U does not depend on the manner in which the configuration is arranged.

(iii) It is natural to ask: where is this energy U stored? In the charges? In the field? Or in both? The answers to these questions are not straightforward and the reader is referred to Ref. [5] for further discussion. For our purposes, we will regard the energy as stored in the electrostatic field; (6) shows that we can define the energy per unit volume (i.e. the energy density) as

$$u = \tfrac{1}{2}\,\epsilon_0 \mathbf{E} \cdot \mathbf{E}. \tag{10}$$

In Question 7.6 we will show that (10) is also the energy density in a *time-dependent* electric field $\mathbf{E}(\mathbf{r}, t)$.

(iv) Also mentioned in Ref. [5] is the apparently infinite energy of point charges[†] present in classical electrodynamics. This difficulty persists even in the quantum theory of electrodynamics where the matter is still not completely resolved.

[‡] In Question 2.20 we find that Φ and $\mathbf{E}$ always fall off with distance at least as fast as $1/R$ and $1/R^2$ respectively. Since $da' \propto R^2$, we have $\Phi\,\mathbf{E} \cdot d\mathbf{a}' \sim R^2/R^3$ which $\to 0$ as $R \to \infty$.

[#] This becomes obvious when (5) is rewritten in the alternative form (6).

[†] Consider the model of an isolated electron as a uniformly charged sphere of radius r_0. From (6) the field energy is

$$U = \tfrac{1}{2}\,\epsilon_0 \int_0^{2\pi} \int_0^{\pi} \int_{r_0}^{\infty} \left(\frac{1}{4\pi\epsilon_0} \frac{-e}{r^2} \right)^2 r^2 \sin\theta\, dr\, d\theta\, d\phi.$$

As $r_0 \to 0$ the electron approaches a 'point' charge and the radial integral clearly diverges.

---

[5]   R. P. Feynman, R. B. Leighton and M. Sands, *The Feynman lectures on physics*, vol. II, Chap. 8, pp. 8–12. Massachusetts: Addison Wesley, 1964.

Question 2.18

Three small particles carrying positive charges q_1, q_2 and q_3 are constrained to move in a circle of radius a. The position vectors of these charges make angles θ_1, θ_2 and θ_3 relative to the x-axis, as shown in the figure. Suppose the system is in equilibrium.

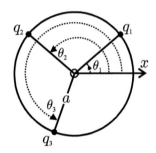

(a) Show that the electrostatic energy U is given by

$$U = \frac{1}{4\pi\epsilon_0} \frac{1}{2a}\left\{ \frac{q_1 q_2}{\sin\frac{1}{2}\alpha} + \frac{q_2 q_3}{\sin\frac{1}{2}\beta} + \frac{q_3 q_1}{\sin\frac{1}{2}\gamma}\right\}, \qquad (1)$$

where $\alpha = \theta_2 - \theta_1$, $\beta = \theta_3 - \theta_2$ and $\gamma = 2\pi - (\theta_3 - \theta_1)$ are the angles *between* neighbouring charges.

(b) Use (1) to derive coupled equations which may be used to calculate α, β and γ.

(c) Hence calculate α, β and γ using *Mathematica's* `FindRoot` function for the case $q_1 = \frac{1}{2}q_2 = \frac{1}{3}q_3$.

Solution

(a) Clearly $U = U_{12} + U_{13} + U_{23}$, and then from (1) of Question 2.17 we have

$$U = \frac{1}{4\pi\epsilon_0}\left[\frac{q_1 q_2}{|\mathbf{r}'_1 - \mathbf{r}'_2|} + \frac{q_1 q_3}{|\mathbf{r}'_1 - \mathbf{r}'_3|} + \frac{q_2 q_3}{|\mathbf{r}'_2 - \mathbf{r}'_3|}\right], \qquad (2)$$

where $|\mathbf{r}'_1 - \mathbf{r}'_2| = 2a\sin\frac{1}{2}\alpha$, $|\mathbf{r}'_2 - \mathbf{r}'_3| = 2a\sin\frac{1}{2}\beta$ and $|\mathbf{r}'_3 - \mathbf{r}'_2| = 2a\sin\frac{1}{2}\gamma$ follow from the sine rule of trigonometry.

(b) The equilibrium positions are found by minimizing the potential energy with respect to α and β, taken as independent variables. We substitute $\gamma = 2\pi - \alpha - \beta$ in (2), and calculate $\partial U/\partial\alpha$ and $\partial U/\partial\beta$. Equating these derivatives to zero yields

$$\left.\begin{array}{l} \dfrac{\cos\frac{1}{2}\alpha}{1 - \cos\alpha} + \dfrac{q_3}{q_2}\dfrac{\cos\frac{1}{2}(\alpha + \beta)}{1 - \cos(\alpha + \beta)} = 0 \\[3mm] \dfrac{\cos\frac{1}{2}\beta}{1 - \cos\beta} + \dfrac{q_1}{q_2}\dfrac{\cos\frac{1}{2}(\alpha + \beta)}{1 - \cos(\alpha + \beta)} = 0 \end{array}\right\}. \qquad (3)$$

(c) Solving (3) simultaneously (see `cell` 1 of the notebook on p. 88) yields $\alpha \simeq 101°$, $\beta \simeq 142°$ and $\gamma \simeq 117°$.

Comments

(i) For $q_1 = q_2 = q_3$ it is obvious from the symmetry that $\alpha = \beta = \gamma = 120°$, and the angles calculated in (c) above may be regarded as a perturbation of this case.

(ii) In Question 2.19, we make a simple numerical study of the dynamics of this system.

Question 2.19

Suppose the charged particles described in Question 2.18 are now set in motion and oscillate without friction.

(a) Calculate the Lagrangian of the system ($\mathsf{L} = T - U$) and hence show that the equations of motion are

$$
\left.
\begin{array}{l}
\ddot{\theta}_1 + \dfrac{1}{16\pi\epsilon_0 m_1 a^3}\left[\dfrac{q_1 q_2 \cos\frac{1}{2}(\theta_2 - \theta_1)}{1 - \cos(\theta_2 - \theta_1)} + \dfrac{q_1 q_3 \cos\frac{1}{2}(\theta_3 - \theta_1)}{1 - \cos(\theta_3 - \theta_1)}\right] = 0 \\[3mm]
\ddot{\theta}_2 - \dfrac{1}{16\pi\epsilon_0 m_2 a^3}\left[\dfrac{q_1 q_2 \cos\frac{1}{2}(\theta_2 - \theta_1)}{1 - \cos(\theta_2 - \theta_1)} - \dfrac{q_2 q_3 \cos\frac{1}{2}(\theta_3 - \theta_2)}{1 - \cos(\theta_3 - \theta_2)}\right] = 0 \\[3mm]
\ddot{\theta}_3 - \dfrac{1}{16\pi\epsilon_0 m_3 a^3}\left[\dfrac{q_1 q_3 \cos\frac{1}{2}(\theta_3 - \theta_1)}{1 - \cos(\theta_3 - \theta_1)} + \dfrac{q_2 q_3 \cos\frac{1}{2}(\theta_3 - \theta_2)}{1 - \cos(\theta_3 - \theta_2)}\right] = 0
\end{array}
\right\}, \qquad (1)
$$

where m_i are the masses of the particles.

(b) Write a *Mathematica* notebook to solve (1) for the following initial conditions:

☞ The particles are released from rest at $\theta_1 = 0°$, $\theta_2 = 120°$ and $\theta_3 = 240°$ at time $t = 0$.

☞ The particles start at the same positions as before, but now have the initial velocities $\dot{\theta}_1 = \dot{\theta}_2 = \dot{\theta}_3 = 1\,\mathrm{s}^{-1}$.

In both cases assume that

$$
\left.
\begin{array}{c}
\frac{1}{2}q_1 = \frac{3}{2}q_2 = \frac{5}{2}q_3 = q \\[2mm]
m_1 = \frac{1}{2}m_2 = \frac{1}{3}m_3 = m \\[2mm]
16\pi\epsilon_0 ma^3/q^2 = 1\,\mathrm{s}^2
\end{array}
\right\},
$$

and plot the $\theta_i(t)$ on the same axes. For the first set of initial conditions take $0 \le t \le 80\,\mathrm{s}$, and for the second set take $0 \le t \le 20\,\mathrm{s}$. Animate the motion of the system using *Mathematica*'s `Manipulate` command. Then run the animation for different combinations of the q_i, m_i and initial conditions.

Solution

(a) Taking $T_i = \frac{1}{2}m_i(a\dot{\theta}_i)^2$ and with U given by (1) of Question 2.18 we obtain

$$
\mathsf{L} = \frac{1}{2}(m_1\dot{\theta}_1{}^2 + m_2\dot{\theta}_2{}^2 + m_3\dot{\theta}_3{}^2)a^2 -
$$

$$
\frac{1}{4\pi\epsilon_0}\frac{1}{2a}\left\{\frac{q_1 q_2}{\sin\frac{1}{2}(\theta_2 - \theta_1)} + \frac{q_1 q_3}{\sin\frac{1}{2}(\theta_3 - \theta_1)} + \frac{q_2 q_3}{\sin\frac{1}{2}(\theta_3 - \theta_2)}\right\}. \qquad (2)
$$

Using (2) in the Euler–Lagrange equation $\dfrac{d}{dt}\left(\dfrac{\partial \mathsf{L}}{\partial \dot{\theta}_i}\right) - \dfrac{\partial \mathsf{L}}{\partial \theta_i} = 0$ gives (1).

(b) Evaluating `cell` 2 of the notebook below gives the graphs shown on p. 89. `Cell` 3 of the notebook produces an animated display of the motion.

```
In[1]:=  q1 = 1;  q2 = 2;  q3 = 3;
```

$$U[\alpha\_, \beta\_] := \frac{q1\,q2}{Sin\left[\frac{\alpha}{2}\right]} + \frac{q2\,q3}{Sin\left[\frac{\beta}{2}\right]} + \frac{q3\,q1}{Sin\left[\frac{\alpha+\beta}{2}\right]};$$

```
f[α_, β_] := D[U[α, β], α];   g[α_, β_] := D[U[α, β], β];

Sol = FindRoot[{f[α, β] == 0, g[α, β] == 0}, {{α, 1, 0, 2 π}, {β, 2, 0, 2 π}}];

α = α /. Sol;   β = β /. Sol;   γ = 2 π - α - β;
```

```
In[2]:=  q1 = 2;   q2 = 3;   q3 = 5;   m1 = 1;   m2 = 2;   m3 = 3;   k = 1;   tmax = 80;
```

$$eqn1 = \theta1''[t] + \frac{k}{m1}\left(\frac{q1\,q2\,Cos\left[\frac{\theta2[t]-\theta1[t]}{2}\right]}{1 - Cos[\theta2[t] - \theta1[t]]} + \frac{q1\,q3\,Cos\left[\frac{\theta3[t]-\theta1[t]}{2}\right]}{1 - Cos[\theta3[t] - \theta1[t]]}\right) == 0;$$

$$eqn2 = \theta2''[t] - \frac{k}{m2}\left(\frac{q1\,q2\,Cos\left[\frac{\theta2[t]-\theta1[t]}{2}\right]}{1 - Cos[\theta2[t] - \theta1[t]]} - \frac{q2\,q3\,Cos\left[\frac{\theta3[t]-\theta2[t]}{2}\right]}{1 - Cos[\theta3[t] - \theta2[t]]}\right) == 0;$$

$$eqn3 = \theta3''[t] - \frac{k}{m3}\left(\frac{q1\,q3\,Cos\left[\frac{\theta3[t]-\theta1[t]}{2}\right]}{1 - Cos[\theta3[t] - \theta1[t]]} + \frac{q2\,q3\,Cos\left[\frac{\theta3[t]-\theta2[t]}{2}\right]}{1 - Cos[\theta3[t] - \theta2[t]]}\right) == 0;$$

```
Sol = NDSolve[{eqn1, eqn2, eqn3, θ1[0] == 0, θ1'[0] == 0,
```
$$\theta2[0] == \frac{2\,\pi}{3},\ \theta2'[0] == 0,\ \theta3[0] == \frac{4\,\pi}{3},\ \theta3'[0] == 0\},$$
```
     {θ1[t], θ2[t], θ3[t]}, {t, 0, tmax}];
```
$$gr1 = Plot\left[(Evaluate[\theta1[t]] /. Sol) \times \left(\frac{180}{\pi}\right), \{t, 0, tmax\}\right];$$
$$gr2 = Plot\left[(Evaluate[\theta2[t]] /. Sol) \times \left(\frac{180}{\pi}\right), \{t, 0, tmax\}\right];$$
$$gr3 = Plot\left[(Evaluate[\theta3[t]] /. Sol) \times \left(\frac{180}{\pi}\right), \{t, 0, tmax\}\right];$$
```
Show[gr1, gr2, gr3, Axes → True]
```

```
In[3]:=  x1[τ_] := Cos[θ1[t]] /. Sol /. t → τ;   y1[τ_] := Sin[θ1[t]] /. Sol /. t → τ;
         x2[τ_] := Cos[θ2[t]] /. Sol /. t → τ;   y2[τ_] := Sin[θ2[t]] /. Sol /. t → τ;
         x3[τ_] := Cos[θ3[t]] /. Sol /. t → τ;   y3[τ_] := Sin[θ3[t]] /. Sol /. t → τ;
         traj[τ_] := Graphics[{Thickness[.01], Black, Circle[{0, 0}, 1],
             PointSize[0.045], Black, Point[{First[x1[τ]], First[y1[τ]]}],
             PointSize[0.045], Red, Point[{First[x2[τ]], First[y2[τ]]}],
             PointSize[0.045], Blue, Point[{First[x3[τ]], First[y3[τ]]}]}]
         Manipulate[traj[τ], {τ, 0, tmax}]
```

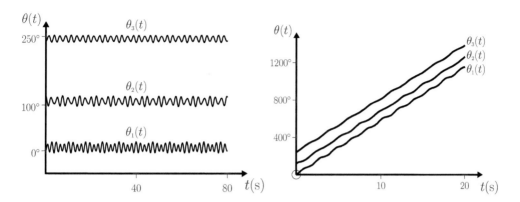

Question 2.20*

Consider a continuous, localized distribution of stationary electric charges in vacuum, having density $\rho(\mathbf{r}')$. Suppose O is an arbitrarily chosen origin (we have taken it to be inside the distribution as shown below). The charge within an infinitesimal volume element dv' is $\rho(\mathbf{r}')\,dv'$, where $\mathbf{r}'$ is the position vector of dv'. Let P be a field point having position vector $\mathbf{r}$. The electrostatic potential is

$$\Phi(\mathbf{r}) = \frac{1}{4\pi\epsilon_0} \int_v \frac{\rho(\mathbf{r}')}{|\mathbf{r} - \mathbf{r}'|}\,dv'. \tag{1}$$

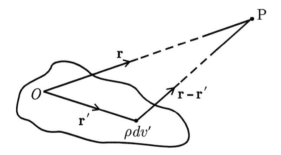

(a) Suppose $r \gg r'$ (i.e. P is a *distant* field point). Using the expansion of $|\mathbf{r} - \mathbf{r}'|^{-1}$ given by (2) of Question 1.29, show that

$$\Phi(\mathbf{r}) = \frac{1}{4\pi\epsilon_0} \left[\frac{1}{r} \int_v \rho(\mathbf{r}')\,dv' + \frac{r_i}{r^3} \int_v r_i'\rho(\mathbf{r}')\,dv' + \frac{(3r_i r_j - r^2\delta_{ij})}{2r^5} \int_v r_i' r_j'\rho(\mathbf{r}')\,dv' + \right.$$

$$\left. \frac{(5r_i r_j r_k - r^2(r_i\delta_{jk} + r_j\delta_{ki} + r_k\delta_{ij}))}{2r^7} \int_v r_i' r_j' r_k'\rho(\mathbf{r}')\,dv' + \cdots \right]. \tag{2}$$

(b) Hence show that the electric field at P is given by

$$E_i(\mathbf{r}) = \frac{1}{4\pi\epsilon_0} \left[\frac{r_i}{r^3} \int_v \rho(\mathbf{r}')\, dv' + \frac{(3r_i r_j - r^2 \delta_{ij})}{r^5} \int_v r_j' \rho(\mathbf{r}')\, dv' + \right.$$

$$\left. \frac{3\left(5r_i r_j r_k - r^2(r_i \delta_{jk} + r_j \delta_{ki} + r_k \delta_{ij})\right)}{2r^7} \int_v r_j' r_k' \rho(\mathbf{r}')\, dv' + \cdots \right]. \quad (3)$$

Solution

(a) The first four terms in the expansion (2) of Question 1.29 are

$$|\mathbf{r} - \mathbf{r}'|^{-1} = \frac{1}{r} + \frac{r_i}{r^3} r_i' + \frac{3r_i r_j - r^2 \delta_{ij}}{2r^5} r_i' r_j' + $$

$$\frac{5r_i r_j r_k - r^2(r_i \delta_{jk} + r_j \delta_{ki} + r_k \delta_{ij})}{2r^7} r_i' r_j' r_k' + \cdots . \quad (4)$$

Substituting (4) in (1) gives (2).

(b) The electric field follows from $\mathbf{E} = -\nabla\Phi(\mathbf{r})$ with

$$E_i = -\frac{\partial \Phi}{\partial r_i}. \quad (5)$$

Equations (3), (7) and (8) of Question 1.1, together with (5), yield (3).

Comments

(i) The various integrals in (2) and (3) are the electric multipole moments of the distribution relative to O, and they are defined as follows:

$$q = \int_v \rho(\mathbf{r}')\, dv' \qquad \text{(zeroth moment or electric monopole moment)}$$

$$p_i = \int_v r_i' \rho(\mathbf{r}')\, dv' \qquad \text{(first moment or electric dipole moment)}$$

$$q_{ij} = \int_v r_i' r_j' \rho(\mathbf{r}')\, dv' \qquad \text{(second moment or electric quadrupole moment)}$$

$$q_{ijk} = \int_v r_i' r_j' r_k' \rho(\mathbf{r}')\, dv' \quad \text{(third moment or electric octopole moment)}$$

$$, \quad (6)$$

and so on. They are Cartesian tensors of rank zero, one, two, three, ... Obviously, q defined by $(6)_1$ is the net charge of the distribution. All of them beyond the dipole are symmetric in their subscripts. This symmetry is immediately evident upon interchange of subscripts, as one can see from (6). Consequently:

☞ q_{ij} has at most six independent components,[‡] since $q_{xy} = q_{yx}$, $q_{xz} = q_{zx}$ and $q_{yz} = q_{zy}$.

☞ q_{ijk} has at most ten independent components, and so on.

(ii) Equations (2) and (3) are known respectively as the multipole expansion of the electrostatic potential and field relative to the origin O.

(iii) Ref. [6] draws six important conclusions from these multipole expansions. Quoting directly from this reference, (2) and (3) show that:

☞ Each multipole contributes separately to the potential and field at an external point.

☞ Each multipole behaves as if it were located at the origin. This is because its contribution to the potential and field contains, apart from the moment itself, only **r**: the displacement of the field point P from the origin O.

☞ The potential and field due to each multipole depend not only on its moment and its distance r from the field point, but in general also on its orientation relative to **r**. For example, the dipole contribution in (2) and (3) contains $r_i p_i = rp\cos\theta$, where θ is the angle between **r** and **p**.

☞ The r-dependences of the potential and field respectively, are charge: r^{-1} and r^{-2}; dipole: r^{-2} and r^{-3}; quadrupole: r^{-3} and r^{-4}, etc.

☞ At sufficiently large distances (distance $\gg$ dimensions of charge distribution), the potential and field are dominated by the leading non-vanishing term in $1/r$, all higher-order terms being negligible. Associated with the leading non-vanishing term is a unique multipole moment independent of choice of origin (see Question 2.23). Thus, at a distant point the SO_4^{2-} ion behaves like a charge, $HC\ell$ like a dipole, CO_2 like a quadrupole, etc., even though each has higher moments. If the distance to the field point is not sufficiently large, then higher-order terms contribute as well.

☞ The potential and field external to a neutral spherically symmetric charge distribution are both zero.

(iv) For a distribution of N stationary point charges $q^{(\alpha)}$ located at $\mathbf{r}'^{(\alpha)}$, where $\alpha = 1, 2, \ldots, N$, the discrete forms of (6) can be derived. So for the electric dipole moment

$$p_i = \int_v r_i' \rho(\mathbf{r}')dv', \tag{7}$$

[‡]Of course, this number may be reduced further by the symmetry of the charge distribution itself. For example, the quadrupole moment tensor of a uniformly charged sphere centred on the origin has only one independent component.

[6]   R. E. Raab and O. L. de Lange, *Multipole theory in electromagnetism*, Chap. 1, p. 4. Oxford: Clarendon Press, 2005.

where $\rho(\mathbf{r}') = \sum\limits_{\alpha=1}^{N} q^{(\alpha)} \delta(\mathbf{r}' - \mathbf{r}'^{(\alpha)})$, it follows that

$$p_i = \sum_{\alpha=1}^{N} \int_v r_i' q^{(\alpha)} \delta(\mathbf{r}' - \mathbf{r}'^{(\alpha)}) \, dv' = \sum_{\alpha=1}^{N} r_i'^{(\alpha)} q^{(\alpha)}, \tag{8}$$

because of (XI) of Appendix E. In a similar way we obtain

$$\left. \begin{array}{l} q_{ij} = \sum\limits_{\alpha=1}^{N} r_i'^{(\alpha)} r_j'^{(\alpha)} q^{(\alpha)} \\[2ex] q_{ijk} = \sum\limits_{\alpha=1}^{N} r_i'^{(\alpha)} r_j'^{(\alpha)} r_k'^{(\alpha)} q^{(\alpha)} \end{array} \right\}, \tag{9}$$

and so on for the higher multipole moments.

(v) ☞ For a point electric charge (at the origin) the scalar potential and electric field are given by the first terms of expansions (2) and (3) respectively:

$$\left. \begin{array}{l} \Phi(\mathbf{r}) = \dfrac{1}{4\pi\epsilon_0}\dfrac{q}{r} \quad \text{and} \\[3ex] E_i(\mathbf{r}) = \dfrac{1}{4\pi\epsilon_0}\dfrac{q r_i}{r^3} \end{array} \right\}. \tag{10}$$

☞ For a point electric dipole (at the origin) the scalar potential and electric field are given by the second terms of expansions (2) and (3) respectively:

$$\left. \begin{array}{l} \Phi(\mathbf{r}) = \dfrac{1}{4\pi\epsilon_0}\dfrac{p_i r_i}{r^3} \quad \text{and} \\[3ex] E_i(\mathbf{r}) = \dfrac{1}{4\pi\epsilon_0}\dfrac{3 r_i r_j - r^2 \delta_{ij}}{r^5} p_j \end{array} \right\}. \tag{11}$$

☞ For a point electric quadrupole (at the origin) the scalar potential and electric field are given by the third terms of expansions (2) and (3) respectively:

$$\left. \begin{array}{l} \Phi(\mathbf{r}) = \dfrac{1}{4\pi\epsilon_0}\dfrac{3 r_i r_j - r^2 \delta_{ij}}{2r^5} q_{ij} \quad \text{and} \\[3ex] E_i(\mathbf{r}) = \dfrac{1}{4\pi\epsilon_0}\dfrac{3\{5 r_i r_j r_k - r^2(r_i \delta_{jk} + r_j \delta_{ki} + r_k \delta_{ij})\}}{2r^7} q_{jk} \end{array} \right\}. \tag{12}$$

This hierarchy continues indefinitely.

Question 2.21

(a) Find the simplest linear arrangement of equispaced point charges for which the leading multipole moment is

☞ the electric quadrupole,

☞ the electric octopole.

Hint: Arrange the charges on the positive z-axis of Cartesian coordinates with q_1 positioned at the origin and neighbouring charges separated by a distance a. Then use the discrete definitions $(9)_1$ and $(9)_2$ of Question 2.20.

(b) For both the quadrupole and octopole arrangements above, plot the leading term in the multipole expansion of Φ and compare with the exact potential. In each case, choose a suitable system of units and use the interval $4 \leq z/a \leq 10$.

Solution

(a) quadrupole

We attempt a solution for an arrangement of three charges: q_1, q_2 and q_3. The conditions of the question require that

$$
\left.
\begin{aligned}
q_1 + q_2 + q_3 &= 0 \quad \text{(the distribution has no net charge)} \\
q_2 + 2q_3 &= 0 \quad \text{(the distribution has no dipole moment)}
\end{aligned}
\right\}.
$$

Solving these equations for q_2 and q_3 gives

$$
\left.
\begin{aligned}
q_1 &= q \\
q_2 &= -2q_1 = -2q \\
q_3 &= q_1 = q
\end{aligned}
\right\}. \tag{1}
$$

With the z-axis of Cartesian coordinates chosen to coincide with the line of charges, all components of q_{ij} are zero except q_{zz}, which is

$$
q_{zz} = \sum_{\alpha=1}^{3} z_\alpha z_\alpha q_\alpha = 0 + (-2q)(a)^2 + (q)(2a)^2 = 2qa^2. \tag{2}
$$

Since $q_{zz} \neq 0$, the required solution is given by (1).

octopole

Now we have

$$
\left.
\begin{aligned}
q_1 + q_2 + q_3 + q_4 &= 0 \quad \text{(the distribution has no net charge)} \\
q_2 + 2q_3 + 3q_4 &= 0 \quad \text{(the distribution has no dipole moment)} \\
q_2 + 4q_3 + 9q_4 &= 0 \quad \text{(the distribution has no quadrupole moment)}
\end{aligned}
\right\}.
$$

Solving these equations for q_2, q_3 and q_4 gives

$$\left.\begin{array}{l} q_1 = q \\ q_2 = -3q_1 = -3q \\ q_3 = 3q_1 = 3q \\ q_4 = -q_1 = -q \end{array}\right\}. \tag{3}$$

With the z-axis of Cartesian coordinates chosen to coincide with the line of charges, all components of q_{ijk} are zero except q_{zzz} which is

$$q_{zzz} = \sum_{\alpha=1}^{4} z_\alpha z_\alpha z_\alpha \, q_\alpha = 0 + (-3q)(a)^3 + (3q)(2a)^3 + (-q)(3a)^3 = -6qa^3. \tag{4}$$

Since $q_{zzz} \neq 0$, the required solution is given by (3).

(b) quadrupole

With the charges located on the positive z-axis, the exact potential obtained by superposition is

$$\Phi_{\text{ex}}(z) = \frac{1}{4\pi\epsilon_0}\left[\frac{q}{z} - \frac{2q}{z-a} + \frac{q}{z-2a}\right]$$

$$= \frac{1}{4\pi\epsilon_0}\frac{q}{a}\left[\frac{1}{z/a} - \frac{2}{z/a-1} + \frac{1}{z/a-2}\right]. \tag{5}$$

Using the quadrupole potential $\Phi(\mathbf{r}) = \frac{1}{4\pi\epsilon_0}\frac{3r_i r_j - r^2\delta_{ij}}{2r^5}q_{ij}$ and taking $q_{zz} = 2qa^2$ yield the approximate potential

$$\Phi_{\approx}(z) = \frac{1}{4\pi\epsilon_0}\frac{3z^3 - z^3}{2z^7}q_{zz} = \frac{1}{4\pi\epsilon_0}\frac{q}{a}\frac{2a^3}{z^3}. \tag{6}$$

Letting $\frac{1}{4\pi\epsilon_0}\frac{q}{a} = 1$ in (5) and (6) leads to the plot shown on p. 95.

octopole

Proceeding as before,

$$\Phi_{\text{ex}}(z) = \frac{1}{4\pi\epsilon_0}\left[\frac{q}{z} - \frac{3q}{z-a} + \frac{3q}{z-2a} - \frac{q}{z-3a}\right]$$

$$= \frac{1}{4\pi\epsilon_0}\frac{q}{a}\left[\frac{1}{z/a} - \frac{3}{z/a-1} + \frac{3}{z/a-2} - \frac{1}{z/a-3}\right]. \tag{7}$$

Using the octopole potential $\Phi(\mathbf{r}) = \frac{15r_i r_j r_k - 3r^2(r_i\delta_{jk} + r_j\delta_{ki} + r_k\delta_{ij})}{2r^7}q_{ijk}$ and taking $q_{zzz} = -6qa^3$ yield the approximate potential

$$\Phi_{\approx}(z) = \frac{1}{4\pi\epsilon_0}\frac{5z^3 - 3z^3}{2z^7}q_{zzz} = -\frac{1}{4\pi\epsilon_0}\frac{q}{a}\frac{6a^4}{z^4}. \tag{8}$$

Substituting $\frac{1}{4\pi\epsilon_0}\frac{q}{a} = -1$ in (7) and (8) leads to the plot shown on p. 95.

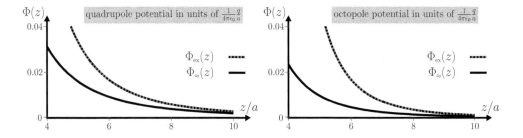

Comments

(i) Equations (1) and (2) are the simplest solutions to the question, in the sense that no solution can be found for fewer than three charges for the quadrupole and fewer than four charges for the octopole.

(ii) As expected, the graphs confirm that the multipole approximations are inadequate as $z/a \to 0$ but become progressively better as $z/a \to \infty$.

(iii) Methane (although not linear in shape) is an example of a molecule that has a leading electric octopole moment.

Question 2.22

Charge q is distributed uniformly

(a) over the surface of a sphere of radius a, and

(b) throughout the volume of a sphere of radius a.

For each of these charge distributions, determine the electric dipole and quadrupole moments about the origin O (chosen at the centre of the sphere).

Solution

(a) Due to the spherical symmetry, it is obvious that the dipole moment about O is zero. For the components of the quadrupole tensor, symmetry again requires that $q_{xx} = q_{yy} = q_{zz}$. Clearly, this latter component is $q_{zz} = \oint_s z^2 dq$, where $dq = \sigma \, da'$ is an element of charge on the surface and $\sigma = \dfrac{q}{4\pi a^2}$. In spherical polar coordinates $da' = a^2 \sin\theta' \, d\theta' \, d\phi'$ and $z = a \cos\theta'$. So

$$q_{zz} = \frac{q \, a^2}{4\pi} \int_0^{2\pi} \int_0^{\pi} \cos^2\theta' \sin\theta' \, d\theta' \, d\phi' = \tfrac{1}{2} q a^2 \int_{-1}^{1} \cos^2\theta' \, d\cos\theta' = \tfrac{1}{3} q a^2.$$

Similarly $q_{xz} = q_{yz} = q_{xy}$, and because $x = a \sin\theta' \cos\phi'$ we have

$$q_{xz} = \frac{q\,a^2}{4\pi}\int_0^{2\pi}\int_0^{\pi}\sin^2\theta'\cos\theta'\cos\phi'\,d\theta'\,d\phi' = 0.$$

Thus all components of q_{ij} are zero except $q_{xx} = q_{yy} = q_{zz} = \frac{1}{3}qa^2$.

(b) As before, the p_i about O are all zero for symmetry reasons, and $q_{xx} = q_{yy} = q_{zz}$ where $q_{zz} = \int_v z^2 dq = \rho\int_v z^2 dv'$ for a charge element inside the volume. Now $\rho = \frac{q}{\frac{4}{3}\pi a^3}$, $dv' = r'^2\sin\theta'\,dr'\,d\theta'\,d\phi'$ and $z = r'\cos\theta'$. Thus,

$$q_{zz} = \frac{3q}{4\pi a^3}\int_0^{2\pi}\int_0^{\pi}\int_0^{a}r'^4\cos^2\theta'\sin\theta'\,dr'\,d\theta'\,d\phi' = \frac{3qa^2}{10}\int_{-1}^{1}\cos^2\theta'\,d(\cos\theta') = \frac{1}{5}qa^2.$$

Similarly $q_{xz} = q_{yz} = q_{xy}$, and because $x = r'\sin\theta'\cos\phi'$, we have

$$q_{xz} = \frac{3q}{4\pi a^3}\int_0^{2\pi}\int_0^{\pi}\int_0^{a}r'^4\sin^2\theta'\cos\theta'\cos\phi'\,dr'\,d\theta'\,d\phi' = 0.$$

Thus all components of q_{ij} are zero except $q_{xx} = q_{yy} = q_{zz} = \frac{1}{5}qa^2$.

Comment

Since this distribution has a net charge, the dipole and quadrupole moments calculated above are origin-dependent quantities. They are not unique properties characterizing the charged sphere, as we consider now in Question 2.23.

Question 2.23

Determine the effect of a change in the coordinate origin on the electric dipole and quadrupole moments.

Hint: Use $(6)_2$ and $(6)_3$ of Question 2.20.

Solution

In the dipole and quadrupole moment definitions let $dq = \rho(\mathbf{r}')dv'$. Then, relative to coordinate origins O and $\bar{O}$, the dipole moments are

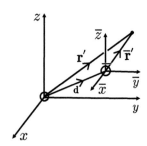

$$p_i = \int_v r_i'\,dq \quad\text{and}\quad \bar{p}_i = \int_v \bar{r}_i'\,d\bar{q} \qquad (1)$$

respectively. From the vector triangle shown in the figure alongside it is clear that $\bar{r}_i' = r_i' - d_i$, where the d_i are constants. Now $dq = d\bar{q}$ since electric charge is an invariant quantity, and $(1)_2$ can be written as

$$\bar{p}_i = \int_v (r'_i - d_i) dq = \int_v r'_i dq - d_i q, \tag{2}$$

where q is the total charge of the distribution. Substituting $(1)_1$ in (2) gives

$$\bar{p}_i = p_i - d_i q. \tag{3}$$

Similarly, for the quadrupole moment we have

$$\bar{q}_{ij} = \int_v \bar{r}'_i \bar{r}'_j \, d\bar{q} = \int_v (r'_i - d_i)(r'_j - d_j) \, dq$$

$$= q_{ij} - p_i d_j - p_j d_i + q d_i d_j. \tag{4}$$

Equations (3) and (4) are the desired results for the origin dependences of p_i and q_{ij}.

Comments

(i) Equations (3) and (4) are also valid for discrete distributions of charge.

(ii) It follows from (3) that the dipole moment p_i is origin-independent only if the charge distribution is neutral ($q = 0$).

(iii) Similarly, from (4) the quadrupole moment q_{ij} is origin-independent only if the total charge q and the dipole moment $\mathbf{p}$ are both zero. So, the electric quadrupole moment of a dipolar molecule (one with $p_i \neq 0$) is origin-dependent. Nevertheless, a value for q_{ij} has been measured relative to a particular molecular origin for certain dipolar molecules.[7,8]

(iv) These conclusions regarding origin dependence of p_i and q_{ij} can be generalized: only the leading non-vanishing electric multipole moment of an arbitrary charge distribution is origin-independent.

(v) In multipole theory, origin independence is a necessary criterion for the validity of an expression for an observable property which is expected to be independent of origin.[9]

(vi) If $q \neq 0$, (3) shows that $\mathbf{d} = \mathbf{p}/q$ gives $\bar{\mathbf{p}} = 0$. It is always possible to find an origin $\bar{O}$ about which the electric dipole moment of a *charged* distribution is zero. By contrast, no origin shift can be made which reduces the higher multipole moments to zero.‡

‡To see this for q_{ij}, put $\bar{q}_{ij} = 0$ in (4) and write an equation for each independent quadrupole component (six in total). One is easily convinced that no solution can be found, since the three d_i cannot satisfy these six independent equations simultaneously.

[7]   A. D. Buckingham and H. C. Longuet-Higgins, 'The quadrupole moments of dipolar molecules', *Molecular Physics*, vol. 14, pp. 63–72, 1968.

[8]   A. D. Buckingham, R. L. Disch and D. A. Dunmur, 'The quadrupole moments of some simple molecules', *Journal of the American Chemical Society*, vol. 90, pp. 3104–7, 1968.

[9]   O. L. de Lange, R. E. Raab and A. Welter, 'On the transition from microscopic to macroscopic electrodynamics', *Journal of Mathematical Physics*, vol. 53, pp. 013513-1–17, 2012.

Question 2.24

Consider, as a model, a point electric quadrupole at the origin O of Cartesian coordinates as the only source of potential and field. Axes are chosen such that all components of q_{ij} are zero except q_{zz}.

(a) Show that the electric potential is given by

$$\Phi(x, y, z) = \frac{q_{zz}}{8\pi\epsilon_0} \frac{(-x^2 - y^2 + 2z^2)}{(x^2 + y^2 + z^2)^{5/2}}. \tag{1}$$

(b) Show that the electric-field components are given by

$$
\left.
\begin{aligned}
E_x(x, y, z) &= \frac{3q_{zz}}{8\pi\epsilon_0} \frac{(-x^3 - xy^2 + 4xz^2)}{(x^2 + y^2 + z^2)^{7/2}} \\[2mm]
E_y(x, y, z) &= \frac{3q_{zz}}{8\pi\epsilon_0} \frac{(-x^2 y - y^3 + 4yz^2)}{(x^2 + y^2 + z^2)^{7/2}} \\[2mm]
E_z(x, y, z) &= \frac{3q_{zz}}{8\pi\epsilon_0} \frac{(-3x^2 z - 3y^2 z + 2z^3)}{(x^2 + y^2 + z^2)^{7/2}}
\end{aligned}
\right\}. \tag{2}
$$

(c) Use *Mathematica* to plot the quadrupole field in the xz-plane at $y = 0$. Suppose $3q_{zz}/8\pi\epsilon_0 = 1$ and take $x, z \in [-4, 4]$ in an arbitrary system of units. On the plot, also show some lines of constant potential.

Solution

(a) Use $(12)_1$ of Question 2.20 with $r^2 = (x^2 + y^2 + z^2)$. The result follows directly, taking $i = j = z$ and contracting subscripts.

(b) Use $(12)_2$ of Question 2.20 with $r^2 = (x^2 + y^2 + z^2)$. The results for $i = x$ or y or z follow directly, taking $j = k = z$ and contracting subscripts.

(c) The notebook below was used to generate the plot shown on p. 99. Lines of electric field given by $E_y = 0$ are drawn with arrows; the dot-dashed lines represent the intersections of the equipotential surfaces with the xz-plane. The two straight lines are for $\Phi = 0$ and they divide the xz-plane into four regions. The upper and lower regions have $\Phi > 0$. The left and right regions have $\Phi < 0$.

```
In[1]:= Φ[x_, z_] := (2 z² - x²) / (x² + z²)^(5/2);

Ex[x_, z_] = -D[Φ[x, z], x]; Ez[x_, z_] = -D[Φ[x, z], z];
gr1 = ContourPlot[Φ[x, z], {x, -5, 5}, {z, -5, 5}];
gr2 = StreamPlot[{Ex[x, z], Ez[x, z]}, {x, -5, 5}, {z, -5, 5}];
Show[{gr1, gr2}]
```

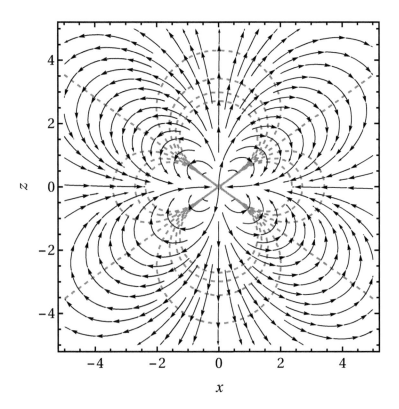

Question 2.25

Two point charges q and $-q$ are a distance $2a$ apart. Choose Cartesian axes such that the coordinates of these charges are $(0, 0, a)$ and $(0, 0, -a)$ respectively.

(a) Use the principle of superposition to show that the (exact) electric potential Φ and field $\mathbf{E}$ in the xz-plane are

$$\Phi = \frac{1}{4\pi\epsilon_0}\frac{q}{a}\left[\frac{1}{\{X^2+(Z-1)^2\}^{1/2}} - \frac{1}{\{X^2+(Z+1)^2\}^{1/2}}\right] \tag{1}$$

and

$$\left.\begin{aligned} E_x &= \frac{1}{4\pi\epsilon_0}\frac{q}{a^2}\left[\frac{X}{\{X^2+(Z-1)^2\}^{3/2}} - \frac{X}{\{X^2+(Z+1)^2\}^{3/2}}\right] \\[2mm] E_y &= 0 \\[2mm] E_z &= \frac{1}{4\pi\epsilon_0}\frac{q}{a^2}\left[\frac{(Z-1)}{\{X^2+(Z-1)^2\}^{3/2}} - \frac{(Z+1)}{\{X^2+(Z+1)^2\}^{3/2}}\right] \end{aligned}\right\} \tag{2}$$

Here $X = x/a$ and $Z = z/a$ are dimensionless coordinates.

(b) Using the multipole expansions of Question 2.20, and retaining terms up to and including quadrupole order only, show that relative to the origin

$$\Phi \simeq \frac{1}{4\pi\epsilon_0}\frac{q}{a}\left[\frac{X^2(-1+2Z) + 2Z^2(1+Z)}{\{X^2+Z^2\}^{5/2}}\right] \tag{3}$$

and

$$\left.\begin{aligned} E_x &\simeq \frac{1}{4\pi\epsilon_0}\frac{q}{a^2}\left[\frac{3X\{X^2(1+2Z) + Z^2(-1+2Z)\}}{\{X^2+Z^2\}^{7/2}}\right] \\[2mm] E_y &= 0 \\[2mm] E_z &\simeq \frac{1}{4\pi\epsilon_0}\frac{q}{a^2}\left[\frac{X^2(-2X^2 - 9Z + 4Z^2) + 2Z^2(3Z + 2Z^2 - X^2)}{\{X^2+Z^2\}^{7/2}}\right] \end{aligned}\right\}. \tag{4}$$

(c) Use *Mathematica* to plot the exact equipotentials and field, (1) and (2). Suppress the terms $q/4\pi\epsilon_0 a$ and $q/4\pi\epsilon_0 a^2$ and take X, $Z \in [-20, 20]$. Repeat for the approximate equipotentials and field, (3) and (4).

Solution

(a) Superposition gives

$$\Phi = \Phi_+ + \Phi_- = \frac{1}{4\pi\epsilon_0}\left(\frac{q}{r_+} - \frac{q}{r_-}\right) \quad \text{and} \quad \mathbf{E} = \mathbf{E}_+ + \mathbf{E}_- = \frac{1}{4\pi\epsilon_0}\left(\frac{q\mathbf{r}_+}{r_+^3} - \frac{q\mathbf{r}_-}{r_-^3}\right), \tag{5}$$

where

$$\mathbf{r}_\pm = x\hat{\mathbf{x}} + (z \mp a)\hat{\mathbf{z}} \tag{6}$$

are the position vectors of the field point at $(x, 0, z)$ from the $\genfrac{}{}{0pt}{}{\text{positive}}{\text{negative}}$ charge. Substituting (6) in (5) and using the definitions of X and Z yield (1) and (2).

(b) The net charge of the distribution is zero. So

$$\Phi \simeq \Phi_{\text{dipole}} + \Phi_{\text{quadrupole}} \qquad \text{and} \qquad \mathbf{E} \simeq \mathbf{E}_{\text{dipole}} + \mathbf{E}_{\text{quadrupole}}, \tag{7}$$

since we neglect contributions from the electric octopole and all higher terms. Substituting $(11)_1$ and $(12)_1$ of Question 2.20 in $(7)_1$ and rearranging give (3). Substituting $(11)_2$ and $(12)_2$ of Question 2.20 in $(7)_2$ and rearranging give (4).

(c) Selecting the appropriate expression for Φ and evaluating the notebook on p. 101 gives the graphs shown:

In[1]:= $\Phi[\mathbf{x\_}, \mathbf{z\_}] := \dfrac{1}{\left(\mathbf{x}^2 + (\mathbf{z} - 1)^2\right)^{1/2}} - \dfrac{1}{\left(\mathbf{x}^2 + (\mathbf{z} + 1)^2\right)^{1/2}};$

```
(* EXACT POTENTIAL *)

(* Φ[x_,z_]:= x²(-1+2z)+2z²(1+z) / (x²+z²)^(5/2) ;      (*  APPROXIMATE  POTENTIAL *) *)

Ex[x_, z_] = -D[Φ[x, z], x];  Ez[x_, z_] = -D[Φ[x, z], z];

gr1 = ContourPlot[Φ[x, z], {x, -20, 20}, {z, -20, 20}];

gr2 = StreamPlot[{Ex[x, z], Ez[x, z]}, {x, -20, 20}, {z, -20, 20}];

Show[{gr1, gr2}]
```

exact field

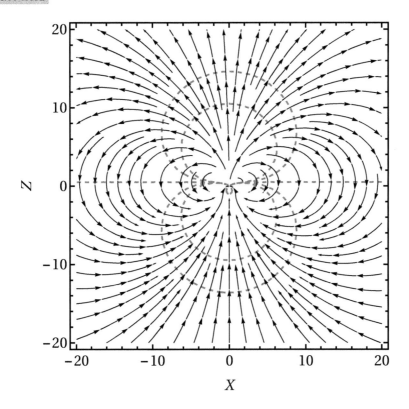

approximate field

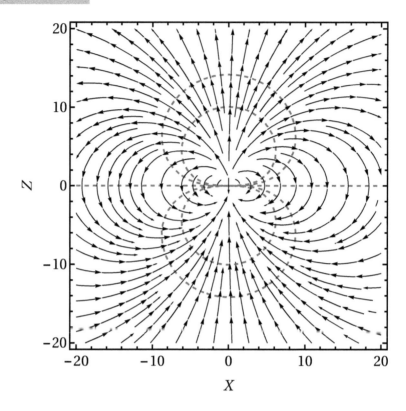

Comments

(i) Changing the plot range of these graphs by a factor of two is an instructive exercise. With X, $Z \in [-10, 10]$ significant differences between the exact and approximate fields become apparent. This is not surprising, since the multipole expansion is not well converged near the origin: the contributions from the higher-order multipoles are significant here. On the other hand, the graphs obtained for X, $Z \in [-40, 40]$ are essentially identical, these being the far field of the dipole.[‡]

(ii) Because the exact forms (1) and (2) for Φ and $\mathbf{E}$ above are not significantly more complicated than the approximate forms (3) and (4), the reader may reasonably ask: why work with the approximate forms at all? The answer lies in the nature of the charge distribution. For simple systems (e.g. continuous arrangements having high symmetry or arbitrary configurations with a small number of point charges) it is often straightforward to derive exact expressions for the potential and field. By contrast, arbitrary distributions involving a large number of charges can usually be studied only by using a suitably truncated multipole expansion. The resulting approximate forms for Φ and $\mathbf{E}$ are then the *only* equations that are available.

[‡]Of course, with this scale the sensitivity of the plot is reduced and the near-field differences are masked to some extent anyway.

Question 2.26

Consider the following model of an atom: a point nucleus (charge q) is surrounded by a spherical electron 'cloud' having radius a and charge $-q$. The atom is placed in a weak uniform electrostatic field $\mathbf{E}_0$ which produces a small displacement $\mathbf{d}$ between the centre of the cloud and the nucleus (taken to be the origin O). Assume the following:

1. the displaced electron cloud remains spherical (i.e. it is not distorted by the field), and

2. the charge $-q$ in the cloud is uniformly distributed throughout the volume[‡] $\frac{4}{3}\pi a^3$.

(a) Show that the field $\mathbf{E}_0$ induces in the atom an electric dipole whose moment is

$$\mathbf{p} = \alpha\,\mathbf{E}_0, \tag{1}$$

where

$$\alpha = 4\pi\epsilon_0 a^3. \tag{2}$$

(b) Suppose that the atom is hydrogen. Take $a =$ the Bohr radius $= 0.59\,\text{Å}$. With $E_0 = 30\,\text{kV cm}^{-1}$ as an upper limit for a laboratory field,

☞ show that $d/a \ll 1$ (which justifies the first assumption above), and

☞ calculate p.

Solution

(a) The displacement of the cloud's centre relative to the nucleus continues until the field $\mathbf{E}$ at O due to the cloud[♯] cancels the external field $\mathbf{E}_0$; that is, $\mathbf{E} + \mathbf{E}_0 = 0$ or

$$\mathbf{E}_0 = \frac{1}{4\pi\epsilon_0}\frac{q\mathbf{d}}{a^3}. \tag{3}$$

But $\mathbf{p} = q\mathbf{d}$, which is (1) with α given by (2).

(b) ☞ Substituting the numerical values in (3) gives

$$d/a = \frac{4\pi\epsilon_0 E_0 a^2}{q} = \frac{3\times 10^6}{9\times 10^9} \times \frac{(0.59\times 10^{-10})^2}{1.6\times 10^{-19}}$$

$$\simeq 7\times 10^{-6} \ll 1.$$

☞ $p = qd = 1.6\times 10^{-19}\times 7\times 10^{-6}\times 0.59\times 10^{-10} \simeq 7\times 10^{-35}\,\text{C m}.$

[‡]This is obviously not a good assumption for real atoms and is the weakest part of the above model.

[♯]The required field also follows directly from Gauss's law (see (6) of Question 2.3). For a volume

density $\rho = \dfrac{-q}{\frac{4}{3}\pi a^3}$, the flux $E\times 4\pi d^2 = \dfrac{1}{\epsilon_0}\displaystyle\int_v \rho\,dv = \dfrac{-q}{\epsilon_0}\left(\dfrac{d}{a}\right)^3$ or $E = -\dfrac{1}{4\pi\epsilon_0}\dfrac{q\,d}{a^3}$.

Comments

(i) The quantity α in (1) is the polarizability. It is a property of a charge distribution (atom, molecule, etc.). According to our atomic model, its value is proportional (through a^3) to the volume of the atom. We expect that, all else being equal, big atoms have larger polarizabilities than small atoms. This reflects the fact that electrons in outer shells, being further from the nucleus, are more easily displaced by an external electric field. Although (2) was derived for a primitive atomic model, it nevertheless gives correct order-of-magnitude results for many simple atoms.

(ii) An exact quantum-mechanical calculation (using static perturbation theory) for the ground-state polarizability of a hydrogen-like atom yields

$$\alpha = \frac{9}{2}\frac{4\pi\epsilon_0 a_0^3}{Z^4},\tag{4}$$

where $a_0 = \dfrac{4\pi\epsilon_0\hbar^2}{m_e e^2}$ is the Bohr radius and Z the atomic number. That (4) is different from (2) is hardly surprising in view of the second assumption in the statement of the question on p. 103.

(iii) Molecules that possess a permanent electric dipole moment p_0 are called polar molecules, and as an order of magnitude

p_0 = electronic charge × bond length

$= 1.6 \times 10^{-19} \times 10^{-10} \simeq 10^{-29}\,\mathrm{C\,m} \gg p$ calculated in (b) above.

This is an instructive result because it shows that an electric dipole moment induced by a field as strong as $30\,\mathrm{kV\,cm^{-1}}$ is very much less than a permanent dipole moment (when the latter exists).

(iv) In a strong external field, one might expect a series expansion for p of the form

$$p = \alpha E_0 + \beta E_0^2 + \gamma E_0^3 + \lambda E_0^4 + \cdots.\tag{5}$$

The coefficients β, γ, λ,... are properties of the atom (molecule, etc.) and are known collectively as the hyperpolarizability coefficients. These non-linear contributions to p are not always negligible and they can produce important physical effects (e.g. the Kerr effect[†] in a gas of spheres[10]).

(v) Because the atom in our model has a centre of symmetry, it turns out that all coefficients of the even powers of E_0 in (5) are necessarily zero, since the induced dipole must reverse when the field reverses; that is, **p** is an odd function of the polarizing field:

$$p = \alpha E_0 + \gamma E_0^3 + \epsilon E_0^5 + \cdots.\tag{6}$$

[†] Electric-field induced birefringence in an isotropic material.

[10]   A. D. Buckingham and J. A. Pople, 'Theoretical studies of the Kerr effect I: Deviations from a linear polarization law', *Proceedings of the Physical Society*, vol. 68, pp. 905–9, 1955.

(vi) For a molecule of arbitrary symmetry in a uniform electrostatic field, a field component E_x (say) could induce a dipole moment p_y or p_z. So in general,

$$\left.\begin{aligned}
p_x &= \alpha_{xx} E_x + \alpha_{xy} E_y + \alpha_{xz} E_z \\
p_y &= \alpha_{yx} E_x + \alpha_{yy} E_y + \alpha_{yz} E_z \\
p_z &= \alpha_{zx} E_x + \alpha_{zy} E_y + \alpha_{zz} E_z
\end{aligned}\right\}, \tag{7}$$

indicating that the polarizability α has nine components (a tensor of rank two). Equation (7) can be expressed in the form

$$p_i = \alpha_{ij} E_j, \tag{8}$$

where α_{ij} is called the polarizability tensor. It is symmetric in its subscripts, as we show in Question 2.29.

(vii) All the property tensors of a spherically symmetric atom are isotropic (this is guaranteed by Neumann's principle; see Comment (vii) of Question 1.11). In particular, its second-rank property tensors are proportional to δ_{ij} (see Comment (iii) of Question 1.1) and so we have

$$\alpha_{ij} = \alpha\,\delta_{ij}, \tag{9}$$

where α is a positive constant.

(viii) It is important to realise that the Cartesian tensor α_{ij} is more than just a tensor, because, in addition to satisfying the tensor transformation laws discussed in Appendix A, it also relates the effect p_i to the cause E_j acting on a sample of matter. Thus, α_{ij} and other tensors entering similar physical cause-and-effect relationships are called *physical property tensors*. The tensors such as E_j and p_i, representing the cause and effect, are called *physical tensors*. Clearly, not all tensors are physical tensors or even physical property tensors. For example, the Kronecker delta and the Levi-Civita tensor are neither. See also Comment (v) of Question 1.11.

Question 2.27

Consider a continuous distribution of electric charge located in an external electrostatic field $\mathbf{E}(\mathbf{r})$. Show that the resultant force $\mathbf{F}$ and torque $\boldsymbol{\Gamma}$ exerted on the distribution are given by

(a) $\quad F_i = q E_i + p_j \nabla_j E_i + \frac{1}{2} q_{jk} \nabla_j \nabla_k E_i + \frac{1}{6} q_{jk\ell} \nabla_j \nabla_k \nabla_\ell E_i + \cdots$, and $\qquad$ (1)

(b) $\quad \Gamma_i = \varepsilon_{ijk} \left(p_j E_k + q_{j\ell} \nabla_\ell E_k + \frac{1}{2} q_{j\ell m} \nabla_\ell \nabla_m E_k + \cdots \right)$. $\qquad$ (2)

In (1) and (2), $\mathbf{E}$ and its derivatives are understood to be evaluated at an arbitrarily chosen origin O.

Solution

(a) The total force on the distribution is $\mathbf{F} = \int_v \mathbf{E}\,dq = \int_v \rho(\mathbf{r}')\mathbf{E}(\mathbf{r}')\,dv'$, and so

$$F_i = \int_v \rho E_i(\mathbf{r}')\,dv'. \tag{3}$$

For a slowly varying field, $E_i(\mathbf{r}')$ above is given by (3) of Question 1.7:

$$E_i(\mathbf{r}') = (E_i)_0 + (\nabla_j E_i)_0 r'_j + \tfrac{1}{2}(\nabla_j \nabla_k E_i)_0 r'_j r'_k + \tfrac{1}{6}(\nabla_j \nabla_k \nabla_\ell E_i)_0 r'_j r'_k r'_\ell + \cdots. \tag{4}$$

Substituting (4) in (3) and using (6) of Question 2.20 yield (1).

(b) The net torque is $\boldsymbol{\Gamma} = \int_v \mathbf{r}' \times d\mathbf{F} = \int_v \mathbf{r}' \times \mathbf{E}\,dq = \int_v \rho(\mathbf{r}')\{\mathbf{r}' \times \mathbf{E}(\mathbf{r}')\}\,dv'$. So

$$\Gamma_i = \varepsilon_{ijk}\int_v \rho\,r'_j E_k(\mathbf{r}')\,dv'$$

$$= \varepsilon_{ijk}\left\{ E_k \int_v \rho\,r'_j\,dv' + (\nabla_\ell E_k)\int_v \rho\,r'_j r'_\ell\,dv' + \tfrac{1}{2}(\nabla_\ell \nabla_m E_k)\int_v \rho\,r'_j r'_\ell r'_m\,dv' + \cdots \right\}. \tag{5}$$

Equations (6) of Question 2.20 give (2).

Comments

(i) Equations (1) and (2) confirm the following elementary results:

☞ In a uniform field, a neutral charge distribution ($q = 0$) experiences no force.

☞ An electric dipole experiences a net force in a non-uniform field where

$$\mathbf{F} = (\mathbf{p} \cdot \boldsymbol{\nabla})\mathbf{E}. \tag{6}$$

☞ An electric dipole experiences a net torque in a uniform field where

$$\boldsymbol{\Gamma} = \mathbf{p} \times \mathbf{E}. \tag{7}$$

(ii) Suppose that the charges q_i of the distribution are able to move[‡] relative to one another. In the presence of an external field, these charges will establish new[♯] equilibrium positions r'_i. It follows from their definitions that the multipole moments of the polarized distribution will be different from those of the unpolarized distribution. The multipole moments in (1) and (2) are understood to be those of the *polarized* distribution.

[‡]The distribution is said to be non-rigid.

[♯]A process usually referred to as polarization.

Question 2.28*

A charge distribution is moved quasi-statically from a field-free region of space into an electrostatic field $\mathbf{E}(\mathbf{r})$. Show that the potential energy U of the final configuration is given by

$$U = \int_i^f \left[q\,(d\Phi)_0 \; - \; p_i\, d(E_i)_0 \; - \; \tfrac{1}{2} q_{ij}\, d(\nabla_j E_i)_0 \; - \; \cdots \right], \tag{1}$$

where the limits i and f denote the initial and final configurations.

Hint: Start by expressing the work W done by the external agent to move the distribution as the line integral of the applied force $\mathbf{F}_e$.

Solution

The potential energy $U = W = \displaystyle\int_i^f \mathbf{F}_e \cdot d\mathbf{r}$. But $\mathbf{F}_e = -\mathbf{F}$, where $\mathbf{F}$ is the force given by (1) of Question 2.27. So

$$U = -\int_i^f F_i\, dr_i = -\int_i^f \left[qE_i + p_j \nabla_j E_i + \tfrac{1}{2} q_{jk} \nabla_j \nabla_k E_i + \cdots \right] dr_i. \tag{2}$$

As in Question 2.27, it is understood that the field in (2) and its derivatives are evaluated at the origin. Using the properties $E_i = -\nabla_i \Phi$ and $\nabla_i E_j = \nabla_j E_i$ for any electrostatic field (see (5) of Question 1.3) yields

$$U = -\int_i^f \left[-q\nabla_i \Phi + p_j \nabla_i E_j + \tfrac{1}{2} q_{jk} \nabla_j \nabla_i E_k + \cdots \right] dr_i$$

$$= \int_i^f q\, \frac{\partial \Phi}{\partial r_i}\, dr_i - \int_i^f p_j\, \frac{\partial E_j}{\partial r_i}\, dr_i - \frac{1}{2} \int_i^f q_{jk}\, \frac{\partial(\nabla_k E_j)}{\partial r_i}\, dr_i - \cdots$$

$$= \int_i^f q\, d\Phi - \int_i^f p_i\, dE_i - \frac{1}{2} \int_i^f q_{ij}\, d(\nabla_j E_i) - \cdots, \tag{3}$$

where, in the last step, we recognize that $\dfrac{\partial f}{\partial r_i}\, dr_i$ is the total differential of the quantity f (i.e. df). This proves the result, since (3) is the same as (1).

Comments

(i) We emphasize the following features of equation (1):

☞ The result applies only to electrostatic fields,

☞ The result applies to both rigid and non-rigid (i.e. polarizable) charge distributions,

☞   Each point multipole at the origin interacts with the potential or gradients
of the potential, evaluated at itself. Thus,

- a monopole (charge) interacts with Φ,
- a dipole interacts with the field $(-\nabla_i\Phi)$,
- a quadrupole interacts with the field gradient $(-\nabla_j\nabla_i\Phi)$, and so on.

(ii) Suppose a neutral spherically symmetric atom (having polarizability α) is placed
in a weak uniform electrostatic field $\mathbf{E}_0$. The energy required to polarize the atom
is given by (3):

$$U = -\int_0^{\mathbf{E}_0} p_i\,dE_i = -\tfrac{1}{2}\alpha E_0^2, \qquad (4)$$

since $p_i = \alpha E_i$. The corresponding result for a charge distribution with arbitrary
symmetry is $U = -\tfrac{1}{2}\,\alpha_{ij}E_{0i}E_{0j}$.

(iii) For a rigid charge distribution the multipole moments in (1) are constants that
are independent of the external field. Their values are equal to the permanent
moments defined in Question 2.20. Integrating (1) gives

$$U = q\Phi_0 - p_i\,(E_i)_0 - \tfrac{1}{2}q_{ij}\,(\nabla_j E_i)_0 - \cdots . \qquad (5)$$

(iv) In the special case of a rigid neutral distribution in a uniform electrostatic field
(3) reduces to

$$U = -\mathbf{p}\cdot\mathbf{E}. \qquad (6)$$

(v) The figure below shows three different orientations of a point electric dipole in an
external electrostatic field $\mathbf{E}$:

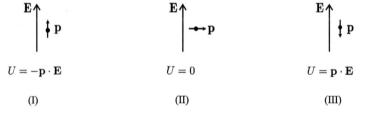

$$U = -\mathbf{p}\cdot\mathbf{E} \qquad\qquad U = 0 \qquad\qquad U = \mathbf{p}\cdot\mathbf{E}$$

(I) \qquad\qquad\qquad (II) \qquad\qquad\qquad (III)

In (I) the configuration is most stable (U is a minimum) and in (III) it is least
stable (U is a maximum).

(vi) It is clear from Comment (v) that the potential energy of an electric dipole is
$U \sim pE$. Because permanent dipole moments in molecules are of the order
$10^{-29}\,\mathrm{C\,m}$ $\big($see Comment (iii) of Question 2.26$\big)$, and with $E = 10^6\,\mathrm{V\,m^{-1}}$ (which
is typical for the maximum field in a gas before electrical breakdown occurs)
we obtain $U \sim 10^{-23}\,\mathrm{J}$. Above standard temperature $(273\,\mathrm{K})$ the mean kinetic
energy $\tfrac{3}{2}k_{\mathrm{B}}T$ of a gas molecule is at least 300 times larger than U, implying
that the random thermal motion is essentially unaffected by the dipole potential
energy: molecules are found in all orientations relative to the field.

Question 2.29**

Consider three parallel-plate capacitors X, Y and Z, each connected via an open switch to a battery. When charged, the capacitors produce mutually orthogonal uniform electrostatic fields (E_{0x}, E_{0y} and E_{0z}, say) in a region v of space. Suppose that, within v, there is a neutral electric-charge distribution whose polarizability is α_{ij}. In the following gedanken‡ experiments, two capacitors are switched on in sequence. The third capacitor remains uncharged.

(a) Use (8) of Question 2.26 to calculate the work required to polarize the distribution when the field from each capacitor is turned on in the following order:

☞ X then Y,

☞ Y then X.

(b) Use the results of these experiments (and the other possible combinations like Y and Z) to deduce an important property of the polarizability tensor.

Solution

(a) All fields are initially zero, and the work required to polarize the distribution$^{\#}$ is given by (3) of Question 2.28:

$$W = -\int_i^f p_i \, dE_i, \tag{1}$$

because $q = 0$ (the distribution is neutral) and the fields are uniform (the field gradients are all zero). Substituting $p_i = \alpha_{ij} E_j$ in (1) gives

$$W = -\alpha_{ij} \int_i^f E_j \, dE_i. \tag{2}$$

☞ The work done by the field of X is

$$W(E_{0x}, 0, 0) = -\alpha_{xj} \int_0^{E_{0x}} E_j \, dE_x = -\alpha_{xx} \int_0^{E_{0x}} E_x \, dE_x = -\tfrac{1}{2}\alpha_{xx} E_{0x}^2.$$

With X fully charged, the work done by the field of Y is

$$W(E_{0x}, E_{0y}, 0) = -\alpha_{yj} \int_0^{E_{0y}} E_j \, dE_y = -\alpha_{yx} E_{0x} E_{0y} - \tfrac{1}{2}\alpha_{yy} E_{0y}^2.$$

So the net energy required to polarize the distribution is

$$W_{XY} = W(E_{0x}, 0, 0) + W(E_{0x}, E_{0y}, 0) = -\tfrac{1}{2}\alpha_{xx} E_{0x}^2 - \alpha_{yx} E_{0x} E_{0y} - \tfrac{1}{2}\alpha_{yy} E_{0y}^2. \tag{3}$$

---

‡From the German word meaning 'thought'.

$^{\#}$The work required to charge up the capacitors is of no interest here.

☞ Repeating the experiment in reverse sequence gives

$$W(0, E_{0y}, 0) = -\alpha_{yj} \int_0^{E_{0y}} E_j \, dE_y = -\alpha_{yy} \int_0^{E_{0y}} E_y \, dE_y = -\tfrac{1}{2}\alpha_{yy} E_{0y}^2.$$

With Y fully charged, the work done by the field of X is

$$W(E_{0x}, E_{0y}, 0) = -\alpha_{xj} \int_0^{E_{0x}} E_j \, dE_x = -\alpha_{xy} E_{0x} E_{0y} - \tfrac{1}{2}\alpha_{xx} E_{0x}^2.$$

In this case, the net energy required to polarize the distribution is

$$W_{YX} = W(0, E_{0y}, 0) + W(E_{0x}, E_{0y}, 0) = -\tfrac{1}{2}\alpha_{yy} E_{0y}^2 - \alpha_{xy} E_{0x} E_{0y} - \tfrac{1}{2}\alpha_{xx} E_{0x}^2. \quad (4)$$

(b) Since electrostatic fields are conservative, the work done in polarizing the charge distribution must be independent of history. Thus $W_{XY} = W_{YX}$ and from (3) and (4) we require that $\alpha_{xy} = \alpha_{yx}$. Repeating the above experiments with capacitors X and Z charged gives $\alpha_{xz} = \alpha_{zx}$. Similarly, with capacitors Y and Z charged we obtain $\alpha_{yz} = \alpha_{zy}$. We therefore deduce that the polarizability tensor is symmetric:

$$\alpha_{ij} = \alpha_{ji}. \quad (5)$$

Comment

Although the result (5) was obtained by considering a uniform electrostatic field, it can also be confirmed using quantum mechanics.

Question 2.30*

(a) Two rigid point electric dipoles $\mathbf{p}_1$ and $\mathbf{p}_2$ are a distance r apart. Suppose the dipoles are spatially fixed relative to each other.

☞ Show that the force $\mathbf{F}_{12}$ exerted by $\mathbf{p}_1$ on $\mathbf{p}_2$ is

$$\mathbf{F}_{12} = \frac{1}{4\pi\epsilon_0}\frac{3}{r^4}\left\{5\left[(\mathbf{p}_1\cdot\mathbf{p}_2) - (\mathbf{p}_1\cdot\hat{\mathbf{r}})(\mathbf{p}_2\cdot\hat{\mathbf{r}})\right]\hat{\mathbf{r}} + (\mathbf{p}_1\cdot\hat{\mathbf{r}})\mathbf{p}_2 + (\mathbf{p}_2\cdot\hat{\mathbf{r}})\mathbf{p}_1\right\}, \quad (1)$$

where $\hat{\mathbf{r}}$ is the unit vector from $\mathbf{p}_1$ to $\mathbf{p}_2$.

☞ State two obvious features of equation (1).

(b) Consider a spherical atom (having polarizability α) that is a large distance r from a polar molecule, whose permanent electric dipole moment is p. Derive an order-of-magnitude expression for the intermolecular force F between the atom–molecule pair in terms of α, p and r.

(c) Consider two identical spherical atoms, each having polarizability α. Suppose $p = p(t)$ is an *instantaneous* dipole moment that arises from a time-dependent fluctuation in the electron density of either atom. What can one deduce about the mean interatomic force $\langle F(t) \rangle$?

Solution

(a) ☞ Suppose $\mathbf{E}_1$ is the field at $\mathbf{p}_2$ due to $\mathbf{p}_1$. Then

$$U = -\mathbf{p}_2 \cdot \mathbf{E}_1 = \frac{1}{4\pi\epsilon_0}\left\{ \frac{r^2(\mathbf{p}_1 \cdot \mathbf{p}_2) - 3(\mathbf{p}_1 \cdot \mathbf{r})(\mathbf{p}_2 \cdot \mathbf{r})}{r^5} \right\}, \qquad (2)$$

because of (2) of Question 2.11. Now $\mathbf{F} = -\nabla U$ or $F_i = -\partial U/\partial r_i$. So we obtain

$$F_i = -\frac{1}{4\pi\epsilon_0}\frac{\partial}{\partial r_i}\left\{ \frac{r^2 p_{1j}p_{2j} - 3p_{1j}p_{2k}r_j r_k}{r^5} \right\}$$

$$= \frac{1}{4\pi\epsilon_0}\left\{ \frac{3(r^2 p_{1j}p_{2j} - 15 p_{1j}p_{2k}r_j r_k)r_i + 3r^2 p_{1j}p_{2k}(r_k\delta_{ij} + r_j\delta_{ik})}{r^7} \right\}. \qquad (3)$$

Contraction of the subscripts yields (1).‡

☞ It is evident from (1) that:

- $\mathbf{F}_{12} = -\mathbf{F}_{21}$, as required by Newton's third law.
- The interaction between these dipoles gives rise to a force that is, in general, non-central.♯ (Contrast this with the Coulomb force between two stationary point charges).

(b) The molecule induces in the atom a temporary electric dipole $p' \sim \alpha E$ where the inducing dipole field is $E \sim \frac{1}{4\pi\epsilon_0}\frac{p}{r^3}$ (see (11)$_2$ of Question 2.20). The leading term in the pair interaction energy is $U \sim -p'E$. This gives

$$U \sim -\alpha E^2 \sim -\left(\frac{1}{4\pi\epsilon_0}\right)^2 \frac{\alpha p^2}{r^6}, \qquad (4)$$

and an attractive force

$$F \sim -\frac{\partial U}{\partial r}\hat{\mathbf{r}} \sim -\left(\frac{1}{4\pi\epsilon_0}\right)^2 \frac{\alpha p^2}{r^7}\hat{\mathbf{r}}. \qquad (5)$$

(c) The order-of-magnitude calculation in (b) also applies here. The mean value $\langle p(t) \rangle$ over a period of time is clearly zero. However, the mean-square dipole moment $\langle p^2(t) \rangle$ is not zero and we conclude that, on average, there will be a net attractive force between the atoms.

‡The reader may easily verify that (1) also follows from (6) of Question 2.27.
♯That is, $\mathbf{F}$ does not act in the direction of $\hat{\mathbf{r}}$.

Comments

(i) Equation (1) shows that the force between two spatially fixed dipoles scales as r^{-4}. In molecules, these dipole–dipole interactions are an important type of inter-molecular force. They are weaker than intramolecular forces (e.g. covalent bonds) and sometimes give rise to effects which can have important consequences. For example, the hydrogen bonds between H_2O molecules, which are due to dipole–dipole interactions, significantly affect the physical and chemical properties of water. In the absence of hydrogen bonding, water would be a gas at temperatures well below $0\,°C$. Ice would be more dense than liquid water, and life on Earth, if it existed at all, would be very different from its present form.

(ii) This question also illustrates two further types of molecular interaction:

☞ dipole–induced dipole interactions, and

☞ instantaneous dipole–induced dipole interactions.

These interactions are present in the atoms and molecules of matter: known as van der Waal's forces they are caused by permanent or temporary polarization, and are characterized by the following features:

☞ the intermolecular potential energy U scales as r^{-6},

☞ the force is attractive, and

☞ U is proportional to the atomic (or molecular) polarizability α.

Van der Waal's forces are important between large molecules and in certain crystals.

(iii) Although (4) is a classical result, the detailed expression must be derived using quantum mechanics.[11] It applies to all molecules, whether polar or not.

(iv) The above solution shows that the atoms in (c) attract each other and U scales as r^{-6}. But as $r \to 0$, the electronic wave functions of both atoms overlap. The Pauli exclusion principle now determines their behaviour, and a short-range repulsive potential ($U \sim r^{-n}$) dominates their interaction. Many properties of matter are insensitive to the exact value of the exponent n, and the choice $n = 12$ is often made 'for mathematical convenience rather than physical significance'.[12] The combination of an attractive r^{-6} van der Waal's potential and a repulsive r^{-12} potential is known as the Lennard-Jones 6-12 potential:

$$U(r) = -\frac{a}{r^6} + \frac{b}{r^{12}}, \tag{6}$$

where a and b are positive constants. Equation (6) is frequently used as a model for an interatomic potential.

---

[11] See, for example, G. Baym, *Lectures on quantum mechanics*. New York: W. A. Benjamin, 1969.

[12] A. P. Thorne, *Spectrophysics*, p. 270. London: Chapman and Hall, 1974.

3

The electrostatics of conductors

Any material containing mobile-charge carriers which can move in the presence of an external electric field is a conductor (e.g. metals, electrolytes and semiconductors). Under these conditions, an electric current is produced because a momentum transfer is established between the charge carriers (these being electrons in a metal, positive and negative ions in an electrolyte and positively charged holes in a semiconductor). In this chapter, we are concerned only with metallic conductors where the charge carriers are electrons. The term *electrostatic equilibrium* implies that there is no movement of charge, except when this equilibrium is disturbed. When this happens, it turns out that the time required to re-establish equilibrium—in a metallic conductor—is unimaginably small (see Question 10.9).

At any arbitrary point inside a conductor the microscopic electric field $\mathbf{E}_{\mathrm{micro}}$ fluctuates rapidly and randomly with time. Here, and for material media in general, at an instant of time these fluctuations are smoothed out by considering macroscopic 'point' properties (this term is defined in the footnote on p. 409). The macroscopic electric field $\mathbf{E}_{\mathrm{macro}}$ at a 'point' in the conductor is defined as the average value of $\mathbf{E}_{\mathrm{micro}}$ over a small but finite (macroscopic) volume Δv, and

$$\mathbf{E}_{\mathrm{macro}} = \frac{1}{\Delta v} \int_{\Delta v} \mathbf{E}_{\mathrm{micro}} \, dv_{\mathrm{micro}}. \tag{I}$$

In (I), the choice of Δv requires some care; the reader is referred to Ref. [1] and Ref. [2] for further discussion.

We begin in Question 3.1 by proving some important properties of conductors in equilibrium. Then in Questions 3.2 and 3.3 we introduce Laplace's equation and discuss some general properties of its solutions. These results underpin most of the remaining questions of this chapter. The coefficients of capacitance for an arbitrary arrangement of conductors are introduced early on, and various numerical calculations then follow in a number of subsequent questions. Some important techniques (both analytical and numerical) for finding solutions to Laplace's equation are considered. These include: the method of separation of variables (the Fourier method), the relaxation method, the method of images and the method of conformal transformation. All of these are discussed in detail, and with appropriate examples.

[1] G. Russakoff, 'A derivation of the macroscopic Maxwell equations', *American Journal of Physics*, vol. 38, pp. 1188–95, 1970.
[2] J. D. Jackson, *Classical electrodynamics*, Chap. 6, pp. 249–58. New York: Wiley, 3 edn, 1998.

Solved Problems in Classical Electromagnetism. J. Pierrus, Oxford University Press (2018).
© J. Pierrus. DOI: 10.1093/oso/9780198821915.001.0001

Question 3.1

Prove the following properties for a conductor of arbitrary shape (see, for example, Fig. (I) below) under conditions of electrostatic equilibrium:

(a) The macroscopic electric field $\mathbf{E}_{\mathrm{macro}}$ (or simply $\mathbf{E}$) is zero everywhere inside the conductor.

(b) Any excess charge resides on, or near,[‡] the surface of the conductor.

(c) The conductor is an equipotential.

(d) Only a normal component of $\mathbf{E}$ can exist at the surface of the conductor.

Solution

(a) Because the conductor is in electrostatic equilibrium, the resultant electric force $\mathbf{F}$ on a (free) charge carrier q inside the medium is zero. Since $\mathbf{F} = q\mathbf{E} = 0$ and $q \neq 0$, we conclude that $\mathbf{E} = 0$ everywhere.

(b) Consider a Gaussian surface G (see Fig. (I) below) lying a small distance δ just beneath the actual surface s of the conductor,[♯] as shown. It follows immediately from (a) that the net flux through G is zero, and Gauss's law guarantees no excess charge of either type inside this surface. Since δ has atomic dimensions, any excess charge can only appear *on* the conductor's surface, or within several atomic layers beneath it.

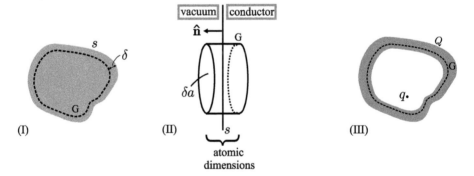

vacuum conductor

$\hat{\mathbf{n}}$ G

δa

s

atomic
dimensions

(I)   (II)   (III)

(c) If different parts of the conductor were at different potentials, finite currents would exist (for example, in a metal electrons would move from low to high potential). But the presence of such currents would contradict the condition of electrostatic equilibrium. Since there can be no such currents, we must conclude that the entire conductor is an equipotential.

[‡]We use the word 'near' to mean within several atomic layers of the surface.

[♯]Suppose that δ is a microscopic quantity that is of the order 10^{-9} m (i.e. several atomic radii).

(d) If the field at the surface of the conductor had a tangential component $\mathbf{E}_t$, the (free) charge carriers in the surface would experience a resultant electric force $q\mathbf{E}_t$. This force would produce surface currents, the presence of which would contradict the condition of electrostatic equilibrium. As previously stated, there can be no such currents. We must therefore conclude that $\mathbf{E}_t = 0$. Hence only a normal component of the electrostatic field may exist at a point on the surface.

Comments

(i) Although most modern textbooks—including this one—usually prove property (b) using Gauss's law (because this is convenient), it is worth remembering the historical facts. Gauss's law was advanced only *after* it was known from a variety of experiments (such as Faraday's ice-pail experiment) that any excess charge on a conductor resides on its surface. In fact, these experiments are sometimes taken as 'proof' of Gauss's law rather than the converse.

(ii) A useful relation exists between the field at a point on the surface of a conductor and the surface-charge density σ at the point. Choose an infinitesimal Gaussian cylinder[†] G straddling the surface with its axis perpendicular to it (see Fig. (II) on p. 114). The flux through the curved part of G is negligible on the microscopic scale; the flux through the end cap of G inside the conductor is zero because $\mathbf{E} = 0$ there, and the flux through the end cap of G on the vacuum side of the boundary is $\mathbf{E} \cdot \delta\mathbf{a} = E_n \delta a$ because of property (d). Thus, by Gauss's law, $E_n \delta a = \sigma \delta a / \epsilon_0$ or

$$E_n = -\frac{\partial \Phi}{\partial n} = \frac{\sigma}{\epsilon_0}. \tag{1}$$

(iii) The electric field is also zero everywhere inside a *hollow* conducting shell carrying excess charge Q on its outer surface. Furthermore, if the shell totally encloses a charge q as shown in Fig. (III) on p. 114, $\mathbf{E}$ remains zero at every point on G. Gauss's law still requires $\psi = 0$ as before, implying that the induced charge on the inner surface of the conducting shell is $-q$.

Question 3.2

(a) Derive Poisson's equation in electrostatics,

$$\nabla^2 \Phi(\mathbf{r}) = -\rho(\mathbf{r})/\epsilon_0, \tag{1}$$

from the differential form of Gauss's law.

(b) Verify that

$$\Phi(\mathbf{r}) = \frac{1}{4\pi\epsilon_0} \int_v \frac{\rho(\mathbf{r}')}{|\mathbf{r} - \mathbf{r}'|} \, dv' \tag{2}$$

(see (4) of Question 2.3) is a solution to (1).

[†]In the vicinity of the infinitesimal cylinder G the surface s is essentially flat.

Solution

(a) Substituting $\mathbf{E} = -\nabla\Phi$ in $\nabla \cdot \mathbf{E} = \rho/\epsilon_0$ yields (1).

(b) Applying the ∇^2 (Laplacian) operator to (2) gives

$$\nabla^2\Phi(\mathbf{r}) = \frac{1}{4\pi\epsilon_0}\int_v \nabla^2\left[\frac{\rho(\mathbf{r}')}{|\mathbf{r}-\mathbf{r}'|}\right]dv' = \frac{1}{4\pi\epsilon_0}\int_v \rho(\mathbf{r}')\,\nabla^2|\mathbf{r}-\mathbf{r}'|^{-1}\,dv'. \tag{3}$$

The last step in (3) follows because ∇^2 does not differentiate the charge density ρ, which is a function of the primed coordinates only. Replacing $\nabla^2|\mathbf{r}-\mathbf{r}'|^{-1}$ with $-4\pi\delta(\mathbf{r}-\mathbf{r}')$ and applying the sifting property of the delta function $\big($see $(\mathrm{XI})_2$ of Appendix E$\big)$ yield $-\rho(\mathbf{r})/\epsilon_0$ as required.

Comments

(i) In regions of space where there are no charges, Poisson's equation reduces to

$$\nabla^2\Phi(\mathbf{r}) = 0. \tag{4}$$

Equation (4) is known as Laplace's equation and it arises in many branches of physics. Its solutions are called harmonic functions, and they possess certain important properties which we discuss in Question 3.3.

(ii) In most practical situations involving charged conductors having arbitrary shapes, the *integral* equation (2) cannot be solved directly, because $\rho(\mathbf{r}')$ is almost always an unknown function of position.[‡] By contrast, we often know and have direct control over the electric potential: consider, for example, a conductor connected to a power supply set to some predetermined voltage. In these situations, we usually wish to find $\Phi(\mathbf{r})$ in regions of space away from the conductors where the charge density is zero. The determination of Φ is now reduced to solving a *differential* equation subject to the particular boundary conditions of our problem.

(iii) Although there are no general mathematical techniques for solving (1) and (4), analytically solutions can sometimes be found for problems with high symmetry. Techniques like the method of images or the method of separation of variables may, in certain circumstances, be used to find a solution. Uniqueness of the solution is guaranteed, provided the boundary conditions have been suitably specified.

(iv) Appropriate boundary conditions required for the solution of (1) and (4) can be stated in several ways. Two of these considered in this book are:

- ☞ Dirichlet boundary conditions which specify the values of the potential on all boundary surfaces.
- ☞ Neumann boundary conditions which specify the normal derivatives[♯] of the potential on all boundary surfaces.

[‡]In practice, we usually cannot influence how a given amount of charge arranges itself over a conducting surface of irregular shape.

[♯]This is equivalent to specifying the surface-charge density σ. See (1) of Question 3.1.

Question 3.3

(a) Suppose Φ_1, Φ_2, ..., Φ_n are solutions of Laplace's equation. Prove that the linear combination

$$\Phi = \alpha_1\Phi_1 + \alpha_2\Phi_2 + \cdots + \alpha_n\Phi_n \tag{1}$$

(where the α_i are constants) is also a solution.

(b) Prove that in a *charge-free* region of space, the value of the potential Φ at an arbitrary point O is the average value over the surface of any sphere in the space that is centred at O.

(c) Prove that the solutions of Laplace's equation can have no maxima or minima (extreme values of the potential may occur only at the boundaries).

(d) Prove that Laplace's equation has a unique solution, assuming that the values of Φ on all boundary surfaces have been specified.

Solution

(a) Substituting (1) in Laplace's equation gives

$$\nabla^2\Phi = \alpha_1\nabla^2\Phi_1 + \alpha_2\nabla^2\Phi_2 + \cdots + \alpha_n\nabla^2\Phi_n = 0,$$

because the α_i are constants and the Φ_i all satisfy $\nabla^2\Phi_i = 0$. Thus Φ is also a solution of Laplace's equation.

(b) Consider the average potential over a spherical surface s of radius r_0 due to a single point charge q located *outside* the sphere. Choose coordinates with origin O at the centre of the sphere and q located on the z-axis, as shown in the figure below.

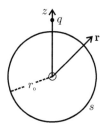

The potential at an arbitrary point on s is given by

$$\Phi(\mathbf{r}_0) = \frac{1}{4\pi\epsilon_0}\frac{q}{|\mathbf{z} - r_0\hat{\mathbf{r}}|}, \tag{2}$$

and the average potential $\overline{\Phi}$ over s is

$$\overline{\Phi} = \frac{1}{4\pi r_0^2} \oint_s \Phi(\mathbf{r}_0) \, da. \tag{3}$$

Substituting (2) in (3) and using (1)$_1$ of Question 1.26 give

$$\overline{\Phi} = \frac{1}{4\pi\epsilon_0} \frac{q}{z}, \tag{4}$$

which is just the potential at O due to q. This result can be generalized to an arbitrary distribution of stationary point charges outside the sphere, as follows. Let q_i be point charges located on z_i-axes which all share a common origin O at the centre of the sphere. The average potential over s due to the distribution is obtained from (4) and the superposition principle

$$\overline{\Phi} = \overline{\Phi}_1 + \overline{\Phi}_2 + \cdots = \frac{1}{4\pi\epsilon_0} \left[\frac{q_1}{z_1} + \frac{q_2}{z_2} + \cdots \right], \tag{5}$$

which proves the result.

(c) This property is proved by contradiction. Suppose that a solution Φ of Laplace's equation attains an extreme value Φ_P at a point P inside some charge-free region v of space. Consider an arbitrary sphere centred on P. The average value of Φ over this sphere would be less than Φ_P for a maximum or greater than Φ_P for a minimum. But this contradicts the mean-value property proved in (b): the extreme values of Φ may occur only at the boundaries of the region.

(d) This is again a proof by contradiction. Let Φ_1 and Φ_2 be two solutions of Laplace's equation which satisfy the *same* boundary conditions. Then the linear combination $\Phi = \Phi_1 - \Phi_2$ is also a solution (see (1) with $\alpha_1 = -\alpha_2 = 1$) and is zero both at infinity ($\Phi_1, \Phi_2 \to 0$ as $r \to \infty$) and on the boundaries ($\Phi_1 = \Phi_2$ there). Because of property (c) the extrema of Φ occur only at the boundaries of the region. Thus $\Phi = 0$ everywhere, implying that $\Phi_1 = \Phi_2$. Hence the solution is unique.

Comments

(i) Variations of the uniqueness property (d) can also be proved for, for example, Laplace's equation with Neumann boundary conditions or Poisson's equation (where, as always, the boundary conditions need to be suitably specified).

(ii) An immediate consequence of property (c) is that a charged particle cannot be maintained in stable equilibrium in an electrostatic field, a result known as Earnshaw's theorem. Of course, it is possible for a charge to be in unstable equilibrium,[‡] but these are saddle points in the field, not extrema.

(iii) The result proved in (d) is an example of a 'uniqueness theorem'. We shall refer to it—in later questions—by this name.

[‡]Consider the stability of a stationary test charge placed midway between two identical charges.

Question 3.4

Consider two long hollow coaxial metal cylinders having radii a and b respectively with $a < b$. Suppose the potential difference between the cylinders is V_0 with the outer cylinder connected to ground. Solve Laplace's equation, and then show that the electric field in the region $a \leq r \leq b$ far away from the ends (i.e. neglect fringing) is

$$\mathbf{E}(\mathbf{r}) = \frac{V_0}{\ln b/a} \frac{\hat{\mathbf{r}}}{r}. \tag{1}$$

Solution

Choose the origin of cylindrical polar coordinates at the centre of the cylinders with their common axis along $\hat{\mathbf{z}}$. Because of the symmetry, Φ is necessarily independent of the azimuthal angle θ and is also independent of z away from the ends. In these coordinates, Laplace's equation reduces to

$$\nabla^2 \Phi = \frac{1}{r}\frac{d}{dr}\left(r\frac{d\Phi}{dr}\right) = 0 \tag{2}$$

because of $(VIII)_4$ of Appendix D. Then (2) implies that $r\, d\Phi/dr$ is a constant (α, say), and so

$$\Phi(r) = \alpha \ln r + \beta, \tag{3}$$

where β is a constant of integration. Applying the boundary conditions $(\Phi(a) = V_0$ and $\Phi(b) = 0)$ gives

$$V_0 = \alpha \ln a + \beta \quad \text{and} \quad 0 = \alpha \ln b + \beta. \tag{4}$$

Solving (4) simultaneously for α and β and substituting these in (3) give

$$\Phi(r) = V_0 \frac{\ln b/r}{\ln b/a}. \tag{5}$$

Now $\mathbf{E} = -\hat{\mathbf{r}}\, \partial\Phi/\partial r$. Hence (1).

Comments

(i) Because (1) satisfies both Laplace's equation and the boundary conditions of the problem, it is guaranteed to be the only solution (see (d) of Question 3.3).

(ii) This question illustrates one of the simplest applications of Laplace's equation for calculating the potential $\Phi(r)$ in the vicinity of charged conductors. Here, this relatively straightforward solution arises because of the high symmetry of the problem. Needless to say, this is not usually the case.

(iii) Apart from the method of separation of variables described in Question 1.18, analytical solutions to Laplace's equation can sometimes be found using:

☞   the method of images, which is particularly useful for problems where the boundaries are straight or circular. See Questions 3.21–3.25 for examples.

☞   conformal transformation. This powerful technique, which is based on the theory of analytic functions, is considered further in Questions 3.30–3.34.

(iv) Numerical examples of solving Laplace's equation are provided in Questions 3.17, 3.19 and 3.20.

Question 3.5

(a) The electrostatic field of an infinite plane conductor, maintained at a constant potential Φ_0, is $\mathbf{E}_0$. Show that the electric potential at an arbitrary point P in space is

$$\Phi(\mathbf{r}) \;=\; \Phi_0 - \mathbf{E}_0\cdot\mathbf{r}, \tag{1}$$

where the origin O of $\mathbf{r}$ is located on the surface of the conductor.

(b) Now suppose that the conductor has a hemispherical bulge of radius a somewhere on its surface. Use (1) and (3) of Question 1.19 to show that

$$\Phi(\mathbf{r}) \;=\; \Phi_0 - (1 - a^3/r^3)\,\mathbf{E}_0\cdot\mathbf{r}, \tag{2}$$

where O is at the centre of the bulge.

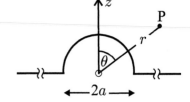

(c) Hence determine the field $\mathbf{E}$ at an arbitrary point $P(r,\theta,\phi)$ in the vicinity of the bulge.

(d) Use *Mathematica* to plot $E(0,\,y,\,z)$ for $-3a \le y \le 3a$ and $0 \le z \le 6a$.

Solution

(a) The electric potential at $\mathbf{r}$, relative to the electric potential at O, is given by

$$\Phi(\mathbf{r}) \;=\; \Phi(0) - \int_0^{\mathbf{r}} \mathbf{E}\cdot d\mathbf{r}. \tag{3}$$

But the field is constant, and so (3) becomes

$$\Phi(\mathbf{r}) \;=\; \Phi(0) - \mathbf{E}_0\cdot\int_0^{\mathbf{r}} d\mathbf{r} \;=\; \Phi(0) - \mathbf{E}_0\cdot\mathbf{r},$$

which is (1) because $\Phi(0) = \Phi_0$.

(b) The potential is clearly symmetric about the z-axis and the general solution is

$$\Phi(r,\theta) = \sum_{n=0}^{\infty} \left[A_n r^n + B_n r^{-(n+1)} \right] P_n(\cos\theta), \tag{4}$$

where the Legendre polynomials are discussed in Appendix F. In the limit $r \to \infty$, (4) must reduce to (1), since the perturbation caused by the bulge is negligible at large distances from it. This is true only if

$$A_2 = A_3 = \cdots = 0,$$

and so

$$\Phi(r,\theta) = A_0 + A_1 r \cos\theta + \frac{B_0}{r} + \frac{B_1}{r^2} \cos\theta + \sum_{n=2}^{\infty} \frac{B_n}{r^{n+1}} P_n(\cos\theta), \tag{5}$$

since $P_0(\cos\theta) = 1$ and $P_1(\cos\theta) = \cos\theta$. The boundary condition $\Phi(a,\theta) = \Phi_0$ gives

$$\Phi_0 = \left(A_0 + \frac{B_0}{a} \right) + \left(A_1 a + \frac{B_1}{a^2} \right) \cos\theta + \sum_{n=2}^{\infty} \frac{B_n}{a^{n+1}} P_n(\cos\theta),$$

which may be written as

$$0 = \left(A_0 + \frac{B_0}{a} - \Phi_0 \right) P_0(\cos\theta) + \left(A_1 a + \frac{B_1}{a^2} \right) P_1(\cos\theta) + \sum_{n=2}^{\infty} \frac{B_n}{a^{n+1}} P_n(\cos\theta). \tag{6}$$

Because the $P_n(\cos\theta)$ are linearly independent functions, the coefficient of each Legendre polynomial in (6) is necessarily zero. Thus,

$$\left. \begin{aligned} A_0 &= \Phi_0 - B_0/a \\ A_1 &= -B_1/a^3 \\ B_2 &= B_3 = \cdots = 0 \end{aligned} \right\}. \tag{7}$$

Substituting (7) in (5) yields $\Phi(r,\theta) = \Phi_0 - B_0 \left(\dfrac{1}{a} - \dfrac{1}{r} \right) - B_1 r \left(\dfrac{1}{a^3} - \dfrac{1}{r^3} \right) \cos\theta.$

Comparing this with (1) in the limit $r \to \infty$ shows that $B_0 = 0$ and $B_1 = E_0 a^3$.

Hence $\Phi(r,\theta) = \Phi_0 - E_0 r \left(1 - a^3/r^3 \right) \cos\theta$, which is (2).

(c) The electric-field components follow from $\mathbf{E} = -\nabla\Phi$, where nabla in spherical polar coordinates is given by $(XI)_1$ of Appendix C. So

$$E_r = -\frac{\partial \Phi}{\partial r} = E_0\left(1 + 2a^3/r^3\right)\cos\theta$$

$$E_\theta = -\frac{1}{r}\frac{\partial \Phi}{\partial \theta} = -E_0\left(1 - a^3/r^3\right)\sin\theta \qquad (8)$$

$$E_\phi = -\frac{1}{r\sin\theta}\frac{\partial \Phi}{\partial \phi} = 0$$

Thus,

$$\mathbf{E} = \hat{\mathbf{r}}E_r + \hat{\boldsymbol{\theta}}E_\theta + \hat{\boldsymbol{\phi}}E_\phi = E_0\left[\hat{\mathbf{r}}\left(1 + 2a^3/r^3\right)\cos\theta - \hat{\boldsymbol{\theta}}\left(1 - a^3/r^3\right)\sin\theta\right]. \qquad (9)$$

(d) The field shown in the figure below was calculated using the notebook on p. 123.

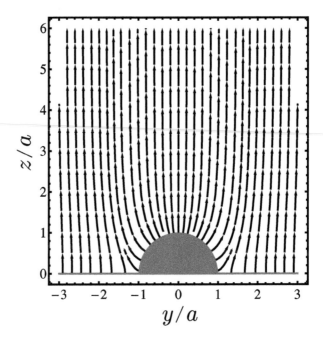

Comments

(i) Although infinite planes (like the above conductor) are an idealization, this question illustrates a noteworthy feature of unbounded charge distributions: namely, $\Phi(\mathbf{r}) \to \infty$ (and not zero) as $r \to \infty$. One also encounters this feature in problems involving infinite lines and infinite cylinders.

(ii) Equation (8) shows that everywhere on the surface of the bulge the electric field is radial ($E_\theta = 0$) and normal to the conductor, as required by property (d) of Question 3.1. Furthermore, E has its largest value at the top of the bulge where $E(a,0) = 3E_0$.

```
In[1]:= (* Mathematica's StreamPlot function requires E in Cartesian form.
        Using (V) of Appendix C to convert r̂ and θ̂ to x̂ and ŷ and putting
        y = rSinθ; z = rCosθ gives the Ey and Ez below. *)
```

$$Ey[y_, z_] := \frac{3\,y\,z}{\left(y^2 + z^2\right)^{5/2}}; \quad Ez[y_, z_] := 1 - \frac{1}{\left(y^2 + z^2\right)^{3/2}} + \frac{3\,z^2}{\left(y^2 + z^2\right)^{5/2}};$$

```
gr1 = StreamPlot[{ Ey[y, z], Ez[y, z]}, {y, -3, 3}, {z, 0, 6},
    RegionFunction → Function[{y, z}, y² + z² ≥ 1 && z ≥ 0], StreamStyle →
    {Black, Thickness[0.0075]}, FrameStyle → Directive[Black,
    Thickness[0.0075]], AxesLabel → {y, z}, LabelStyle →
    Directive[FontSize → 18]];

gr2 = Graphics[{Gray, Disk[{0, 0}, 1, {0, π}]}];
gr3 = Graphics[{Thickness[0.0075], Gray, Line[{{-3, 0}, {3, 0}}]}];
Show[gr1, gr2, gr3]
```

Question 3.6

A thin conducting disc of radius a is maintained at a potential V_0. The electric potential at an arbitrary point P on the symmetry axis (z, say) of the disc is given by (1) of Question 2.5: $\Phi(z) = \frac{\sigma}{2\epsilon_0}\left(\mp z + \sqrt{a^2 + z^2}\right)$ where the upper (lower) sign is for $z > 0$ ($z < 0$). Now $\sigma/2\epsilon_0 = V_0/a$, and so

$$\Phi(z) = V_0\left(\mp \alpha + \sqrt{1 + \alpha^2}\right), \tag{1}$$

where $\alpha = z/a$.

(a) Show that (1) can be expressed as a series expansion

$$\Phi(z) = V_0\left[1 \mp \left(\frac{z}{a}\right) + \frac{1}{2}\left(\frac{z}{a}\right)^2 - \frac{1}{8}\left(\frac{z}{a}\right)^4 - \cdots\right]. \tag{2}$$

(b) Use (2) and the hint below to determine the potential at an arbitrary point P(r,θ), where $r < a$.

Hint: The solution of Laplace's equation with azimuthal symmetry is given by[‡]

$$\Phi(r, \theta) = \sum_{n=0}^{\infty}\left[A_n r^n + B_n r^{-(n+1)}\right]P_n(\cos\theta), \tag{3}$$

[‡]See (3) of Question 1.19.

where the coefficients A_n and B_n, determined from the boundary conditions, are unique. Use the fact that on the symmetry axis, $\theta = 0$ and the $P_n(\cos\theta)$ become $P_n(1) = 1$.

(c) Repeat (b) for $r > a$.

Solution

(a) Using *Mathematica*'s series function to expand (1) about $\alpha = 0$ gives (2).[#]

(b) Because of the hint,

$$\Phi(z) = \left(A_0 + A_1 z^1 + A_2 z^2 + A_3 z^3 + \cdots\right) + \left(B_0 + B_1/z + B_2/z^2 + B_3/z^3 + \cdots\right). \quad (4)$$

Comparing (4) with (2) shows that

$$\left. \begin{array}{llll} A_0 = V_0, & A_1 = \mp V_0/a, & A_2 = V_0/2a^2, & A_3 = -V_0/8a^4, & \cdots \\ B_0 = B_1 = B_2 = B_3 = \cdots = 0 \end{array} \right\}. \quad (5)$$

Substituting (5) in (3) yields

$$\Phi(r,\theta) = V_0\left[P_0(\mu) \mp \left(\frac{r}{a}\right)P_1(\mu) + \frac{1}{2}\left(\frac{r}{a}\right)^2 P_2(\mu) - \frac{1}{8}\left(\frac{r}{a}\right)^4 P_4(\mu) + \cdots\right], \quad (6)$$

where $\mu = \cos\theta$.

(c) The simplest way of obtaining the potential for $r > a$ is to replace $(r/a)^n$ in (6) by $(a/r)^{n+1}$ (see Appendix F). Then

$$\Phi(r,\theta) = V_0\left[\left(\frac{a}{r}\right)P_0(\mu) \mp \left(\frac{a}{r}\right)^2 P_1(\mu) + \frac{1}{2}\left(\frac{a}{r}\right)^3 P_2(\mu) - \frac{1}{8}\left(\frac{a}{r}\right)^5 P_4(\mu) + \cdots\right]. \quad (7)$$

Comment

The uniqueness of the solution of Laplace's equation discussed in Question 3.3 ensures that (6) and (7) are unique series expansions for the potential.

[#]The notebook is:

```
In[1]:=  (*Let α=z/a*)        VO = Series[-α + √(1 + α²) , {α, 0, 10}]
```

Question 3.7*

Consider two eccentric[‡] spherical conducting shells having radii a and b with $a < b$. The point O (O') is at the centre of the inner (outer) shell, and let d be the distance between these two origins. See the figure below, but assume that $d \ll a$. The potential difference between the conductors is V_0 with the outer shell connected to ground.

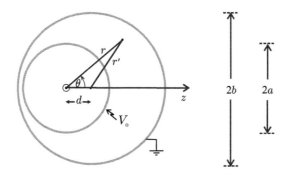

(a) Show that in the region between the conductors

$$\Phi(r,\theta) \simeq V_0 \frac{ab}{b-a}\left[\frac{1}{r} - \frac{1}{b} + \frac{d(r - a^3/r^2)\cos\theta}{(b^3 - a^3)}\right]. \tag{1}$$

Hint: Neglect powers of d/b greater than one.

(b) Hence show that the charge density $\sigma(\theta)$ induced on the outer surface of the inner shell is

$$\sigma(\theta) \simeq \frac{q}{4\pi a^2}\left[1 + \frac{3a^2 d \cos\theta}{(b^3 - a^3)}\right], \tag{2}$$

where q is the charge.

Solution

(a) The potential is symmetric about the z-axis and is given by (3) of Question 1.19:

$$\Phi(r,\theta) = \sum_{n=0}^{\infty}\left[A_n r^n + B_n r^{-(n+1)}\right]P_n(\mu), \tag{3}$$

where $P_n(\mu)$ (with $\mu = \cos\theta$) are Legendre polynomials of order n (see Appendix F). Before proceeding, we must first express r' in terms of r. Applying the cosine rule gives $r'^2 = r^2 + d^2 - 2rd\cos\theta$, and because $d/r \ll 1$ for all values of r between the conductors we obtain $r' \simeq (r - d\cos\theta)$. Now when $r' = b$ let $r = R$. Then

[‡]Eccentric spheres, unlike concentric spheres, do not share a common centre.

$R = b + d\cos\theta$. The boundary conditions of the problem are: $\Phi(a) = V_0$ and $\Phi(R) = 0$. These, together with (3), give

$$
\left.\begin{array}{l}
V_0 = \left(A_0 + \dfrac{B_0}{a}\right)P_0(\mu) + \left(A_1 a + \dfrac{B_1}{a^2}\right)P_1(\mu) + \displaystyle\sum_{n=2}^{\infty}\left[A_n a^n + B_n a^{-(n+1)}\right]P_n(\mu) \\[4mm]
0 = \left(A_0 + \dfrac{B_0}{R}\right)P_0(\mu) + \left(A_1 R + \dfrac{B_1}{R^2}\right)P_1(\mu) + \displaystyle\sum_{n=2}^{\infty}\left[A_n R^n + B_n R^{-(n+1)}\right]P_n(\mu)
\end{array}\right\}. \quad (4)
$$

The constants A_n and B_n are evaluated as follows. In doing this we draw on the results of Appendix F, using specifically (VI) and (VII)$_5$.

$n = 0$

Multiply both sides of (4) by $P_0(\mu)$ and integrate with respect to μ from $\mu = -1$ to $\mu = 1$. Then making the approximation $R^{-1} = b^{-1}(1 - (d/b)\cos\theta)$ gives

$$
\left.\begin{array}{ll}
V_0 \displaystyle\int_{-1}^{1} P_0(\mu)\,d\mu = \left(A_0 + \dfrac{B_0}{a}\right)\displaystyle\int_{-1}^{1} P_0(\mu)\,P_0(\mu)\,d\mu & \Rightarrow \quad A_0 + \dfrac{B_0}{a} = V_0 \\[4mm]
0 = \displaystyle\int_{-1}^{1}\left[A_0 + \dfrac{B_0}{b}\left(1 - \dfrac{d}{b}P_1(\mu)\right)\right]P_0(\mu)\,d\mu & \Rightarrow \quad A_0 + \dfrac{B_0}{b} = 0
\end{array}\right\}.
$$

Solving these equations simultaneously gives

$$
A_0 = -\frac{aV_0}{b-a} \quad \text{and} \quad B_0 = \frac{abV_0}{b-a}. \quad (5)
$$

$n = 1$

Multiply both sides of (4) by $P_1(\mu)$ and integrate with respect to μ from $\mu = -1$ to $\mu = 1$. Then making the approximation $R^{-2} = b^{-2}(1 - 2(d/b)\cos\theta)$ gives

$$
V_0 \int_{-1}^{1} P_1(\mu)\,d\mu = \left(A_1 a + \frac{B_1}{a^2}\right)\int_{-1}^{1} P_1(\mu)\,P_1(\mu)\,d\mu \quad \Rightarrow \quad A_1 a + \frac{B_1}{a^2} = 0 \quad (6)
$$

and,

$$
\begin{aligned}
0 = \int_{-1}^{1}\Bigg(&\frac{B_0}{b}\left\{1 - \frac{d}{b}P_1(\mu)\right\}P_0(\mu) + \left\{A_1\left[b + d\,P_1(\mu)\right] + \right. \\
&\left. \frac{B_1}{b^2}\left[1 - \frac{2d}{b}P_1(\mu)\right]\right\}P_1(\mu)\Bigg)P_1(\mu)\,d\mu
\end{aligned},
$$

which reduces to

$$
0 = -\frac{B_0 d}{b^2}\int_{-1}^{1}\mu^2\,d\mu + \left(A_1 b + \frac{B_1}{b^2}\right)\int_{-1}^{1}\mu^2\,d\mu + \left(A_1 d - 2\frac{B_1 d}{b^3}\right)\int_{-1}^{1}\mu^3\,d\mu.
$$

Now the integral involving μ^3 is obviously zero, and so

$$A_1 b + \frac{B_1}{b^2} = \frac{B_0 d}{b^2} = \frac{a d V_0}{b(b-a)}, \tag{7}$$

because of $(5)_2$. Solving (6) and (7) simultaneously yields

$$A_1 = \frac{a b d V_0}{(b-a)(b^3-a^3)} \quad \text{and} \quad B_1 = -\frac{a^4 b d V_0}{(b-a)(b^3-a^3)}. \tag{8}$$

higher-order coefficients

We proceed order by order. For $n = 2$, multiply both sides of (4) by $P_2(\mu)$ and integrate with respect to μ from $\mu = -1$ to $\mu = 1$. Then

$$\left.\begin{aligned}
0 &= \left(A_2 a^2 + \frac{B_2}{a^3}\right) \int_{-1}^{1} P_2(\mu) P_2(\mu)\, d\mu \quad \Rightarrow \quad A_2 a^2 + \frac{B_2}{a^3} = 0 \\[2mm]
0 &= \left(A_2 b^2 + \frac{B_2}{b^3}\right) \int_{-1}^{1} P_2(\mu) P_2(\mu)\, d\mu \quad \Rightarrow \quad A_2 b^2 + \frac{B_2}{b^3} = 0
\end{aligned}\right\}.$$

These two equations can only be satisfied if $A_2 = 0$ and $B_2 = 0$. By proceeding in this way, one can show that

$$A_2 = A_3 = \cdots = 0 \quad \text{and} \quad B_2 = B_3 = \cdots = 0. \tag{9}$$

Equations (3), (5), (8) and (9) give (1).

(b) The surface-charge density $\sigma = \epsilon_0 E_r$ follows from (1) of Question 3.1. At the surface of the conducting shell, $E_r = -\partial\Phi/\partial r$. Now with Φ given by (1) we obtain

$$\sigma = -\epsilon_0 \frac{\partial\Phi}{\partial r}\bigg|_{r=a} = \frac{\epsilon_0 V_0 a b}{b-a}\left[\frac{1}{a^2} + \frac{3 d \cos\theta}{(b^3-a^3)}\right]. \tag{10}$$

The total charge appearing on the outer surface of the inner shell is

$$q = \int_s \sigma\, da = \int_0^{2\pi}\int_0^{\pi} \sigma a^2 \sin\theta\, d\theta\, d\phi = 2\pi a^2 \int_0^{\pi} \sigma \sin\theta\, d\theta. \tag{11}$$

Substituting (10) in (11) gives

$$q = \frac{2\pi a^3 b V_0 \epsilon_0}{(b-a)}\left[\frac{1}{a^2}\int_{-1}^{1} d\mu + \frac{3d}{(b^3-a^3)}\int_{-1}^{1}\mu\, d\mu\right] = \frac{4\pi a b V_0 \epsilon_0}{(b-a)}. \tag{12}$$

Using (12) to eliminate $\epsilon_0 V_0 a b/(b-a)$ from (10) yields (2).

Question 3.8

A conducting sphere of radius a has charge q distributed uniformly over its surface. The electric field is given by

$$\mathbf{E}(\mathbf{r}) = \begin{cases} 0 & \text{for } r < a \\ \dfrac{1}{4\pi\epsilon_0}\dfrac{q}{r^2}\,\hat{\mathbf{r}} & \text{for } r \geq a. \end{cases} \tag{1}$$

(a) Integrate the energy density $u = \frac{1}{2}\epsilon_0 \mathbf{E}\cdot\mathbf{E}$ (see (10) of Question 2.17), and show that the potential energy stored in the field is

$$U = \frac{1}{8\pi\epsilon_0}\frac{q^2}{a}. \tag{2}$$

(b) Re-derive (2) from first principles by starting with $dW = \Phi\,dq$.

Solution

(a) Clearly $U = \displaystyle\int u\,dv = \frac{\epsilon_0}{2}\int_{\substack{\text{all} \\ \text{space}}} \mathbf{E}\cdot\mathbf{E}\,dv$ where $dv = 4\pi r^2 dr$ because of the spherical symmetry. Thus, $U = 2\pi\epsilon_0\displaystyle\int_0^a E^2 r^2 dr + 2\pi\epsilon_0\int_a^\infty E^2 r^2 dr$. Now since $\mathbf{E} = 0$ in the first integral we have

$$U = 2\pi\epsilon_0\left(\frac{q}{4\pi\epsilon_0}\right)^2\int_a^\infty \frac{dr}{r^2} = \frac{q^2}{8\pi\epsilon_0}\left[-\frac{1}{r}\right]_a^\infty,$$

which is (2).

(b) Suppose we start with a neutral sphere and add charge in incremental amounts until the final charge q is reached. At some point during the charging process, the instantaneous charge on the sphere is q' and its potential is $\Phi' = q'/4\pi\epsilon_0 a$ (this being a necessary consequence of $(1)_2$). The work done dW in moving an incremental amount of charge dq' from infinity (the zero of potential) to the surface of the sphere at $r = a$ is $\Phi' dq'$. Thus

$$W = \int dW = \int_0^q \Phi' dq' = \frac{1}{4\pi\epsilon_0}\frac{1}{a}\int_0^q q' dq',$$

which—as before—is (2).

Comment

The potential energy given by (2) may also be written in the alternative form $U = \frac{1}{2}q\Phi$ (this is, of course, an expected result because of (2) of Question 2.17).

Question 3.9

Consider two conducting spheres having radii r_1 and r_2 whose centres are separated by a distance d that is large compared to their radii. Suppose an amount of charge q is distributed arbitrarily between the conductors with q_1 on sphere 1 and $q_2 = q - q_1$ on sphere 2.

(a) Neglecting the perturbation caused by each sphere on the potential of the other sphere, determine the condition for which the electrostatic energy U of the system is a minimum. Then express q_1 and q_2 in terms of q.

(b) Repeat (a) using the hint below, and include the perturbing effect of each sphere on the other.

Hint: Take the potential of the distant sphere to be that of a *point* charge.

Solution

(a) Because $d \gg r_1, r_2$ the spheres are effectively non-interacting. The charge density on each sphere may be regarded as uniform and the potential is due only to the charge on it. Then it follows from (2) of Question 3.8 that

$$U = U_1 + U_2 = \frac{1}{8\pi\epsilon_0}\frac{q_1^2}{r_1} + \frac{1}{8\pi\epsilon_0}\frac{q_2^2}{r_2}. \tag{1}$$

Now suppose q_1 and q_2 are variables with q constant. Then

$$\frac{\partial U}{\partial q_1} = \frac{1}{8\pi\epsilon_0}\left[\frac{q_1}{r_1} - \frac{q_2}{r_2}\right], \tag{2}$$

because $\partial q_2/\partial q_1 = -1$. The electrostatic energy has an extremum if $\partial U/\partial q_1 = 0$, or, as (2) shows, when

$$\frac{1}{4\pi\epsilon_0}\frac{q_1}{r_1} = \frac{1}{4\pi\epsilon_0}\frac{q_2}{r_2}. \tag{3}$$

So when both spheres have the same potential (Φ_0, say), U is a minimum (see Comment (ii) below). Solving $q = q_1 + q_2$ and (3) simultaneously gives

$$q_1 = \left(\frac{r_1}{r_1 + r_2}\right)q \quad \text{and} \quad q_2 = \left(\frac{r_2}{r_1 + r_2}\right)q. \tag{4}$$

Substituting (4) in (3) yields

$$\Phi_0 = \frac{1}{4\pi\epsilon_0} \frac{q}{r_1 + r_2}. \tag{5}$$

(b) Because of the hint, we now have

$$\Phi_1 = \frac{1}{4\pi\epsilon_0}\left(\frac{q_1}{r_1} + \frac{q_2}{d}\right) \quad \text{and} \quad \Phi_2 = \frac{1}{4\pi\epsilon_0}\left(\frac{q_2}{r_2} + \frac{q_1}{d}\right), \tag{6}$$

and so

$$U = U_1 + U_2 = \tfrac{1}{2}q_1\Phi_1 + \tfrac{1}{2}q_2\Phi_2$$

$$= \frac{1}{8\pi\epsilon_0}\left(\frac{q_1^2}{r_1} + \frac{q_2^2}{r_2} + \frac{2q_1q_2}{d}\right). \tag{7}$$

Calculating $\partial U/\partial q_1$ from (7) and solving $\partial U/\partial q_1 = 0$ gives

$$q_1\left(\frac{1}{r_1} - \frac{1}{d}\right) = q_2\left(\frac{1}{r_2} - \frac{1}{d}\right). \tag{8}$$

Proceeding as in (a) we obtain

$$q_1 = \frac{r_2(d - r_1)\,q}{(r_1 + r_2)d - 2r_1r_2} \quad \text{and} \quad q_2 = \frac{r_1(d - r_2)\,q}{(r_1 + r_2)d - 2r_1r_2}. \tag{9}$$

Substituting (9) in (6) shows that U is a minimum when both spheres have a common potential (which is the same condition as before), and that now

$$\Phi_0 = \frac{1}{4\pi\epsilon_0}\frac{(d - r_1r_2/d)\,q}{(r_1 + r_2)d - 2r_1r_2}. \tag{10}$$

Comments

(i) We emphasize that both the solutions for Φ_0 in (a) and (b) given above are approximate. The solution for (a) is crude yet simple and may be regarded as the zeroth approximation. The first approximation given by (b) improves on (a) in that it attempts to correct the perturbing effect of each sphere on the other. But because the distant sphere is not strictly a point charge, the correction is still inexact. See also the numerical method of Questions 3.25 and 3.26 which applies for any $d > r_1 + r_2$.

(ii) It follows from (2) that

$$\frac{\partial^2 U}{\partial q_1^2} = \frac{1}{8\pi\epsilon_0}\left[\frac{1}{r_1} + \frac{1}{r_2}\right] > 0,$$

and so the extreme value of U occurs when the electrostatic energy of the system is a minimum and not a maximum.

(iii) The first approximation (10) reduces, as one might expect, to the zeroth approximation (5) by letting $d \to \infty$.

(iv) This question illustrates a particular case of Thomson's theorem which states that: electric charge distributes itself over conducting surfaces in such a way that the electrostatic energy of the system is a minimum.

Question 3.10*

(This question and its solution are based on Purcell's approach.[3])

An arrangement of three conductors c_1, c_2 and c_3 of arbitrary shape is depicted in the figure[‡] below.

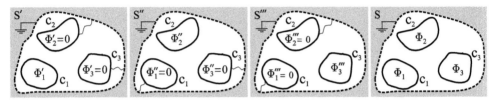

Consider a state of the system S' in which c_2 and c_3 are connected by a fine wire to ground and c_1 is at a potential Φ'_1 (measured relative to ground). Suppose q'_1, q'_2 and q'_3 are the charges on the various conductors in this state. The electric field in the system and the q'_i are uniquely determined by Φ'_1. With $\Phi'_2 = \Phi'_3 = 0$, doubling Φ'_1 would double the field and also the q'_i; that is, each of the three charges is directly proportional to Φ'_1, and so

$$\left.\begin{array}{l} \text{State S'} \\ \Phi'_2 = \Phi'_3 = 0 \end{array}\right\} \quad q'_1 = C_{11}\,\Phi'_1; \qquad q'_2 = C_{21}\,\Phi'_1; \qquad q'_3 = C_{31}\,\Phi'_1. \tag{1}$$

Here C_{11}, C_{21} and C_{31} are constants that depend on the size, shape and spatial orientation of the conductors.

(a) Consider a second state S'' in which c_1 and c_3 are grounded and c_2 is at potential Φ''_2. Express the charges q''_i in terms of Φ''_2.

(b) Consider a third state S''' in which c_1 and c_2 are grounded and c_3 is at potential Φ'''_3. Express the charges q'''_i in terms of Φ'''_3.

(c) Consider a general state S in which c_1, c_2 and c_3 are at potentials Φ_1, Φ_2 and Φ_3 respectively. Use superposition to express the charges q_i in terms of the Φ_i and the C_{ij}.

(d) Generalize your answer to (c) for an arbitrary number of conductors, n say.

‡The figure above is based on Figure 3.18 of Ref. [3]. It is reproduced here courtesy of Cambridge University Press.

[3] E. M. Purcell and D. J. Morin, *Electricity and magnetism*, Chap. 3, pp. 147–8. New York: Cambridge University Press, 3 edn, 2013.

Solution

(a) By analogy with (1), we have for

$$\left.\begin{array}{c} \text{State S''} \\ \Phi_1'' = \Phi_3'' = 0 \end{array}\right\} \quad q_1'' = C_{12}\,\Phi_2''; \qquad q_2'' = C_{22}\,\Phi_2''; \qquad q_3'' = C_{32}\,\Phi_2'', \tag{2}$$

where C_{12}, C_{22} and C_{32} are constants.

(b) Similarly, for

$$\left.\begin{array}{c} \text{State S'''} \\ \Phi_1''' = \Phi_2''' = 0 \end{array}\right\} \quad q_1''' = C_{13}\,\Phi_3'''; \qquad q_2''' = C_{23}\,\Phi_3'''; \qquad q_3''' = C_{33}\,\Phi_3''', \tag{3}$$

where C_{13}, C_{23} and C_{33} are constants.

(c) The superposition of states S', S'' and S''' is also a possible state of the system. The electric field at any point in the space between the conductors is the vector sum of the fields at that point in these three states, whilst the charge q_i on conductor i is $q_i' + q_i'' + q_i'''$. Thus,

$$\left.\begin{array}{c} q_1 = C_{11}\,\Phi_1 + C_{12}\,\Phi_2 + C_{13}\,\Phi_3 \\[4pt] q_2 = C_{21}\,\Phi_1 + C_{22}\,\Phi_2 + C_{23}\,\Phi_3 \\[4pt] q_3 = C_{31}\,\Phi_1 + C_{32}\,\Phi_2 + C_{33}\,\Phi_3 \end{array}\right\}. \tag{4}$$

(d) Generalizing (4) gives

$$q_i \;=\; \sum_{j=1}^{n} C_{ij}\,\Phi_j \qquad (i = 1,2,\dots,n), \tag{5}$$

where the constants C_{ij} depend (apart from ϵ_0) only on the size, shape and spatial orientation of the conductors.

Comments

(i) In the preceding discussion, we see that for a given arrangement of n conductors maintained at potentials Φ_i the charges q_i are determined by the set of n linear equations (5). Conversely, if the same configuration of conductors carry charges q_i, then the potentials Φ_i are determined by the set of equations

$$\Phi_i \;=\; \sum_{j=1}^{n} P_{ij}\,q_j \qquad (i = 1,2,\dots,n), \tag{6}$$

where the constants P_{ij} also depend (apart from ϵ_0) only on the size, shape and spatial orientation of the conductors.

(ii) The matrix forms of (5) and (6) are

$$
\left.\begin{array}{c}
Q = C\Phi \\[6pt]
\text{and} \\[6pt]
\Phi = PQ
\end{array}\right\}, \tag{7}
$$

where

$$
Q = \begin{pmatrix} q_1 \\ q_2 \\ \vdots \\ q_n \end{pmatrix}, \qquad\qquad \Phi = \begin{pmatrix} \Phi_1 \\ \Phi_2 \\ \vdots \\ \Phi_n \end{pmatrix},
$$

$$
\left.\begin{array}{l}
\end{array}\right\}. \tag{8}
$$

$$
C = \begin{pmatrix} C_{11} & C_{12} & \cdots & C_{1n} \\ C_{21} & C_{22} & \cdots & C_{2n} \\ \vdots & \vdots & & \vdots \\ C_{n1} & C_{n2} & \cdots & C_{nn} \end{pmatrix}, \quad \text{and} \quad P = \begin{pmatrix} P_{11} & P_{12} & \cdots & P_{1n} \\ P_{21} & P_{22} & \cdots & P_{2n} \\ \vdots & \vdots & & \vdots \\ P_{n1} & P_{n2} & \cdots & P_{nn} \end{pmatrix}
$$

(iii) The diagonal elements C_{ii} of the matrix C are the coefficients of capacitance of the system. The off-diagonal elements C_{ij} are the coefficients of induction. The elements P_{ij} of the matrix P are the coefficients of potential. In Question 3.11, we prove some useful properties of the matrices C and P.

(iv) Two adjacent conductors that are electrically isolated from each other constitute a device known as a capacitor. Suppose $V = \Phi_1 - \Phi_2$ is the potential difference between the conductors when they carry equal and opposite charges $\pm q$. The capacitance of the system is a purely geometrical quantity and is defined as

$$
C = \frac{q}{V}. \tag{9}
$$

Dimensionally $C = \epsilon_0 \times$ a characteristic length scale of the system, and it is positive by definition. The SI unit of capacitance—the farad (F)—is named in honour of Michael Faraday, where $1\,\mathrm{F} = 1\,\mathrm{C\,V^{-1}}$.

(v) It is sometimes useful to express the capacitance C of a two-conductor system in terms of the potential coefficients P_{ij}. Since $V = \Phi_1 - \Phi_2$ and because of (6) we have

$$
V = (P_{11}q_1 + P_{12}q_2) - (P_{21}q_1 + P_{22}q_2) = (P_{11} - P_{21})q_1 + (P_{12} - P_{22})q_2.
$$

But $q_1 = -q_2 = q$, and so

$$
C = \frac{1}{P_{11} + P_{22} - P_{12} - P_{21}}. \tag{10}
$$

Some simple applications involving capacitance and the coefficients P_{ij} and C_{ij} are considered elsewhere in this chapter.

Question 3.11**

(a) Consider a configuration of n conductors carrying charges $q_1, q_2, \ldots, q_n$ distributed with surface-charge densities $\sigma_1, \sigma_2, \ldots, \sigma_n$ over the surfaces $s_1, s_2, \ldots, s_n$. Show that the electrostatic potential energy U of this system can be written as

$$U = \tfrac{1}{2} Q^t P Q, \tag{1}$$

where the matrices P and Q are given by (8) of Question 3.10. Here the row matrix $Q^t = (q_1, q_2, \ldots, q_n)$ is the transpose of the column matrix Q.

Hint: Start with (5) of Question 2.17.

(b) Prove the following properties:

☞ the matrices C and P are inverses of each other. That is,

$$C^{-1} = P. \tag{2}$$

☞ C and P are both symmetric matrices. That is,

$$\left. \begin{aligned} C &= C^t & \text{and} && P &= P^t, \\ &\text{or equivalently} \\ C_{ij} &= C_{ji} & \text{and} && P_{ij} &= P_{ji} \end{aligned} \right\}. \tag{3}$$

Hint: To prove (3), begin by calculating the work required to transfer an amount of charge dq_k from a charge reservoir at infinity to the surface of the kth conductor (keeping the charge on all the other conductors constant).

Solution

(a) Making the replacement $\rho\, dv \to \sigma\, da$ in (5) of Question 2.17 and summing over all conductors give

$$U = \tfrac{1}{2} \sum_{i=1}^{n} \int_{s_i} \sigma_i(\mathbf{r}')\Phi_i(\mathbf{r}')\, da = \tfrac{1}{2} \sum_{i=1}^{n} \Phi_i \int_{s_i} \sigma_i\, da, \tag{4}$$

since each surface is an equipotential. By definition, $q_i = \int_{s_i} \sigma_i\, da$, and so

$$U = \tfrac{1}{2} \sum_{i=1}^{n} q_i \Phi_i. \tag{5}$$

Substituting (6) of Question 3.10 in (5) gives

$$U = \tfrac{1}{2} \sum_{i,j=1}^{n} q_i P_{ij} q_j, \tag{6}$$

which is the same as (1).

(b) ☞ Substituting $(7)_2$ into $(7)_1$ of Question 3.10 gives $Q = \mathcal{C}PQ$ which implies that $I = \mathcal{C}P$, the unit matrix. It then follows that $\mathcal{C}^{-1}I = \mathcal{C}^{-1}\mathcal{C}P$ or $\mathcal{C}^{-1} = P$, which is (2).

☞ Because of the hint, the work dW expended in transferring dq_k from infinity to s_k is

$$dW = \Phi_k\, dq_k = \sum_{i=1}^{n} P_{ki}\, q_i\, dq_k, \tag{7}$$

where the last step follows from (6) of Question 3.10. This work manifests as a change in the electrostatic potential energy dU of the system. Thus

$$dW = dU = \frac{\partial U}{\partial q_k}\, dq_k$$

$$= \tfrac{1}{2}\left[\sum_{j=1}^{n} P_{kj}\, q_j + \sum_{i=1}^{n} q_i P_{ik}\right] dq_k$$

$$= \tfrac{1}{2}\sum_{i=1}^{n}(P_{ki} + P_{ik})\, q_i\, dq_k. \tag{8}$$

Comparing (7) and (8) shows that $P_{ki} = \tfrac{1}{2}(P_{ki} + P_{ik})$ or $P_{ki} = P_{ik}$. That is, the matrix P is equal to its transpose and is therefore symmetric.

☞ Since $P = P^t$, it follows from (2) that $\mathcal{C}^{-1} = (\mathcal{C}^{-1})^t$. Taking the transpose of $I = \mathcal{C}^{-1}\mathcal{C}$ gives $I^t = I = (\mathcal{C}^{-1})^t\mathcal{C}^t$ and $I = \mathcal{C}^{-1}\mathcal{C}^t$. So $\mathcal{C} = \mathcal{C}^t$, which proves that $\mathcal{C}_{ij} = \mathcal{C}_{ji}$.

Question 3.12

Calculate the coefficients of potential, and hence the capacitance C for a system comprising:

(a) Two concentric spherical conducting shells having radii r_1 and r_2.

(b) Three concentric spherical conducting shells having radii r_1, r_2 and r_3 with the innermost and outermost shells connected by a fine, insulated copper wire.

In both cases, assume the shells have negligible thickness and that $r_1 < r_2 < r_3$.

Solution

(a) Suppose the shells carry charges q_1 and q_2. The results of Question 2.4 and the principle of superposition give

$$\left.\begin{aligned}\Phi_1 &= \frac{1}{4\pi\epsilon_0}\left(\frac{q_1}{r_1} + \frac{q_2}{r_2}\right) \\ \Phi_2 &= \frac{1}{4\pi\epsilon_0}\left(\frac{q_1}{r_2} + \frac{q_2}{r_2}\right)\end{aligned}\right\}.$$

Comparing these equations with

$$\left.\begin{aligned}\Phi_1 &= P_{11}q_1 + P_{12}q_2 \\ \Phi_2 &= P_{21}q_1 + P_{22}q_2\end{aligned}\right\} \quad \text{yields} \quad P = \frac{1}{4\pi\epsilon_0}\begin{pmatrix} r_1^{-1} & r_2^{-1} \\ r_2^{-1} & r_2^{-1} \end{pmatrix}.$$

Substituting the P_{ij} coefficients above in (10) of Question 3.10 gives

$$C = \frac{4\pi\epsilon_0}{r_1^{-1} - r_2^{-1}} = 4\pi\epsilon_0 \frac{r_1 r_2}{r_2 - r_1}. \tag{1}$$

(b) Proceeding as before we obtain:

$$\left.\begin{aligned}\Phi_1 &= \frac{1}{4\pi\epsilon_0}\left(\frac{q_1}{r_1} + \frac{q_2}{r_2} + \frac{q_3}{r_3}\right) \\ \Phi_2 &= \frac{1}{4\pi\epsilon_0}\left(\frac{q_1}{r_2} + \frac{q_2}{r_2} + \frac{q_3}{r_3}\right) \\ \Phi_3 &= \frac{1}{4\pi\epsilon_0}\left(\frac{q_1}{r_3} + \frac{q_2}{r_3} + \frac{q_3}{r_3}\right)\end{aligned}\right\},$$

where q_3 is the charge on the outermost shell. Comparing these equations with

$$\left.\begin{aligned}\Phi_1 &= P_{11}q_1 + P_{12}q_2 + P_{13}q_3 \\ \Phi_2 &= P_{21}q_1 + P_{22}q_2 + P_{23}q_3 \\ \Phi_3 &= P_{31}q_1 + P_{32}q_2 + P_{33}q_3\end{aligned}\right\} \quad \text{yields} \quad P = \frac{1}{4\pi\epsilon_0}\begin{pmatrix} r_1^{-1} & r_2^{-1} & r_3^{-1} \\ r_2^{-1} & r_2^{-1} & r_3^{-1} \\ r_3^{-1} & r_3^{-1} & r_3^{-1} \end{pmatrix}. \tag{2}$$

Now

$$V = \Phi_1 - \Phi_2 = (P_{11} - P_{21})q_1, \tag{3}$$

because $P_{12} = P_{22}$ for this system of conductors, and $P_{13} = P_{23}$. Also, since $\Phi_1 = \Phi_3$ it is easily shown that

$$q_1 = \frac{(P_{12} - P_{32})}{(P_{11} - P_{31})}q. \tag{4}$$

From (2)–(4) we obtain

$$C = \frac{q}{V} = \frac{(P_{11} - P_{31})}{(P_{11} - P_{21})(P_{12} - P_{32})} = \frac{4\pi\epsilon_0(r_1^{-1} - r_3^{-1})}{(r_1^{-1} - r_2^{-1})(r_2^{-1} - r_3^{-1})}$$

$$= 4\pi\epsilon_0 \frac{r_2^2(r_3 - r_1)}{(r_2 - r_1)(r_3 - r_2)}. \tag{5}$$

Comments

(i) Equation (1) is a well-known elementary result for the capacitance of two concentric spherical conductors.

(ii) If the two conductors in (a) above carry different charges (q and q', say), their capacitance C is still given by (1). That this must be so follows from Gauss's law: the charge induced on the inner surface of the outer shell is always $-q$. This charge and the charge q on the inner shell form the 'plates' of a capacitor, carrying equal and opposite charges. The charge $q + q'$ on the outer surface of the outer shell has no effect on C, but of course it determines the field and potential of the system for $r > r_2$. Similarly, for the conductors of (b), any excess charge on the outer surface of shell 3 will not affect the capacitance (5).

(iii) This question illustrates how the coefficients P_{ij} (and hence C) may be determined analytically for an arrangement of conductors that possesses high symmetry. In Questions 3.13 and 3.14, we determine approximate values of the P_{ij} for systems of conductors having lower symmetry.

Question 3.13

Consider two conducting spheres having radii r_1 and r_2 whose centres are separated by a distance $d \gg r_1, r_2$.

(a) Show that, to a first approximation, the capacitance of the system is given by

$$C = \frac{4\pi\epsilon_0 r_1 r_2 d}{(r_1 + r_2)d - 2r_1 r_2}. \tag{1}$$

Hint: Start with (6) of Question 3.9.

(b) Suppose $r_1 = 3\,\text{cm}$, $r_2 = 4\,\text{cm}$ and $d = 10\,\text{cm}$.[‡] Calculate the coefficients of the matrix C.

Solution

(a) From (6) of Question 3.9

$$\left.\begin{aligned}
\Phi_1 &= \frac{1}{4\pi\epsilon_0}\left(\frac{q_1}{r_1} + \frac{q_2}{d}\right) \\[2mm]
\Phi_2 &= \frac{1}{4\pi\epsilon_0}\left(\frac{q_1}{r_2} + \frac{q_2}{d}\right)
\end{aligned}\right\} \quad \text{yields} \quad P = \frac{1}{4\pi\epsilon_0}\begin{pmatrix} r_1^{-1} & d^{-1} \\ d^{-1} & r_2^{-1} \end{pmatrix}. \tag{2}$$

Substituting the coefficients P_{ij} from (2) in (10) of Question 3.10 gives

[‡]Here we intentionally violate the condition $d \gg r_2$. See the comment on p. 138.

$$C = \frac{4\pi\epsilon_0}{r_1^{-1} + r_2^{-1} - 2d^{-1}},$$

which is (1).

(b) Inverting$^\sharp$ the matrix (2) gives

$$C = \frac{4\pi\epsilon_0 d}{d^2 - r_1 r_2}\begin{pmatrix} r_1 d & -r_1 r_2 \\ -r_1 r_2 & r_2 d \end{pmatrix}, \tag{3}$$

and substituting the numerical values yields

$$C = \frac{10\times10^{-2}}{(9.0\times10^9)(100-12)}\begin{pmatrix} 30 & -12 \\ -12 & 40 \end{pmatrix} \simeq \begin{pmatrix} 3.788 & -1.515 \\ -1.515 & 5.051 \end{pmatrix}\times10^{-12}\,\text{F}. \tag{4}$$

Comment

Comparing (4) with the (essentially) exact answer $(3)_2$ of Question 3.25 shows that the error in the coefficients C_{ij} is less than 5%. This is a surprising result, because the approximation which treats the field of the distant conductor like a point charge is expected to be valid for $d/r_2 \gg 1$ only; in the above calculation $d/r_2 \sim 3$.

Question 3.14

Consider three spherical conductors having radii r_1, r_2 and r_3 centred at the corners of an equilateral triangle of side d. The potentials of these spheres are Φ_1, Φ_2 and Φ_3 (relative to $\Phi = 0$ at infinity). Using *Mathematica* to perform all appropriate algebraic/numerical calculations, answer the following questions.

(a) Write down, to a first approximation, the potential coefficients P_{ij}. Then show that the capacitance matrix is given by

$$C = k\begin{pmatrix} r_1(d^2 - r_2 r_3)d & r_1 r_2(r_3 - d)d & r_1 r_3(r_2 - d)d \\ r_1 r_2(r_3 - d)d & r_2(d^2 - r_1 r_3)d & r_2 r_3(r_1 - d)d \\ r_1 r_3(r_2 - d)d & r_2 r_3(r_1 - d)d & r_3(d^2 - r_1 r_2)d \end{pmatrix}, \tag{1}$$

where $k = \dfrac{4\pi\epsilon_0}{d^3 + 2r_1 r_2 r_3 - (r_2 r_3 + r_1(r_2 + r_3))d}.$

(b) Suppose $d = 12r_1 = 8r_2 = 6r_3 = 24\,\text{cm}$ and $9\Phi_1 = 10\Phi_2 = 15\Phi_3 = 18\,000\,\text{V}$. Use (1) to calculate the charge on each sphere.

$^\sharp$See the notebook given in Question 3.14 where we use *Mathematica*'s Inverse function to invert a matrix.

Solution

(a) We proceed as in (a) of Question 3.13, and obtain

$$P = \frac{1}{4\pi\epsilon_0} \begin{pmatrix} r_1^{-1} & d^{-1} & d^{-1} \\ d^{-1} & r_2^{-1} & d^{-1} \\ d^{-1} & d^{-1} & r_3^{-1} \end{pmatrix}. \tag{2}$$

The inverse of (2), evaluated in `cell 1` of the notebook below, gives (1).

(b) From $(7)_1$ of Question 3.10 we have $Q = C\Phi$. Calculating the elements of matrix Q in `cell 2` of the *Mathematica* notebook yields the values

$$q_1 = 3.70\,\text{nC}, \qquad q_2 = 5.05\,\text{nC} \quad \text{and} \quad q_3 = 3.87\,\text{nC}. \tag{3}$$

```
In[1]:= PotCoeffs = ───── × {{r1⁻¹, d⁻¹, d⁻¹}, {d⁻¹, r2⁻¹, d⁻¹}, {d⁻¹, d⁻¹, r3⁻¹}};
                    4 π ε0
        CapCoeffs = MatrixForm[FullSimplify[Inverse[PotCoeffs]]]

In[2]:= SpherePotentials = {2000, 1800, 1200};
                                                                           1
        NumValues = {r1 → 0.02, r2 → 0.03, r3 → 0.04, d → 0.24, ε0 → ───────────────};
                                                                     4 π (9.0 × 10⁹)
        CapCoeffValues = Inverse[PotCoeffs] /. NumValues;
        Charges = CapCoeffValues /. SpherePotentials
```

Comment

The charges given in (3) are accurate only to within this first approximation. However, these values may, in fact, be better than we might expect in view of the comment on p. 138.

Question 3.15

A long conducting wire having radius a is bent into a circle of radius $r_0 \gg a$. Choose Cartesian coordinates such that the wire is centred on the origin O and lies in the xy-plane. This geometry, known as an 'anchor ring', is illustrated in cross-section in the figure below.

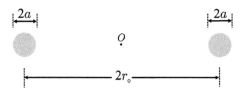

(a) Explain why the circular line-charge formula

$$\Phi(r,\theta) = \frac{q}{2\pi^2\epsilon_0}\frac{K\left[\frac{-4rr_0\sin\theta}{r^2+r_0^2-2rr_0\sin\theta}\right]}{\sqrt{r^2+r_0^2-2rr_0\sin\theta}},\tag{1}$$

derived in Question 2.10, is a good approximation for the potential of the anchor ring carrying charge q.

(b) Hence show that the capacitance of the anchor ring is given by

$$C = \frac{4\pi^2\epsilon_0 r_0}{\ln(8\gamma^{-1})},\tag{2}$$

where $\gamma = a/r_0 \ll 1$.

Hint: At the last step, use *Mathematica* to make an appropriate series expansion involving the EllipticK function.

Solution

(a) The condition $a/r_0 \ll 1$ ensures that the equipotentials have an essentially circular cross-section in the immediate vicinity of the line charge.[‡] Consequently, one of these equipotentials can be made to coincide almost exactly with the surface of the wire.[#] Because the line charge replicates the boundary conditions of the actual physical problem, (1) will give the correct potential in the space outside the anchor ring.

(b) Consider a second, larger anchor ring of radius R_0 (made from the same wire as the first) lying in the xy-plane and also centred on O. Suppose charges q $(-q)$ are placed on the inner (outer) anchor rings. Then $C = q/V$, where V is the potential difference between these two conductors. Now in the limit $R_0 \to \infty$ neither ring makes a contribution to the potential of the other, and V is given by (1) taking $r = r_0 \pm a$; $\theta = \frac{1}{2}\pi$. Hence

$$C = \frac{2\pi^2\epsilon_0\sqrt{(r_0\pm a)^2+r_0^2-2(r_0\pm a)r_0}}{K\left[\frac{-4(r_0\pm a)r_0}{(r_0\pm a)^2+r_0^2-2(r_0\pm a)r_0}\right]},$$

where it is easily shown that $\sqrt{(r_0\pm a)^2+r_0^2-2(r_0\pm a)r_0} = a = \gamma r_0$. So

$$C = \frac{2\pi^2\epsilon_0 r_0\gamma}{K[-4(1\pm\gamma)\gamma^{-2}]} = 2\pi^2\epsilon_0 r_0 f(\gamma),\tag{3}$$

[‡]See Comment (i) of Question 2.10.

[#]The conductor is an equipotential.

where $f(\gamma) = \gamma/K[-4(1 \pm \gamma)\gamma^{-2}]$. Using the *Mathematica* notebook below, we make a series expansion of $f(\gamma)$ and find that

$$f(\gamma) = \frac{2}{\ln(8\gamma^{-1})} + \text{higher-order terms that are negligible for } \gamma \ll 1. \qquad (4)$$

Substituting (4) in (3) gives (2).

```
In[1]:= Series[ ------Y------ , {Y, 0, 1}, Assumptions → 0 ≤ Y < 1]
              EllipticK[-4 (1 - Y) Y⁻²]
```

Question 3.16

Consider a closed metal box having rectangular cross-section $a \times b \times c$. The lid (shown shaded in the figure below) is located in the xy-plane at $z = c$, and is maintained at a potential V_0 relative to the other five sides of the box which are connected to ground.[‡]

(a) Using the boundary conditions that apply to the fives sides of the box except the lid, show that

$$\Phi_{mn} = \Phi_0 \sin\left(\frac{m\pi x}{a}\right) \sin\left(\frac{n\pi y}{b}\right) \sinh \gamma_{mn} z. \qquad (1)$$

Here m and n are integers having the values $1, 2, 3 \ldots$ and γ_{mn} is a constant which should be defined in the answer.

Hint: Use the results of Question 1.18.

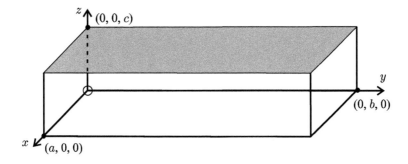

[‡]Suppose a rubber seal electrically insulates the lid from the rest of the box.

(b) The final boundary condition ($\Phi = V_0$ at $z = c$) can be satisfied if we make a series expansion of Φ of the form:

$$\Phi = \sum_{m,n=1}^{\infty} A_{mn}\Phi_{mn} = \sum_{m,n=1}^{\infty} A_{mn}\sin\left(\frac{m\pi x}{a}\right)\sin\left(\frac{n\pi y}{b}\right)\sinh\gamma_{mn}z. \quad (2)$$

Use (2) and the orthonormality condition

$$\frac{2}{L}\int_0^L \sin\left(\frac{m\pi v}{L}\right)\sin\left(\frac{n\pi v}{L}\right)dv = \delta_{mn}, \quad (3)$$

to show that the coefficients A_{mn} are given by

$$A_{mn} = \frac{1}{mn\sinh\gamma_{mn}c}\frac{4V_0/\pi^2}{}[1-(-1)^m][1-(-1)^n]. \quad (4)$$

(c) Hence write down an expression for the potential $\Phi(x, y, z)$ inside the box.

Solution

(a) The boundary condition $\Phi = 0$ anywhere $x = 0$ or $y = 0$ or $z = 0$ vetoes any combination in (8) of Question 1.18 involving cos or cosh functions. The only remaining possibility is thus

$$\Phi = \Phi_0\sin\alpha x\,\sin\beta y\,\sinh\gamma z. \quad (5)$$

The boundary conditions $\Phi = 0$ at $x = a$ and $\Phi = 0$ at $y = b$ impose restrictions on α and β: namely, that $\alpha a = m\pi$ and $\beta b = n\pi$. So

$$\alpha = \alpha_m = \frac{m\pi}{a}, \quad \beta = \beta_n = \frac{n\pi}{b} \quad \text{and} \quad \gamma = \gamma_{mn} = \pi\sqrt{\frac{m^2}{a^2}+\frac{n^2}{b^2}} \quad (6)$$

follows from (5) of Question 1.18. Substituting (6) in (5) yields (1).

(b) Putting $z = c$ in (2) and using the boundary condition $\Phi = V_0$ for the lid give

$$V_0 = \sum_{p,q=1}^{\infty} A_{pq}\sin\left(\frac{p\pi x}{a}\right)\sin\left(\frac{q\pi y}{b}\right)\sinh\gamma_{pq}c. \quad (7)$$

The coefficients A_{pq} are extracted from (7) by multiplying both sides of this equation by $\sin\frac{m\pi x}{a}\sin\frac{n\pi y}{b}$ and integrating over the rectangle $0 \le x \le a$, $0 \le y \le b$. This gives

☞ for the left-hand side of (7):

$$\int_0^b \int_0^a V_0 \sin\left(\frac{m\pi x}{a}\right) \sin\left(\frac{n\pi y}{b}\right) dx\, dy = \frac{V_0}{\pi^2} \frac{ab}{mn}\left[1-(-1)^m\right]\left[1-(-1)^n\right] \quad (8)$$

because $\int_0^L \sin\left(\frac{m\pi v}{L}\right) dv = -\frac{\cos(m\pi v/L)}{m\pi/L}\Big|_0^L = \frac{L}{m\pi}\left[1-(-1)^m\right].$

☞ for the right-hand side of (7):

$$\sum_{p,q=1}^{\infty} A_{pq} \sinh\gamma_{pq}c \int_0^b \int_0^a \sin\left(\frac{m\pi x}{a}\right) \sin\left(\frac{p\pi x}{a}\right) \sin\left(\frac{n\pi y}{b}\right) \sin\left(\frac{q\pi y}{b}\right) dx\, dy$$

$$= \sum_{p,q} A_{pq}\left(\tfrac{1}{2}a\delta_{mp}\right)\left(\tfrac{1}{2}b\delta_{nq}\right) \sinh\gamma_{pq}c$$

$$= \tfrac{1}{4}ab\sum_{p,q} A_{pq}\delta_{mp}\delta_{nq}\sinh\gamma_{pq}c = \tfrac{1}{4}ab A_{mn}. \quad (9)$$

(In the second step, we use the orthonormality condition (3).) Equating (8) and (9) yields (4).

(c) Substituting (4) in (2) and recognizing that the coefficients A_{mn} are non-zero only if m and n are odd integers give

$$\Phi(x,y,z) = \frac{16V_0}{\pi^2} \sum_{\substack{\text{odd}\\m,n}}^{\infty} \frac{1}{mn} \sin\left(\frac{m\pi x}{a}\right) \sin\left(\frac{n\pi y}{b}\right) \frac{\sinh\gamma_{mn}z}{\sinh\gamma_{mn}c}. \quad (10)$$

Comments

(i) This question is an example of an electrostatic potential expressed in terms of a Fourier series. The quantities A_{mn} are the Fourier coefficients and they determine the 'weight' of the Fourier components. It is clear from (4) that the $A_{mn} \to 0$ as $m, n \to \infty$. The calculation of these coefficients hinges on the crucial steps leading to (9): the orthogonality of the sine functions on the interval $[0, \pi]$ enables all but one of the A_{mn} to be eliminated from the infinite sum. Because the integral in (8) can be evaluated in closed form, an analytical expression for this sole surviving Fourier coefficient then follows.

(ii) The method used here is known as the Fourier method (or the method of eigenfunction expansion). It usually succeeds in solving Laplace's equation when the boundary surfaces of the problem coincide with coordinate surfaces, as illustrated in this question for a system possessing rectangular symmetry.

(iii) In Question 3.17, we consider a numerical application of (10).

Question 3.17

Consider the metal box described in Question 3.16. Suppose that $a = b = c = 10\,\text{cm}$ and $\Phi_0 = 1000\,\text{V}$. Approximate solutions for $\Phi(x, y, z)$ may be obtained by truncating the infinite series at specified upper limits (m_0 and n_0, say) for both of the summations in (10) above.

(a) Write a *Mathematica* notebook to calculate $\Phi(x, y_0, z_0)$ for $0 \le x \le 10\,\text{cm}$ and $2y_0 = z_0 = 10\,\text{cm}$. Use a step size $\Delta x = 0.01\,\text{cm}$. Take

   ☞ the first five Fourier coefficients ($m_0 = n_0 = 9$), and

   ☞ the first twenty-five Fourier coefficients ($m_0 = n_0 = 49$).

Store the data in an array. Then, on the same set of axes, plot graphs of $\Phi(x, y_0, z_0)$ vs x.

(b) Repeat (a) for $z_0 = 9.5\,\text{cm}$.

(c) Add a second `cell` to your notebook in which you let $m_0 = n_0 = 149$. Then calculate the lines of constant potential:

   ☞ in the xy-plane at $z = 5\,\text{cm}$,

   ☞ in the xz-plane at $y = 5\,\text{cm}$ (use the symbol $*$ to indicate electrical insulation at the points in the plot where the lid joins the rest of the box).

(d) We now wish to calculate the potential at discrete points inside the box on a lattice having unit cell size $\Delta x = \Delta y = \Delta z = 1\,\text{cm}$. For this purpose, take $1 \le x \le 9\,\text{cm}$, $1 \le y \le 9\,\text{cm}$ and $0 \le z \le 9\,\text{cm}$. At each point, terminate the Fourier series when successive terms in the sum make a relative contribution to Φ of less than 10^{-6}. In doing this, you will need to write *Mathematica* code for a third `cell` where the computer should decide suitable values of m_0 and n_0 (these must be able to vary from one lattice point to the next). Produce a grid displaying the potentials (quoted to the first decimal place) at twenty evenly spaced points in the yz-plane for $x = 5\,\text{cm}$.

Solution

Using the notebook below, we obtain for the graphs (a) and (b):

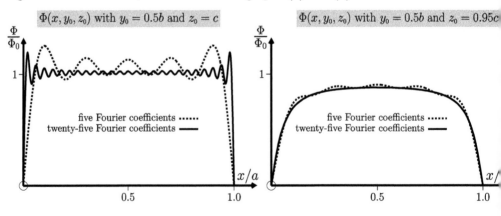

the contour plots for (c):

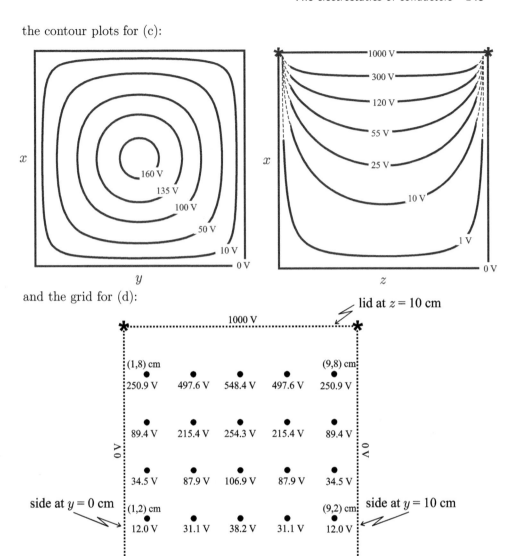

and the grid for (d):

Comments

(i) It turns out that the rate at which the infinite series ((10) of Question 3.16) converges is much faster in certain regions of parameter space than in others. This important aspect can be explored further using cell 3 of the notebook where a tolerance of 10^{-6} has been specified. Here, one can establish that convergence at the majority of points inside the box on a grid comprising $\Delta x = \Delta y = \Delta z = 1\,\text{cm}$ occurs for values of m and n equal to nine. The electric potential converges at most other points well before these integers reach thirty-nine. By contrast, the convergence of Φ at points on the box's lid (where $z = 10\,\mathrm{cm}$) is extremely slow. Here the m, n values required to reach the desired level of accuracy can exceed one million. Clearly, repetitive cycling around the Do loop at these points consumes a significant fraction of the total CPU time. Readers interested in the computational aspects of this problem may be tempted to refine or rewrite the code provided below. (For example, symmetry features of the problem may be used to eliminate duplicate calculations.)

(ii) In Question 3.20 we calculate the potentials on a grid inside the box using the relaxation method. This important technique is discussed in Question 3.18.

```
In[1]:= Φ0 = 1000.0; a = b = c = 10; max1 = 9; max2 = 49;
```
$$\gamma[m\_,\ n\_] := \pi \sqrt{\left(\frac{m}{a}\right)^2 + \left(\frac{n}{b}\right)^2}\ ;\quad \alpha[m\_,\ n\_] := \frac{16\ \Phi0}{\pi^2\,m\,n}\ \frac{1}{\mathrm{Sinh}[\gamma[m,\ n]\ c]}\ ;$$
$$\mathrm{pot}[x\_,\ y\_,\ z\_] := \alpha[m,\ n]\ \mathrm{Sin}\!\left[\frac{m\,\pi\,x}{a}\right]\ \mathrm{Sin}\!\left[\frac{n\,\pi\,y}{b}\right]\ \mathrm{Sinh}[\gamma[m,\ n]\ z]$$
```
Φ1[x_, y_, z_] := Sum[pot[x, y, z], {n, 1, max1, 2}, {m, 1, max1, 2}]
Φ2[x_, y_, z_] := Sum[pot[x, y, z], {n, 1, max2, 2}, {m, 1, max2, 2}]
```
$$\mathrm{dat1} = \mathrm{Table}\!\left[\left\{x,\ \Phi1\!\left[x,\ \frac{b}{2},\ a\right]\right\},\ \left\{x,\ 0,\ a,\ \frac{a}{1000}\right\}\right];$$
$$\mathrm{dat2} = \mathrm{Table}\!\left[\left\{x,\ \Phi2\!\left[x,\ \frac{b}{2},\ a\right]\right\},\ \left\{x,\ 0,\ a,\ \frac{a}{1000}\right\}\right];$$
$$\mathrm{ListLinePlot}\!\left[\{\mathrm{dat1},\ \mathrm{dat2}\},\ \mathrm{PlotRange} \to \mathrm{All}\right]$$

```
In[2]:= max = 149;
```
```
Φ[x_, y_, z_] := Sum[pot[x, y, z], {n, 1, max, 2}, {m, 1, max, 2}]
```
$$\mathrm{dat3} = \mathrm{Table}\!\left[\left\{x,\ y,\ \Phi\!\left[x,\ y,\ \frac{c}{2}\right]\right\},\ \left\{x,\ 0,\ a,\ \frac{a}{100}\right\},\ \left\{y,\ 0,\ b,\ \frac{b}{100}\right\}\right];$$
$$\mathrm{ListContourPlot}\!\left[\mathrm{Flatten}\!\left[\mathrm{dat3},\ 1\right],\ \mathrm{ContourShading} \to \mathrm{None}\right]$$

```
In[3]:=
    tolerance = 1 × 10^-6; Δa = 1;
    ε = 10^-6 (*an infinitesimal quantity to avoid possible division by zero*);
    dat4 = {{0, 0, 0, 0, 0, 0}};
    Do[Converged = False; m0 = 5; n0 = 5; Φ = 0;
```
$$\mathrm{While}\!\left[\mathrm{Not}\!\left[\mathrm{Converged}\right],\ \mathrm{Old}\Phi = \Phi;\ m0 = m0 + 2;\ n0 = n0 + 2;\right.$$
```
     Φ = Sum[pot[x, y, z], {n, 1, n0, 2}, {m, 1, m0, 2}];
```
$$\mathrm{error} = \mathrm{Abs}\!\left[\frac{2\ (\Phi - \mathrm{Old}\Phi)}{(\Phi + \mathrm{Old}\Phi + \varepsilon)}\right];\ \mathrm{Converged} = \left(\mathrm{error} < \mathrm{tolerance}\right)\Big];$$
```
    AppendTo[dat4, {m0, n0, x, y, z, Φ}],
    {x, 1, a - 1, Δa}, {y, 1, a - 1, Δa}, {z, 1, a - 1, Δa}]
    dat4 = Delete[dat4, 1]; MatrixForm[dat4]
```

Question 3.18

Consider a two-dimensional electrostatic field that exists in a charge-free region of space ($\mathcal{R}$, say).

(a) Write Taylor-series expansions for the electrostatic potential $\Phi(x,y)$ about the following points in $\mathcal{R}$: ☞ $(x{+}h,\ y)$; ☞ $(x{-}h,\ y)$; ☞ $(x,\ y{+}h)$; ☞ $(x,\ y{-}h)$, up to and including derivatives of third order in x and y.

(b) Hence show that

$$\Phi(x,y) \;=\; \frac{1}{4}\Big[\Phi(x{+}h,y)+\Phi(x{-}h,y)+\Phi(x,y{+}h)+\Phi(x,y{-}h)\Big]+\mathcal{O}(h^4). \quad (1)$$

Solution

(a) Using (1) of Question 1.7 gives

$$\left.\begin{aligned}
\Phi(x\pm h,y) &= \Phi(x,y)\pm\frac{\partial\Phi}{\partial x}h+\frac{1}{2!}\frac{\partial^2\Phi}{\partial x^2}h^2\pm\frac{1}{3!}\frac{\partial^3\Phi}{\partial x^3}h^3+\mathcal{O}(h^4)\\[2mm]
\Phi(x,y\pm h) &= \Phi(x,y)\pm\frac{\partial\Phi}{\partial y}h+\frac{1}{2!}\frac{\partial^2\Phi}{\partial y^2}h^2\pm\frac{1}{3!}\frac{\partial^3\Phi}{\partial y^3}h^3+\mathcal{O}(h^4)
\end{aligned}\right\}. \quad (2)$$

(b) From the four equations in (2) we obtain

$$\Phi(x{+}h,y)+\Phi(x{-}h,y)+\Phi(x,y{+}h)+\Phi(x,y{-}h)=4\Phi(x,y)+\tfrac{1}{2}h^2\nabla^2\Phi+\mathcal{O}(h^4).$$

But $\nabla^2\Phi=0$, and so (1) follows.

Comments

(i) For a field that also varies in the z-direction, the equivalent form of (1) is

$$\begin{aligned}
\Phi(x,y,z) \;=\; \frac{1}{6}\Big[&\Phi(x+h,y,z)+\Phi(x-h,y,z)+\Phi(x,y+h,z)+\Phi(x,y-h,z)\\
&+\Phi(x,y,z+h)+\Phi(x,y,z-h)\Big]+\mathcal{O}(h^4). \quad (3)
\end{aligned}$$

(ii) Equations (1) and (3) provide the basis for a numerical algorithm for solving Laplace's equation. We outline this algorithm for a two-dimensional problem in Comments (iii) and (iv) on p. 148.

(iii) Suppose that a square lattice is superimposed on the region $\mathcal{R}$ as shown in the figure below. With h representing the side of a unit cell, (1) suggests that the potential at any node[‡] can be approximated by the average of the four nearest-neighbour potentials. So, for the labelled node n,

$$\Phi(x, y) \simeq \tfrac{1}{4}(\Phi_1 + \Phi_2 + \Phi_3 + \Phi_4), \tag{4}$$

since terms of $\mathcal{O}(h^4)$ are negligible in the limit $h \to 0$ (see the next comment).

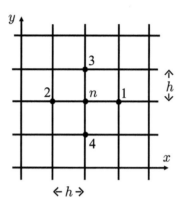

(iv) We begin the numerical calculation for $\Phi(x, y)$ by first choosing the unit cell size h and assigning values to Φ at all boundary nodes. Next, we make sensible guesses for the potential elsewhere on the lattice. These guesses are now refined through the repetitive application of (4). This process continues until successive values of Φ at corresponding nodes are less than some specified tolerance. Finally, we repeat the calculation for smaller and smaller values of the unit cell size. By extrapolating h to zero, $\Phi(x, y)$ relaxes[♯] to the correct solution for the specified boundary conditions. Purcell describes the process as follows:

> The relaxation of the values toward an eventually unchanging distribution is closely related to the physical phenomenon of *diffusion*. If you start with much too high a value at one point, it will 'spread' to its nearest neighbors, then to its next nearest neighbors, and so on, until the bump is smoothed out.[4]

Obviously, rapid convergence to the correct solution is achieved through a judicious choice of initial potentials.

(v) Questions 3.19 and 3.20 use the relaxation method to solve Laplace's equation for a two-dimensional and three-dimensional field respectively.

[‡]Defined here as a corner of a unit cell.

[♯]Numerical procedures for solving Laplace's equation based on (1) are known as 'relaxation methods'.

---

[4]   E. M. Purcell and D. J. Morin, *Electricity and magnetism*, Chap. 3, pp. 175–6. New York: Cambridge University Press, 3 edn, 2013.

Question 3.19 **

Two identical metal cylinders c_1 and c_2 having diameter a are mounted inside a cylindrical metal pipe whose internal diameter is $5a$. The origin of Cartesian axes is chosen at the centre of the pipe; the axes of all three conductors are parallel to $\hat{z}$ and those of c_1 and c_2 pass through $(x_{01}, y_{01}, 0)$ and $(x_{02}, y_{02}, 0)$ respectively. A cross-section of this configuration is shown in the figure below, together with the two-dimensional lattice required for the numerical calculations that follow. The cylinders are maintained at potentials Φ_1 and Φ_2 relative to the pipe which is at zero potential.

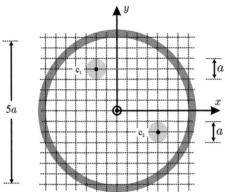

(a) Write a *Mathematica* program to find the solution of Laplace's equation everywhere inside the pipe and away from the ends, for the following cases:

☞ $\Phi_1 = \Phi_2 = 100\,\text{V}$ and $x_{01} = -y_{01} = -x_{02} = y_{02} = -2a$.

☞ $\Phi_1 = -\Phi_2 = 100\,\text{V}$ and $x_{01} = -y_{01} = -x_{02} = y_{02} = -2a$.

☞ $\Phi_1 = 100\,\text{V}, \quad \Phi_2 = 150\,\text{V}$ and $x_{01} = -y_{01} = -x_{02} = y_{02} = -2a$.

☞ $\Phi_1 = 100\,\text{V}, \quad \Phi_2 = -150\,\text{V}$ and $x_{01} = -y_{01} = -2x_{02} = y_{02} = -2a$.

Suppose the length scale $a = 1$ (in some system of units).

Hint: Adopt the algorithm outlined in the comments of Question 3.18. Choose an initial lattice comprising 12×12 unit cells (or 13×13 nodes). Stop iterating when the values of Φ between successive calculations agree to within $10^{-6}\,\text{V}$ at all nodes. Repeat for lattices with 24×24, 48×48 and 96×96 unit cells. For each of these four lattices, extract the values of $\Phi(i, j)$ at the initial set of nodes $(i, j = 1, 2, \ldots, 13)$. The result is a 13×13 matrix, each element of which is itself a 4×1 matrix. Taking these elements in turn, make a linear extrapolation to $h = 0$ and obtain the relaxed Φ-value for that node. Store these values in a data file.

(b) Use a Do loop to associate an (x, y) coordinate pair with each entry in the data file. Then use *Mathematica*'s ListContourPlot command to draw the equipotentials for the four cases outlined in (a).

Solution

(a) The notebook below is for the last set of potentials ($\Phi_1 = 100\,\text{V}$, $\Phi_2 = -150\,\text{V}$):

(b)

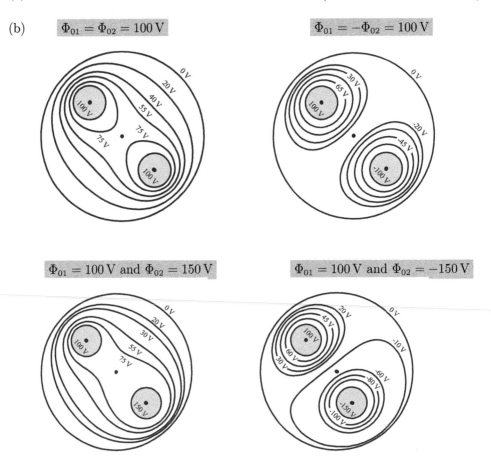

$\Phi_{01} = \Phi_{02} = 100\,\text{V}$

$\Phi_{01} = -\Phi_{02} = 100\,\text{V}$

$\Phi_{01} = 100\,\text{V}$ and $\Phi_{02} = 150\,\text{V}$

$\Phi_{01} = 100\,\text{V}$ and $\Phi_{02} = -150\,\text{V}$

Comment

The notebook may be readily adapted to calculate equipotentials for various arrangements of conductors held at different potentials relative to each other, such as:

☞ three conducting cylinders inside a cylindrical pipe,

☞ a cylinder inside a square pipe,

☞ a cylindrical pipe split (through its axis of symmetry) into two halves, and so on.

```
In[3]:= Φ01 = 100.0; (* CYLINDER c1 POTENTIAL *)

Φ02 = -150.0; (* CYLINDER c2 POTENTIAL *)

Φ03 = 0.0; (* PIPE POTENTIAL *)


con1a[i_, j_] := (low1i ≤ i ≤ up1i && low1j ≤ j ≤ up1j );


con1b[i_, j_] := (Ceiling[c01j - √((r01² - (i - c01i)²))] ≤ j ≤
    Floor[c01j + √((r01² - (i - c01i)²))]);


con2a[i_, j_] := (low2i ≤ i ≤ up2i && low2j ≤ j ≤ up2j );


con2b[i_, j_] := (Ceiling[c02j - √((r02² - (i - c02i)²))] ≤ j ≤
    Floor[c02j + √((r02² - (i - c02i)²))]);


con3a[i_] := (i < 1 + 2^m || i > 1 + 11 × 2^m);


con3b[i_, j_] := (Ceiling[c03 - √(r03² - (i - c03)²)] ≤ j ≤
    Floor[c03 + √(r03² - (i - c03)²)]);

con3c[i_, j_] := (((i == low3 || i == up3) && j == c03) ||
    (i == c03 && (j == low3 || j == up3)));

f[i_, j_] := 1/4 (table0[[i, j - 1]] + table0[[i, j + 1]] + table0[[i + 1, j]] +
    table0[[i - 1, j]]);

tolerance = 1 × 10^-6;

mMax = 3;
(* m above is an integer which effectively sets the grid size.
   As m→ ∞ the grid size → 0 *)


hValues = {};
```

```
In[2]:=  (* ITERATE THE POTENTIALS AND STOP WHEN THE ERROR IS LESS THAN TOL *)

        Do[counter = 0; TolFlag = True; n0 = 13; n = 1 + (n0 - 1) × 2^m;

          c01j = 1 + 4 × 2^m;    low1j = 1 + 3 × 2^m;   up1j = 1 + 5 × 2^m;   c01i = 1 + 4 × 2^m;
          low1i = 1 + 3 × 2^m;   up1i = 1 + 5 × 2^m;    r01 = (up1i - low1i)/2;


          c02j = 1 + 7 × 2^m;    low2j = 1 + 6 × 2^m;   up2j = 1 + 8 × 2^m;   c02i = 1 + 8 × 2^m;
          low2i = 1 + 7 × 2^m;   up2i = 1 + 9 × 2^m;    r02 = (up2i - low2i)/2;


          c03 = 1 + 6 × 2^m;    low3 = 1 + 1 × 2^m;   up3 = 1 + 11 × 2^m;   r03 = (up3 - low3)/2;


          AppendTo[hValues, 1.0/2^m];
          table0 =
           Table[
            If[con3a[i], Φ03,
             If[con3b[i, j], If[con3c[i, j], Φ03, If[con1a[i, j],
                If[con1b[i, j], Φ01, (Φ01 + Φ02 + Φ03)/3], If[con2a[i, j],
                 If[con2b[i, j], Φ02, 1/3 (Φ01 + 2 Φ02 + Φ03)], (2 Φ01 + Φ02 + Φ03)/3]]], Φ03]],
            {i, 1, n}, {j, 1, n}];


          While[TolFlag,
            table1 = Table[If[con3a[i], Φ03, If[con3b[i, j], If[con3c[i, j], Φ03,
                    If[con1a[i, j], If[con1b[i, j], Φ01, f[i, j]], If[con2a[i, j],
                      If[con2b[i, j], Φ02, f[i, j]], f[i, j]]]], Φ03]],
                {i, 1, n}, {j, 1, n}] ;; difference = table1 - table0;
            e = Max[Abs[difference]];
            p = First[First[Position[Abs[difference], e]]];
            q = Last[First[Position[Abs[difference], e]]];
            BiggestError = table1[[p, q]]; error = Abs[e/BiggestError];
            TolFlag = (error > tolerance); table0 = table1; counter++];

          If[m == 0,
           array0 = table1,
           array1 = Table[table1[[(i - 1) 2^m + 1, (j - 1) 2^m + 1]], {i, 1, n0}, {j, 1, n0}];
           array2 = Partition[Partition[Riffle[Flatten[array0, 1],
               Flatten[array1]], 2], n0]; array0 = array2],
          {m, 0, mMax, 1}]
```

```
In[3]:= (* RELAX THE POTENTIALS AND STORE THESE VALUES IN THE ARRAY final *)
        array3 = {}; final = {};


        Do[array4 = Flatten[array2[[i, j]]]; AppendTo[array3, array4],
           {i, 1, n0}, {j, 1, n0}];        array3 = Partition[array3, n0];


        Do[array5 = Flatten[array3[[i, j]]];
           array6 = Partition[Riffle[hValues, array5], 2];
           array3[[i, j]] = array6, {i, 1, n0}, {j, 1, n0}];
        Partition[array3, n0];


        Do[int = InterpolatingPolynomial[array3[[i, j]], x] /. x -> 0;
           AppendTo[final, int], {i, 1, n0}, {j, 1, n0}];
        final = Partition[final, n0];


        (* IT IS CONVENIENT FOR THE PURPOSE OF DRAWING EQUIPOTENTIALS TO USE
           THE DATA FROM THE LAST ITERATION STORED IN table1. TO EACH OF THESE
           POTENTIALS WE FIRST ASSIGN (x,y) COORDINATES AND THEN PLOT. *)
```

$$a = 1.0; \quad n = 97; \quad data = \{\}; \quad x[k\_] := \frac{2\,(k - c03)}{n - 17}\,a; \quad y[k\_] := \frac{2\,(c03 - k)}{n - 17}\,a;$$

```
        Do[AppendTo[data, {x[j], y[i], table1[[i, j]]}], {i, 9, n - 8}, {j, 9, n - 8}];


        gr = ListContourPlot[data, PlotRange -> {{-1.2, 1.2}, {-1.2, 1.2}},
             ContourShading -> None];
```

Question 3.20 **

Consider the metal box described previously in Question 3.16. Take $a = b = c = 10\,\text{cm}$.

(a) Suppose, as in Question 3.17, that the potential on the lid is $\Phi_0 = 1000\,\text{V}$. Write a *Mathematica* notebook based on the relaxation method (see (3) of Question 3.18) to calculate the electric potential at discrete points inside the box on a lattice having unit cell size $\Delta_1 = 1\,\text{cm}$. For this purpose, take $1 \leq x, y \leq 9\,\text{cm}$ and $0 \leq z \leq 9\,\text{cm}$. Produce a grid, similar to that for (d) of Question 3.17, displaying the potentials at twenty evenly spaced points in the yz-plane for $x = 5\,\text{cm}$. Quote your answers, as before, to the first decimal place. Take the following into account:

☞ Set the tolerance to 1×10^{-6} (don't be too impatient: *Mathematica* is slow).

☞ For the relaxation part of the program, re-run *Mathematica* taking $\Delta_2 = 0.5\,\text{cm}$. Use the results from Δ_1 and Δ_2 to extrapolate to zero grid size. Two points for each (x, y, z) is far from ideal, but once again it is necessary to reach a compromise between 'accuracy' and 'calculation time'.

(b) For each of the grid points calculated in (a) and (d) of Question 3.17, determine the following percentage difference and illustrate these values on their own grid:

$$\left(\frac{\Phi_{\text{relaxation}} - \Phi_{\text{Fourier}}}{\Phi_{\text{Fourier}}}\right) \times 100.$$

Solution

We show below the grid for (a) derived from the relaxation method:

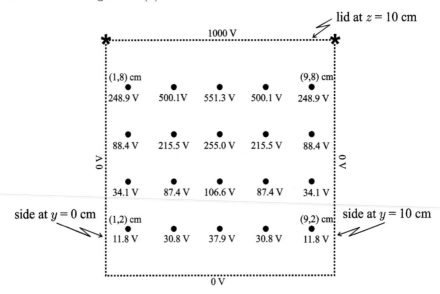

and the grid for (b):

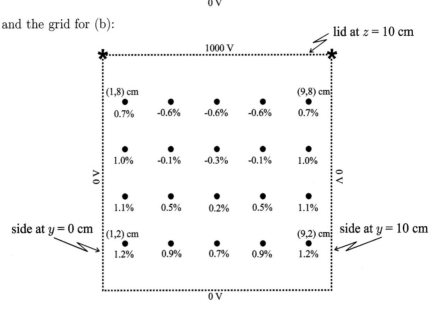

Comments

(i) The two techniques (Fourier and relaxation) give values for the potential that agree to within ∼1% of each other. Each method has its own distinct convergence issues which interested readers may readily explore.

(ii) The notebook below can be used in its present form (albeit with some small modifications to `cell 4`) for the problem of a metal box, where each face is assigned a different potential. This is a distinct, generalizing advantage of the relaxation method over the Fourier approach.

```
In[1]:=  (*Assume box is a rectangular cuboid so a=b=c*) a = 10;  p = 10;


  (*NOTE: Program requires that p=a/Δa is an INTEGER≥3*)


  (*Allow different sides of the box to assume different potentials*)


  V1 = 0 (*yz plane @ x=0*); V2 = 0 (*yz plane @ x=a*);
  V3 = 0 (*xz plane @ y=0*); V4 = 0 (*xz plane @ y=b*);
  V5 = 0 (*xy plane @ z=0*); V6 = 1000.0 (*xy plane @ z=c*);
```

$$\text{weight}[x\_, y\_, z\_] := \text{Exp}\left[\frac{-\text{Abs}[x-5]-\text{Abs}[y-5]}{8}\right]\left(\frac{z}{a}\right)^2 \times V6;$$

```
  (*The weight function is a starting potential for interior grid points.
    Be guided by the starting potentials on each face in deciding the form
    of weight[x,y,z]*)


  (*Points on the face which are neither corner nor edge. Their
    potentials will remain unchanged during the calculation*)
  con1[i_, j_, k_] := (i == 0 && (0 < j < m) && (0 < k < m))
  con2[i_, j_, k_] := (i == m && (0 < j < m) && (0 < k < m))
  con3[i_, j_, k_] := ((0 < i < m) && j == 0 && (0 < k < m))
  con4[i_, j_, k_] := ((0 < i < m) && j == m && (0 < k < m))
  con5[i_, j_, k_] := ((0 < i < m) && (0 < j < m) && k == 0)
  con6[i_, j_, k_] := ((0 < i < m) && (0 < j < m) && k == m)


  (*Exclude points at a corner*)
  con7[i_, j_, k_] := ((i == 0 && j == 0 && (k == 0 || k == m)))
  con8[i_, j_, k_] := ((i == 0 && j == m && (k == 0 || k == m)))
  con9[i_, j_, k_] := ((i == m && j == 0 && (k == 0 || k == m)))
  con10[i_, j_, k_] := ((i == m && j == m && (k == 0 || k == m)))


  (*Exclude points along an edge*)
  con11[i_, j_, k_] := ((i == 0 || i == m) && (j ≠ 0 && j ≠ m) && (k == 0 || k == m))
  con12[i_, j_, k_] := ((i ≠ 0 && i ≠ m) && (j == 0 || j == m) && (k == 0 || k == m))
  con13[i_, j_, k_] := ((i == 0 || i == m) && (j == 0 || j == m) && (0 < k < m))
```

```
In[2]:= Do[Δa = a / m;
    (*ASSIGN INITIAL VALUES TO DATA ARRAY*)
    data0 = {{0, 0, 0, 0, 0, 0, 0}};
    Do[x = i × Δa; y = j × Δa; z = k × Δa;
     If[con1[i, j, k], Φ = V1,
      If[con2[i, j, k], Φ = V2,
       If[con3[i, j, k], Φ = V3,
        If[con4[i, j, k], Φ = V4,
         If[con5[i, j, k], Φ = V5,
          If[con6[i, j, k], Φ = V6,
           If[con7[i, j, k], Φ = V,
            If[con8[i, j, k], Φ = V,
             If[con9[i, j, k], Φ = V,
              If[con10[i, j, k], Φ = V,
               If[con11[i, j, k], Φ = V,
                If[con12[i, j, k], Φ = V,
                 If[con13[i, j, k], Φ = V,
                  Φ = weight[x, y, z]]]]]]]]]]]]]];
     AppendTo[data0, {i, j, k, x, y, z, Φ}], {i, 0, m}, {j, 0, m}, {k, 0, m}];
    data0 = Delete[data0, 1];

    (* IMPLEMENT THE ALGORITH & ITERATE*)
    Converged = False; n = 1; olderror = 0; tolerance = 1.0 × 10^-6;

    While[Not[Converged], If[n > 1, counter = 0; data1 = {{0, 0, 0, 0, 0, 0, 0}};
     Do[counter++;
      If[(((i ≠ 0 && i ≠ m) && (j ≠ 0 && j ≠ m) && (k ≠ 0 && k ≠ m)),
       Φ = (data0[[counter + (m + 1)^2, 7]] + data0[[counter - (m + 1)^2, 7]] +
          data0[[counter + (m + 1), 7]] + data0[[counter - (m + 1), 7]] +
          data0[[counter + 1, 7]] + data0[[counter - 1, 7]]) / 6.0;
      AppendTo[data1, {data0[[counter, 1]], data0[[counter, 2]],
         data0[[counter, 3]], data0[[counter, 4]], data0[[counter, 5]],
         data0[[counter, 6]], Φ}], AppendTo[data1, data0[[counter]]]],
      {i, 0, m}, {j, 0, m}, {k, 0, m}]; data1 = Delete[data1, 1];
     difference = Abs[data0 - data1]; e = Max[difference];
     difference1 = Flatten[Position[difference, {0, 0, 0, 0, 0, 0, e}]];
     L = Length[difference1];
     Do[value = difference1[[1]]; error = e / data1[[value, 7]];
      If[error ≥ olderror, olderror = error], {1, 1, L}];
     Converged = (error ≤ tolerance); data0 = data1;
     n++, n = 2]]; If[m == p, Listp = data1, If[m == 2 p, List2p = data1,
      If[m == 4 p, List4p = data1]]]; Print[m], {m, {p, 2 p}}]
```

```
In[3]:=  (*SOME LIST MANIPULATION*)
         Lp = Length[Listp];  L2p = Length[List2p];
         con14[i_, j_] := ((Listp[[i, 4]] == List2p[[j, 4]]) &&
             (Listp[[i, 5]] == List2p[[j, 5]]) && (Listp[[i, 6]] == List2p[[j, 6]]))


         NewList = {{0, 0, 0, 0, 0}};

         Do[If[con14[i, j], AppendTo[NewList, {Listp[[i, 4]], Listp[[i, 5]],
             Listp[[i, 6]], Listp[[i, 7]], List2p[[j, 7]]}]],
          {i, 1, Lp}, {j, 1, L2p}]
         NewList = Delete[NewList, 1] /. V → π;
```

```
In[4]:=  NewestList = {{0, 0, 0, 0}}; max = 59;
```

$$\gamma[m\_, n\_] := \pi \sqrt{\left(\frac{m}{a}\right)^2 + \left(\frac{n}{a}\right)^2} \; ; \; \alpha[m\_, n\_] := \frac{16\,V6}{\pi^2} \; \frac{1}{m\,n\,\text{Sinh}[\gamma[m, n]\,a]} \; ;$$

$$V[x\_, y\_, z\_] := \text{Sum}\left[\alpha[m, n]\,\text{Sin}\left[\frac{m\,\pi\,x}{a}\right]\,\text{Sin}\left[\frac{n\,\pi\,y}{a}\right]\,\text{Sinh}[\gamma[m, n]\,(z)],\right.$$

$$\left.\{n, 1, max, 2\}, \{m, 1, max, 2\}\right]$$

```
         (*PERFORM THE RELAXATION OF THE GRID*)
         Do[If[NewList[[i, 4]] ≠ π && NewList[[i, 4]] ≠ π && NewList[[i, 4]] ≠ π,
           Φ1 = NewList[[i, 4]]; Φ2 = NewList[[i, 5]];
```

$$\text{data} = \left\{\left\{\frac{a}{p}, \Phi1\right\}, \left\{\frac{a}{2\,p}, \Phi2\right\}\right\};$$

```
           Φ = Fit[data, {1, h}, h] /. h → 0;
           x = NewList[[i, 1]];
           y = NewList[[i, 2]];
           z = NewList[[i, 3]];
           AppendTo[NewestList, {x, y, z, V[x, y, z], Φ,
```

$$\frac{100\,(V[x, y, z] - \Phi)}{V[x, y, z] + 10^{-6}}\}]], \{i, 1, Lp\}]$$

```
         (*THE TERM 10^-6 ABOVE AVOIDS POSSIBLE DIVISION BY ZERO*)
         NewestList = Delete[NewestList, 1];
```

```
In[5]:=  Data = {{0, 0, 0, 0, 0, 0}};
         Do[x = NewestList[[i, 1]]; y = NewestList[[i, 2]]; z = NewestList[[i, 3]];
          If[1 ≤ 1 x ≤ 9 && 1 ≤ 1 y ≤ 9 && 1 ≤ 1 z ≤ 9,
            AppendTo[Data, {x, y, z, NewestList[[i, 4]], NewestList[[i, 5]],
```

$$\frac{1}{10.0} \times \text{Floor}\left[\frac{(1000\,(\text{NewestList}[[i, 4]] - \text{NewestList}[[i, 5]]))}{\text{NewestList}[[i, 4]]}\right]\}]],$$

```
          {i, 1, 1215}]
         Data = Delete[Data, 1];
```

Question 3.21

Two point charges q and q' lie on a straight line (the z-axis, say, of Cartesian coordinates) passing through an origin O. Consider a sphere having radius a and surface s centred on O. The figure alongside shows a field point P having the spherical polar coordinates (r, θ, ϕ).

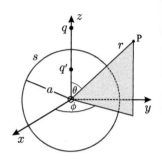

(a) Write down an expression for the electric potential Φ at P.

(b) Is it possible, and if so under what circumstances, for the potential to be zero everywhere on s (i.e. $\Phi(a\hat{\mathbf{r}}) = 0$)?

Solution

(a) By superposition

$$\Phi_P = \frac{1}{4\pi\epsilon_0}\left[\frac{q}{|\mathbf{r}-\mathbf{z}|} + \frac{q'}{|\mathbf{r}-\mathbf{z}'|}\right] = \frac{1}{4\pi\epsilon_0}\left[\frac{q}{|r\hat{\mathbf{r}}-d\hat{\mathbf{z}}|} + \frac{q'}{|r\hat{\mathbf{r}}-b\hat{\mathbf{z}}|}\right], \quad (1)$$

where d and b are the distances of q and q' from O respectively.

(b) Putting $r = a$ in (1) and rearranging give

$$\Phi(a) = \frac{1}{4\pi\epsilon_0}\left[\frac{q}{|a\hat{\mathbf{r}}-d\hat{\mathbf{z}}|} + \frac{q'}{|a\hat{\mathbf{r}}-b\hat{\mathbf{z}}|}\right]$$

$$= \frac{1}{4\pi\epsilon_0}\frac{q}{a}\left[\frac{a/d}{|\frac{a}{d}\hat{\mathbf{r}}-\hat{\mathbf{z}}|} + \frac{q'/q}{|\hat{\mathbf{r}}-\frac{b}{a}\hat{\mathbf{z}}|}\right]. \quad (2)$$

It is evident from (2) that $\Phi(a)$ is zero everywhere on s for all possible values of $\theta = \cos^{-1}(\hat{\mathbf{r}}\cdot\hat{\mathbf{z}})$ and $\phi \in [0, 2\pi]$ if we choose $q'/q = -a/d$ and $a/d = b/a$. That is,

$$q' = -q\frac{a}{d} \quad \text{and} \quad b = \frac{a^2}{d}. \quad (3)$$

Comments

(i) In the field of a *single* point charge q, all equipotential surfaces are necessarily spheres centred on the charge. Intuitively, one expects this spherical symmetry to be destroyed completely when a second point charge $-\alpha q$ is introduced.[‡] And it does. Except, astonishingly, for a single equipotential surface that is located in the

‡Here the arbitrary parameter α can have any value greater than zero.

immediate vicinity of the two charges. The radius and position of this sphere can be calculated from (3). For example, if $q = 4\,\mathrm{nC}$ and $q' = -3\,\mathrm{nC}$ are $14\,\mathrm{cm}$ apart, then the $\Phi = 0$ equipotential has radius $a = 24\,\mathrm{cm}$, with centre O a distance $b = 18\,\mathrm{cm}$ from q' and a distance $d = 32\,\mathrm{cm}$ from q (where O, q' and q lie on a straight line).

(ii) This property of the electrostatic field of two point charges having opposite signs, outlined in (i) above, can be exploited in an ingenious way to calculate the field of a point charge near a *spherical conductor*. See Question 3.22.

Question 3.22*

Consider a spherical[‡] conductor having radius a centred on an origin O and connected to ground with a fine wire. A point charge q is placed a distance d from O. Use the uniqueness theorem (see Comment (iii) of Question 3.3) and the results of Question 3.21 to determine the electric field $\mathbf{E}(\mathbf{r})$ everywhere in space for:

(a) $d > a$, and

(b) $d < a$.

(Choose a Cartesian coordinate system with origin at O and q on the z-axis.)

Solution

(a) We consider the regions $r \geq a$ and $r < a$ separately.

    ☞ $r \geq a$

The uniqueness of the solutions of Poisson's equation guarantees that, provided we can find a solution satisfying

- $\nabla^2 \Phi = -\rho/\epsilon_0$ everywhere and
- the boundary conditions,

this solution is the only possible one for the problem. With the results of Question 3.21 in mind, we replace the conductor with a point charge $q' = -q\dfrac{a}{d}$ located a distance $b = \dfrac{a^2}{d}$ from O on the z-axis. This preserves the boundary conditions of the original problem:

$$\left.\begin{array}{l} \Phi(a) = 0, \text{ and} \\ \Phi(r) \to 0 \text{ as } r \to \infty \end{array}\right\}. \tag{1}$$

Since q' is *inside* the region $r < a$, Poisson's equation *outside* this region remains unchanged. So for $r \geq a$, (1) of Question 3.21 yields

---

[‡]Either a solid conducting sphere or a hollow conducting shell.

$$\Phi(\mathbf{r}) = \frac{1}{4\pi\epsilon_0}\left[\frac{q}{|\mathbf{r} - d\hat{\mathbf{z}}|} + \frac{q'}{|\mathbf{r} - b\hat{\mathbf{z}}|}\right]$$

$$= \frac{1}{4\pi\epsilon_0}\left[\frac{q}{|\mathbf{r} - d\hat{\mathbf{z}}|} - \frac{q\frac{a}{d}}{|\mathbf{r} - \frac{a^2}{d}\hat{\mathbf{z}}|}\right]$$

$$= \frac{q}{4\pi\epsilon_0}\left[\frac{1}{\sqrt{r^2 + d^2 - 2rd\cos\theta}} - \frac{a}{\sqrt{r^2 d^2 + a^4 - 2rda^2\cos\theta}}\right], \qquad (2)$$

where $\theta = \cos^{-1}(\hat{\mathbf{r}} \cdot \hat{\mathbf{z}})$. The field has components $E_r = -\dfrac{\partial\Phi}{\partial r}$, $E_\theta = -\dfrac{1}{r}\dfrac{\partial\Phi}{\partial\theta}$ and $E_\phi = -\dfrac{1}{r\sin\theta}\dfrac{\partial\Phi}{\partial\phi}$, where $\Phi = \Phi(r, \theta)$. Differentiating (2) gives

$$\left.\begin{array}{l} E_r = \dfrac{q}{4\pi\epsilon_0}\left[\dfrac{(r - d\cos\theta)}{(r^2 + d^2 - 2rd\cos\theta)^{3/2}} - \dfrac{ad(rd - a^2\cos\theta)}{(r^2 d^2 + a^4 - 2rda^2\cos\theta)^{3/2}}\right] \\[4mm] E_\theta = \dfrac{qd\sin\theta}{4\pi\epsilon_0}\left[\dfrac{1}{(r^2 + d^2 - 2rd\cos\theta)^{3/2}} - \dfrac{a^3}{(r^2 d^2 + a^4 - 2rda^2\cos\theta)^{3/2}}\right] \\[4mm] E_\phi = 0 \end{array}\right\} \qquad (3)$$

in this region.

☞ $r < a$

Since the conductor is an equipotential, $\Phi = 0$ for $r \leq a$. Hence $\mathbf{E}(\mathbf{r}) = -\nabla\Phi = 0$ everywhere.

(b) We again divide the problem into two regions and consider each in turn.

☞ $r \geq a$

Assume that the conductor is a spherical shell. Because E is necessarily zero inside the walls of this shell, Gauss's law requires that a charge $-q$ be induced on its inner surface. The connection to ground ensures that the charge on the outer surface is zero and $\mathbf{E} = 0$ everywhere.

☞ $r < a$

Interchanging q with q' and b with d leads to a potential Φ given by (2), and so (3) is also the field in this region.

Comments

(i) Let us reflect for a moment on the method used to solve this question. Guided by the results of Question 3.21, we began by creating an 'image problem' in which the conducting surface was removed and *replaced* by a single image charge. This image charge was chosen to have just the right magnitude and position in space so that the pair of charges q and q' duplicated the boundary conditions of the actual problem. The uniqueness theorem was then invoked to assert that the potential given by the image problem was also a unique solution to the actual problem, but only in the region of space not containing q'. This approach is known as 'the method of images', and it is described elegantly as follows:

> the method of images concerns itself with the problem of one or more point charges in the presence of boundary surfaces, for example, conductors either grounded or held at fixed potentials. Under favourable conditions it is possible to infer from the geometry of the situation that a small number of suitably placed charges of appropriate magnitudes, external to the region of interest, can simulate the required boundary conditions. These charges are called *image charges*, and the replacement of the actual problem with boundaries by an enlarged region with image charges but not boundaries is called the *method of images*. The image charges must be external to the volume of interest, since their potentials must be solutions of the Laplace equation inside the volume; the 'particular integral' (i.e., solution of the Poisson equation) is provided by the sum of the potentials of the charges inside the volume.[5]

(ii) It is clear from the quotation at the end of Comment (i) above that the method of images works only in certain cases. The skill is to find the right combination and placement of image charges because 'there's as much art as science in the method of images, for you must somehow think up the right "auxiliary problem" to look at'.[6] The simplest application of this technique solves the problem of a point charge placed a known distance from an infinite plane conductor at zero potential (see the footnote on p. 173). Although we do not consider this standard example any further, we will explore related problems in Questions 3.28 and 9.10.

(iii) Substituting $r = a$ in $(3)_1$ gives $E_r(a)$, which is the normal component of the field at the surface of the conductor (here $E_\theta = E_\phi = 0$). The induced surface-charge density is $\sigma(a) = \pm\epsilon_o E_r(a)$ where the upper (lower) sign[†] is for $d > a$ ($d < a$). It then follows directly that

$$\sigma(\theta) = -\frac{1}{4\pi}\frac{q}{a^2}\frac{(a/d)\,|1 - a^2/d^2|}{[1 + a^2/d^2 - 2(a/d)\cos\theta]^{3/2}}, \tag{4}$$

which is clearly a non-uniform distribution.

(iv) The total charge induced on the surface of the sphere is

$$q_{\text{ind}} = \int \sigma(\theta)\,da = \int_0^{2\pi}\int_0^\pi \sigma(\theta)\,a^2\sin\theta\,d\theta\,d\phi = 2\pi a^2\int_{-1}^1 \sigma(\theta)\,d(\cos\theta). \tag{5}$$

[†]These are determined by the outward normal to s in Gauss's law.

---

[5]   J. D. Jackson, *Classical electrodynamics*, Chap. 2, p. 57. New York: Wiley, 3 edn, 1998.
[6]   D. J. Griffiths, *Introduction to electrodynamics*, Chap. 3, p. 125. New York: Prentice Hall, 3 edn, 1999.

Substituting (4) in (5) and evaluating the integral give

$$q_{ind} = \begin{cases} -q\dfrac{a}{d} & \text{for } d > a \\[2mm] -q & \text{for } d < a. \end{cases} \tag{6}$$

We note the following:

☞ q_{ind} is equal to the image charge both in magnitude and sign if $d > a$.

☞ q_{ind} is equal in magnitude but opposite in sign to the actual charge if $d < a$. Furthermore, q_{ind} appears on the inner surface of the shell, whilst the outer surface is uncharged.

(v) We can also use the image charge to calculate the force F exerted by q on the conductor. Consider the case $d > a$. Because q cannot distinguish between the conductor in the real problem and the image charge q' in the auxiliary problem, the force exerted by q on q' is also the force exerted by q on the conductor. This force follows immediately from Coulomb's law and is

$$F = \frac{1}{4\pi\epsilon_0}\frac{qq'}{(d-b)^2} = -\frac{q^2}{4\pi\epsilon_0}\frac{a}{d^3}\frac{1}{(1-a^2/d^2)^2}, \tag{7}$$

where the minus sign implies a force of attraction.

(vi) The method of images can also be used to solve variations of the above problem. See Questions 3.24 and 3.25.

Question 3.23

Consider a conducting thin-walled spherical shell of radius a centred on the origin O of Cartesian coordinates which is held at zero potential. Suppose that a point charge q is located near the shell at $(d, 0, 0)$.

(a) Use the results of Question 3.22 to show that the electric potential Φ at an arbitrary point $P(x, y, 0)$ is given by

$$\Phi(x, y) = \frac{1}{4\pi\epsilon_0}\frac{q}{a}\left[\frac{1}{\sqrt{(X-D)^2 + Y^2}} - \frac{1}{\sqrt{(XD-1)^2 + D^2Y^2}}\right], \tag{1}$$

where $X = x/a$, $Y = y/a$ and $D = d/a$ are dimensionless coordinates.

(b) Write a *Mathematica* notebook to calculate the electric field in the xy-plane passing through O, taking $\dfrac{1}{4\pi\epsilon_0}\dfrac{q}{a} = 1\,\text{V}$ and $D = 4$ (i.e. q located outside the shell). Plot this field and also show some lines of constant potential.

(c) Repeat (b) for $D = 0.6$ (i.e. q located inside the shell).

Solution

(a) Superposition of the potentials of q and its image charge q' give

$$\Phi(x, y) = \frac{1}{4\pi\epsilon_0} \frac{q}{\sqrt{(x-d)^2 + y^2}} + \frac{1}{4\pi\epsilon_0} \frac{q'}{\sqrt{(x-b)^2 + y^2}} ,$$

where $q' = -qa/d$ and $b = a^2/d$. Then

$$\Phi(x, y) = \frac{q}{4\pi\epsilon_0} \left[\frac{1}{\sqrt{(x-d)^2 + y^2}} - \frac{a/d}{\sqrt{(x-a^2/d)^2 + y^2}} \right], \tag{2}$$

together with the definitions of X, Y and D yields (1).

(b, c) The *Mathematica* notebook below produces the following figures, in which the • represents the point charge q. The surface of the shell is shown in grey.

```
In[1]:=  (* USE THIS SET OF PARAMETERS FOR (a) *)


     D0 = 4; Xmin = -3; Xmax = 7; Ymin = -5; Ymax = 5;
     Φ1 = 1.5; Φ2 = 0.6; Φ3 = 0.3; Φ4 = 0.05;


     (* USE THIS SET OF PARAMETERS FOR (b) *)
     D0 = 0.6; Xmin = -1; Xmax = 1; Ymin = -1; Ymax = 1;
     Φ1 = 3.0; Φ2 = 0.5; Φ3 = 0.1; Φ4 = -0.05;

                     ⎛        1                   1            ⎞
     Φ[X_, Y_] :=    ⎜ ─────────────────  -  ──────────────────── ⎟
                     ⎝ √(X-D0)² + Y²       √(X D0-1)² + Y² D0²   ⎠
     EX[X_, Y_] = -D[Φ[X, Y], X];   EY[X_, Y_] = -D[Φ[X, Y], Y];


In[2]:=  gr1 = StreamPlot[{{EX[X, Y], EY[X, Y]}}, {X, Xmin, Xmax},
            {Y, Ymin, Ymax}, StreamStyle → Directive[Thickness[0.006], Black],
            AspectRatio → 1, StreamPoints → 50,
            RegionFunction → Function[{X, Y}, If[D0 > 1, X² + Y² ≥ 1, X² + Y² ≤ 1]],
            Epilog → {PointSize[Large], Point[{{D0, 0}}]}];
         gr2 = ContourPlot[{Φ[X, Y] == Φ1, Φ[X, Y] == Φ2, Φ[X, Y] == Φ3,
            Φ[X, Y] == Φ4}, {X, Xmin, Xmax}, {Y, Ymin, Ymax},
            ContourStyle → Directive[Thickness[0.006], Black, Dashed],
            AspectRatio → 1, PlotPoints → 100,
            RegionFunction → Function[{X, Y}, If[D0 > 1, X² + Y² ≥ 1, X² + Y² ≤ 1]]];
         gr3 = Graphics[{Thickness[0.015], Gray, Circle[{0, 0}, 1.0]}];
         Show[{gr1, gr2, gr3}, FrameStyle → Directive[Thickness[0.004], 12],
          FrameLabel → {TraditionalForm[X], TraditionalForm[Y]},
          LabelStyle → Directive[Large]]
```

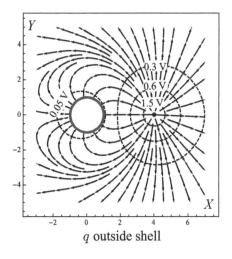

q outside shell

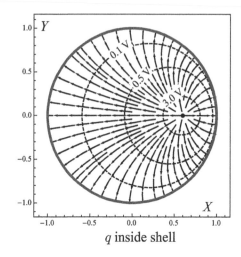

q inside shell

Question 3.24

Consider a point charge q a distance d from the centre O of a spherical conductor (radius a; surface s). Use the method of images to answer the following questions, assuming that $d > a$.

(a) Suppose the conductor is connected via a variable power supply (output is V_0) to ground. The role of the power supply is to raise (lower) the potential of the conductor above (below) ground potential. Determine the charge Q induced on the conductor.

(b) Suppose the conductor is uncharged and 'floating' (i.e. isolated from ground). Determine the potential Φ of the conductor.

(c) Suppose the conductor carries a net charge equal to q. Calculate the value of d such that the resultant force acting on the conductor is zero.

Solution

(a) It follows from Questions 3.21 and 3.22 that q induces on s a non-uniform charge density $\sigma'(\theta)$ and a net charge q'. The remaining charge $Q - q'$ delivered by the power supply is *uniformly* distributed over s, because an electrostatic equilibrium already exists between q and q'. The image problem therefore consists of three charges: q at $z = d$, q' at $z = b$ and $Q - q'$ at the origin. But q and q' give $\Phi(a) = 0$, and so $V_0 = \dfrac{1}{4\pi\epsilon_0}\dfrac{Q - q'}{a}$, or

$$Q = 4\pi\epsilon_0 V_0 a + q' = 4\pi\epsilon_0 V_0 a - qa/d \qquad (1)$$

because of $(3)_1$ of Question 3.21.

(b) Substituting $Q = 0$ (the conductor is uncharged) in (1), we calculate the floating potential

$$V_0 = \frac{1}{4\pi\epsilon_0}\frac{q}{d}. \tag{2}$$

(c) With the point charge q located at $z = d$, we also require an image charge q' at $z = b$ and a second image charge $q - q'$ at $z = 0$. The magnitude of the resultant force exerted by q on both q' and $q - q'$ is also the force exerted by q on the conductor (see Comment (v) of Question 3.22). So

$$F = \frac{1}{4\pi\epsilon_0}\left[\frac{qq'}{(d-b)^2} + \frac{q(q-q')}{d^2}\right]$$

$$= \frac{1}{4\pi\epsilon_0}\left[\frac{-q^2\frac{a}{d}}{(d-\frac{a^2}{d})^2} + \frac{q(q+q\frac{a}{d})}{d^2}\right]$$

$$= \frac{1}{4\pi\epsilon_0}\frac{q^2}{a^2}\frac{\gamma^5 - 2\gamma^3 - 2\gamma^2 + \gamma + 1}{\gamma^3(\gamma^2-1)^2}, \tag{3}$$

where $\gamma = d/a$ is a real number greater than one. Using *Mathematica*'s NSolve command to find the roots of $\gamma^5 - 2\gamma^3 - 2\gamma^2 + \gamma + 1 = 0$ gives $\gamma \simeq 1.618$. Thus

$$F = 0 \qquad \text{for } d = 1.618a.$$

Question 3.25 **

Consider two spherical conductors c_1 and c_2 having radii r_1 and r_2 whose centres O_1 and O_2 are a distance d apart. We wish to determine the electrostatic field in the vicinity of the spheres when they are maintained at potentials Φ_1 and Φ_2. Based on the results of Questions 3.21 and 3.22, we attempt a solution using image charges and proceed through a sequence of successive approximations.

first approximation

Regard c_2 as a point charge of magnitude $q_2 = 4\pi\epsilon_0\Phi_2 r_2$ located at O_2 and replace c_1 with two image charges:

- $q_1 = 4\pi\epsilon_0\Phi_1 r_1$ located at O_1, and
- $q_2' = -q_2 r_1/d$ a distance $b_1' = r_1^2/d$ from O_1.

These three charges establish surface s_1 as an equipotential but not surface s_2. This is corrected as follows:

second approximation

Place an image charge of magnitude $q'_1 = -q_1 r_2/d$ a distance $b'_2 = r_2^2/d$ from O_2. Whilst this has made s_2 an equipotential, it has destroyed the equipotential s_1.

third approximation

Another image charge of magnitude $q''_1 = -q'_1 r_1/(d - b'_2)$ a distance $b''_1 = r_1^2/(d - b'_2)$ from O_1 re-establishes s_1 as an equipotential but destroys the equipotential s_2.

By continuing in this way, it is evident that the solution to the problem requires an infinite set of image charges for each sphere: it turns out that these series converge rapidly. The figure below illustrates the notation used and the first few image charges for each conductor.

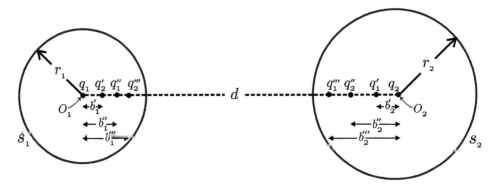

In what follows, assume that $r_1 = 3$ cm, $r_2 = 4$ cm and $d = 10$ cm.

(a) For the following values of Φ_1 and Φ_2, use *Mathematica* to calculate the first ten[‡] image charges for each conductor. Then plot the field in a plane containing O_1 and O_2 (the xy-plane say) and draw somes lines of constant potential.

   ☞ $\Phi_1 = 1500$ V and $\Phi_2 = 1800$ V

   (Show the equipotentials: 500 V, 600 V, 800 V and 1400 V.)

   ☞ $\Phi_1 = -1500$ V and $\Phi_2 = 1800$ V

   (Show the equipotentials: -250 V, -500 V, -750 V, 350 V, 500 V, 750 V and 1250 V.)

   Hint: Use the results of Question 3.22.

(b) Modify your *Mathematica* notebook to calculate the coefficients of potential and the coefficients of capacitance for this two-conductor system.

   Hint: Use the results of Questions 3.10 and 3.13.

[‡]Convince yourself that the error in the equipotentials s_1 and s_2 is of the order 0.01%.

Solution

(a) Evaluating `cells` 1 and 2 of the notebook below gives the field diagrams:

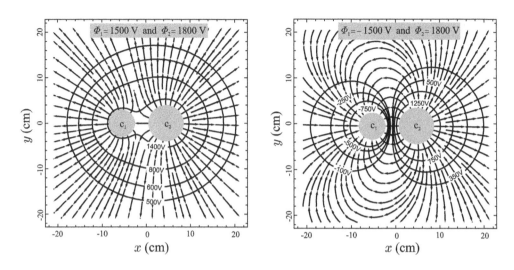

```
In[1]:= (* CALCULATE THE IMAGES *)

      Clear[q1, q2];  r1 = 3/100;  r2 = 4/100;  d = 1/10;  counter = 15;

      x1[dist_] := -dist/2 + r1/2 - r2/2;   x2[dist_] := dist/2 + r1/2 - r2/2;

      charges1 = {q1};  pos1 = {x1[d]};  charges2 = {q2};  pos2 = {x2[d]};

      ch1 = q1; d1 = d; ch2 = q2; d2 = d;

      Do[
         im1 = -ch2 r1/d1; im2 = -ch1 r2/d2; b1 = r1^2/d1; b2 = r2^2/d2; d1 = d - b2;
         d2 = d - b1; charges1 = Append[charges1, im1];
         charges2 = Append[charges2, im2]; pos1 = Append[pos1, (x1[d] + b1)];
         pos2 = Append[pos2, (x2[d] - b2)]; ch1 = im1; ch2 = im2,
         {i, counter}]
```

```
In[2]:=  (* NUMERICAL & PLOTTING *)
         Φ1 = -1500;   Φ2 = 1800;   k = 9.0 × 10⁹;   q1 = (Φ1 × r1)/k;   q2 = (Φ2 × r2)/k;

         FieldPoint = {x, y};   Charges = Join[charges1, Reverse[charges2]];
         Positions = Join[pos1, Reverse[pos2]];   n = Length[Charges];

         Φ[q_, p_, r_] := k ∑ᵢ₌₁ⁿ q[[i]]/(√((r[[1]] - p[[i]])² + (r[[2]])²));


In[3]:=  f[x_, y_] = -D[Φ[Charges, Positions, FieldPoint], x];
         g[x_, y_] = -D[Φ[Charges, Positions, FieldPoint], y];
         range = 0.21;
         InsideSpheres = Function[{x, y}, (x - x1[d])² + y² > r1² &&
             (x - x2[d])² + y² > r2²];
         gr1 = StreamPlot[{f[x, y], g[x, y]}, {x, -range, range},
             {y, -range, range}, RegionFunction → InsideSpheres,
             StreamStyle → Directive[Thickness[0.006], Black], AspectRatio → 1,
             Epilog → {Gray, Disk[{x1[d], 0}, r1], Disk[{x2[d], 0}, r2]}];
         gr2 = ContourPlot[V[Charges, Positions, FieldPoint], {x, -range, range},
             {y, -range, range}, Contours → {-100, -250, -500, -750, 350, 500,
                 750, 1250}, ContourStyle → Directive[Thickness[0.006], Black],
             AspectRatio → 1, ContourShading → None];
         Show[{gr1, gr2}, Frame → True, FrameStyle → Directive[Thickness[0.004],
             12], FrameLabel → {TraditionalForm[x], TraditionalForm[y]},
             LabelStyle → Directive[Large]]

In[4]:=  Clear[q1, q2, ch1, ch2];   k = 9.0 × 10⁹;
         Sol = Flatten[NSolve[{Q1 == Total[charges1],
             Q2 == Total[charges2]}, {q1, q2}]];
         ch1 = q1 /. Sol;   ch2 = q2 /. Sol;
         P11 = k/r1 ch1 /. {Q1 → 1, Q2 → 0};   P12 = k/r1 ch1 /. {Q1 → 0, Q2 → 1};

         P21 = k/r2 ch2 /. {Q1 → 1, Q2 → 0};   P22 = k/r2 ch2 /. {Q1 → 0, Q2 → 1};
         Pot = {{P11, P12}, {P21, P22}};   Cap = Inverse[Pot];
         MatrixForm[Pot]
         MatrixForm[Cap]
```

(b) Let Q_1 and Q_2 be the net charge on c_1 and c_2 respectively. Then

$$Q_1 = q_1 + q_2' + q_1'' + q_2''' + \cdots \qquad \text{and} \qquad Q_2 = q_2 + q_1' + q_2'' + q_1''' + \cdots . \quad (1)$$

Summing the series in (1) and solving for q_1 and q_2 in terms of Q_1 and Q_2 (see cell 4 of the notebook) give

$$q_1 \simeq 0.977\, Q_1 + 0.302\, Q_2 \Bigg\}$$
$$q_2 \simeq 0.402\, Q_1 + 0.988\, Q_2 \Bigg\}.$$

But $q_1 = 4\pi\epsilon_0 \Phi_1 r_1$ and $q_2 = 4\pi\epsilon_0 \Phi_2 r_2$, and so

$$\left. \begin{aligned} \Phi_1 &\simeq \frac{1}{4\pi\epsilon_0}\left[\frac{0.977}{r_1}Q_1 + \frac{0.302}{r_1}Q_2\right] \\ \Phi_2 &\simeq \frac{1}{4\pi\epsilon_0}\left[\frac{0.402}{r_2}Q_1 + \frac{0.988}{r_2}Q_2\right] \end{aligned} \right\}. \tag{2}$$

The coefficients of the Q_i in (2) are the potential coefficients P_{ij}. Substituting the values of r_1, r_2 and ϵ_0 in (2) yields

$$P = \begin{pmatrix} 2.931 & 0.905 \\ 0.905 & 2.223 \end{pmatrix} \times 10^{11}\, \mathrm{F}^{-1} \quad \text{and} \quad C = \begin{pmatrix} 3.903 & -1.589 \\ -1.589 & 5.145 \end{pmatrix} \times 10^{-12}\, \mathrm{F}. \tag{3}$$

Comments

(i) The electrostatic states of this two-conductor system are determined by

$$\left. \begin{aligned} \Phi_1 &= P_{11}Q_1 + P_{12}Q_2 \\ \Phi_2 &= P_{21}Q_1 + P_{22}Q_2 \end{aligned} \right\} \quad \text{or, equivalently,} \quad \left. \begin{aligned} Q_1 &= C_{11}\Phi_1 + C_{12}\Phi_2 \\ Q_2 &= C_{21}\Phi_1 + C_{22}\Phi_2 \end{aligned} \right\}. \tag{4}$$

Here the coefficients P_{ij} and C_{ij} are given by $(3)_1$ and $(3)_2$ respectively. See Question 3.26 for some applications.

(ii) The above calculation confirms a general result that was proved in Question 3.10: $P_{12} = P_{21}$ and $C_{12} = C_{21}$.

Question 3.26

Consider two spherical conductors c_1 and c_2 having radii 3 cm and 4 cm respectively, whose centres are 10 cm apart. The coefficients P_{ij} and C_{ij} for this system are given by (3) of Question 3.25. Use these results to answer the following:

(a) Under what conditions will the potentials of the conductors be equal in magnitude but opposite in sign?

(b) Calculate the capacitance C of this two-conductor system.

(c) Suppose c_1 and c_2 carry charges of $-5\,\mathrm{nC}$ and $8\,\mathrm{nC}$ respectively. Calculate their potentials. The spheres are now brought into contact and then returned to their original positions. Determine the final charge on each conductor and also calculate their common potential.

Solution

(a) Putting $\Phi_1 = -\Phi_2$ in $(4)_2$ of Question 3.25 and solving for Q_1/Q_2 give

$$\frac{Q_1}{Q_2} = \frac{C_{11} - C_{12}}{C_{21} - C_{22}}$$

$$= -0.816.$$

The conditions of the question are therefore satisfied when the charges carried by the conductors are in the above ratio.

(b) Suppose the conductors carry equal and opposite charges $Q_1 = -Q_2 = Q$. Then by definition $C = Q/V$, where the potential difference $V = \Phi_1 - \Phi_2$ follows from $(4)_2$ of Question 3.25. So

$$C = \frac{Q}{P_{11}Q - P_{12}Q - P_{21}Q + P_{22}Q}$$

$$= \frac{1}{P_{11} + P_{22} - 2P_{12}}$$

because $P_{21} = P_{12}$. Substituting the values of P_{ij} from (3) of Question 3.25 yields

$$C = 2.991\,\mathrm{pF}.$$

(Alternatively, expressing C in terms of the C_{ij} gives

$$C = \frac{C_{11}C_{22} - C_{12}^2}{C_{11} + C_{22} + 2C_{12}}$$

which returns the same answer.)

(c) Substituting $Q_1 = -5\,\mathrm{nC}$ and $Q_2 = 8\,\mathrm{nC}$ in $(4)_1$ of Question 3.25 and using $(3)_1$ yield

$$\Phi_1 = -741\,\mathrm{V} \quad \text{and} \quad \Phi_2 = 1326\,\mathrm{V}.$$

Bringing the conductors into contact equalizes their potentials, $\Phi_1' = \Phi_2'$. The net charge of the system is of course conserved, and so $Q_1' + Q_2' = -5 + 8 = 3\,\mathrm{nC}$. Also, $P_{11}Q_1' + P_{12}Q_2' = P_{21}Q_1' + P_{22}Q_2'$. Solving these two equations simultaneously and inserting the numerical values calculated in Question 3.25 give

$$Q_1' = 1.183\,\mathrm{nC} \quad \text{and} \quad Q_2' = 1.817\,\mathrm{nC}.$$

It then follows directly from $(4)_1$ that

$$\Phi_1' = \Phi_2' = 511\,\mathrm{V}.$$

Question 3.27*

Consider two circular conducting cylinders each of radius r_0 and length L lying parallel to the z-axis of Cartesian coordinates. Suppose the cylinders are separated by a distance $2d$ (measured from centre to centre) and that $r_0 < d \ll L$. Each conductor is maintained at a constant potential $\pm \frac{1}{2} V_0$ where the upper (lower) sign is for the right-hand (left-hand) cylinder. In answering the following questions, write *Mathematica* code to perform any cumbersome algebra (see the notebook on p. 172).

(a) Show that the electric potential at an arbitrary point P in the space surrounding the cylinders is given by

$$\Phi(x, y) = \frac{1}{2} V_0 \frac{\ln \left[\dfrac{(x + \sqrt{d^2 - r_0^2})^2 + y^2}{(x - \sqrt{d^2 - r_0^2})^2 + y^2} \right]}{\ln \left[\dfrac{(d - r_0 + \sqrt{d^2 - r_0^2})^2}{(d - r_0 - \sqrt{d^2 - r_0^2})^2} \right]}. \tag{1}$$

Hint: Start with (1) of Question 2.16.

(b) Translate the origin to the centre of the right-hand cylinder and introduce the polar coordinates (r, θ). Show that the radial electric field at the surface of this conductor is

$$E_r(\theta) = \frac{V_0 \sqrt{d^2 - r_0^2}}{r_0 (d + r_0 \cos \theta) \ln \left[2d^2/r_0^2 - 1 + \sqrt{d^2/r_0^2 - 1} \right]}. \tag{2}$$

(c) Hence prove that the capacitance of this two-cylinder configuration is given by

$$C = \frac{2\pi \epsilon_0 L}{\ln \left[2d^2/r_0^2 - 1 + \sqrt{d^2/r_0^2 - 1} \right]}. \tag{3}$$

Solution

(a) With the parallel line charges passing through the points $(\mp a, 0, 0)$, the electrostatic potential at P is given by

$$\Phi(x, y) = \frac{\lambda}{4\pi \epsilon_0} \ln \left[\frac{(x + a)^2 + y^2}{(x - a)^2 + y^2} \right]. \tag{4}$$

Now because the surfaces of the conducting cylinders are equipotentials, we require for the right-hand conductor that

$$r_0 = \frac{2\alpha a}{\alpha^2 - 1} \quad \text{and} \quad d = \frac{\alpha^2 + 1}{\alpha^2 - 1} a \tag{5}$$

(see (4) of Question 2.16). Using $(5)_2$ to eliminate α from $(5)_1$ gives $a = \sqrt{d^2 - r_0^2}$ which when substituted in (4) yields

$$\Phi = \frac{\lambda}{4\pi\epsilon_0} \ln\left[\frac{(x + \sqrt{d^2 - r_0^2})^2 + y^2}{(x - \sqrt{d^2 - r_0^2})^2 + y^2}\right]. \tag{6}$$

The line density is determined by choosing P to lie at some convenient point on the surface of either conductor, say $(x, y) = (d - r_0, 0)$. Now $\Phi = \frac{1}{2}V_0$ here, and solving for λ gives

$$\lambda = \frac{2\pi\epsilon_0 V_0}{\ln\left[\dfrac{(d - r_0 + \sqrt{d^2 - r_0^2})^2}{(d - r_0 - \sqrt{d^2 - r_0^2})^2}\right]}. \tag{7}$$

Equation (1) follows immediately from (6) and (7).

(b) Making the change of variable

$$\left.\begin{aligned} x &= d + r\cos\theta \\ y &= r\sin\theta \end{aligned}\right\}$$

in (1), and evaluating $E_r = -\partial\Phi/\partial r$ at the surface $r = r_0$ yield (2).

(c) The surface-charge density on the right-hand cylinder is given by $\sigma(\theta) = \epsilon_0 E_r$. Hence the net charge induced on this conductor is

$$q = \int_0^L \int_0^{2\pi} \sigma(\theta) r_0 \, d\theta \, dz = \frac{\epsilon_0 V_0 L \sqrt{d^2 - r_0^2}}{\ln\left[2d^2/r_0^2 - 1 + \sqrt{d^2/r_0^2 - 1}\right]} \int_0^{2\pi} \frac{d\theta}{d + r_0\cos\theta}, \tag{8}$$

where the definite integral has the value $2\pi/\sqrt{d^2 - r_0^2}$. Equation (3) then follows immediately from (8) using the definition of capacitance $C = q/V_0$.

```
In[1]:=  (* FIELD COMPONENTS *)

             V0   Log[((x + √(d² - r0²))² + y²) / ((x - √(d² - r0²))² + y²)]
Φ[r_, θ_] := ── ──────────────────────────────────────────────────────────── /.
             2        Log[(d - r0 + √(d² - r0²))² / (d - r0 - √(d² - r0²))²]

   {x → d + r Cos[θ], y → r Sin[θ]}

   Er[r_, θ_] = FullSimplify[-D[Φ[r, θ], r] /. {r → r0}]

   Eθ[r_, θ_] = FullSimplify[-1/r D[Φ[r, θ], θ] /. {r → r0}]

In[2]:=  (* CAPACITANCE *)
   Integrate[Er[r, θ], {θ, 0, 2π}, Assumptions → 0 < r0 < d]
```

Comments

(i) In this problem, each line charge serves as an image charge for the other conductor. Being located 'outside the region of interest' and offset from the cylinder axes, these line charges produce equipotentials that coincide exactly with the actual conducting surfaces. Uniqueness of the solution is therefore guaranteed.

(ii) The charge density induced on the surfaces of the cylinders is given by

$$\sigma(\theta) \; = \; \pm\epsilon_0 E_r(\theta) \; = \; \frac{\pm\epsilon_0 V_0 \sqrt{d^2 - r_0^2}}{r_0(d + r_0\cos\theta)\ln\left[2d^2/r_0^2 - 1 + \sqrt{d^2/r_0^2 - 1}\right]}.$$

Clearly $\sigma(\theta)$ is non-uniform and has a minimum (maximum) value at $\theta = 0$ ($\theta = \pi$), which is intuitively obvious.

(iii) In the notebook on p. 172 we also confirm that E_θ is zero at $r = r_0$ (no tangential component of electric field at a conducting surface).

Question 3.28*

An electric dipole (comprising charges q and $-q$ separated by a distance $2a$) is located a distance x from an infinite grounded conducting plane. The dipole (which is not free to move) makes an angle θ with the x-axis, as indicated in the figure below.

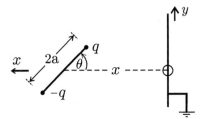

(a) Using image charges,[‡] show that the resultant force and torque about the centre of mass (exerted on the dipole) are

$$\left.\begin{aligned}
\mathbf{F} &= -\frac{1}{4\pi\epsilon_0}\frac{q^2}{4}\left[\frac{1}{(x + a\cos\theta)^2} + \frac{1}{(x - a\cos\theta)^2} - \frac{x}{(x^2 + a^2\sin^2\theta)^{3/2}}\right]\hat{\mathbf{x}} \\[2mm]
\boldsymbol{\Gamma} &= \frac{1}{4\pi\epsilon_0}\frac{q^2 a\sin\theta}{4}\left[\frac{1}{(x - a\cos\theta)^2} - \frac{1}{(x + a\cos\theta)^2}\right]\hat{\mathbf{z}}
\end{aligned}\;\right\} . \quad (1)$$

[‡] Assume the following elementary result: for a *single* point charge Q placed a distance d from a grounded conducting plane, the image problem comprises a pair of point charges $\pm Q$ separated by a distance $2d$.

(b) Show that for a *point* electric dipole having moment **p**, equation (1) reduces to

$$
\left.
\begin{aligned}
\mathbf{F} &= -\frac{1}{4\pi\epsilon_0}\frac{3p^2(2-\sin^2\theta)}{16x^4}\hat{\mathbf{x}} \\[2mm]
\mathbf{\Gamma} &= \frac{1}{4\pi\epsilon_0}\frac{p^2\sin 2\theta}{8x^3}\hat{\mathbf{z}}
\end{aligned}
\right\}.
\tag{2}
$$

Hint: Use *Mathematica's* `Series` function.

Solution

(a) In the image problem illustrated below, the quantities

$$
R = 2\sqrt{x^2 + a^2\sin^2\theta} \quad \text{and} \quad \cos\alpha = 2x/R
\tag{3}
$$

follow from simple trigonometry. The force between charge q and the conducting plane is the same as the force between q and its image. Then Coulomb's law and the definition of torque $(\mathbf{\Gamma} = \mathbf{r} \times \mathbf{F})$ give

$$
\left.
\begin{aligned}
\mathbf{F} &= -\frac{1}{4\pi\epsilon_0}\left[\frac{q^2}{[2(x+a\cos\theta)]^2} + \frac{q^2}{[2(x-a\cos\theta)]^2} - \frac{2q^2}{R^2}\cos\alpha\right]\hat{\mathbf{x}} \\[2mm]
\mathbf{\Gamma} &= \frac{1}{4\pi\epsilon_0}\left[\frac{q^2 a\sin\theta}{[2(x-a\cos\theta)]^2} - \frac{q^2 a\sin\theta}{[2(x+a\cos\theta)]^2}\right]\hat{\mathbf{z}}
\end{aligned}
\right\}.
\tag{4}
$$

Substituting (3) in (4) yields (1).

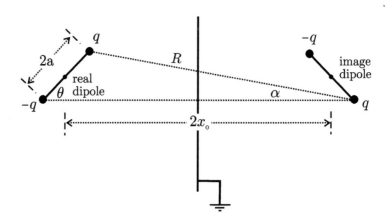

(b) In the limit $a \to 0$; $q \to \infty$ the physical dipole in (a) becomes a point dipole. Expanding (1) in powers of (a/x) gives (2) in this limit, since $p = 2aq$ is constant by construction (see Comment (ii) of Question 2.11).

Question 3.29

(a) Consider the physical dipole described in (a) of Question 3.28 where each of the two charges have the same mass m. Suppose the dipole is free to move and that $x = x_0$ at time $t = 0$. Use Newton's second law to show that the equations of motion of the dipole (in dimensionless form) are

$$
\left.
\begin{aligned}
&\frac{d^2 X}{d\tau^2} + \frac{1}{4\gamma^2}\left[\frac{1}{(X + \gamma\cos\theta)^2} + \frac{1}{(X - \gamma\cos\theta)^2} - \frac{2X}{(X^2 + \gamma^2\sin^2\theta)^{3/2}}\right] = 0 \\
&\frac{d^2\theta}{d\tau^2} + \frac{1}{4\gamma^2}\frac{X\sin 2\theta}{(X^2 - \gamma^2\cos^2\theta)^{3/2}} = 0
\end{aligned}
\right\}. \quad (1)
$$

Here $\gamma = a/x_0$, $X = x/x_0$ and $\tau = t/T$ where the characteristic time

$$
T = \sqrt{\frac{16\pi\epsilon_0 m x_0^5}{p^2}}.
$$

(b) Taking $\gamma = \frac{1}{50}$ and the initial conditions $(X(0) = 1, \dot{X}(0) = 0, \theta(0) = \frac{\pi}{6}, \dot{\theta}(0) = 0)$, use *Mathematica* to solve (1) numerically. Then plot graphs of $X(\tau)$ and $\theta(\tau)$ in the interval $0 \le \tau \le \tau_{\max}$. (Here $\tau_{\max}$, the time taken for the dipole to reach the conducting plane, should be determined by trial and error.)

Solution

(a) We begin by substituting Newton's second law $\mathbf{F} = m\ddot{\mathbf{x}}$ and $\boldsymbol{\Gamma} = I\dfrac{d\boldsymbol{\omega}}{dt}$‡ in

$$
\left.
\begin{aligned}
\mathbf{F} &= -\frac{1}{4\pi\epsilon_0}\frac{q^2}{4}\left[\frac{1}{(x + a\cos\theta)^2} + \frac{1}{(x - a\cos\theta)^2} - \frac{x}{(x^2 + a^2\sin^2\theta)^{3/2}}\right]\hat{\mathbf{x}} \\
\boldsymbol{\Gamma} &= \frac{1}{4\pi\epsilon_0}\frac{q^2 a\sin\theta}{4}\left[\frac{1}{(x - a\cos\theta)^2} - \frac{1}{(x + a\cos\theta)^2}\right]\hat{\mathbf{z}}
\end{aligned}
\right\}. \quad (2)
$$

Equation (1) then follows from (2) and the definitions of γ, X and τ.

(b) The notebook below yields the graphs, with the value of $\tau_{\max} \simeq 0.765$.

Comment

The graphs reveal that

☞ in the final stages of its motion, the dipole moves rapidly towards the conducting plane;

‡The moment of inertia about the dipole's centre of mass is $I = 2ma^2$, and $\boldsymbol{\omega} = -d\theta/dt\,\hat{\mathbf{z}}$.

☞ the net torque causes the dipole to oscillate with increasing frequency as $X \to 0$.

In[1]:= $\gamma = 0.01;$ τmax $= 0.765;$

$$\text{eqn1} = X''[\tau] + \frac{1}{4\,\gamma^2} \left(\frac{1}{(X[\tau] - \gamma \cos[\theta[\tau]])^2} + \right.$$

$$\left. \frac{1}{(X[\tau] + \gamma \cos[\theta[\tau]])^2} - \frac{2\,X[\tau]}{(X[\tau]^2 + \gamma^2 \sin[\theta[\tau]]^2)^{\frac{3}{2}}} \right) == 0;$$

$$\text{eqn2} = \theta''[\tau] + \frac{1}{2\,\gamma^2} \frac{X[\tau] \cos[\theta[\tau]] \sin[\theta[\tau]]}{(X[\tau]^2 - \gamma^2 \cos[\theta[\tau]]^2)^{\frac{3}{2}}} == 0;$$

$$\text{Sol} = \text{NDSolve}\left[\left\{\text{eqn1, eqn2, } X[0] == 1, \theta[0] == \frac{\pi}{6}, X'[0] == 0, \theta'[0] == 0\right\},\right.$$

$$\left. \{X[\tau], \theta[\tau], X'[\tau], \theta'[\tau]\}, \{\tau, 0, \tau\text{max}\}, \text{MaxSteps} \to 500\,000\right];$$

$$\text{Plot}\left[X[\tau] \text{ /. Sol, } \{\tau, 0, \tau\text{max}\}, \text{PlotPoints} \to 1000\right]$$

$$\text{Plot}\left[\theta[\tau] \text{ /. Sol, } \{\tau, 0, \tau\text{max}\}, \text{PlotPoints} \to 1000\right]$$

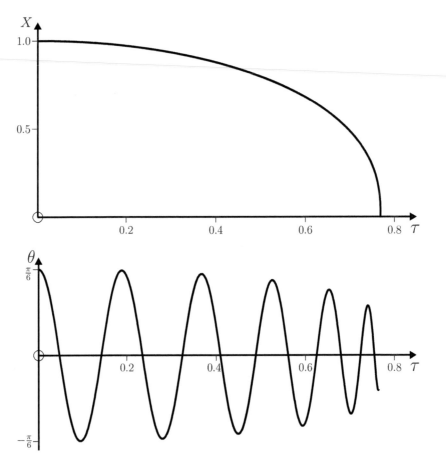

Question 3.30*

(a) Suppose

$$w(z) = u(x, y) + iv(x, y), \tag{1}$$

is an analytic function of a complex variable $z = x + iy$. Prove that $u(x, y)$ and $v(x, y)$ satisfy Laplace's equation in two dimensions.

Hint: Use the Cauchy–Riemann equations derived in Appendix I.

(b) Hence show that

$$\left.\begin{array}{c} (\nabla u) \cdot (\nabla v) = 0 \\ \text{and} \\ |\nabla u| = |\nabla v| \end{array}\right\}. \tag{2}$$

Solution

(a) Differentiating the Cauchy–Riemann equations

$$\frac{\partial u}{\partial x} = \frac{\partial v}{\partial y} \qquad \text{and} \qquad \frac{\partial v}{\partial x} = -\frac{\partial u}{\partial y} \tag{3}$$

gives,

$$\frac{\partial^2 u}{\partial x^2} = \frac{\partial^2 v}{\partial x \partial y} \qquad \text{and} \qquad \frac{\partial^2 u}{\partial y^2} = -\frac{\partial^2 v}{\partial y \partial x}. \tag{4}$$

For well-behaved functions $\dfrac{\partial^2 v}{\partial y \partial x} = \dfrac{\partial^2 v}{\partial x \partial y}$, and so

$$\frac{\partial^2 u}{\partial x^2} + \frac{\partial^2 u}{\partial y^2} = 0 \qquad \text{or} \qquad \boxed{\nabla^2 u = 0}, \tag{5}$$

where ∇^2 is the two-dimensional Laplacian. In a similar way, we obtain

$$\frac{\partial^2 v}{\partial x^2} + \frac{\partial^2 v}{\partial y^2} = 0 \qquad \text{or} \qquad \boxed{\nabla^2 v = 0}. \tag{6}$$

(b) We begin by expressing (3) in the form

$$\nabla u = \nabla v \times \hat{\mathbf{z}}, \tag{7}$$

where $\nabla = \hat{\mathbf{x}} \partial/\partial x + \hat{\mathbf{y}} \partial/\partial y$. Using (7) and the cyclic properties of the scalar triple product then gives

$$(\nabla u) \cdot (\nabla v) = (\nabla v \times \hat{\mathbf{z}}) \cdot (\nabla v) = (\nabla v \times \nabla v) \cdot \hat{\mathbf{z}} = 0, \tag{8}$$

which is $(2)_1$. Next, we note that

$$(\nabla u) \cdot (\nabla u) = \left(\frac{\partial u}{\partial x}\right)^2 + \left(\frac{\partial u}{\partial y}\right)^2$$

$$= \left(\frac{\partial v}{\partial y}\right)^2 + \left(-\frac{\partial v}{\partial x}\right)^2 \quad \text{(because of (3))}$$

$$= (\nabla v) \cdot (\nabla v). \tag{9}$$

Hence $(2)_2$.

Comments

(i) Both u and v, being solutions of Laplace's equation, are harmonic functions.

(ii) The condition described by (2) is that the gradients ∇u and ∇v are perpendicular and have the same magnitude. This leads to the following important geometric interpretation: the set of curves $u(x, y) = $ constant is mutually orthogonal to the set of curves $v(x, y) = $ constant. In electrostatics we may either regard u as the equipotential lines and v the field lines, or vice versa.

Question 3.31**

(a) Consider the analytic function $\Phi(x, y) = \Psi(u(x, y), v(x, y))$. Show that

$$\left.\begin{array}{l} \dfrac{\partial^2 \Phi}{\partial x^2} = \left(\dfrac{\partial u}{\partial x}\right)^2 \dfrac{\partial^2 \Psi}{\partial u^2} + \left(\dfrac{\partial v}{\partial x}\right)^2 \dfrac{\partial^2 \Psi}{\partial v^2} + 2\dfrac{\partial u}{\partial x}\dfrac{\partial v}{\partial x}\dfrac{\partial^2 \Psi}{\partial u \partial v} + \dfrac{\partial^2 u}{\partial x^2}\dfrac{\partial \Psi}{\partial u} + \dfrac{\partial^2 v}{\partial x^2}\dfrac{\partial \Psi}{\partial v} \\[4mm] \dfrac{\partial^2 \Phi}{\partial y^2} = \left(\dfrac{\partial u}{\partial y}\right)^2 \dfrac{\partial^2 \Psi}{\partial u^2} + \left(\dfrac{\partial v}{\partial y}\right)^2 \dfrac{\partial^2 \Psi}{\partial v^2} + 2\dfrac{\partial u}{\partial y}\dfrac{\partial v}{\partial y}\dfrac{\partial^2 \Psi}{\partial u \partial v} + \dfrac{\partial^2 u}{\partial y^2}\dfrac{\partial \Psi}{\partial u} + \dfrac{\partial^2 v}{\partial y^2}\dfrac{\partial \Psi}{\partial v} \end{array}\right\}. \tag{1}$$

(b) Suppose that, in the z-plane, $\Phi(x, y)$ satisfies Laplace's equation in two dimensions. Use (1) to show that, in the w-plane, $\Psi(u, v)$ is also a solution to this equation.

Solution

(a) Begin with

$$\frac{\partial \Phi}{\partial x} = \frac{\partial \Psi}{\partial u}\frac{\partial u}{\partial x} + \frac{\partial \Psi}{\partial v}\frac{\partial v}{\partial x} \quad \text{and} \quad \frac{\partial \Phi}{\partial y} = \frac{\partial \Psi}{\partial u}\frac{\partial u}{\partial y} + \frac{\partial \Psi}{\partial v}\frac{\partial v}{\partial y}. \tag{2}$$

Then

$$\frac{\partial^2 \Phi}{\partial x^2} = \frac{\partial}{\partial u}\left(\frac{\partial \Psi}{\partial u}\frac{\partial u}{\partial x} + \frac{\partial \Psi}{\partial v}\frac{\partial v}{\partial x}\right)\frac{\partial u}{\partial x} + \frac{\partial}{\partial v}\left(\frac{\partial \Psi}{\partial u}\frac{\partial u}{\partial x} + \frac{\partial \Psi}{\partial v}\frac{\partial v}{\partial x}\right)\frac{\partial v}{\partial x}, \tag{3}$$

which leads to $(1)_1$. Similarly for $(1)_2$.

(b) We begin by adding $(1)_1$ and $(1)_2$. In the resulting sum of ten terms, six yield zero because the real and imaginary parts u and v of the analytic function $w(z)$ satisfy the Laplace and Cauchy–Riemann equations. Thus

$$\frac{\partial^2 \Phi}{\partial x^2} + \frac{\partial^2 \Phi}{\partial y^2} = \left[\left(\frac{\partial u}{\partial x}\right)^2 + \left(\frac{\partial v}{\partial x}\right)^2\right]\left(\frac{\partial^2 \Psi}{\partial u^2} + \frac{\partial^2 \Psi}{\partial v^2}\right). \tag{4}$$

Now from (III) of Appendix I we have

$$\frac{dw}{dz} = \frac{\partial u}{\partial x} + i\frac{\partial v}{\partial x} \quad \text{and so} \quad \left|\frac{dw}{dz}\right|^2 = \left(\frac{\partial u}{\partial x}\right)^2 + \left(\frac{\partial v}{\partial x}\right)^2. \tag{5}$$

Substituting (5) in (4) gives

$$\frac{\partial^2 \Phi}{\partial x^2} + \frac{\partial^2 \Phi}{\partial y^2} = \left|\frac{dw}{dz}\right|^2\left(\frac{\partial^2 \Psi}{\partial u^2} + \frac{\partial^2 \Psi}{\partial v^2}\right). \tag{6}$$

Clearly, $\dfrac{\partial^2 \Phi}{\partial x^2} + \dfrac{\partial^2 \Phi}{\partial y^2} = 0 \Rightarrow \dfrac{\partial^2 \Psi}{\partial u^2} + \dfrac{\partial^2 \Psi}{\partial v^2} = 0$ as required.

Comments

(i) It follows that a curve in the z-plane, along which Φ is constant, maps into a new equipotential in the w-plane along which Ψ is constant. This suggests that if a potential problem can be solved in uv-space, then by transforming back to xy-space we will obtain a solution to our original problem. The trick is to find a complex-variable transformation which maps a difficult problem into a simpler problem which can be more easily solved. Although this is usually a matter of trial and error, it is often helpful to consult a compilation of mappings (available in some specialized books) to assist in this regard.

(ii) The idea outlined in Comment (i) forms the basis of a powerful technique known as conformal[‡] transformation. This method can sometimes be used to solve a range of potential problems which involve a two-dimensional geometry with infinite uniform extent in the third direction. Based on the theory of analytic functions, the technique is capable of dealing with boundaries having a more complicated shape than other analytical approaches.[7] A simple application of this method is presented in Question 3.32.

[‡] So named because the mapping preserves the angles between intersecting lines (except at points where $w'(z)$ is singular or zero).

[7]   K. J. Binns and P. J. Lawrenson, *Analysis and computation of electric and magnetic field problems.* New York: Pergamon Press, 1973.

Question 3.32

(a) Consider the mapping

$$z = \frac{d}{2\pi}(1 + w + e^w),\tag{1}$$

where d is a positive constant. Show that

$$x = \frac{d}{2\pi}(1 + u + e^u \cos v) \quad \text{and} \quad y = \frac{d}{2\pi}(v + e^u \sin v).\tag{2}$$

(b) Fig. (I) below shows the cross-section between two infinite parallel plates in the w-plane (an infinite capacitor). The top plate at $v = \pi$ is at a potential $\frac{1}{2}V_0$; the bottom plate at $v = -\pi$ is at a potential $-\frac{1}{2}V_0$. Briefly explain, using the conformal transformation (1), why the plates map into the semi-infinite capacitor shown in Fig. (II).

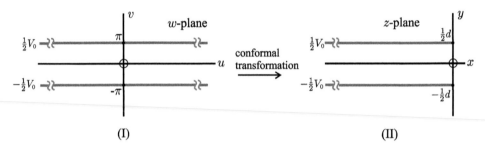

(I)   (II)

(c) It is well known from elementary physics that the electric potential within the infinite capacitor varies linearly between the plates. That is,

$$\Psi = \left(\frac{V_0}{2\pi}\right)v.\tag{3}$$

Using (2) and (3), write a *Mathematica* notebook to draw some electric-field lines and equipotentials near the end of the semi-infinite capacitor at $x = 0$. For this purpose, take $V_0 = 1000\,\text{V}$.

Hint: This is most easily done using *Mathematica*'s `ParametricPlot` function.

Solution

(a) Substituting $z = x + iy$ and $w = u + iv$ in (1) gives

$$x + iy = \frac{d}{2\pi}\left[1 + u + iv + e^u e^{iv}\right] = \frac{d}{2\pi}\left[(1 + u + e^u \cos v) + i(v + e^u \sin v)\right].$$

Hence (2).

(b) For $v = \pm\pi$ in the w-plane, it follows that $x = (d/2\pi)(1 + u - e^u)$; $y = \pm\frac{1}{2}d$ in the u-plane. Therefore, the points $u = 0$; $v = \pm\pi$ map to $x = 0$; $y = \pm\frac{1}{2}d$. Also, as $u \to \pm\infty$, $x \to -\infty$. The plates of the infinite capacitor map into the semi-infinite plates with their right-hand edges ending at $x = 0$; $y = \pm\frac{1}{2}d$. The transformation effectively folds the top and bottom plates back onto themselves, making the capacitor into a device half the original size (of course, this is not literally true, since other scaling factors are also involved).

(c) The transformation which changes the infinite plates to the semi-infinite plates may also be applied to map the field lines and equipotentials. In the notebook below, we assign to u a set of constant values and make a parametric plot of the corresponding lines of electric field near $x = 0$. Similarly, for the equipotentials. This results in the following field diagram:

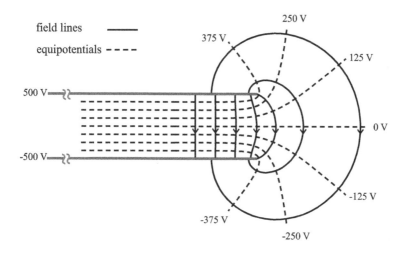

```
In[1]:=  X[u_, v_] := 1 + u + e^u Cos[v];        Y[u_, v_] := v + e^u Sin[v];
         u1 = {-7, -5, -3, -1.2, 0, 1, 2};        v1 = Range[ (-3 π)/4, (3 π)/4, π/4 ];
         xf = X[u, v] /. u → u1;                   yf = Y[u, v] /. u → u1;
         xp = X[u, v] /. v → v1;                   yp = Y[u, v] /. v → v1;
         FieldDat = Partition[Riffle[xf, yf], 2];
         PotDat = Partition[Riffle[xp, yp], 2];
         field = ParametricPlot[FieldDat, {v, -π, π}];
         potentials = ParametricPlot[PotDat, {u, -9, 2.1}];
         Show[field, potentials]
```

Comment

The field and potential deep inside the semi-infinite capacitor will (essentially) be equal to the corresponding values within the infinite device. However, near the edges of the semi-infinite plates a fringing field is present which bulges outwards from the space between them. It is in this region where significant differences between the fields of the two capacitors are to be found.

Question 3.33

Consider the conformal mapping

$$z = k \cosh w, \tag{1}$$

where k is a real constant.

(a) Use (1) to show that the lines $u = $ constant transform into the confocal ellipses

$$\frac{x^2}{a^2} + \frac{y^2}{b^2} = 1 \tag{2}$$

in the z-plane. Here $2a$ and $2b$ are the lengths of the major and minor axes respectively, where

$$\left. \begin{array}{l} a = k \cosh u \\ b = k \sinh u \end{array} \right\}. \tag{3}$$

Hint: The distance from the origin O to either focus F is $\sqrt{a^2 - b^2}$ (see, for example, Ref. [8]).

(b) Use (1) to show that the lines $v = $ constant transform into the confocal hyperbolas

$$\frac{x^2}{a'^2} - \frac{y^2}{b'^2} = 1 \tag{4}$$

in the z-plane. Here a' and b', where

$$\left. \begin{array}{l} a' = k \cos v \\ b' = k \sin v \end{array} \right\}, \tag{5}$$

determine the asymptotes of the hyperbolas according to the equation

$$y = \pm \frac{b'}{a'} x . \tag{6}$$

Hint: The distance from the origin O to either focus F is $\sqrt{a'^2 + b'^2}$ (see, for example, Ref. [8]).

Solution

(a) Expanding $z = k \cosh(u + iv)$ into its real and imaginary parts gives

$$(x + iy) = \tfrac{1}{2} k (e^u e^{iv} + e^{-u} e^{-iv}) = \tfrac{1}{2} k \left[(e^u + e^{-u}) \cos v + i (e^u - e^{-u}) \sin v \right]$$

$$= k \left[\cosh u \cos v + i \sinh u \sin v \right],$$

[8]   O. L. de Lange and J. Pierrus, *Solved problems in classical mechanics: Analytical and numerical solutions with comments*, Chap. 8, pp. 231–2. Oxford: Oxford University Press, 2010.

and so

$$\left.\begin{array}{l} x = k\cosh u \cos v \\ y = k\sinh u \sin v \end{array}\right\} \tag{7}$$

or

$$\left.\begin{array}{l} x/a = \cos v \\ y/b = \sin v \end{array}\right\}. \tag{8}$$

Hence (2). Now, since $\cosh^2 u - \sinh^2 u = 1$ and because of the hint, we calculate the focal distance

$$\sqrt{a^2 - b^2} = \sqrt{(k\cosh u)^2 - (k\sinh u)^2} = k.$$

Thus the curves $u = $ constant are a family of *confocal* ellipses: their foci are all the same distance k from O. See Fig. (I) below.

(b) Substituting (5) in (7) gives

$$\left.\begin{array}{l} x/a' = \cosh u \\ y/b' = \sinh u \end{array}\right\}. \tag{9}$$

Hence (4). As before, the focal distance follows immediately and is

$$\sqrt{a'^2 + b'^2} = \sqrt{(k\cos v)^2 + (k\sin v)^2} = k,$$

showing that the curves $v = $ constant are a family of confocal hyperbolas sharing the same foci as the family of confocal ellipses. See Fig. (II) below.

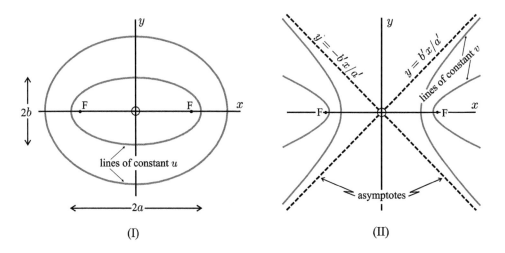

(I)   (II)

Comment

The above results are required for solving the electrostatic problems outlined in Question 3.34.

Question 3.34

Use *Mathematica* and the results of Question 3.33 to draw the electric field and some lines of constant potential for the configurations illustrated below. In each case, use the dimensions and potentials shown in the relevant figure.

(a) Two infinitely long confocal cylindrical conducting shells having elliptical cross-sections. See Fig. (I) below.

(b) Two coplanar semi-infinite conducting plates separated by an 8 cm gap. See Fig. (II) below.

(c) An infinite conducting plate perpendicular to a semi-infinite conducting plate with a 5 cm gap between them. See Fig. (III) below.

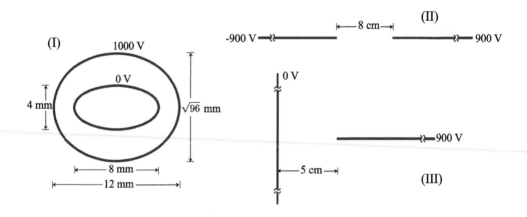

Solution

The notebook required to plot the field lines and equipotentials is similar to that for Question 3.32 with appropriate modifications. Some points to keep in mind are:

(a) The equipotentials are a family of confocal ellipses, and the electric-field lines are a family of confocal hyperbolas. Suppose the surface of the inner (outer) conductor is given by $u_i = $ constant ($u_o = $ constant). Then for a line of constant potential $\Phi_i \leq \Phi \leq \Phi_o$ corresponding to $u = $ constant it follows that

$$\Phi = \alpha u + \beta, \qquad (1)$$

where α and β are constants which we determine from the boundary conditions as follows:

$$\Phi_i = \alpha u_i + \beta \qquad \text{and} \qquad \Phi_o = \alpha u_o + \beta.$$

Solving for α and β and eliminating these from (1) give

$$\Phi(u) = \frac{(\Phi_i - \Phi_o)u + \Phi_o u_i - \Phi_i u_o}{(u_i - u_o)}. \tag{2}$$

Substituting $\Phi_i = 0\,\mathrm{V}$, $\Phi_o = 1000\,\mathrm{V}$, $u_i = \tanh^{-1}(b/a) = \tanh^{-1}(2/4)$ and $u_o = \tanh^{-1}(12/\sqrt{96})$ in (2) gives an equation for finding the value of u corresponding to a given potential. It is this value of u which is required in calculating the ellipse of constant potential. We obtain the figure:

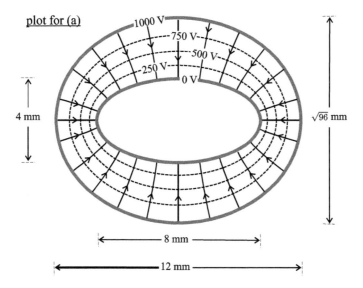

(b) In this problem, the equipotentials are a family of confocal hyperbolas, and the electric-field lines are a family of confocal ellipses. With $k = 4\,\mathrm{cm}$ and $v = 0$ ($v = \pi$) in (7) of Question 3.33, we generate the right-hand (left-hand) plate. From the equation

$$\Phi(v) = \left(\frac{\pi/2 - v}{\pi}\right)(\Phi_{\text{right plate}} - \Phi_{\text{left plate}}),$$

we establish the value of v for a particular equipotential Φ. The field lines shown in the first figure on p. 186 were obtained for $0 \le v \le 2\pi$ and $u = 0.3$, $u = 0.8$, $u = 1.2$ and $u = 1.5$.

(c) As in (b), the equipotentials are a family of confocal hyperbolas, and the electric-field lines are a family of confocal ellipses. With $k = 3\,\mathrm{cm}$ and $v = 0$ ($v = \frac{1}{2}\pi$) in (7) of Question 3.33, we generate the horizontal (vertical) plate. From the equation

$$\Phi(v) = \left(\frac{\pi/2 - v}{\pi/2}\right)\Phi_{\text{horizontal}},$$

we establish the value of v for a particular equipotential Φ. The field lines shown in the second figure on p. 186 were obtained for $0 \le v \le 2\pi$ and $u = 0.25$, $u = 0.5$, $u = 0.75$ and $u = 1.0$.

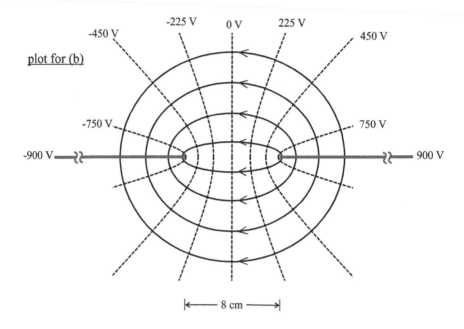

plot for (b)

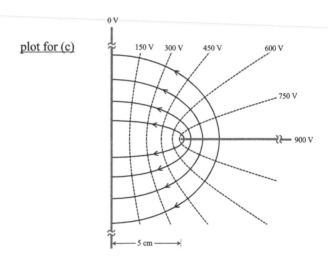

plot for (c)

Comments

(i) As the gap → 0 in (b) and (c) above (and by introducing suitable insulation between the plates), the field lines become circles and the hyperbolas become straight lines.

(ii) In addition to the properties described in Question 3.3, another useful feature of harmonic functions is their scalability. Suppose $\dfrac{\partial^2 \Phi}{\partial x^2} + \dfrac{\partial^2 \Phi}{\partial y^2} = 0$, then

$$\frac{\partial^2 (\alpha\Phi)}{\partial(\beta y)^2} + \frac{\partial^2 (\alpha\Phi)}{\partial(\beta y)^2} = 0,$$

where α and β are arbitrary constants not equal to zero. Evidently, both electric potential and distance can be scaled linearly without affecting a solution to Laplace's equation. This property ensures that we can always find appropriate solutions in problems such as (a)–(c) when the potentials of the conductors and their physical dimensions are changed.

(iii) It is sometimes necessary to calculate the electric-field components, E_x and E_y. This is demonstrated below for the configuration described in (a). We begin by eliminating v from (7) of Question 3.33 to obtain

$$x^2 = k^2 \cosh^2 u - y^2 \coth^2 u. \tag{3}$$

Then using (3) to calculate the partial derivatives $\partial u / \partial x$ and $\partial u / \partial y$ yields, after a little algebra,

$$\frac{\partial u}{\partial x} = \frac{x \tanh u \sinh^2 u}{k^2 \sinh^4 u + y^2} \quad \text{and} \quad \frac{\partial u}{\partial y} = \frac{y \coth u \cosh^2 u}{k^2 \cosh^4 u - x^2}. \tag{4}$$

Now $E_x = -\dfrac{\partial \Phi}{\partial x} = -\dfrac{\partial \Phi}{\partial u}\dfrac{\partial u}{\partial x}$ and $E_y = -\dfrac{\partial \Phi}{\partial y} = -\dfrac{\partial \Phi}{\partial u}\dfrac{\partial u}{\partial y}$, where $\dfrac{\partial \Phi}{\partial u} = \dfrac{(\Phi_o - \Phi_i)}{(u_o - u_i)}$

follows from (3) of Question 3.33. Substituting (4) in these results gives

$$\left.\begin{aligned}
E_x &= -\left(\frac{\Phi_o - \Phi_i}{u_o - u_i}\right)\frac{x \tanh u \sinh^2 u}{k^2 \sinh^4 u + y^2} \\[2mm]
E_y &= -\left(\frac{\Phi_o - \Phi_i}{u_o - u_i}\right)\frac{y \coth u \cosh^2 u}{k^2 \cosh^4 u - x^2}
\end{aligned}\right\},$$

where, for given values of x and y, the parameter u can be found from (3) using *Mathematica*'s NSolve function.

4

Static magnetic fields in vacuum

In 1799, Count Alessandro Volta from Italy invented the voltaic pile (a battery) and produced for the first time a source of *continuous* electric current in a laboratory. This was a crucial milestone in science, because the connection between electricity and magnetism (which had long been suspected by some) was now waiting to be found. Surprisingly, it took as long as twenty years for this to happen. Then in 1820, Oersted made an interesting observation whilst delivering a public lecture. He noticed that a compass needle in one of his lecture demonstrations deflected briefly when the current in a nearby electric circuit was turned on or off. This crucial observation, although not explained by Oersted at the time, confirmed the suspicions of the enlightened few: a link between electrical and magnetic phenomena did indeed exist. Of course, prior to this, certain factual information about electrostatics and magnetism (in particular) was known; namely, that iron objects were attracted to stones containing magnetite, and that compasses were used to detect the presence of what we now call a magnetic field. Within a short period following Oersted's observation, Ampère announced the results of his experiments which suggested that two parallel currents attract each other, and anti-parallel currents repel. A few months later, Biot and Savart reported to a meeting of the French Académie des sciences that they too had conducted experiments similar to those of Oersted. Furthermore, they laid claim to having deduced how the force depends on key experimental variables such as angle and distance. The culmination of the work of these four men (and others)[‡] is central to the questions of this chapter.

An electric charge q, moving in a magnetic field $\mathbf{B}$, is found to experience a velocity-dependent force,

$$\mathbf{F} = q\mathbf{v} \times \mathbf{B}, \tag{I}$$

where $\mathbf{v}$ is the velocity of q. This force is quite unlike an electric force $q\mathbf{E}$, which does not depend on $\mathbf{v}$ and is present even for charges at rest. As it turns out, the study of magnetostatics is algebraically more complex than the corresponding study of electrostatics. This is partly because the source of $\mathbf{B}$ (the current density $\mathbf{J}(\mathbf{r})$) is a vector and partly because cross-products arise in the ensuing equations. Now although (I) provides the definition for magnetic field, it cannot be solved explicitly for $\mathbf{B}$: a matter we deal with in Question 4.1. The Biot–Savart law, as it is now known, is the magnetostatic counterpart of Coulomb's law in electrostatics. We apply it to calculate $\mathbf{B}$ for various current distributions. Furthermore, in attempting to develop this chapter

---

[‡]This period in science was marked by major controversy. Put simplistically, the issue was who discovered what and when.

Solved Problems in Classical Electromagnetism. J. Pierrus, Oxford University Press (2018).
© J. Pierrus. DOI: 10.1093/oso/9780198821915.001.0001

along similar lines to Chapter 2, we will use the Biot–Savart law to deduce Maxwell's magnetostatic equations,

$$\nabla \cdot \mathbf{B} = 0 \quad \text{and} \quad \nabla \times \mathbf{B} = \mu_0 \, \mathbf{J}(\mathbf{r}), \quad \text{(II)}$$

which apply at a point in vacuum. In deriving (II), the magnetic vector potential $\mathbf{A}(\mathbf{r})$ emerges quite naturally and features prominently in questions throughout this book. The magnetic scalar potential Φ_m also makes an appearance, and is used as an effective device for solving certain types of problem (see, for example, Questions 4.18, 4.19 and 9.21). The chapter concludes with a multipole expansion of the vector potential for a bounded distribution of stationary currents,[‡] the associated magnetic multipole moments and various applications involving magnetic dipoles and quadrupoles.

Question 4.1

A charge q moving with velocity $\mathbf{v}$ in a magnetic field $\mathbf{B}$ experiences a force per unit charge given by (I) above:

$$\mathbf{f} = \frac{\mathbf{F}}{q} = \mathbf{v} \times \mathbf{B}. \quad (1)$$

(a) Suppose that $\mathbf{f}$ and $\mathbf{v}$ are known at a point P in space. Prove that this information is insufficient to determine $\mathbf{B}$ at P. Do this in two ways:

  ☞ Consider the (coordinate-free) definition of a cross-product.

  ☞ Write out the Cartesian components of (1) and then show that the resulting set of equations has no solution.

(b) Use (1) to derive the equation

$$\mathbf{B} = \frac{\mathbf{f} \times \hat{\mathbf{v}}}{v} + (\mathbf{B} \cdot \hat{\mathbf{v}})\hat{\mathbf{v}}. \quad (2)$$

(c) Hence show that *two* separate measurements of $\mathbf{f}$ and $\mathbf{v}$ (with $\mathbf{v}_1$ and $\mathbf{v}_2$ in different directions) are required to determine the magnetic field at P.

  Hint: Choose the vectors $\mathbf{v}_1$ and $\mathbf{v}_2$ to be mutually perpendicular.

Solution

(a) ☞ The force per unit charge $f = vB \sin \theta$, where θ is the angle between $\mathbf{v}$ and $\mathbf{B}$, follows from (1). Clearly a single equation cannot be used to solve for the two unknowns, B and θ.

  ☞ The Cartesian components of (1) are:

[‡] A current $i(t)$ is quasi-stationary if at time t it produces effects, in some region of space, that are indistinguishable from those of a direct current having the same instantaneous magnitude. A formal definition is given in the second paragraph on p. 249.

$$f_x = v_y B_z - v_z B_y, \qquad f_y = v_z B_x - v_x B_z, \qquad f_z = v_x B_y - v_y B_x, \qquad (3)$$

which may be written in matrix form as

$$\begin{pmatrix} f_x \\ f_y \\ f_z \end{pmatrix} = \begin{pmatrix} 0 & -v_z & v_y \\ v_z & 0 & -v_x \\ -v_y & v_x & 0 \end{pmatrix} \begin{pmatrix} B_x \\ B_y \\ B_z \end{pmatrix}. \qquad (4)$$

It is easy to show that the determinant of the matrix $\begin{pmatrix} 0 & -v_z & v_y \\ v_z & 0 & -v_x \\ -v_y & v_x & 0 \end{pmatrix}$ is zero,

and so (3) cannot be inverted to obtain the (column) vector $\mathbf{B}$.

(b) Taking the cross-product of both sides of (1) with $\hat{\mathbf{v}}$ and using the **BAC–CAB** rule $\big($see (3) of Question 1.8$\big)$ yield (2).

(c) We obtain from (2):

$$\mathbf{B} = \frac{\mathbf{f}_1 \times \hat{\mathbf{v}}_1}{v_1} + (\mathbf{B} \cdot \hat{\mathbf{v}}_1)\hat{\mathbf{v}}_1 \qquad \text{and} \qquad \mathbf{B} = \frac{\mathbf{f}_2 \times \hat{\mathbf{v}}_2}{v_2} + (\mathbf{B} \cdot \hat{\mathbf{v}}_2)\hat{\mathbf{v}}_2. \qquad (5)$$

With $\mathbf{v}_1$ and $\mathbf{v}_2$ orthogonal, $\mathbf{v}_1 \cdot \mathbf{v}_2 = 0$ and $(5)_2$ gives $\mathbf{B} \cdot \hat{\mathbf{v}}_1 = \dfrac{(\mathbf{f}_2 \times \hat{\mathbf{v}}_2) \cdot \hat{\mathbf{v}}_1}{v_2}$.

Substituting this equation in $(5)_1$ and using the properties of the scalar triple product yield

$$\mathbf{B} = \frac{\mathbf{f}_1 \times \hat{\mathbf{v}}_1}{v_1} - \left[\frac{\mathbf{f}_2 \cdot (\hat{\mathbf{v}}_1 \times \hat{\mathbf{v}}_2)}{v_2} \right] \hat{\mathbf{v}}_1, \qquad (6)$$

which is the result we seek.

Comments

(i) It follows from $q(\mathbf{v} \times \mathbf{B})$ that a magnetic force $\mathbf{F}$ acting on a moving charge is always perpendicular to its velocity. Consequently, a magnetic force can change the direction of $\mathbf{v}$, but not its magnitude. Since the kinetic energy of q is constant, the work done by $\mathbf{F}$ is zero.[‡] This may be stated as a general result:

$$\boxed{\text{magnetic forces do no work}}. \qquad (7)$$

The statement (7) is also true for time-dependent magnetic fields as well.

[‡] As required by the work-energy theorem: $W = \Delta E_{\text{k}}$.

(ii) In the presence of both an electric and a magnetic field, the force on a charge q is

$$\mathbf{F} = q[\mathbf{E} + \mathbf{v} \times \mathbf{B}]. \tag{8}$$

Now three independent measurements of $\mathbf{f}$ are needed to determine $\mathbf{E}$ and $\mathbf{B}$ at the point P. In addition to the two measurements required to calculate $\mathbf{B}$, the electric field at the point is obtained with q at rest ($\mathbf{v} = 0$). Hence the definition of $\mathbf{E}$ at a point is the force per unit *stationary* test charge (see the first footnote on p. 48).

(iii) The fields in (8) may depend on both position and time, and the electromagnetic force $\mathbf{F}(\mathbf{r}, t)$ on q is known as the Lorentz force.

Question 4.2

A particle having mass m and charge q moves with velocity $\mathbf{v}(t) = d\mathbf{r}/dt$ in static electric and magnetic fields. Prove that

$$\tfrac{1}{2}mv^2 + q\Phi(\mathbf{r}) = \tfrac{1}{2}mv_0^2 + q\Phi(\mathbf{r}_0), \tag{1}$$

where Φ is the electric potential and $\mathbf{r}_0$ and $\mathbf{v}_0$ are the initial position and velocity of the particle.

Solution

Power is the rate of doing work, and so

$$\frac{dW}{dt} = \mathbf{F} \cdot \mathbf{v} = m\frac{d\mathbf{v}}{dt} \cdot \mathbf{v} = \tfrac{1}{2}m\frac{d}{dt}(\mathbf{v} \cdot \mathbf{v}) = \tfrac{1}{2}m\frac{dv^2}{dt}. \tag{2}$$

With $\mathbf{F}$ given by (8) of Question 4.1 we have

$$\mathbf{F} \cdot \mathbf{v} = q[\mathbf{E} + \mathbf{v} \times \mathbf{B}] \cdot \mathbf{v} = q\mathbf{E} \cdot \mathbf{v}, \tag{3}$$

since $(\mathbf{v} \times \mathbf{B}) \cdot \mathbf{v} = 0$. Equating (2) and (3) gives $\tfrac{1}{2}m\,d\mathbf{v}^2 = q\mathbf{E} \cdot d\mathbf{r}$. Now $\mathbf{E} = -\nabla\Phi$, and so

$$\tfrac{1}{2}m \int_{\mathbf{v}_0}^{\mathbf{v}} d\mathbf{v}^2 = -q \int_{\mathbf{r}_0}^{\mathbf{r}} \nabla\Phi \cdot d\mathbf{r} = -q \int_{\mathbf{r}_0}^{\mathbf{r}} d\Phi. \tag{4}$$

Integrating (4) yields

$$\tfrac{1}{2}m(v^2 - v_0^2) = -q[\Phi(\mathbf{r}) - \Phi(\mathbf{r}_0)],$$

which is (1).

Comments

(i) Equation (1) is a statement of energy conservation: for a particle moving in static $\mathbf{E}$- and $\mathbf{B}$-fields, the quantity $\tfrac{1}{2}mv^2 + q\Phi$ is a constant of the motion.

(ii) Consider the particle (mass m; charge q) moving in the uniform magnetostatic and electrostatic fields $\mathbf{B} = (0, 0, B)$ and $\mathbf{E} = (0, E, 0)$ described in Question 7.18 of Ref. [1]. Suppose $\mathbf{v}_0 = 0$ and $\Phi(\mathbf{r}_0) = 0$. The reader can easily verify that $\frac{1}{2}mv^2 + q\Phi(\mathbf{r})$, with $\Phi(\mathbf{r}) = \Phi(\mathbf{r}_0) - \mathbf{E} \cdot \mathbf{r}$ (see (1) of Question 3.5), remains zero throughout the motion of the particle.

Question 4.3

Consider a conducting medium in which m different species of charge carrier (electrons, protons, ions, ...) coexist and which move through space in some way. Show that the current I through a finite surface s is

$$I = \int_s \mathbf{J} \cdot d\mathbf{a}, \tag{1}$$

where the current density

$$\mathbf{J} = \sum_{i=1}^{m} \rho_i \overline{\mathbf{v}}_i. \tag{2}$$

Here ρ_i and $\overline{\mathbf{v}}_i$ are the charge density and average velocity of the ith species respectively.

Solution

Consider charge carriers of a particular species (j, say) having N members in a macroscopic‡ volume element dv and number density $n = N/dv$. Suppose $\overline{\mathbf{v}}$ is the average velocity of this species within dv. Then $\overline{\mathbf{v}} = \frac{1}{N} \sum_{k=1}^{N} \mathbf{v}_k$ where $\mathbf{v}_k$ is the velocity of the kth member. In a time dt the number of carriers of this species which (on average) cross an infinitesimal area $d\mathbf{a}$ is $n(\overline{\mathbf{v}} \, dt) \cdot d\mathbf{a}$. Hence the rate at which charge flows is

$$\frac{dq}{dt} = nq\overline{\mathbf{v}} \cdot d\mathbf{a},$$

where q is the charge of the jth species. Now the *net* rate at which charge crosses the infinitesimal area $d\mathbf{a}$ is the sum over all species. If we represent this current by dI then

$$dI = \sum_{j=1}^{m} n_j q_j \overline{\mathbf{v}}_j \cdot d\mathbf{a} = \sum_{j=1}^{m} \rho_j \overline{\mathbf{v}}_j \cdot d\mathbf{a}, \tag{3}$$

where $\rho_j = n_j q_j$. Integrating (3) over the surface s yields (1) with $\mathbf{J}$ given by (2).

‡This term is defined in the footnote on p. 409.

[1] O. L. de Lange and J. Pierrus, *Solved problems in classical mechanics: Analytical and numerical solutions with comments*, Chap. 7, p. 190–1. Oxford: Oxford University Press, 2010.

Comments

(i) An important case of (2) occurs for a single species ($m = 1$). Then $\mathbf{J} = \rho \bar{\mathbf{v}}$ with $\rho = nq$, and so

$$\mathbf{J} = nq\bar{\mathbf{v}}. \tag{4}$$

(ii) In some circumstances the flow of charge is restricted to a *surface*. Then it is usually convenient to introduce a surface-current density $\mathbf{K}$. At an arbitrary point P in the surface, $\mathbf{K}$ is defined to be the current per unit width perpendicular to the flow. Consider a ribbon-like strip lying in the surface having infinitesimal width $\delta\ell_\perp$ as shown in the figure.

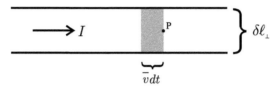

In a time dt all the charge contained within the shaded region will pass P. This charge is $dq = \sigma(\bar{v}\,dt)\delta\ell_\perp$ where σ is the surface density of mobile-charge carriers whose mean velocity is $\bar{\mathbf{v}}$. Hence $K = \dfrac{1}{\delta\ell_\perp}\dfrac{dq}{dt} = \sigma\bar{v}$ or

$$\mathbf{K} = \sigma\bar{\mathbf{v}}. \tag{5}$$

Equation (5) is the surface analogue of (2), and it can be generalized to include surface-charge carriers of more than one species.

(iii) For an element of charge dq moving with velocity $\mathbf{v}$ we may express the quantity $\mathbf{v}\,dq$ in the following equivalent forms:

$$\mathbf{v}\,dq = \mathbf{J}\,dv = \mathbf{K}\,da = I\,dl. \tag{6}$$

The relationships between these equivalent forms, illustrated in the figure below, turn out to be very useful.

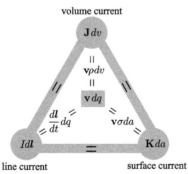

For example, a volume integral involving $\mathbf{J}\,dv$ can be transformed into a surface integral involving $\mathbf{K}\,da$, or a line integral involving $I\,dl$. That is,

$$\int_v (\cdots)\mathbf{J}\,dv' \;\leftrightarrow\; \oint_s (\cdots)\mathbf{K}\,da' \;\leftrightarrow\; \oint_c (\cdots)I\,dl' \tag{7}$$

(see Comment (v) of Question 4.4 for an application).

(iv) Kirchhoff's junction rule is a familiar result from elementary circuit theory. It states that for stationary or quasi-stationary currents, charge cannot accumulate anywhere in a circuit (this being a necessary consequence of charge conservation). Students often remember this rule as: 'current into a node = current out of the node', and because of (1) it can be expressed mathematically as $\oint_s \mathbf{J}\cdot d\mathbf{a} = 0$. The differential form of this integral, which follows from Gauss's theorem, is:

$$\nabla\cdot\mathbf{J} = 0. \tag{8}$$

Equation (8) is an important result which we will use later in this chapter. It is the magnetostatic form of the continuity equation considered in Question 7.1.

Question 4.4*

(a) The figure below shows two rigid circuits c and c' of arbitrary shape carrying currents I and I' respectively. Suppose that the relative orientation and separation of these circuits remains constant. Experimental measurements on circuits of many different sizes and shapes suggests that the force exerted by c' on c is of the form

$$\mathbf{F} = \frac{\mu_0}{4\pi}II'\oint_c \oint_{c'} \frac{d\mathbf{l}\times[d\mathbf{l}'\times(\mathbf{r}-\mathbf{r}')]}{|\mathbf{r}-\mathbf{r}'|^3}. \tag{1}$$

Here $d\mathbf{l}$ and $d\mathbf{l}'$ are infinitesimal line elements of c and c', and their displacements (relative to an arbitrary origin O) are $\mathbf{r}$ and $\mathbf{r}'$ respectively.

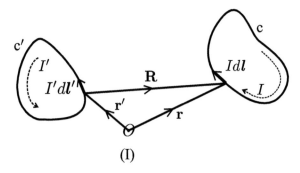

(I)

Express the force given by (1) in the symmetrical form

$$\mathbf{F} = -\frac{\mu_0}{4\pi}II'\oint_c \oint_{c'} \frac{(\mathbf{r}-\mathbf{r}')\,d\mathbf{l}\cdot d\mathbf{l}'}{|\mathbf{r}-\mathbf{r}'|^3}. \tag{2}$$

(b) Now suppose that the relative separation of the two circuits c and c' can be varied. It is convenient to choose separate origins O and O' for each circuit. Let $\mathbf{r}$ and $\mathbf{r}'$ be the position vectors of current elements Idl and $I'dl'$ relative to O and O' respectively. The vector $\mathcal{R}$ shown in the figure below denotes the displacement of origin O relative to O'.

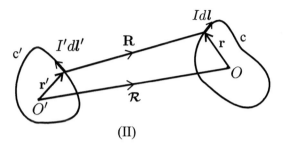

(II)

Hence show that

$$\mathbf{F} = II'\,\nabla_{\mathcal{R}}\left(\frac{\mu_0}{4\pi}\oint_c\oint_{c'}\frac{dl\cdot dl'}{|\mathbf{r}-\mathbf{r}'+\mathcal{R}|}\right),\tag{3}$$

where $\nabla_{\mathcal{R}}$ comprises derivatives of the form $\partial/\partial\mathcal{R}$.

Solution

(a) From Fig. (I) on p. 194 we see that $\mathbf{R}=\mathbf{r}-\mathbf{r}'$. Then applying the **BAC–CAB** rule to the vector triple product $dl\times(dl'\times\mathbf{R})$ gives $dl'(dl\cdot\mathbf{R})-\mathbf{R}(dl\cdot dl')$. Substituting this result in (1) yields

$$\mathbf{F} = \frac{\mu_0}{4\pi}II'\oint_c\oint_{c'}\left[\frac{(dl\cdot\mathbf{R})\,dl'}{R^3}-\frac{\mathbf{R}(dl\cdot dl')}{R^3}\right].$$

Now $\mathbf{R}/R^3=-\nabla R^{-1}$ (see $(1)_1$ of Question 1.6), and so

$$\mathbf{F} = -\frac{\mu_0}{4\pi}II'\oint_c\oint_{c'}\left[(dl\cdot\nabla R^{-1})\,dl'+\frac{\mathbf{R}(dl\cdot dl')}{R^3}\right]$$

$$= -\frac{\mu_0}{4\pi}II'\left\{\oint_{c'}\left[\oint_c d\left(\frac{1}{R}\right)\right]dl'+\oint_c\oint_{c'}\frac{\mathbf{R}(dl\cdot dl')}{R^3}\right\},\tag{4}$$

where, in the last step, we interchange the order of integration and express the term $dl\cdot\nabla R^{-1}$ as the perfect differential $d(R^{-1})$. The integral in square brackets in (4) is therefore zero. Hence (2).

(b) Refer to Fig. (II) on p. 195. Then, making the replacement $\mathbf{r} \to \mathcal{R} + \mathbf{r}$, (2) becomes

$$\mathbf{F} = -\frac{\mu_0}{4\pi} I I' \oint_c \oint_{c'} \frac{(\mathbf{r} - \mathbf{r}' + \mathcal{R}) \, dl \cdot dl'}{|\mathbf{r} - \mathbf{r}' + \mathcal{R}|^3}. \tag{5}$$

Now the vectors shown in the figure satisfy $\mathbf{R} = \mathbf{r} - \mathbf{r}' + \mathcal{R}$, and so $-\dfrac{(\mathbf{r} - \mathbf{r}' + \mathcal{R})}{|\mathbf{r} - \mathbf{r}' + \mathcal{R}|^3} =$

$-\dfrac{\mathbf{R}}{R^3} = \nabla_{\mathcal{R}} R^{-1}$. Substituting this result in (5) gives

$$\mathbf{F} = \frac{\mu_0}{4\pi} I I' \oint_c \oint_{c'} \nabla_{\mathcal{R}} R^{-1} (dl \cdot dl')$$

$$= \frac{\mu_0}{4\pi} I I' \oint_c \oint_{c'} \left\{ \nabla_{\mathcal{R}} \left(\frac{dl \cdot dl'}{R} \right) - \frac{\nabla_{\mathcal{R}} (dl \cdot dl')}{R} \right\}, \tag{6}$$

where the last step follows from the identity $f\nabla g = \nabla(fg) - g\nabla f$ (see (4) of Question 1.8). Now $\nabla_{\mathcal{R}} (dl \cdot dl') = 0$ because dl and dl' are independent of $\mathcal{R}$. Hence the result, since the order of differentiation and integration in (6) may be interchanged.

Comments

(i) In the introduction to this chapter, we mentioned that Ampère, Biot, Savart and others conducted a series of ingenious experiments in the first twenty-five years of the nineteenth century with current-carrying circuits of various shapes. After extensive investigations, and not without controversy, Ampère published (c. 1827) a '*Memoir on the mathematical theory of electrodynamic phenomena uniquely deduced from experiment*'. This seminal body of work was the first serious attempt to develop a physical and mathematical theory that explained the relationship between electricity and magnetism. Maxwell, writing more than half a century later, described the Frenchman's contribution in these terms:

> The experimental investigation by which Ampère established the laws of the mechanical action between electric currents is one of the most brilliant achievements in science. The whole, theory and experiment, seems as if it had leaped, full grown and full armed, from the brain of the 'Newton of electricity'. It is perfect in form, and unassailable in accuracy, and it is summed up in a formula from which all the phenomena may be deduced, and which must always remain the cardinal formula of electro-dynamics.[2]

Ampère's work is today remembered for his law of force ((see (7)$_1$ below), for his circuital law $\left(\oint_c \mathbf{B} \cdot dl = \mu_0 I_{\text{net}} \right)$ and for the SI unit of electric current.

[2]   J. Clerk Maxwell, *A treatise on electricity and magnetism*, vol. 2, Chap. III, p. 528. Cambridge: Cambridge University Press, 1873.

(ii) Equation (1) may be rearranged as follows:

$$\mathbf{F} = \oint_c I\, dl \times \mathbf{B}, \quad \text{where} \quad \mathbf{B} = \frac{\mu_0}{4\pi} \oint_{c'} \frac{I'\, dl' \times \mathbf{R}}{R^3} \tag{7}$$

is the magnetic field at circuit c due to the current in c' (it is understood that $\mathbf{R}$ is the vector *from* the current source *to* the field point). Equations $(7)_1$ and $(7)_2$ are known as Ampère's law of force and the Biot–Savart law respectively (see also Comment (v)). The latter is reminiscent of Coulomb's law for the electrostatic field: both laws are inverse square and their associated forces both involve a product of source terms ($\rho\, dv$ and $\rho'\, dv'$ for the force between two static point charges; $I\, dl$ and $I'\, dl'$ for the force between two current elements).

(iii) Since these are stationary currents, it is reasonable to expect that the force between the two circuits c and c' obeys Newton's third law: $\mathbf{F}_{c'c} = -\mathbf{F}_{cc'}$. Unlike (1), equation (2) shows explicitly that it does.

(iv) The quantity in brackets in (3) is the mutual inductance of the two circuits and is represented by the symbol $M(\mathcal{R})$. It is an important physical property and is considered in more detail in Chapter 5 (see Questions 5.14 and 5.19). Suppose circuits c, c' comprise N, N' turns of wire. Then their mutual inductance is

$$M(\mathcal{R}) = \frac{\mu_0}{4\pi} N N' \oint_c \oint_{c'} \frac{dl \cdot dl'}{|\mathbf{r} - \mathbf{r}' + \mathcal{R}|}, \tag{8}$$

and the force between them is

$$\mathbf{F} = II' \nabla_{\mathcal{R}} M(\mathcal{R}). \tag{9}$$

(v) For a volume distribution of electric current having density $\mathbf{J}(\mathbf{r}')$, the corresponding forms of Ampère's law and the Biot–Savart law follow from (7) of Question 4.3, and are

$$\mathbf{F} = \int_v \mathbf{J}(\mathbf{r}') \times \mathbf{B}(\mathbf{r}')\, dv' \quad \text{and} \quad \mathbf{B} = \frac{\mu_0}{4\pi} \int_v \frac{\mathbf{J}(\mathbf{r}') \times (\mathbf{r} - \mathbf{r}')}{|\mathbf{r} - \mathbf{r}'|^3}\, dv'. \tag{10}$$

(vi) Consider a single charge q moving with velocity $\mathbf{v}$ in a magnetostatic field $\mathbf{B}$. Taking $\mathbf{J}(\mathbf{r}') = \rho(\mathbf{r}')\mathbf{v}$ with $\rho(\mathbf{r}') = q\,\delta(\mathbf{r} - \mathbf{r}')$ and substituting in $(10)_1$ give

$$\mathbf{F} = \int_v q\,\delta(\mathbf{r}-\mathbf{r}')\,\mathbf{v} \times \mathbf{B}(\mathbf{r}')\, dv' = q\mathbf{v} \times \int_v \delta(\mathbf{r}-\mathbf{r}')\mathbf{B}(\mathbf{r}')\, dv' = q\mathbf{v} \times \mathbf{B}(\mathbf{r}), \tag{11}$$

where the last step follows because of the sifting property of a delta function (see Appendix E). Equation (11) should be regarded as an experimental result, being a direct consequence of Ampère's force law.[‡]

[‡]Of course, Ampère had no means of measuring forces on *individual* charges: recall that J. J. Thomson discovered the electron only in 1897, more than fifty years after Ampère died.

Question 4.5 **

Use the Biot–Savart law to obtain the magnetostatic equations:

$$\nabla \cdot \mathbf{B} = 0 \quad \text{and} \quad \nabla \times \mathbf{B} = \mu_0 \, \mathbf{J}(\mathbf{r}), \tag{1}$$

where $\mathbf{J}(\mathbf{r})$ is the current density.

Solution

In $(10)_2$ of Question 4.4 we replace $(\mathbf{r} - \mathbf{r}')/|\mathbf{r} - \mathbf{r}'|^3$ with $-\nabla|\mathbf{r} - \mathbf{r}'|^{-1}$ to obtain

$$\mathbf{B} = -\frac{\mu_0}{4\pi} \int_v \mathbf{J}(\mathbf{r}') \times \nabla|\mathbf{r} - \mathbf{r}'|^{-1} \, dv'$$

$$= \frac{\mu_0}{4\pi} \int_v \nabla \times \left[\frac{\mathbf{J}(\mathbf{r}')}{|\mathbf{r} - \mathbf{r}'|} \right] dv' - \frac{\mu_0}{4\pi} \int_v \frac{\nabla \times \mathbf{J}(\mathbf{r}')}{|\mathbf{r} - \mathbf{r}'|} \, dv', \tag{2}$$

where in the last step we use the identity $\nabla \times (f\mathbf{a}) = f\nabla \times \mathbf{a} + \nabla f \times \mathbf{a}$ (see (6) of Question (1.8)). Now $\nabla \times \mathbf{J}(\mathbf{r}') = 0$ (since $\mathbf{J}(\mathbf{r}')$ is a function of primed coordinates only), and (2) becomes

$$\mathbf{B}(\mathbf{r}) = \nabla \times \left(\frac{\mu_0}{4\pi} \int_v \frac{\mathbf{J}(\mathbf{r}')}{|\mathbf{r} - \mathbf{r}'|} \, dv' \right), \tag{3}$$

where the order of differentiation and integration has been interchanged.

☞ Equation $(1)_1$ follows immediately from (3) because the divergence of the curl of any vector is zero.

☞ Next we take the curl of (3) and use the identity $\nabla \times (\nabla \times \mathbf{a}) = \nabla(\nabla \cdot \mathbf{a}) - \nabla^2 \mathbf{a}$ (see (11) of Question 1.8). This gives

$$\nabla \times \mathbf{B} = \nabla \times \left(\nabla \times \frac{\mu_0}{4\pi} \int_v \frac{\mathbf{J}(\mathbf{r}')}{|\mathbf{r} - \mathbf{r}'|} \, dv' \right)$$

$$= \nabla \frac{\mu_0}{4\pi} \int_v \nabla \cdot \left[\frac{\mathbf{J}(\mathbf{r}')}{|\mathbf{r} - \mathbf{r}'|} \right] dv' - \frac{\mu_0}{4\pi} \int_v \mathbf{J}(\mathbf{r}') \nabla^2 |\mathbf{r} - \mathbf{r}'|^{-1} dv', \tag{4}$$

where, as before, we interchange the order of differentiation and integration. Now the identity $\nabla \cdot (f\mathbf{a}) = f\nabla \cdot \mathbf{a} + \mathbf{a} \cdot \nabla f$ (see (5) of Question 1.8) allows us to express the first integral in (4) as

$$\frac{\mu_0}{4\pi} \int_v \mathbf{\nabla} \cdot \left[\frac{\mathbf{J}(\mathbf{r}')}{|\mathbf{r} - \mathbf{r}'|} \right] dv' = \frac{\mu_0}{4\pi} \int_v \left[\frac{\mathbf{\nabla} \cdot \mathbf{J}(\mathbf{r}')}{|\mathbf{r} - \mathbf{r}'|} + \mathbf{J}(\mathbf{r}') \cdot \mathbf{\nabla} |\mathbf{r} - \mathbf{r}'|^{-1} \right] dv'$$

$$= \frac{\mu_0}{4\pi} \int_v \mathbf{J}(\mathbf{r}') \cdot \mathbf{\nabla} |\mathbf{r} - \mathbf{r}'|^{-1} dv' \quad \text{(because } \mathbf{\nabla} \cdot \mathbf{J}(\mathbf{r}') = 0 \text{, as before)}$$

$$= -\frac{\mu_0}{4\pi} \int_v \mathbf{J}(\mathbf{r}') \cdot \mathbf{\nabla}' |\mathbf{r} - \mathbf{r}'|^{-1} dv' \quad \text{(because } \mathbf{\nabla} R^{-1} = -\mathbf{\nabla}' R^{-1})$$

$$= -\frac{\mu_0}{4\pi} \int_v \left\{ \mathbf{\nabla}' \cdot \left[\frac{\mathbf{J}(\mathbf{r}')}{|\mathbf{r} - \mathbf{r}'|} \right] - \frac{\mathbf{\nabla}' \cdot \mathbf{J}(\mathbf{r}')}{|\mathbf{r} - \mathbf{r}'|} \right\} dv',$$

where in the final step we again use the identity $\mathbf{a} \cdot \mathbf{\nabla}' f = \mathbf{\nabla}' \cdot (f\mathbf{a}) - f\mathbf{\nabla}' \cdot \mathbf{a}$. In magnetostatics $\mathbf{\nabla}' \cdot \mathbf{J}(\mathbf{r}') = 0$, because of (8) of Question 4.3. Hence

$$\frac{\mu_0}{4\pi} \int_v \mathbf{\nabla} \cdot \left[\frac{\mathbf{J}(\mathbf{r}')}{|\mathbf{r} - \mathbf{r}'|} \right] dv' = -\frac{\mu_0}{4\pi} \int_v \mathbf{\nabla}' \cdot \left[\frac{\mathbf{J}(\mathbf{r}')}{|\mathbf{r} - \mathbf{r}'|} \right] dv' = -\frac{\mu_0}{4\pi} \oint_s \frac{\mathbf{J}(\mathbf{r}') \cdot d\mathbf{a}'}{|\mathbf{r} - \mathbf{r}'|}, \tag{5}$$

as a result of Gauss's theorem. Now for a bounded distribution[‡] $\mathbf{J}(\mathbf{r}') \cdot d\mathbf{a}'$ in (5) is zero everywhere on s. This proves that the first term on the right-hand side of (4) is identically zero. Using $\nabla^2 |\mathbf{r} - \mathbf{r}'|^{-1} = -4\pi\delta(\mathbf{r} - \mathbf{r}')$ then yields

$$\mathbf{\nabla} \times \mathbf{B} = -\frac{\mu_0}{4\pi} \int_v \mathbf{J}(\mathbf{r}') \nabla^2 |\mathbf{r} - \mathbf{r}'|^{-1} dv' = \mu_0 \int_v \mathbf{J}(\mathbf{r}') \delta(\mathbf{r} - \mathbf{r}') dv'$$

$$= \mu_0 \mathbf{J}(\mathbf{r}), \tag{6}$$

because of the sifting property of delta functions (see $(XI)_2$ of Appendix E). The desired result is given by (6).

Comments

(i) The result $\mathbf{\nabla} \cdot \mathbf{B} = 0$ implies that

$$\mathbf{B} = \mathbf{\nabla} \times \mathbf{A}(\mathbf{r}), \tag{7}$$

because of (2) of Question 1.15. Comparing (3) and (7) shows that

$$\mathbf{A}(\mathbf{r}) = \frac{\mu_0}{4\pi} \int_v \frac{\mathbf{J}(\mathbf{r}')}{|\mathbf{r} - \mathbf{r}'|} dv', \tag{8}$$

which is the vector potential for an arbitrary current distribution. It is given by (8) up to an arbitrary additive gradient $\mathbf{\nabla}\chi$, say,[♯] where $\chi(\mathbf{r})$ is any suitably differentiable scalar field.

[‡]See the footnote on p. 45.

[♯]Notice that $\mathbf{B}$ remains unaffected by the $\mathbf{\nabla}\chi$ term, because the curl of any gradient is always zero. This is discussed further in Chapter 8.

(ii) It follows from (7) of Question 4.3 that the equivalent of (8) for a filamentary current I is

$$\mathbf{A}(\mathbf{r}) = \frac{\mu_0}{4\pi} \oint_c \frac{I dl'}{|\mathbf{r} - \mathbf{r}'|}, \tag{9}$$

which is a useful result.

(iii) ☞ Integrating $(1)_1$ over an arbitrary volume and applying Gauss's theorem give

$$\int_v \boldsymbol{\nabla} \cdot \mathbf{B} \, dv' = \oint_s \mathbf{B} \cdot d\mathbf{a}' = 0, \tag{10}$$

which is Gauss's law for a magnetostatic field $\mathbf{B}(\mathbf{r})$.

☞ Integrating $(1)_2$ over an arbitrary surface and applying Stokes's theorem yield

$$\int_s (\boldsymbol{\nabla} \times \mathbf{B}) \cdot d\mathbf{a}' = \oint_c \mathbf{B} \cdot dl' = \mu_0 \int_s \mathbf{J} \cdot d\mathbf{a}' = \mu_0 I_{\text{net}}, \tag{11}$$

where $I_{\text{net}} = \int_s \mathbf{J} \cdot d\mathbf{a}'$ is the net current passing through s (see (1) of Question 4.3). This result known as Ampère's circuital law[†] may be stated in words as follows: the line integral of magnetic field around the perimeter of any area in vacuum equals $\mu_0 \times$ the net current crossing that area.

(iv) The differential equations (1) apply at a point in vacuum, whereas the integral equations (10) and (11) apply over a finite region of space.

(v) Because of the asymmetry in the equations $(\boldsymbol{\nabla} \cdot \mathbf{B} = 0;\ \boldsymbol{\nabla} \times \mathbf{B} = \mu_0\, \mathbf{J}(\mathbf{r})$ on the one hand, $\boldsymbol{\nabla} \cdot \mathbf{E} = \epsilon_0^{-1} \rho(\mathbf{r});\ \boldsymbol{\nabla} \times \mathbf{E} = 0$ on the other), the mathematical development of magnetostatics is different from electrostatics. However, in regions of space where $\mathbf{J} = 0$ the curl of $\mathbf{B}$ is zero, and here the magnetic field may be expressed as the gradient of a magnetic scalar potential Φ_{m} (see (2) of Question 1.14). That is,

$$\mathbf{B} = -\mu_0 \boldsymbol{\nabla} \Phi_{\text{m}}. \tag{12}$$

Clearly, Φ_{m} satisfies Laplace's equation because $\boldsymbol{\nabla} \cdot \mathbf{B}$ is always zero, and so the techniques for solving electrostatic boundary-value problems can sometimes be adapted to magnetostatic boundary-value problems. However, care must be exercised in applying the boundary conditions. For example, it is necessary that $\mathbf{J}$ be zero not only inside v where (12) holds, but *also* on the boundaries of v. This is unlike electrostatics where the charge density is not *required* to vanish anywhere.

(vi) Applications involving the magnetic scalar potential are given in Questions 4.18, 4.19 and 9.21.

[†]Henceforth we omit the word 'circuital', and refer to this result as Ampère's law.

Question 4.6

Show that the vector potential given by (8) of Question 4.5 satisfies the condition:

$$\nabla \cdot \mathbf{A} = 0. \tag{1}$$

Solution

Taking the divergence of $\mathbf{A}(\mathbf{r}) = \dfrac{\mu_0}{4\pi} \displaystyle\int_v \dfrac{\mathbf{J}(\mathbf{r}')}{|\mathbf{r} - \mathbf{r}'|}\, dv'$ yields

$$\nabla \cdot \mathbf{A} = \nabla \cdot \left[\frac{\mu_0}{4\pi} \int_v \frac{\mathbf{J}(\mathbf{r}')}{|\mathbf{r} - \mathbf{r}'|}\, dv' \right] = \frac{\mu_0}{4\pi} \int_v \nabla \cdot \left[\frac{\mathbf{J}(\mathbf{r}')}{|\mathbf{r} - \mathbf{r}'|} \right] dv'. \tag{2}$$

Now the right-hand side of (2) is zero as we showed in the solution to Question 4.5. Hence (1). See also Question 8.10.

Comment

Equation (1) shows that the vector potential always has zero divergence in magneto-statics. However, in electrodynamics this condition is not *required* to be true. Specifying the divergence of $\mathbf{A}$ is referred to as 'choosing a gauge' and the particular choice (1) is known variously as the Coulomb, radiation or transverse gauge. We return to this important topic again in Chapter 8.

Question 4.7

Consider a long (assume infinite) cylindrical conductor of radius a lying along the z-axis of cylindrical polar coordinates. Suppose the conductor carries a current I and that the current density $\mathbf{J} = \dfrac{I \hat{\mathbf{z}}}{\pi a^2}$ is constant over its cross-section. Use Ampère's law, discussed in Question 4.5, to show that the magnetic field $\mathbf{B}(r)$ is given by

$$\mathbf{B}(r) = \begin{cases} \dfrac{\mu_0}{2} \mathbf{J} \times \mathbf{r} & \text{for } r \leq a \\[2ex] \dfrac{\mu_0}{2} \left(\dfrac{a}{r}\right)^2 \mathbf{J} \times \mathbf{r} & \text{for } r > a. \end{cases} \tag{1}$$

Solution

Since this is a cylindrically symmetric current, the magnetic-field lines are necessarily circles centred on the conductor. Around any one of these circles the magnitude of **B** is constant. Thus the line integral of **B** around an arbitrary circular contour of radius r is $B \times 2\pi r$, and Ampère's law gives $B \times 2\pi r = \mu_0 I_{\text{net}}$ where

$$I_{\text{net}} = \begin{cases} J\pi r^2 & \text{for } r \leq a \\ J\pi a^2 & \text{for } r > a. \end{cases}$$

It then follows that

$$B(r) = \begin{cases} \dfrac{\mu_0}{2} Jr & \text{for } r \leq a \\ \dfrac{\mu_0}{2} \dfrac{Ja^2}{r} & \text{for } r > a. \end{cases} \tag{2}$$

The direction of **B** is related to **J** by the right-hand rule; the vector form of (2) is (1).

Comments

(i) Equation (1) is an elementary result which is often encountered in the alternative form

$$\mathbf{B}(r) = \begin{cases} \dfrac{\mu_0}{2\pi} \dfrac{Ir}{a^2} \hat{\boldsymbol{\theta}} & \text{for } r \leq a \\ \dfrac{\mu_0}{2\pi} \dfrac{I}{r} \hat{\boldsymbol{\theta}} & \text{for } r > a, \end{cases} \tag{3}$$

since $I = J \times \pi a^2$. The graph below of B/B_0 vs r/a is a plot of (3).[‡] Inside the conductor, B increases linearly with distance from $r = 0$ to $r = a$ and then decreases inversely with distance for $r > a$.

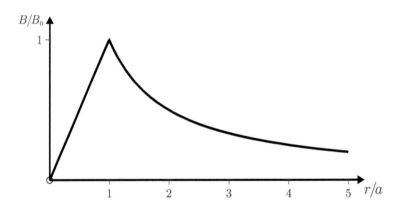

(ii) Equation $(3)_2$ is sometimes used to calculate **B** in the vicinity of a finite conductor of length ℓ. This is a reasonable approximation, provided the field point $P(\mathbf{r})$ is sufficiently far from the ends of the conductor with $r \ll \ell$.

[‡] Here $B_0 = \dfrac{\mu_0}{2\pi} \dfrac{I}{a}$ is the magnitude of the magnetic field at the surface of the conductor.

(iii) In the solution and comments above, we have assumed that the conductor is non-ferromagnetic and that its permeability has the vacuum value μ_0 (the permeability of ferromagnetic materials is discussed briefly in Chapter 9).

Question 4.8

Current I is present in a wire having the shapes described below. Use the Biot–Savart law to calculate the magnetic field at an arbitrary point P

(a) on the symmetry axis (z, say) of a circular loop of wire having radius a.

(b) due to a straight wire of length ℓ.[‡] Suppose the wire is centred on the origin and lies along the z-axis of cylindrical polar coordinates.

Solution

(a) Refer to the cylindrical polar coordinates shown in the figure below. Clearly, the distance from the current element $I\,d\mathbf{l}'$ to P is $R = \sqrt{a^2 + z^2}$ where $I\,d\mathbf{l}' = I\,a\,d\theta'\,\hat{\boldsymbol{\theta}} = I\,a(-\hat{\mathbf{x}}\sin\theta'+\hat{\mathbf{y}}\cos\theta')\,d\theta'$, because of (III) of Appendix D. Furthermore, the unit vector $\hat{\mathbf{R}} = -\hat{\mathbf{r}}\cos\chi + \hat{\mathbf{z}}\sin\chi$ where $\hat{\mathbf{r}} = \hat{\mathbf{x}}\cos\theta' + \hat{\mathbf{y}}\sin\theta'$. Therefore $\hat{\mathbf{R}} = -\hat{\mathbf{x}}\cos\chi\cos\theta' - \hat{\mathbf{y}}\cos\chi\sin\theta' + \hat{\mathbf{z}}\sin\chi$.

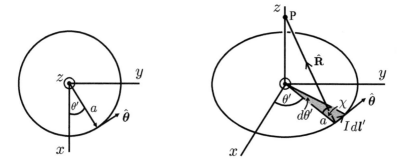

Substituting these results in the Biot–Savart law $\big($see $(7)_2$ of Question 4.4$\big)$ gives

$$
\begin{aligned}
d\mathbf{B} &= \frac{\mu_0}{4\pi}\,I\,a\,\frac{(-\hat{\mathbf{x}}\sin\theta' + \hat{\mathbf{y}}\cos\theta') \times (-\hat{\mathbf{x}}\cos\chi\cos\theta' - \hat{\mathbf{y}}\cos\chi\sin\theta' + \hat{\mathbf{z}}\sin\chi)}{a^2 + z^2}\,d\theta' \\
&= \frac{\mu_0}{4\pi}\,I\,a\,\frac{\sin\chi\cos\theta'\,\hat{\mathbf{x}} + \sin\chi\sin\theta'\,\hat{\mathbf{y}} + \cos\chi(\sin^2\theta' + \cos^2\theta')\,\hat{\mathbf{z}}}{a^2 + z^2}\,d\theta'.
\end{aligned}
$$

But $\sin\chi = \dfrac{z}{\sqrt{a^2 + z^2}}$, $\quad \cos\chi = \dfrac{a}{\sqrt{a^2 + z^2}}$, and so

[‡]Although this wire is necessarily part of a larger circuit, we are concerned here only with the contribution which *this* wire makes to the magnetic field at P.

$$dB = \frac{\mu_0}{4\pi} I a \frac{(\hat{\mathbf{x}}\cos\theta' + \hat{\mathbf{y}}\sin\theta')z + \hat{\mathbf{z}}a}{(a^2 + z^2)^{3/2}} d\theta'.$$

Hence

$$\mathbf{B} = \int_0^{2\pi} d\mathbf{B} = \frac{\mu_0}{4\pi} \frac{I a}{(a^2 + z^2)^{3/2}} \int_0^{2\pi} [(\hat{\mathbf{x}}\cos\theta' + \hat{\mathbf{y}}\sin\theta')z + \hat{\mathbf{z}}a] d\theta'$$

$$= \frac{\mu_0 I}{2} \frac{a^2 \hat{\mathbf{z}}}{(a^2 + z^2)^{3/2}} = \frac{B_0 a^3 \hat{\mathbf{z}}}{(a^2 + z^2)^{3/2}}, \tag{1}$$

where $B_0 = \mu_0 I / 2a$.

(b) Let the Cartesian coordinates of P be $(r\cos\theta, r\sin\theta, z)$. Then the vector $\mathbf{R}$ from the current element $I\,d\mathbf{z}'$ to P is $\mathbf{R} = (r\cos\theta, r\sin\theta, z-z')$ with $R^2 = r^2 + (z-z')^2$. Thus $d\mathbf{z}' \times \mathbf{R} = \hat{\mathbf{z}} \times \mathbf{R}\,dz' = r(-\hat{\mathbf{x}}\sin\theta + \hat{\mathbf{y}}\cos\theta)dz'$, and the Biot–Savart law gives

$$\mathbf{B} = \frac{\mu_0}{4\pi} I r(-\hat{\mathbf{x}}\sin\theta + \hat{\mathbf{y}}\cos\theta) \int_{-\frac{1}{2}\ell}^{\frac{1}{2}\ell} \frac{dz'}{[r^2 + (z-z')^2]^{3/2}}$$

$$= \frac{\mu_0}{4\pi} \frac{I}{r} \left[\frac{(\ell + 2z)}{\sqrt{4r^2 + (\ell + 2z)^2}} + \frac{(\ell - 2z)}{\sqrt{4r^2 + (\ell - 2z)^2}} \right] \hat{\boldsymbol{\theta}}, \tag{2}$$

where, in the last step, we use *Mathematica*'s Integrate function to evaluate the integral.

Comments

(i) Putting $z = 0$ in (1) gives the magnetic field at the centre of a ring of current: $\mathbf{B} = \frac{\mu_0}{2} \frac{I}{a} \hat{\mathbf{z}}$, which is an elementary result.

(ii) In the limit $\ell \to \infty$ it is clear that (2) gives the familiar field of an infinitely long straight wire: $\mathbf{B} = \frac{\mu_0}{2\pi} \frac{I}{r} \hat{\boldsymbol{\theta}}$.

Question 4.9*

(a) Consider a solenoid having length ℓ, cross-sectional radius a and N turns of wire wound as a single layer, carrying a current I. Suppose the solenoid lies along the z-axis of cylindrical polar coordinates with its centre at the origin. Use (1) of Question 4.8 to show that the axial magnetic field is given by

$$\mathbf{B}(z) = \frac{1}{2} B_0 \left[\frac{(z + \frac{1}{2}\ell)}{\sqrt{a^2 + (z + \frac{1}{2}\ell)^2}} - \frac{(z - \frac{1}{2}\ell)}{\sqrt{a^2 + (z - \frac{1}{2}\ell)^2}} \right] \hat{\mathbf{z}}, \tag{1}$$

where $B_0 = \mu_0 n I$ and $n = N/\ell$ is the number of turns per unit length.

(b) Now suppose that the solenoid has a toroidal shape; all relevant dimensions and coordinates are indicated in the figure below. Use Ampère's law to show that the magnetic field inside the solenoid is given by

$$\mathbf{B}(r,\phi) = \frac{\mu_0}{2\pi} \frac{NI}{R + r\cos\phi} \hat{\boldsymbol{\theta}}, \tag{2}$$

where $\hat{\boldsymbol{\theta}}$ is a unit vector perpendicular to, and out of, the plane of the page.

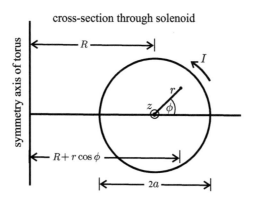

cross-section through solenoid

Solution

(a) Consider a single turn whose centre is displaced along the z-axis from $z = 0$ to $z = z'$. The field due to this turn (call it $\mathbf{b}(z)$) follows from (1) of Question 4.8 if we make the substitution $z \rightarrow z - z'$. This gives

$$\mathbf{b}(z) = \frac{\mu_0 I}{2} \frac{a^2}{[a^2 + (z - z')^2]^{3/2}} \hat{\mathbf{z}}.$$

Now in an interval dz' the number of turns is Ndz'/ℓ, and their contribution to the net magnetic field is

$$d\mathbf{B}(z) = \frac{N\mathbf{b}(z)dz'}{\ell} = \frac{\mu_0 NI}{2\ell} \frac{a^2 dz'}{[a^2 + (z - z')^2]^{3/2}} \hat{\mathbf{z}}$$

or

$$\mathbf{B}(z) = \frac{\mu_0 NIa^2}{2\ell} \int_{-\frac{1}{2}\ell}^{\frac{1}{2}\ell} \frac{dz'}{[a^2 + (z - z')^2]^{3/2}} \hat{\mathbf{z}}. \tag{3}$$

If we let $z - z' = a\tan\theta$ then $dz' = -a\sec^2\theta\,d\theta$ and (3) becomes

$$\mathbf{B}(z) = \frac{\mu_0 NI}{2\ell} \int_{\theta_-}^{\theta_+} \cos\theta\,d\theta\,\hat{\mathbf{z}} \quad \text{where} \quad \sin\theta_\pm = \frac{(z \pm \frac{1}{2}\ell)}{\sqrt{a^2 + (z \pm \frac{1}{2}\ell)^2}},$$

which is (1).

(b) Because of the symmetry, the circulation of **B** around a circular contour of radius $R + r\cos\phi$ (see the figure on p. 205) is $2\pi(R + r\cos\phi)B$. It then follows from Ampère's law that $2\pi(R+r\cos\phi)B = \mu_0 I_{net} = \mu_0 NI$. Hence (2) with the direction of magnetic field given by the right-hand rule.

Comments

(i) In the limit $\ell \to \infty$ equation (1) reduces to the usual result $\mathbf{B} = B_0\hat{\mathbf{z}}$ for the field of an ideal infinitely long solenoid. This result is, of course, valid anywhere inside the solenoid and not just on the axis itself.

(ii) The graph below shows a plot of B/B_0 vs z/a for $\ell/a = 10$ and $\ell/a = 5$. Notice that for $\ell/a = 10$, the magnetic field on the axis is essentially constant from one end of the solenoid to the other.

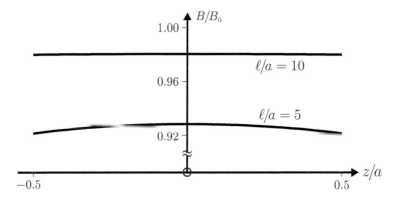

Question 4.10*

Consider a circular current loop of radius a lying in the xy-plane and centred on the origin as shown in the figure on p. 207. A power supply (not drawn) maintains a steady current I in the loop. Suppose P represents a distant field point $(r \gg a)$ that is conveniently chosen to lie in the xz-plane where the spherical polar coordinate ϕ is zero. This choice involves no loss of generality, because the current loop is axially symmetric.

(a) Show that the vector potential at P is given by

$$\mathbf{A}(\mathbf{r}) = \frac{\mu_0}{4\pi} \frac{\mathbf{m} \times \hat{\mathbf{r}}}{r^2}, \tag{1}$$

where $\mathbf{m} = (\pi a^2)I\hat{\mathbf{z}}$.

(b) Use (1) to derive the field

$$\mathbf{B}(\mathbf{r}) = \frac{\mu_0}{4\pi}\left[\frac{(3\mathbf{m}\cdot\hat{\mathbf{r}})\hat{\mathbf{r}} - \mathbf{m}}{r^3}\right]. \tag{2}$$

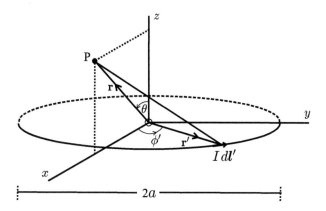

Solution

(a) We start with the vector potential (9) of Question 4.5:

$$\mathbf{A}(\mathbf{r}) = \frac{\mu_0}{4\pi} \oint_c \frac{I d\mathbf{l}'}{|\mathbf{r} - \mathbf{r}'|}. \tag{3}$$

Here $I d\mathbf{l}'$ is an arbitrary current element of the loop where

$$d\mathbf{l}' = (-\hat{\mathbf{x}} \sin \phi' + \hat{\mathbf{y}} \cos \phi') a\, d\phi'. \tag{4}$$

The distance from $I d\mathbf{l}'$ to the field point P follows from (VII) of Appendix C (taking $\theta' = \frac{1}{2}\pi$; $\phi = 0$) and is $|\mathbf{r} - \mathbf{r}'| = \sqrt{r^2 + a^2 - 2ra \sin \theta \cos \phi'}$. Expanding $|\mathbf{r} - \mathbf{r}'|$ in powers of a/r and retaining terms up to first order only (recall that $r \gg a$) yield

$$|\mathbf{r} - \mathbf{r}'| \simeq r\left(1 - \frac{a}{r} \sin \theta \cos \phi'\right) \quad \text{or} \quad |\mathbf{r} - \mathbf{r}'|^{-1} \simeq r^{-1}\left(1 + \frac{a}{r} \sin \theta \cos \phi'\right). \tag{5}$$

Substituting (4) and (5) in (3) gives

$$\begin{aligned}
\mathbf{A}(\mathbf{r}) &= \frac{\mu_0}{4\pi} I \frac{a}{r} \oint_c (-\hat{\mathbf{x}} \sin \phi' + \hat{\mathbf{y}} \cos \phi')\left(1 + \frac{a}{r} \sin \theta \cos \phi'\right) d\phi' \\
&= \frac{\mu_0}{4\pi} \frac{I a}{r} \int_0^{2\pi} (-\hat{\mathbf{x}} \sin \phi' + \hat{\mathbf{y}} \cos \phi')\left(1 + \frac{a}{r} \sin \theta \cos \phi'\right) d\phi' \qquad (6) \\
&= \frac{\mu_0}{4\pi} \frac{I a^2}{r^2} \sin \theta \int_0^{2\pi} \cos^2 \phi'\, d\phi'\, \hat{\mathbf{y}},
\end{aligned}$$

because the other three integrals in (6) are clearly zero. Now $\int_0^{2\pi} \cos^2 \phi'\, d\phi' = \pi$, and so

$$\mathbf{A}(\mathbf{r}) = \frac{\mu_0}{4\pi} \frac{I \pi a^2}{r^2} \sin \theta\, \hat{\mathbf{y}} = \frac{\mu_0}{4\pi} \frac{I \pi a^2}{r^2} \sin \theta\, \hat{\boldsymbol{\phi}},$$

since $\hat{\mathbf{y}} = \hat{\boldsymbol{\phi}}$ when P lies in the xz-plane. But $\sin \theta\, \hat{\boldsymbol{\phi}} = \hat{\mathbf{z}} \times \hat{\mathbf{r}}$. Hence (1).

(b) From $\mathbf{B} = \nabla \times \mathbf{A}$ we have $B_i = (\nabla \times \mathbf{A})_i = \varepsilon_{ijk}\nabla_j A_k$ where $A_k = \dfrac{\mu_0}{4\pi}\varepsilon_{klm}m_l r_m r^{-3}$ follows from (1). So $4\pi B_i = \mu_0\varepsilon_{ijk}\varepsilon_{klm}\nabla_j(m_l r_m r^{-3}) = \mu_0 m_l \varepsilon_{kij}\varepsilon_{klm}\nabla_j(r_m r^{-3})$ because of the cyclic properties of the Levi-Civita subscripts. Using (1) of Question 1.3 and the product rule of differentiation give

$$B_i = \frac{\mu_0}{4\pi}m_l\left(\delta_{il}\delta_{jm} - \delta_{im}\delta_{jl}\right)\left(r^{-3}\nabla_j r_m + r_m\nabla_j r^{-3}\right)$$

$$= \frac{\mu_0}{4\pi}m_l\left(\delta_{il}\delta_{jm} - \delta_{im}\delta_{jl}\right)\left(r^{-3}\delta_{jm} - 3r_j r_m r^{-5}\right), \tag{7}$$

where in the last step we use (1) and (3) of Question 1.2. Multiplying out the brackets in (7) and contracting subscripts yield $4\pi B_i = \mu_0 m_l(3r_i r_l r^{-5} - \delta_{il}r^{-3}) = \mu_0(3m_l r_l r_i - r^2 m_i)r^{-5} = \mu_0(3r_i\mathbf{m}\cdot\mathbf{r} - r^2 m_i)r^{-5}$. Equation (2) follows immediately since this last result is true for $i = x,\ y$ and z.

Comments

(i) The above current loop is an example of a magnetic dipole and the vector $\mathbf{m}$ is its magnetic dipole moment. The magnitude of $\mathbf{m}$ is given by $I \times$ the cross-sectional area of the loop, and its direction is determined by the right-hand rule: with the fingers of the right hand curling in the direction of I the thumb points along $\mathbf{m}$.

(ii) Equations (1) and (2) are the dipole potential and field respectively. Because of the approximation made in deriving $\mathbf{A}$, these equations are valid only at points for which $r \gg a$. Now suppose we let $a \to 0$ and $I \to \infty$ in such a way that $\mathbf{m}$ remains finite and constant. Then this current loop becomes a 'point' magnetic dipole located at the origin and we may regard (1) and (2) as being valid at any $r > 0$.

(iii) Strictly speaking, the dipole field (2) is incomplete because it omits a term that is zero everywhere except at the origin itself. Suppose that, in calculating the magnetic field in (b) above, we use (5) of Question 1.21 (instead of (7) of Question 1.1) to differentiate $r_m r^{-3}$. Then

$$B_i = -\frac{\mu_0}{4\pi}m_l\left(\delta_{il}\delta_{jm} - \delta_{im}\delta_{jl}\right)\nabla_j\nabla_m r^{-1}$$

$$= -\frac{\mu_0}{4\pi}m_l\left(\delta_{il}\delta_{jm} - \delta_{im}\delta_{jl}\right)\left[(3r_j r_m r^{-5} - r^{-3}\delta_{jm}) - \frac{4\pi}{3}\delta_{jm}\delta(\mathbf{r})\right]$$

$$= -\frac{\mu_0}{4\pi}m_l\left[(\delta_{il}\delta_{jm} - \delta_{im}\delta_{jl})(3r_j r_m r^{-5} - r^{-3}\delta_{jm}) - \frac{8\pi}{3}\delta_{il}\delta(\mathbf{r})\right]$$

$$= \frac{\mu_0}{4\pi}\left[\frac{(3m_j r_j r_i - r^2 m_i)}{r^5} + \frac{8\pi}{3}m_i\delta(\mathbf{r})\right],$$

or

$$\mathbf{B} = \frac{\mu_0}{4\pi} \left[\frac{3(\mathbf{m} \cdot \hat{\mathbf{r}})\hat{\mathbf{r}} - \mathbf{m}}{r^3} + \frac{8\pi}{3} \mathbf{m} \, \delta(\mathbf{r}) \right]. \tag{8}$$

Omission of the delta-function term in (8) is usually inconsequential; but not always so, as Ref. [3] explains. See Comment (iii) of Question 2.11 and also Question 4.21.

(iv) Magnetic dipoles are an important member of a class of entities called magnetic multipoles. This topic is discussed further in Question 4.24.

Question 4.11 **

Charge q is uniformly distributed throughout a sphere of radius a. The sphere spins with constant angular velocity $\boldsymbol{\omega}$ about an axis which we choose to be the z-axis of Cartesian coordinates.

(a) Show that the vector potential at an arbitrary point P outside the sphere is given by

$$\mathbf{A} = \frac{\mu_0 q a^2}{20\pi} \frac{(\boldsymbol{\omega} \times \mathbf{r})}{r^3}. \tag{1}$$

Hint: The integration leading to (1) can be simplified by temporarily orienting the spin axis in the xz-plane and positioning P on the z-axis. Evaluate the resulting integral with *Mathematica*.

(b) Hence show that the magnetic field at P is

$$\mathbf{B} = \frac{\mu_0 q \omega a^2}{20\pi} \frac{(2\hat{\mathbf{r}} \cos\theta + \hat{\boldsymbol{\theta}} \sin\theta)}{r^3}, \tag{2}$$

where r and θ are spherical polar coordinates.
Hint: Use the curl operator $(\mathrm{XI})_3$ of Appendix C.

Solution

(a) Because of the hint, $\mathbf{r} = (0, 0, z)$. Then letting Θ be the angle between the spin axis and $\hat{\mathbf{z}}$ yields $\boldsymbol{\omega} = \omega(\sin\Theta, 0, \cos\Theta)$. The vector potential given by (8) of Question 4.5 is

$$\mathbf{A}(\mathbf{r}) = \frac{\mu_0}{4\pi} \int_v \frac{\mathbf{J}(\mathbf{r}') \, dv'}{|\mathbf{r} - \mathbf{r}'|}, \tag{3}$$

[3] D. J. Griffiths, 'Hyperfine splitting in the ground state of hydrogen', *American Journal of Physics*, vol. 50, pp. 698–703, 1982.

where, for an arbitrary volume element dv' of the sphere, the source vector $\mathbf{r}' = r'(\sin\theta'\cos\phi', \sin\theta'\sin\phi', \cos\theta')$ and $\mathbf{J}(\mathbf{r}') = \rho\mathbf{v}(\mathbf{r}')$. Here $\mathbf{v}(\mathbf{r}')$ is the velocity of dv' and it is related to $\boldsymbol{\omega}$ by $\mathbf{v} = \boldsymbol{\omega}\times\mathbf{r}'$.[4] Also $|\mathbf{r}-\mathbf{r}'| = \sqrt{z^2 + r'^2 - 2zr'\cos\theta'}$, and so

$$\mathbf{A} = \frac{\mu_0\rho}{4\pi}\int_v \frac{(\boldsymbol{\omega}\times\mathbf{r}')\,dv'}{\sqrt{z^2 + r'^2 - 2zr'\cos\theta'}} = \frac{\mu_0\rho}{4\pi}\int_0^{2\pi}\int_0^\pi\int_0^a \frac{(\boldsymbol{\omega}\times\mathbf{r}')\,r'^2\sin\theta'\,dr'\,d\theta'\,d\phi'}{\sqrt{z^2 + r'^2 - 2zr'\cos\theta'}}. \quad (4)$$

Next we substitute

$$(\boldsymbol{\omega}\times\mathbf{r}') = \omega r'(-\cos\Theta\sin\theta'\sin\phi', \cos\Theta\sin\theta'\cos\phi' - \sin\Theta\cos\theta', \sin\Theta\sin\theta'\sin\phi')$$

in (4), and because $\int_0^{2\pi}\cos\phi'\,d\phi' = \int_0^{2\pi}\sin\phi'\,d\phi' = 0$ the only non-zero contribution to $\mathbf{A}$ arises from the y-component of $\boldsymbol{\omega}\times\mathbf{r}'$. Thus,

$$A_y = -\frac{\mu_0\rho\omega\sin\Theta}{4\pi}\int_0^{2\pi}\int_0^\pi\int_0^a \frac{\cos\theta'\sin\theta'\,r'^3\,dr'\,d\theta'\,d\phi'}{\sqrt{z^2 + r'^2 - 2zr'\cos\theta'}}$$

$$= -\frac{\mu_0\rho\omega\sin\Theta}{2}\int_0^\pi\int_0^a \frac{\cos\theta'\sin\theta'\,r'^3\,dr'\,d\theta'}{\sqrt{z^2 + r'^2 - 2zr'\cos\theta'}}. \quad (5)$$

The integral in (5) evaluates to $\dfrac{2a^5}{15z^2}$ (see the *Mathematica* notebook on p. 211).

Therefore $A_y = -\dfrac{\mu_0\rho\omega a^5\sin\Theta}{15z^2}$ where $\rho = \dfrac{q}{\frac{4}{3}\pi a^3}$, and we obtain

$$\mathbf{A} = -\frac{\mu_0 q\omega a^2\sin\Theta}{20\pi z^2}\hat{\mathbf{y}}. \quad (6)$$

Replacing $\omega\sin\Theta\,\hat{\mathbf{y}}$ in (6) with $-\boldsymbol{\omega}\times\hat{\mathbf{z}}$ gives

$$\mathbf{A} = \frac{\mu_0 q a^2}{20\pi}\frac{(\boldsymbol{\omega}\times\hat{\mathbf{z}})}{z^2}. \quad (7)$$

Now we must revert to the original configuration where $\boldsymbol{\omega}$ is along the z-axis and the field point P has the arbitrary coordinates (r,θ,ϕ). This is achieved by making the transformation in (7): $\Theta\to\theta$ and $\mathbf{z}\to\mathbf{r}$. Hence (1).

(b) With $\mathbf{A}$ given by (1), the field components are obtained from $\mathbf{B} = \nabla\times\mathbf{A}$ and are

$$\left.\begin{aligned}
B_r &= \frac{\mu_0 q\omega}{20\pi}\frac{a^2}{r^2}\frac{1}{r\sin\theta}\frac{\partial}{\partial\theta}\sin^2\theta = \frac{\mu_0 q\omega a^2}{20\pi}\frac{2\cos\theta}{r^3} \\[2mm]
B_\theta &= -\frac{\mu_0 q\omega}{20\pi}\frac{a^2\sin\theta}{r}\frac{\partial}{\partial r}\frac{1}{r} = \frac{\mu_0 q\omega a^2}{20\pi}\frac{\sin\theta}{r^3}
\end{aligned}\right\},$$

which is (2).

---

[4]   O. L. de Lange and J. Pierrus, *Solved problems in classical mechanics: Analytical and numerical solutions with comments*, Chap. 12, p. 402. Oxford: Oxford University Press, 2010.

Comment

Comparing (2) with (2) of Question 4.10 shows that for all $r \geq a$, the magnetic field of this charged spinning sphere is that of a magnetic dipole at the origin whose dipole moment **m** is

$$\mathbf{m} = \tfrac{1}{5}q\omega a^2 \hat{\mathbf{z}}. \tag{8}$$

See also Question 4.26 for an alternative calculation of **m**.

```
In[1]:= Asp = {a > 0 && z ≥ a};

        Integrate[ (rp³ Sin[θp] Cos[θp]) / √(rp² + z² - 2 rp z Cos[θp]), {rp, 0, a}, {θp, 0, π}, Assumptions → Asp]
```

Question 4.12

Consider a long (assume infinite) cylindrical conductor of radius a that has a cylindrical hole of radius b bored parallel to and offset from the axis by a displacement **d**, where $d < a - b$. The conductor carries a steady current I, and the current density **J** is constant everywhere (except obviously inside the hole, where it is zero). Use (1) of Question 4.7 and the principle of superposition to calculate **B(r)** everywhere.

Solution

Consider uniform current densities **J** and **J'** that are confined to infinitely long cylindrical regions having radii a and b respectively. Suppose that these currents are centred on parallel axes passing through origins O and O' which are separated by a displacement **d**. Take $\mathbf{J}' = -\mathbf{J}$. Superposition of these currents produces a distribution whose density is **J** everywhere for $r \leq a$ (except in a cylindrical region centred on O' and having radius b where it is zero). This may be represented schematically[‡] in the following 'picture equation':

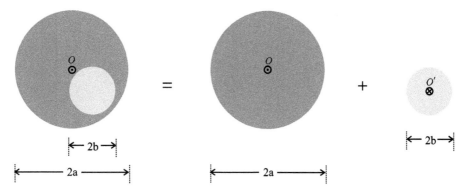

[‡]In the figure above, the symbols $\otimes$ ($\odot$) represent current directions into (out of) the page.

The superposition of $\mathbf{J}$ and $\mathbf{J}'$ produces, in each of the three regions (see the figure below), the following magnetic field:

Region 1: inside the hole

$$
\begin{aligned}
\mathbf{B}(\mathbf{r}) &= \frac{\mu_0}{2}\,\mathbf{J}\times\mathbf{r} + \frac{\mu_0}{2}\,\mathbf{J}'\times\mathbf{r}' \\
&= \frac{\mu_0}{2}\,\mathbf{J}\times\mathbf{r} - \frac{\mu_0}{2}\,\mathbf{J}\times\mathbf{r}' \\
&= \frac{\mu_0}{2}\,\mathbf{J}\times(\mathbf{r}-\mathbf{r}') \\
&= \frac{\mu_0}{2}\,\mathbf{J}\times\mathbf{d}.
\end{aligned}
\tag{1}
$$

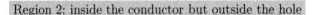

Region 2: inside the conductor but outside the hole

$$
\mathbf{B}(\mathbf{r}) = \frac{\mu_0}{2}\,\mathbf{J}\times\mathbf{r} + \frac{\mu_0}{2}\left(\frac{b}{r'}\right)^2\mathbf{J}'\times\mathbf{r}' = \frac{\mu_0}{2}\,\mathbf{J}\times\left[\mathbf{r} - \left(\frac{b}{r'}\right)^2\mathbf{r}'\right].
\tag{2}
$$

Region 3: outside the conductor

$$
\mathbf{B}(\mathbf{r}) = \frac{\mu_0}{2}\left(\frac{a}{r}\right)^2\mathbf{J}\times\mathbf{r} + \frac{\mu_0}{2}\left(\frac{b}{r'}\right)^2\mathbf{J}'\times\mathbf{r}' = \frac{\mu_0}{2}\,\mathbf{J}\times\left[\left(\frac{a}{r}\right)^2\mathbf{r} - \left(\frac{b}{r'}\right)^2\mathbf{r}'\right].
\tag{3}
$$

Comment

Equation (1) shows that the magnetic field $\mathbf{B}$ is uniform and perpendicular to $\mathbf{d}$ inside the hole. Compare this with the corresponding electrostatic problem discussed in Question 2.7.

Question 4.13

Suppose that the current-carrying conductor with a hole, described in Question 4.12, is oriented with its central axis along the z-axis and origin O' located on the x-axis of Cartesian coordinates. Let $b = \frac{2}{5}a$ and $\mathbf{d} = \left(\frac{1}{2}a, 0, 0\right)$.

(a) Use *Mathematica* to draw the field lines as a function of x/a and y/a.

(b) On the same set of axes, plot graphs of $B(x, 0, 0)$ vs x/a and $B(0, y, 0)$ vs y/a in the interval $[-3, 3]$. For comparative purposes, show also the graph of a conductor with no hole.

Solution

(a) We first derive the Cartesian form of the field by substituting $\mathbf{r} = (x, y, 0)$ and $\mathbf{r}' = (x - \frac{1}{2}a, y, 0)$ in (1)–(3) of Question 4.12. This gives

$$
\mathbf{B} = B_0 \times
\begin{cases}
(0, \frac{1}{2}, 0) & \text{Region 1} \\[2mm]
(-Y, X, 0) - \dfrac{(b^2/a^2)}{(X - \frac{1}{2})^2 + Y^2}(-Y, X - \frac{1}{2}, 0) & \text{Region 2} \\[2mm]
\dfrac{(-Y, X, 0)}{X^2 + Y^2} - \dfrac{(b^2/a^2)}{(X - \frac{1}{2})^2 + Y^2}(-Y, X - \frac{1}{2}, 0) & \text{Region 3}
\end{cases}
\quad , \qquad (1)
$$

where $X = x/a$ and $Y = y/a$ are dimensionless coordinates. Then implementing (1) in the following notebook yields the field[‡] diagram shown on p. 214.

```
In[1]:= a = 5;  b = 2;  d = a/2;  x0 = y0 = 1.5 a;  z = 2/3;

R[X_, Y_] := Sqrt[X^2 + Y^2] ;   Rp[X_, Y_] := Sqrt[(X - 0.5)^2 + Y^2] ;

B1x[X_, Y_] := 0;  B1y[X_, Y_] := 0.5;

B2x[X_, Y_] := -(Y - (b^2/a^2 Y)/Rp[X, Y]^2) ;   B2y[X_, Y_] := (X - (b^2/a^2 (X - 0.5))/Rp[X, Y]^2) ;

B3x[X_, Y_] := -(Y/R[X, Y]^2 - (b^2/a^2 Y)/Rp[X, Y]^2) ;

B3y[X_, Y_] := (X/R[X, Y]^2 - (b^2/a^2 (X - 0.5))/Rp[X, Y]^2) ;

(* Streamlines to be forced thru the points pts1 & pts2 *)
pts1 = {{d - 3 z, 0}, {d - 2 z, 0}, {d - z, 0}, {d, 0}, {d + z, 0},
    {d + 2 z, 0}, {d + 3 z, 0}};
pts2 = {{-0.75, 0}, {-0.2, 0}, {d, 0}, {d - 2 z, Sqrt[b^2 - 4 z^2]},
    {d - z, Sqrt[b^2 - z^2]}, {d, 2}, {d + z, Sqrt[b^2 - z^2]}, {d + 2 z, Sqrt[b^2 - 4 z^2]},
    {d + z, -Sqrt[b^2 - z^2]}, {d + 2 z, -Sqrt[b^2 - 4 z^2]}, {4.5, 0}};
```

[‡]Let $B_0 = 1$ in some arbitrary system of units.

```
In[2]:=  (* RF1, RF2 and RF3 are regions to be excluded from the plot *)
         RF1 = Function[{x, y}, (x - d)² + y² ≤ b²];
         RF2 = Function[{x, y}, (x - d)² + y² ≥ b² && x² + y² ≤ a²];
         RF3 = Function[{x, y}, x² + y² ≥ a²];


         gr1 = StreamPlot[{B1x[x, y], B1y[x, y]}, {x, -x0, x0}, {y, -y0, y0},
            StreamPoints → pts1, RegionFunction → RF1];
         gr2 = StreamPlot[{B2x[x, y], B2y[x, y]}, {x, -x0, x0}, {y, -y0, y0},
            StreamPoints → pts2, RegionFunction → RF2];
         gr3 = StreamPlot[{B3x[x, y], B3y[x, y]}, {x, -x0, x0}, {y, -y0, y0},
            RegionFunction → RF3];

         di0 = Graphics[{Gray, Disk[{0, 0}, a]}];
         di1 = Graphics[{Black, Disk[{0, 0}, a/50]}];
         di2 = Graphics[{White, Disk[{d, 0}, b]}];
         Show[di0, di1, di2, gr1, gr2, gr3, Frame → True]
```

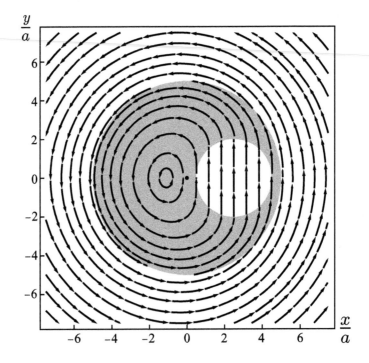

(b) We obtain the graph:

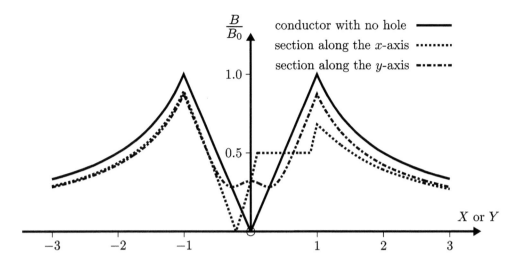

Comment

The position of the neutral point in the field, obtained by putting $B_y = 0$ in $(1)_2$ and solving for X, occurs at $X = -0.222$.

Question 4.14*

(a) Show that the vector potential **A** and magnetic field **B** are related by the equation

$$\oint_c \mathbf{A} \cdot d\mathbf{l} = \int_s \mathbf{B} \cdot d\mathbf{a}. \tag{1}$$

Hint: Start with the definition of magnetic flux $\phi = \int_s \mathbf{B} \cdot d\mathbf{a}$.

(b) A long (assume infinite) cylindrical conductor of radius a carrying current I lies along the z-axis of cylindrical polar coordinates. Suppose that the current density is constant inside the conductor. Use (1) with **B** given by (3) of Question 4.7, to show that

$$A_z = \begin{cases} \dfrac{\mu_0 I}{4\pi}\left(1 - \left[\dfrac{r}{a}\right]^2\right) & \text{for } r \leq a \\[2ex] -\dfrac{\mu_0 I}{2\pi}\ln\left[\dfrac{r}{a}\right] & \text{for } r > a. \end{cases} \tag{2}$$

Hint: It is convenient to choose the zero of potential at the surface of the conductor $(r = a)$.

Solution

(a) Substituting $\mathbf{B} = \nabla \times \mathbf{A}$ in the definition of flux gives

$$\int_s \mathbf{B} \cdot d\mathbf{a} = \int_s (\nabla \times \mathbf{A}) \cdot d\mathbf{a}.$$

Then (1) follows immediately from Stokes's theorem.

(b) Consider the rectangular contour (having sides ℓ and r) whose corners are labelled 1, 2, 3, 4 as shown in the figure below.

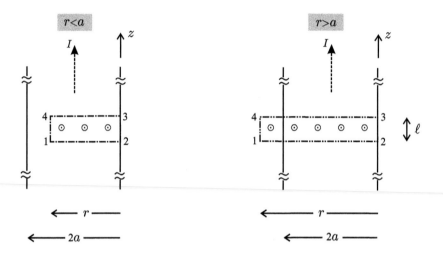

Suppose the origin of coordinates is at corner 2 of the rectangle. The magnetic flux through the contour is $\int_s \mathbf{B} \cdot d\mathbf{a}$, where s is the area of the rectangle and

$$\mathbf{B}(r) = \begin{cases} \dfrac{\mu_0}{2\pi} \dfrac{Ir}{a^2} \hat{\boldsymbol{\theta}} & \text{for } r \leq a \\[3mm] \dfrac{\mu_0}{2\pi} \dfrac{I}{r} \hat{\boldsymbol{\theta}} & \text{for } r > a. \end{cases} \tag{3}$$

From (1) and $d\mathbf{a} = \ell\, dr\, \hat{\boldsymbol{\theta}}$ we obtain

$$\ell \int_0^r B_{\lessgtr}\, dr = \oint_c \mathbf{A} \cdot d\mathbf{l}, \tag{4}$$

where $B_{\lessgtr}$ is the field inside (outside) the conductor. We consider each of the two regions separately.

$r \leq a$

Substituting $(3)_1$ in (4) and integrating yield

$$\frac{\mu_0 I}{4\pi} \frac{r^2 \ell}{a^2} = \oint_c \mathbf{A} \cdot d\mathbf{l}, \tag{5}$$

where the contour integral is

$$\oint_c \mathbf{A} \cdot d\mathbf{l} = \int_1^2 \mathbf{A} \cdot d\mathbf{l} + \int_2^3 \mathbf{A} \cdot d\mathbf{l} + \int_3^4 \mathbf{A} \cdot d\mathbf{l} + \int_4^1 \mathbf{A} \cdot d\mathbf{l}.$$

Now the first and third terms on the right-hand side cancel. Furthermore, symmetry requires that $\mathbf{A}$ is constant and independent of z along path segments 23 and 41. So

$$\oint_c \mathbf{A} \cdot d\mathbf{l} = A_z(0) \int_2^3 dz - A_z(r) \int_1^4 dz$$

$$= [A_z(0) - A_z(r)] \ell. \tag{6}$$

Equating (5) and (6) gives $\dfrac{\mu_0 I}{4\pi} \dfrac{r^2 \ell}{a^2} = [A_z(0) - A_z(r)] \ell$, or

$$A_z(r) = A_z(0) - \frac{\mu_0 I}{4\pi} \frac{r^2}{a^2}. \tag{7}$$

Substituting $r = a$ in (7) and using the boundary condition $A_z(a) = 0$ yield

$$A_z(0) = \frac{\mu_0 I}{4\pi}. \tag{8}$$

Then from (7) and (8) we obtain

$$A_z(r) = \frac{\mu_0 I}{4\pi} - \frac{\mu_0 I}{4\pi} \frac{r^2}{a^2}, \tag{9}$$

which is $(2)_1$.

$r > a$

Proceeding as above, (4) becomes

$$\ell \int_0^a B_< dr + \ell \int_a^r B_> dr = \oint_c \mathbf{A} \cdot d\mathbf{l}. \tag{10}$$

Substituting (3) in (10) and integrating give

$$\frac{\mu_0 I \ell}{4\pi} + \frac{\mu_0 I \ell}{2\pi} \ln \left[\frac{r}{a} \right] = [A_z(0) - A_z(r)] \ell.$$

Now $A_z(0)$ is given by (8). Hence $(2)_2$.

Question 4.15**

Consider a circular loop of wire having radius a centred on the origin of cylindrical polar coordinates. Suppose there is a current I in the loop.

(a) Use (8) of Question 4.5 to show that the vector potential at an arbitrary point $P(r, \theta, z)$ is

$$A_\theta = \frac{\mu_0 I a}{2\pi} \int_0^\pi \frac{\cos\theta' d\theta'}{\sqrt{r^2 + z^2 + a^2 - 2ar\cos\theta'}}. \tag{1}$$

(b) Express (1) in the form

$$A_\theta = \frac{\mu_0 I}{2\pi} \frac{(a^2 + r^2 + z^2)}{r\sqrt{(a+r)^2 + z^2}} \left\{ K\left(\frac{2\gamma^2}{1+\gamma^2}\right) - (1+\gamma^2)E\left(\frac{2\gamma^2}{1+\gamma^2}\right) \right\}, \tag{2}$$

where the functions E and K are elliptic integrals (in calculations we will use the EllipticE and EllipticK functions as they are defined in *Mathematica*) and

$$\gamma^2 = \frac{2ar}{r^2 + z^2 + a^2}. \tag{3}$$

(c) Use *Mathematica* to calculate $\mathbf{B}(r, z)$ and then plot this magnetic field in the vicinity of the loop. Take $B_0 = \dfrac{\mu_0 I}{2a} = 1\,\mathrm{T}$.

Solution

(a) Because of the cylindrical symmetry, it is convenient to choose P in the plane where $\theta = 0$. The current density has a component only in the $\hat{\theta}$ direction. Thus

$$J_\theta = I\delta(r' - a)\,\delta(z'), \tag{4}$$

which can be written as

$$\mathbf{J} = J_\theta(-\hat{\mathbf{x}}\sin\theta' + \hat{\mathbf{y}}\cos\theta'), \tag{5}$$

because of (III) of Appendix D. Substituting (4) and (5) in (8) of Question 4.5 gives

$$\mathbf{A} = \frac{\mu_0 I}{4\pi} \int_v \frac{\delta(r' - a)\,\delta(z')(-\hat{\mathbf{x}}\sin\theta' + \hat{\mathbf{y}}\cos\theta')}{|\mathbf{r} - \mathbf{r}'|}\,dv',$$

where $dv' = r'dr'd\theta'dz'$ and $|\mathbf{r} - \mathbf{r}'| = \sqrt{r^2 + r'^2 - 2rr'\cos\theta' + (z - z')^2}$. So

$$\mathbf{A} = \frac{\mu_0 I}{4\pi} \int_{-\infty}^{\infty} \int_0^{2\pi} \int_0^{\infty} \frac{\delta(r'-a)\,\delta(z')(-\hat{\mathbf{x}}\sin\theta' + \hat{\mathbf{y}}\cos\theta')}{\sqrt{r^2 + r'^2 - 2rr'\cos\theta' + (z-z')^2}}\,r'\,dr'\,d\theta'\,dz'$$

$$= 2 \times \frac{\mu_0 I}{4\pi} \int_{-\infty}^{\infty} \int_0^{\pi} \int_0^{\infty} \frac{\hat{\mathbf{y}}\,\delta(r'-a)\,\delta(z')\cos\theta'}{\sqrt{r^2 + r'^2 - 2rr'\cos\theta' + (z-z')^2}}\,r'\,dr'\,d\theta'\,dz',$$

where the factor of two and limit change arise because the azimuthal integration is symmetric about $\theta' = 0$. Then (1) follows immediately because of the sifting property of delta functions.

(b) Expressed in terms of γ, the vector potential is

$$A_\theta = \frac{\mu_0 I a}{2\pi} \frac{1}{\sqrt{r^2 + z^2 + a^2}} \int_0^\pi \frac{\cos\theta'\,d\theta'}{\sqrt{1 - \gamma^2\cos\theta'}}. \tag{6}$$

Evaluating the integral in (6) with the aid of *Mathematica* yields (2), after some rearrangement of terms. See `cell 1` in the notebook on p. 220.

(c) In cylindrical polar coordinates, the components of $\nabla \times \mathbf{A}$ are

$$\left.\begin{aligned} B_r &= -\frac{\partial A_\theta}{\partial z} \\[2mm] B_\theta &= 0 \\[2mm] B_z &= \frac{1}{r}\frac{\partial}{\partial r}(rA_\theta) \end{aligned}\right\}. \tag{7}$$

Now because the geometry is cylindrically symmetric, it is convenient to plot this field in the xz-plane where $B_x = B_r$ and $B_y = B_\theta = 0$. Furthermore, the computation is simplified if we introduce the dimensionless coordinates $X = x/a$, $Y = y/a$, $Z = z/a$ and $R = r/a$. Then

$$\left.\begin{aligned} B_x &= -\frac{B_0}{\pi}\frac{\partial}{\partial Z}\left[\frac{1}{X}\sqrt{\frac{(1+X^2+Z^2)^2}{(1+X)^2+Z^2}}\left\{K\left(\frac{2\gamma^2}{1+\gamma^2}\right) - (1+\gamma^2)\,E\left(\frac{2\gamma^2}{1+\gamma^2}\right)\right\}\right] \\[3mm] B_z &= \frac{B_0}{\pi}\frac{1}{X}\frac{\partial}{\partial X}\left[\sqrt{\frac{(1+X^2+Z^2)^2}{(1+X)^2+Z^2}}\left\{K\left(\frac{2\gamma^2}{1+\gamma^2}\right) - (1+\gamma^2)\,E\left(\frac{2\gamma^2}{1+\gamma^2}\right)\right\}\right] \end{aligned}\right\}, \tag{8}$$

where γ^2 in (8) is now understood to equal $\dfrac{2|X|}{1+X^2+Z^2}$, putting $r = x$ in (3).

Implementing (8) in the notebook below produces the field diagram shown on p. 220.

In[1]:= $\gamma[r\_, z\_] := \left(\dfrac{2 a \sqrt{r^2}}{r^2 + z^2 + a^2}\right)^{1/2}$

$\dfrac{a}{2\pi\left(r^2 + z^2 + a^2\right)^{1/2}}$ Integrate$\left[\dfrac{\text{Cos}[\theta p]}{\sqrt{1 - \gamma[r, z]^2 \, \text{Cos}[\theta p]}}, \{\theta p, 0, \pi\},\right.$

   Assumptions $\to \{0 \le \gamma[r, z] < 1 \,\&\&\, r \ge 0\}\Big]$

In[2]:= $\gamma[u\_, v\_] := \left(\dfrac{2 \sqrt{u^2}}{1 + u^2 + v^2}\right)^{1/2}$

BX[X_, Z_] = $-D\left[\dfrac{1}{\pi u}\left(\dfrac{\left(1 + u^2 + v^2\right)^2}{(1 + u)^2 + v^2}\right)^{1/2}\left(\text{EllipticK}\left[\dfrac{2\,\gamma[u, v]^2}{1 + \gamma[u, v]^2}\right] - \right.\right.$

   $\left.\left.\left(1 + \gamma[u, v]^2\right) \text{EllipticE}\left[\dfrac{2\,\gamma[u, v]^2}{1 + \gamma[u, v]^2}\right]\right), v\right] \,/.\, \{u \to X, v \to Z\};$

BZ[X_, Z_] = $\dfrac{1}{u} D\left[u \dfrac{1}{\pi u}\left(\dfrac{\left(1 + u^2 + v^2\right)^2}{(1 + u)^2 + v^2}\right)^{1/2}\left(\text{EllipticK}\left[\dfrac{2\,\gamma[u, v]^2}{1 + \gamma[u, v]^2}\right] - \right.\right.$

   $\left.\left.\left(1 + \gamma[u, v]^2\right) \text{EllipticE}\left[\dfrac{2\,\gamma[u, v]^2}{1 + \gamma[u, v]^2}\right]\right), u\right] \,/.\, \{u \to X, v \to Z\};$

In[3]:= max = 3.8; min = -max;
   StreamPlot$[\{$BX[X, Z], BZ[X, Z]$\}, \{$X, min, max$\}, \{$Z, min, max$\},$
   StreamPoints $\to$ 100, StreamStyle $\to$ {Directive[Black, Thickness[0.005]]},
   FrameStyle $\to$ Directive[Black, Thickness[0.00575]],
   LabelStyle $\to$ Directive[FontSize $\to$ 14, FontFamily $\to$ "Times"],
   Prolog $\to$ {Gray, Disk[{1, 0}, .1], Disk[{-1, 0}, .1]}]

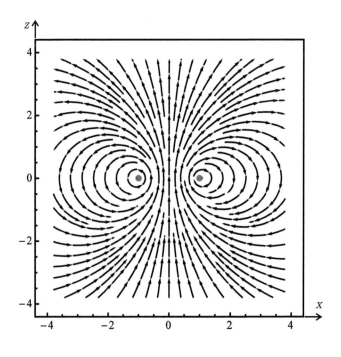

Question 4.16

Two plane circular coils of radius a have a common axis of symmetry (the z-axis of cylindrical polar coordinates, say) and are centred at $z = \pm d$ as shown. Each coil comprises N turns and carries a current I in the same direction. The axial magnetic field between the coils, which follows from (1) of Question 4.8 and the principle of superposition, is

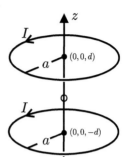

$$B(z) = B_0 \left\{ \frac{a^3}{[a^2 + (z - d)^2]^{3/2}} + \frac{a^3}{[a^2 + (z + d)^2]^{3/2}} \right\}, \quad (1)$$

where now $B_0 = \mu_0 N I / 2a$.

(a) In the vicinity of the origin $B(z)$ may be replaced by the series expansion

$$B(z) = \sum_{n=0}^{\infty} \frac{d^n B(z)}{dz^n} \frac{z^n}{n!}, \quad (2)$$

where all the derivatives $d^n B / dz^n$ are evaluated at $z = 0$. Using a symmetry argument, decide which values of n are absent from (2).

(b) With the aid of *Mathematica*, expand (1) about $z = 0$ and establish the value of d/a which produces the most nearly-uniform axial magnetic field.

(c) For the value of d/a established in (b) above, plot a graph of $B(z)/B_0$ vs z/a. Show also the separate contribution made by each coil to the field.

Solution

(a) Since this current configuration is symmetric about the origin, the magnetic field is an even function of z. That is, $B(z) = B(-z)$ and therefore all *odd* values of n in (2) are vetoed.

(b) Using *Mathematica*'s `Series` function we obtain

$$B(z) = \frac{2a^3 B_0}{(a^2 + d^2)^{3/2}} \left[1 - \frac{3 (a^2 - 4d^2)}{2 (a^2 + d^2)^4} z^2 + \frac{15 (a^4 - 12a^2 d^2 + 8d^4)}{8 (a^2 + d^2)^4} z^4 - \cdots \right]. \quad (3)$$

The choice $d = \frac{1}{2} a$ eliminates the quadratic term in z and the remainder

$$B(z) = \frac{16 B_0}{5\sqrt{5}} \left[1 - \frac{144}{125} \left(\frac{z}{a} \right)^4 + \cdots \right] \quad (4)$$

is an almost-uniform field near the origin.

(c) We obtain the graph:

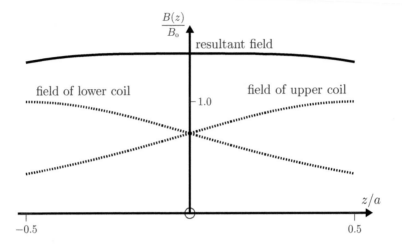

Comment

With $d = \frac{1}{2}a$, the arrangement of coils described above is known as a Helmholtz pair. The image alongside shows a unit which is marketed commercially for teaching laboratories. Helmholtz coils are often used in experiments where uniform magnetic fields are required in open and relatively unconfined spaces.[‡]

Question 4.17

Consider the coil configuration described in Question 4.16, but now suppose that the currents circulate in *opposite* directions. Show that the axial magnetic field is given by

$$B(z) = B_0 \left\{ \frac{a^3}{[a^2 + (z - d)^2]^{3/2}} - \frac{a^3}{[a^2 + (z + d)^2]^{3/2}} \right\}. \tag{1}$$

(a) Again using a symmetry argument, decide which values of n are absent from the series expansion (2) of Question 4.16.

(b) Expand (1) about $z = 0$ (using *Mathematica*) and establish the value of d/a which produces an almost-uniform axial field *gradient*.

(c) For the value of d/a established in (b) above, plot a graph of $B(z)/B_0$ vs z/a.

[‡]Solenoids also produce magnetic fields that are uniform away from their ends. However, the working area near the centre of a current-carrying solenoid is often relatively inaccessible.

Solution

(a) Here the magnetic field is an odd function of z: $B(z) = -B(-z)$. This requires that all *even* values of n in expansion (2) of Question 4.16 are vetoed.

(b) Using *Mathematica's* **Series** function we obtain

$$B(z) = \frac{6da^3 B_0}{(a^2 + d^2)^{5/2}} \left[z - \frac{5(3a^2 - 4d^2)}{6(a^2 + d^2)^2} z^3 + \frac{21(5a^4 - 20a^2 d^2 + 8d^4)}{24(a^2 + d^2)^4} z^5 - \cdots \right].$$

The choice $d = \frac{1}{2}\sqrt{3}a$ eliminates the cubic term in z and the remainder

$$B(z) = \frac{96}{49}\sqrt{\frac{3}{7}} B_0 \left[\left(\frac{z}{a}\right) - \frac{176}{343}\left(\frac{z}{a}\right)^5 + \cdots \right], \tag{2}$$

gives a nearly-uniform field gradient in the vicinity of the origin.

(c) We obtain the graph:

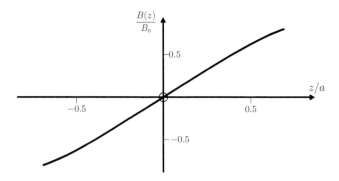

Comment

The arrangement described above with $d = \frac{1}{2}\sqrt{3}a$ is known as an anti-Helmholtz pair. The nearly-uniform axial field gradient makes coils like these particularly useful in a variety of experiments (for example, magneto-optical traps).

Question 4.18*

In the figure alongside, c is a simple closed loop and P is an arbitrary point in space. The solid angle Ω subtended by c at P is given by

$$\Omega = \int_s \frac{\mathbf{R} \cdot d\mathbf{a}}{R^3}, \tag{1}$$

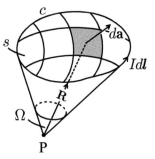

where s is *any* surface spanning c and $\mathbf{R}$ is the vector from P to area element $d\mathbf{a}$ (shown shaded). If P undergoes an infinitesimal displacement $\delta\mathbf{r}$, the change in solid angle $d\Omega = \nabla\Omega \cdot \delta\mathbf{r}$.

(a) Show that

$$\boldsymbol{\nabla}\Omega = -\oint_c \frac{dl \times \hat{\mathbf{R}}}{R^3}.\qquad (2)$$

Hint: For the purposes of this calculation, displacing P by $\delta\mathbf{r}$ is equivalent to displacing the circuit by $-\delta\mathbf{r}$.

(b) Now suppose c is a conducting circuit carrying a current I. Show that the magnetic scalar potential Φ_{m} at P is related to Ω by

$$\Phi_{\mathrm{m}} = -\frac{I\Omega}{4\pi}.\qquad (3)$$

Hint: Use (12) of Question 4.5.

Solution

(a) The displacement $-\delta\mathbf{r}$ of c generates a ribbon-like surface (shown shaded in the adjacent figure) whose area we now calculate. Consider the parallelogram spanned by the vectors $-\delta\mathbf{r}$ and dl. Its area is $-\delta\mathbf{r} \times dl$. The projection of this area onto the surface of a sphere of radius R and centred at P is $-\delta\mathbf{r}\times dl\cdot\hat{\mathbf{R}} = -\delta\mathbf{r}\cdot dl\times\hat{\mathbf{R}}$ (in the last step we use the cyclic property of a scalar triple product). Therefore the change in solid angle $\delta\Omega$ subtended by the parallelogram at P is $\delta\Omega = (-\delta\mathbf{r}\cdot dl \times \hat{\mathbf{R}})\div R^2$, and the change in solid angle $d\Omega$ subtended by the circuit at P—obtained by summing all the $\delta\Omega$s around the ribbon—is

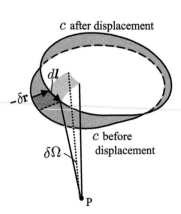

c after displacement

c before displacement

$$d\Omega = \oint_c \delta\Omega = \oint_c \frac{-\delta\mathbf{r}\cdot dl \times \hat{\mathbf{R}}}{R^2} = -\delta\mathbf{r}\cdot\oint_c \frac{dl \times \hat{\mathbf{R}}}{R^2}.\qquad (4)$$

Comparing (4) with $d\Omega = \delta\mathbf{r}\cdot\boldsymbol{\nabla}\Omega$ yields (2).

(b) In the Biot–Savart law we replace $\mathbf{R}$ in $(7)_2$ of Question 4.4 with $-\mathbf{R}$ (because here $\mathbf{R}$ is the vector *from* the field point *to* the source). This gives

$$\mathbf{B} = -\frac{\mu_0 I}{4\pi}\oint_c \frac{dl \times \mathbf{R}}{R^3}.\qquad (5)$$

Substituting (2) in (5) yields

$$\mathbf{B} = \frac{\mu_0 I}{4\pi}\boldsymbol{\nabla}\Omega.\qquad (6)$$

Comparing (6) with $\mathbf{B} = -\mu_0\boldsymbol{\nabla}\Phi_{\mathrm{m}}$ gives (3).

Comments

(i) There is a sign convention implicit in (1). The direction of $d\mathbf{a}$ is taken to bear the same relationship to the sense of circulation around c as the curling fingers of the right hand have with the thumb. The direction of the magnetic field at P is determined by the sign of $\nabla\Omega$, as is evident from (6).[‡]

(ii) The solid angle subtended by a point P inside a closed surface of any shape is 4π. So the solid angle subtended at the centre of a cube by any one of its six faces is $4\pi/6 = 2\pi/3$. If the cube is now compressed in such a way that the area of one pair of opposite faces approaches infinity whilst the area of the remaining four faces tends to zero, then (with P at the centre of this flat rectangular cuboid) the solid angle subtended by either of the large faces approaches $4\pi/2 = 2\pi$. It follows that the solid angle subtended by an *arbitrary* surface s of finite shape is also 2π in the limit P$\rightarrow s$ (any surface is effectively an infinite plane when P is close enough to it). Because of the sign convention outlined in (i) above, it is clear that Ω changes *discontinuously* from 2π to -2π (or vice versa) when P crosses s.

(iii) The solid angle subtended by c at P is not uniquely defined. Solid angles differing from each other by integral multiples of 4π are equivalent, in much the same way as the plane angles $\theta + 2m\pi$ (where m is an integer) correspond to multiple rotations about an axis normal to the plane. This of course has no effect on $\mathbf{B}$ because the gradient is independent of an additive constant. But it does lead to a multi-valued magnetic scalar potential (see (3)). This is quite unlike the electrostatic potential Φ which, although not unique, is always single-valued.

(iv) The magnetic scalar potential is also unlike Φ in another significant way: Φ_m plays no role in determining either the work required to move an electric charge or the energy stored in a magnetostatic field (see Question 5.15). This is not surprising given that magnetostatic forces cannot change the kinetic energy of an electric charge (see (7) of Question 4.1).

(v) In Question 4.19 we use (3) to determine the magnetic field on the axis of a circular current loop.

Question 4.19

Consider a plane circular conducting loop of radius a carrying a current I. Suppose Cartesian coordinates are chosen with the z-axis normal to the plane of the loop and passing through its centre, taken to be the origin.

(a) Show that the solid angle subtended by the loop at an arbitrary point P on the z-axis is

$$\Omega = 2\pi\left[\frac{z}{\sqrt{a^2 + z^2}} \mp 1\right],\tag{1}$$

where the upper (lower) sign is for $z > 0$ ($z < 0$).

[‡]Clearly, the direction of $\mathbf{B}$ must change when the current in c reverses.

(b) Use (1) to calculate the magnetic scalar potential $\Phi_m(z)$.

(c) Hence determine the magnetic field $\mathbf{B}(z)$.

Solution

(a) Consider the annular ring of width dr and area $d\mathbf{a} = 2\pi r\, dr\, \hat{\mathbf{z}}$ shown shaded in the figure below. The differential solid angle $d\Omega$ subtended by $d\mathbf{a}$ at P is $\hat{\mathbf{R}} \cdot d\mathbf{a}/R^2 = 2\pi r\, dr\, \hat{\mathbf{R}} \cdot \hat{\mathbf{z}}/R^2 = -2\pi r z\, dr/R^3$ since $\hat{\mathbf{R}} \cdot \hat{\mathbf{z}} = -z/R$ is positive (negative) for $z < 0$ ($z > 0$). Thus

$$\Omega = \int d\Omega = -2\pi z \int_0^a \frac{r\, dr}{(r^2 + z^2)^{3/2}}.$$

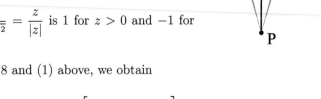

Making the change of variable $u^2 = r^2 + z^2$ yields

$$\Omega = -2\pi z \int_{\sqrt{z^2}}^{\sqrt{a^2+z^2}} \frac{du}{u^2} = 2\pi \left[\frac{z}{\sqrt{a^2 + z^2}} - \frac{z}{\sqrt{z^2}} \right],$$

which is (1) because $\dfrac{z}{\sqrt{z^2}} = \dfrac{z}{|z|}$ is 1 for $z > 0$ and -1 for $z < 0$.

(b) From (3) of Question 4.18 and (1) above, we obtain

$$\Phi_m(z) = -\tfrac{1}{2}I \left[\frac{z}{\sqrt{a^2 + z^2}} \mp 1 \right]. \tag{2}$$

(c) Substituting (2) in (12) of Question 4.5 gives

$$\mathbf{B}(z) = -\mu_0 \frac{\partial \Phi_m}{\partial z} \hat{\mathbf{z}} = \frac{\mu_0 I}{2} \frac{a^2}{(a^2 + z^2)^{3/2}} \hat{\mathbf{z}}. \tag{3}$$

Comments

(i) The graph on p. 227 is a plot of (1): it reveals the discontinuity in Ω discussed in Comment (ii) of Question 4.18. The magnetic field $\mathbf{B}(z)$, which is proportional to the gradient of $\Omega(z)$, is continuous and positive everywhere.

(ii) As one would expect, (3) agrees with (1) of Question 4.8 determined from the Biot–Savart law.

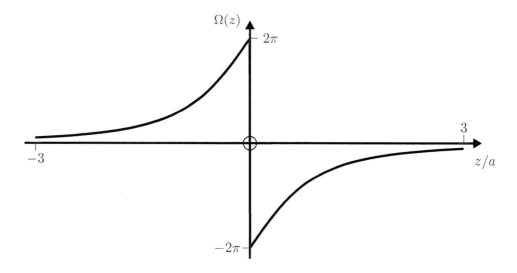

Question 4.20**

(a) Consider an arbitrary distribution of currents having density $\mathbf{J}(\mathbf{r}')$. Let v be a spherical region of space of radius r_0 centred on an origin O. The average magnetic field $\overline{\mathbf{B}}$ inside v is given by

$$\overline{\mathbf{B}} \;=\; \frac{1}{\frac{4}{3}\pi r_0^3}\int_v \mathbf{B}(\mathbf{r})\,dv \;=\; \overline{\mathbf{B}}_{\text{int}} + \overline{\mathbf{B}}_{\text{ext}}, \tag{1}$$

where $\overline{\mathbf{B}}_{\text{int}}$ is the average field due to all the currents inside v and $\overline{\mathbf{B}}_{\text{ext}}$ is the average field due to all the currents outside v. Prove that

$$\left.\begin{aligned}
\overline{\mathbf{B}}_{\text{int}} &= \frac{\mu_0}{4\pi}\frac{2\mathbf{m}}{r_0^3} \\[2ex]
\overline{\mathbf{B}}_{\text{ext}} &= -\frac{\mu_0}{4\pi}\int_V \frac{\mathbf{J}(\mathbf{r}')\times\mathbf{r}'}{r'^3}\,dv'
\end{aligned}\right\}, \tag{2}$$

where $\mathbf{m}$ is the magnetic dipole moment (about O) of the internal currents, and V is the region of space excluding v where $\mathbf{J}(\mathbf{r}') \neq 0$.

(b) A circular conducting loop of radius a lying in the xy-plane of Cartesian coordinates and centred on O carries a steady current I. Use (2) to determine the average magnetic field $\overline{\mathbf{B}}$ inside the spherical region referred to above, assuming that

☞ $a < r_0$, and

☞ $a > r_0$.

Solution

(a) Substituting $\mathbf{B} = \nabla \times \mathbf{A}$ in (1) and using (2) of Question 1.22 give

$$\overline{\mathbf{B}} \; = \; \frac{1}{\frac{4}{3}\pi r_0^3} \int_v \mathbf{B}(\mathbf{r})\, dv \; = \; \frac{1}{\frac{4}{3}\pi r_0^3} \int_v \left[\nabla \times \mathbf{A}(\mathbf{r}) \right] dv \; = \; -\frac{1}{\frac{4}{3}\pi r_0^3} \oint_s \mathbf{A}(\mathbf{r}) \times d\mathbf{a}. \quad (3)$$

Now $\mathbf{A}(\mathbf{r}) = \dfrac{\mu_0}{4\pi} \displaystyle\int_v \dfrac{\mathbf{J}(\mathbf{r}')}{|\mathbf{r} - \mathbf{r}'|}\, dv'$, and so (3) becomes

$$\overline{\mathbf{B}} \; = \; -\frac{1}{\frac{4}{3}\pi r_0^3} \frac{\mu_0}{4\pi} \int_v \mathbf{J}(\mathbf{r}') \times \left(\oint_s \frac{d\mathbf{a}}{|\mathbf{r} - \mathbf{r}'|} \right) dv' \quad (4)$$

where the order of integration has been interchanged. But the surface integral in (4) is given by (8) of Question 1.26, and so

$$\left. \begin{aligned} \overline{\mathbf{B}}_{\text{int}} &= -\frac{\mu_0}{4\pi} \frac{1}{r_0^3} \int_v \mathbf{J}(\mathbf{r}') \times \mathbf{r}'\, dv' \\[2mm] \overline{\mathbf{B}}_{\text{ext}} &= -\frac{\mu_0}{4\pi} \int_V \frac{\mathbf{J}(\mathbf{r}') \times \mathbf{r}'}{r'^3}\, dv' \end{aligned} \right\}, \quad (5)$$

which proves (1), because $\mathbf{m} = \frac{1}{2} \displaystyle\int_v \mathbf{r}' \times \mathbf{J}(\mathbf{r}')\, dv'$ by definition $\big($see $(2)_1$ of Question 4.24$\big)$.

(b) ☞ $a < r_0$

Since the loop is inside the sphere, $\overline{\mathbf{B}}$ is given by $(2)_1$ with $\mathbf{m} = I\pi a^2 \hat{\mathbf{z}}$ and

$$\overline{\mathbf{B}}_{\text{int}} \; = \; \frac{\mu_0 I a^2 \, \hat{\mathbf{z}}}{2 r_0^3}.$$

☞ $a > r_0$

We begin by using delta functions in spherical polar coordinates to confine the current to a ring of radius a.[‡] Then with $\mathbf{J}(\mathbf{r}') = \dfrac{I}{a}\, \delta(\cos\theta')\, \delta(r' - a)\, \hat{\boldsymbol{\phi}}$ we obtain $\mathbf{J} \times \mathbf{r}' = \dfrac{Ir'}{a}\, \delta(\cos\theta')\, \delta(r' - a)\, \hat{\boldsymbol{\theta}}$. Substituting this last result in $(2)_2$ yields

$$\overline{\mathbf{B}}_{\text{ext}} \; = \; -\frac{\mu_0 I}{4\pi a} \int_0^{2\pi} \int_{-1}^{1} \int_{r_0}^{\infty} \left[(\hat{\mathbf{x}} \cos\theta' \cos\phi' + \hat{\mathbf{y}} \cos\theta' \sin\phi' - \hat{\mathbf{z}} \sin\theta') \right] \times$$
$$r' \delta(\cos\theta')\, \delta(r' - a)\, \frac{r'^2 dr'\, d(\cos\theta')\, d\phi'}{r'^3}, \quad (6)$$

[‡]Use the technique of Question 2.2. Here $\int_v \mathbf{J}\, dv' = \oint_c I\, d\mathbf{l} = 2\pi I a \hat{\boldsymbol{\phi}}$. Hence $2\pi I a = \alpha \int_0^{2\pi} \int_{-1}^{1} \times \int_0^{\infty} \delta(\cos\theta')\, \delta(r' - a)\, r'^2 dr'\, d(\cos\theta')\, d\phi' = 2\pi \alpha a^2 \Rightarrow \alpha = I/a$. So $\mathbf{J}(\mathbf{r}') = \dfrac{I}{a}\, \delta(\cos\theta')\, \delta(r' - a)\, \hat{\boldsymbol{\phi}}$.

where $\hat{\boldsymbol{\theta}}$ and dv' are given by $(V)_2$ and $(VIII)_3$ of Appendix C respectively. Now the integrals in (6) involving $\cos\phi'$ and $\sin\phi'$ are both obviously zero, and so

$$\overline{\mathbf{B}}_{\text{ext}} = \frac{\mu_0 I}{2a} \int_{-1}^{1} \int_{r_0}^{\infty} \sin\theta' \, \delta(\cos\theta') \, \delta(r'-a) \, dr' \, d(\cos\theta') \, \hat{\mathbf{z}}. \tag{7}$$

Applying the sifting property of delta functions gives $\int_{r_0}^{\infty} \delta(r'-a) \, dr' = 1$, since $r_0 < a$ and $\int_{-1}^{1} \sqrt{1 - \cos^2\theta'} \, \delta(\cos\theta') \, d(\cos\theta') = 1$. Hence

$$\overline{\mathbf{B}}_{\text{ext}} = \frac{\mu_0 I}{2a} \hat{\mathbf{z}}.$$

Comment

Equation $(2)_2$ is the Biot–Savart law $\big(\text{see } (10)_2 \text{ of Question 4.4 with } \mathbf{r} = 0\big)$. For the dipole of (b), we see that $\overline{\mathbf{B}}_{\text{ext}}$ has the following simple interpretation: the average magnetic field inside the sphere due to the external current loop equals the field which it produces at O.

Question 4.21**

$\big($This question and its solution are based on §II and §III of Ref. [3].$\big)$

Consider a point magnetic dipole located at the origin O of Cartesian coordinates with the z-axis chosen along the direction of $\mathbf{m}$.

(a) Express the field of this dipole in the form

$$\mathbf{B}(r,\theta,\phi) = \frac{\mu_0}{4\pi} \frac{m\big[3\sin\theta\cos\theta(\hat{\mathbf{x}}\cos\phi + \hat{\mathbf{y}}\sin\phi) + (3\cos^2\theta - 1)\hat{\mathbf{z}}\big]}{r^3}, \tag{1}$$

where r, θ and ϕ are spherical polar coordinates.

(b) The average magnetic field inside a sphere of radius r_0 centred on O is

$$\overline{\mathbf{B}} = \frac{1}{\frac{4}{3}\pi r_0^3} \int_v \mathbf{B}(r,\theta,\phi) \, dv. \tag{2}$$

With $\mathbf{B}(r,\theta,\phi)$ given by (1) and using dv for spherical polar coordinates, show that—surprisingly—$\overline{\mathbf{B}}$ is indeterminate. $\Big($Recall that the correct average field calculated in Question 4.20 is $\dfrac{\mu_0}{4\pi}\dfrac{2\mathbf{m}}{r_0^3}.\Big)$

(c) The difficulty exposed in (b) can be resolved as follows: choose an infinitesimal sphere (radius ϵ) centred on O and take

$$\mathbf{B}(\mathbf{r}, \theta, \phi) = \begin{cases} \text{using (1) above} & \text{for } r \geq \epsilon \text{ (region I)}, \\[2mm] \dfrac{2\mu_0}{3}\, \mathbf{m}\, \delta(\mathbf{r}) & \text{for } r < \epsilon \text{ (region II)}. \end{cases} \tag{3}$$

Show that (3) now gives the correct average field.[‡]

Solution

(a) The result follows immediately when $\hat{\mathbf{r}}$ in (2) of Question 4.10 is expressed in terms of the Cartesian unit vectors (see (V) of Appendix C).

(b) Taking $dv = r^2 \sin\theta\, dr\, d\theta\, d\phi$ and evaluating the angular integrals give zero. But the integral of $1/r$ between 0 and r_0 is infinite, and so the result is indeterminate.

(c) With $\mathbf{B}(\mathbf{r}, \theta, \phi)$ given by (3), we obtain from (2)

$$\overline{\mathbf{B}} = \frac{1}{\frac{4}{3}\pi r_0^3}\left[\int_{\mathrm{I}} \mathbf{B}\, dv + \int_{\mathrm{II}}\left(\frac{2\mu_0}{3}\, \mathbf{m}\, \delta(\mathbf{r})\right) dv\right]. \tag{4}$$

The first term on the right-hand side of (4) is now unambiguously zero, since $\ln r_0/\epsilon$ remains finite (the angular integrals are zero as before). The second term gives the expected answer because $\displaystyle\int_{\mathrm{II}} \delta(\mathbf{r})\, dv = 1$ (see Appendix E).

Comments

(i) The calculation in (b) fails because the correct dipole field is given by (8), and not (2), of Question 4.10. This is an example where the delta-function contribution to the field plays a crucial role.

(ii) The entire contribution to the average magnetic field in a spherical region centred on a point dipole arises from the delta-function term at the origin. Although this term is infinite, its integral over a finite region of space is not.

(iii) The delta-function term in the interaction energy of two magnetic dipoles accounts for the hyperfine splitting in the spectrum of atomic hydrogen.[3]

(iv) A similar analysis may also be carried out for the field of a point electric dipole. See Question 2.12.

[‡]See Comment (iii) of Question 4.10 if you are puzzled by the odd-looking equation $(3)_2$.

Question 4.22

Consider a particle having mass M and charge q moving in the field of a magnetic dipole located at the origin O of Cartesian coordinates and having moment $\mathbf{m} = m_0 \hat{\mathbf{z}}$.

(a) Show that

$$\mathbf{B} = B_0 \frac{(3xz, \, 3yz, \, -x^2 - y^2 + 2z^2) r_0^3}{r^5} \tag{1}$$

where $B_0 = \dfrac{\mu_0}{4\pi} \dfrac{m_0}{r_0^3}$ is the field, in the equatorial plane of the dipole, at a distance r_0 from O.

(b) Show that the equation of motion of the particle has components

$$\left.\begin{aligned}
(x^2 + y^2 + z^2)^{5/2} \ddot{x} + \omega\big[(x^2 + y^2 - 2z^2)\dot{y} + 3yz\dot{z}\big] &= 0 \\
(x^2 + y^2 + z^2)^{5/2} \ddot{y} - \omega\big[(x^2 + y^2 - 2z^2)\dot{x} + 3xz\dot{z}\big] &= 0 \\
(x^2 + y^2 + z^2)^{5/2} \ddot{z} + \omega\big[3z(x\dot{y} - y\dot{x})\big] &= 0
\end{aligned}\right\}, \tag{2}$$

where $\omega = |q|B_0/M$.

(c) Hence prove that the kinetic energy of the particle is conserved.

Solution

(a) The result follows immediately from (2) of Question 4.10, since $\mathbf{r} = (x, y, z)$ and $\mathbf{m} \cdot \mathbf{r} = m_0 z$.

(b) Newton's second law and $\mathbf{F} = q\mathbf{v} \times \mathbf{B}$ give

$$M(\ddot{x}, \ddot{y}, \ddot{z}) = \frac{qB_0}{r^5}(\dot{x}, \dot{y}, \dot{z}) \times (3xz, \, 3yz, \, -x^2 - y^2 + 2z^2)$$

which has the component equations:

$$\left.\begin{aligned}
r^5 M\ddot{x} &= -qB_0\big[(x^2 + y^2 - 2z^2)\dot{y} + 3yz\dot{z}\big] \\
r^5 M\ddot{y} &= -qB_0\big[(x^2 + y^2 - 2z^2)\dot{x} + 3xz\dot{z}\big] \\
r^5 M\ddot{z} &= -qB_0 \, 3z(x\dot{y} - y\dot{x})
\end{aligned}\right\}. \tag{3}$$

Equation (2) follows immediately from (3) and the definition of ω (for positive q).

(c) By definition $E_k = \frac{1}{2}m(\dot{x}^2 + \dot{y}^2 + \dot{z}^2)$, and so $\dot{E}_k = m(\dot{x}\ddot{x} + \dot{y}\ddot{y} + \dot{z}\ddot{z})$. Using (3) to eliminate $\ddot{x}$, $\ddot{y}$ and $\ddot{z}$ gives $\dot{E}_k = 0$, which proves the result.

Comments

(i) Since magnetostatic forces do no work $\big($see (7) of Question 4.1$\big)$, the kinetic energy of the particle is necessarily constant.

(ii) In Question 4.23, we consider a numerical solution of the equation of motion.

Question 4.23**

The equation of motion of the charged particle described in Question 4.22 may be readily expressed as

$$
\left.\begin{array}{l}
(X^2 + Y^2 + Z^2)^{5/2}\ddot{X} + (X^2 + Y^2 - 2Z^2)\dot{Y} + 3YZ\dot{Z} = 0 \\
(X^2 + Y^2 + Z^2)^{5/2}\ddot{Y} - (X^2 + Y^2 - 2Z^2)\dot{X} - 3XZ\dot{Z} = 0 \\
(X^2 + Y^2 + Z^2)^{5/2}\ddot{Z} + 3Z(X\dot{Y} - Y\dot{X}) = 0
\end{array}\right\} , \tag{1}
$$

where the dimensionless coordinates X, Y, Z and τ are defined as follows: $X = x/r_0$, $Y = y/r_0$, $Z = z/r_0$ and $\tau = \omega t$.‡ Suppose that the initial conditions are

$$
\left.\begin{array}{l}
X(0) = 1; \quad \dot{X}(0) = 0.005 \\
Y(0) = 1; \quad \dot{Y}(0) = 0.005 \\
Z(0) = 0; \quad \dot{Z}(0) = 0.010
\end{array}\right\} . \tag{2}
$$

(a) Write a *Mathematica* notebook to solve (1) for the initial conditions given in (2), then plot graphs of $X(\tau)$, $Y(\tau)$ and $Z(\tau)$ for $0 \leq \tau \leq \tau_{\max}$. (For each graph, choose a value of $\tau_{\max}$ which gives a suitable representation of the particle's motion.)

(b) Taking $\tau_{\max} = 25\,850$, make a parametric plot of the particle's trajectory.

(c) There are three principal oscillations present in the motion of y whose periods are labelled τ_1, τ_2 and τ_3 in the graphs shown below. Estimate the ratio $\tau_1 : \tau_2 : \tau_3$.

(d) Give a brief qualitative description of the particle's motion, making reference to the three types of oscillation described in (c).

Solution

(a) Using the notebook given on p. 234, we obtain the following graphs:

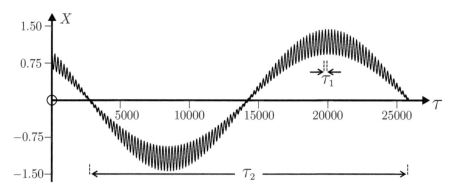

‡Here r_0 and ω are defined in Question 4.22 and the dot notation in (1) denotes differentiation with respect to the dimensionless time.

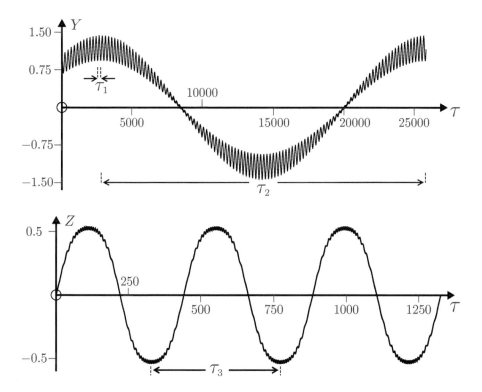

(b) The trajectory below follows from *Mathematica*'s `ParametricPlot3D` function:

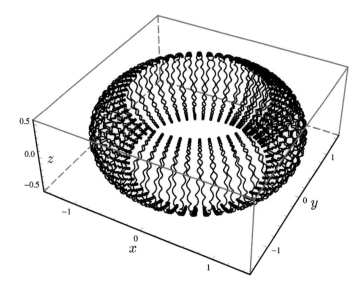

(c) The period τ_1 varies during the motion. Using the average value of τ_1 we find that $\tau_1 : \tau_2 : \tau_3 \simeq 1 : 100 : 2$.

(d) Gyration, drift and bounce are the three main components which characterize the motion of the particle. We discuss each of these briefly below:

☞ The gyration is a rapid oscillation about a magnetic field line and occurs on a time scale of order τ_1.

☞ The drift is a lateral motion about the z-axis with a period of order τ_2. For the given initial conditions, this time is typically two orders of magnitude greater than τ_1.

☞ The bounce motion occurs between the 'North' and 'South' poles assuming that the dipole field is produced by a bar magnet aligned along the z-axis. The particle is effectively reflected by the field, which acts as a magnetic mirror. This motion has a period of order τ_3.

```
In[1]:= τmax = 1328;

      f1[τ_] := (X[τ]² + Y[τ]² + Z[τ]²)^(5/2);   f2[τ_] := X[τ]² + Y[τ]² - 2 Z[τ]²;
      Ekin[τ_] := X'[τ]² + Y'[τ]² + Z'[τ]²;
      eqn1 = f1[τ] X''[τ] + f2[τ] Y'[τ] + 3 Z[τ] Z'[τ] Y[τ] == 0;
      eqn2 = f1[τ] Y''[τ] - f2[τ] X'[τ] - 3 Z[τ] Z'[τ] X[τ] == 0;
      eqn3 = f1[τ] Z''[τ] + 3 (X[τ] Y'[τ] - Y[τ] X'[τ]) Z[τ] == 0;
      eqn4 = X[0] == 1;         eqn6 = Y[0] == 1;          eqn8 = Z[0] == 0;
      eqn5 = X'[0] == 0.005;  eqn7 = Y'[0] == 0.005;  eqn9 = Z'[0] == 0.01;
```

```
In[2]:= Sol = NDSolve[{eqn1, eqn2, eqn3, eqn4, eqn5, eqn6, eqn7, eqn8, eqn9},
          {X[τ], X'[τ], Y[τ], Y'[τ], Z[τ], Z'[τ]}, {τ, 0, τmax}];
```

```
In[3]:= Plot[Evaluate[X[τ] /. Sol], {τ, 0, τmax}]
      Plot[Evaluate[Y[τ] /. Sol], {τ, 0, τmax}]
      Plot[Evaluate[Z[τ] /. Sol], {τ, 0, τmax}]
      Plot[Evaluate[Ekin[τ] /. Sol], {τ, 0, τmax}]

      ParametricPlot3D[{{X[τ], Y[τ], Z[τ]} /. Sol}, {τ, 0, τmax}, PlotPoints → 200]
```

Comment

Charged particles from the solar wind, once trapped in the magnetic field of the Earth, undergo a motion qualitatively similar to the behaviour described in (d) above. Typical order-of-magnitude values for electrons (having energies of a few keVs) are: milliseconds for the gyration time, hours for the equatorial drift time and seconds for the time to bounce between the poles. The interaction between these energetic particles (mostly electrons) results in the excitation of gas molecules (nitrogen, oxygen and traces of other elements) in the upper atmosphere. The visible light emitted when these molecules relax to their ground state is responsible for the 'northern lights' (*aurora borealis*) and 'southern lights' (*aurora australis*). These aurorae must surely rank amongst the most spectacular phenomena in nature.

Question 4.24**

Consider a bounded distribution of steady electric currents in vacuum having density $\mathbf{J}(\mathbf{r}')$. As in Question 2.20, we let $\mathbf{r}$ and $\mathbf{r}'$ denote the position vectors of a field point P and a source point relative to an arbitrarily chosen origin O respectively.

(a) Make a multipole expansion of the vector potential (8) of Question 4.5 and show that

$$A_i(\mathbf{r}) \;=\; \frac{\mu_0}{4\pi}\left[\varepsilon_{ijk}\, m_j \frac{r_k}{r^3} + \varepsilon_{ijl}\, m_{jk}\frac{(3 r_k r_l - r^2 \delta_{kl})}{2 r^5} + \cdots\right], \tag{1}$$

where the magnetic dipole moment m_i and magnetic quadrupole moment m_{ij} are defined as follows:

$$m_i \;=\; \frac{1}{2}\int_v (\mathbf{r}' \times \mathbf{J})_i \, dv' \qquad \text{and} \qquad m_{ij} \;=\; \frac{2}{3}\int_v (\mathbf{r}' \times \mathbf{J})_i\, r'_j \, dv'. \tag{2}$$

Hints:

1. $|\mathbf{r} - \mathbf{r}'|^{-1} = \dfrac{1}{r} + \dfrac{r_i}{r^3} r'_i + \dfrac{3 r_i r_j - r^2 \delta_{ij}}{2 r^5} r'_i r'_j + \cdots$ $\left(\text{see (2) of Question 1.29}\right)$.

2. Use the integral transforms (2)–(4) of Question 1.30 with $\dot{\rho} = \partial \rho / \partial t = 0$.

(b) Show that the magnetic field $\mathbf{B}(\mathbf{r})$ of the distribution is given by

$$B_i \;=\; \frac{\mu_0}{4\pi}\left[\frac{(3 r_i r_j - r^2 \delta_{ij})}{r^5}\, m_j + \frac{3\{5 r_i r_j r_k - r^2 (r_i \delta_{jk} + r_j \delta_{ki} + r_k \delta_{ij})\}}{2 r^7}\, m_{jk} + \cdots\right]. \tag{3}$$

Solution

(a) Substituting the expansion for $|\mathbf{r} - \mathbf{r}'|^{-1}$ in (1) gives

$$A_i(\mathbf{r}) = \frac{\mu_0}{4\pi}\left[\frac{1}{r}\int_v J_i\, dv' + \frac{r_j}{r^3}\int_v r'_j J_i\, dv' + \frac{(3 r_j r_k - r^2 \delta_{jk})}{2 r^5}\int_v r'_j r'_k J_i\, dv' + \cdots\right]. \tag{4}$$

The first three terms in (4) are considered separately below.

☞ **first term**

Equation (2) of Question 1.30 shows that

$$\int_v J_i\, dv' = 0. \tag{5}$$

☞ **second term**

Applying the integral transform (3) of Question 1.30 yields

$$\frac{\mu_0 \, r_j}{4\pi r^3} \int_v r'_j J_i \, dv' = -\frac{\mu_0 \, r_j}{4\pi r^3} \frac{\varepsilon_{ijk}}{2} \int_v (\mathbf{r}' \times \mathbf{J})_k \, dv'. \tag{6}$$

Substituting $(2)_1$ in (6) then gives

$$\frac{\mu_0 \, r_j}{4\pi r^3} \int_v r'_j J_i \, dv' = -\frac{\mu_0 \, r_j}{4\pi r^3} \frac{\varepsilon_{ijk}}{2} \int_v (\mathbf{r}' \times \mathbf{J})_k \, dv'$$

$$= -\frac{\mu_0 \, r_j}{4\pi r^3} \varepsilon_{ijk} m_k = \frac{\mu_0}{4\pi} \varepsilon_{ikj} m_k \frac{r_j}{r^3} = \frac{\mu_0}{4\pi} \varepsilon_{ijk} m_j \frac{r_k}{r^3}, \tag{7}$$

where the last step follows since the subscripts are arbitrary.

☞ **third term**

Applying the integral transform (4) of Question 1.30 yields

$$\frac{\mu_0}{4\pi} \frac{(3r_j r_k - r^2 \delta_{jk})}{2r^5} \int_v r'_j r'_k J_i \, dv'$$

$$= -\frac{\mu_0}{4\pi} \frac{(3r_j r_k - r^2 \delta_{jk})}{2r^5} \int_v \frac{\left[\varepsilon_{ijl}(\mathbf{r}' \times \mathbf{J})_l r'_k + \varepsilon_{ikl}(\mathbf{r}' \times \mathbf{J})_l r'_j \right]}{3} \, dv'$$

$$= \frac{\mu_0}{4\pi} \frac{(\varepsilon_{ilj} m_{lk} + \varepsilon_{ilk} m_{lj})}{2} \underbrace{\frac{(3r_j r_k - r^2 \delta_{jk})}{2r^5}}_{T_{jk}}, \tag{8}$$

where in the last step we use the definition $(2)_2$. Since the subscripts are arbitrary, we let

- $j \rightarrow l$; $k \rightarrow k$; $l \rightarrow j$. Then $\varepsilon_{ilj} m_{lk} T_{jk} = \varepsilon_{ijl} m_{jk} T_{kl}$,

- $j \rightarrow k$; $k \rightarrow l$; $l \rightarrow j$. Then $\varepsilon_{ilk} m_{lj} T_{jk} = \varepsilon_{ijl} m_{jk} T_{kl}$,

because the second-rank tensor labelled T_{jk} in (8) is symmetric in its subscripts. This proves that

$$\frac{\mu_0}{4\pi} \frac{(3r_j r_k - r^2 \delta_{jk})}{2r^5} \int_v r'_j r'_k J_i \, dv' = \frac{\mu_0}{4\pi} \varepsilon_{ijl} m_{jk} \frac{(3r_k r_l - r^2 \delta_{kl})}{2r^5}. \tag{9}$$

Equations (5), (7) and (9) yield (1).

(b) The magnetic field follows from $B_i = (\boldsymbol{\nabla} \times \mathbf{A})_i = \varepsilon_{ijk} \nabla_j A_k$, with A_k given by (1). We showed in Question 4.10 that the dipole potential gives rise to the dipole field (the first terms of (1) and (3) respectively). Here we differentiate the quadrupole potential and show that it produces the quadrupole field (the second terms of (1) and (3) respectively). So

$$B_i = \frac{\mu_0}{4\pi} \varepsilon_{ijk} \nabla_j \varepsilon_{klm} m_{lp} \frac{(3r_m r_p - r^2 \delta_{mp})}{2r^5}$$

$$= \frac{\mu_0}{4\pi} (\delta_{il}\delta_{jm} - \delta_{im}\delta_{jl}) m_{lp} \nabla_j \frac{(3r_m r_p - r^2\delta_{mp})}{2r^5} \qquad (\text{see (1) of Question 1.3})$$

$$= \frac{\mu_0}{4\pi} (\delta_{il}\delta_{jm} - \delta_{im}\delta_{jl}) m_{lp} \frac{3r^2 (r_j\delta_{mp} + r_m\delta_{pj} + r_p\delta_{jm}) - 15 r_j r_m r_p}{2r^7}, \qquad (10)$$

using (8) of Question 1.1. Multiplying out the terms in (10) and contracting subscripts give the desired result.

Comments

(i) Notice the following differences between the multipole expansions of the scalar and vector potentials (compare (2) of Question 2.20 with (1) above):

☞ The zeroth order term is always absent from the expansion of **A**; a feature which reflects the fact that there is no magnetic equivalent of electric charge (i.e. magnetic monopoles do not exist).[‡]

☞ In the expansion of Φ we include terms to electric octopole order, whereas in the expansion of **A** we retain terms to magnetic quadrupole order only. This apparent anomaly is discussed further in Comment (i) of Question 11.1.

(ii) The electric multipoles are defined in terms of integrals arising directly in the multipole expansion of Φ. But for the magnetic multipoles the procedure is less straightforward, in that the integrals in the multipole expansion of **A** must first be transformed using the results of Question 1.30 (see Questions 8.20 and 11.1).

(iii) All the electric multipole moments are symmetric in their subscripts, whereas the magnetic multipole moments are generally not. So, for example, $q_{ij} = q_{ji}$ but $m_{ij} \neq m_{ji}$ (except, of course, if the system has a particular symmetry).

(iv) The leading non-vanishing term in the expansion (1) dominates at large distances. Associated with this term is a unique multipole moment independent of the choice of origin. As in the electric case, all higher multipoles are then expected to be origin-dependent. So the magnetostatic dipole moment (unlike its electrostatic counterpart discussed in Question 2.23) is *always* origin-independent. See also Question 10.5 where we consider the time-dependent dipole moments.

(v) The vector potential and magnetic field of a point magnetic dipole at the origin (given by the first term of the expansions (1) and (3) respectively) have the familiar forms:

$$\mathbf{A}(\mathbf{r}) = \frac{\mu_0}{4\pi} \frac{\mathbf{m} \times \mathbf{r}}{r^3} \qquad \text{and} \qquad \mathbf{B}(\mathbf{r}) = \frac{\mu_0}{4\pi} \left[\frac{3(\mathbf{m} \cdot \hat{\mathbf{r}})\hat{\mathbf{r}} - \mathbf{m}}{r^3} \right]. \qquad (11)$$

See also (1) and (2) of Question 4.10.

[‡] For a long time, physicists have speculated that at least one magnetic monopole might exist somewhere in the universe. So far the search has proved to be highly elusive, but it continues nevertheless.

Question 4.25

(a) Consider a closed plane conducting loop c in which there is a filamentary current I. Show that the magnetic dipole moment is

$$\mathbf{m} = I\mathbf{a}, \tag{1}$$

where $\mathbf{a}$ is the vector area of the loop (the direction of $\mathbf{a}$ is determined by the right-hand rule).

(b) Determine $\mathbf{m}$ for a circular current loop of radius r_0 with its axis along $\hat{\mathbf{z}}$.

Solution

(a) Substituting $\mathbf{J}\,dv' = I\,d\mathbf{l}'$ (see (7) of Question (4.3)) in the definition $(2)_1$ of Question 4.24 gives

$$\mathbf{m} = \frac{I}{2}\oint_c \mathbf{r}' \times d\mathbf{l}'.$$

Then (1) follows immediately from (5) of Question 1.22.

(b) Assume the current circulates anticlockwise as seen from above, then $\mathbf{a} = \pi r_0^2\hat{\mathbf{z}}$ and $\mathbf{m} = I\pi r_0^2\hat{\mathbf{z}}$.

Comments

(i) The higher magnetic multipole moments may also be formulated in terms of a contour integral: for example, the magnetic quadrupole moment m_{ij} (see $(2)_2$ of Question 4.24) is

$$m_{ij} = \frac{2I}{3}\varepsilon_{ikl}\oint_c r'_j r'_k\,dl'_l. \tag{2}$$

(ii) It is evident from (2) that the magnetic quadrupole moment is traceless:

$$m_{xx} + m_{yy} + m_{zz} = 0. \tag{3}$$

(Proof: let $j = i$ in (2) and use (4) of Question 1.5.)

Question 4.26**

Charge q is distributed uniformly

(a) over the surface s of a sphere of radius a, and

(b) throughout the volume of a sphere of radius a.

Suppose the sphere spins about an axis (the z-axis, say) with constant angular velocity $\boldsymbol{\omega}$. For each of these charge distributions, calculate the magnetic dipole moment about the origin (chosen at the centre of the sphere).

Hint: Start with $(2)_1$ of Question 4.24.

Solution

(a) Because of the remarks leading to (7) of Question 4.3, we note that

$$\mathbf{m} = \frac{1}{2} \oint_s (\mathbf{r}' \times \mathbf{K}) \, da', \tag{1}$$

where $\mathbf{K}$ is the surface-current density and da' is an element of the spherical surface at $\mathbf{r}'$. Now $\mathbf{K} = \sigma \mathbf{v}(\mathbf{r}')$ (see (5) of Question 4.3) where $\sigma = q/4\pi a^2$ and $\mathbf{v} = \boldsymbol{\omega} \times \mathbf{r}'$.[4] Then

$$\mathbf{m} = \frac{\sigma}{2} \oint_s \mathbf{r}' \times (\boldsymbol{\omega} \times \mathbf{r}') \, da' = \frac{\sigma a^2}{2} \int_0^{2\pi} \int_0^{\pi} \mathbf{r}' \times (\boldsymbol{\omega} \times \mathbf{r}') \sin\theta' d\theta' d\phi'. \tag{2}$$

Taking $\boldsymbol{\omega} = (0, 0, \omega)$ and $\mathbf{r}' = a(\sin\theta'\cos\phi', \sin\theta'\sin\phi', \cos\theta')$, we obtain for the cross-product $\boldsymbol{\omega} \times \mathbf{r}' = \omega r'(-\sin\theta'\sin\phi', \sin\theta'\cos\phi', 0)$. Thus,

$$\mathbf{r}' \times (\boldsymbol{\omega} \times \mathbf{r}') = \omega r'^2(-\sin\theta'\cos\theta'\cos\phi', \sin\theta'\cos\theta'\sin\phi', \sin^2\theta'). \tag{3}$$

Substituting (3) in (2) and putting $r' = a$ yield

$$\mathbf{m} = \frac{\sigma \omega a^4}{2} \int_0^{2\pi} \int_0^{\pi} (-\sin\theta'\cos\theta'\cos\phi', \sin\theta'\cos\theta'\sin\phi', \sin^2\theta') \sin\theta' d\theta' d\phi'. \tag{4}$$

Because $\int_0^{2\pi} \cos\phi' \, d\phi' = \int_0^{2\pi} \sin\phi' \, d\phi' = 0$, the only non-zero contribution to (4) comes from the z-component of $\mathbf{r}' \times (\boldsymbol{\omega} \times \mathbf{r}')$. Therefore,

$$m_z = \frac{\sigma \omega a^4}{2} \int_0^{2\pi} \int_0^{\pi} \sin^3\theta' d\theta' d\phi' = \pi\sigma\omega a^4 \int_0^{\pi} \sin^3\theta' d\theta'$$

$$= \pi\sigma\omega a^4 \int_{-1}^{1} (1 - \cos^2\theta') d(\cos\theta').$$

Now this definite integral is easily shown to equal $\frac{4}{3}$, and so

$$\mathbf{m} = \frac{4\pi}{3} \times \frac{q}{4\pi a^2} \times wa^4 \hat{\mathbf{z}}$$

$$= \tfrac{1}{3} q w a^2 \hat{\mathbf{z}}. \tag{5}$$

(b) Proceeding as for (a),

$$\mathbf{m} = \frac{1}{2} \int_v (\mathbf{r}' \times \mathbf{J}) \, dv', \tag{6}$$

where dv' is a volume element at $\mathbf{r}'$ and $\mathbf{J} = \rho \mathbf{v}(\mathbf{r}') = \rho \boldsymbol{w} \times \mathbf{r}'$ because of (4) of Question 4.3 (here $\rho = q/\frac{4}{3}\pi a^3$). So

$$\mathbf{m} = \frac{\rho}{2} \int_v \mathbf{r}' \times (\boldsymbol{w} \times \mathbf{r}') \, dv'$$

$$= \frac{\rho}{2} \int_0^{2\pi} \int_0^{\pi} \int_0^{a} \mathbf{r}' \times (\boldsymbol{w} \times \mathbf{r}') \, r'^2 \sin\theta' \, dr' \, d\theta' \, d\phi'. \tag{7}$$

Substituting (3) in (7) and recognizing that, as before, there is only a z-component of $\mathbf{m}$ gives

$$m_z = \frac{\rho w}{2} \int_0^{2\pi} \int_0^{\pi} \int_0^{a} r'^4 \sin^3\theta' \, dr' \, d\theta' \, d\phi'$$

$$= \pi \rho w \int_0^{\pi} \int_0^{a} r'^4 \sin^3\theta' \, dr' \, d\theta'. \tag{8}$$

Both integrations in (8) are trivial and the result is

$$\mathbf{m} = \pi \rho w \times \frac{a^5}{5} \times \frac{4}{3} \hat{\mathbf{z}}$$

$$= \frac{4\pi w a^5}{15} \frac{3q}{4\pi a^3} \hat{\mathbf{z}}$$

$$= \tfrac{1}{5} q w a^2 \hat{\mathbf{z}}. \tag{9}$$

Comment

The dipole moment (9) agrees with that calculated earlier by a different approach; see (8) of Question 4.11.

Question 4.27**

Two plane circular current loops of radius r_0 have a common axis of symmetry (the z-axis, say). Suppose they are centred at $(0, 0, \pm z_0)$ and that the current in each loop is $\pm I$ respectively. Here the sign of I denotes the sense of circulation. As we discovered in Question 4.17, this arrangement is known as an anti-Helmholtz pair.

(a) Explain why the magnetic dipole moment of the pair of coils is zero.

(b) Calculate the components of the magnetic quadrupole tensor m_{ij}.

   Hint: Use (2) of Question 4.25.

Solution

(a) For the loop centred at $\pm z_0$ the corresponding dipole moment is $\pm m\hat{z}$ where $m = I\pi r_0^2$. The net dipole moment, being the sum of the moments of the two current loops, is thus obviously zero.

(b) Step 1: Consider the loop at $z = z_0$.

The components of m_{ij} are calculated using (2) of Question 4.25 and symmetry. By inspection, some of the contour integrals below are seen to be zero; the others are conveniently evaluated in polar coordinates with $d\boldsymbol{l}' = r_0 d\theta' \hat{\boldsymbol{\theta}} = r_0(-\hat{\boldsymbol{x}}\sin\theta' + \hat{\boldsymbol{y}}\cos\theta') d\theta'$. In what follows, we omit the primes on the r_i for the sake of clarity.

$m_{xy} = m_{yx}$

$$\tfrac{2}{3}I\left(\varepsilon_{xyz}\oint_c y^2 dz + \varepsilon_{xzy} z_0 \oint_c y\, dy\right) = 0.$$

$m_{xz} = m_{yz}$

$$\tfrac{2}{3}I\left(\varepsilon_{xyz} z_0\oint_c y\, dz + \varepsilon_{xzy} z_0^2 \oint_c dy\right) = 0.$$

$m_{zx} = m_{zy}$

$$\tfrac{2}{3}I\left(\varepsilon_{zxy}\oint_c x^2 dy + \varepsilon_{zyx}\oint_c xy\, dx\right) = \tfrac{2}{3}Ir_0^3\int_0^{2\pi}(\cos^2\theta' + \sin^2\theta')\cos\theta'\,d\theta' = 0.$$

$m_{xx} = m_{yy}$

$$\tfrac{2}{3}I\left(\varepsilon_{xyz}\oint_c xy\,dz + \varepsilon_{xzy} z_0\oint_c x\,dy\right) = -\tfrac{2}{3}Iz_0\oint_c x\,dy = -\tfrac{2}{3}Iz_0 r_0^2\int_0^{2\pi}\cos^2\theta\,d\theta = -\tfrac{2}{3}I\pi r_0^2 z_0.$$

m_{zz}

One may proceed as above and evaluate the contour integrals, but it is quicker to use the traceless property given by (3) of Question 4.25: $m_{zz} = -(m_{xx} + m_{yy}) = -2m_{xx} = \tfrac{4}{3}I\pi r_0^2 z_0$.

Step 2: Consider both loops together.

The non-zero m_{ij} for the two-loop configuration are obtained by summing individual contributions

$$\left.\begin{aligned} -m_{xx} = -m_{yy} &= \tfrac{2}{3}I\pi r_0^2 z_0 + \tfrac{2}{3}(-I)\pi r_0^2(-z_0) = \tfrac{4}{3}m_0 z_0 \\ m_{zz} &= \tfrac{4}{3}I\pi r_0^2 z_0 + \tfrac{4}{3}(-I)\pi r_0^2(-z_0) = \tfrac{8}{3}m_0 z_0 \end{aligned}\right\}. \tag{1}$$

Step 3: Use (1) and (3) of Question 4.24 to obtain B.

The calculation of B_x, B_y and B_z involves the contraction of tensors and is straightforward but tedious. As an example, we outline the calculation for B_x below.

$$\begin{aligned} B_x &= \frac{\mu_0}{4\pi}\left\{\frac{3(5x^3 - 3r^2x)\,m_{xx} + 3(5xy^2 - r^2x)\,m_{yy} + 3(5xz^2 - r^2x)\,m_{zz}}{2r^2}\right\} \\ &= \frac{\mu_0}{4\pi}\left\{\frac{-(5x^3 - 3r^2x) - 3(5xy^2 - r^2x) + 2(5xz^2 - r^2x)}{2r^2}\right\}4m_0 z_0 \\ &= \frac{3\mu_0}{2}\frac{x(4z^2 - x^2 - y^2)}{(x^2 + y^2 + z^2)^{7/2}}I\,r_0^2 z_0. \tag{2} \end{aligned}$$

Similarly,

$$B_y = \frac{3\mu_0}{2}\frac{y(4z^2 - x^2 - y^2)}{(x^2 + y^2 + z^2)^{7/2}}I\,r_0^2 z_0 \tag{3}$$

$$B_z = \frac{3\mu_0}{2}\frac{z(2z^2 - 3x^2 - 3y^2)}{(x^2 + y^2 + z^2)^{7/2}}I\,r_0^2 z_0. \tag{4}$$

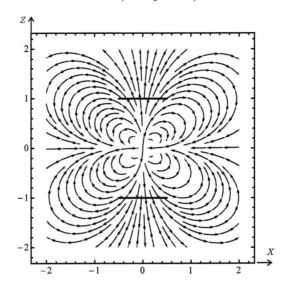

Comment

The diagram on p. 242 shows the quadrupole field of the coils in the xz-plane and is drawn for $z_0/r_0 = 1$. Here X and Z are the dimensionless coordinates, $X = x/r_0$ and $Z = z/r_0$. Each coil is represented in cross-section by a solid horizontal line. See Question 2.24 for a sample *Mathematica* notebook.

Question 4.28 **

Consider a bounded distribution of steady currents in an external magnetostatic field $\mathbf{B}(\mathbf{r})$. Show that the force $\mathbf{F}$ and torque $\boldsymbol{\Gamma}$ on the distribution are given by

(a) $F_i = m_j \nabla_j B_i + \frac{1}{2} m_{jk} \nabla_j \nabla_k B_i + \cdots,$ and$\qquad\qquad$ (1)

(b) $\Gamma_i = \varepsilon_{ijk} \big[m_j B_k + \frac{1}{2} (m_{jl} + m_{lj}) \nabla_l B_k + \cdots \big].$ $\qquad\qquad$ (2)

Solution

(a) The total force on the distribution is obtained from $(10)_1$ of Question 4.4:

$$F_i = \varepsilon_{ijk} \int_v J_j(\mathbf{r}') B_k(\mathbf{r}') \, dv'. \qquad\qquad (3)$$

Assuming that the spatial variation of $B_k(\mathbf{r}')$ varies sufficiently slowly, we make the Taylor-series expansion $\big($see, for example, (3) of Question 1.7$\big)$

$$B_k(\mathbf{r}') = B_k + (\nabla_l B_k) r'_l + \frac{1}{2} (\nabla_l \nabla_m B_k) r'_l r'_m + \cdots, \qquad (4)$$

where the field and its derivatives are understood to be evaluated at the origin O. Substituting (4) in (3) and recalling that $\int_v \mathbf{J}(\mathbf{r}') \, dv' = 0$ $\big($see (5) of Question (4.24)$\big)$ give

$$F_i = \varepsilon_{ijk} (\nabla_l B_k) \int_v r'_l J_j \, dv' + \frac{1}{2} \varepsilon_{ijk} (\nabla_l \nabla_m B_k) \int_v r'_l r'_m J_j \, dv' + \cdots. \qquad (5)$$

The first two terms in (5) are considered separately below.

☞ **first term**

Applying the integral transform (3) of Question 1.30 with $\dot{\rho} = 0$ gives

$$\varepsilon_{ijk}(\nabla_l B_k) \int_v r'_l J_j \, dv' = -\tfrac{1}{2} \varepsilon_{ijk}(\nabla_l B_k) \varepsilon_{jlm} \int_v (\mathbf{r}' \times \mathbf{J})_m dv' = \varepsilon_{jki} \varepsilon_{jml} m_m (\nabla_l B_k),$$

because of the definition $(2)_1$ of Question 4.24. Contracting subscripts yields

$$\varepsilon_{jki}\varepsilon_{jml}\,m_m(\nabla_l B_k) \;=\; (\delta_{km}\delta_{il} - \delta_{kl}\delta_{im})\,m_m(\nabla_l B_k)$$

$$= \; m_k\nabla_i B_k - m_i\nabla_k B_k = m_j\nabla_i B_j, \tag{6}$$

where in the last step we use $\nabla\cdot\mathbf{B}=0$ and invoke the arbitrary nature of the subscripts.

☞ second term

Applying the integral transform (4) of Question 1.30 with $\dot\rho=0$ gives

$$\tfrac{1}{2}\varepsilon_{ijk}(\nabla_l\nabla_m B_k)\int_v r_i' r_m' J_j\,dv'$$

$$= \; -\tfrac{1}{2}\,\varepsilon_{ijk}(\nabla_l\nabla_m B_k)\int_v \frac{[\varepsilon_{jln}(\mathbf{r}'\times\mathbf{J})_n r_m' + \varepsilon_{jmn}(\mathbf{r}'\times\mathbf{J})_n r_l']}{3}\,dv'$$

$$= \; \tfrac{1}{2}\,\varepsilon_{jki}\frac{(\varepsilon_{jnl}m_{nm} + \varepsilon_{jnm}m_{nl})}{2}(\nabla_l\nabla_m B_k),$$

because of the definition $(2)_2$ of Question 4.24. Contracting subscripts, interchanging the order of differentiation, $\nabla_i\nabla_j \leftrightarrow \nabla_j\nabla_i$ and using $\nabla\cdot\mathbf{B}=0$ give

$$\tfrac{1}{4}(\varepsilon_{jki}\varepsilon_{jnl}m_{nm} + \varepsilon_{jki}\varepsilon_{jnm}m_{nl})(\nabla_l\nabla_m B_k)$$

$$= \; \tfrac{1}{4}[(\delta_{kn}\delta_{il} - \delta_{kl}\delta_{in})m_{nm} + (\delta_{kn}\delta_{im} - \delta_{km}\delta_{in})m_{nl}](\nabla_l\nabla_m B_k)$$

$$= \; \tfrac{1}{2}m_{jk}(\nabla_k\nabla_i B_j) \;=\; \tfrac{1}{2}m_{jk}(\nabla_k\nabla_j B_i). \tag{7}$$

In the last step we recognize that the sources of the magnetostatic field $\mathbf{B}$ are external to the current distribution. So $\nabla\times\mathbf{B}=0$ everywhere and $\nabla_i B_j = \nabla_j B_i$. Equations (6) and (7) in (5) yield (1).

(b) For the total torque on the distribution we have

$$\boldsymbol{\Gamma} \;=\; \int \mathbf{r}'\times d\mathbf{F} \;=\; \int_v \mathbf{r}'\times(\mathbf{J}\times\mathbf{B})\,dv' \tag{8}$$

(see $(10)_1$ of Question 4.4). Thus

$$\Gamma_i \;=\; \varepsilon_{ijk}\,\varepsilon_{klm}\int_v r_j' J_l B_m\,dv' \;=\; \varepsilon_{kij}\,\varepsilon_{klm}\int_v r_j' J_l\,[B_m + (\nabla_n B_m)\,r_n' + \cdots]dv', \tag{9}$$

using (4) for a slowly varying magnetostatic field. As before, we consider the first two terms in (9) separately.

☞ **first term**

Equation (3) of Question 1.30 with $\dot{\rho} = 0$ and the definition $(2)_1$ give

$$\varepsilon_{kij}\varepsilon_{klm}B_m\int_v r'_j J_l\, dv' = -\tfrac{1}{2}\varepsilon_{kij}\varepsilon_{klm}\varepsilon_{ljp}B_m\int_v (\mathbf{r}'\times\mathbf{J})_p\, dv'$$

$$= \varepsilon_{kij}\varepsilon_{klm}\varepsilon_{jlp}m_p B_m$$

$$= (\delta_{il}\delta_{jm} - \delta_{im}\delta_{jl})\varepsilon_{jlp}m_p B_m = \varepsilon_{ijk}m_j B_k, \qquad (10)$$

using the properties of the Levi-Civita tensor and the arbitrary nature of the subscripts.

☞ **second term**

Equation (4) of Question 1.30 with $\dot{\rho} = 0$ and the definition $(2)_2$ give

$$\varepsilon_{ijk}\varepsilon_{klm}(\nabla_n B_m)\int_v r'_j r'_n J_l\, dv' = -\tfrac{1}{2}\varepsilon_{ijk}\varepsilon_{klm}[\varepsilon_{ljp}m_{pn} + \varepsilon_{lnp}m_{pj}](\nabla_n B_m)$$

$$= \tfrac{1}{2}\varepsilon_{ijk}[(\delta_{jk}\delta_{mp} - \delta_{kp}\delta_{jm})m_{pn} + (\delta_{kn}\delta_{mp} - \delta_{kp}\delta_{mn})m_{pj}](\nabla_n B_m)$$

$$= \tfrac{1}{2}\varepsilon_{ijk}(m_{jl} + m_{lj})(\nabla_l B_k), \qquad (11)$$

since $\nabla_i B_i = 0$, $\nabla_i B_j = \nabla_j B_i$ and subscripts are arbitrary.

Equations (10) and (11) in (9) yield (2).

Comment

Equations (1) and (2) confirm an elementary result: a magnetic dipole experiences a resultant force

$$\mathbf{F} = (\mathbf{m}\cdot\nabla)\mathbf{B} \qquad (12)$$

in a non-uniform magnetic field, and a net torque

$$\mathbf{\Gamma} = \mathbf{m}\times\mathbf{B}. \qquad (13)$$

Question 4.29

Consider a circular coil of wire having radius a and N turns centred on the origin of Cartesian coordinates with its symmetry axis along $\hat{\mathbf{z}}$. A steady current I is maintained in the coil. A small cylindrical permanent magnet (having magnetic dipole moment $\mathbf{m}$ with $|\mathbf{m}| = m_0$) pivoted about its centre is located at $(0, 0, z)$. Assume that the magnet is free to rotate (without friction) about the x-axis.

(a) Show that in the position of stable equilibrium (when $\mathbf{m} = m_0\hat{\mathbf{z}}$), the magnet exerts a force on the coil given by

$$\mathbf{F} = \frac{3\mu_0 NImo_0 a^2 z}{2(a^2+z^2)^{5/2}}\hat{\mathbf{z}}. \tag{1}$$

Hint: Use the magnetic field of the coil calculated in Question 4.8(a) which has been adjusted for N turns of wire:

$$B_z = \frac{\mu_0 NIa^2}{2(a^2+z^2)^{3/2}}. \tag{2}$$

(b) The magnet is now displaced from its equilibrium position and performs small-amplitude oscillations. Show that the motion is simple harmonic with angular frequency

$$\omega = \sqrt{\frac{6\mu_0 NImo_0 a^2}{M(3b^2+\ell^2)(a^2+z^2)^{3/2}}}, \tag{3}$$

where b, ℓ and M are the radius, length and mass of the cylindrical magnet respectively.

Hint: The moment of inertia of a circular cylinder of radius b, length ℓ and mass M about a perpendicular axis through the centre of mass is

$$I_\perp = \tfrac{1}{12}M(3a^2+\ell^2) \tag{4}$$

(see, for example, Ref. [5]).

(c) For what values of z are $\mathbf{F}$ and ω a maximum?

Solution

(a) The force exerted by the coil on the magnet, given by (12) of Question 4.28, is $(\mathbf{m}\cdot\nabla)\mathbf{B}$. So by Newton's third law, the force we seek is

$$\mathbf{F} = -(\mathbf{m}\cdot\nabla)\mathbf{B} = -m_z\frac{\partial B_z}{\partial z}\hat{\mathbf{z}} = -\frac{\mu_0 NImo_0 a^2}{2}\frac{\partial}{\partial z}(a^2+z^2)^{-3/2}\hat{\mathbf{z}}, \tag{5}$$

where B_z is given by (2). Performing the differentiation in (5) yields (1).

(b) The torque exerted by the coil on the magnet, given by (13) of Question 4.28, is $\mathbf{\Gamma} = \mathbf{m}\times\mathbf{B} = m_0 B\sin\theta\,\hat{\mathbf{x}}$, where θ is the angular displacement measured from the

[5]   O. L. de Lange and J. Pierrus, *Solved problems in classical mechanics: Analytical and numerical solutions with comments*, Chap. 12, p. 413. Oxford: Oxford University Press, 2010.

z-axis. Newton's second law in the form: net torque = rate of change of angular momentum = moment of inertia × angular acceleration, gives

$$\mathbf{\Gamma} = I_\perp \frac{d\omega}{dt} = -I_\perp \frac{d^2\theta}{dt^2}\,\hat{\mathbf{x}}, \qquad \text{and so} \qquad m_0 B \sin\theta = -I_\perp \frac{d^2\theta}{dt^2}, \tag{6}$$

where the minus sign in (6) is required for a restoring torque. Now for small-amplitude oscillations we make the approximation $\sin\theta \approx \theta$. Then

$$\ddot{\theta} + \frac{m_0 B}{I_\perp}\theta = 0, \tag{7}$$

which is the equation of motion of an undamped, simple harmonic oscillator having angular frequency $\omega = \sqrt{m_0 B/I_\perp}$. Substituting (2) and (4) in this last equation yields (3).

(c) The maximum value of $\mathbf{F}$ follows immediately from $d\mathbf{F}/dz = 0$ and occurs at $z = \frac{1}{2}a$. It is obvious, by inspection, that the maximum value of ω occurs at $z = 0$.

Comment

With $z = 0$, the period of oscillation of the magnet follows from $T = 2\pi/\omega$ and is

$$T = \sqrt{\frac{2Ma(3b^2 + \ell^2)}{3\mu_0 N I m_0}}. \tag{8}$$

Equation (8) suggests a simple experimental technique for determining the dipole moment of a permanent magnet (pivoted at the centre of a current-carrying coil) from measurements of its period.

Question 4.30*

(a) Consider a discrete distribution of N point charges $q^{(\alpha)}$ at $\mathbf{r}'^{(\alpha)}$ moving with velocities $\mathbf{u}^{(\alpha)}$. Show that the magnetic dipole moment of these charges may be expressed in the form

$$\mathbf{m} = \frac{1}{2}\sum_{\alpha=1}^{N} q^{(\alpha)}\mathbf{r}'^{(\alpha)} \times \mathbf{u}^{(\alpha)}. \tag{1}$$

Hint: Begin with the definition

$$\mathbf{m} = \frac{1}{2}\int_v \mathbf{r}' \times \mathbf{J}(\mathbf{r}')\,dv', \tag{2}$$

given by (2)₁ of Question 4.24.

(b) Consider a classical model of an atom where the $q^{(\alpha)}$ in (a) are electrons orbiting the nucleus. Use (1) to show that

$$\mathbf{m} = -\frac{e\hbar}{2m_e}\left(\frac{\mathbf{L}}{\hbar}\right), \tag{3}$$

where e/m_e is the electron charge-to-mass ratio, $\mathbf{L}$ is the total angular momentum of the electrons about the nucleus and $\hbar$ is the reduced Planck constant.

Solution

(a) The current density $\mathbf{J}$ given by (2) of Question (4.3) may be written as

$$\mathbf{J}(\mathbf{r}') = \sum_{\alpha=1}^{N} q^{(\alpha)}\mathbf{u}^{(\alpha)}\delta(\mathbf{r}' - \mathbf{r}'^{(\alpha)}). \tag{4}$$

Substituting (4) in (2) and using the sifting property of the delta function gives (1).

(b) The result follows directly from (1) and the definition $\mathbf{L} = \sum_{\alpha=1}^{N} \mathbf{r}'^{(\alpha)} \times m_e\mathbf{u}^{(\alpha)}$, since the electrons have a common charge-to-mass ratio.

Comments

(i) It is easy to check that $\mathbf{L}/\hbar$ is dimensionless and that the quantity $\mu_B = \dfrac{e\hbar}{2m_e}$, known as the Bohr magneton, has the value

$$\mu_B = \frac{(1.60 \times 10^{-19}) \times (1.06 \times 10^{-34})}{2 \times (9.11 \times 10^{-31})}$$

$$\sim 9.3 \times 10^{-24}\,\mathrm{A\,m^2}. \tag{5}$$

Now in quantum mechanics we learn that any component of $\mathbf{L}/\hbar$ (say, $L_z/\hbar$) may only assume the discrete values $0, \pm1, \pm2, \dots$. This suggests that the Bohr magneton is a natural measure of the magnetic dipole moment of an atom.

(ii) As well as the moment due to their orbital motion, electrons are also known to have an intrinsic magnetic moment of the order μ_B aligned with their spin. It is this magnetic moment which is primarily responsible for ferromagnetism in matter (a topic which is briefly discussed in Chapter 9).

5

Quasi-static electric and magnetic fields in vacuum

Following on from Ampère's research which showed that electric currents resulted in magnetic forces, the picture remained incomplete (at least for a few more years). One might expect on grounds of symmetry, if nothing else, that the converse must also be true: a magnetic field should generate an electric current in a closed conducting loop (known as a circuit). At this point enter Michael Faraday, who in 1831 performed his now-famous experiments on electromagnetic induction. This important phenomenon, whereby a changing magnetic field induces an electric field (whose curl is no longer zero), forms the basis of most of the questions which follow. In this chapter we begin the transition from time-independent to time-dependent source densities and fields, and some new and interesting examples, not usually encountered in other textbooks, are introduced. These are treated from both an analytical and numerical point of view.

However, before proceeding it is worth clarifying the meaning of the term *quasi-static* used in the title of this chapter. To do this, consider a circuit having a typical linear dimension d carrying current $i(t)$.[‡] Electromagnetic effects (known to travel in vacuum with speed $c = 3.0 \times 10^8 \, \mathrm{m \, s^{-1}}$; see Chapter 7) traverse the circuit in a time of order d/c. If the current changes on a time scale $\tau \gg d/c$ (in circuit theory τ is called the time constant), the current and consequently the associated magnetic field are said to be *quasi-static* (or *quasi-stationary*). In this approximation the laws of Ampère and Biot–Savart, along with the results of Chapter 4, remain applicable (see also Question 5.1).

The study of quasi-static fields involves the concept of magnetic flux ϕ, whose definition (as we already know; see, for example, Question 1.10) is

$$\phi = \int_s \mathbf{B} \cdot d\mathbf{a}, \tag{I}$$

where s is any surface spanning a fixed contour c. This contour may be a conducting circuit, or any closed geometrical path in space (not necessarily conducting). Faraday discovered in his experiments that a time-dependent current could be induced in an electric circuit in different ways, including:

[‡]In the last chapter, the symbol I was used to denote a stationary current. Here we introduce the notation $i(t)$ or simply i to emphasize the time dependence.

Solved Problems in Classical Electromagnetism. J. Pierrus, Oxford University Press (2018).
© J. Pierrus. DOI: 10.1093/oso/9780198821915.001.0001

☞   changing the current in a neighbouring circuit,

☞   changing the size, shape, orientation or position of a neighbouring circuit,

☞   moving a permanent magnet in the vicinity of the circuit.

Faraday attributed the source of this current to a change in the magnetic flux linking the circuit. This changing flux induces in the circuit an electric field $\mathbf{E}$ whose line integral is the *electromotive force* $\mathcal{E}$ (or emf), where

$$\mathcal{E} = \oint_c \mathbf{E} \cdot d\mathbf{l}. \tag{II}$$

Electromotive forces are clearly non-electrostatic in nature (recall from Chapter 2 that the circulation of an electrostatic field around any closed path is always zero) and often have a mechanical or chemical origin. Faraday's observations may be stated simply as follows: the emf induced in an electric circuit is proportional to the rate of change of magnetic flux linking the circuit. In mathematical form

$$\mathcal{E} = -k\frac{d\phi}{dt},$$

where the constant of proportionality k—in the SI system of units—has the assigned value one. A lucid account of this important point is provided in Ref. [1]. Then

$$\mathcal{E} = -\frac{d\phi}{dt}, \tag{III}$$

where the negative sign appearing in (III) is a manifestation of Lenz's law (see Question 5.3). We choose to adopt the approach of Ref. [2], and refer to $\mathcal{E} = -d\phi/dt$ as Faraday's flux rule. Faraday's law, which follows from (I)–(III), is then

$$\oint_c \mathbf{E} \cdot d\mathbf{l} = -\frac{d}{dt} \int_s \mathbf{B} \cdot d\mathbf{a}. \tag{IV}$$

The magnetic energy associated with the quasi-stationary fields surrounding current-carrying wires is an important topic also considered in this chapter. The concept of inductance, which may be regarded as the magnetic analogue of mechanical inertia, arises from considerations of discrete electric circuits coupled by their magnetic flux. Because of the practical importance of inductance, several questions analysing some simple circuits are included towards the end.

[1]   J. D. Jackson, *Classical electrodynamics*, Chap. 5, pp. 208–11. New York: Wiley, 3 edn, 1998.
[2]   D. J. Griffiths, *Introduction to electrodynamics*, Chap. 7, pp. 296–8. New York: Prentice Hall, 3 edn, 1999.

Question 5.1

Suppose the linear dimension of a typical laboratory circuit is $d \sim 30\,\text{cm}$. A current varying harmonically with frequency f is present in the circuit. Estimate a lower bound for the value of $f_\approx$ where the quasi-static approximation (see the second paragraph of the introduction) begins to fail.

Solution

Assume changes in the current occur on a time scale τ. Then, because of our earlier remarks, this approximation starts to break down when $\tau \lesssim d/c$, or

$$f_\approx \gtrsim \frac{c}{d} \sim \frac{3 \times 10^8}{0.3} \sim 10 \times 10^9 \,\text{Hz}. \tag{1}$$

Comment

For many experiments in physics the condition $f \ll 1\,\text{GHz}$ is often satisfied, and (1) shows that the quasi-static approximation is not particularly restrictive in practice. Of course, as the circuit size decreases (increases), the value of $f_\approx$ increases (decreases).

Question 5.2

Suppose c is an arbitrary closed contour, and that s is any open surface spanning c. Show that the magnetic flux through c is independent of the choice of s.

Solution

Let s_1 and s_2 be two surfaces spanning c, as shown in the adjacent figure. Suppose the volume enclosed by $s = s_1 + s_2$ is v. Then it follows from the definition of flux that

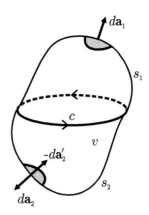

$$\oint_s \mathbf{B} \cdot d\mathbf{a} = \int_{s_1} \mathbf{B} \cdot d\mathbf{a}_1 + \int_{s_2} \mathbf{B} \cdot d\mathbf{a}_2. \tag{1}$$

Now the magnetic flux through any closed surface is always zero (see (10) of Question 4.5), and so it follows that

$$\int_{s_1} \mathbf{B} \cdot d\mathbf{a}_1 = -\int_{s_2} \mathbf{B} \cdot d\mathbf{a}_2 = \int_{s_2} \mathbf{B} \cdot d\mathbf{a}_2', \tag{2}$$

where $d\mathbf{a}_2' = -d\mathbf{a}_2$. Hence the result.

Question 5.3

Consider a closed conducting loop c moving with velocity $\mathbf{u}$ in a magnetic field (created, say, by a bar magnet). Use the Lorentz force law (see (1) of Question 4.1) to verify that the emf induced in the loop is $-d\phi/dt$.

Solution

Consider an element $d\mathbf{l}$ of the loop. The conduction electrons in this element experience a force per unit charge $\mathbf{F}/q = \mathbf{u} \times \mathbf{B}$. From the definition (II) in the introduction, we have

$$\mathcal{E} = \oint_c \frac{\mathbf{F}}{q} \cdot d\mathbf{l} = \oint_c (\mathbf{u} \times \mathbf{B}) \cdot d\mathbf{l}. \tag{1}$$

Now suppose that in a time interval dt, the element $d\ell$ sweeps out an area $d\mathbf{a} = \mathbf{u}\,dt \times d\ell$ (shown shaded in the figure below). The corresponding change in magnetic flux $d\phi$ through the loop during this time is

$$d\phi = \sum_{\text{elements}} \mathbf{B} \cdot (\mathbf{u}\,dt \times d\mathbf{l}) \xrightarrow{\text{limit}} \oint_c \mathbf{B} \cdot (\mathbf{u} \times d\mathbf{l})\,dt = -\oint_c (\mathbf{u} \times \mathbf{B}) \cdot d\mathbf{l}\,dt, \tag{2}$$

where in the last step we use the anti-cyclic property of a scalar triple product. The flux rule then follows from (1) and (2).

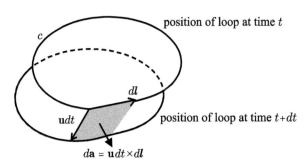

Comments

(i) Experiments such as those performed by Faraday showed that the velocity of the bar magnet relative to the stationary loop is the same as the velocity of the loop relative to the stationary bar magnet. Now the relativity principle[‡] ensures that the emf induced will be the same in both cases, although the physical processes involved are quite different. An observer in the rest frame (S′, say) of the loop cannot explain the source of the induced current in terms of a Lorentz force (which an observer at rest in the lab frame (S, say) is able to do). The S′ observer

---

[‡] Einstein's relativity principle asserts that the laws of physics are equally valid in all inertial reference frames.

must therefore conclude that there is an induced electric field $\mathbf{E}'$ which exerts an electrical force $q\mathbf{E}'$ on the conduction electrons.

(ii) The negative sign in $\mathcal{E} = -d\phi/dt$ implies Lenz's law: the emf $\mathcal{E}$ causes an induced current $i(t)$ whose magnetic flux acts to oppose the change in flux which produced it. See the figure below.

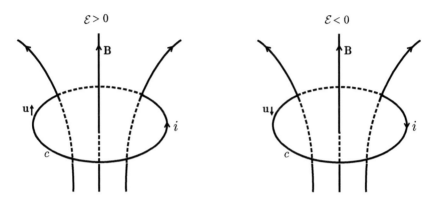

(iii) Unlike many of the best scientists of the eighteenth and nineteenth centuries, Faraday received little formal schooling. Born into a poor family, he was the son of a blacksmith and by the age of fourteen was working as an apprentice to a London bookbinder. This exposure to books enabled Faraday to read on a wide range of scientific topics, which effectively provided him with an education. Later on he became a chemical assistant to Sir Humphry Davy, and during this period of his life met many influential scientists. Despite his lack of rigorous mathematical training, Faraday made fundamental and lasting contributions to science. This is revealed in these closing quotations about the man:

> ☞ Faraday was the greatest physicist of the nineteenth century and the greatest of all experimental investigators of physical nature. He is a member of the small class of supreme scientists, which includes Archimedes, Galileo, Newton, Lavoisier and Darwin. Einstein has said that the history of physical science contains two couples of equal magnitude: Galileo and Newton, and Faraday and Clerk Maxwell. This is not one of the less interesting of Einstein's equations. From it one can deduce an instructive result. No one would allow that the wonderful Clerk Maxwell was as great a scientist as Newton. If the Faraday–Maxwell couple is to equal the Galileo–Newton couple Faraday must be accounted a greater scientist than Galileo. This deduction indicates his place in the history of science.[3]

> ☞ Whatever work Michael Faraday might have done he would have done well but it was in his character to do scientific work supremely well.[4]

[3]   J. G. Crowther, *British scientists of the nineteenth century*, vol. 1, p. 85. Middlesex: Penguin, 1935.

[4]   Source unknown: possibly Sir Humphry Davy.

Question 5.4

(a) Consider a stationary circuit c, or more generally a medium. Show that the differential form of Faraday's law,

$$\oint_c \mathbf{E} \cdot d\mathbf{l} = -\frac{d}{dt} \int_s \mathbf{B} \cdot d\mathbf{a}, \tag{1}$$

is

$$\nabla \times \mathbf{E} = -\frac{\partial \mathbf{B}}{\partial t}. \tag{2}$$

(b) Now suppose that c (or the medium) moves with velocity $\mathbf{v}(t)$ where $v \ll c$. Again, show that the differential form of (1) is (2).

Hint: Use the convective derivative $\dfrac{d}{dt} = \dfrac{\partial}{\partial t} + (\mathbf{v} \cdot \nabla)$ discussed in Question 1.9.

Solution

(a) At an arbitrary fixed point on the stationary surface s, we are concerned only with the explicit variation of $\mathbf{B}$ with time. So in Faraday's law a partial time derivative is required, and

$$\oint_c \mathbf{E} \cdot d\mathbf{l} = -\int_s \frac{\partial \mathbf{B}}{\partial t} \cdot d\mathbf{a}. \tag{3}$$

Then, using Stokes's theorem, we obtain

$$\oint_c \mathbf{E} \cdot d\mathbf{l} = \int_s (\nabla \times \mathbf{B}) \cdot d\mathbf{a} = -\int_s \frac{\partial \mathbf{B}}{\partial t} \cdot d\mathbf{a}$$

or

$$\int_s \left[(\nabla \times \mathbf{E}) + \frac{\partial \mathbf{B}}{\partial t} \right] \cdot d\mathbf{a} = 0. \tag{4}$$

Now since (4) is true for any surface spanning c, the term in square brackets is zero. Hence (2).

(b) For a moving circuit or medium, we must replace the partial time derivative in (3) with a total derivative. Then

$$\oint_c \mathbf{E} \cdot d\mathbf{l} = -\int_s \frac{d\mathbf{B}}{dt} \cdot d\mathbf{a}. \tag{5}$$

In (5) we recognize that $\dfrac{d\mathbf{B}}{dt} = \dfrac{\partial \mathbf{B}}{\partial t} + (\mathbf{v} \cdot \nabla)\mathbf{B}$. Then using (8) of Question 1.8 to re-write the convective derivative and putting $\nabla \cdot \mathbf{B} = 0$ give

$$\nabla \times (\mathbf{v} \times \mathbf{B}) = -\mathbf{B}(\nabla \cdot \mathbf{v}) + (\mathbf{B} \cdot \nabla)\mathbf{v} - (\mathbf{v} \cdot \nabla)\mathbf{B}. \tag{6}$$

The first and second terms on the right-hand side of (6) are zero because $\mathbf{v}$ is a function of t only, and so the convective derivative becomes

$$\frac{d\mathbf{B}}{dt} = \frac{\partial \mathbf{B}}{\partial t} + (\mathbf{v} \cdot \nabla)\mathbf{B} = \frac{\partial \mathbf{B}}{\partial t} - \nabla \times (\mathbf{v} \times \mathbf{B}). \tag{7}$$

Substituting (7) in (5) yields

$$\oint_c \mathbf{E}' \cdot d\mathbf{l} = -\int_s \left[\frac{\partial \mathbf{B}}{\partial t} - \nabla \times (\mathbf{v} \times \mathbf{B}) \right] \cdot d\mathbf{a},$$

where $\mathbf{E}'$ is the electric field in the rest frame of the circuit (see Comment (i) of Question 5.3). Proceeding as for (a) above then gives

$$\int_s \nabla \times \left[\mathbf{E}' - (\mathbf{v} \times \mathbf{B}) \right] \cdot d\mathbf{a} = -\int_s \frac{\partial \mathbf{B}}{\partial t} \cdot d\mathbf{a},$$

implying, as before, that

$$\nabla \times \left[\mathbf{E}' - (\mathbf{v} \times \mathbf{B}) \right] = -\frac{\partial \mathbf{B}}{\partial t}. \tag{8}$$

Now the term in square brackets in (8) is the electric field $\mathbf{E}$ measured in the rest frame of the *observer*, and so

$$\mathbf{E} = \mathbf{E}' - \mathbf{v} \times \mathbf{B}, \tag{9}$$

giving

$$\nabla \times \mathbf{E} = -\frac{\partial \mathbf{B}}{\partial t}.$$

This is the same equation as (2).

Comments

(i) Equation (9) is the non-relativistic transformation of the electric field for two inertial observers moving relative to each other. The $\mathbf{v} \times \mathbf{B}$ term produces an emf

$$\mathcal{E}_{\text{mot}} = \oint_c (\mathbf{v} \times \mathbf{B}) \cdot d\mathbf{l}, \tag{10}$$

which, for obvious reasons, is named *motional emf*.

(ii) In Question 5.5, we use Galilean invariance to derive the (non-relativistic) transformation of electric and magnet fields for observers in relative motion. Questions 14.8–14.10 of Ref. [5] offer a brief introduction to Galilean relativity and will be helpful to readers who wish to revise this topic.

[5]   O. L. de Lange and J. Pierrus, *Solved problems in classical mechanics: Analytical and numerical solutions with comments.* Oxford: Oxford University Press, 2010.

Question 5.5

A charge q moves with velocity $\mathbf{u}$ (relative to an inertial reference frame S) in an electric field $\mathbf{E}$ and a magnetic field $\mathbf{B}$. Suppose $E \ll cB$. Show that in an inertial frame S′, moving with velocity $\mathbf{v}$ relative to S, the fields are

$$\left.\begin{array}{c} \mathbf{E}' = \mathbf{E} + \mathbf{v} \times \mathbf{B} \\ \mathbf{B}' = \mathbf{B} \end{array}\right\}, \tag{1}$$

assuming $v \ll c$.

Hint: Use the force transformation $\mathbf{F}' = \mathbf{F}$ of Galilean relativity discussed in Comment (ii) of Question 12.4.

Solution

The force on q is $\mathbf{F} = q(\mathbf{E} + \mathbf{u} \times \mathbf{B})$ in S and $q(\mathbf{E}' + \mathbf{u}' \times \mathbf{B}')$ in S′ where $\mathbf{u}' = \mathbf{u} - \mathbf{v}$ (note that electric charge is an invariant, and so has the same value in any reference frame). Now because of the hint,

$$\mathbf{E} + \mathbf{u} \times \mathbf{B} = \mathbf{E}' + (\mathbf{u} - \mathbf{v}) \times \mathbf{B}',$$

which is true for all $\mathbf{u}$ if $\mathbf{B}' = \mathbf{B}$, and then $\mathbf{E}' = \mathbf{E} + \mathbf{v} \times \mathbf{B}$. Hence (1).

Comment

The transformation (1), derived from Galilean relativity, is valid at speeds $v \ll c$. It is a limiting case of a more general result which is based on the Lorentz transformation. The relativistically correct transformation of the electromagnetic field between inertial reference frames is considered in Chapter 12; see (2) of Question 12.8.

Question 5.6

The figure on p. 257 shows a stationary rectangular wire track abcd lying in the vertical plane and perpendicular to a uniform magnetic field $\mathbf{B}$. The horizontal rail b′c′ having length ℓ is free to move in the vertical direction without friction. At time $t = 0$ the rail is released from rest. Throughout its fall, suppose that the rail maintains good electrical contact with the track. Stating any assumptions which you make, show that the current induced in the circuit is

$$i(t) = \frac{\alpha t}{1 + \beta t^2}, \tag{1}$$

where α and β are positive constants.

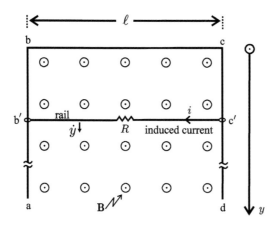

Solution

The instantaneous flux through the circuit b'bcc' at time t is $\phi(t) = B\ell y$, where y is the distance of the rail (from the origin) at this time. Faraday's flux rule gives the induced emf $\mathcal{E} = d\phi/dt = B\ell dy/dt$. According to Lenz's law, the direction of the induced current $i(t)$ must be clockwise, as shown. Assuming the resistance R of the circuit is proportional to y, we have $R = R_0 + k(y - y_0)$. Here, R_0 is the resistance at $t = 0$ and the constant $k = 2 \times$ the resistivity of the wire $\div$ its cross-sectional area. Now $i = \mathcal{E}/R$, and so

$$i(t) = \frac{B\ell dy/dt}{R_0 + k(y - y_0)}.$$ (2)

Furthermore, if we assume that B is weak, the retarding force exerted on the rail is very much less than its weight (the rail is in free fall). Then $y = y_0 + \frac{1}{2}gt^2$ with $dy/dt = gt$, and (2) becomes

$$i(t) = \frac{B\ell gt}{R_0 + \frac{1}{2}kgt^2}.$$

But this is (1), if we let $\alpha = B\ell g/R_0$ and $\beta = kg/2R_0$.

Question 5.7

A rectangular coil having area A, resistance R, self-inductance L and N turns rotates with constant angular velocity ω about an axis perpendicular to a uniform magnetic field **B**.

(a) Show that the current induced in the coil is given by

$$i(t) = \frac{BAN\omega}{(R^2 + \omega^2 L^2)^{1/2}} \sin(\theta_0 + \omega t - \phi),$$ (1)

where $\phi = \tan^{-1} \omega L/R$ and θ_0 is the angle between $\mathbf{B}$ and the normal $\hat{n}$ to the coil at time $t = 0$.

(b) Suppose that at $t = 0$ the plane of the coil is perpendicular to $\mathbf{B}$ (i.e. $\theta_0 = 0$). Calculate the torque acting at time t, and then show that the time-average torque required to keep the coil turning is

$$\langle \Gamma \rangle = \frac{B^2 A^2 N R}{2(R^2 + \omega^2 L^2)} \omega, \tag{2}$$

assuming that frictional forces are negligible.

Solution

(a) Because the field is uniform, the total magnetic flux linking the coil is $\phi = N\mathbf{B} \cdot \hat{n} A = BAN \cos\theta$ where $\theta = \theta_0 + \omega t$ is the instantaneous angle between $\mathbf{B}$ and $\hat{n}$. The induced emf follows from Faraday's flux rule and is $\mathcal{E} = -d\phi/dt = BAN\omega \sin(\theta_0 + \omega t)$. Now a back emf $-Ldi/dt$ drops the potential of an electric charge moving through it, just like a resistor, and so may be regarded as a potential drop $v = Ldi/dt$ in the direction of i. Then Kirchhoff's loop rule (net emf in circuit = sum of voltage drops; see p. 286) gives

$$\mathcal{E} = L\frac{di}{dt} + Ri,$$

or

$$L\frac{di}{dt} + Ri = BAN\omega \sin(\theta_0 + \omega t). \tag{3}$$

We attempt a solution to (3) of the form $i(t) = i_0 \sin(\theta_0 + \omega t - \phi)$ where i_0 and ϕ are constants which must be determined. Substituting this trial solution in (3) and using elementary trigonometric identities yield the equation

$$i_0 \left[\omega\tau \cos\phi - \sin\phi \right] \cos(\theta_0 + \omega t) + \left[i_0\omega\tau \sin\phi + i_0 \cos\phi - \frac{BAN\omega}{R} \right] \sin(\theta_0 + \omega t) = 0,$$

where $\tau = L/R$ is a characteristic time. Now since this last equation must be true for arbitrary t, we require that each of the coefficients in square brackets is zero. That is,

$$\left. \begin{array}{l} \omega\tau \cos\phi - \sin\phi = 0 \\ i_0\omega\tau \sin\phi + i_0 \cos\phi - BAN\omega/R = 0 \end{array} \right\}. \tag{4}$$

Solving $(4)_1$ and $(4)_2$ simultaneously gives

$$\left. \begin{array}{l} i_0 = \dfrac{BAN\omega}{R(1 + \omega^2\tau^2)^{1/2}} \\[2mm] \phi = \tan^{-1} \omega\tau \end{array} \right\}. \tag{5}$$

Hence $i = \dfrac{BAN\omega\sin(\theta_0 + \omega t - \phi)}{R(1 + \omega^2\tau^2)^{1/2}}$, which is (1).

(b) The uniform nature of **B** allows us to calculate the torque acting on the coil using the formula for a point magnetic dipole $\big($see (13) of Question 4.28$\big)$

$$\mathbf{\Gamma} = \mathbf{m} \times \mathbf{B}, \tag{6}$$

where the dipole moment **m**, given by (1) of Question 4.25, equals the current per turn × the number of turns × the vector area. That is,

$$\mathbf{m} = iA\hat{\mathbf{n}} \tag{7}$$

with i given by (1). Then from (1), (6) and (7) we obtain

$$\mathbf{\Gamma} = \frac{BA^2N\omega\sin(\theta_0 + \omega t - \phi)}{(R^2 + \omega^2L^2)^{1/2}}\,\hat{\mathbf{n}} \times \mathbf{B}.$$

Now $\hat{\mathbf{n}} \times \mathbf{B}$ is a vector along the axis of rotation of the coil (z, say). Then $\hat{\mathbf{n}} \times \mathbf{B} = B\sin(\theta_0 + \omega t)\hat{\mathbf{z}}$, and so

$$
\begin{aligned}
\mathbf{\Gamma} &= \frac{B^2A^2N\omega\sin(\theta_0 + \omega t - \phi)\sin(\theta_0 + \omega t)}{(R^2 + \omega^2L^2)^{1/2}}\,\hat{\mathbf{z}} \\[2mm]
&= \frac{B^2A^2N\sin(\theta_0 + \omega t - \phi)\sin(\theta_0 + \omega t)}{(R^2 + \omega^2L^2)^{1/2}}\,\boldsymbol{\omega},
\end{aligned}
\tag{8}
$$

where, in the last step, we recognize that $\boldsymbol{\omega} = \omega\hat{\mathbf{z}}$. Using trigonometric identities, and for the initial condition $\theta_0 = 0$, (8) becomes

$$\mathbf{\Gamma}(t) = \frac{B^2A^2N(\sin^2\omega t\cos\phi - \cos\omega t\sin\omega t\sin\phi)}{(R^2 + \omega^2L^2)^{1/2}}\,\boldsymbol{\omega}. \tag{9}$$

The time average of (9) follows immediately from (3) of Question 1.27 and is

$$\langle\mathbf{\Gamma}\rangle = \frac{B^2A^2N\cos\phi}{2(R^2 + \omega^2L^2)^{1/2}}\,\boldsymbol{\omega}. \tag{10}$$

Now $\cos\phi = \dfrac{1}{\sqrt{1 + \tan^2\phi}}$, and because of (5)$_2$ we obtain $\cos\phi = \dfrac{R}{(R^2 + \omega^2L^2)^{1/2}}$. Hence (2).

Comment

We see from (1) that the current in the coil lags the induced emf in phase by $\phi = \tan^{-1}\omega L/R$. This is a standard result in circuit theory; see Question 6.1.

Question 5.8[*]

A circular coil of wire has radius a, resistance R, inductance L and N turns. It lies in the xy-plane of Cartesian coordinates centred on the origin with its symmetry axis along $\hat{\mathbf{z}}$. A small cylindrical bar magnet having moment $\mathbf{m} = m_0\hat{\mathbf{z}}$ is released from rest on the z-axis above the plane of the coil. The magnet accelerates under gravity, passes through the centre of the coil, and continues to fall indefinitely.

(a) Show that the current induced in the coil satisfies the equation

$$L\frac{di}{dt} + Ri = \frac{3\mu_0 m_0 N a^2 z v}{2(a^2 + z^2)^{5/2}}, \tag{1}$$

where $v = dz/dt$ is the instantaneous velocity of the magnet.

Hint: Assume that the field of the bar magnet is that of a point magnetic dipole.

(b) Show that the equation of motion of the magnet (having mass M) is

$$\frac{d^2z}{dt^2} + \frac{3\mu_0 m_0 N i a^2 z}{2M(a^2 + z^2)^{5/2}} + g = 0. \tag{2}$$

(c) Calculate the total charge Δq that is displaced in the circuit when the bar magnet, released from $z = z_0$, moves through the coil and then travels towards negative infinity.

Solution

(a) The total magnetic flux linking the coil $= N\times$ the flux linkage per turn $= N\oint_c \mathbf{A}\cdot d\mathbf{l}$ (see (1) of Question 4.14), where, for a point magnetic dipole at the origin, the vector potential is given (1) of Question 4.10. If the dipole is displaced from the origin by $\mathbf{r}'$, then

$$\mathbf{A}(\mathbf{r}) = \frac{\mu_0}{4\pi}\frac{\mathbf{m}\times(\mathbf{r}-\mathbf{r}')}{|\mathbf{r}-\mathbf{r}'|^3},$$

and

$$\Phi = \frac{\mu_0 N}{4\pi}\oint_c \frac{[\mathbf{m}\times(\mathbf{r}-\mathbf{r}')]\cdot d\mathbf{l}}{|\mathbf{r}-\mathbf{r}'|^3}. \tag{3}$$

In this problem $\mathbf{r}' = z\hat{\mathbf{z}}$, $\mathbf{r} = a\hat{\mathbf{r}}$ and $(\mathbf{r}-\mathbf{r}') = a\hat{\mathbf{r}} - z\hat{\mathbf{z}}$. So $\mathbf{m}\times(\mathbf{r}-\mathbf{r}') = m_0 a\hat{\boldsymbol{\theta}}$ (a cylindrical polar unit vector), $d\mathbf{l} = ad\theta\hat{\boldsymbol{\theta}}$ and $[\mathbf{m}\times(\mathbf{r}-\mathbf{r}')]\cdot d\mathbf{l} = m_0 a^2 d\theta$. Substituting this last result in (3) gives

$$\Phi = \frac{\mu_0 N m_0 a^2}{4\pi}\oint_c \frac{d\theta}{|\mathbf{r}-\mathbf{r}'|^3} = \frac{\mu_0 N m_0 a^2}{4\pi}\oint_c \frac{d\theta}{(a^2 + z^2)^{3/2}} = \frac{\mu_0 N m_0 a^2}{2(a^2 + z^2)^{3/2}}. \tag{4}$$

The emf induced in the coil follows from (4) and the flux rule:

$$\mathcal{E} = -\frac{d\phi}{dt} = -\frac{d\phi}{dz}\frac{dz}{dt} = -\frac{\mu_0 N m_0 a^2 v}{2}\frac{d}{dz}(a^2 + z^2)^{-3/2} = \frac{3\mu_0 N m_0 a^2 z v}{2(a^2 + z^2)^{5/2}}. \quad (5)$$

As in Question 5.7, we apply Kirchhoff's loop rule to the coil and obtain the equation $L\dfrac{di}{dt} + Ri = \mathcal{E}$, which is (1).

(b) The induced current $i(t)$ produces a magnetic field which tends to repel (attract) the bar magnet when it is above (below) the coil. On the z-axis, this field is given by (1) of Question 4.8: $\mathbf{B} = \dfrac{\mu_0 N i a^2 \hat{\mathbf{z}}}{2(a^2 + z^2)^{3/2}}$, where, for obvious reasons, we include a factor N. Then, because of (12) of Question 4.28, the repulsive (attractive) force exerted on the magnet is

$$\mathbf{F} = m_0\frac{\partial \mathbf{B}}{\partial z} = \frac{\mu_0 m_0 N i a^2}{2}\frac{\partial}{\partial z}(a^2 + z^2)^{-3/2}\hat{\mathbf{z}} = -\frac{3\mu_0 m_0 N i a^2 z}{2(a^2 + z^2)^{5/2}}\hat{\mathbf{z}}, \quad (6)$$

where the signs of both i and z determine the direction of $\mathbf{F}$. From Newton's second law, net force = mass × acceleration, we obtain

$$\mathbf{F} + M\mathbf{g} = \mathbf{F} - Mg\hat{\mathbf{z}} = M\frac{d^2\mathbf{z}}{dt^2}. \quad (7)$$

Substituting (6) in (7) and rearranging yield (2).

(c) Replacing Ri in (1) with Rdq/dt gives $Ldi + Rdq = \dfrac{3\mu_0 m_0 N a^2 z\,dz}{2(a^2 + z^2)^{5/2}}$. Then

$$L\int_{z_0}^{-\infty} di + R\int_{z_0}^{-\infty} dq = \frac{3\mu_0 m_0 N a^2}{2}\int_{z_0}^{-\infty}\frac{z\,dz}{(a^2 + z^2)^{5/2}}.$$

Now $i(z_0) = 0$ and $i(-\infty) = 0$ (the transient current decays to zero as the magnet recedes from the coil). So

$$R\Delta q = \frac{3\mu_0 m_0 N a^2}{2}\int_{z_0}^{-\infty}\frac{z\,dz}{(a^2 + z^2)^{5/2}}. \quad (8)$$

The integral in (8) is trivial, and equals $-\frac{1}{3}(a^2 + z_0^2)^{-3/2}$. Hence

$$\Delta q = \frac{\mu_0 m_0 N a^2}{2R(a^2 + z_0^2)^{3/2}}. \quad (9)$$

Comments

(i) With the dipole moment along $\hat{\mathbf{z}}$, the north pole of the magnet is above its south pole. Lenz's law determines the direction of i, which is shown in the following figure when the magnet is above (below) the coil, $z > 0$ ($z < 0$).

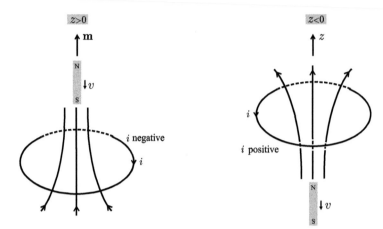

(ii) In Question 5.9, we use *Mathematica* to find numerical solutions of (1) and (2).

(iii) Equation (9) provides a convenient way for determining the dipole moment of a bar magnet dropped through a coil. A measurement of Δq (current-integrating electronic devices are usually available in undergraduate laboratories) immediately yields a value for m_0.

Question 5.9

Consider the bar magnet moving through the coil described in Question 5.8. Suppose that $a = 8\,\text{cm}$, $R = 20\,\Omega$, $L = 80\,\text{mH}$, $N = 1000$, $M = 50\,\text{g}$, $m_0 = 0.25\,\text{A m}^2$ and $g = 10\,\text{m s}^{-2}$. The initial conditions are $v(0) = 0$, $z(0) = 2a$ and $i(0) = 0$.

(a) Write a *Mathematica* notebook to solve the differential equations (1) and (2) of Question 5.8. Then plot a graph of $i(t)$ vs t. On the same axes, also show the graph of $i(t)$ vs t, assuming $L = 0$.

(b) Use the solution for $i(t)$ from (a) to calculate the energy dissipated in the coil.

(c) Use (9) of Question 5.8 to calculate the total charge Δq that moves through the coil.

Solution

(a) The graph shown on p. 263 was produced by `cell 1` of the notebook on p. 264.

(b) From $\mathcal{P} = Ri^2(t)$, we determine the energy converted to heat

$$E_{\text{heat}} = R \int_0^{t_{\text{max}}} i^2(t)\, dt, \tag{1}$$

where the time t_{max} should be sufficiently long enough to ensure that $i(t)$ has decayed (essentially) to zero. Using the notebook to evaluate (1) with $t_{\text{max}} = 3\,\text{s}$ gives $E_{\text{heat}} = 4.5\,\mu\text{J}$ (see `cell 2` on p. 264).

(c) For the coil having the properties stated above, we obtain the value $\Delta q = 8.78\,\mu\text{C}$.

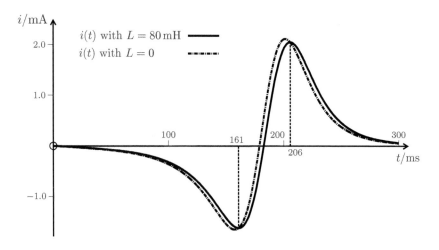

Comments

(i) The graph of $i(t)$ reveals the following:

☞ At $t = 161$ ms the current magnitude is smaller than the current magnitude at $t = 206$ ms. Obviously the speed of the magnet increases in this time interval, resulting in a greater rate of change of flux, and consequently a bigger induced current as the magnet enters and then recedes from the coil.

☞ Evidently, the inductance of the coil has only a small perturbing effect on $i(t)$ for the given parameters.

(ii) Because of energy conservation, the decrease in the total mechanical energy of the bar magnet should equal E_{heat}. That is,

$$E_{\text{heat}} = \tfrac{1}{2}Mv_{\text{initial}}^2 - M(gz + \tfrac{1}{2}v^2)_{\text{final}}, \tag{2}$$

where the final values of z and v must be taken as $t \to \infty$ (say 3 s, as with the integration in (b)). In cell 3 of the notebook on p. 264, we confirm that (2) also gives $E_{\text{heat}} = 4.5\,\mu\text{J}$.

(iii) The force F (defined in Question 5.8) acting on the magnet is plotted as a function of time in units of Mg (see p. 264). It turns out that F is several orders of magnitude smaller than the weight of the magnet. (This confirms an earlier assumption that the bar magnet is essentially in free fall throughout). Because of Lenz's law, F always acts in the positive z-direction except, curiously, for a very brief period of time when $z(t)$ changes sign and before $i(t)$ changes sign (there is a slight time lag between the displacement and the current; $z(t)$ passes through zero first). During this short time interval, the coil effectively pushes the magnet downwards, albeit very slightly. Once $i(t)$ has reversed direction, the direction of F is again upward. This effect is due to the inductance of the coil and it disappears when $L = 0$, as the reader can verify by experimenting with the notebook below.

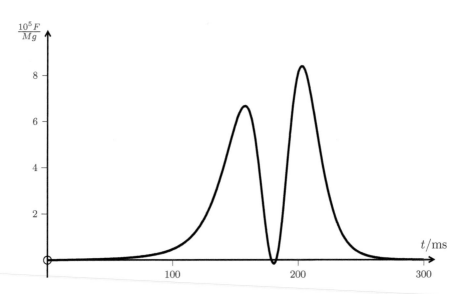

```
In[1]:=   (* In this notebook, we represent the current by Y(t)
            since i is a protected symbol in Mathematica *)

          L = 80 × 10⁻³;  M = 50 × 10⁻³;  a = 8 × 10⁻²;  μ0 = 4 π × 10⁻⁷;
          R = 20;  g = 10;  m0 = 0.25;  No = 1000;
          I0 = 0;  z0 = 2 a;  vz0 = -0;  tmax = 3;
```

$$eqn1 = \left\{L\,Y'[t] + R\,Y[t] - \frac{3\,\mu 0\,m0\,a^2\,No\,z[t]\,z'[t]}{2\,\left(a^2 + z[t]^2\right)^{\frac{5}{2}}} == 0\right\};$$

$$eqn2 = \left\{z''[t] + g + \frac{3\,\mu 0\,No\,m0\,a^2\,Y[t]\,z[t]}{2\,M\,\left(a^2 + z[t]^2\right)^{\frac{5}{2}}} == 0\right\};$$

```
          Sol = NDSolve[{eqn1, eqn2, Y[0] == I0, z[0] == z0, z'[0] == vz0}, {Y[t],
              Y'[t], z[t], z'[t], z''[t]}, {t, 0, tmax}, AccuracyGoal → 20];
```

$$\text{Plot}\left[\{1000\,\text{Evaluate}[Y[t]\ /.\ \text{Sol}]\},\ \left\{t,\ 0,\ \frac{tmax}{10}\right\},\ \text{PlotRange} \to \text{All}\right]$$

In[2]:= $\text{Eheat} = \text{NIntegrate}\left[\text{Evaluate}\left[R\,Y[t]^2\ /.\ \text{Sol}\right],\ \{t,\ 0,\ tmax\}\right]$

In[3]:= $\text{ChangeInMechEnergy} = \left(\text{Evaluate}\left[M\,g\,z[t]\ /.\ \text{Sol}\right]\ /.\ t \to 0\right) -$
$\qquad \text{Evaluate}\left[M\left(\frac{1}{2}\,z'[t]^2 + g\,z[t]\right)\ /.\ \text{Sol}\right]\ /.\ t \to tmax$

Question 5.10

A small permanent magnet centred on an origin O and having magnetic dipole moment $\mathbf{m}$ rotates about a perpendicular axis with constant angular velocity $\boldsymbol{\omega}$.

(a) Show that the electric field induced at a field point P is

$$\mathbf{E}(\mathbf{r}, t) = \frac{\mu_0}{4\pi} \frac{\mathbf{r} \times (\boldsymbol{\omega} \times \mathbf{m})}{r^3}, \tag{1}$$

where $\mathbf{r}$ is the position vector of point P relative to O.

Hint: Use (2) of Question 5.4.

(b) Calculate $\mathbf{E}(\mathbf{r}, t)$ at the point $P(x, y, z)$, assuming that $\mathbf{m} = m\hat{\mathbf{x}}$ at time $t = 0$.

Solution

(a) Substituting $\mathbf{B} = \nabla \times \mathbf{A}$ in the differential form of Faraday's law gives

$$\nabla \times \mathbf{E} = -\frac{\partial \mathbf{B}}{\partial t} = -\frac{\partial}{\partial t} \nabla \times \mathbf{A} = -\nabla \times \frac{\partial \mathbf{A}}{\partial t}.$$

Now it follows from this equation and (2) of Question 1.14 that $\nabla \times \left(\mathbf{E} + \dfrac{\partial \mathbf{A}}{\partial t} \right) = 0$ implies that $\mathbf{E} + \dfrac{\partial \mathbf{A}}{\partial t} = -\nabla \Phi(\mathbf{r})$, where $\Phi(\mathbf{r})$ is a scalar field and the minus sign is conventional. So

$$\mathbf{E} = -\frac{\partial \mathbf{A}}{\partial t} - \nabla \Phi(\mathbf{r}).$$

In Question 8.1 we discover that Φ is the electric scalar potential. Because of this, and since the magnet is electrically neutral (a reasonable assumption), we let Φ equal zero. Thus

$$\mathbf{E} = -\frac{\partial \mathbf{A}}{\partial t} \quad \text{where} \quad \mathbf{A} = \frac{\mu_0}{4\pi} \frac{\mathbf{m} \times \mathbf{r}}{r^3} \quad \text{for a point magnetic dipole at } O,$$

and so

$$\mathbf{E} = -\frac{\mu_0}{4\pi} \frac{\partial \mathbf{m}/\partial t \times \mathbf{r}}{r^3}. \tag{2}$$

If $\mathbf{m}$ lies in the xy-plane we have $\mathbf{m} = m(\hat{\mathbf{x}} \cos \omega t + \hat{\mathbf{y}} \sin \omega t)$ and $\boldsymbol{\omega} = \omega\hat{\mathbf{z}}$. Then $\partial \mathbf{m}/\partial t = m\omega(-\hat{\mathbf{x}} \sin \omega t + \hat{\mathbf{y}} \cos \omega t) = \boldsymbol{\omega} \times \mathbf{m}$. Substituting this last result in (2) gives (1).

(b) The initial condition is satisfied taking $\mathbf{m} = m(\hat{\mathbf{x}} \cos \omega t + \hat{\mathbf{y}} \sin \omega t)$. Then

$$\mathbf{r} \times (\boldsymbol{\omega} \times \mathbf{m}) = (\hat{\mathbf{x}}x + \hat{\mathbf{y}}y + \hat{\mathbf{z}}z) \times m\omega(-\hat{\mathbf{x}} \sin \omega t + \hat{\mathbf{y}} \cos \omega t)$$

$$= m\omega(\hat{\mathbf{x}}z \cos \omega t - \hat{\mathbf{y}}z \sin \omega t + \hat{\mathbf{z}}(x \cos \omega t + y \sin \omega t)). \tag{3}$$

Substituting (3) in (1) yields

$$\mathbf{E}(\mathbf{r},t) \;=\; \frac{\mu_0}{4\pi}\,\frac{m\omega\big[(\hat{\mathbf{z}}x - \hat{\mathbf{x}}z)\cos\omega t \;+\; (\hat{\mathbf{z}}y - \hat{\mathbf{y}}z)\sin\omega t\big]}{(x^2 + y^2 + z^2)^{3/2}}. \tag{4}$$

Comments

(i) Under normal laboratory conditions, the electric field given by (4) is weak. For instance, at a distance 1 cm from a bar magnet ($m=0.3\,\mathrm{A\,m^2}$) rotating on a wood lathe at 3000 r.p.m. ($\omega = 100\pi\,\mathrm{rad\,s^{-1}}$), the maximum value of E is of the order

$$E \sim \frac{4\pi \times 10^{-7} \times 0.3 \times 100\pi}{4\pi(0.01)^2} \sim 0.1\,\mathrm{V\,m^{-1}}.$$

(ii) It is interesting to plot (4) in the $z = 0$ plane, and view the field dynamically. This can be done with the following *Mathematica* notebook:

```
In[1]:= Tmax = 2;

        EX[X_, Y_, T_] := (X Cos[2 π T])/(X² + Y²)^(3/2);   EY[X_, Y_, T_] := (Y Sin[2 π T])/(X² + Y²)^(3/2);

        Field[T_] := StreamPlot[{EX[X, Y, T], EY[X, Y, T]}, {X, -2, 2}, {Y, -2, 2}]

In[2]:= Manipulate[Field[T], {T, 0, Tmax}]
```

Question 5.11

A small permanent magnet having magnetic dipole moment $\mathbf{m}$ moves with constant velocity $\mathbf{v}$ where $v \ll c$. Suppose that at time $t = 0$ the magnet passes through the origin O.

(a) Show that the electric field induced at a field point P is

$$\mathbf{E}(\mathbf{r},t) \;=\; \frac{\mu_0}{4\pi}\left[\frac{(r^2 - 2v^2t^2 + \mathbf{r}\cdot\mathbf{v}t)(\mathbf{m}\times\mathbf{v}) \;+\; 3(v^2t - \mathbf{r}\cdot\mathbf{v})(\mathbf{m}\times\mathbf{r})}{(r^2 + v^2t^2 - 2\mathbf{r}\cdot\mathbf{v}t)^{5/2}}\right], \tag{1}$$

where $\mathbf{r}$ is the position vector of P relative to O.

(b) Suppose the magnet (moment $\mathbf{m} = m\hat{\mathbf{x}}$) moves with velocity $\mathbf{v} = v\hat{\mathbf{y}}$. Calculate $\mathbf{E}(\mathbf{r},t)$ at the point $\mathrm{P}(x_0, 0, 0)$.

(c) Suppose the magnet (moment $\mathbf{m} = m\hat{\mathbf{z}}$) moves with velocity $\mathbf{v} = v\hat{\mathbf{y}}$. Calculate $\mathbf{E}(\mathbf{r},t)$ at the point $\mathrm{P}(x_0, 0, 0)$.

Solution

(a) Let $\mathbf{R} = \mathbf{r} - \mathbf{v}t$ be the vector from the instantaneous position of the magnet to P. Then the vector potential is given by

$$\mathbf{A} = \frac{\mu_0}{4\pi} \frac{\mathbf{m} \times \mathbf{R}}{R^3}$$

(see (1) of Question 4.10), where $R = (r^2 + v^2 t^2 - 2\mathbf{r} \cdot \mathbf{v}t)^{1/2}$. Now, as in Question 5.10, $\mathbf{E} = -\partial \mathbf{A}/\partial t$, and so

$$\mathbf{E}(\mathbf{r}, t) = -\frac{\mu_0}{4\pi} \mathbf{m} \times \frac{\partial}{\partial t} \frac{\mathbf{R}}{R^3}$$

$$= -\frac{\mu_0}{4\pi} \left[\frac{R^3 \left(\mathbf{m} \times \frac{\partial \mathbf{R}}{\partial t} \right) - 3 \left(\mathbf{m} \times \mathbf{R} \right) R^2 \frac{\partial R}{\partial t}}{R^6} \right], \tag{2}$$

where $\dfrac{\partial \mathbf{R}}{\partial t} = -\mathbf{v}$, $2R\dfrac{\partial R}{\partial t} = (2v^2 t - 2\mathbf{r} \cdot \mathbf{v})$ or $\dfrac{\partial R}{\partial t} = \dfrac{v^2 t - \mathbf{r} \cdot \mathbf{v}}{R}$, and $(\mathbf{m} \times \mathbf{R}) = (\mathbf{m} \times \mathbf{r}) - (\mathbf{m} \times \mathbf{v}t)$. Substituting these results in (2) gives

$$\mathbf{E}(\mathbf{r}, t) = -\frac{\mu_0}{4\pi} \left[\frac{-R^2 \left(\mathbf{m} \times \mathbf{v} \right) - 3 \left(\mathbf{m} \times (\mathbf{r} - \mathbf{v}t) \right) \left(v^2 t - \mathbf{r} \cdot \mathbf{v} \right)}{R^5} \right]. \tag{3}$$

Replacing R in (3) with $(r^2 + v^2 t^2 - 2\mathbf{r} \cdot \mathbf{v}t)^{1/2}$ gives (1) after a little algebra.

(b) Here $\mathbf{m} \times \mathbf{v} = mv\hat{\mathbf{z}}$, $\mathbf{m} \times \mathbf{r} = 0$ and $\mathbf{r} \cdot \mathbf{v} = 0$. Substituting these results in (1) gives

$$\mathbf{E}(x_0, 0, 0, t) = \frac{\mu_0 mv}{4\pi} \frac{(x_0^2 - 2v^2 t^2)\hat{\mathbf{z}}}{(x_0^2 + v^2 t^2)^{5/2}}. \tag{4}$$

(c) Now $\mathbf{m} \times \mathbf{v} = -mv\hat{\mathbf{x}}$, $\mathbf{m} \times \mathbf{r} = mx_0\hat{\mathbf{y}}$, $\mathbf{r} \cdot \mathbf{v} = 0$ and (1) becomes

$$\mathbf{E}(x_0, 0, 0, t) = \frac{\mu_0 mv}{4\pi} \left[\frac{(2v^2 t^2 - x_0^2)\hat{\mathbf{x}} + 3x_0 vt\,\hat{\mathbf{y}}}{(x_0^2 + v^2 t^2)^{5/2}} \right]. \tag{5}$$

Question 5.12

Consider a cylindrical region of space (radius r_0) where there is a uniform magnetic field $\mathbf{B}(t) = B(t)\,\hat{\mathbf{z}}$ (see Fig. (I) below). Use Faraday's law to calculate the induced electric field $\mathbf{E}(r, t)$ everywhere.

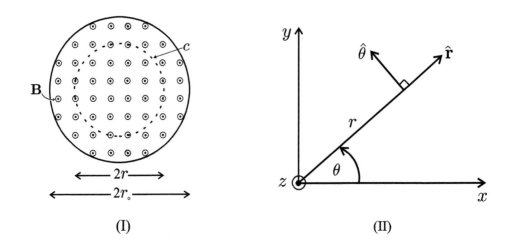

(I)   (II)

Solution

Because of the symmetry about the z-axis, there will be an induced electric field $\mathbf{E}(r, t) = E(r, t)\hat{\boldsymbol{\theta}}$ where $\hat{\boldsymbol{\theta}}$ is a cylindrical polar unit vector (see Fig. (II) above). This electric field is related to $\mathbf{B}$ by Faraday's law $\oint_c \mathbf{E} \cdot d\mathbf{l} = -\dfrac{d}{dt}\displaystyle\int_s \mathbf{B} \cdot d\mathbf{a}$. For the contour of integration c, we choose a circle of radius r lying in the xy-plane. The magnitude of $\mathbf{E}$ around this circle is momentarily constant, although its value changes from one instant to the next. Taking $d\mathbf{l} = rd\theta\hat{\boldsymbol{\theta}}$, $d\mathbf{a} = da\hat{\mathbf{z}}$ and $\mathbf{B} = B\hat{\mathbf{z}}$ gives

$$2\pi r E = -\frac{dB}{dt}\int_s da, \tag{1}$$

where the integral in (1) is either πr_0^2 for $r > r_0$ or πr^2 for $r < r_0$. It then follows that

$$\mathbf{E}(r, t) = \begin{cases} -\dfrac{r}{2}\dfrac{dB}{dt}\hat{\boldsymbol{\theta}} & \text{for } r < r_0 \\[2ex] -\dfrac{r_0^2}{2r}\dfrac{dB}{dt}\hat{\boldsymbol{\theta}} & \text{for } r > r_0. \end{cases} \tag{2}$$

Comments

(i) If the path c coincides with an ohmic conductor, then an induced current having density $\mathbf{J} = \sigma \mathbf{E}$ will be present in the wire (here σ is the electrical conductivity of the metal). Clearly, the direction of this current is determined by (2) and accords with Lenz's law.

(ii) This question illustrates yet again an important feature of the electromagnetic field: a time-dependent magnetic field induces an electric field. The converse of this statement is also true, and an example with the roles of $\mathbf{E}$ and $\mathbf{B}$ reversed can also be constructed.

Question 5.13

A particle with mass m and charge $q \, (> 0)$ moves in the magnetic field $\mathbf{B}(t) = B(t) \, \hat{\mathbf{z}}$ described in Question 5.12. In what follows, consider motion in the xy-plane only.

(a) Show that the equation of motion of the particle is

$$\left.\begin{aligned} m(\ddot{r} - r\dot{\theta}^2) &= qBr\dot{\theta} \\ m(r\ddot{\theta} + 2\dot{r}\dot{\theta}) &= q(E_\theta - B\dot{r}) \end{aligned}\right\}, \tag{1}$$

where r and θ are cylindrical polar coordinates and E_θ is the induced electric field given by (2) of Question 5.12.

(b) Suppose $B(t) = B_0(1 - e^{-t/t_0})$ where B_0 and t_0 are positive constants. Express (1) in the dimensionless form

$$\left.\begin{aligned} \frac{d^2 R}{d\tau^2} - R\left(\frac{d\theta}{d\tau}\right)^2 - R\left(\frac{d\theta}{d\tau}\right)(1 - e^{-\tau/\tau_0})\, f(R) &= 0 \\ R\frac{d^2\theta}{d\tau^2} + 2\frac{dR}{d\tau}\left(\frac{d\theta}{d\tau}\right) + g(R)\, e^{-\tau/\tau_0} + \frac{dR}{d\tau}(1 - e^{-\tau/\tau_0})\, f(R) &= 0 \end{aligned}\right\}, \tag{2}$$

where $R = r/r_0$ is a dimensionless coordinate (r_0 was defined in Question 5.12) and $\tau = \omega_B t$ is a dimensionless time (here $\omega_B = qB_0/m$ is the cyclotron frequency). The definitions of $f(R)$ and $g(R)$ are:

$$f(R) = \begin{cases} 1 & \text{for } R < 1 \\ 0 & \text{for } R \geq 1, \end{cases}$$

and

$$g(R) = \frac{1}{2\tau_0} \begin{cases} R & \text{for } R < 1 \\ R^{-1} & \text{for } R \geq 1. \end{cases}$$

(c) Taking $\tau_0 = 50$, write a *Mathematica* notebook to solve (2) for $R(\tau)$ and $\theta(\tau)$. Then plot the trajectory of the particle for $0 \le \tau \le 600$ using the following initial conditions:

1. $R(0) = \dfrac{2}{5}$; $\dot{R}(0) = 0$; $\theta(0) = 0$ and $\dot{\theta}(0) = 0$.

2. $R(0) = \dfrac{9}{10}$; $\dot{R}(0) = 0$; $\theta(0) = \frac{1}{6}\pi$ and $\dot{\theta}(0) = -\frac{1}{4}$.

Solution

(a) The velocity and acceleration of the particle, expressed in terms of cylindrical polar coordinates, are

$$\mathbf{v} = \dot{r}\hat{\mathbf{r}} + r\dot{\theta}\hat{\boldsymbol{\theta}} + \dot{z}\hat{\mathbf{z}} \qquad \text{and} \qquad \mathbf{a} = (\ddot{r} - r\dot{\theta}^2)\hat{\mathbf{r}} + (r\ddot{\theta} + 2\dot{r}\dot{\theta})\hat{\boldsymbol{\theta}} + \ddot{z}\hat{\mathbf{z}}.$$

Then with Newton's second law $m\mathbf{a} = q[\mathbf{E} + \mathbf{v} \times \mathbf{B}]$ we obtain (1), since $\dot{z} = 0$, $\ddot{z} = 0$, $\mathbf{E} = E_\theta\hat{\boldsymbol{\theta}}$, $\hat{\boldsymbol{\theta}} \times \hat{\mathbf{z}} = \hat{\mathbf{r}}$ and $\hat{\mathbf{r}} \times \hat{\mathbf{z}} = -\hat{\boldsymbol{\theta}}$.

(b) Substituting $B(t) = B_0(1 - e^{-t/t_0})f(R)$ in (1) yields (2) after some trivial algebra.

(c) In the following figures, the solid line gives the trajectory for $\tau \le 60$ and the dotted line shows the path followed by the particle in the limit $\tau \to \infty$. In the interval $60 < \tau < 594$ the trajectory has been suppressed for the purpose of presentation. The notebook on p. 271 is for the first set of initial conditions.

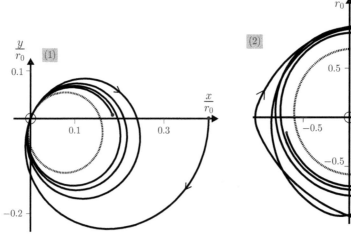

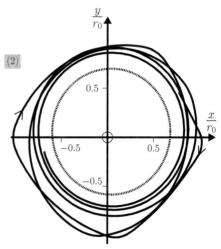

Comments

(i) The notebook below may readily be modified to study the dynamics of the particle in various other magnetic fields, such as

$$B(t) = B_0 \times \begin{cases} (1 + t/t_0), \\ e^{-t/t_0}, \\ (1 + t/t_0)^{-1} \\ \sin \omega t. \end{cases}$$

(ii) For $r < r_0$ the equation of motion $(1)_2$ can be written in terms of the angular momentum $L = mr^2 \dot{\theta}$ as follows: $\dfrac{1}{r}\dfrac{d}{dt}(mr^2\dot{\theta}) = \dfrac{1}{r}\dfrac{dL}{dt} = q(E_\theta - B\dot{r})$. From $(2)_1$ of Question 5.12, we have $E_\theta = -\frac{1}{2}r\dot{B}$, and so $\dfrac{d}{dt}(L + \frac{1}{2}qr^2B) = 0$. The quantity $(L + \frac{1}{2}qr^2B)$ is a constant of the motion; a fact which can easily be verified numerically by suitably adapting the given notebook.

(iii) This question illustrates an important principle. The kinetic energy of a charged particle moving in a time-dependent magnetic field changes because of the work done on it by the *induced electric* field. We already know that the magnetic field itself can do no work on the particle (see (7) of Question 4.1).

(iv) The principle whereby an induced electric field increases the kinetic energy of a charged particle was first applied successfully in the 1940s to produce high-energy electrons (beta particles) in accelerator devices known as betatrons.

```
In[1]:=  t0 = 50; tmax = 600; steps = 400 000; R0 = 0.4; θ0 = 0; dotR0 = 0; dotθ0 = 0;

In[2]:=  f[t_] := Piecewise[{{1, R[t] < 1}, {0, R[t] ≥ 1}}]

         g[t_] := Piecewise[{{R[t]/(2 t0), R[t] < 1}, {1/(2 t0 R[t]), R[t] ≥ 1}}]

         eqn1 = R''[t] - R[t] (θ'[t]² + θ'[t] (1 - Exp[-t/t0]) f[t])  == 0;

         eqn2 = R[t] θ''[t] + 2 R'[t] θ'[t] +
                g[t] Exp[-t/t0] + R'[t] (1 - Exp[-t/t0]) f[t] == 0;

         Sol = NDSolve[{eqn1, eqn2, R[0] == R0, R'[0] == dotR0, θ[0] == θ0,
                θ'[0] == dotθ0}, {R[t], θ[t]}, {t, 0, tmax}, MaxSteps → steps];

         ParametricPlot[{{R[t] Cos[θ[t]], R[t] Sin[θ[t]]} /. Sol}, {t, 0, tmax/10},
            PlotRange → {{-0.02, 0.42}, {-0.34, 0.1}}, PlotPoints → 1000]
```

Question 5.14**

(Questions 5.14–5.15 and their solutions adopt the approach of Ref. [6].)

Consider n rigid current-carrying circuits having arbitrary size, shape and orientation at rest in vacuum. Assume that the circuits are magnetically coupled to each other. The flux ϕ_j through the jth circuit is

$$\phi_j = \phi_{j1} + \phi_{j2} + \cdots + \phi_{jj} + \cdots + \phi_{jn} = \sum_{k=1}^{n} \phi_{jk}, \tag{1}$$

where ϕ_{jk} is the flux through circuit j due to current i_k in circuit k. The emf $\mathcal{E}_j$ induced in the jth circuit is

$$\mathcal{E}_j = -\frac{d\phi_j}{dt} = -\sum_{k=1}^{n} \frac{d\phi_{jk}}{dt} = -\sum_{k=1}^{n} \frac{d\phi_{jk}}{di_k}\frac{di_k}{dt}. \tag{2}$$

For quasi-stationary currents

$$M_{jk} = \frac{d\phi_{jk}}{di_k} \quad \text{and} \quad L_j = M_{jj} = \frac{d\phi_{jj}}{di_j}. \tag{3}$$

Here M_{jk} is the *mutual inductance* between the jth and kth circuits;‡ L_j is the *self-inductance* of circuit j.

(a) Consider the important case where the fluxes ϕ_{jk} are directly proportional to the currents i_k. Use (3) to establish alternative definitions for mutual and self-inductance.

(b) Consider two circuits, 1 and 2. Show that their mutual inductance is

$$M_{12} = \frac{\mu_0}{4\pi} \oint_{c_1} \oint_{c_2} \frac{dl_1 \cdot dl_2}{|r_1 - r_2|}. \tag{4}$$

Hint: Introduce the vector potential in the flux integral and use (9) of Question 4.5.

Solution

(a) From (3) we obtain $d\phi_{jk} = M_{jk}di_k$ and $d\phi_{jj} = L_j di_j$. Integrating these equations gives $\phi_{jk} = M_{jk}i_k$ and $\phi_{jj} = L_j i_j$ where the constraint of direct proportionality requires that both constants of integration are zero. It then follows that

‡The M_{jk}, known also as the coefficients of inductance, may be regarded as magnetostatic counterparts of the C_{jk} in the corresponding electrostatic problem (see Question 3.10).

[6]   J. R. Reitz and F. J. Milford, *Foundations of electromagnetic theory*, Chap. 12, pp. 231–5. Massachusetts: Addison-Wesley, 1960.

$$M_{jk} = \frac{\Phi_{jk}}{i_k} \qquad \text{and} \qquad L_j = \frac{\Phi_{jj}}{i_j} \quad \left(\text{or } L = \frac{\Phi}{i}\right). \tag{5}$$

(b) Using the definition $(5)_1$ gives $M_{12} = \dfrac{\Phi_{12}}{i_2}$ where $\Phi_{12} = \displaystyle\int_{s_1} \mathbf{B}_2 \cdot d\mathbf{a}_1$. So

$$M_{12}\, i_2 = \int_{s_1} \mathbf{B}_2 \cdot d\mathbf{a}_1 = \int_{s_1} (\nabla \times \mathbf{A}_2) \cdot d\mathbf{a}_1 = \oint_{c_1} \mathbf{A}_2 \cdot d\mathbf{l}_1, \tag{6}$$

because of Stokes's theorem. Substituting $\mathbf{A}_2 = \dfrac{\mu_0}{4\pi} \displaystyle\oint_{c_2} \dfrac{i_2\, d\mathbf{l}_2}{|\mathbf{r}_1 - \mathbf{r}_2|}$ in (6) yields (4).

Comments

(i) The following aspects of the foregoing discussion require clarification:

- ☞ The assumption of 'rigid circuits at rest' ensures that no flux changes occur because of a change in either shape or position. Flux changes can occur only through a change in one or more of the currents $(i_1, i_2, \ldots, i_k)$.

- ☞ Expressing the flux in terms of the circulation of $\mathbf{A}$ around a closed path $\big($see (6)$\big)$ assumes single-loop circuits. The flux in a multiloop circuit is enhanced by a factor N (the number of turns of wire).

(ii) The integral in (4) depends only on the circuit geometry and it represents an average separation of the two circuits. This average is weighted, through the dot product, in favour of parallel circuit elements. The symmetry of the subscripts in (4) shows that $M_{12} = M_{21}$, meaning that the two coefficients are identical. Because of this, we may represent M_{12} and M_{21} by the symbol M (without subscripts) and call it simply *the* mutual inductance between circuits c_1 and c_2.

(iii) The result (4) is known as Neumann's formula. It can be generalized as follows. Suppose c_1 (c_2) comprises N_1 (N_2) turns of wire, then

$$M = \frac{\mu_0}{4\pi} N_1 N_2 \oint_{c_1} \oint_{c_2} \frac{d\boldsymbol{\ell}_1 \cdot d\boldsymbol{\ell}_2}{|\mathbf{r}_1 - \mathbf{r}_2|}, \tag{7}$$

and is valid when the distance between the nearest part of c_1 and c_2 is much greater than the mean diameter of the wires. Clearly this condition is not satisfied for a single circuit (e.g. a coil) and Neumann's formula,

$$L = \frac{\mu_0}{4\pi} \oint_c \oint_{c'} \frac{d\boldsymbol{\ell} \cdot d\boldsymbol{\ell}'}{|\mathbf{r} - \mathbf{r}'|},$$

results in a divergent integral (note the singularity at $\mathbf{r} = \mathbf{r}'$) and an infinite L.

(iv) Neumann's formula is usually difficult to apply in practice, except in cases where the symmetry of the circuits is high.

Question 5.15 **

Consider a rigid circuit of arbitrary size, shape and orientation at rest in vacuum. Suppose that the circuit has a seat of emf $\mathcal{E}_b$ (a battery, say) connected in series with a switch S which is closed at time $t = 0$. This results in a time-dependent current[‡] $i(t)$ and a time-dependent magnetic flux in the circuit that induces a back emf $\mathcal{E}(t) = -d\phi/dt$. Kirchhoff's loop rule (see p. 286) requires that

$$\mathcal{E}(t) + \mathcal{E}_b = R\,i(t),$$

where R is the total resistance to the flow of charge. The work done by $\mathcal{E}_b$ in moving an increment of charge $dq = i\,dt$ around the circuit is

$$dW_b = \mathcal{E}_b\,dq = (-\mathcal{E} + R\,i)dq = \underbrace{i\,d\phi}_{dU} + \underbrace{R\,i^2 dt}_{dU'}.$$

The term labelled dU represents the energy expended by $\mathcal{E}_b$ in moving dq against the induced emf, whilst the term dU' represents the irreversible conversion of electrical energy into heat (joule losses). In this question, we are concerned only with changes in the *magnetic* energy of the circuit

$$dU = i\,d\phi. \tag{1}$$

(a) Use (1) to show that the magnetic energy U of a system comprising n rigid circuits of arbitrary size, shape and orientation, at rest in vacuum, is

$$U = \tfrac{1}{2}\sum_{j=1}^{n} i_j\,\phi_j. \tag{2}$$

Hint: Assume that, since the circuits are in vacuum, there is no hysteresis and so U does not depend on how the currents i_j reach their steady-state values (i_{0j}, say). Because of this, it is convenient to consider a process whereby the currents are increased from zero to i_{0j} in the *same time.*

(b) Show that (2) may be expressed as

$$U = \tfrac{1}{2}\sum_{j=1}^{n} \oint_{c_j} i_j\,\mathbf{A}\cdot d\mathbf{l}'_j, \tag{3}$$

where $\mathbf{A}$ is the value of the vector potential at $d\mathbf{l}'_j$ (assume single-loop circuits).

(c) We now wish to generalize (3) and derive the magnetic energy U in an extended region of space v (e.g. a conducting medium that supports quasi-stationary currents). Ref. [6] explains how to proceed:

> suppose that we do not have well-defined current circuits, but instead each 'circuit' is a closed path in the medium (which we take to be conducting). Equation (3) may be made to approximate this situation very closely by choosing a large

---

[‡] $i(t)$ grows from zero to its steady-state value i_0.

number of contiguous circuits c_j, replacing $i_j dl'_j \rightarrow \mathbf{J} dv'$, and, finally, by the substitution of $\int_v$ for $\sum_j \oint_{c_j}$.

Then

$$U = \tfrac{1}{2} \int_v \mathbf{J} \cdot \mathbf{A} \, dv'. \tag{4}$$

Use (4) to derive the result

$$U = \frac{1}{2\mu_0} \int_v \mathbf{B} \cdot \mathbf{B} \, dv', \tag{5}$$

where the permeability of the medium is that of vacuum.

Hint: The derivation is similar to the corresponding electrostatic case given in Question 2.17.

Solution

(a) It follows from (1) that $dU = \sum_j^n i_j \, d\phi_j$. Adopting the procedure suggested in the hint, the currents (and hence the fluxes) are raised from zero to their final values in the same time. Thus, at any instant, all currents and fluxes will be the *same* fraction of their final values; α, say. We therefore write $i_j = \alpha i_{0j}$ and $\phi_j = \alpha \phi_{0j}$. Hence $dU = \sum_j^n i_{0j} \phi_{0j} \, \alpha \, d\alpha$, which may be integrated over the interval $0 \le \alpha \le 1$ to give

$$U = \sum_{j=1}^n i_{0j} \phi_{0j} \int_0^1 \alpha \, d\alpha = \tfrac{1}{2} \sum_{j=1}^n i_{0j} \phi_{0j}. \tag{6}$$

This is the result we seek: the first subscript on i and ϕ in (6) is implicit in (2) where the currents and fluxes are understood to represent steady-state values.

(b) From the definition of flux and Stokes's theorem we obtain

$$\phi_j = \int_{s_j} \mathbf{B} \cdot d\mathbf{a}'_j = \int_{s_j} (\nabla \times \mathbf{A}) \cdot d\mathbf{a}'_j = \oint_{c_j} \mathbf{A} \cdot dl'_j. \tag{7}$$

Substituting (7) in (2) yields (3).

(c) With the current density $\mathbf{J}(\mathbf{r}')$ in (4) given by the Maxwell equation $\nabla \times \mathbf{B} = \mu_0 \mathbf{J}$, we have $U = \dfrac{1}{2\mu_0} \int_v (\nabla \times \mathbf{B}) \cdot \mathbf{A} \, dv'$ where the region of integration v is a sphere of radius R that is large enough to include all the currents. Transforming the integrand using (7) of Question 1.8 gives

$$U = \frac{1}{2\mu_0} \left[\int_v (\nabla \times \mathbf{A}) \cdot \mathbf{B} \, dv' - \int_v \nabla \cdot (\mathbf{A} \times \mathbf{B}) \, dv' \right]$$

$$= \frac{1}{2\mu_0} \left[\int_v \mathbf{B} \cdot \mathbf{B} \, dv' - \oint_s (\mathbf{A} \times \mathbf{B}) \cdot d\mathbf{a}' \right], \tag{8}$$

where in the last step we use $\mathbf{B} = \nabla \times \mathbf{A}$ and Gauss's theorem. But the surface integral in (8) is zero in the limit[#] $R \to \infty$ and (5) follows.

Comments

(i) Substituting (1) and (5)$_1$ of Question 5.14 in (2) gives

$$U = \tfrac{1}{2}\sum_{j=1}^{n} i_j \sum_{k=1}^{n} \Phi_{jk} = \tfrac{1}{2}\sum_{j=1}^{n} i_j \sum_{k=1}^{n} i_k M_{jk} = \tfrac{1}{2}\sum_{j=1}^{n}\sum_{k=1}^{n} M_{jk} i_j i_k, \qquad (9)$$

which is necessarily a positive quantity.[†] Two special cases of (9) are:

☞ For a single circuit $U = \tfrac{1}{2}Li^2$ which is an elementary result.

☞ For two coupled circuits, $U = \tfrac{1}{2}L_1 i_1^2 + M i_1 i_2 + \tfrac{1}{2}L_2 i_2^2$ follows from the properties of M described in Question 5.14. Expressing this equation in the form

$$U = \tfrac{1}{2}L_1(i_1 + M/L_1\, i_2)^2 + \tfrac{1}{2}(L_2 - M^2/L_1)i_2^2 \qquad (10)$$

shows that U (which is always positive) can be minimized by choosing $i_2 = -i_1 L_1/M$. Imposing the condition $U > 0$ requires that $M^2 < L_1 L_2$ or $M < \sqrt{L_1 L_2}$ (the mutual inductance of two interacting coils is always less than the geometric mean of their self-inductances).

(ii) We may regard U as energy stored in the magnetic field, with density

$$u = \frac{1}{2\mu_0}\mathbf{B} \cdot \mathbf{B}. \qquad (11)$$

It turns out that (11) is also the energy density for a time-dependent magnetic field, as we will find in Question 7.6. (Note that (2) and (4) are correct for quasi-stationary currents only.)

(iii) We end with the following quote:

> You might find it strange that it takes energy to set up a magnetic field—after all, magnetic fields *themselves* do no work. The point is that producing a magnetic field, where previously there was none, requires *changing* the field, and a changing **B**-field, according to Faraday, induces an *electric* field. The latter, of course, *can* do work. In the beginning there is no **E**, and at the end there is no **E**; but in between, while **B** is building up, there *is* an **E**, and it is against *this* that the work is done.[7]

[#] **A** and **B** decrease with distance at least as fast as $1/R$ and $1/R^2$ respectively, whilst $da' \propto R^2$. So $(\mathbf{A} \times \mathbf{B}) \cdot da'$ scales as R^2/R^3 which $\to 0$ as $R \to \infty$.

[†] The sources of emf do work to establish the currents.

---

[7] D. J. Griffiths, *Introduction to electrodynamics*, Chap. 7, p. 319. New York: Prentice Hall, 3 edn, 1999.

Question 5.16

Show that the self-inductance of a square single-turn plane conducting loop having side 4ℓ is

$$L \simeq \frac{2\mu_0\ell}{\pi}\left[\ln\left(\frac{\ell-b}{b}\right) - 0.77\right], \tag{1}$$

assuming that the radius of the wire $b \ll \ell$.

Hint: Use the magnetic field $\mathbf{B}$ for the wire of length ℓ calculated in (2) of Question 4.8, and evaluate the flux integral with *Mathematica*.

Solution

Choose the origin of cylindrical planar coordinates at the centre of one side of the loop with the wire lying along the z-axis. Because of the symmetry, each of the four sides of the square make an equal contribution to the flux ϕ through the current-carrying loop. Then

$$\phi = 4 \times \int_s \mathbf{B} \cdot d\mathbf{a} = 4 \times \int_{-\frac{\ell}{2}}^{\frac{\ell}{2}} \int_b^{\ell-b} \mathbf{B} \cdot (dr\,dz\,\hat{\boldsymbol{\theta}})$$

$$= \frac{\mu_0 I}{\pi} \int_{-\frac{\ell}{2}}^{\frac{\ell}{2}} \int_b^{\ell-b} [f_+(r, z) + f_-(r, z)]\,dr\,dz,$$

where $f_\pm(r, z) = \dfrac{(\ell \pm 2z)}{r\sqrt{4r^2 + (\ell \pm 2z)^2}}$. Now by definition $L = \phi/I$, and so

$$L = \frac{\mu_0}{\pi} \int_{-\frac{1}{2}\ell}^{\frac{1}{2}\ell} \int_b^{\ell-b} [f_+(r, z) + f_-(r, z)]\,dr\,dz. \tag{2}$$

Evaluating the integral in (2) using *Mathematica*'s `Integrate` function (see `cell 1` in the notebook on p. 280) and applying the approximation $b \ll \ell$ give (1).

Comment

The self-inductance (1) applies at low frequencies where the current in the loop is distributed uniformly across the cross-sectional area πb^2 of the wire. At high frequencies, where the skin effect[‡] changes the current distribution, corrections to (1) are required.

[‡]See (11) of Question 10.10.

Question 5.17

Show that the self-inductance of the toroidal solenoid described in Question 4.9 is

$$L = \mu_0 N^2 \left[R - \sqrt{R^2 - a^2} \right], \qquad (1)$$

where the radii a and R are defined in the figure on p. 205.

Hint: Evaluate the flux integral with Mathematica.

Solution

Using the notation of Question 4.9, the flux per turn is $\phi = \displaystyle\int_s \mathbf{B} \cdot d\mathbf{a} = \int_s \mathbf{B} \cdot (r \, dr \, d\phi) \hat{\boldsymbol{\theta}}$

where $\mathbf{B} = \dfrac{\mu_0 N I}{2\pi} \dfrac{\hat{\boldsymbol{\theta}}}{R + r \cos \phi}$. Now $L = \dfrac{\Phi_{\text{total}}}{I} = \dfrac{N\phi}{I}$, and so

$$L = \frac{\mu_0 N^2}{2\pi} \int_0^{2\pi} \int_0^a \frac{r \, dr \, d\phi}{R + r \cos \phi}$$

$$= \frac{\mu_0 N^2}{\pi} \int_0^\pi \int_0^a \frac{r \, dr \, d\phi}{R + r \cos \phi} \qquad (2)$$

because the integrand is symmetric about $\phi = 0$. In (2), we make the change of variable $u = r/R$ which gives

$$L = \frac{\mu_0 N^2 R}{\pi} \int_0^\pi \int_0^\alpha \frac{u \, du \, d\phi}{1 + u \cos \phi}, \qquad (3)$$

where $\alpha = a/R < 1$. *Mathematica's* `Integrate` function (see `cell 2` in the notebook on p. 280) then yields

$$\int_0^\pi \int_0^\alpha \frac{u \, du \, d\phi}{1 + u \cos \phi} = \pi - \pi\sqrt{1 - \alpha^2}. \qquad (4)$$

Hence (1).

Question 5.18

Show that the self-inductance L of the single-turn circular conducting loop considered in Question 4.15 is

$$L \simeq \mu_0 a \left[\ln\left(\frac{8a}{b}\right) - 2 \right], \qquad (1)$$

assuming $a \gg b$ (where b is the radius of the wire).

Hint: Use Mathematica whenever appropriate.

Solution

In terms of the vector potential, the flux through a surface s spanned by a contour c is $\Phi = \oint_c \mathbf{A} \cdot d\mathbf{l}$ (see (1) of Question 4.14). For the contour of integration, it is convenient to choose a circle of radius $r = a - b$ centred on the origin and lying in the $z = 0$ plane which clearly coincides with the inner circumference of the loop. Now with $\mathbf{A}$ given by (1) of Question 4.15 and $d\mathbf{l} = r\,d\theta\,\hat{\boldsymbol{\theta}}$ we have $\mathbf{A} \cdot d\mathbf{l} = (a - b)A_\theta\,d\theta$. Thus

$$\Phi = \frac{\mu_0 I a}{2\pi} \frac{(a - b)}{\sqrt{(a - b)^2 + a^2}} \int_0^{2\pi} \int_0^\pi \frac{\cos u}{\sqrt{1 - \gamma^2 \cos u}} \, du \, d\theta,$$

where u is a dummy variable and $\gamma^2 = \dfrac{2a(a - b)}{(a - b)^2 + a^2}$. By definition $L = \Phi/I$, and so

$$L = \frac{\mu_0 a(a - b)}{\sqrt{(a - b)^2 + a^2}} \int_0^\pi \frac{\cos u}{\sqrt{1 - \gamma^2 \cos u}} \, du, \tag{2}$$

since the θ-integration is 2π. Evaluating the integral in (2) with *Mathematica* (see cell 3 in the notebook on p. 280) yields

$$L = \frac{2\mu_0 a(a - b)}{\sqrt{(a - b)^2 + a^2}} \left[\frac{\text{EllipticK}\left(\frac{2\gamma^2}{1+\gamma^2}\right) - (1 + \gamma^2)\text{EllipticE}\left(\frac{2\gamma^2}{1+\gamma^2}\right)}{\gamma^2\sqrt{1 + \gamma^2}} \right], \tag{3}$$

where the elliptic integrals EllipticE and EllipticK are *Mathematica*-defined functions. Now in order to make progress, we define $\alpha = b/a$ (then $\gamma^{-2} = 1 - \frac{1}{2}\alpha^2(1 - \alpha)^{-1}$) and introduce the approximation $\alpha \ll 1$. Expanding (3) about $\alpha = 0$ using *Mathematica*'s `Series` function produces the infinite series

$$L \simeq \mu_0 a \left\{ \left[\ln(8\gamma^{-1}) - 2 \right] - \frac{1}{2} \left[\ln(8\gamma^{-1}) - 1 \right] \gamma + \mathcal{O}[\gamma^2] \right\}, \tag{4}$$

whose lowest-order term is (1).

Comments

(i) The approximation sign in (4) arises partly from truncating the infinite series and partly because the contour integration at $r = a - b$ omits the small contribution to the flux inside the conductor itself.

(ii) The comment for Question 5.16 also applies here.

In[1]:= (* INDUCTANCE OF A SQUARE LOOP *)
(* Below we use h because Mathematica's 1 looks like 1 *)

$$\text{Integrate}\left[\frac{1}{r}\left(\frac{(h+2z)}{\sqrt{4\,r^2+(h+2z)^2}}+(h-2z)\Big/\left(\sqrt{\big(4\,r^2+(h-2z)^2\big)}\right)\right)\right],$$

$$\{r, b, h-b\}, \left\{z, \frac{-h}{2}, \frac{h}{2}\right\}, \text{Assumptions} \to \{0 < 2b < h\}\right]$$

In[2]:= (* INDUCTANCE OF A TOROIDAL SOLENOID *)

$$\text{Integrate}\left[\frac{u}{1+u\,\text{Cos}\,[\phi]}, \{u, 0, \alpha\}, \{\phi, 0, \pi\}, \text{Assumptions} \to 0 < \alpha < 1\right]$$

In[3]:= (* INDUCTANCE OF A CIRCULAR CURRENT LOOP *)

$$\left(\frac{(1-\alpha)}{\sqrt{2-2\,\alpha+\alpha^2}}\,\text{Integrate}\left[\frac{\text{Cos}\,[u]}{\sqrt{1-\gamma^2\,\text{Cos}\,[u]}}, \{u, 0, \pi\}, \text{Assumptions} \to 0 < \gamma < 1\right]\right)\Big/.$$

$$\gamma \to \sqrt{\frac{1-\alpha}{1-\alpha+\frac{\alpha^2}{2}}}$$

In[4]:= W[α_] := (2 (1-α)) $\Big/\left(\left(1-\alpha+\frac{\alpha^2}{2}\right)\left(1+\frac{1-\alpha}{1-\alpha+\frac{\alpha^2}{2}}\right)\right);$

$$\text{Series}\left[\left(\left(1-\alpha+\frac{\alpha^2}{2}\right)\left(-2\left(1+\frac{1-\alpha}{1-\alpha+\frac{\alpha^2}{2}}\right)\text{EllipticE}\,[w[\alpha]] + 2\,\text{EllipticK}\,[w[\alpha]]\right)\right)\Big/\right.$$

$$\left.\left(\sqrt{2-2\,\alpha+\alpha^2}\,\sqrt{1+\frac{1-\alpha}{1-\alpha+\frac{\alpha^2}{2}}}\right), \{\alpha, 0, 2\}, \text{Assumptions} \to 0 < \alpha < 1\right]$$

Question 5.19*

Two circular coaxial coils c_1 and c_2 having radii r_1 and r_2 carry currents I_1 and I_2 respectively. The origin of cylindrical polar coordinates is chosen at the centre of c_1 and the z-axis along the symmetry axis of the coils. The centre of c_2 is at $(0, 0, z)$.

(a) Show that the mutual inductance of these coils is

$$M(z) = \frac{\mu_0 N_1 N_2 r_1 r_2}{2} \int_0^{2\pi} \frac{\cos\theta\,d\theta}{\sqrt{z^2 + (r_2 - r_1)^2 + 2r_1 r_2(1 - \cos\theta)}}, \tag{1}$$

where N_1 (N_2) is the number of turns of coil 1 (2) and θ is a variable of integration.

(b) Show that the force exerted by c_1 on c_2 is

$$\mathbf{F}_{12} = I_1 I_2 \frac{\partial M}{\partial z} \hat{\mathbf{z}}. \tag{2}$$

(c) Suppose $r_1 = \frac{1}{4} r_2 = r$ and that the currents, which are of equal magnitude, circulate in opposite directions (i.e. $I_1 = -I_2 = I_0$). Use *Mathematica* to plot graphs of $M(z)$ vs z/r and $F(z)$ vs z/r. Determine the value of z/r for which the force of repulsion between the coils is a maximum.

Hint: For (a) and (b) use (8) and (9) of Question 4.4 respectively.

Solution

(a) Choose an origin O_1 or O_2 at the centre of each coil. Then the line elements dl_1 and dl_2 of c_1 and c_2 have the polar coordinates (r_1, θ_1) and (r_2, θ_2) respectively, with corresponding Cartesian coordinates $(r_1 \cos \theta_1, r_1 \sin \theta_1)$ and $(r_2 \cos \theta_2, r_2 \sin \theta_2)$. Substituting $\mathcal{R} = z\hat{\mathbf{z}}$ in the denominator of (8) of Question 4.4 gives

$$
\begin{aligned}
|\mathbf{r}_1 - \mathbf{r}_2 + \mathcal{R}| &= \sqrt{(r_1 \cos \theta_1 - r_2 \cos \theta_2)^2 + (r_1 \sin \theta_1 - r_2 \sin \theta_2)^2 + z^2} \\
&= \sqrt{z^2 + r_1^2 + r_2^2 - 2 r_1 r_2 (\cos \theta_1 \cos \theta_2 + \sin \theta_1 \sin \theta_2)} \\
&= \sqrt{z^2 + (r_2 - r_1)^2 + 2 r_1 r_2 (1 - \cos \theta)}, \tag{3}
\end{aligned}
$$

where $\theta = \theta_2 - \theta_1$. Furthermore, the line elements in the numerator of (8) of Question 4.4 are $dl_1 = r_1 \, d\theta_1 \hat{\boldsymbol{\theta}}_1$ and $dl_2 = r_2 \, d\theta_2 \hat{\boldsymbol{\theta}}_2$. It can be seen from the figure below that the angle between dl_1 and dl_2 is also θ. So

$$dl_1 \cdot dl_2 = r_1 r_2 \cos \theta \, d\theta_1 \, d\theta_2. \tag{4}$$

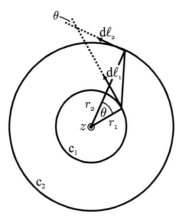

It then follows that

$$M(z) = \frac{\mu_0 N_1 N_2 r_1 r_2}{4\pi} \int_0^{2\pi} \int_0^{2\pi} \frac{\cos\theta \, d\theta_1 \, d\theta_2}{\sqrt{z^2 + (r_2 - r_1)^2 + 2r_1 r_2(1 - \cos\theta)}}. \tag{5}$$

Now with θ_1 constant, $d\theta_2 = d\theta$ and (5) is

$$M(z) = \frac{\mu_0 N_1 N_2 r_1 r_2}{4\pi} \int_0^{2\pi} d\theta_1 \int_{-\theta}^{2\pi-\theta} \frac{\cos\theta \, d\theta}{\sqrt{z^2 + (r_2 - r_1)^2 + 2r_1 r_2(1 - \cos\theta)}}$$

$$= \frac{\mu_0 N_1 N_2 r_1 r_2}{2} \int_{-\theta}^{2\pi-\theta} \frac{\cos\theta \, d\theta}{\sqrt{z^2 + (r_2 - r_1)^2 + 2r_1 r_2(1 - \cos\theta)}}. \tag{6}$$

The integrand in (6) is a periodic function in θ having periodicity 2π, and so we can replace the lower (upper) limit of the integral with 0 (2π). Hence (1).

(b) The force exerted by c_1 on c_2 is $I_1 I_2 \nabla_z M$, which is (2) because $\nabla_z = \dfrac{\partial}{\partial z}\hat{z}$. Clearly, **F** is attractive (repulsive) when the currents have the same (opposite) sense of circulation.

(c) Introducing the dimensionless variable $u = z/2r$ and rearranging (1) and (2) give

$$M(u) = M_0 \int_0^{2\pi} \frac{\cos\theta \, d\theta}{\sqrt{u^2 + \beta^2 + 2(1 - \cos\theta)}} \quad \text{and} \quad \mathbf{F}(u) = -F_0 \frac{\partial M}{\partial u}\hat{z}, \tag{7}$$

where $\beta = 3/2$, $M_0 = \mu_0 N_1 N_2 r$ and $F_0 = \frac{1}{2}\mu_0 N_1 N_2 I_0^2$. Implementing (7) in the notebook on p. 283 yields the following graphs:

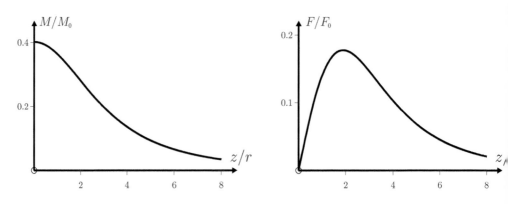

The maximum value of F occurs at $z \sim 1.9r$.

```
In[1]:=  β = 3/2;          asp = {β > 0 && u ≥ 0};

         f[β_, u_] = Integrate[ Cos[θ] / √(u² + β² + 2 (1 - Cos[θ])) , {θ, 0, 2 π}, Assumptions → asp];

In[2]:=  Plot[f[β, u], {u, 0, 4}]
         Plot[-D[f[β, u], u] /. {u → x}, {x, 0, 4}]

In[3]:=  FindMaximum[-D[f[β, u], u] /. {u → x}, x]
```

Question 5.20

Consider two coils having self-inductances L_1 and L_2 respectively, and mutual inductance M. The relationship between L_1, L_2 and M is

$$M^2 = L_1 L_2. \tag{1}$$

Use a suitable plausibility argument to justify (1).

Hint: Consider how each of these inductances depends on the number of turns.

Solution

Because of the results of Questions 5.17 and 5.19 we let

$$L_1 = N_1^2 \phi_0, \qquad L_2 = N_2^2 \phi_0 \qquad \text{and} \qquad M = N_1 N_2 \phi_0, \tag{2}$$

where ϕ_0 is the magnetic flux linking any single turn of wire carrying a unit current. Equation (1) then follows from (2).

Comment

Equation (1) is an idealization: in practice, because of the leakage of magnetic-field lines,

$$M = k\sqrt{L_1 L_2}, \tag{3}$$

where $k < 1$ $\left(\text{as we have already seen in Comment (i) of Question 5.15}\right)$.

Question 5.21

Adopting Maxwell's notation where $[Q]$ denotes the dimensions of some quantity Q, express the dimensions of the following quantities in terms of the fundamental units of mass M, length L, time T and current I:

(a) magnetic field, (b) magnetic flux, (c) inductance and (d) permeability.

Solution

(a) magnetic field

From $\mathbf{F} = q\mathbf{v} \times \mathbf{B}$ we have $[B] = \dfrac{[F]}{[q][v]} = \dfrac{\mathtt{MLT^{-2}}}{\mathtt{ITLT^{-1}}} = \mathtt{MT^{-2}I^{-1}}.$ \hfill (1)

(b) magnetic flux

From $\phi = \displaystyle\int \mathbf{B} \cdot d\mathbf{a}$ and (1) we have $\phi = [B][a] = \mathtt{ML^2T^{-2}I^{-1}}.$ \hfill (2)

(c) inductance

From $L = \dfrac{\phi}{i}$ and (2) we have $[L] = \dfrac{[\phi]}{[i]} = \dfrac{\mathtt{ML^2T^{-2}I^{-1}}}{\mathtt{I}} = \mathtt{ML^2T^{-2}I^{-2}}.$ \hfill (3)

(d) permeability

From $\displaystyle\oint_c \mathbf{B} \cdot d\mathbf{l} = \mu_0 i$ and (1) we have $[\mu_0] = \dfrac{\mathtt{MLT^{-2}I^{-1}}}{\mathtt{I}} = \mathtt{MLT^{-2}I^{-2}}.$ \hfill (4)

Comments

(i) In the International System (SI):

- ☞ the unit of magnetic field is the tesla where $1\,\mathrm{T} = 1\,\mathrm{kg\,s^{-2}\,A^{-1}}$.
- ☞ the unit of magnetic flux is the weber where $1\,\mathrm{Wb} = 1\,\mathrm{kg\,m^2\,s^{-2}\,A^{-1}}$.
- ☞ the unit of inductance is the henry where $1\,\mathrm{H} = 1\,\mathrm{kg\,m^2\,s^{-2}\,A^{-2}}$.
- ☞ the unit of permeability may be conveniently written as $1\,\mathrm{H\,m^{-1}}$.

(ii) The tesla is named in honour of the Serbian-born physicist–engineer–inventor Nikola Tesla (1856–1943), who made significant contributions to the design of modern ac electrical supply systems.

(iii) The weber is named in honour of the German physicist Wilhelm Weber (1804–91) who, together with Gauss, invented electromagnetic telegraphy.

(iv) The henry is named in honour of the famous American scientist Joseph Henry (1797–1878) whose work in electromagnetic induction led to the design and construction of early prototypes of the dc motor. Faraday and Henry, working at about the same time, independently recognized the property of self-inductance.

(v) Other famous scientists are also honoured in different systems of units. For example, in the cgs system magnetic field is measured in gauss (G) and magnetic flux in maxwell (Mx).

6
Ohm's law and electric circuits

In this chapter we consider various dc and ac circuits which contain at least one active element (here always a voltage source) and passive elements—resistors, capacitors and inductors arranged in different combinations to form a bilateral[‡] network. The questions below have been selected to illustrate some of the basic techniques used in circuit theory. Before attempting these, we first provide some background information that may be useful.

The goal of circuit theory is to find the current in each element and the potential (or voltage) at certain points in the circuit relative to some reference point (usually earth). For all passive elements used in the circuits below, we assume the following:

☞ the associated resistance R, capacitance C and inductance L are concentrated or *lumped* in a particular element. Situations in which this is not the case and where the parameters are *distributed* along the inter-connecting wires are not considered here.

☞ they are *ideal* (unless otherwise stated) and possess only a single property: either R, C or L. Thus, for example, a capacitor is assumed to be pure capacitance with neither resistance nor inductance. In real circuits, this assumption is often not valid.

☞ they are *linear* devices where the voltage v across an element is proportional to the first power of the current i, or its differential or its integral. Specifically:

Resistors

For any resistance $R = v/i$ by definition, and for those resistors which satisfy Ohm's law (R is *constant* and independent of i at constant temperature) a linear relationship between v and i is implied:

$$v = Ri \qquad \text{or} \qquad i = \frac{v}{R}. \tag{I}$$

With conventional current there is a potential drop (commonly referred to as the 'voltage') across the resistance in the direction of the flow; if the current reverses then so does v—resistors are bidirectional circuit components. In the circuits of this chapter, we consider ohmic resistors only.

[‡]The current remains the same if the connections to the device terminals are interchanged.

Solved Problems in Classical Electromagnetism. J. Pierrus, Oxford University Press (2018).
© J. Pierrus. DOI: 10.1093/oso/9780198821915.001.0001

Capacitors

By definition $C = q/v$ and with $q = \int i\, dt$ we have

$$v = \frac{1}{C}\int i\, dt \qquad \text{or} \qquad i = C\frac{dv}{dt}, \qquad\qquad (\text{II})$$

which is again a linear relationship between voltage and current. All capacitors are assumed to be bidirectional and lossless.

Inductors

Using the definition of inductance $L = \phi/i$ and Faraday's flux rule $\mathcal{E} = -d\phi/dt$, we obtain the emf across an inductor: $\mathcal{E} = -L\,di/dt$. Here the negative sign implies that $\mathcal{E}$ opposes the current change and is known as a *back emf*. A positive emf raises the potential of a conventional charge carrier but a back emf $-L\,di/dt$ drops the potential like a resistor, and so may be regarded as a voltage drop $v = L\,di/dt$ in the direction of i. Thus we have a third important linear relationship between voltage and current:

$$v = L\frac{di}{dt} \qquad \text{or} \qquad i = \frac{1}{L}\int v\, dt. \qquad\qquad (\text{III})$$

In purely resistive dc circuits the branch[#] currents are sometimes calculated using Kirchhoff's rules. These are:

☞ Kirchhoff's junction rule: the sum of the currents at any node in a circuit is zero ('current in = current out').

☞ Kirchhoff's loop rule: the sum of the emfs around any closed loop in a circuit equals the sum of the voltage drops around that loop.

The term 'ac theory' generally refers to quasi-stationary currents (see the introduction to Chapter 5), where we assume that at some instant the current is the same throughout any branch of the circuit. Then the basic laws for dc circuits (e.g. Kirchhoff) remain valid with appropriate modifications. The applicability of these laws requires that the circuit dimensions should be very much less than the wavelength λ of electromagnetic radiation at the frequency of the ac. Suppose d represents a typical physical dimension of the circuit. Then at 50 Hz we have $\lambda = 3 \times 10^8 \div 50 \sim 6000\,\text{km} \Rightarrow d \lesssim 60\,\text{km}$. But at 10 MHz the wavelength decreases to 30 m and now $d \lesssim 30\,\text{cm}$ (see also Question 5.1). Of course, this approximation sometimes fails, as for example in a national power grid for which d may be several thousand kilometres; such systems are studied using the theory of transmission lines.

[#]We use the following terminology: a *branch* is a section of a circuit with two terminals between which connections may be made; a *node* is a point where more than two branches meet; a *loop* is any closed path formed by connecting branches.

We now consider a harmonically varying current i and voltage v. The analysis of ac circuits is often greatly simplified (at least algebraically) if we use complex exponentials instead of sines and cosines. Then

$$i = I_0 e^{j\omega t} \quad \text{and} \quad v = V_0 e^{j(\omega t + \phi)}, \tag{IV}$$

where I_0 and V_0 are real magnitudes, ϕ is the phase difference between current and voltage and $j = \sqrt{-1}$.[‡] The complex impedance z is defined as v/i, and so from (IV)

$$z = \frac{V_0 e^{j(\omega t + \phi)}}{I_0 e^{j\omega t}} = Z e^{j\phi}, \tag{V}$$

where $Z = V_0/I_0$ is the impedance magnitude. Using this last equation to express z as the sum of a real (resistive) part R and an imaginary (reactive) part X gives

$$\left. \begin{aligned} z &= Z e^{j\phi} = Z\cos\phi + jZ\sin\phi = R + jX \quad \text{where} \\ R &= Z\cos\phi, \quad X = Z\sin\phi, \quad Z^2 = R^2 + X^2, \quad \phi = \tan^{-1}(X/R) \end{aligned} \right\}. \tag{VI}$$

Measurable quantities (like current and voltage) may be obtained from these complex quantities in the usual way by recovering either a real or an imaginary part at the end of a calculation.

We now state (without proof) two important 'network theorems' which will be used repeatedly in the questions which follow.

Thévenin's theorem

Suppose A and B are the ends of a branch of some linear active network. Thévenin's theorem states that as far as this branch is concerned, the remainder of the network may be replaced by a single active source consisting of an ideal voltage generator $\mathcal{E}_{\text{Th}}$ in series with internal impedance z_{Th}. The value of $\mathcal{E}_{\text{Th}}$ is the open-circuit voltage appearing between A and B with the branch removed, and z_{Th} is the impedance between A and B with the branch removed and with the active sources replaced by their internal impedances.

Superposition theorem

This theorem states that the response in any element of a linear bilateral network containing two or more active sources is the sum of the responses due to each source acting independently and with all other sources replaced by their internal impedances.

Various simple applications of these theorems appear throughout this chapter, which contains questions on bridges, filters, audio amplifiers and coupled circuits (e.g. transformers). Important topics such as series and parallel resonance in LRC circuits are treated along the way. Much of the laborious algebra, involving the manipulation of complex quantities, is avoided by relegating this task to *Mathematica*.

[‡]Throughout this book we adopt the notation $i = \sqrt{-1}$, which is traditional in physics. However, in this chapter we will break with tradition and use $j = \sqrt{-1}$ to avoid possible confusion with the complex current i.

Question 6.1

Consider a resistor, inductor and capacitor in turn. For each of these single circuit elements, write down the resistive and reactive parts of the complex impedance z and also the angle ϕ by which the voltage leads (lags) the current in phase.

Hint: If necessary, refer to equations (IV)–(VI) of the introduction.

Solution

resistor
$$v = Ri, \quad \boxed{R = R}, \quad \boxed{X = 0}, \quad \boxed{\phi = 0}. \tag{1}$$

inductor
$$v = L\frac{di}{dt} = L\frac{d}{dt}(I_0 e^{j\omega t}) = j\omega L I_0 e^{j\omega t} = j\omega L i.$$

Hence $z = v/i = j\omega L$ with $\boxed{R = 0}$, $\boxed{X = \omega L}$, $\boxed{\phi = \pi/2}$. $\tag{2}$

capacitor
$$v = \frac{1}{C}\int i\,dt = \frac{1}{C}\int I_0 e^{j\omega t}dt = \frac{I_0}{j\omega C}e^{j\omega t} = \frac{i}{j\omega C}.$$

Hence $z = v/i = \dfrac{1}{j\omega C} = \dfrac{-j}{\omega C}$ with $\boxed{R = 0}$, $\boxed{X = \dfrac{1}{\omega C}}$, $\boxed{\phi = -\pi/2}$. $\tag{3}$

Comments

(i) It is useful to summarize the results (1)–(3) above. In ac circuits:

☞ the voltage across a resistor is in phase with the current, and the resistance $= R$.

☞ the voltage across an inductor (with no internal resistance) leads the current in phase by $\frac{1}{2}\pi$, and the inductive reactance $= j\omega L$.

☞ the voltage across a (lossless) capacitor lags the current in phase by $\frac{1}{2}\pi$, and the capacitative reactance $= \dfrac{1}{j\omega C} = \dfrac{-j}{\omega C}$.

(ii) In most of the questions which now follow, reactances are shown as a magnitude only (i.e. as ωL or $1/\omega C$). The reader must remember to insert the $\pm j$ (as appropriate) in any subsequent calculations.

Question 6.2

Suppose z_1 and z_2 are two complex impedances. Show that when they are connected in

(a) series, their equivalent impedance is $z = z_1 + z_2$.   (1)

(b) parallel, their equivalent impedance is $z = \dfrac{z_1 z_2}{(z_1 + z_2)}$.   (2)

Solution

(a) The net voltage v across these resistors connected in series is the sum of the separate voltages v_1 and v_2. That is, $zi = v = v_1 + v_2 = (z_1 + z_2) i$, since the current i is the same in each impedance. Hence (1).

(b) The net current i flowing towards the node where these resistors are connected in parallel is the sum of the separate branch currents i_1 and i_2. That is, $v/z = i = i_1 + i_2 = v/z_1 + v/z_2$, since the voltage v is the same across each impedance. Then $z^{-1} = z_1^{-1} + z_2^{-1}$, which is (2).

Comments

(i) For two impedances in parallel, (2) is the familiar 'product-over-sum' rule: $z = $ the product of the two impedances $\div$ their sum.

(ii) The above results can be generalized. For n impedances connected in series, or in parallel,

$$z_{\text{series}} = z_1 + z_2 + \cdots + z_n \quad \text{and} \quad z_{\text{parallel}}^{-1} = z_1^{-1} + z_2^{-1} + \cdots + z_n^{-1}. \quad (3)$$

(iii) Three cases of particular interest are for two resistances (R_1, R_2), capacitances $(C_1, C_2$ with $X_i = 1/j\omega C_i)$ or inductances $(L_1, L_2$ with $X_i = j\omega L_i)$ connected either in series or in parallel. The following standard results then follow immediately from (1) and (2):

<table>
<tr><td align="center">series</td><td align="center">parallel</td></tr>
<tr><td>$R_{\text{eq}} = R_1 + R_2$</td><td>$R_{\text{eq}} = \dfrac{R_1 R_2}{R_1 + R_2}$</td></tr>
<tr><td>$C_{\text{eq}} = \dfrac{C_1 C_2}{C_1 + C_2}$</td><td>$C_{\text{eq}} = C_1 + C_2$</td></tr>
<tr><td>$L_{\text{eq}} = L_1 + L_2$</td><td>$L_{\text{eq}} = \dfrac{L_1 L_2}{L_1 + L_2}$</td></tr>
</table>

(In the last two equations we assume negligible coupling between the inductors.)

(iv) Two impedances in series (parallel) constitute a voltage (current) divider.

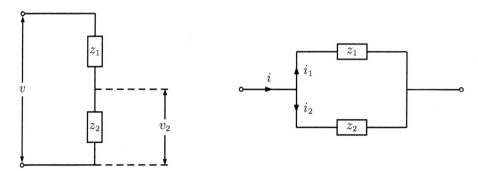

With reference to the diagrams above, we see that

$$v_2 = \left(\frac{z_2}{z_1 + z_2}\right)v \qquad \text{and} \qquad i_2 = \left(\frac{z_1}{z_1 + z_2}\right)i, \tag{4}$$

where the quantities in (4) are, in general, complex.

Question 6.3

Consider a voltage source $v = V_0 e^{j\omega t}$ having internal impedance $z_s = R_s + jX_s$ that is connected to a load impedance $z_\ell = R_\ell + jX_\ell$. Establish the conditions for maximizing power transfer to the load.

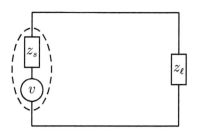

Solution

The net circuit impedance is $z = z_s + z_\ell = (R_s + R_\ell) + j(X_s + X_\ell)$ having magnitude $Z = [(R_s + R_\ell)^2 + (X_s + X_\ell)^2]^{1/2}$. The rms current in the circuit is thus $I = V/Z$ where $V = V_0/\sqrt{2}$. Now the average power dissipated in the load is

$$P = R_\ell I^2 = \frac{R_\ell V^2}{(R_s + R_\ell)^2 + (X_s + X_\ell)^2}. \tag{1}$$

Suppose we adjust the load by altering R_ℓ and X_ℓ. With respect to variation in X_ℓ, the load power is clearly a maximum when $\boxed{X_\ell = -X_s}$ (so if the source impedance is inductive we make the load impedance capacitive, and vice versa). Once the condition $X_\ell = -X_s$ has been satisfied, (1) becomes

$$P = \frac{R_\ell V^2}{(R_s + R_\ell)^2}.$$

This is a maximum with respect to variation in R_ℓ if $dP/dR_\ell = 0$. That is,

$$\frac{(R_s + R_\ell)^2 - 2R_\ell(R_s + R_\ell)}{(R_s + R_\ell)^4} = 0 \quad \Rightarrow \quad \boxed{R_\ell = R_s}.$$

The conditions we seek are highlighted above.

Comment

When $R_\ell = R_s$ and $X_\ell = -X_s$ we say that the load is 'matched to the source'. Because of the requirement $X_\ell = -X_s$, this matching condition is obviously frequency-dependent in general.

Question 6.4

Suppose a time-harmonic voltage $V = V_0\cos(\omega t + \phi)$ applied across the terminals of an impedance z produces the current $I = I_0\cos\omega t$. Show that the average power dissipated in the impedance is

$$P = \tfrac{1}{2}V_0 I_0 \cos\phi. \tag{1}$$

Solution

By definition, the instantaneous power delivered to z is voltage × current. So

$$\begin{aligned}
\mathcal{P}_{\text{inst}} &= V_0 I_0 \cos\omega t \cos(\omega t + \phi)\\
&= V_0 I_0 \cos\omega t\left(\cos\omega t \cos\phi - \sin\omega t \sin\phi\right)\\
&= V_0 I_0\left(\cos^2\omega t \cos\phi - \tfrac{1}{2}\sin 2\omega t \sin\phi\right). \tag{2}
\end{aligned}$$

Now we require the time-average power $\langle\mathcal{P}_{\text{inst}}\rangle$. So from (2),

$$\langle\mathcal{P}_{\text{inst}}\rangle = V_0 I_0\left(\langle\cos^2\omega t\rangle \cos\phi - \tfrac{1}{2}\langle\sin 2\omega t\rangle \sin\phi\right). \tag{3}$$

From (3) of Question 1.27 we know that $\langle\cos^2\omega t\rangle = \tfrac{1}{2}$ and $\langle\sin 2\omega t\rangle = 0$. Therefore, $\langle\mathcal{P}_{\text{inst}}\rangle = \tfrac{1}{2}V_0 I_0 \cos\phi$. In circuit theory, the term 'power' usually implies 'average power' and the angular brackets are often omitted. Hence (1).

Comments

(i) The term $\cos\phi$ in (1) is called the power factor of the circuit. In an ac circuit the only power dissipation is that which occurs in the resistive part of z. On average, no power is consumed in the reactive part of z. So, for example, a capacitor connected to an ac source will remove energy from the mains supply as it charges up, and then return this energy to the source during discharge.

(ii) For harmonically varying voltages and currents the root-mean-square values are related to the peak values by $V_{\mathrm{rms}} = V_0/\sqrt{2}$ and $I_{\mathrm{rms}} = I_0/\sqrt{2}$. Then (1) is sometimes written as

$$\mathcal{P} = V_{\mathrm{rms}} I_{\mathrm{rms}} \cos\phi \quad \text{or more simply} \quad \mathcal{P} = VI\cos\phi. \tag{4}$$

(iii) It is often convenient to use the complex voltage and current: $v = V_0 e^{j(\omega t+\phi)}$ and $i = I_0 e^{j\omega t}$. Then it is easily seen that (1) can be expressed in the equivalent forms

$$\mathcal{P} = \tfrac{1}{2}Re(vi^*) = \tfrac{1}{2}Re(v^*i) \quad \text{or for rms quantities} \quad \mathcal{P} = Re(vi^*) = Re(v^*i). \tag{5}$$

Question 6.5

An arrangement of resistors which connect three nodes of a network in a 'delta' configuration is shown in Fig. (I) below. In circuit analysis it is sometimes convenient to replace the delta by the 'star' configuration illustrated in Fig. (II). Corresponding node voltages and node currents are identical in both circuits.

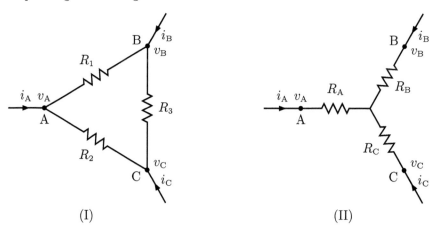

(I) (II)

Prove that the resistance of each star branch equals (the product of adjacent delta resistances) ÷ (the sum of delta resistances). That is,

$$R_{\mathrm{A}} = \frac{R_1 R_2}{R}, \qquad R_{\mathrm{B}} = \frac{R_1 R_3}{R}, \qquad \text{and} \qquad R_{\mathrm{C}} = \frac{R_2 R_3}{R}, \tag{1}$$

where $R = R_1 + R_2 + R_3$.

Solution

In the circuit diagram below, we assign branch currents as shown and then apply Kirchhoff's junction rule (see p. 286) to the delta configuration. This gives:

$$i_A = i_1 - i_2, \qquad i_B = i_3 - i_1, \qquad \text{and} \qquad i_C = i_2 - i_3. \qquad (2)$$

Equivalence requires that corresponding voltage drops are the same in both configurations, and so

$$\left. \begin{aligned} v_{AB} &= R_1 i_1 = R_A i_A - R_B i_B \\ v_{CA} &= R_2 i_2 = R_C i_C - R_A i_A \\ v_{BC} &= R_3 i_3 = R_B i_B - R_C i_C \end{aligned} \right\}. \qquad (3)$$

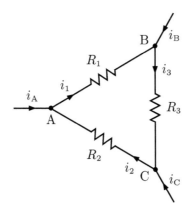

Using (2) to eliminate i_A, i_B and i_C from (3) yields

$$\left. \begin{aligned} (R_1 - R_A - R_B)i_1 + R_A i_2 + R_B i_3 &= 0 \\ R_A i_1 + (R_2 - R_A - R_C)i_2 + R_C i_3 &= 0 \\ R_B i_1 + R_C i_2 + (R_3 - R_B - R_C)i_3 &= 0 \end{aligned} \right\}. \qquad (4)$$

The set of equations (4) has non-trivial solutions if the determinant of the coefficients is zero, that is

$$\begin{vmatrix} (R_1 - R_A - R_B) & R_A & R_B \\ R_A & (R_2 - R_A - R_C) & R_C \\ R_B & R_C & (R_3 - R_B - R_C) \end{vmatrix} = 0. \qquad (5)$$

Expanding (5) gives

$$\begin{aligned} R_1 R_2 R_3 &- R_1 R_3 R_A - R_2 R_3 R_A - R_1 R_2 R_B - R_2 R_3 R_B + R_1 R_A R_B \\ &+ R_2 R_A R_B + R_3 R_A R_B - R_1 R_2 R_C - R_1 R_3 R_C + R_1 R_A R_C \\ &+ R_2 R_A R_C + R_3 R_A R_C + R_1 R_B R_C + R_2 R_B R_C + R_3 R_B R_C = 0, \qquad (6) \end{aligned}$$

which may be rearranged in the form

$$\left[R_{\mathrm{A}} - \frac{R_1 R_2}{R}\right] R_{\mathrm{B}} + \left[R_{\mathrm{B}} - \frac{R_1 R_3}{R}\right] R_{\mathrm{C}} + \left[R_{\mathrm{C}} - \frac{R_2 R_3}{R}\right] R_{\mathrm{A}} + \frac{R_1 R_2 R_3}{R} -$$

$$\left[\frac{R_{\mathrm{A}} R_1 R_3 + R_{\mathrm{B}} R_2 R_3 + R_{\mathrm{C}} R_1 R_2}{R}\right] = 0. \tag{7}$$

With R_{A}, R_{B} and R_{C} given by (1), it is evident that (7) is satisfied.

Comments

(i) In circuit theory the transformation discussed above has various names. We shall refer to it as the delta–star transformation. The inverse transformation (star–delta) is sometimes also required, and is

$$\left.\begin{aligned} R_1 &= R_{\mathrm{A}} + R_{\mathrm{B}} + R_{\mathrm{A}} R_{\mathrm{B}}/R_{\mathrm{C}} \\ R_2 &= R_{\mathrm{A}} + R_{\mathrm{C}} + R_{\mathrm{A}} R_{\mathrm{C}}/R_{\mathrm{B}} \\ R_3 &= R_{\mathrm{B}} + R_{\mathrm{C}} + R_{\mathrm{B}} R_{\mathrm{C}}/R_{\mathrm{A}} \end{aligned}\right\}. \tag{8}$$

Stated in words: the resistance between any two nodes in the delta configuration is the sum of:

☞ the resistances at the corresponding nodes in the star configuration, and

☞ the product of these two resistances divided by the resistance at the remaining star node.

(The proof of (8) follows from a different rearrangement of the terms in (6)).

(ii) The delta–star transformation and its inverse also hold for ac circuits containing reactive circuit elements. Then each resistance is replaced by a corresponding *complex* impedance, with

$$\left.\begin{aligned} z_{\mathrm{A}} &= \frac{z_1 z_2}{z_1 + z_2 + z_3} \\ z_{\mathrm{B}} &= \frac{z_1 z_3}{z_1 + z_2 + z_3} \\ z_{\mathrm{C}} &= \frac{z_2 z_3}{z_1 + z_2 + z_3} \end{aligned}\right\} \quad \text{and} \quad \left.\begin{aligned} z_1 &= z_{\mathrm{A}} + z_{\mathrm{B}} + \frac{z_{\mathrm{A}} z_{\mathrm{B}}}{z_{\mathrm{C}}} \\ z_2 &= z_{\mathrm{A}} + z_{\mathrm{C}} + \frac{z_{\mathrm{A}} z_{\mathrm{C}}}{z_{\mathrm{B}}} \\ z_3 &= z_{\mathrm{B}} + z_{\mathrm{C}} + \frac{z_{\mathrm{B}} z_{\mathrm{C}}}{z_{\mathrm{A}}} \end{aligned}\right\}. \tag{9}$$

(iii) In Question 6.6 we consider some simple applications of these transformations.

Question 6.6

For each of the following circuits, use the results of Question 6.5 to calculate the equivalent resistance R_{AC}.

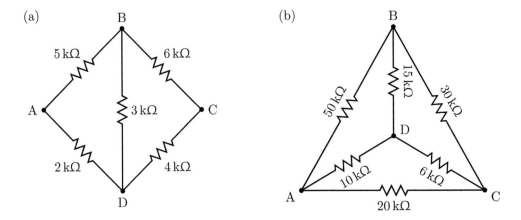

(a)

(b)

Solution

(a) Either delta may be transformed to a star. However, in this example, it is simpler arithmetically to transform the 5–3–2 delta, since $\Sigma R = 10\,\mathrm{k\Omega}$. The equivalent circuit (I) below follows from a straightforward application of (1) of Question 6.5. The series-connected resistors in this arrangement then leads to equivalent circuit (II), and so

$$R_{AC} = \text{series} + \text{parallel} = 1 + \frac{(7.5)(4.6)}{(7.5 + 4.6)} = 3.85\,\mathrm{k\Omega}.$$

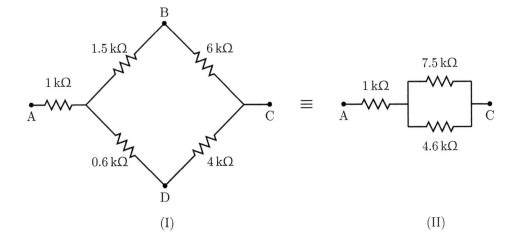

(I)

(II)

(b) The simplest solution involves transforming the inner star to a delta using (8) of Question 6.5. This produces the equivalent circuit (I) shown below.

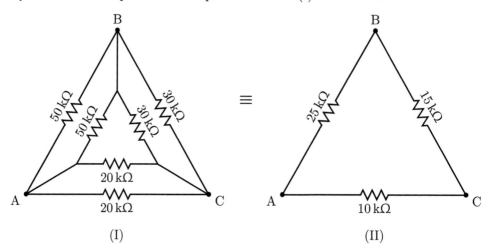

(I)   (II)

Each pair of identical resistors in Fig. (I) forms a parallel combination, which leads immediately to the equivalent circuit (II) following an application of the product-over-sum rule. Then $(25 + 15)\|10$ gives $R_{AC} = (40)(10)/(40 + 10) = 8\,\mathrm{k\Omega}$.

Question 6.7

In the circuit shown below, the galvanometer G and power source (emf $\mathcal{E}$) have negligible internal resistance.

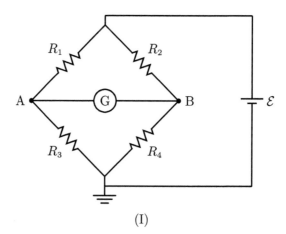

(I)

Use Thévenin's theorem to show that the current i in the galvanometer is

$$i = \frac{(R_2 R_3 - R_1 R_4)\,\mathcal{E}}{R_1 R_2 R_3 + R_1 R_2 R_4 + R_1 R_3 R_4 + R_2 R_3 R_4}. \tag{1}$$

Solution

Thévenin equivalent resistance

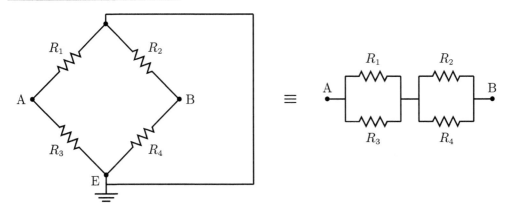

$$R_{\text{Th}} = R_1 \| R_3 + R_2 \| R_4 = \frac{R_1 R_3}{R_1 + R_3} + \frac{R_2 R_4}{R_2 + R_4}$$

$$= \frac{R_1 R_3 (R_2 + R_4) + R_2 R_4 (R_1 + R_3)}{(R_1 + R_3)(R_2 + R_4)}. \tag{2}$$

Thévenin equivalent emf

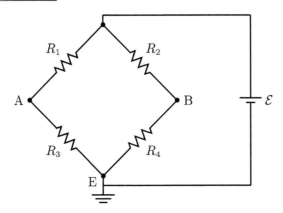

We use the voltage-divider rule (see Comment (iv) of Question 6.2):

$$\mathcal{E}_{\text{Th}} = (V_A - V_E) - (V_B - V_E) = \left[\left(\frac{R_3}{R_1 + R_3} \right) - \left(\frac{R_4}{R_2 + R_4} \right) \right] \mathcal{E}$$

$$= \frac{(R_2 R_3 - R_1 R_4)\, \mathcal{E}}{(R_1 + R_3)(R_2 + R_4)}. \tag{3}$$

Thévenin equivalent circuit

Equation (1) follows immediately from
$i = \mathcal{E}_{\mathrm{Th}}/R_{\mathrm{Th}}$ and with R_{Th} and $\mathcal{E}_{\mathrm{Th}}$
given by (2) and (3).

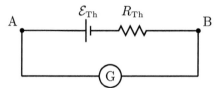

Comment

Circuit (I) above shows a configuration of resistors known as a Wheatstone bridge. It is sometimes used to determine an unknown resistance (R_4, say). Suppose R_1, R_2 and R_3 are known resistances and that one of these is variable (R_3, say). The bridge is said to be 'balanced' when the galvanometer registers no deflection. This is achieved by adjusting the variable resistance. Whilst the bridge is unbalanced, the sense of deflection of G indicates whether R_3 is too low or too high. Clearly, at balance $i = 0$, and it is evident from (1) that this corresponds to

$$R_4 = \frac{R_2 R_3}{R_1}. \tag{4}$$

The unknown resistance follows directly from (4).

Question 6.8

Consider the circuit shown below.

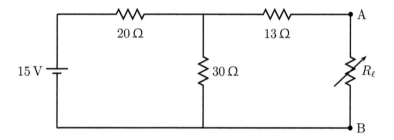

(a) Calculate the values of R_{Th} and $\mathcal{E}_{\mathrm{Th}}$ for the branch labelled AB.

(b) Hence deduce the value of the load resistance R_ℓ for which the power $\mathcal{P}$ dissipated in the load is a maximum. Then calculate $\mathcal{P}$.

Solution

(a) Thévenin equivalent resistance

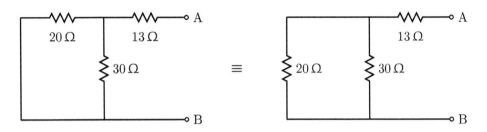

$$R_{Th} = 13 + \frac{(20)(30)}{20 + 30} = 13 + 12 = 25 \,\Omega. \tag{1}$$

Thévenin equivalent emf

With R_L removed from the branch AB, the Thévenin equivalent emf equals the voltage across the $30\,\Omega$ resistor. So

$$\mathcal{E}_{Th} = \left(\frac{30}{30 + 20}\right) \times 15 = 9 \,\text{V}.$$

(b) From the results of Question 6.3 we obtain the following Thévenin equivalent circuit:

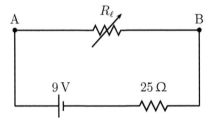

The power dissipated in the load is a maximum when $R_\ell = 25\,\Omega$, and then

$$\mathcal{P}_{\max} = R_\ell \left(\frac{9}{25 + R_\ell}\right)^2 = 25 \times \left(\frac{9}{50}\right)^2 = 0.81 \,\text{W}. \tag{2}$$

Comment

Suppose $R_\ell = 25\alpha\,\Omega$ where α is a mismatching factor $(0 < \alpha < \infty)$. Then it follows immediately from (2) that the load power $\mathcal{P} = 4\alpha\,\mathcal{P}_{\max}/(1 + \alpha)^2$. For $\alpha = 1 \pm 0.2$, the power $\mathcal{P} \not< 0.99\mathcal{P}_{\max}$. A mismatch between the source and load resistances of up to 20% results in a less than 1% decrease in power dissipated in R_ℓ. Evidently, the maximum power dissipated is rather insensitive to an exact matching between the load and source resistances.

Question 6.9

The switch S shown in the circuit below has been open for a 'long time'. Suppose that it is closed at time $t = 0$.

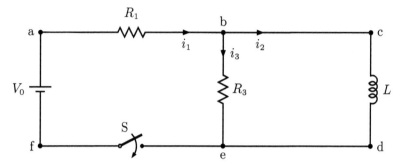

(a) Use Kirchhoff's rules (see p. 286) to show that the branch currents satisfy the equations

$$
\left.
\begin{array}{l}
\dfrac{di_1}{dt} + \dfrac{i_1}{\tau} = \dfrac{V_0}{\tau R_1} \\[2mm]
i_2 - \dfrac{(R_1 + R_3) i_1}{R_3} - \dfrac{V_0}{R_3} \\[2mm]
i_3 = i_1 - i_2
\end{array}
\right\},
\tag{1}
$$

where

$$
\tau = \frac{L(R_1 + R_3)}{R_1 R_3}.
\tag{2}
$$

(b) Solve (1) and express i_1, i_2 and i_3 in terms of the circuit parameters.

(c) Suppose $R_1 = 3\,\Omega$, $R_3 = 2\,\Omega$, $L = 300\,\mathrm{mH}$ and $V_0 = 1.0\,\mathrm{V}$. Show graphically how $i_1(t)$, $i_2(t)$ and $i_3(t)$ vary with time.

Solution

(a) Applying Kirchhoff's loop rule to loops abef and acdf gives

$$
\left.
\begin{array}{l}
V_0 = R_1 i_1 + R_3 i_3 \\
\text{and} \\
V_0 = R_1 i_1 + L\dfrac{di_2}{dt}
\end{array}
\right\}.
\tag{3}
$$

At node b, Kirchhoff's junction rule yields $i_1 = i_2 + i_3$. Hence $(1)_3$. Substituting $i_3 = i_1 - i_2$ in $(3)_1$ and rearranging give $(1)_2$. Equation $(1)_1$ then follows immediately from $(1)_2$ and $(3)_2$.

(b) At time $t = 0$ the large back emf in the inductor ensures that the current in this branch is instantaneously zero, and so we have the initial conditions

$$\left. \begin{array}{c} i_1(0) = i_3(0) = \dfrac{V_0}{(R_1 + R_3)} \\[2mm] i_2(0) = 0 \end{array} \right\}. \tag{4}$$

Rearranging $(1)_1$ gives

$$R_1 \int_{i_1(0)}^{i_1(t)} \frac{di_1}{R_1 i_1 - V_0} = -\int_0^t \frac{dt}{\tau},$$

where $i_1(0)$ is given by $(4)_1$. Integration yields

$$i_1(t) = \frac{V_0 \left(R_1 + R_3 (1 - e^{-t/\tau})\right)}{R_1 (R_1 + R_3)}. \tag{5}$$

Substituting (5) in $(1)_2$ and rearranging give

$$i_2(t) = \frac{V_0}{R_1} \left(1 - e^{-t/\tau}\right), \tag{6}$$

and then from $(1)_3$ we have

$$i_3(t) = \frac{V_0}{R_1 + R_3} e^{-t/\tau}. \tag{7}$$

(c) We obtain the following graph:

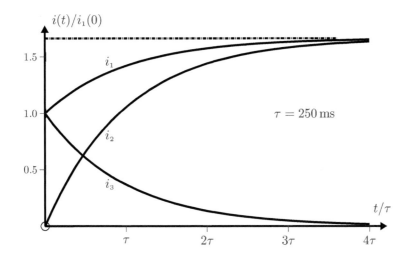

Comments

(i) The characteristic time τ given by (2) is called the time constant of the circuit. It effectively determines the rate at which the currents grow (decay) to their equilibrium values. The term 'long time' used in the statement of the question, can now be quantified to mean for 'times greater than about 10τ'.

(ii) The transient response of circuits can also be determined—elegantly—through the use of an integral transform (e.g. Laplace). Readers interested in this approach should consult an appropriate textbook.[1]

Question 6.10

In the series LRC circuit below, the switch S is closed at time $t = 0$.

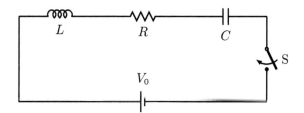

(a) Show that the charge $q(t)$ on the capacitor satisfies the equation

$$\frac{d^2q}{dt^2} + \frac{2}{\tau}\frac{dq}{dt} + \omega_0^2 q = \frac{V_0}{L}, \tag{1}$$

where $\omega_0 = 1/\sqrt{LC}$ and $\tau = 2L/R$.

(b) Suppose that the capacitor is initially uncharged and that $R < \sqrt{L/C}$. Use (1) to show that

$$q(t) = CV_0\left(1 - e^{-t/\tau}\left(\cos\omega t + \frac{1}{\omega\tau}\sin\omega t\right)\right) \tag{2}$$

and

$$i(t) = CV_0\left(\frac{1+\omega^2\tau^2}{\omega\tau^2}\right)e^{-t/\tau}\sin\omega t, \tag{3}$$

where $\omega = \sqrt{\omega_0^2 - \tau^{-2}}$.

(c) Plot a graph showing the voltages $V_L(t)$ and $V_C(t)$, taking $V_0 = 1.0\,\text{V}$, $L = 2.0\,\text{mH}$, $C = 3.0\,\mu\text{F}$ and $R = 10.0\,\Omega$.

---

[1] See, for example, R. N. Bracewell, *The Fourier transform and its applications*. New York: Mc-Graw-Hill, 2 edn, 1978.

Solution

(a) At time t, the voltages appearing across the inductor, resistor and capacitor are
$V_L = L\dfrac{di}{dt} = L\dfrac{d^2q}{dt^2}$, $V_R = Ri = R\dfrac{dq}{dt}$ and $V_C = \dfrac{q}{C}$. Then by Kirchhoff's loop
rule, $V_L + V_R + V_C = V_0$ or

$$\frac{d^2q}{dt^2} + \frac{R}{L}\frac{dq}{dt} + \frac{q}{LC} = \frac{V_0}{L}, \tag{4}$$

which is (1).

(b) From the theory of differential equations, we know that the general solution of
(1) consists of a complementary function q_c (which is the general solution of (1)
with $V_0 = 0$), and a particular integral q_p (which is a particular solution of (1)):

$$q(t) = q_c(t) + q_p(t). \tag{5}$$

complementary function

Here the equation to solve is

$$\frac{d^2q}{dt^2} + \frac{R}{L}\frac{dq}{dt} + \frac{q}{LC} = 0, \tag{6}$$

so we attempt a solution of the form $q = e^{\alpha t}$ where α satisfies the characteristic
equation $\alpha^2 + R\alpha/L + 1/LC = 0$, or

$$\alpha^2 + 2\alpha/\tau + \omega_0^2 = 0. \tag{7}$$

Consequently, the general solution of (6) is

$$q_c(t) = q_1 e^{\alpha_1 t} + q_2 e^{\alpha_2 t}, \tag{8}$$

where the roots of (7) are

$$\alpha_1 = -\tau^{-1}\left(1 + \sqrt{1 - \omega_0^2\tau^2}\right) \quad \text{and} \quad \alpha_2 = -\tau^{-1}\left(1 - \sqrt{1 - \omega_0^2\tau^2}\right).$$

Now because $R < \sqrt{L/C}$ it follows that $\omega_0\tau > 1$, and α_1 and α_2 are complex:
$\alpha_1 = -\tau^{-1} - i\omega_d$, $\alpha_2 = -\tau^{-1} + i\omega_d$ where

$$\omega_d = \sqrt{\omega_0^2 - \tau^{-2}} > 0.$$

Then from (8)

$$q_c(t) = e^{-t/\tau}\left(q_1 e^{i\omega_d t} + q_2 e^{-i\omega_d t}\right).$$

The term in brackets $(q_1 e^{i\omega_d t} + q_2 e^{-i\omega_d t})$ can be expressed as $Q\cos(\omega_d t - \varphi)$ where
Q and φ are constants, and so

$$q_c(t) = Q e^{-t/\tau} \cos(\omega_d t - \varphi). \tag{9}$$

particular integral

It is obvious by inspection that a particular solution of (1) is

$$q_p(t) = C V_0. \tag{10}$$

general solution

Combining (5), (9) and (10) yields

$$q(t) = C V_0 + Q e^{-t/\tau} \cos(\omega_d t - \varphi). \tag{11}$$

We must now determine the constants Q and φ from the initial conditions

$$\left. \begin{array}{l} q(0) = 0 \\ i(0) = 0 \end{array} \right\}. \tag{12}$$

Differentiating (11) gives

$$i(t) = -Q e^{-t/\tau} \left(\tau^{-1} \cos(\omega_d t - \varphi) + \omega_d \sin(\omega_d t - \varphi) \right), \tag{13}$$

then from (11)–(13) we obtain

$$\left. \begin{array}{l} C V_0 + Q \cos\varphi = 0 \\ -\tau^{-1} \cos\varphi + \omega_d \sin\varphi = 0 \end{array} \right\}. \tag{14}$$

Solving $(14)_1$ and $(14)_2$ simultaneously yields

$$\left. \begin{array}{l} Q \cos\varphi = -C V_0 \\ Q \sin\varphi = \dfrac{-C V_0}{\omega_d \tau} \end{array} \right\}, \tag{15}$$

then substituting (15) in (11) gives (2) and substituting (15) in (13) gives (3).

(c) The voltages $V_L(t)$ and $V_C(t)$ follow directly from (3) and (2) and are given by

$$\left. \begin{array}{l} V_L(t) = L\dfrac{di}{dt} = V_0 \left(\cos\omega_d t - \dfrac{1}{\omega_d \tau} \sin\omega_d t \right) e^{-t/\tau} \\[3mm] V_C(t) = \dfrac{q(t)}{C} = V_0 \left(1 - e^{-t/\tau} \left(\cos\omega_d t + \dfrac{1}{\omega_d \tau} \sin\omega_d t \right) \right) \end{array} \right\}. \tag{16}$$

For the given circuit parameters, we obtain $\omega_d = 12\,666 \text{ rad s}^{-1}$ and $\tau = 0.4 \text{ ms}$. Using these values in (16) leads to the following graphs:

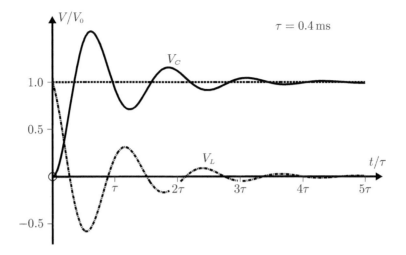

Comments

(i) When the switch is closed at time $t = 0$, the current $i(0) = 0$ and $V_R = 0$. Also, $q(0) = 0$ and so $V_C = 0$. However, at the instant the switch closes, $di/dt \neq 0$ and the voltage of the battery necessarily appears across the inductor. That is, $V_L(0) = V_0$. As the capacitor charges up, the voltage across it increases rapidly and oscillates about V_0 with decreasing amplitude. Meanwhile, di/dt reduces and $V_L \to 0$. The voltage across the resistor also oscillates and decays to zero after a long time ($\gg \tau$), at which point there is no further movement of charge in the circuit and $i = 0$.

(ii) The behaviour described in (i) above is the transient response of the circuit. As this name implies, the response is short-lived, lasting for a period of time $\lesssim 5\tau$. For times greater than this, the behaviour of the circuit is governed by the steady-state solution to (1): $V_C = V_0$, $V_L = V_R = 0$ and $i = 0$.

(iii) If the battery is replaced by an oscillator producing a sinusoidal output $v = v_0 e^{j\omega t}$, the transient response does not change, but the steady-state behaviour becomes more interesting. Now $z = R + j(\omega L - 1/\omega C)$ and

$$i(t) = \frac{v_0\, e^{j(\omega t - \phi)}}{\sqrt{R^2 + (\omega L - 1/\omega C)^2}}, \quad \text{where } \phi = \tan^{-1}\left(\frac{\omega^2 LC - 1}{\omega RC}\right). \tag{17}$$

At low driving frequencies the circuit behaves capacitively and v lags i in phase. As ω approaches ω_0 the circuit starts to resonate, and at resonance ($\omega = \omega_0$) the current in the circuit is in phase with v. For driving frequencies above ω_0 the circuit behaves inductively, and then v leads i in phase. The theory of a driven LRC circuit is the same as its mechanical analogue: the damped, driven harmonic oscillator (see, for example, Question 4.7 of Ref. [2]).

[2] O. L. de Lange and J. Pierrus, *Solved problems in classical mechanics: Analytical and numerical solutions with comments.* Oxford: Oxford University Press, 2010.

Question 6.11

A coil and a capacitor are connected in parallel, as shown in the circuit below. Assume that the ammeter is ideal and that the current in the capacitor is 5 A.

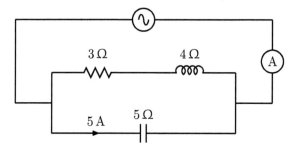

Calculate:

(a) the reading i on the ammeter,

(b) the average power dissipated in the parallel branch,

(c) the phase angle between the current in the capacitor and the current in the coil.

Solution

(a) Because of the current-divider rule (see Comment (iv) of Question 6.2)

$$i_{cap} = \frac{z_{coil}}{z_{coil} + z_{cap}} \times i, \tag{1}$$

where $z_{coil} = (3 + 4j)\Omega$ and $z_{cap} = -5j\Omega$. So

$$5 = \frac{3 + 4j}{3 - j} \times i \quad \Rightarrow \quad i = \frac{15 - 5j}{3 + 4j} \times \frac{3 - 4j}{3 - 4j} = \frac{25 - 75j}{25} = (1 - 3j)\text{A}.$$

The reading on the ammeter is $I = \sqrt{i\,i^*} = \sqrt{(1 - 3j)(1 + 3j)} = \sqrt{10} = \boxed{3.16\ \text{A}}$.

(b) From Kirchhoff's junction rule: $i_{coil} = i - i_{cap} = (1 - 3j) - 5 = (-4 - 3j)\text{A}$ or $I_{coil} = \sqrt{(-4)^2 + (-3)^2} = 5\ \text{A}$, and so $\mathcal{P} = RI^2 = 3 \times 5^2 = \boxed{75\ \text{W}}$.

(c) With reference to the phasor diagram below, we see that i_{coil} lags i_{cap} in phase by $90° + \alpha = 90° + \tan^{-1}(4/3) = 90° + 53.1° = \boxed{143.1°}$.

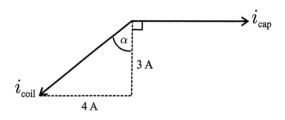

Question 6.12

Consider the circuit shown below.

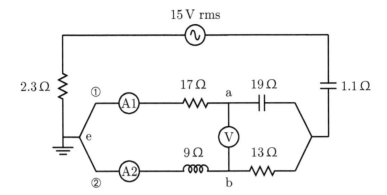

15 V rms

(a) Calculate the readings on the (ideal) ammeters A1 and A2.

(b) Hence calculate the reading on the (ideal) voltmeter V.

Solution

(a) Let $z = (2.3-1.1j)\Omega$, $z_1 = (17-19j)\Omega$ and $z_2 = (13+9j)\Omega$. Then the impedance presented to the oscillator is $z_{\text{tot}} = z + z_1 z_2/(z_1 + z_2)$, or

$$z_{\text{tot}} = (2.3 - 1.1j) + \frac{(17 - 19j)(13 + 9j)}{(17 - 19j) + (13 + 9j)} = (2.3 - 1.1j) + (12.7 + 1.1j)$$

$$= 15\,\Omega.$$

The current output from the oscillator is $i = \dfrac{v}{z_{\text{tot}}} = \dfrac{30}{15} = 2.0\,\text{A}$, and so

$$i_1 = \frac{z_2}{z_1 + z_2}\,i = \frac{(13 + 9j)}{(30 - 10j)} \times 2 = \frac{(13 + 9j)\,(3 + j)}{5(3 - j)\,(3 + j)}$$

$$= (0.6 + 0.8j)\,\text{A}.$$

Then $i_2 = i - i_1 = 2 - 0.6 - 0.8j = (1.4 - 0.8j)\,\text{A}$. The readings on the ammeters are thus $I_1 = \sqrt{0.6^2 + 0.8^2} = \boxed{0.50\,\text{A}}$ and $I_1 = \sqrt{1.4^2 + (-0.8)^2} = \boxed{1.61\,\text{A}}$.

(b) The voltages of points a and b relative to ground are: $v_{\text{ea}} = v_{\text{a}} - v_{\text{e}} = Ri_1 = (17)(0.6 + 0.8j) = (10.2 + 13.6j)\,\text{V}$ and $v_{\text{eb}} = v_{\text{b}} - v_{\text{e}} = X_L i_2 = (9j)(1.4 - 0.8j) = (7.2 + 12.6j)\,\text{V}$. So $v_{\text{a}} - v_{\text{b}} = v_{\text{ea}} - v_{\text{eb}} = (10.2 + 13.6j) - (7.2 + 12.6j) = (3 + j)\,\text{V}$. Hence $V_{\text{a}} - V_{\text{b}} = \sqrt{10} = \boxed{3.16\,\text{V}}$.

Question 6.13

Consider the RC network shown below. Let $\omega_0 = \dfrac{1}{RC}$.

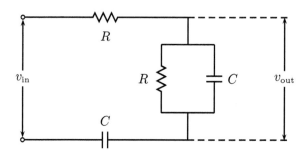

(a) Show that the output voltage v_{out} is given by

$$v_{\text{out}} = \frac{v_{\text{in}}\,e^{-j\phi}}{\sqrt{9 + \left(\omega/\omega_0 - \omega_0/\omega\right)^2}} \qquad \text{and} \qquad \phi = \tan^{-1}\left(\frac{\omega^2 - \omega_0^2}{3\,\omega\omega_0}\right). \tag{1}$$

(b) Draw graphs of $V_{\text{out}}/V_{\text{in}}$ and ϕ as a function of ω/ω_0.

Solution

(a) The voltage-divider rule gives: $v_{\text{out}} = v_{\text{in}}\dfrac{z_{\|}}{z_{\|} + z_{\text{total}}}$ where $z_{\|} = \dfrac{\frac{R}{j\omega C}}{R + \frac{1}{j\omega C}}$

$= \dfrac{R}{1 + j\omega RC}$ and $z_{\text{total}} = R + \dfrac{1}{j\omega C} + z_{\|}$. So

$$v_{\text{out}} = \frac{v_{\text{in}}\,R}{\left(1 + j\omega RC\right)\left(R + \dfrac{1}{j\omega C} + \dfrac{R}{1 + j\omega RC}\right)}$$

$$= \frac{v_{\text{in}}\,R}{\left(3R + j(\omega R^2 C - 1/\omega C)\right)}$$

$$= \frac{v_{\text{in}}\,R}{\left(3R + j(\omega R^2 C - 1/\omega C)\right)} \times \frac{\left(3R - j(\omega R^2 C - 1/\omega C)\right)}{\left(3R - j(\omega R^2 C - 1/\omega C)\right)}$$

$$= \frac{v_{\text{in}}\,R\,e^{-j\phi}}{\sqrt{9R^2 + (\omega R^2 C - 1/\omega C)^2}}.$$

Hence (1).

(b) We obtain the following graphs:

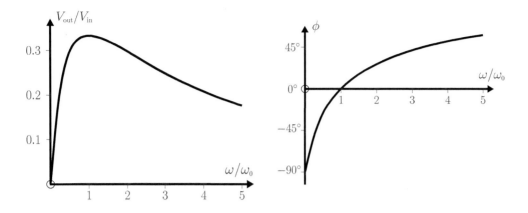

Comment

At high frequencies X_C is small, $v_{\text{out}} \to 0$ and the circuit is a type of low-pass filter. From (1), we see that at $\omega = \omega_0$ the voltages v_{out} and v_{in} are in phase and $V_{\text{out}} = \frac{1}{3} V_{\text{in}}$.

Question 6.14

In the circuit shown below, the output from the oscillator is 15 V rms. Suppose that $R_1 = 12\,\Omega$, $R_2 = 8\,\Omega$, $L_1 = 2\,\text{mH}$, $L_2 = 1\,\text{mH}$, $C_1 = 2\,\mu\text{F}$ and $C_3 = 1\,\mu\text{F}$.

$$\mathcal{E} = 15\,\text{V rms}$$

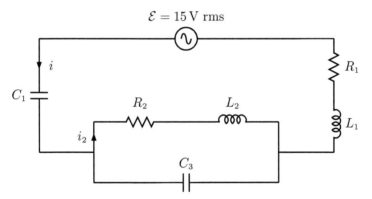

(a) Show that i and i_2 are given by

$$ i = \frac{(z_2 + z_3)\mathcal{E}}{(z_1 z_2 + z_1 z_3 + z_2 z_3)} \quad \text{and} \quad i_2 = \frac{z_3 \mathcal{E}}{(z_1 + z_2 + z_1 z_2 / z_3)}. \tag{1} $$

(b) Use *Mathematica* to calculate the power $\mathcal{P}$ dissipated in this circuit at the (cyclic) frequency f, and plot a graph of $\mathcal{P}(f)$ for $0 \le f \le 11\,\text{kHz}$.

Solution

(a) The individual branch impedances are $z_1 = R_1 + j\omega L_1 + \dfrac{1}{j\omega C_1}$, $z_2 = R_2 + j\omega L_2$ and

$z_3 = \dfrac{1}{j\omega C_3}$ resulting in a net impedance $z = z_1 + \dfrac{z_2 z_3}{z_2 + z_3} = \dfrac{z_1 z_2 + z_1 z_3 + z_2 z_3}{z_2 + z_3}$.

Then $i = \mathcal{E}/z$. Because of the current-divider rule, $i_2 = i z_3/(z_2 + z_3)$. Hence (1).

(b) Using the following notebook, we obtain the graph below:

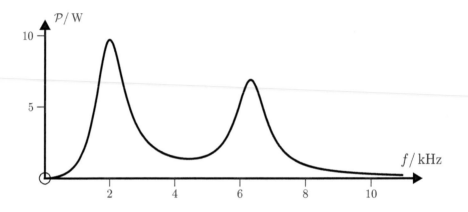

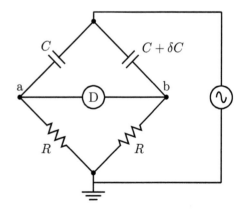

Question 6.15

Two arms of the bridge circuit shown along-side have equal resistance R. The other two arms have the capacitances shown with $\delta C \ll C$. Assume that both the detector D and the oscillator have negligible internal impedance. Use Thévenin's theorem to show that the current i in the detector is

$$i = \dfrac{-\omega\,\delta C\,\mathcal{E}\,e^{j\phi}}{2\sqrt{1 + \omega^2 R^2 C^2}}, \qquad (1)$$

where $\phi = \tan^{-1}\left(\dfrac{1}{\omega RC}\right)$.

Solution

Thévenin equivalent impedance

$$z_{\text{Th}} = R \| C + R \| (C + \delta C) = \frac{R}{1 + j\omega RC} + \frac{R}{1 + j\omega R(C + \delta C)}$$

$$\simeq \frac{2R}{1 + j\omega RC} \quad \text{since } \delta C \ll C. \tag{2}$$

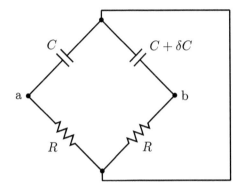

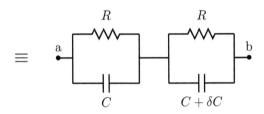

Thévenin equivalent emf

Applying the voltage-divider rule we calculate the potentials of v_{a} and v_{b} relative to ground:

$$v_{\text{a}} = \frac{R}{R + \dfrac{1}{j\omega C}} \cdot \mathcal{E} = \frac{j\omega RC\mathcal{E}}{1 + j\omega RC} \quad \text{and similarly} \quad v_{\text{b}} = \frac{j\omega R(C + \delta C)\mathcal{E}}{1 + j\omega R(C + \delta C)}.$$

Then

$$\mathcal{E}_{\text{Th}} = v_{\text{a}} - v_{\text{b}} = \frac{j\omega RC\mathcal{E}}{1 + j\omega RC} - \frac{j\omega R(C + \delta C)\mathcal{E}}{1 + j\omega R(C + \delta C)} \simeq \frac{-j\omega R\,\delta C\mathcal{E}}{(1 + j\omega RC)^2}, \tag{3}$$

where in the final step we make the approximation $(C + \delta C) \to C$ in the denominator.

Thévenin equivalent circuit

With the detector connected in series with $\mathcal{E}_{\text{Th}}$ and z_{Th}, we obtain from (2) and (3)

$$i = \frac{\mathcal{E}_{\text{Th}}}{z_{\text{Th}}} = \frac{-j\omega\,\delta C\,\mathcal{E}}{2(1 + j\omega RC)} = \frac{-j(1 - j\omega RC)\omega\,\delta C\,\mathcal{E}}{2(1 + \omega^2 R^2 C^2)} = \frac{-(\omega RC + j)\omega\,\delta C\,\mathcal{E}}{2(1 + \omega^2 R^2 C^2)}.$$

Hence (1).

Comment

Notice that when the bridge is balanced ($\delta C = 0$), the current in the detector is zero (as one would expect on grounds of symmetry).

Question 6.16*

In the circuit shown below, assume that both oscillators have negligible output impedance.

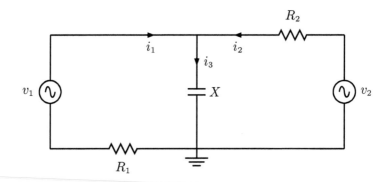

(a) Use the superposition theorem to show that the branch currents are given by

$$
\left.
\begin{aligned}
i_1 &= \frac{(R_2 - jX)v_1 + jX v_2}{R_1 R_2 - jX(R_1 + R_2)} \\[2mm]
i_2 &= \frac{-jX v_1 - (R_1 - jX)v_2}{R_1 R_2 - jX(R_1 + R_2)} \\[2mm]
i_3 &= \frac{R_2 v_1 + R_1 v_2}{R_1 R_2 - jX(R_1 + R_2)}
\end{aligned}
\right\}. \tag{1}
$$

(b) Suppose

$$
\left.
\begin{aligned}
R_1 &= 4\,\Omega, \qquad R_2 = 3\,\Omega, \qquad X = 5\,\Omega \\
v_1 &= 15 e^{j\omega t}\,\text{V} \\
v_2 &= 12 e^{j(\omega t + \pi/6)}\,\text{V}
\end{aligned}
\right\}.
$$

Calculate i_1, i_2 and i_3 and express your answers in polar form.

Hint: Use *Mathematica* to assist with the algebra.

Solution

(a) In the two circuits below, we calculate the net impedance z which leads directly to the current supplied by the oscillator. The branch currents then follow from a straightforward application of the current-divider rule. In the final step we make use of the superposition theorem.

oscillator #1 acting independently

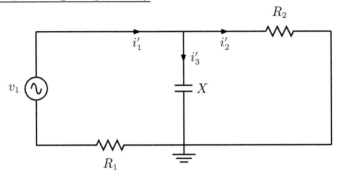

$$z = R_1 + \frac{-jR_2X}{R_2 - jX} = \frac{R_1R_2 - jX(R_1 + R_2)}{R_2 - jX} \Rightarrow i'_1 = \frac{v_1}{z} = \frac{(R_2 - jX)v_1}{R_1R_2 - jX(R_1 + R_2)}.$$

$$\text{Then } i'_2 = \frac{-jX}{(R_2 - jX)}\frac{v_1(R_2 - jX)}{R_1R_2 - jX(R_1 + R_2)} = \frac{-jXv_1}{R_1R_2 - jX(R_1 + R_2)}, \text{ and}$$

$$i'_3 = \frac{R_2}{(R_2 - jX)}\frac{v_1(R_2 - jX)}{R_1R_2 - jX(R_1 + R_2)} = \frac{R_2v_1}{R_1R_2 - jX(R_1 + R_2)}.$$

oscillator #2 acting independently

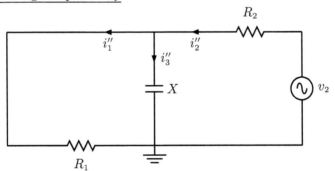

$$z = R_2 + \frac{-jR_1X}{R_1 - jX} = \frac{R_1R_2 - jX(R_1 + R_2)}{R_1 - jX} \Rightarrow i_2'' = \frac{v_2}{z} = \frac{(R_1 - jX)v_2}{R_1R_2 - jX(R_1 + R_2)}.$$

$$\text{Then } i_1'' = \frac{-jX}{(R_1 - jX)} \frac{v_2(R_1 - jX)}{R_1R_2 - jX(R_1 + R_2)} = \frac{-jXv_2}{R_1R_2 - jX(R_1 + R_2)}, \text{ and}$$

$$i_3'' = \frac{R_1}{(R_1 - jX)} \frac{v_2(R_1 - jX)}{R_1R_2 - jX(R_1 + R_2)} = \frac{R_1v_2}{R_1R_2 - jX(R_1 + R_2)}.$$

apply the superposition theorem

For the current directions assigned to the three circuits above, we have

$$i_1 = i_1' - i_1'', \qquad i_2 = i_2' - i_2'', \qquad i_3 = i_3' + i_3''. \tag{2}$$

Substituting the currents calculated above in (2) yields (1).

(b) Implementing (1) in the notebook below gives:

$$\left. \begin{aligned} i_1 &= 0.74e^{j(\omega t + 14.14°)} \\ i_2 &= 2.31e^{j(\omega t \ 75.61°)} \\ i_3 &= 2.43e^{j(\omega t + 86.57°)} \end{aligned} \right\} \text{A.} \tag{3}$$

```
In[1]:= R1 = 4.0;  R2 = 3;  X = 5;  v1 = 15;  v2 = 12 e^{i π/6};
        i1 = ComplexExpand[ (R2 - i X) v1 + i X v2
                            ─────────────────────── ];
                            R1 R2 - i X (R1 + R2)
        i2 = ComplexExpand[ -i X v1 - (R1 - i X) v2
                            ─────────────────────── ];
                            R1 R2 - i X (R1 + R2)
        i3 = ComplexExpand[ R2 v1 + R1 v2
                            ───────────────────── ];
                            R1 R2 - i X (R1 + R2)
        I1 = Abs[i1]; φ1 = 180 ArcTan[ Im[i1] ];
                          ───          ──────
                           π           Re[i1]
        I2 = Abs[i2]; φ2 = 180 ArcTan[ Im[i2] ];
                          ───          ──────
                           π           Re[i2]
        I3 = Abs[i3]; φ3 = 180 ArcTan[ Im[i3] ];
                          ───          ──────
                           π           Re[i3]
```

Comment

It is left as an exercise for the reader to confirm that the currents calculated in (b) satisfy Kirchhoff's junction rule: $i_1 = i_2 + i_3$.

Question 6.17

A 'coil' having inductance L and internal resistance R is connected in series with a capacitor, as shown.

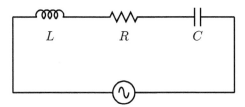

Determine the frequency at which the impedance of this series LRC circuit is real.

Solution

The net impedance of the combination is $z = R + j\omega L + \dfrac{1}{j\omega C} = R + j\left(\omega L - \dfrac{1}{\omega C}\right)$

which is clearly real at the frequency ω_0 when $\omega_0 L - \dfrac{1}{\omega_0 C} = 0$, or

$$\omega_0 = (LC)^{-1/2}. \tag{1}$$

Comments

(i) Resonance is usually regarded as the condition for which z is purely resistive. For the above combination of a coil in series with a capacitor it occurs at the frequency given by (1). It is sometimes useful, when discussing resonance, to introduce the complex admittance y defined as the reciprocal of the complex impedance:

$$y = \frac{1}{z} = \frac{1}{Ze^{j\phi}} = \frac{1}{Z}e^{-j\phi} = Ye^{j\phi'}, \qquad \text{where} \qquad Y = Z^{-1} \text{ and } \phi' = -\phi.$$

(ii) At resonance the impedance of a series LRC circuit is a minimum ($Z = R$), the admittance is a maximum ($Y = 1/R$) and the current is in phase with the applied voltage ($\phi = 0$). Of course, it follows from (2) and (3) of Question 6.1 that the voltage across the inductor (capacitor) leads (lags) i in phase by $90°$.

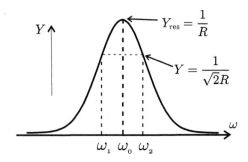

(iii) The sharpness of resonance can be described by how rapidly Y falls off from its maximum value R^{-1} as ω varies in an interval around ω_0. Suppose that ω_1 and ω_2 are frequencies at which $Y = Y_{\mathrm{res}}/\sqrt{2}$. Then $Y = \dfrac{1}{Z} = \dfrac{1}{\sqrt{2}R} \Rightarrow Z^2 = 2R^2$.

Now at ω_1 and ω_2 the reactance $X^2 = Z^2 - R^2 = R^2$ where $X = (\omega L - 1/\omega C)$. So

$$\left.\begin{aligned}\text{at the higher frequency } \omega_2 \text{ we have: } \omega_2 L - \frac{1}{\omega_2 C} &= R\\[2mm]\text{at the lower frequency } \omega_1 \text{ we have: } \omega_1 L - \frac{1}{\omega_1 C} &= -R\end{aligned}\right\}. \tag{2}$$

From $\omega_2 \times (2)_1 - \omega_1 \times (2)_2$ it follows that $(\omega_2^2 - \omega_1^2)L = (\omega_1 + \omega_2)R$, or

$$\Delta\omega = \omega_2 - \omega_1 = \frac{R}{L}. \tag{3}$$

The resonance is regarded as being of high quality if the interval $\Delta\omega = \omega_2 - \omega_1$ is narrow relative to ω_0 (here $\Delta\omega$ is sometimes called the 'bandwidth' of the circuit). With this in mind, we define the quality factor $Q = \omega_0/\Delta\omega$. Then with (3) we have

$$Q = \frac{\omega_0 L}{R}, \tag{4}$$

and, because of (1) we have the alternative forms

$$\left.\begin{aligned}Q &= \frac{1}{\omega_0 RC}\\[2mm]Q &= \frac{1}{R}\sqrt{\frac{L}{C}}\end{aligned}\right\}. \tag{5}$$

(iv) For high Q the resonant impedance of the coil is $Z_{\mathrm{coil}} = \omega_0 L(1+1/Q^2)^{1/2} \approx \omega_0 L$, and then $V_L = Z_{\mathrm{coil}}I = \omega_0 LI = I/(\omega_0 C) = V_{\mathrm{cap}}$. Now at resonance $V_{\mathrm{tot}} = RI$, and so

$$\frac{V_L}{V_{\mathrm{tot}}} = \frac{\omega_0 L}{R} = Q \quad \text{and} \quad \frac{V_C}{V_{\mathrm{tot}}} = \frac{1}{\omega_0 RC} = Q. \tag{6}$$

These equations shows that the voltages across L and C at resonance have magnitudes Q times the net circuit voltage. Evidently, the resonant series LRC circuit acts like a voltage transformer.

(v) The following figure is the resonance curve for a series LRC circuit for which $L = 1\,\mathrm{mH}$, $R = 2\,\Omega$ and $C = 3\,\mu\mathrm{F}$.

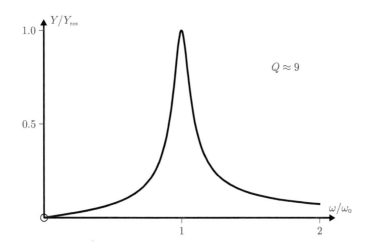

(vi) The reader should be aware that current resonance, voltage resonance and power resonance do not occur at the same frequency. However, for high-Q circuits these different resonant frequencies are essentially so close to one another that there is usually no reason to distinguish between them in practice.

Question 6.18

Suppose that the coil of Question 6.17 is now connected in parallel with the capacitor, as shown below.

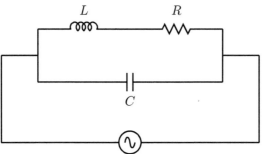

(a) Show that resonance occurs at

$$\omega_0 = \frac{1}{\sqrt{LC}}\left(1 - \frac{1}{Q^2}\right) \qquad \text{where } Q = \frac{1}{R}\sqrt{\frac{L}{C}}. \tag{1}$$

(b) Prove that the impedance of this circuit at resonance is

$$Z_{\text{res}} = Q^2 R. \tag{2}$$

(c) Suppose that $Q \gtrsim 10$. Show that at resonance the individual branch currents are almost equal in magnitude and opposite in phase.

Solution

(a) The branch impedances are $z_1 = R + j\omega L$ and $z_2 = \dfrac{1}{j\omega C}$. Then for two impedances in parallel, (2) of Question 6.2 gives

$$z = \frac{z_1 z_2}{z_1 + z_2} = \frac{(R + j\omega L)\dfrac{1}{j\omega C}}{(R + j\omega L) + \dfrac{1}{j\omega C}} = \frac{(R + j\omega L)}{(1 - \omega^2 LC) + j\omega RC}$$

$$= \frac{(R + j\omega L)}{[(1 - \omega^2 LC) + j\omega RC]} \frac{[(1 - \omega^2 LC) - j\omega RC]}{[(1 - \omega^2 LC) - j\omega RC]}$$

$$= \frac{R + j[\omega L(1 - \omega^2 LC) - \omega R^2 C]}{[(1 - \omega^2 LC)^2 + \omega^2 R^2 C^2]}. \tag{3}$$

Resonance occurs when the reactive part of $z = 0$, or $\omega_0 L(1 - \omega_0^2 LC) = \omega_0 R^2 C$. The non-trivial root of this equation is

$$\omega_0 = \sqrt{\frac{1}{LC} - \frac{R^2}{L^2}}. \tag{4}$$

Substituting the definition of Q in (4) gives (1).

(b) At resonance the denominator in (3) is $(1 - \omega_0^2 LC)^2 + \omega_0^2 R^2 C^2 = R^2 C/L = 1/Q^2$ because of (4), and the definition of Q. Hence (2).

(c) Suppose $v = Ve^{j\omega t}$. At resonance the branch currents are $i_1 = v/z_1$ and $i_2 = v/z_2$, where $z_1 = (R + j\omega_0 L)$ and $z_2 = 1/j\omega_0 C$. So

$$i_1 = \frac{Ve^{j\omega t}}{R + j\omega_0 L} = \frac{Ve^{j\omega t}(R - j\omega_0 L)}{R^2 + \omega_0^2 L^2} = \frac{Ve^{j(\omega t - \phi)}}{\sqrt{R^2 + \omega_0^2 L^2}}$$

$$i_2 = \frac{Ve^{j\omega t}}{1/(j\omega_0 C)} = jV\omega_0 C\, e^{j\omega t} = V\omega_0 C\, e^{j(\omega t + \frac{1}{2}\pi)}$$

where $\tan\phi = \dfrac{\omega_0 L}{R} = Q$. Now $\tan(\frac{1}{2}\pi - \phi) = \cot\phi = Q^{-1}$. So $\phi = \frac{1}{2}\pi - \tan^{-1} Q^{-1}$.

For $Q \gtrsim 10$ we have $\tan^{-1} Q^{-1} \lesssim 6°$, and so $\phi \gtrsim 84° \approx \frac{1}{2}\pi$. Then

$$i_1 = V\omega_0 C\, e^{j(\omega_0 t - \frac{1}{2}\pi)}$$
$$i_2 = V\omega_0 C\, e^{j(\omega_0 t + \frac{1}{2}\pi)} \tag{5}$$

where $I_1 = I_2 = V\omega_0 C$. Equation (5) shows also that the two currents are essentially in anti-phase, since i_1 lags v by $\sim \pi/2$ and i_2 leads v by $\sim \pi/2$.

Comments

(i) For high Q, the current I delivered by the oscillator is $V/Z_{\text{res}} = V/(Q^2 R)$. Then

$$\frac{I_1}{I} = \frac{I_2}{I} = Q^2 \omega_0 R C = Q, \tag{6}$$

because of $(5)_1$ of Question 6.17. We see from (6) that the branch currents in the coil and capacitor at resonance have magnitudes Q times the feed current from the oscillator. Evidently, the parallel LRC circuit at resonance acts like a current transformer.

(ii) Because i_1 and i_2 are approximately equal in magnitude and $180°$ out of phase at resonance, the feed current $i = i_1 + i_2$ from the oscillator is almost zero. For readers who enjoy algebra, it is possible to show that the current minimum occurs not at ω_0, but at the nearby frequency $\omega_0' = \omega_0(1 - 2/Q^2)$. Clearly, for $Q \gtrsim 10$, these two frequencies are practically indistinguishable from one another. This means that the resonant impedance of the circuit $Z_{\text{res}} = V/I_{\text{res}}$ is a maximum, and is equal to $Q^2 R$, as we found in (b).

Question 6.19

(a) In the bridge circuit shown in Fig. (I) below, the z_i represent arbitrary impedances (here $i = 1, \ldots, 4$). Show that the balance condition for the bridge (recall from Question 6.7 that the bridge is balanced when $i_D = 0$) is

$$z_1 z_4 = z_2 z_3. \tag{1}$$

(b) Hence show that the balance conditions for the bridge shown in Fig. (II) below are

$$\left. \begin{array}{l} R_2 R_3 = R_1(R_4 + r_4) \\ L_4 = C_1 R_2 R_3 \end{array} \right\}. \tag{2}$$

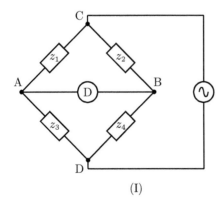

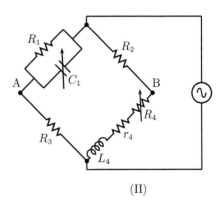

(I)   (II)

Solution

(a) At balance, the potential difference across the detector is zero. This requires that $V_{CA} = V_{CB}$ and $V_{AD} = V_{BD}$, or $i_1 z_1 = i_2 z_2$ and $i_3 z_3 = i_4 z_4$. With no current in the galvanometer $i_1 = i_3$ and $i_2 = i_4$. So

$$\frac{i_1 z_1}{i_1 z_3} = \frac{i_2 z_2}{i_2 z_4}.$$

Hence (1).

(b) Here $z_1 = R_1/(1 + j\omega R_1 C_1)$, $z_2 = R_2$, $z_3 = R_3$ and $z_4 = (R_4 + r_4 + j\omega L_4)$. Then from (1)

$$R_1(R_4 + r_4) + jR_1\omega L_4 = R_2 R_3(1 + j\omega R_1 C_1),$$

or

$$\left[R_1(R_4 + r_4) - R_2 R_3\right] + j\omega R_1\left[L_4 - R_2 R_3 C_1\right] = 0. \qquad (3)$$

This equation can only be satisfied if its real and imaginary parts are each zero. Hence (2).

Comment

The circuit shown in Fig. (II) on p. 319, known as Maxwell's bridge, can be used to determine an unknown inductance. In the decades preceding digital electronics, ac bridges were widely used in laboratories to determine unknown capacitances, inductances and frequencies. In this regard, their usefulness these days has largely been surpassed by the advance of technology: the ubiquitous handheld digital multimeter can now measure quantities like these both quickly and accurately.

Question 6.20

Two loudspeakers, each of resistance R and negligible inductance, are connected to the output stage of an audio amplifier via the network shown below.

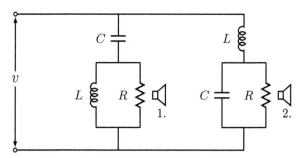

(a) Explain which speaker is the 'woofer' (driven strongly at low frequencies) and which is the 'tweeter' (driven strongly at high frequencies).

(b) Derive a relation between L and C for a given R such that the network presents a purely resistive load to the amplifier at all frequencies. What is the value of this load?

(c) At the crossover angular frequency w_c, each speaker receives half the power delivered by the amplifier. Determine w_c in terms of L and C. What is the phase relation between speaker currents at this frequency?

Solution

(a) At low frequencies $X_C \to \infty$ and $X_L \to 0$. The output voltage v from the amplifier effectively appears across each capacitor and speaker 2 is being driven strongly, whereas speaker 1 is not. At high frequencies the converse is true because now $X_C \to 0$ and $X_L \to \infty$. Hence speaker 1 is the 'tweeter' and speaker 2 is the 'woofer'.

(b) We use the subscript 1 to indicate impedances in the left-hand branch and the subscript 2 to indicate impedances in the right-hand branch. Then

$$\left.\begin{array}{l} z_1 = \dfrac{1}{jwC} + \dfrac{R \times jwL}{R + jwL} = \dfrac{R(1 - w^2 LC) + jwL}{-w^2 LC + jwRC}, \\[4mm] \text{and} \\[4mm] z_2 = jwL + \dfrac{\frac{1}{jwC} \times R}{R + \frac{1}{jwC}} = jwL + \dfrac{R}{1 + jwRC} = \dfrac{R(1 - w^2 LC) + jwL}{1 + jwRC} \end{array}\right\} \qquad (1)$$

Now z_1 and z_2 are in parallel and the net impedance z is thus

$$\frac{1}{z} = \frac{1}{z_1} + \frac{1}{z_2} = \frac{-w^2 LC + 1 + 2jwRC}{R(1 - w^2 LC) + jwL} = \frac{1}{R}\left\{\frac{(1 - w^2 LC) + 2jwRC}{(1 - w^2 LC) + jwL/R}\right\},$$

which is resistive at all frequencies if $2wRC = wL/R$ or

$$R = \sqrt{\frac{L}{2C}}. \qquad (2)$$

(c) Clearly the crossover condition corresponds to the same current in each speaker. These currents are given by the current-divider rule:

$$i_{sp1} = \frac{v}{z_1} \times \frac{jwL}{R + jwL} \qquad \text{and} \qquad i_{sp2} = \frac{v}{z_2} \times \frac{\frac{1}{jwC}}{R + \frac{1}{jwC}}. \qquad (3)$$

Substituting (1) in (3) yields

$$i_{sp1} = \frac{-v\,w^2 LC}{R(1 - w^2 LC) + jwL} \qquad \text{and} \qquad i_{sp2} = \frac{v}{R(1 - w^2 LC) + jwL}. \qquad (4)$$

These are equal in magnitude if $\omega^2 LC = 1$, or

$$\omega_c = \frac{1}{\sqrt{LC}} \, . \tag{5}$$

Recalling that $e^{-j\pi} = -1$, it is evident that the speaker currents are out of phase, with i_{sp1} lagging i_{sp2} by 180°.

Question 6.21

A 'bridged-T' filter is shown in the circuit below.

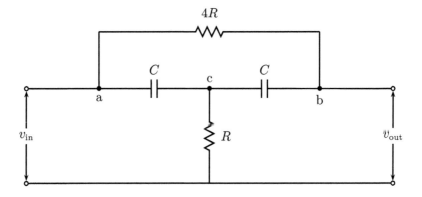

(a) Explain how the circuit behaves in the limit of very low and very high frequencies.

(b) Transform the delta combination in the circuit to a star (see Question 6.5), and show that the open-circuit output voltage is

$$\left.\begin{array}{c} V_{\text{out}} = V_{\text{in}} \sqrt{\dfrac{1 - 4\omega^2 R^2 C^2 + 16\omega^4 R^4 C^4}{1 + 28\omega^2 R^2 C^2 + 16\omega^4 R^4 C^4}} \; e^{j\phi} \\[2em] \text{where} \\[1em] \phi = \tan^{-1}\!\left[\dfrac{\omega R C (4\omega^2 R^2 C^2 - 1)}{12\omega^2 R^2 C^2 + (4\omega^2 R^2 C^2 - 1)^2}\right] \end{array}\right\} . \tag{1}$$

(c) Draw graphs of $V_{\text{out}}/V_{\text{in}}$ and ϕ as a function of the cyclic frequency f in the interval $0 \le f \le 12\,\text{kHz}$.

Solution

(a) At low (high) frequencies $X_C \to \infty$ ($X_C \to 0$), and we obtain the following equivalent circuits:

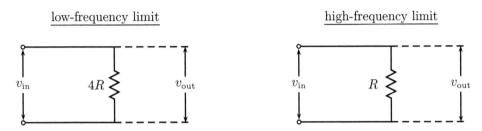

In both cases v_{out} is in phase with v_{in} and $v_{\mathrm{out}} = v_{\mathrm{in}}$.

(b) Transforming the delta in this circuit to a star yields:

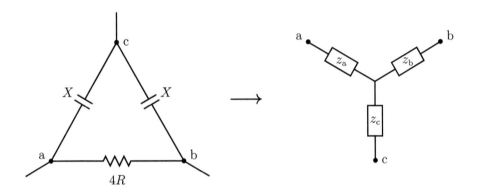

Using (9) of Question 6.5 and putting $X = \dfrac{1}{j\omega C}$; $z = 4R + \dfrac{2}{j\omega C}$ give

$$
\left.
\begin{aligned}
z_a = z_b &= \frac{4R}{2 + 4j\omega RC} \\[1em]
\text{and} & \\[1em]
z_c &= \frac{-j}{\omega C(2 + 4j\omega RC)}
\end{aligned}
\right\}. \tag{2}
$$

Then the current i shown in the circuit below is $i = \dfrac{v_{\mathrm{in}}}{R + z_a + z_c}$, along with an output voltage $v_{\mathrm{out}} = (R + z_c)i$. So

$$
v_{\mathrm{out}} = \frac{(R + z_c)}{R + z_a + z_c} v_{\mathrm{in}}. \tag{3}
$$

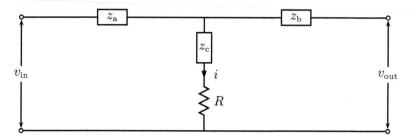

Substituting (2) in (3) and simplifying yield

$$v_{out} = \frac{[2\omega RC + j(4\omega^2 R^2 C^2 - 1)]}{[6\omega RC + j(4\omega^2 R^2 C^2 - 1)]} v_{in} \tag{4}$$

$$= \frac{[2\omega RC + j(4\omega^2 R^2 C^2 - 1)]\,[6\omega RC - j(4\omega^2 R^2 C^2 - 1)]}{[6\omega RC + j(4\omega^2 R^2 C^2 - 1)]\,[6\omega RC - j(4\omega^2 R^2 C^2 - 1)]} v_{in}$$

$$= \frac{(12\omega^2 R^2 C^2 + (4\omega^2 R^2 C^2 - 1)^2) + 4j\omega RC(4\omega^2 R^2 C^2 - 1)}{36\omega^2 R^2 C^2 + (4\omega^2 R^2 C^2 - 1)^2} v_{in}.$$

Hence (1).

(c) We obtain the graphs:

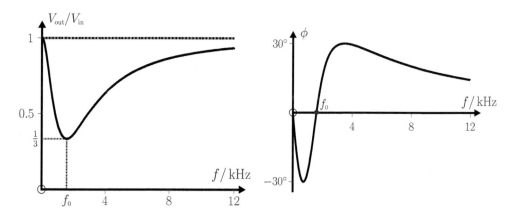

Comment

From (4) we see that v_{in} and v_{out} are in phase at

$$f = f_0 = \frac{1}{4\pi RC},$$

with $V_{out} = \frac{1}{3} V_{in}$.

Question 6.22*

The two circuits shown below are coupled via their mutual inductance M. Assume that $\mathcal{E}_1$ and $\mathcal{E}_2$ are emfs that vary harmonically with the same frequency ω.

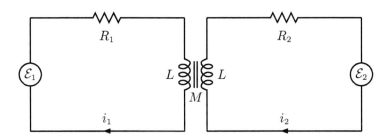

(a) Show that

$$\left.\begin{aligned}\mathcal{E}_1 &= z_1 i_1 - j\omega M i_2 \\ \mathcal{E}_2 &= -j\omega M i_1 + z_2 i_2\end{aligned}\right\}, \tag{1}$$

where $z_1 = R_1 + j\omega L_1$ and $z_2 = R_2 + j\omega L_2$.

(b) Suppose $\mathcal{E}_2 = 0$. Show that

$$\left.\begin{aligned}\mathcal{E}_1 &= \left(z_1 + \frac{\omega^2 M^2}{z_2}\right) i_1 \\ \frac{j\omega M \mathcal{E}_1}{z_1} &= \left(z_2 + \frac{\omega^2 M^2}{z_1}\right) i_2\end{aligned}\right\}. \tag{2}$$

Solution

(a) We must be careful about assigning a polarity to the induced emfs. Suppose that i_1 is producing an increasing magnetic flux ϕ_1. The current induced in circuit 2 through the coupling creates a flux ϕ_2 which *opposes* ϕ_1. The changing flux ϕ_2 in turn induces an emf in circuit 1 which *reinforces* ϕ_1. Hence this emf has the same polarity as $\mathcal{E}_1$, and similarly for $\mathcal{E}_2$. Applying Kirchhoff's loop rule then gives

$$\left.\begin{aligned}\mathcal{E}_1 + M\frac{di_2}{dt} &= R_1 i_1 + L_1\frac{di_1}{dt} \\ \mathcal{E}_2 + M\frac{di_1}{dt} &= R_2 i_2 + L_2\frac{di_2}{dt}\end{aligned}\right\}. \tag{3}$$

For time-harmonic currents $d/dt \rightarrow j\omega$ and (3) becomes

$$\left.\begin{array}{l} \mathcal{E}_1 = (R_1 + j\omega L_1)i_1 - j\omega M i_2 \\[2mm] \mathcal{E}_2 = -j\omega M i_1 + (R_2 + j\omega L_2)i_2 \end{array}\right\}, \qquad (4)$$

which is (1).

(b) Here we are required to solve the equations

$$\left.\begin{array}{l} \mathcal{E}_1 = z_1 i_1 - j\omega M i_2 \\[2mm] 0 = -j\omega M i_1 + z_2 i_2 \end{array}\right\}, \qquad (5)$$

and so from $(5)_2$ we obtain

$$i_1 = -\frac{j\,z_2}{\omega M}\,i_2 \qquad \text{and} \qquad i_2 = \frac{j\omega M}{z_2}\,i_1. \qquad (6)$$

Eliminating i_2 and i_1, in turn, from $(5)_1$ gives (2).

Comments

(i) With $\mathcal{E}_2 = 0$ we have the model of a simple transformer, where 1 and 2 represent the primary and secondary circuits respectively.

(ii) The quantity in brackets in $(2)_1$ is the effective impedance of the primary circuit, made up of z_1 and $\omega^2 M/z_2$ 'reflected from the secondary into the primary circuit'. We see from $(2)_2$ that $\omega^2 M/z_1$ is also a reflected impedance, and that there is an effective emf $\mathcal{E}'_2 = j\omega M\mathcal{E}_1/z_1$. Here $\mathcal{E}'_2$ is given as if by a primary current $i' = \mathcal{E}_1/z_1$, which would be the actual primary current without coupling, producing an emf $M\,di'/dt$.

(iii) We consider the case of an ideal transformer[‡] where $\mathcal{E}_2 = 0$ and $R_1 = 0$. Substituting $z_1 = j\omega L_1$ and $z_2 = R_2 + j\omega L_2$ in (4) gives

$$\left.\begin{array}{l} \mathcal{E}_1 = j\omega L_1 i_1 - j\omega M i_2 \\[2mm] 0 = -j\omega M i_1 + z_2 i_2 \end{array}\right\} \quad\Rightarrow\quad \left.\begin{array}{l} \mathcal{E}_1 L_2 = j\omega L_1 L_2 i_1 - j\omega M L_2 i_2 \\[2mm] -R_2 i_2 = -j\omega M i_1 + j\omega L_2 i_2 \end{array}\right\}. \qquad (7)$$

For perfect coupling $L_1 L_2 = M^2$, and so

$$\mathcal{E}_1 L_2 = M(j\omega M i_1 - j\omega L_2 i_2) = M R_2 i_2, \qquad (8)$$

because of $(7)_2$. It follows from (2) of Question 5.20 that $L_2/M = N_2/N_1$ and then (8) becomes

$$V_1 \frac{N_2}{N_1} = V_2, \qquad (9)$$

[‡]That is, no dissipation in the windings or in the core and with perfect coupling ($k = 1$ in (3) of Question 5.20).

where V_1 and V_2 are voltage magnitudes. In the absence of dissipation, energy conservation requires that $V_1 I_1 = V_2 I_2$ where I_1 and I_2 are current magnitudes. Hence

$$\frac{V_1}{V_2} = \frac{I_2}{I_1} = \frac{N_1}{N_2}, \tag{10}$$

with the last step following from (9). Equation (10) is a well-known result which holds for ideal transformers.

Question 6.23*

Consider the coupled circuits of Question 6.22 and suppose that $\mathcal{E}_1 = V_0 \sin \omega t$ and $\mathcal{E}_2 = 0$.

(a) Show that the current in the secondary circuit satisfies the differential equation

$$(L_1 L_2 - M^2) \frac{d^2 i_2}{dt^2} + (L_1 R_2 + L_2 R_1) \frac{d i_2}{dt} + R_1 R_2 i_2 = \omega M V_0 \cos \omega t. \tag{1}$$

(b) Solve (1) and show that the steady-state amplitude of the current in the secondary is given by

$$I_2 = \frac{\omega M V_0}{\sqrt{\omega^2 (L_1 R_2 + L_2 R_1)^2 + (R_1 R_2 - \omega^2 (L_1 L_2 - M^2))^2}}. \tag{2}$$

Solution

(a) Substituting $\mathcal{E}_1 = V_0 \sin \omega t$ and $\mathcal{E}_2 = 0$ in (3) of Question 6.22 gives

$$\left. \begin{aligned} V_0 \sin \omega t + M \frac{d i_2}{dt} &= R_1 i_1 + L_1 \frac{d i_1}{dt} \\[2mm] M \frac{d i_1}{dt} &= R_2 i_2 + L_2 \frac{d i_2}{dt} \end{aligned} \right\}. \tag{3}$$

Differentiating $(3)_1$ and using $(3)_2$ to eliminate derivatives of i_1 give (1).

(b) It is convenient to write (1) in the form

$$a_1 \frac{d^2 i_2}{dt^2} + a_2 \frac{d i_2}{dt} + a_3 i_2 = a_4 \cos \omega t, \tag{4}$$

where

$$a_1 = (L_1 L_2 - M^2), \quad a_2 = (L_1 R_2 + L_2 R_1), \quad a_3 = R_1 R_2 \quad \text{and} \quad a_4 = \omega M V_0. \tag{5}$$

Since the transient response of the secondary circuit is of no interest here, we seek a particular integral of (4) of the form $i_2(t) = I_0 \sin(\omega t + \phi)$, with ϕ the phase difference between the primary voltage and the secondary current. Substituting this trial solution in (4) yields

$$\beta I_0 \sin(\omega t + \phi) + a_2 \omega I_0 \cos(\omega t + \phi) = a_4 \cos \omega t, \tag{6}$$

where $\beta = (a_3 - a_1\omega^2)$. Expanding (6) using standard trigonometric identities gives

$$I_0 \Big[\beta \cos \phi - a_2 \omega \sin \phi \Big] \sin \omega t + \Big[I_0 \beta \sin \phi + I_0 a_2 \omega \cos \phi - a_4 \Big] \cos \omega t = 0. \tag{7}$$

Now this equation is satisfied at arbitrary t if each of the bracketed coefficients in (7) is zero. Then

$$\beta \cos \phi - a_2 \omega \sin \phi = 0 \quad \text{and} \quad I_0 \beta \sin \phi + I_0 a_2 \omega \cos \phi = a_4. \tag{8}$$

Solving (8) simultaneously yields

$$
\left.
\begin{aligned}
\sin \phi &= \frac{\beta}{\sqrt{a_2^2 \omega^2 + \beta^2}} & \Rightarrow & \quad \sin \phi = \frac{a_3 - a_1\omega^2}{\sqrt{a_2^2 \omega^2 + (a_3 - a_1\omega^2)^2}} \\[2mm]
\cos \phi &= \frac{a_2 \omega}{\sqrt{a_2^2 \omega^2 + \beta^2}} & \Rightarrow & \quad \cos \phi = \frac{a_2 \omega}{\sqrt{a_2^2 \omega^2 + (a_3 - a_1\omega^2)^2}} \\[2mm]
I_0 &= \frac{a_4}{\sqrt{a_2^2 \omega^2 + \beta^2}} & \Rightarrow & \quad I_0 = \frac{a_4}{\sqrt{a_2^2 \omega^2 + (a_3 - a_1\omega^2)^2}}
\end{aligned}
\right\}.
$$

Substituting (5) in this last expression for I_0 gives (2).

Comment

We consider the case of a transformer for which $M^2 \approx L_1 L_2$ (a condition which is usually met in practice). Then

$$I_2 = \frac{\omega M V_0}{\sqrt{R_1^2 R_2^2 + \omega^2 (L_1 R_2 + L_2 R_1)^2}} \simeq \frac{M V_0}{L_1 R_2} \left(1 - \frac{R_1^2}{2\omega^2 L_1^2} \right), \tag{9}$$

for $R_1 \ll \omega L_1$. Now $M/L_1 = N_2/N_1$ and (9) becomes

$$I_2 = \frac{V_0}{R_2} \left(\frac{N_2}{N_1} \right) \left[1 - \frac{R_1^2}{2\omega^2 L_1^2} \right], \tag{10}$$

where $\dfrac{V_0}{R_2} \left(\dfrac{N_2}{N_1} \right)$ is current in the secondary windings of an ideal transformer ($R_1 = 0$).

Question 6.24*

In the coupled circuit shown below, $M = kL$ with $k < 1$.

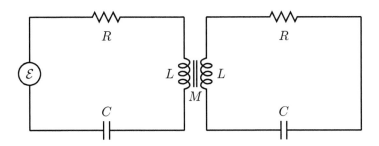

(a) Use the results of Question 6.22 to show that the effective impedance of the secondary circuit is

$$z_{\text{eff}} = \frac{R(R^2 + X^2 + \omega^2 M^2) + jX(R^2 + X^2 - \omega^2 M^2)}{R^2 + X^2}, \qquad (1)$$

where $X = \left(\omega L - \dfrac{1}{\omega C}\right)$.

(b) Hence determine the resonant frequencies of the secondary circuit.

(c) Show that the magnitude of the current in the secondary circuit is

$$I_{02} = \frac{k\omega L \mathcal{E}}{\sqrt{4R^2(\omega L - 1/\omega C)^2 + \left((\omega L - 1/\omega C)^2 - R^2 - k^2\omega^2 L^2\right)^2}}. \qquad (2)$$

Hint: Use *Mathematica* to assist with the algebra.

Solution

(a) The impedance reflected from the primary into the secondary circuit is $\omega^2 M^2 / z$, where $z = R + jX$. So

$$z_{\text{eff}} = R + jX + \frac{\omega^2 M^2}{R + jX}$$

$$= R + jX + \frac{\omega^2 M^2}{R + jX}\frac{R - jX}{R - jX}$$

$$= \frac{(R + jX)(R^2 + X^2) + \omega^2 M^2(R - jX)}{R^2 + X^2}.$$

Hence (1).

(b) Equating the imaginary part of the complex impedance in (1) to zero (this being our condition for resonance) gives

$$\frac{X(R^2 + X^2 - \omega^2 M^2)}{R^2 + X^2} = 0 \quad \Rightarrow \quad \boxed{X = 0} \text{ or } \boxed{X^2 = \omega^2 M^2 - R^2}. \quad (3)$$

From the first condition above we see that $X = \omega_0 L - \dfrac{1}{\omega_0 C} = 0$, which occurs

when $\omega = \omega_0 = \dfrac{1}{\sqrt{LC}}$. Examining the second condition shows that

$$\left(\omega L - \frac{1}{\omega C}\right)^2 = \omega^2 M^2 - R^2, \quad (4)$$

which can be rearranged to yield the quadratic equation

$$(1 - k^2)\omega^4 + (\omega_0^4 \tau^2 - 2\omega_0^2)\omega^2 + \omega_0^4 = 0, \quad (5)$$

where $\tau = RC$. The roots of (5) are

$$\omega_\pm = \sqrt{\frac{(2\omega_0^2 - \omega_0^4 \tau^2) \pm \sqrt{\omega_0^8 \tau^4 - 4\omega_0^6 \tau^2 + 4k^2 \omega_0^4}}{2(1 - k^2)}}, \quad (6)$$

and possible resonant frequencies are thus ω_0, ω_+ and ω_-.

(c) The effective emf in the secondary circuit is $\mathcal{E}_2' = j\omega M \mathcal{E}/z$ (see Comment (ii) of Question 6.22). Then $i_2 = \mathcal{E}_2'/z_{\text{eff}}$ where $z_{\text{eff}} = z + \omega^2 M^2/z$ is given by (1). So

$$i_2 = \frac{j\omega M \mathcal{E}}{z^2 + \omega^2 M^2} = \frac{j\omega M \mathcal{E}}{(R + jX)^2 + \omega^2 M^2}$$

$$= \frac{j\omega M \mathcal{E}}{R^2 - (\omega L - 1/\omega C)^2 + \omega^2 M^2 + 2jR(\omega L - 1/\omega C)}. \quad (7)$$

The magnitude of i_2 in (7) is most easily calculated using *Mathematica* (see cell 1 in the notebook on p. 331) and the result is (2).

Comment

(i) The second condition in (3) can be satisfied only if $\omega^2 M^2 > R^2$. At frequencies for which $\omega M < R$ (*under-coupling*), there is only one real root of the equation, that for which $X = 0$. When $\omega M > R$ (*over-coupling*), there are three roots. At *critical coupling*, when $\omega M = R$, the three roots coincide. As the coupling is increased beyond critical, the single maximum splits into two, both with the same amplitude as that which occurs at critical coupling. At the three frequencies

ω_0, ω_- and ω_+ which satisfy $X^2 = \omega^2 M^2 - R^2$, the primary impedance is purely resistive, but only at the latter two is it equal to $2R$ (see (1)),[‡] thus matching and giving maximum power (and hence maximum current) in the secondary.

(ii) The graph below shows I_{02} plotted as a function of ω, taking

$$L = 1\,\text{mH}, \ C = 1\,\mu\text{F}, \ R = 2\,\Omega, \ \mathcal{E} = 1\,\text{V}, \ k_{\text{critical}} = R\sqrt{\frac{C}{L}}, \ k_{\text{under}} = 0.5 \times k_{\text{critical}}$$

and $k_{\text{over}} = 1.8 \times k_{\text{critical}}$.

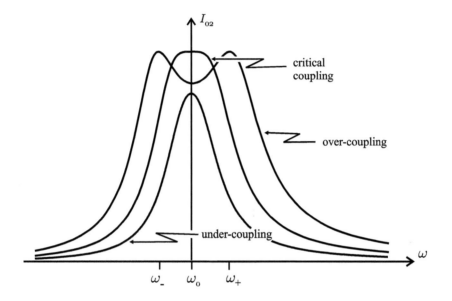

```
In[1]:=  z = R1 + i (ω L1 - 1/(ω C1));  zeff = (z² + ω² L1² k²)/z;

         I0[ω_] := ComplexExpand[Abs[(i ω L1 k v1)/(z zeff)]]

         v1 = 1.0;  L1 = 0.001;  C1 = 0.000001;  R1 = 2;

         kc = R1 √(C1/L1);  ko = R1 √(C1/L1) × 1.8;  ku = R1 √(C1/L1) × 0.5;

         Plot[{I0[ω] /. k → ku, I0[ω] /. k → kc, I0[ω] /. k → ko},
              {ω, 25 000, 40 000}, PlotRange → All]
         Clear[v1, L1, C1, R1, k]
```

[‡]The effective impedance of the primary circuit is also given by (1).

Question 6.25

(a) Two electrodes of arbitrary shape are embedded in an infinite medium having uniform permittivity and conductivity (ϵ and σ respectively).[‡] Show that

$$RC = \epsilon/\sigma, \tag{1}$$

where R is the resistance between the electrodes (both assumed to be perfect conductors) and C is the capacitance.

(b) Calculate the resistance R between the electrodes for the following configurations. In each case, assume that the medium is an electrolyte (having conductivity σ) contained in a very large tank.

☞ **two spheres**

The electrodes are two spheres having radii a and b whose centres are separated by a distance $d \gg a, b$.

Hint: Use (1) of Question 3.13.

☞ **two rods**

The electrodes are two parallel cylindrical rods having radius a and length L separated by a distance d where $a \ll d \ll L$.

Hint: Use (3) of Question 3.27.

Solution

(a) Suppose that each electrode is maintained at a constant potential and carries a charge $\pm q$, as shown. The closed surfaces drawn with dashed lines in the figure below serve as Gaussian surfaces in what follows.

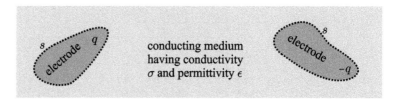

With V equal to the potential difference between the electrodes and assuming quasi-stationary currents, we have

$$RC = \frac{V}{i}\frac{q}{V} = \frac{q}{i}. \tag{2}$$

[‡]These material properties of a medium are discussed further in Chapter 9.

Now $q = \epsilon \oint_s \mathbf{E} \cdot d\mathbf{a}$ (Gauss's law; see Question 9.2) and $i = \oint_s \mathbf{J} \cdot d\mathbf{a} = \sigma \oint_s \mathbf{E} \cdot d\mathbf{a}$

for an ohmic conductor. Substituting these results in (2) gives (1).

(b) ☞ two spheres

Substituting $C = \dfrac{4\pi\epsilon}{\dfrac{1}{a} + \dfrac{1}{b} - \dfrac{2}{d}}$ in (1) gives

$$R = \frac{1}{4\pi\sigma}\left(\frac{1}{a} + \frac{1}{b} - \frac{2}{d}\right). \tag{3}$$

☞ two rods

We begin by making the replacements $r_0 \to a$ and $d \to \frac{1}{2}d$ in the capacitance formula (3) of Question 3.27. Then for $a \ll d$ we obtain $C \simeq \dfrac{2\pi\epsilon L}{\ln(d^2/2a^2)}$, and so

$$R \simeq \frac{\ln(d^2/2a^2)}{2\pi\sigma L}. \tag{4}$$

Comment

Dimensionally RC is a time and, as we have indicated previously, this quantity serves as a characteristic time scale for changes to occur in the circuit.

7

Electromagnetic fields and waves in vacuum

In earlier chapters of this book, we encountered the following experimental laws of electromagnetism:

$$\left. \begin{array}{ll} \nabla \cdot \mathbf{E} = \epsilon_0{}^{-1}\rho(\mathbf{r}) & \text{(Gauss's law in electrostatics)} \\[1.5ex] \nabla \times \mathbf{E} = -\partial \mathbf{B}/\partial t & \text{(Faraday's law of electromagnetic induction)} \\[1.5ex] \nabla \cdot \mathbf{B} = 0 & \text{(Gauss's law in magnetostatics)} \\[1.5ex] \nabla \times \mathbf{B} = \mu_0 \mathbf{J}(\mathbf{r}) & \text{(Ampère's law for stationary currents)} \end{array} \right\}. \quad \text{(I)}$$

We now seek to generalize these equations (where appropriate) for time-dependent charge and current densities $\rho(\mathbf{r}, t)$ and $\mathbf{J}(\mathbf{r}, t)$ respectively. The result is a set of four coupled differential equations known as Maxwell's equations for a vacuum, and they form the theoretical framework of classical electrodynamics. One of the most important features emerging from Maxwell's theory is the prediction of electromagnetic waves, and indeed an entire spectrum of electromagnetic radiation. Topics considered include solutions of the wave equation, some important properties of waves travelling in free space, polarized waves, energy and momentum conservation in the electromagnetic field and various simple applications of waves propagating along transmission lines and in wave guides. Before doing any of this, we begin in Question 7.1 with a fundamental result: the continuity equation for electric charge.

Question 7.1

Use the law of conservation of charge and the definition of electric current to derive the equation

$$\nabla \cdot \mathbf{J} + \frac{\partial \rho}{\partial t} = 0, \quad (1)$$

relating the time-dependent charge density $\rho(\mathbf{r}, t)$ and current density $\mathbf{J}(\mathbf{r}, t)$ at an arbitrary point in space.

Solved Problems in Classical Electromagnetism. J. Pierrus, Oxford University Press (2018).
© J. Pierrus. DOI: 10.1093/oso/9780198821915.001.0001

Solution

Consider a region of space having volume v bounded by the closed surface s. Let $q(t) = \int_v \rho \, dv'$ be the charge contained within v at time t. The rate at which $q(t)$ changes with time is given by

$$\frac{dq}{dt} = \frac{d}{dt} \int_v \rho(\mathbf{r}', t) \, dv', \tag{2}$$

and the current entering v is the flux of $\mathbf{J}(\mathbf{r}', t)$:

$$I_{\text{in}} = -\oint_s \mathbf{J}(\mathbf{r}', t) \cdot d\mathbf{a}' \tag{3}$$

(here the minus sign is necessary because of the convention that $d\mathbf{a}'$ is chosen along an *outward* normal). Charge conservation requires that (2) and (3) are equal and so

$$\frac{d}{dt} \int_v \rho(\mathbf{r}', t) \, dv' = -\oint_s \mathbf{J}(\mathbf{r}', t) \cdot d\mathbf{a}' = -\int_v \nabla \cdot \mathbf{J} \, dv', \tag{4}$$

where, in the last step, we use Gauss's theorem. Now the total derivative on the left-hand side of (4) can be written as $\int_v \frac{\partial \rho}{\partial t} \, dv'$ because at a fixed volume element at $\mathbf{r}'$ we are concerned only with explicit variations of $\rho(\mathbf{r}', t)$ in time. Thus (4) becomes $\int_v \frac{\partial \rho}{\partial t} \, dv' = -\int_v \nabla \cdot \mathbf{J} \, dv'$, or

$$\int_v \left[\nabla \cdot \mathbf{J} + \frac{\partial \rho}{\partial t} \right] dv' = 0. \tag{5}$$

Since the volume element v in (5) is arbitrary, we conclude that $\nabla \cdot \mathbf{J} + \dfrac{\partial \rho}{\partial t} = 0$ which is (1).

Comments

(i) Charge conservation is an empirical law that was first formulated by Benjamin Franklin in the eighteenth century. No convincing experiment has ever been performed which indicates a violation of this law. The conservation of electric charge of an isolated system is associated with the global gauge invariance of the electromagnetic field. This is an example of Noether's theorem, which states that each invariance of a system implies the existence of a corresponding conserved quantity.

(ii) Equation (1) is the continuity equation for electric charge. Like other continuity equations it expresses the local form of a conservation law (see the next comment).

(iii) There are other examples from physics which describe the transport of a conserved quantity (mass, energy, momentum, etc.) and all of them have the form

$$\nabla \cdot \mathbf{f} + \frac{\partial \psi}{\partial t} = 0,$$

where the vector field $\mathbf{f}$ describes the flux of some conserved quantity ψ.

(iv) For stationary and quasi-stationary currents $\partial \rho / \partial t = 0$, and then (1) reduces to

$$\nabla \cdot \mathbf{J} = 0 \tag{6}$$

which is the continuity equation in magnetostatics (see (8) of Question 4.3).

Question 7.2

Explain how the following equations:

$$\nabla \cdot \mathbf{E} = \frac{\rho(\mathbf{r})}{\epsilon_0} \qquad \text{(Gauss's law in electrostatics)}$$

$$\nabla \times \mathbf{E} = -\frac{\partial \mathbf{B}}{\partial t} \qquad \text{(Faraday's law of electromagnetic induction)}$$

$$\nabla \cdot \mathbf{B} = 0 \qquad \text{(Gauss's law in magnetostatics)}$$

$$\nabla \times \mathbf{B} = \mu_0 \mathbf{J}(\mathbf{r}) \qquad \text{(Ampère's law for stationary currents)}$$

, (1)

can be generalized for time-dependent charge and current densities $\rho(\mathbf{r}, t)$ and $\mathbf{J}(\mathbf{r}, t)$ respectively.

Solution

Of these four equations, only $(1)_2$ was obtained for time-dependent fields. The remaining three equations apply—as we discovered in Chapters 2 and 4—to time-independent electric and magnetic fields and charge and current densities. So we must now consider the matter of generalizing $(1)_1$, $(1)_3$ and $(1)_4$ to include time-dependent phenomena. For this purpose, we proceed in two stages. First we *assume* that Gauss's law for the electrostatic and magnetostatic fields (equations $(1)_1$ and $(1)_3$) apply also to time-dependent phenomena. That is, if $\rho = \rho(\mathbf{r}, t)$ and $\mathbf{J} = \mathbf{J}(\mathbf{r}, t)$ then $\mathbf{E} = \mathbf{E}(\mathbf{r}, t)$ and $\mathbf{B} = \mathbf{B}(\mathbf{r}, t)$. There are no obvious reasons against this assumption, and one can proceed to check whether the resulting theory agrees with observed phenomena. It does.[‡]

[‡] For example, suppose the closed surface in Gauss's law now contains moving charges. The number of field lines from these charges which pass through the surface at an instant, being related to the outward flux of $\mathbf{E}$, is the same as if the charges were at rest. See also Question 12.15.

Second, we consider $(1)_4$. Taking the divergence of both sides of this equation and recalling that the curl of any vector has zero divergence, it follows that $(1)_4$ requires $\nabla \cdot \mathbf{J} = 0$. That is a steady state. For non-steady flow, $\nabla \times \mathbf{B} = \mu_0 \mathbf{J}$ would violate charge conservation, a crucial fact recognized by Maxwell. To generalize $(1)_4$ to time-dependent charge and current densities we must construct a vector which has the following properties:

☞ zero divergence,
☞ reduces to $\mathbf{J}$ in a steady state, and
☞ does not violate charge conservation.

Then we replace $\mathbf{J}(\mathbf{r})$ in $(1)_4$ with this new vector. Starting with the continuity equation $\nabla \cdot \mathbf{J} + \partial\rho/\partial t = 0$ and using Gauss's law to express ρ as a divergence (viz. $\rho = \epsilon_0 \nabla \cdot \mathbf{E}$) give

$$\nabla \cdot \left(\mathbf{J} + \epsilon_0 \frac{\partial \mathbf{E}}{\partial t} \right) = 0.$$

So the desired vector is the quantity in parenthesis[#] and the generalization of $(1)_4$ to time-dependent phenomena is

$$\nabla \times \mathbf{B} = \mu_0 \left(\mathbf{J} + \epsilon_0 \frac{\partial \mathbf{E}}{\partial t} \right). \tag{2}$$

Thus the four differential equations of the electromagnetic field at an arbitrary point in vacuum are:

$$\nabla \cdot \mathbf{E} = \frac{\rho(\mathbf{r}, t)}{\epsilon_0} \qquad \text{(Maxwell–Gauss)}$$

$$\nabla \times \mathbf{E} = -\frac{\partial \mathbf{B}}{\partial t} \qquad \text{(Maxwell–Faraday)}$$

$$\nabla \cdot \mathbf{B} = 0 \qquad \text{(Maxwell–Gauss)} \tag{3}$$

$$\nabla \times \mathbf{B} = \mu_0 \left[\mathbf{J}_\mathrm{f}(\mathbf{r}, t) + \epsilon_0 \frac{\partial \mathbf{E}}{\partial t} \right] \qquad \text{(Maxwell–Ampère)}$$

Here $\mathbf{J}_\mathrm{f}(\mathbf{r}, t)$ is the free or conduction current density (sometimes also written as $\mathbf{J}_\mathrm{c}(\mathbf{r}, t)$) and the term $\epsilon_0 \dfrac{\partial \mathbf{E}}{\partial t}$ is the displacement current density $\mathbf{J}_\mathrm{d}(\mathbf{r}, t)$.

Comment

(i) Equations $(3)_2$ and $(3)_3$—which contain no source terms—are known as the homogeneous Maxwell equations. The remaining two—which contain the source terms $\rho(\mathbf{r}, t)$, $\mathbf{J}_\mathrm{c}(\mathbf{r}, t)$ and $\mathbf{J}_\mathrm{d}(\mathbf{r}, t)$—are the inhomogeneous Maxwell equations.

[#]This vector obviously reduces to $\mathbf{J}$ in a steady state where $\rho = \rho(\mathbf{r})$ and $\mathbf{J} = \mathbf{J}(\mathbf{r})$.

(ii) Ref. [1] observes that:

> It required the genius of J. C. Maxwell, spurred on by Faraday's observations, to see the inconsistency in equations (1) and to modify them into a consistent set that implied new physical phenomena, at the time unknown but subsequently verified in all details by experiment. For this brilliant stroke in 1865, the modified set of equations (3) is justly known as the *Maxwell equations*. ... Maxwell called the added term in (2) the *displacement current*. Its presence means that a changing *electric* field causes a magnetic field, even without a current—the converse of Faraday's law. This necessary addition to Ampère's law is of crucial importance to rapidly fluctuating fields. Without it there would be no electromagnetic radiation, and the greatest part of the remainder of this book would have to be omitted. It was Maxwell's prediction that light was an electromagnetic wave phenomenon, and that electromagnetic waves of all frequencies could be produced, that drew the attention of all physicists and stimulated so much theoretical and experimental research into electromagnetism during the last part of the nineteenth century. ... the Maxwell equations form the basis of all classical electromagnetic phenomena. When combined with the Lorentz force equation and Newton's second law of motion, these equations provide a complete description of the classical dynamics of interacting charged particles and electromagnetic fields.

(iii) An important hallmark of a good physical theory is its ability to accurately predict effects or phenomena *before* they are known or discovered experimentally. In this regard, Maxwell's electrodynamics and Einstein's general relativity are two famous examples. Consider the case of the German physicist Heinrich Hertz, who, through a series of ingenious experiments, was the first person to produce and detect radio waves in the laboratory. This discovery provided irrefutable empirical evidence that the theory of electrodynamics was correct. Hertz, expressing himself on the topic of Maxwell's equations, observed that 'one cannot escape the feeling that these mathematical formulas have an independent existence and an intelligence of their own, that they are wiser than we are, wiser even than their discoverers, that we get more out of them than was originally put in to them'.[2]

(iv) For time-dependent fields $\mathbf{E}(\mathbf{r}, t)$ and $\mathbf{B}(\mathbf{r}, t)$, the integral forms of (1) are:

$$\left. \begin{aligned} \oint_s \mathbf{E} \cdot d\mathbf{a} &= \frac{1}{\epsilon_0} \int_v \rho(\mathbf{r}, t) \, dv && \text{(Gauss's law)} \\ \oint_c \mathbf{E} \cdot d\mathbf{l} &= -\frac{d}{dt} \int_s \mathbf{B} \cdot d\mathbf{a} && \text{(Faraday's law)} \\ \oint_s \mathbf{B} \cdot d\mathbf{a} &= 0 && \text{(Gauss's law)} \\ \oint_c \mathbf{B} \cdot d\mathbf{l} &= \mu_0 \int_s \left[\mathbf{J}_f(\mathbf{r}, t) + \epsilon_0 \frac{\partial \mathbf{E}}{\partial t} \right] \cdot d\mathbf{a} && \text{(modified Ampère's law)} \end{aligned} \right\}. \quad (4)$$

[1]   J. D. Jackson, *Classical electrodynamics*, Chap. 6, pp. 238–9. New York: Wiley, 3rd edn, 1998.
[2]   E. T. Bell, *Men of mathematics*, vol. 1, Chap. 1, p. 16. London: Pelican, 1937.

(v) We note, in passing, that Maxwell's equations for a source-free vacuum ($\rho = 0$; $\mathbf{J} = 0$) are invariant under the transformation

$$\left.\begin{array}{c} \mathbf{E} \rightarrow \pm c\mathbf{B} \\ \mathbf{B} \rightarrow \mp \mathbf{E}/c \end{array}\right\}. \tag{5}$$

This symmetry (known as duality) was first discovered by Heaviside over one hundred years ago. It has created much speculation about its meaning and today plays an important role in modern gauge theories.

Question 7.3

Consider two concentric spherical conducting shells having radii a and b (with $a < b$). This arrangement is a capacitor having capacitance[‡] $C = 4\pi\epsilon_0 ab/(b-a)$. When it is connected in series with a resistor and an oscillator, a quasi-stationary current[#] $i = i_0 e^{-i\omega t}$ is present in the circuit. Find the magnetic field $\mathbf{B}(\mathbf{r}, t)$ between the plates of the capacitor for $a < r < b$.

Solution

We use Ampère's law in the form

$$\oint_c \mathbf{B} \cdot d\mathbf{l} = \epsilon_0 \mu_0 \int_s \frac{\partial \mathbf{E}}{\partial t} \cdot d\mathbf{a} \tag{1}$$

(see $(4)_4$ of Question 7.2 with $\mathbf{J}_f = 0$). Because of the spherical symmetry, it is convenient to choose a circular contour c centred on the origin and lying in the xy-plane of Cartesian coordinates. Let s be the area of a hemispherical surface of radius r spanning c. Clearly, the circulation of $\mathbf{B}$ around c is $B \times 2\pi r$, and so (1) becomes

$$B_\phi = \frac{\epsilon_0 \mu_0}{2\pi r} \int_s \frac{\partial \mathbf{E}}{\partial t} \cdot d\mathbf{a}. \tag{2}$$

Within the quasi-static approximation, we use the electrostatic result $\mathbf{E} = \dfrac{1}{4\pi\epsilon_0} \dfrac{q}{r^2} \hat{\mathbf{r}}$ for the time-dependent electric field between the spheres. So

$$\frac{\partial \mathbf{E}}{\partial t} = \frac{1}{4\pi\epsilon_0} \frac{\partial q/\partial t}{r^2} \hat{\mathbf{r}} = \frac{i_0}{4\pi\epsilon_0} \frac{e^{-i\omega t}}{r^2} \hat{\mathbf{r}}. \tag{3}$$

[‡]See (1) of Question 3.12.

[#]See the second paragraph on p. 249.

Substituting (3) in (2) and integrating over the hemispherical surface s yields

$$B_\phi = \frac{\epsilon_0 \mu_0}{2\pi r} \frac{i_0}{4\pi\epsilon_0} \frac{e^{-i\omega t}}{r^2} \times 2\pi r^2 = \frac{\mu_0}{4\pi} \frac{i_0}{r} e^{-i\omega t},$$

or

$$\mathbf{B}(\mathbf{r}, t) = \frac{\mu_0}{4\pi} \frac{i_0}{r} e^{-i\omega t} \hat{\boldsymbol{\phi}}. \tag{4}$$

Comment

The conduction current in the wires of the circuit disappears at one spherical shell and reappears at the other. Between the plates of the capacitor (where the conduction current is zero) there is a displacement current $\epsilon_0 \oint \frac{\partial \mathbf{E}}{\partial t} \cdot d\mathbf{a}$, and Kirchhoff's junction rule of 'current in = current out' remains satisfied at all times.

Question 7.4

Use Maxwell's equations to show that the $\mathbf{E}$ and $\mathbf{B}$ fields in a source-free vacuum (i.e. $\rho = 0$; $\mathbf{J}_f = 0$) satisfy

$$\nabla^2 \mathbf{E} - \epsilon_0 \mu_0 \frac{\partial^2 \mathbf{E}}{\partial t^2} = 0 \quad \text{and} \quad \nabla^2 \mathbf{B} - \epsilon_0 \mu_0 \frac{\partial^2 \mathbf{B}}{\partial t^2} = 0. \tag{1}$$

Solution

Here we use the Maxwell equations:

$$\nabla \cdot \mathbf{E} = 0, \quad \nabla \times \mathbf{E} = -\frac{\partial \mathbf{B}}{\partial t}, \quad \nabla \cdot \mathbf{B} = 0, \quad \nabla \times \mathbf{B} = \epsilon_0 \mu_0 \frac{\partial \mathbf{E}}{\partial t}. \tag{2}$$

The two curl equations can be decoupled by taking the curl of $(2)_2$ and interchanging the order of space and time differentiation. Then

$$\nabla \times (\nabla \times \mathbf{E}) + \frac{\partial}{\partial t}(\nabla \times \mathbf{B}) = 0. \tag{3}$$

Substituting $(2)_4$ in (3) and using the vector identity (11) of Question 1.8 yield

$$-\nabla^2 \mathbf{E} + \nabla(\nabla \cdot \mathbf{E}) + \epsilon_0 \mu_0 \frac{\partial^2 \mathbf{E}}{\partial t^2} = 0,$$

which is $(1)_1$ since $\nabla \cdot \mathbf{E} = 0$. The proof for $(1)_2$ begins with $(2)_4$ and the ensuing algebra is similar.

Comments

(i) We see from (1) that each Cartesian component of the electromagnetic field satisfies the homogeneous wave equation

$$\nabla^2\psi - \frac{1}{v^2}\frac{\partial^2\psi}{\partial t^2} = 0, \tag{4}$$

where ψ represents one of E_x, E_y, E_z, B_x, B_y or B_z, and $v = (\epsilon_0\mu_0)^{-1/2}$, which has the dimensions of speed, is the phase velocity of the wave. The simplest solutions of the wave equation, plane waves (having angular frequency ω and wave vector **k**) travelling in free space, have the form

$$\psi = \psi_0 e^{\pm i(\mathbf{k}\cdot\mathbf{r}\mp\omega t)}, \tag{5}$$

where ψ_0 is the amplitude and $v = \omega/k$.

(ii) Maxwell was the first person to recognize that light was an electromagnetic wave and that its speed c was identical to $(\epsilon_0\mu_0)^{-1/2}$, as the following quotation from Ref. [3] shows:

> The velocity of light deduced from experiment agrees sufficiently well with the value of v deduced from the only set of experiments we as yet possess. The value of v was determined by measuring the electromotive force with which a condenser of known capacity was charged, and then discharging the condenser through a galvanometer, so as to measure the quantity of electricity in it in electromagnetic measure. The only use made of light in the experiment was to see the instruments. The value of c found by M. Foucault was obtained by determining the angle through which a revolving mirror turned, while the light reflected from it went and returned along a measured course. No use whatever was made of electricity or magnetism. The agreement of the results seems to show that light and magnetism are affections of the same substance, and that light is an electromagnetic disturbance propagated through the field according to electromagnetic laws.

This conclusion, together with the prediction of an entire spectrum of electromagnetic radiation (which incorporates visible light) is one of the most important discoveries in physics. In this bold step, Maxwell succeeded in unifying the hitherto unrelated fields of electricity, magnetism and optics. Thus

$$v = c = \frac{1}{\sqrt{\epsilon_0\mu_0}}, \tag{6}$$

the vacuum speed of all electromagnetic waves. Hence we have the dispersion relation for electromagnetic waves in a vacuum:

$$\frac{\omega}{k} = c. \tag{7}$$

---

[3]   J. Clerk Maxwell, 'A dynamical theory of the electromagnetic field', *Philosophical Transactions of the Royal Society*, vol. VIII, pp. 459–512, 1864.

Question 7.5

Consider the plane wave solutions (5) of Question 7.4. Prove the following properties:

(a) The field is a transverse wave.

(b) The fields $\mathbf{E}$ and $\mathbf{B}$ are mutually perpendicular and satisfy

$$\mathbf{B} = \frac{\hat{\mathbf{k}} \times \mathbf{E}}{c}, \tag{1}$$

where $\hat{\mathbf{k}}$ is a unit vector in the direction of propagation.

Solution

(a) Consider the following plane wave solutions of (1) of Question 7.4:

$$\mathbf{E}(\mathbf{r}, t) = \hat{\epsilon}_1 E_0 e^{i(\mathbf{k} \cdot \mathbf{r} - \omega t)} \quad \text{and} \quad \mathbf{B}(\mathbf{r}, t) = \hat{\epsilon}_2 B_0 e^{i(\mathbf{k} \cdot \mathbf{r} - \omega t)}. \tag{2}$$

Here $\hat{\epsilon}_1$ and $\hat{\epsilon}_2$ are two constant real unit vectors and E_0, B_0 are complex amplitudes which vary neither in space nor in time. Substituting (2) in the Maxwell equations $(3)_1$ and $(3)_3$ of Question 7.2 gives

$$E_0 \nabla \cdot \left[\hat{\epsilon}_1 e^{i(\mathbf{k} \cdot \mathbf{r} - \omega t)} \right] = 0 \quad \text{and} \quad B_0 \nabla \cdot \left[\hat{\epsilon}_2 e^{i(\mathbf{k} \cdot \mathbf{r} - \omega t)} \right] = 0. \tag{3}$$

Applying (5) of Question 1.8 and $\nabla e^{i\mathbf{k} \cdot \mathbf{r}} = i\mathbf{k}\, e^{i\mathbf{k} \cdot \mathbf{r}}$ (see (9) of Question 1.1) to (3) yields

$$\hat{\epsilon}_i \cdot \mathbf{k} = 0 \quad (i = 1, 2). \tag{4}$$

We see from (4) that both $\mathbf{E}$ and $\mathbf{B}$ are perpendicular to $\mathbf{k}$. This is therefore a transverse wave.

(b) Substituting (2) into the Maxwell equation $(3)_2$ of Question 7.2 and using (6) of Question 1.8 give $E_0 \nabla e^{i(\mathbf{k} \cdot \mathbf{r} - \omega t)} \times \hat{\epsilon}_1 - \hat{\epsilon}_2 i \omega B_0 e^{i(\mathbf{k} \cdot \mathbf{r} - \omega t)} = 0$, or

$$i \left[(\hat{\mathbf{k}} \times \hat{\epsilon}_1) k E_0 - \hat{\epsilon}_2 \omega B_0 \right] e^{i(\mathbf{k} \cdot \mathbf{r} - \omega t)} = 0, \tag{5}$$

where we again make use of the result $\nabla e^{i\mathbf{k} \cdot \mathbf{r}} = i\mathbf{k}\, e^{i\mathbf{k} \cdot \mathbf{r}}$. Equation (5) can be satisfied for arbitrary $\mathbf{r}$ and t only if the term in square brackets is zero. This requires that $\hat{\epsilon}_2 = \hat{\mathbf{k}} \times \hat{\epsilon}_1$ and $k E_0 = \omega B_0$. Thus

$$\mathbf{B} = \frac{\mathbf{k} \times \mathbf{E}}{\omega}. \tag{6}$$

Now for a plane wave in vacuum $\omega = ck$ (see (7) of Question 7.4). Hence (1).

Comments

(i) For an unbounded plane wave travelling in vacuum, we have shown the following:

  ☞ **E**, **B** and **k** are mutually orthogonal vectors. Note the cyclic permutations of (1), which are often needed, are:

$$\mathbf{E} = c\mathbf{B} \times \hat{\mathbf{k}} \qquad \text{and} \qquad \hat{\mathbf{k}} = \frac{c\mathbf{E} \times \mathbf{B}}{E^2}. \tag{7}$$

  ☞ $kE_0 = \omega B_0 \Rightarrow$ the fields **E** and **B** are in phase and in constant ratio.

  ☞ The wave described is a transverse wave propagating in the direction **k**. It represents a flux of energy given by the real part of the complex Poynting vector,[‡] which is defined as

$$\mathbf{S} = \frac{\mathbf{E} \times \mathbf{B}}{\mu_0}. \tag{8}$$

For 'rapidly' varying fields we are usually concerned with the time-averaged energy flux $\langle \mathbf{S} \rangle$, which for harmonic waves (see Question 1.27) is

$$\langle \mathbf{S} \rangle = \frac{1}{2\mu_0} \text{Re}\{\mathbf{E} \times \mathbf{B}^*\} = \frac{1}{2\mu_0} \text{Re}\{\mathbf{E}^* \times \mathbf{B}\} \tag{9}$$

$$= \frac{1}{2} \sqrt{\frac{\epsilon_0}{\mu_0}} |E_0|^2 \,\hat{\mathbf{k}}. \tag{10}$$

(ii) It is easy to show that the ratio $\sqrt{\mu_0/\epsilon_0}$ in (10) has the units of ohms. We therefore define the impedance Z_0 presented to a plane electromagnetic wave in vacuum as

$$Z_0 = \sqrt{\frac{\mu_0}{\epsilon_0}}, \tag{11}$$

which has the approximate value

$$Z_0 = \sqrt{\frac{\mu_0}{\epsilon_0}} = \sqrt{\frac{4\pi\mu_0}{4\pi\epsilon_0}} \approx \sqrt{9 \times 10^9 \times 16\pi^2 \times 10^{-7}} = 120\pi \simeq 377\,\Omega.$$

(iii) The force exerted by an electromagnetic wave on a charge q moving with velocity **v** comprises an electric component $F_e = qE$ and a magnetic component $F_m \sim qvB$. Thus $F_m/F_e \sim vB/E \sim v/c \ll 1$ for non-relativistic charges. Since the speed of electrons in atoms[#] is of order $v = c/137$, the interaction between matter and electromagnetic radiation can usually be explained in terms of F_e alone.

(iv) The results (4), (6) and (7) are valid for unbounded plane waves only. They do not apply to all solutions of the wave equation, for example, bounded waves (see Questions 7.12 and 7.16).

[‡] **S** has the dimensions of energy per unit area per unit time.

[#] In Bohr's theory of the atom $v/c = \alpha$, the fine structure constant.

Question 7.6

Use Maxwell's equations to prove that energy conservation at a macroscopic point in the vacuum electromagnetic field can be expressed in terms of the continuity equation

$$\nabla \cdot \mathbf{S} + \frac{\partial u}{\partial t} = -\mathbf{E} \cdot \mathbf{J}_{\mathrm{f}}, \tag{1}$$

where $\mathbf{S}$ is Poynting's vector given by (8) of Question 7.5,

$$u = \tfrac{1}{2}\epsilon_0 \mathbf{E}^2 + \tfrac{1}{2}\mu_0^{-1}\mathbf{B}^2 \tag{2}$$

is the energy density in the electromagnetic field (see Comment (i) on p. 345) and $\mathbf{J}_{\mathrm{f}}$ is the free (conduction) current density.

Hint: Begin with the Lorentz force (see (8) of Question 4.1) acting on a system of N identical free charges at a macroscopic point in vacuum. Show that the rate at which the field expends energy on the charges is $\displaystyle\int_v (\mathbf{E} \cdot \mathbf{J}_{\mathrm{f}}) \, dv$, and then use Maxwell's equations.

Solution

Suppose there are N identical free charges at a macroscopic point in vacuum having velocities $\mathbf{v}_1, \mathbf{v}_2, \mathbf{v}_3, \ldots, \mathbf{v}_N$. The rate at which an electric field[‡] $\mathbf{E}(\mathbf{r}, t)$ does work on these charges is $q\mathbf{E} \cdot (\mathbf{v}_1 + \mathbf{v}_2 + \cdots + \mathbf{v}_N) = qN\mathbf{E} \cdot \bar{\mathbf{v}}$, where $\bar{\mathbf{v}}$ is their average velocity. Then the rate at which the field expends energy per unit volume on these charges at the point is $qn\mathbf{E} \cdot \bar{\mathbf{v}}$, where n is their number density. But $nq\bar{\mathbf{v}} = \rho(\mathbf{r}, t)\bar{\mathbf{v}}$ is the current density $\mathbf{J}(\mathbf{r}, t)$; if at this point there are different types of free charge $q_1, q_2, \ldots, q_k$, then the total rate at which the field expends energy per unit volume on all the free charges at the point is $\mathbf{E} \cdot \mathbf{J}_1 + \mathbf{E} \cdot \mathbf{J}_2 + \cdots + \mathbf{E} \cdot \mathbf{J}_k = \mathbf{E} \cdot \mathbf{J}_{\mathrm{f}}$. It is this energy per unit volume per unit time that is converted into kinetic energy, and so is lost or dissipated from the field at the point. Thus the total rate of doing work by the field in a finite region v is

$$\int_v (\mathbf{E} \cdot \mathbf{J}_{\mathrm{f}}) \, dv = \int_v \mathbf{E} \cdot \left[\left(\nabla \times \frac{\mathbf{B}}{\mu_0} \right) - \epsilon_0 \frac{\partial \mathbf{E}}{\partial t} \right] dv, \tag{3}$$

where we use Maxwell's equation $(3)_4$ of Question 7.2 to express $\mathbf{J}_{\mathrm{f}}$ in terms of $\mathbf{E}$ and $\mathbf{B}$. Now $\nabla \cdot \mathbf{S} = \nabla \cdot \left(\mathbf{E} \times \dfrac{\mathbf{B}}{\mu_0} \right)$ and by the vector identity (7) of Question 1.8 we have

$$\int_v (\nabla \cdot \mathbf{S}) \, dv = \int_v \left[\frac{\mathbf{B}}{\mu_0} \cdot (\nabla \times \mathbf{E}) - \mathbf{E} \cdot \left(\nabla \times \frac{\mathbf{B}}{\mu_0} \right) \right] dv. \tag{4}$$

Adding (3) and (4) gives

[‡]The magnetic field does no work on the charges since $\mathbf{v} \cdot (\mathbf{v} \times \mathbf{B}) = 0$.

$$\int_v (\mathbf{E} \cdot \mathbf{J}_f)\, dv = \int_v \left[\frac{\mathbf{B}}{\mu_0} \cdot (\nabla \times \mathbf{E}) - \nabla \cdot \mathbf{S} - \epsilon_0 \mathbf{E} \cdot \frac{\partial \mathbf{E}}{\partial t} \right] dv$$

$$= - \int_v \left[\nabla \cdot \mathbf{S} + \frac{\partial}{\partial t} \left(\frac{\epsilon_0 \mathbf{E}^2}{2} + \frac{\mathbf{B}^2}{2\mu_0} \right) \right] dv, \tag{5}$$

where, in the last step, we use the Maxwell–Faraday law. But this is (1), since the volume v in (5) is arbitrary.

Comments

(i) Equation (5) is Poynting's theorem for a vacuum. It can be expressed in the alternative form

$$- \int_v (\nabla \cdot \mathbf{S})\, dv = \int_v (\mathbf{E} \cdot \mathbf{J}_f)\, dv + \frac{d}{dt} \int_v \left(\frac{\epsilon_0 \mathbf{E}^2}{2} + \frac{\mathbf{B}^2}{2\mu_0} \right) dv, \tag{6}$$

where we interpret the three main terms as follows:

☞ From the definition that $\nabla \cdot \mathbf{F}$ is the net outward flux of $\mathbf{F}$ per unit volume, and since $\nabla \cdot \mathbf{S}$ has the dimensions energy per unit *volume* per unit time, it is obvious that $- \int_v (\nabla \cdot \mathbf{S})\, dv$ represents the electromagnetic energy per unit time passing *into* the volume v.

☞ $\int_v (\mathbf{E} \cdot \mathbf{J}_f)\, dv$ is the rate at which the field does work on the charges as shown.

☞ We have seen in Chapters 2 and 4 that $\frac{1}{2}\epsilon_0 \mathbf{E}^2$ and $\frac{1}{2}\mu_0^{-1} \mathbf{B}^2$ are the energy densities in the vacuum electrostatic and magnetostatic fields respectively. We must now assume that they are more general than this. Their sum represents the energy density of any time-dependent field and so $\frac{d}{dt} \int_v \left(\frac{1}{2}\epsilon_0 \mathbf{E}^2 + \frac{\mathbf{B}^2}{2\mu_0} \right) dv$ is the rate of change of electromagnetic field energy within v.

Thus the interpretation of Poynting's theorem in the form (6) is as follows:

$$\left. \begin{array}{c} \text{The rate at which electromagnetic field energy enters } v \\ = \\ \text{the rate at which field energy is dissipated} \\ + \\ \text{the rate at which energy is stored in the field} \end{array} \right\}. \tag{7}$$

(ii) The definition of $\mathbf{S}$ (see (8) of Question 7.5) is arbitrary to the extent that the curl of any vector field can be added to the right-hand side without affecting the continuity equation (1), because of (9) of Question 1.8. Relativistic considerations, however, show that the definition $\mathbf{S} = \mu_0^{-1} (\mathbf{E} \times \mathbf{B})$ is unique.[4]

---

[4] J. D. Jackson, *Classical electrodynamics*, Chap. 6, p. 259. New York: Wiley, 3rd edn, 1998.

Question 7.7*

Let ρ and $\mathbf{J}$ be time-dependent charge and current densities describing a distribution of charged particles inside a region of space having volume v.

(a) Use the Lorentz force (see (8) of Question 4.1) and Maxwell's equations to show that the total electromagnetic force exerted on all particles inside v is

$$\mathbf{F} = \int_v \mathbf{f} \, dv = \int_v \left[\epsilon_0 \mathbf{E}(\nabla \cdot \mathbf{E}) + \mu_0^{-1}(\nabla \times \mathbf{B}) \times \mathbf{B} - \epsilon_0 \frac{\partial \mathbf{E}}{\partial t} \times \mathbf{B} \right] dv, \qquad (1)$$

where $\mathbf{f}$ is the force per unit volume.

(b) Hence show that

$$\int_v \left[\mathbf{f} + \epsilon_0 \frac{\partial}{\partial t}(\mathbf{E} \times \mathbf{B}) \right]_i dv = \int_v \nabla_j \left[\epsilon_0 (E_i E_j - \tfrac{1}{2} E^2 \delta_{ij}) + \mu_0^{-1}(B_i B_j - \tfrac{1}{2} B^2 \delta_{ij}) \right] dv. \qquad (2)$$

Solution

(a) The force $d\mathbf{F}$ acting on an element of charge $dq = \rho \, dv$ is

$$d\mathbf{F} = dq \left[\mathbf{E} + \mathbf{v} \times \mathbf{B} \right] = \rho \left[\mathbf{E} + \mathbf{v} \times \mathbf{B} \right] dv = \left[\epsilon_0 \mathbf{E}(\nabla \cdot \mathbf{E}) + (\mathbf{J} \times \mathbf{B}) \right] dv,$$

where, in the last step, we use $\mathbf{J} = \rho \mathbf{v}$ and the Maxwell–Gauss equation. So

$$\mathbf{F} = \int_v \mathbf{f} \, dv = \int_v \left[\epsilon_0 \mathbf{E}(\nabla \cdot \mathbf{E}) + (\mathbf{J} \times \mathbf{B}) \right] dv. \qquad (3)$$

Now, from the Maxwell–Ampère equation, $\mathbf{J} = \mu_0^{-1}(\nabla \times \mathbf{B}) - \epsilon_0 \dfrac{\partial \mathbf{E}}{\partial t}$. Substituting this result in (3) gives (1).

(b) Using the Maxwell–Faraday equation, we can express the last term on the right-hand side of (1) as

$$\epsilon_0 \frac{\partial \mathbf{E}}{\partial t} \times \mathbf{B} = \epsilon_0 \frac{\partial}{\partial t}(\mathbf{E} \times \mathbf{B}) - \epsilon_0 \mathbf{E} \times \frac{\partial \mathbf{B}}{\partial t} = \epsilon_0 \frac{\partial}{\partial t}(\mathbf{E} \times \mathbf{B}) + \epsilon_0 \mathbf{E} \times (\nabla \times \mathbf{E}).$$

Then

$$\int_v \left[\mathbf{f} + \epsilon_0 \frac{\partial}{\partial t}(\mathbf{E} \times \mathbf{B}) \right] dv = \int_v \left[\epsilon_0 \mathbf{E}(\nabla \cdot \mathbf{E}) - \mu_0^{-1} \mathbf{B} \times (\nabla \times \mathbf{B}) - \epsilon_0 \mathbf{E} \times (\nabla \times \mathbf{E}) \right] dv.$$

Now, because $\nabla \cdot \mathbf{B}$ is always zero, we can improve the symmetry of this equation by adding the term $\mu_0 \mathbf{B}(\nabla \cdot \mathbf{B})$ to the right-hand side. Then

$$\int_v \left[\mathbf{f} + \epsilon_0 \frac{\partial}{\partial t} (\mathbf{E} \times \mathbf{B}) \right] dv = \epsilon_0 \int_v \left[\mathbf{E}(\boldsymbol{\nabla} \cdot \mathbf{E}) - \mathbf{E} \times (\boldsymbol{\nabla} \times \mathbf{E}) \right] dv +$$

$$\mu_0^{-1} \int_v \left[\mathbf{B}(\boldsymbol{\nabla} \cdot \mathbf{B}) - \mathbf{B} \times (\boldsymbol{\nabla} \times \mathbf{B}) \right] dv. \qquad (4)$$

Taking the ith component of (4) and using tensors gives

$$\int_v \left[\mathbf{f} + \epsilon_0 \frac{\partial}{\partial t} (\mathbf{E} \times \mathbf{B}) \right]_i dv = \epsilon_0 \int_v \left[E_i \nabla_j E_j - \varepsilon_{ijk} E_j (\boldsymbol{\nabla} \times \mathbf{E})_k \right] dv +$$

$$\mu_0^{-1} \int_v \left[B_i \nabla_j B_j - \varepsilon_{ijk} B_j (\boldsymbol{\nabla} \times \mathbf{B})_k \right] dv. \qquad (5)$$

The integrand involving $\mathbf{E}$ on the right-hand side of (5) can be written as:

$$
\begin{aligned}
E_i \nabla_j E_j - \varepsilon_{ijk} E_j (\boldsymbol{\nabla} \times \mathbf{E})_k &= E_i \nabla_j E_j - \varepsilon_{ijk} \varepsilon_{klm} E_j \nabla_l E_m \\
&= E_i \nabla_j E_j - (\delta_{il} \delta_{jm} - \delta_{im} \delta_{jl}) E_j \nabla_l E_m \\
&= E_i \nabla_j E_j - E_j \nabla_i E_j + E_j \nabla_j E_i \\
&= E_i \nabla_j E_j - E_k \delta_{ij} \nabla_j E_k + E_j \nabla_j E_i \\
&= \nabla_j \left[E_i E_j - \tfrac{1}{2} E^2 \delta_{ij} \right], \qquad (6)
\end{aligned}
$$

and similarly, $\quad B_i \nabla_j B_j - \varepsilon_{ijk} B_j (\boldsymbol{\nabla} \times \mathbf{B})_k = \nabla_j \left[B_i B_j - \tfrac{1}{2} B^2 \delta_{ij} \right]. \qquad (7)$

Equations (5)–(7) yield (2).

Comments

(i) The term in square brackets on the right-hand side of (2) represents a symmetric, second-rank tensor T_{ij} known as the *Maxwell stress tensor*:

$$T_{ij} = \epsilon_0 \left(E_i E_j - \tfrac{1}{2} E^2 \delta_{ij} \right) + \mu_0^{-1} \left(B_i B_j - \tfrac{1}{2} B^2 \delta_{ij} \right). \qquad (8)$$

In terms of T_{ij} we have

$$\int_v \left[\mathbf{f} + \epsilon_0 \frac{\partial}{\partial t} (\mathbf{E} \times \mathbf{B}) \right]_i dv = \int_v \nabla_j T_{ij} \, dv. \qquad (9)$$

(ii) In Question 7.8, we use (9) to introduce the momentum and angular momentum of the electromagnetic field.

Question 7.8**

Consider again the charge distribution of Question 7.7. Show that

$$\frac{d}{dt}\left[\mathbf{p}_m + \epsilon_0 \int_{\substack{\text{all}\\\text{space}}} (\mathbf{E} \times \mathbf{B})\, dv\right] = 0, \tag{1}$$

where $\mathbf{p}_m$ is the momentum of all the charges (the *matter*) within v.

Hint: Introduce the vectors $\mathbf{X} = \hat{\mathbf{x}}T_{xx} + \hat{\mathbf{y}}T_{xy} + \hat{\mathbf{z}}T_{xz}$, $\mathbf{Y} = \hat{\mathbf{x}}T_{yx} + \hat{\mathbf{y}}T_{yy} + \hat{\mathbf{z}}T_{yz}$ and $\mathbf{Z} = \hat{\mathbf{x}}T_{zx} + \hat{\mathbf{y}}T_{zy} + \hat{\mathbf{z}}T_{zz}$. Then consider each component of (9) of Question 7.7 in turn.

Solution

Because of the hint, let $i = x$. Then $\nabla_j T_{xj} = \left(\dfrac{\partial T_{xx}}{\partial x} + \dfrac{\partial T_{xy}}{\partial y} + \dfrac{\partial T_{xz}}{\partial z}\right) = \nabla \cdot \mathbf{X}$, and so

$$\int_v \left[\mathbf{f} + \epsilon_0 \frac{\partial}{\partial t}(\mathbf{E} \times \mathbf{B})\right]_x dv = \int_v \nabla \cdot \mathbf{X}\, dv = \oint_s \mathbf{X} \cdot d\mathbf{a}, \tag{2}$$

using Gauss's theorem. The region of integration v in (2) must be large enough to include all the charges. Clearly, it does not matter if v is expanded to include regions of space where ρ and $\mathbf{J}$ are zero, as this will have no effect on the integral. Extending the domain of integration over all space then gives

$$\int_{\substack{\text{all}\\\text{space}}} \left[\mathbf{f} + \epsilon_0 \frac{\partial}{\partial t}(\mathbf{E} \times \mathbf{B})\right]_x dv = \oint_s \mathbf{X} \cdot d\mathbf{a}. \tag{3}$$

Now the components of the stress tensor involve products of field components that fall off at least as fast as r^{-2} for $\mathbf{E}$ and r^{-3} for $\mathbf{B}$ (these being the electric charge and magnetic dipole dependences). So X falls off faster than r^{-3}, and since da scales as r^2 the surface integral in (3) tends to zero as $r \to \infty$. Thus $\displaystyle\int_{\substack{\text{all}\\\text{space}}}\left[\mathbf{f}+\epsilon_0\frac{\partial}{\partial t}(\mathbf{E}\times\mathbf{B})\right]_x dv = 0$. A similar analysis can be made for the y and z components, and then

$$\int_{\substack{\text{all}\\\text{space}}} \left[\mathbf{f} + \epsilon_0 \frac{\partial}{\partial t}(\mathbf{E} \times \mathbf{B})\right] dv = 0. \tag{4}$$

By definition $\mathbf{F} = \displaystyle\int_{\substack{\text{all}\\\text{space}}} \mathbf{f}\, dv$ and because force is the rate of change of momentum, we have

$$\frac{d\mathbf{p}_m}{dt} + \epsilon_0 \int_{\substack{\text{all}\\\text{space}}} \frac{\partial}{\partial t}(\mathbf{E} \times \mathbf{B})\, dv = 0. \tag{5}$$

The partial time derivative in (5) is evaluated at a fixed volume element dv. When taken outside the integral, it becomes a total derivative d/dt. Hence (1).

Comments

(i) Equation (1) implies that the dynamical quantity $\mathbf{p}_m + \epsilon_0 \int_{\substack{\text{all} \\ \text{space}}} (\mathbf{E} \times \mathbf{B}) \, dv$ is conserved. We can identify the volume integral as the momentum of the *electromagnetic field*:

$$\mathbf{p}_{\text{field}} = \epsilon_0 \int_{\substack{\text{all} \\ \text{space}}} (\mathbf{E} \times \mathbf{B}) \, dv. \tag{6}$$

The result $d(\mathbf{p}_m + \mathbf{p}_{\text{field}})/dt = 0$ is a statement of momentum conservation: the total momentum of the matter plus the field is constant.

(ii) The integrand in (6) may be regarded as a momentum density $\mathbf{g}$, where

$$\mathbf{g} = \epsilon_0 (\mathbf{E} \times \mathbf{B}) = \epsilon_0 \mu_0 \mathbf{S} = \mathbf{S}/c^2, \tag{7}$$

with a corresponding angular momentum density

$$\mathcal{L} = \mathbf{r} \times \mathbf{g} = (\mathbf{r} \times \mathbf{S})/c^2. \tag{8}$$

Question 7.9

A plane monochromatic electromagnetic wave is travelling in vacuum along a direction taken to be the z-axis of Cartesian coordinates. The electric vector of the wave is of the form $\mathbf{E} = \left(E_{0x} \cos(kz - \omega t), \; E_{0y} \cos(kz - \omega t - \varphi), \; 0 \right)$ where E_{0x} and E_{0y} are real amplitudes and φ is a constant phase factor. At an arbitrarily chosen origin $z = 0$ in the xy-plane, we have:

$$\mathbf{E} = \left(E_{0x} \cos \omega t, \; E_{0y} \cos(\omega t + \varphi), \; 0 \right). \tag{1}$$

Use (1) to draw graphs[‡] of E_y vs E_x for the following values of E_{0x}, E_{0y} and φ. In each case, describe how the resultant electric field varies with time in the xy-plane at $z = 0$:

(a) ☞ $E_{0x} = 2E_{0y} = E_0$ and $\varphi = 0$;   ☞ $E_{0x} = 2E_{0y} = E_0$ and $\varphi = \pi$.

(b) ☞ $E_{0x} = E_{0y} = E_0$ and $\varphi = \frac{1}{2}\pi$;   ☞ $E_{0x} = E_{0y} = E_0$ and $\varphi = -\frac{1}{2}\pi$.

(c) ☞ $E_{0x} = 2E_{0y} = E_0$ and $\varphi = \frac{1}{4}\pi$;   ☞ $E_{0x} = 2E_{0y} = E_0$ and $\varphi = -\frac{1}{4}\pi$.

[‡]These graphs are similar to the Lissajous figures seen on oscilloscope screens in undergraduate laboratories. See Comment (i) on p. 351.

Solution

It is convenient to plot the dimensionless quantities: $\dfrac{E_{0y}}{E_0}$ vs $\dfrac{E_{0x}}{E_0}$.

(a) For $\varphi = 0$ (or π), the resultant electric field lies on a straight line inclined at an angle $\tan^{-1}\left(\dfrac{E_{0y}}{E_{0x}}\right) = \tan^{-1}(\frac{1}{2})$ with respect to $\hat{\mathbf{x}}$ (or $-\hat{\mathbf{x}}$). See the figures below.

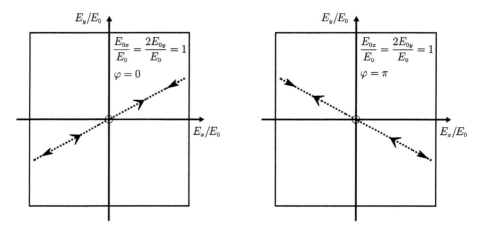

(b) For $\varphi = \frac{1}{2}\pi$, the tip of the electric-field vector moves in a circle, rotating clockwise, as 'seen' by an observer located on the z-axis and facing the approaching electromagnetic wave. Such waves are said to have a negative helicity. For $\varphi = -\frac{1}{2}\pi$, the rotation is anticlockwise (positive helicity). See the figures below.

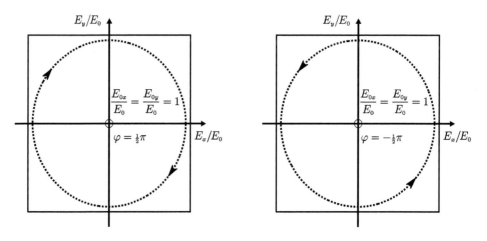

(c) For $\varphi = \frac{1}{4}\pi$, the tip of the electric-field vector moves in an ellipse, rotating clockwise, as 'seen' by an observer located on the z-axis and facing the approaching electromagnetic wave (negative helicity). For $\varphi = -\frac{1}{4}\pi$, the rotation is anticlockwise (positive helicity). See the figures below.

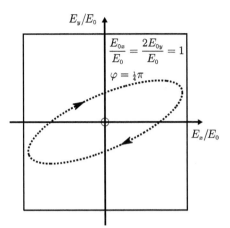
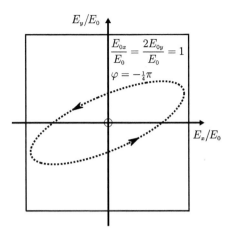

Comments

(i) It is sometimes convenient to describe an electromagnetic wave at a single instant in time by a curve that 'may be thought of as a smooth line joining the tips of a large number of vectors that indicate the directions and magnitudes of the electric field at various positions along the center line of the beam'.[5] Such a figure is known as a *snapshot* pattern. For the wave described in (a), the snapshot pattern is a sinusoidally varying curve lying in a plane inclined with respect to the xz-plane, whereas for (b) and (c), this curve is a helix centred on the z-axis. An alternative representation of a wave is the *sectional* pattern: it may be regarded as an end-on view of the snapshot pattern, as seen by an observer located on the z-axis, facing the oncoming wave. The six figures in the solution above are sectional patterns, and the first four of them are all special cases of (c). Thus the sectional pattern of an ellipse is

- ☞ a straight line when its semi-minor axis tends to zero $\big($see the figures for (a)$\big)$, and
- ☞ a circle when its minor and major axes are equal $\big($see the figures for (b)$\big)$.

(ii) Both pictorial representations referred to in Comment (i) are useful for describing the *polarization state* of an electromagnetic wave, which is said to be:

- ☞ Linearly polarized if its **E**-field has orthogonal components that are either in phase $(\varphi = 0)$ or out of phase $(\varphi = \pi)$. So for the wave considered above, (1) becomes

$$\mathbf{E} = \big(E_{0x},\ \pm E_{0y},\ 0\big)\cos\omega t. \tag{2}$$

---

[5]   W. A. Shurcliff, *Polarized light: production and use*, Chap. 1, pp. 3–4. Massachusetts: Harvard University Press, 1962.

☞ Circularly polarized if its **E**-field has orthogonal components that have equal amplitudes and which are in quadrature ($\varphi = \pm\frac{1}{2}\pi$). Now (1) becomes

$$\mathbf{E} = E_{0x}(\cos\omega t, \mp\sin\omega t, 0). \tag{3}$$

☞ Elliptically polarized if none of the above conditions apply.

As stated above, linearly and circularly polarized waves are special cases of the general elliptical polarization state.

(iii) In optics, the positive and negative helicities are often referred to as 'left-elliptically polarized' ($\mathcal{L}$) and 'right-elliptically polarized' ($\mathcal{R}$) respectively.

(iv) It is sometimes convenient to regard a linearly polarized wave as a superposition of the $\mathcal{R}$ and $\mathcal{L}$ circular states. For example, we have from (3):

$$\mathbf{E}_\mathcal{R} + \mathbf{E}_\mathcal{L} = E_{0x}(\hat{\mathbf{x}}\cos\omega t - \hat{\mathbf{y}}\sin\omega t) + E_{0x}(\hat{\mathbf{x}}\cos\omega t + \hat{\mathbf{y}}\sin\omega t) = 2\hat{\mathbf{x}}E_{0x}\cos\omega t, \tag{4}$$

which is (2) with $E_{0x} \to 2E_{0x}$ and $E_{0y} = 0$. This concept can be extended:

☞ An elliptically polarized wave may be regarded as a superposition of $\mathcal{R}$ and $\mathcal{L}$ circularly polarized states having different amplitudes, and

☞ an unpolarized wave can be represented in terms of two arbitrary incoherent[#] orthogonal linearly polarized states of equal amplitude (see, for example, Ref. [6]).

(v) In this question, the known values of E_{0x}, E_{0y} and φ were used to determine the state of polarization of the wave. But in practice, the converse problem usually presents itself. Consequently, various techniques have been devised for determining a state of polarization. One of these, based on a set of four quantities, was introduced by Stokes in 1854 as a result of his work in optics; these four quantities are now known as the Stokes parameters. In terms of the three independent quantities E_{0x}, E_{0y} and φ introduced earlier, the Stokes parameters for quasi-monochromatic waves are:[7]

$$\left.\begin{aligned}
s_0 &= \langle E_{0x}^2\rangle + \langle E_{0y}^2\rangle \\
s_1 &= \langle E_{0x}^2\rangle - \langle E_{0y}^2\rangle \\
s_2 &= \langle 2E_{0x}E_{0y}\cos\varphi\rangle \\
s_3 &= \langle 2E_{0x}E_{0y}\sin\varphi\rangle
\end{aligned}\right\}, \tag{5}$$

where the angular brackets denote a macroscopic time average. These four parameters may be measured experimentally in a series of operational steps.[7] The quantities E_{0x}, E_{0y}, φ and the helicity are then calculated using (5), thereby establishing the polarization state of the wave.

[#]That is, waves for which φ varies rapidly and randomly in an interval of time.

[6]   E. Hecht and A. Zajac, *Optics*, Chap. 8, p. 223. Massachusetts: Addison Wesley, 1974.
[7]   M. Born and E. Wolf, *Principles of optics*, Chap. X, pp. 550–2. London: Pergamon Press, 1959.

(vi) It is immaterial that the **E**-field (instead of the **B**-field) is used to determine a state of polarization, because, for electromagnetic waves travelling in vacuum (or in an isotropic medium), the two fields maintain a fixed relationship with each other. The choice of **E** is conventional and 'pays tribute to the dominant role of the electric vector in the more familiar absorption processes'.[5]

(vii) An anisotropic medium can change the polarization state of a wave. For example, linearly polarized light entering a calcite crystal will, in general, emerge elliptically polarized. This phenomenon is known as birefringence: in Questions 10.18–10.20 we consider the birefringence of a tenuous plasma in the presence of a uniform magnetostatic field.

Question 7.10

The electric field of a plane electromagnetic wave propagating in vacuum is given by

$$\mathbf{E} = E_0(\hat{\mathbf{x}} - i\hat{\mathbf{y}}) \, e^{i(kz - \omega t)}, \tag{1}$$

where all symbols have their usual meaning.

(a) Explain how to determine the polarization state of this wave.

(b) Using the results of Question 7.5, determine the magnetic field **B**.

(c) Calculate the time-average Poynting vector $\langle \mathbf{S} \rangle$.

Solution

(a) The real part of (1) is $\mathbf{E} = E_0(\hat{\mathbf{x}} \cos(kz - \omega t) + \hat{\mathbf{y}} \sin(kz - \omega t))$, and with the wave propagating along the positive z-axis, the electric vector lies along $\hat{\mathbf{x}}$ at the origin at time $t = 0$. One quarter of a cycle later, this vector is along $-\hat{\mathbf{y}}$, and is therefore 'seen' to be rotating clockwise by an observer looking towards the origin from a position $z > 0$. The wave is therefore right-circularly polarized.

(b) For this wave

$$\mathbf{B} = \frac{\hat{\mathbf{k}} \times \mathbf{E}}{c} = \frac{E_0}{c} \left[\hat{\mathbf{z}} \times (\hat{\mathbf{x}} - i\hat{\mathbf{y}}) \right] e^{i(kz - \omega t)} = \frac{E_0}{c} \left(-i\hat{\mathbf{x}} + \hat{\mathbf{y}} \right) e^{i(kz - \omega t)},$$

because of (6) of Question 7.5.

(c) The electric-field amplitude is $\sqrt{2}E_0$. So from (10) of Question 7.5:

$$\langle \mathbf{S} \rangle = \sqrt{\frac{\epsilon_0}{\mu_0}} \, E_0^2 \, \hat{\mathbf{z}}.$$

Question 7.11

Consider two concentric metal cylinders of length ℓ having radii a and b. Suppose that $\ell \gg b > a$. The space between the cylinders is filled with a material whose electromagnetic properties are essentially those of a vacuum. This arrangement, known as a coaxial (or coax) cable, is a simple example of a transmission line. We assume that the cable lies along the z-axis and that the cylinders are *perfect* conductors. The end of the cable at $z = \ell$ is terminated by a load impedance Z_L. The other end at $z = 0$ is connected to a source of emf that maintains a steady potential difference V between the conductors and which results in a current I.[‡] In the region $a < r < b$ we know that:

$$\mathbf{E}(r) = \frac{V}{\ln(b/a)}\frac{\hat{\mathbf{r}}}{r} \qquad \text{and} \qquad \mathbf{B}(r) = \frac{\mu_0}{2\pi}\frac{I}{r}\hat{\boldsymbol{\theta}}, \tag{1}$$

where r and θ are cylindrical polar coordinates (see (1) of Question 3.4 and (3)$_2$ of Question 4.7 respectively). For the case of time-dependent fields we speculate that both V and I in (1) become functions of z and t, then

$$\mathbf{E}(r,t) = \frac{V(z,t)}{\ln(b/a)}\frac{\hat{\mathbf{r}}}{r} \qquad \text{and} \qquad \mathbf{B}(r,t) = \frac{\mu_0}{2\pi}\frac{I(z,t)}{r}\hat{\boldsymbol{\theta}}. \tag{2}$$

(a) Use Maxwell's equations and (2) to determine the restrictions on $V(z,t)$ and $I(z,t)$.

(b) Hence show that $V(z,t)$ and $I(z,t)$ satisfy the one-dimensional wave equation

$$\frac{\partial^2 V}{\partial z^2} - \epsilon_0\mu_0\frac{\partial^2 V}{\partial t^2} = 0 \qquad \text{and} \qquad \frac{\partial^2 I}{\partial z^2} - \epsilon_0\mu_0\frac{\partial^2 I}{\partial t^2} = 0. \tag{3}$$

(c) The general solution of (3) is

$$\left. \begin{array}{l} V(z,t) = V_+(z-ct) + V_-(z+ct) \\ I(z,t) = I_+(z-ct) + I_-(z+ct) \end{array} \right\}, \tag{4}$$

where V_+, V_-, I_+ and I_- are arbitrary functions.[♯] Use (4) to derive the results

$$\left. \begin{array}{l} V(z,t) = Z_0\left[I_+(z-ct) - I_-(z+ct)\right] \\ I(z,t) = Z_0^{-1}\left[V_+(z-ct) - V_-(z+ct)\right] \end{array} \right\}, \tag{5}$$

where $Z_0 = \dfrac{1}{2\pi}\sqrt{\dfrac{\mu_0}{\epsilon_0}}\ln\left[\dfrac{b}{a}\right]$ is the characteristic impedance of the cable.

[‡] Here we use I for the current (and not i) to avoid possible confusion with $i = \sqrt{-1}$.

[♯] Clearly V_+, I_+ are for waves travelling along the positive z-axis; and V_-, I_- are for waves travelling along the negative z-axis.

(d) Suppose the load impedance is purely resistive, $Z_L = R$. Show that forward-travelling waves are partially reflected in general, and establish the condition for no reflection to occur.

(e) Suppose waves of frequency ω propagate in the cable. Prove that the input impedance $Z_i = V(0,t)/I(0,t)$ is given by

$$Z_i = \frac{(Z_L \cos k\ell - iZ_0 \sin k\ell)}{(Z_0 \cos k\ell - iZ_L \sin k\ell)} Z_0, \tag{6}$$

where $k = \omega/c$.

Solution

(a) Substituting (2) in $\nabla \times \mathbf{E} = -\dfrac{\partial \mathbf{B}}{\partial t}$ and $\nabla \times \mathbf{B} = \epsilon_0 \mu_0 \dfrac{\partial \mathbf{E}}{\partial t}$ gives

$$\frac{\partial V}{\partial z} = -\frac{\mu_0}{2\pi} \ln\left[\frac{b}{a}\right] \frac{\partial I}{\partial t} \qquad \text{and} \qquad \frac{\partial V}{\partial t} = -\frac{1}{2\pi\epsilon_0} \ln\left[\frac{b}{a}\right] \frac{\partial I}{\partial z} \tag{7}$$

respectively. Maxwell's divergence equations confirm that $\nabla \cdot \mathbf{E} = 0$ and $\nabla \cdot \mathbf{B} = 0$, and so provide no further information. Equations $(7)_1$ and $(7)_2$ are therefore the restrictions we seek.

(b) Taking $\dfrac{\partial}{\partial z}$ of $(7)_1 - \epsilon_0\mu_0\dfrac{\partial}{\partial t}$ of $(7)_2$ gives $(3)_1$ because $\dfrac{\partial^2}{\partial z \partial t} = \dfrac{\partial^2}{\partial t \partial z}$.[†] Equation $(3)_2$ may be derived in a similar way.

(c) Substituting (4) in $(7)_1$[b] gives

$$\frac{\partial V}{\partial z} = \frac{\partial}{\partial z}(V_+ + V_-) = -\frac{\mu_0}{2\pi} \ln\left[\frac{b}{a}\right] \frac{\partial}{\partial t}(I_+ + I_-). \tag{8}$$

Changing variables to $u = z - ct$ and $v = z + ct$ enables (8) to be expressed as

$$\frac{dV_+}{du}\frac{\partial u}{\partial z} + \frac{dV_-}{dv}\frac{\partial v}{\partial z} = -\frac{\mu_0}{2\pi} \ln\left[\frac{b}{a}\right]\left(\frac{dI_+}{du}\frac{\partial u}{\partial t} + \frac{dI_-}{dv}\frac{\partial v}{\partial t}\right). \tag{9}$$

Now $\partial u/\partial z = \partial v/\partial z = 1$ and $\partial u/\partial t = -\partial v/\partial t = -c$, and so (9) becomes

$$\frac{d}{du}(V_+ - Z_0 I_+) = -\frac{d}{dv}(V_- + Z_0 I_-). \tag{10}$$

[†] If the first- and second-order derivatives of $f(x,y)$ are continuous, then $\dfrac{\partial^2 f}{\partial x \partial y} = \dfrac{\partial^2 f}{\partial y \partial x}$.

[b] Or alternatively begin with $(7)_2$.

The left-hand side of (10) is a function of u; the right-hand side is a function of v. But u and v are independent variables and (10) can therefore be satisfied only if

$$\frac{d}{du}(V_+ - Z_0 I_+) = \alpha \quad \text{and} \quad -\frac{d}{dv}(V_- + Z_0 I_-) = \alpha, \tag{11}$$

where α is an arbitrary constant. Integrating[‡] (11) yields $(V_+ - Z_0 I_+) = \alpha u$ and $(V_- + Z_0 I_-) = -\alpha v$. Therefore

$$\frac{V_+ - Z_0 I_+}{V_- + Z_0 I_-} = -\frac{z - ct}{z + ct}, \tag{12}$$

which can be rearranged to give

$$z\left\{(V_+ + V_-) - Z_0(I_+ - I_-)\right\} = ct\left\{Z_0(I_+ + I_-) - (V_+ - V_-)\right\}. \tag{13}$$

But z and t are also independent variables and (13) is true only if each term in braces is identically zero. Now $V = (V_+ + V_-)$ and $I = (I_+ + I_-)$. Hence (5).

(d) By definition $Z_L = V(\ell, t)/I(\ell, t)$. Then

$$R = Z_0 \frac{I_+(\ell - ct) - I_-(\ell + ct)}{I_+(\ell - ct) + I_-(\ell + ct)}, \tag{14}$$

because of $(5)_1$. Rearranging gives

$$I_- = \left(\frac{Z_0 - R}{Z_0 + R}\right) I_+, \tag{15}$$

at $z = \ell$. So there is a reflected wave

$$I_-(z + ct) = \left(\frac{Z_0 - R}{Z_0 + R}\right) I_+(z - ct), \tag{16}$$

which is zero for all z and t if the load impedance equals the characteristic impedance (i.e. $R = Z_0$).

(e) Consider waves of the form

$$V_\pm = \left.\begin{array}{c} V_{0_+} e^{ikz} \\ V_{0_-} e^{-ikz} \end{array}\right\} e^{-i\omega t} \quad \text{and} \quad I_\pm = \left.\begin{array}{c} I_{0_+} e^{ikz} \\ I_{0_-} e^{-ikz} \end{array}\right\} e^{-i\omega t}. \tag{17}$$

Then $(4)_2$ and $(5)_1$ give

$$Z_i = \frac{V(0, t)}{I(0, t)} = Z_0 \frac{I_{0_+} - I_{0_-}}{I_{0_+} + I_{0_-}} \quad \text{and} \quad Z_L = \frac{V(\ell, t)}{I(\ell, t)} = Z_0 \frac{I_{0_+} e^{ik\ell} - I_{0_-} e^{-ik\ell}}{I_{0_+} e^{ik\ell} + I_{0_-} e^{-ik\ell}}. \tag{18}$$

Eliminating I_{0_+} and I_{0_-} from $(18)_1$ and $(18)_2$ gives (6) after some manipulation.

[‡]The constant of integration—being a static voltage—has no effect on the wave motion. It is therefore of no interest here, and so we equate it to zero.

Comments

(i) Equation (3) shows that plane electromagnetic waves may propagate with speed c, in both directions along the cable, without attenuation.[‡] Both resistive losses in the conductors of real cables and absorption in the dielectric material (see Chapter 9) limit their use to low frequencies. Between microwave and optical frequencies, wave guides are more appropriate conduits for transmitting electromagnetic energy (see Questions 7.12–7.16).

(ii) The dielectric material used in coax cable usually has $\epsilon_r > 1$ and $\mu_r = 1$, these values often remaining constant over a wide frequency range. Then in the above equations $c \rightarrow c/\sqrt{\epsilon_r(0)}$ and $Z_0 \rightarrow Z_0/\sqrt{\epsilon_r(0)}$, where $\epsilon_r(0)$ is the dc relative permittivity.

(iii) The power transmitted along this coaxial cable is given by $\mathcal{P} = \dfrac{1}{\mu_0} \displaystyle\int_s (\mathbf{E} \times \mathbf{B}) \cdot d\mathbf{a}$.

Using the fields given by (2), we then obtain:

$$\mathcal{P} = \frac{1}{2\pi} \frac{V(z,t)I(z,t)}{\ln(b/a)} \int_a^b \frac{2\pi r dr}{r^2} = V(z,t)\, I(z,t) = Z_0 \big[I_+^2(z-ct) - I_-^2(z+ct)\big],$$

because of (4). This last equation is reminiscent of the power dissipated in a resistance. For a lossless cable there is, however, no dissipation and the wave propagates without attenuation.

(iv) A coax cable terminated by a load equal to its characteristic impedance is said to be *matched*. All the energy propagating in the cable is absorbed and, as (16) shows, there is no reflected wave. The voltage and current (V_+ and I_+) are always in phase and the forward-travelling wave behaves as though the cable were infinitely long. Many items of electronic equipment used in laboratories have input and output impedances of $50\,\Omega$, often connected together with coax cable. Choosing cable with a $50\,\Omega$[♯] characteristic impedance avoids unwanted reflected signals between these various pieces of apparatus.

(v) There are three cases of (6) which are of special interest:

$Z_L = 0$; the short-circuit line

The input impedance $Z_i = -iZ_0 \tan k\ell$ is purely reactive; no power is dissipated in the load.

$Z_L \rightarrow \infty$; the open-circuit line

Here $Z_i = iZ_0 \cot k\ell = -iZ_0 \tan(k\ell + \tfrac{1}{2}\pi)$ which is equivalent to a short-circuited line differing in length by one-quarter wavelength.

$k\ell = \pi/2$; the quarter-wave line

For any resistive load $Z_i = Z_0^2/Z_L$ is real (R say), and $Z_i \rightarrow 0$ as $R \rightarrow \infty$.

[‡]The cable may be regarded as 'lossless', because of the assumed infinite conductivities.

[♯]A $50\,\Omega$ impedance can be achieved using a teflon dielectric ($\epsilon_r(0) = 2.1$) and with $b/a = 3.4$.

Question 7.12

Electromagnetic waves propagate along a hollow wave guide of constant but arbitrary cross-section. Suppose the walls of the guide are perfectly conducting ($\sigma \to \infty$) and that the permittivity and permeability of the enclosed medium (air, say) have the vacuum values ϵ_0 and μ_0 respectively. Choose the z-axis of Cartesian coordinates along the guide.

(a) Seek solutions to the wave equation of the form[‡]

$$\mathbf{E}(x, y, z, t) = \mathbf{E}_0(x, y)\, e^{i(\pm kz - \omega t)} \quad \text{and} \quad \mathbf{B}(x, y, z, t) = \mathbf{B}_0(x, y)\, e^{i(\pm kz - \omega t)} \quad (1)$$

and show that

$$\left[\frac{\partial^2}{\partial x^2} + \frac{\partial^2}{\partial y^2} + \gamma^2 \right] \psi = 0, \quad (2)$$

where $\gamma^2 = \omega^2/c^2 - k^2$. In (2), ψ represents either E_z or B_z and

$$k = 2\pi \div \text{the wavelength of the radiation inside the guide.}$$

(The quantity ω/c is the magnitude of the wave vector in *free space*; k_0, say.)

(b) Use Maxwell's curl equations to show that the field components for forward-travelling waves are given by

$$
\left.
\begin{aligned}
E_x &= \frac{i}{\gamma^2} \left(k \frac{\partial E_z}{\partial x} + \omega \frac{\partial B_z}{\partial y} \right) \\[2mm]
E_y &= \frac{i}{\gamma^2} \left(k \frac{\partial E_z}{\partial y} - \omega \frac{\partial B_z}{\partial x} \right) \\[2mm]
B_x &= \frac{i}{\gamma^2} \left(k \frac{\partial B_z}{\partial x} - \frac{\omega}{c^2} \frac{\partial E_z}{\partial y} \right) \\[2mm]
B_y &= \frac{i}{\gamma^2} \left(k \frac{\partial B_z}{\partial y} + \frac{\omega}{c^2} \frac{\partial E_z}{\partial x} \right)
\end{aligned}
\right\}, \quad (3)
$$

where $\mathbf{k} = k\hat{\mathbf{z}}$.[♯]

(c) Show that (3) may be expressed in the coordinate-free form

$$
\left.
\begin{aligned}
\mathbf{E}_t &= \frac{i}{\gamma^2} \left(k \nabla_t E_\| - \omega \hat{\mathbf{k}} \times \nabla_t B_\| \right) \\[2mm]
\mathbf{B}_t &= \frac{i}{\gamma^2} \left(k \nabla_t B_\| + \frac{\omega}{c^2} \hat{\mathbf{k}} \times \nabla_t E_\| \right)
\end{aligned}
\right\}. \quad (4)
$$

Here $\hat{\mathbf{k}}$ is a unit vector in the direction of propagation, and the subscripts t and $\|$ are components transverse and parallel to $\hat{\mathbf{k}}$.

[‡]These are 'bounded waves' since they are confined by the walls of the wave guide; they are not plane waves because the **E**- and **B**-field amplitudes vary over the guide's cross-section. See (1).

[♯]For waves travelling in the opposite direction, let $k \to -k$ in (3) and (4).

Solution

(a) Substituting (1) in the wave equation (1) of Question 7.4 gives (2).

(b) Inside the wave guide

$$\nabla \times \mathbf{E} = i\omega\mathbf{B} \qquad \text{and} \qquad \nabla \times \mathbf{B} = -i\omega\epsilon_0\mu_0\mathbf{E}. \tag{5}$$

Then from (1) and (5) we obtain, for a wave travelling along positive $\hat{\mathbf{z}}$, the equations

$$\left.\begin{aligned}
\frac{\partial E_z}{\partial y} - ikE_y &= i\omega B_x & -\frac{\partial E_z}{\partial x} + ikE_x &= i\omega B_y \\[2mm]
\frac{\partial B_z}{\partial y} - ikB_y &= -i\omega\epsilon_0\mu_0 E_x & -\frac{\partial B_z}{\partial x} + ikB_x &= -i\omega\epsilon_0\mu_0 E_y
\end{aligned}\right\}.$$

Solving this system of equations simultaneously for E_x, E_y, B_x and B_y gives (3).

(c) Clearly $\mathbf{E}_t = \hat{\mathbf{x}}E_x + \hat{\mathbf{y}}E_y$ follows immediately from $(3)_1$ and $(3)_2$ since $\hat{\mathbf{k}} = (0, 0, 1)$ and $\nabla_t = \hat{\mathbf{x}}\dfrac{\partial}{\partial x} + \hat{\mathbf{y}}\dfrac{\partial}{\partial y}$. Similarly for $\mathbf{B}_t$.

Comments

(i) An alternative derivation of (2) follows by substituting (3) in Maxwell's divergence equations: $\nabla \cdot \mathbf{E} = 0$ and $\nabla \cdot \mathbf{B} = 0$.

(ii) Suitable linear combinations of (1), which satisfy the appropriate boundary conditions at the walls of the guide, give rise to travelling or standing waves along the z-axis.

(iii) For propagating waves, the wavelength of the radiation inside the guide is greater than the free-space wavelength and $\gamma^2 = \omega^2/c^2 - k^2 > 0$. We consider two distinct types of wave:

☞ transverse electric (TE) characterized by $E_z = 0$ everywhere, and
☞ transverse magnetic (TM) characterized by $B_z = 0$ everywhere.

Despite the name, these are not transverse waves (reason: the magnetic (electric) field of the TE (TM) wave has a component in the direction of propagation).

(iv) If $\gamma^2 = \omega^2/c^2 - k^2 = 0$ both E_z and B_z are necessarily zero (see (3)). These are transverse electromagnetic (TEM) waves that propagate as though they were in an infinite medium having no walls. TEM waves cannot exist in the hollow wave guide considered in this question.

(v) Inside a medium of infinite conductivity both $\mathbf{E}$ and $\mathbf{B}$ vanish and the matching conditions (see $(2)_2$ and $(4)_1$ of Question 10.6) become $\hat{\mathbf{n}} \times \mathbf{E} = 0$ and $\hat{\mathbf{n}} \cdot \mathbf{B} = 0$ on the walls. These results require

$$\left.\begin{aligned}
&\text{TE waves with } E_z = 0: & \frac{\partial B_z}{\partial n}\bigg|_{\text{walls}} &= 0 \\[2mm]
&\text{TM waves with } B_z = 0: & E_z|_{\text{walls}} &= 0
\end{aligned}\right\}, \tag{6}$$

where $\partial/\partial n$ is the usual normal derivative. (Equation $(6)_1$ can be understood by considering a rectangular wave guide. Then $\hat{\mathbf{n}} \cdot \mathbf{B} = 0 \Rightarrow B_i = 0$, where $i = x$ or y. Hence it follows from $(3)_3$ and $(3)_4$ that $\partial B_z/\partial n$ must be zero for TE waves since $E_z = 0$ by definition.)

(vi) The wave equation $\partial^2/\partial x^2 + \partial^2/\partial y^2 + \gamma^2 \psi = 0$ together with the boundary condition (6) specify an eigenvalue problem. It is easy to see why 'the constant γ^2 must be nonnegative. Roughly speaking, it is because ψ must be oscillatory to satisfy boundary conditions (6) on opposite sides of the guide. There will be a spectrum of eigenvalues γ_λ^2 and corresponding solutions ψ_λ, $\lambda = 1, 2, 3, \ldots$, which form an orthogonal set. These different solutions are called the *modes of the guide'.[8]*

(vii) Since k must be real for propagating waves it follows that $\gamma_\lambda^2 = \omega^2/c^2 - k_\lambda^2$ or $k_\lambda^2 = \omega^2/c^2 - \gamma_\lambda^2$. Corresponding to each mode is a cut-off frequency ω_λ given by $\omega_\lambda = c\gamma_\lambda$. Frequencies below ω_λ are attenuated and cannot exist inside the wave guide.

(viii) In Question 7.13 we consider TE waves travelling in a guide having a rectangular cross-section. Question 7.16 deals with waves in a circular wave guide.

(ix) The finite conductivity of the walls in a real wave guide leads to the dissipation of energy. See Question 7.14.

Question 7.13

Suppose the wave guide described in Question 7.12 has a rectangular cross-section and that its inside walls coincide with the surfaces $x = 0$, $x = a$, $y = 0$, $y = b$. Consider the propagation of TE waves in the guide and assume that

$$B_z(x, y, t) = X(x)\, Y(y)\, e^{i(kz - \omega t)} \quad \text{with the boundary condition } \left.\frac{\partial B_z}{\partial n}\right|_{\text{walls}} = 0, \quad (1)$$

where the functions $X(x)$ and $Y(y)$ are to be determined.

(a) Using (1) and the method of separation of variables described in Question 1.18, solve (2) of Question 7.12 and obtain the dispersion relation

$$\frac{\omega^2}{c^2} = \frac{m^2\pi^2}{a^2} + \frac{n^2\pi^2}{b^2} + k^2 \quad \text{or} \quad \gamma^2 = \frac{m^2\pi^2}{a^2} + \frac{n^2\pi^2}{b^2}, \quad (2)$$

where m, n are integers having the values $0, 1, 2, \ldots$.

(b) Use (2) to obtain the cut-off frequencies for the wave guide.

(c) Determine all components of the **E**- and **B**-fields inside the wave guide.

[8]   J. D. Jackson, *Classical electrodynamics*, Chap. 8, p. 360. New York: Wiley, 3 edn, 1998.

(d) Show, by integrating the time-average Poynting vector over the cross-section of the wave guide, that the power transmitted in the TE_{mn} mode is

$$\mathcal{P} = \frac{a^3 b^3 \omega k B_0^2}{8\pi^2 \mu_0 (m^2 b^2 + n^2 a^2)} \times q \tag{3}$$

where the integer q has the values

$$q = \begin{cases} 1 & \text{if neither } n \text{ nor } m \text{ is zero,} \\ 2 & \text{if either } n \text{ or } m \text{ is zero.} \end{cases} \tag{4}$$

Solution

(a) Substituting (1) in the wave equation (2) of Question 7.12 gives

$$Y \frac{d^2 X}{dx^2} + X \frac{d^2 Y}{dy^2} + \gamma^2 XY = 0,$$

or

$$\frac{1}{X(x)} \frac{d^2 X(x)}{dx^2} + \frac{1}{Y(y)} \frac{d^2 Y(y)}{dy^2} + \gamma^2 = 0. \tag{5}$$

The first term in (5) is a function of x only; the second term is a function of y only; and the third term is independent of both x and y. So[‡]

$$\frac{1}{X} \frac{d^2 X}{dx^2} = -\alpha^2 \quad \text{and} \quad \frac{1}{Y} \frac{d^2 Y}{dy^2} = -\beta^2, \tag{6}$$

where α and β are the separation constants which satisfy

$$\alpha^2 + \beta^2 = \gamma^2. \tag{7}$$

The general solution of (6) is

$$\left. \begin{array}{l} X(x) = C_1 \sin \alpha x + C_2 \cos \alpha x \\ Y(y) = C_3 \sin \beta y + C_4 \cos \beta y \end{array} \right\}, \tag{8}$$

where the C_i are arbitrary constants. So from (1)

$$B_z = (C_1 \sin \alpha x + C_2 \cos \alpha x)(C_3 \sin \beta y + C_4 \cos \beta y) e^{i(kz - \omega t)}. \tag{9}$$

Differentiating (9) yields

$$\left. \begin{array}{l} \dfrac{\partial B_z}{\partial x} = \alpha (C_1 \cos \alpha x - C_2 \sin \alpha x)(C_3 \sin \beta y + C_4 \cos \beta y) \\[2mm] \dfrac{\partial B_z}{\partial y} = \beta (C_1 \sin \alpha x + C_2 \cos \alpha x)(C_3 \cos \beta y - C_4 \sin \beta y) \end{array} \right\} e^{i(kz - \omega t)}. \tag{10}$$

[‡]See Question 1.18 for a discussion on the signs of the separation constants.

The boundary conditions $\dfrac{\partial B_z}{\partial x}\bigg|_{x=0} = 0$ and $\dfrac{\partial B_z}{\partial y}\bigg|_{y=0} = 0$ can only be satisfied for all x, y and z if

$$C_1 = C_3 = 0. \tag{11}$$

Then (10) becomes

$$\left. \begin{aligned} \frac{\partial B_z}{\partial x} &= -\alpha B_0 \sin \alpha x \cos \beta y \\[2mm] \frac{\partial B_z}{\partial y} &= -\beta B_0 \cos \alpha x \sin \beta y \end{aligned} \right\} e^{i(kz - \omega t)}, \tag{12}$$

where $B_0 = C_2 C_4$ is a constant. The boundary condition $\dfrac{\partial B_z}{\partial x}\bigg|_{x=a} = 0$ requires that

$$\sin \alpha a = 0. \tag{13}$$

Similarly, $\dfrac{\partial B_z}{\partial y}\bigg|_{y=b} = 0$ is only possible if

$$\sin \beta b = 0. \tag{14}$$

We therefore conclude that α and β can only assume the discrete values

$$\left. \begin{aligned} \alpha &= m\pi/a \\ \beta &= n\pi/b \end{aligned} \right\} \quad \text{where } m, n = 0, 1, 2, \dots. \tag{15}$$

Substituting (15) in (7) yields (2). Furthermore, (9), (11) and (15) yield

$$B_z(x, y, t) = B_0 \cos\left(\frac{m\pi x}{a}\right) \cos\left(\frac{n\pi y}{b}\right) e^{i(kz - \omega t)}. \tag{16}$$

(b) For travelling waves k must be real. It then follows that for any pair of integers m and n, the cut-off frequency ω_{mn} of the TE_{mn} mode is

$$\gamma_{mn}^2 = \frac{\omega_{mn}^2}{c^2} = \frac{m^2 \pi^2}{a^2} + \frac{n^2 \pi^2}{b^2}. \tag{17}$$

(c) Recovering the real part of (16) gives

$$B_z = B_0 \cos\left(\frac{m\pi x}{a}\right) \cos\left(\frac{n\pi y}{b}\right) \cos(kz - \omega t). \tag{18}$$

Substituting (18) in (3) of Question 7.12 with $E_z = 0$ (by definition) and again taking real parts yield

$$
\left.
\begin{aligned}
E_x &= \frac{n\pi/b}{\gamma^2}\, \omega B_0 \cos\left(\frac{m\pi x}{a}\right) \sin\left(\frac{n\pi y}{b}\right) \sin(kz - \omega t) \\[2mm]
E_y &= -\frac{m\pi/a}{\gamma^2}\, \omega B_0 \sin\left(\frac{m\pi x}{a}\right) \cos\left(\frac{n\pi y}{b}\right) \sin(kz - \omega t) \\[2mm]
B_x &= \frac{m\pi/a}{\gamma^2}\, k B_0 \sin\left(\frac{m\pi x}{a}\right) \cos\left(\frac{n\pi y}{b}\right) \sin(kz - \omega t) \\[2mm]
B_y &= \frac{n\pi/b}{\gamma^2}\, k B_0 \cos\left(\frac{m\pi x}{a}\right) \sin\left(\frac{n\pi y}{b}\right) \sin(kz - \omega t)
\end{aligned}
\right\}.
\tag{19}
$$

(d) The Poynting vector $\mathbf{S} = S_x \hat{\mathbf{x}} + S_y \hat{\mathbf{y}} + S_z \hat{\mathbf{z}}$, but only S_z survives time-averaging. Substituting (19) in $\langle \mathbf{S} \rangle = \mu_0^{-1} \langle E_x B_y - E_y B_x \rangle \hat{\mathbf{z}}$ and because $\langle \cos^2(kz - \omega t) \rangle = \langle \sin^2(kz - \omega t) \rangle = \frac{1}{2}$ (see (3) of Question 1.27), we obtain

$$
\langle \mathbf{S}(x,y) \rangle = \frac{\pi^2 \omega k B_0^2 (m^2 b^2 \sin^2 \alpha x \, \cos^2 \beta y + n^2 a^2 \cos^2 \alpha x \, \sin^2 \beta y)}{2a^2 b^2 \mu_0 \gamma^4}\, \hat{\mathbf{z}}
\tag{20}
$$

with α and β given by (15). Using trigonometric identities and the dispersion relation (2), we can express (20) as:

$$
\langle \mathbf{S} \rangle = \frac{\omega k B_0^2}{8\pi^2 \mu_0} \left[\frac{(1 - \cos 2\alpha x \cos 2\beta y)}{(m^2/a^2 + n^2/b^2)} - \frac{(m^2/a^2 - n^2/b^2)(\cos 2\alpha x - \cos 2\beta y)}{(m^2/a^2 + n^2/b^2)^2} \right] \hat{\mathbf{z}}.
\tag{21}
$$

The average power $\mathcal{P}$ is

$$
\mathcal{P} = \int_s \langle \mathbf{S} \rangle \cdot d\mathbf{a} = \int_0^b \int_0^a \langle S \rangle \, dx \, dy,
\tag{22}
$$

since $d\mathbf{a} = dx \, dy \, \hat{\mathbf{z}}$. Substituting (21) in (22) and integrating yield

$$
\mathcal{P} = \frac{a^3 b^3 \omega k B_0^2}{8\pi^2 \mu_0 (m^2 b^2 + n^2 a^2)} \left\{ 1 - \frac{m^2 b^2 - n^2 a^2}{m^2 b^2 + n^2 a^2} \left(\frac{\sin 4m\pi}{4m\pi} - \frac{\sin 4n\pi}{4n\pi} \right) \right\}.
\tag{23}
$$

If neither m nor n is zero then the term q represented by the curly brackets is one, but if either integer is zero then q is two because $\lim_{\theta \to 0} \sin \theta / \theta = 1$. Hence (3).

Comments

(i) The wall currents in the wave guide follow from $\mathbf{K}_f = -\hat{\mathbf{n}} \times \mathbf{H}$,[#] where the unit vector $\hat{\mathbf{n}}$ points from a wall towards the interior of the guide and $\mathbf{H} = \mu_0^{-1}\mathbf{B}$ is the field inside the wave guide at the wall surface.[†] For TE modes there will always

---

[#] See (4) of Question 10.6.

[†] The **H**-field referred to above is discussed more generally in Chapters 9 and 10.

be a non-zero circumferential surface current $\mathbf{K}_f$ because of the axial magnetic field.

(ii) For TE modes either m or n can be zero, but not both. For TM modes neither m nor n can be zero because $E_z(x, y, t)$ is of the form

$$E_z(x, y, t) = E_{0z} \sin\left(\frac{m\pi x}{a}\right) \sin\left(\frac{n\pi y}{b}\right) e^{i(kz - \omega t)},$$

and the lowest cut-off frequency is TM_{11}. Assuming $a > b$, the cut-off frequency of TM_{11} exceeds the cut-off frequency of TE_{10} by the factor $\sqrt{1 + a^2/b^2}$. TE_{10} is called the dominant mode and is often used in practical applications.

(iii) Waves generated inside a guide (with some suitable source) will usually excite several modes in varying proportions. If only the dominant mode can propagate at the operating frequency, then all other modes will be attenuated. In applications where several modes are required simultaneously *mode coupling* can occur, which may cause one mode to generate another.

(iv) From the dispersion relation (2) it follows that the phase velocity $v_\phi = \omega/k$ and the group velocity $v_g = d\omega/dk$ are given by

$$v_\phi = \frac{c}{\sqrt{1 - \omega_{mn}^2/\omega^2}} \qquad \text{and} \qquad v_g = c\sqrt{1 - \omega_{mn}^2/\omega^2}, \qquad (24)$$

where $v_g v_\phi = c^2$. The energy travels along the guide at speed $v_g < c$ whilst $v_\phi > c$. (Why should you not be concerned by a phase velocity greater than c?)

Question 7.14*

Suppose the rectangular wave guide described in Question 7.13 is made from a material whose conductivity σ is taken to be *finite* (e.g. copper). Within the conductor the fields decay exponentially as $e^{-\xi/\delta}$, where ξ is the inward normal coordinate at a wall and $\delta = \sqrt{\dfrac{2}{\mu_0 \sigma \omega}}$ is the skin depth[‡] at frequency ω. As a first approximation, we calculate the Poynting vector $\mathbf{S}$ from

$$\mathbf{S} = \sqrt{\frac{\omega \mu_0}{2\sigma}} H^2 e^{-2\xi/\delta} \hat{\mathbf{n}} \qquad (1)$$

(see Comment (vii) of Question 10.10). Here $\mathbf{H} = \mu_0^{-1} \mathbf{B}$ and the unit normal $\hat{\mathbf{n}}$ points from within the wave guide towards a wall.

[‡]With perfectly conducting walls the fields decrease abruptly to zero at the surface. In materials having a finite conductivity, there is a transition zone characterized by a characteristic length scale (called the skin depth) over which the fields decay smoothly. See Question 10.10.

(a) Show that for TE modes in this guide the time-average of (1) inside a wall at $\xi = 0$ is

$$
\langle \mathbf{S} \rangle = \sqrt{\frac{\omega \mu_0}{8\sigma}}\, H_0^2\, \hat{\mathbf{n}}
\begin{cases}
\left[\dfrac{m^2 \pi^2 k^2}{a^2 \gamma^4} + \left(1 - \dfrac{m^2 \pi^2 k^2}{a^2 \gamma^4} \right) \cos^2 \dfrac{m \pi x}{a} \right] \\[2ex]
\left[\dfrac{n^2 \pi^2 k^2}{b^2 \gamma^4} + \left(1 - \dfrac{n^2 \pi^2 k^2}{b^2 \gamma^4} \right) \cos^2 \dfrac{n \pi y}{b} \right],
\end{cases}
\tag{2}
$$

where the first (second) equation in (2) is for the walls lying in the xz-plane (yz-plane).

Hint: Assume that $\mathbf{H}$ is continuous across all air–metal interfaces inside the guide,[‡] and use $\mathbf{B}$ calculated in Question 7.13.

(b) Hence show that the power decays exponentially along the guide with an attenuation length

$$
d_{mn} = \frac{ab\,\gamma^2 \sqrt{\omega^2/c^2 - \gamma^2}}{[\gamma^4 (a+b) + \pi^2 k^2 (m^2/a + n^2/b)]}\, \frac{q}{\delta},
\tag{3}
$$

for the TE$_{mn}$ mode. In (3), the integer q is given by (4) of Question 7.13.

Solution

(a) The parallel components of $\mathbf{H}$ inside the guide at $\xi = 0$ are

$$
\mathbf{H}_{g_{\parallel}} = \frac{1}{\mu_0}
\begin{cases}
\hat{\mathbf{x}} B_x + \hat{\mathbf{z}} B_z & \text{at a wall lying in the } xz\text{-plane} \\[1ex]
\hat{\mathbf{y}} B_y + \hat{\mathbf{z}} B_z & \text{at a wall lying in the } yz\text{-plane},
\end{cases}
\tag{4}
$$

and because of the matching condition for $\mathbf{H}$ this is also the field inside the conducting *walls* at $\xi = 0$. Substituting (4) in (1) and using the components of $\mathbf{B}$ given by (18) and (19) of Question 7.13 give

$$
\mathbf{H}_{w_{\parallel}} = H_0
\begin{cases}
\dfrac{m \pi}{a \gamma^2} k \sin\left(\dfrac{m \pi x}{a} \right) \sin(kz - \omega t)\, \hat{\mathbf{x}} \pm \cos\left(\dfrac{m \pi x}{a} \right) \cos(kz - \omega t)\, \hat{\mathbf{z}} \\[2ex]
\dfrac{n \pi}{b \gamma^2} k \sin\left(\dfrac{n \pi y}{b} \right) \sin(kz - \omega t)\, \hat{\mathbf{y}} \pm \cos\left(\dfrac{n \pi y}{b} \right) \cos(kz - \omega t)\, \hat{\mathbf{z}}.
\end{cases}
\tag{5}
$$

In $(5)_1$ the term involving y has been equated to ± 1 at the walls; in $(5)_2$ the term involving x has been equated to ± 1 at the two other walls. Substituting (5) in (1) yields

[‡] Chapter 10 gives a general discussion of matching conditions in electromagnetism. See, for example, Question 10.6.

$$\mathbf{S} = \sqrt{\frac{\omega\mu_0}{2\sigma}}\, H_0^2 \hat{\mathbf{n}} \begin{cases} \dfrac{m^2\pi^2}{a^2\gamma^4} k^2 \sin^2\left(\dfrac{m\pi x}{a}\right)\sin^2(kz-\omega t) + \cos^2\left(\dfrac{m\pi x}{a}\right)\cos^2(kz-\omega t) \\[3mm] \dfrac{n^2\pi^2}{b^2\gamma^4} k^2 \sin^2\left(\dfrac{n\pi y}{b}\right)\sin^2(kz-\omega t) + \cos^2\left(\dfrac{n\pi y}{b}\right)\cos^2(kz-\omega t), \end{cases}$$

and (2) then follows immediately because $\langle\cos^2(kz-\omega t)\rangle = \langle\sin^2(kz-\omega t)\rangle = \frac{1}{2}$.

(b) Consider the wall lying in the xz-plane at $y=0$. An element of area on this wall is $da = dx\,dz\,\hat{\mathbf{n}}$ (here $\hat{\mathbf{n}} = \hat{\mathbf{y}}$). The energy per unit time dP 'flowing' into this area is removed from the wave propagating along the guide where $dP = -\langle\mathbf{S}\rangle\cdot da = -\langle S\rangle\,dx\,dz$. Integrating this over a strip of 'width' a along x and 'breadth' δz along z gives

$$\delta P = \int dP = -\int_0^{\delta z}\int_0^a \langle S\rangle\,dx\,dz,$$

or

$$\frac{\delta P}{\delta z} = -\int_0^a \langle S\rangle\,dx. \tag{6}$$

Substituting $(2)_1$ in (6) and integrating over x yield

$$\frac{\delta P}{\delta z} = -\sqrt{\frac{\omega\mu_0}{8\sigma}}\, H_0^2 \left[\frac{m^2\pi^2 k^2}{a\gamma^4} + \frac{a}{2}\left(1 - \frac{m^2\pi^2 k^2}{a^2\gamma^4}\right)\right].$$

There is a identical contribution from the opposite wall at $y=b$, and so

$$\frac{\delta P}{\delta z} \rightarrow -\sqrt{\frac{\omega\mu_0}{8\sigma}}\, H_0^2 \left[\frac{2m^2\pi^2 k^2}{a\gamma^4} + a\left(1 - \frac{m^2\pi^2 k^2}{a^2\gamma^4}\right)\right].$$

A similar calculation from the remaining two walls gives the contribution

$$-\sqrt{\frac{\omega\mu_0}{8\sigma}}\, H_0^2 \left[\frac{2n^2\pi^2 k^2}{b\gamma^4} + b\left(1 - \frac{n^2\pi^2 k^2}{b^2\gamma^4}\right)\right],$$

and so the *total* power dissipated per unit distance of travel δz is

$$\frac{\delta P}{\delta z} = -\sqrt{\frac{\omega\mu_0}{8\sigma}}\, H_0^2 \left[(a+b) + \frac{\pi^2 k^2}{\gamma^4}\left(\frac{m^2}{a} + \frac{n^2}{b}\right)\right]. \tag{7}$$

Dividing (7) by (3) of Question 7.13 yields $\delta P/P = -2\,\delta z/d_{mn}$, with d_{mn} given by (3). This shows that the power decays exponentially along the guide with an attenuation distance d_{mn}.

Comments

(i) For practical wave guides operating at microwave frequencies d_{10} is typically $\sim 10^2$ m (see Question 7.15).

(ii) The mode structure of a wave guide constructed from a 'good' conductor is hardly perturbed by the finite σ (which an analysis of the exact boundary-value problem would show).

Question 7.15

The wave guide described in Question 7.13 has dimensions $a = 3b = 3$ cm and is made of copper.

(a) Determine the range of frequencies for which only the TE_{10} mode will propagate.

(b) Suppose the operating frequency $f = 8$ GHz. How does the wavelength in the guide compare with the free-space wavelength λ_0? Calculate also the phase and group velocities of the wave.

(c) Taking the breakdown electric field in air as 30 kV cm^{-1}, calculate the maximum power which this (air-filled) guide can transmit in the TE_{10} mode at 8 GHz.

(d) Calculate the attenuation distance d_{10} for the TE_{10} mode at 8 GHz. Assume that $\sigma_{Cu} = 6.5 \times 10^7 \, \Omega^{-1} \, m^{-1}$.

(e) Sketch a 'snapshot' representation of the field distribution inside the wave guide for the TE_{10} mode.

Solution

(a) For travelling waves k must be real. It then follows that given any pair of integers (m, n), the cut-off frequency ω_{mn} for the TE_{mn} mode is

$$\frac{\omega_{mn}^2}{c^2} = \frac{m^2 \pi^2}{a^2} + \frac{n^2 \pi^2}{b^2}.$$

Since $a > b$ the lowest cut-off frequency is given by $m = 1$ and $n = 0$. Therefore $\omega_{10} = \dfrac{\pi c}{a}$ or $f_{10} = \dfrac{c}{2a} = \dfrac{3 \times 10^8}{0.06} = 5 \times 10^9$ Hz. The cut-off frequencies for the TE_{20} mode and the TE_{01} mode may be similarly calculated and are 10 GHz and 15 GHz respectively. So if $\boxed{5 \text{ GHz}} < f < 10$ GHz, only the TE_{10} mode will be excited.

(b) Only the TE_{10} mode is present at a frequency of 8 GHz. Now $\dfrac{\lambda}{\lambda_0} = \dfrac{k_0}{k}$ where $k_0 = \dfrac{\omega}{c} = \dfrac{16\pi \times 10^9}{3 \times 10^8} = 167.6 \, m^{-1}$ and $k = \sqrt{k_0^2 - \pi^2/a^2} = \sqrt{167.6^2 - \pi^2/0.03^2}$ $= 130.8 \, m^{-1}$. Thus $\dfrac{\lambda}{\lambda_0} = \dfrac{167.6}{130.8} \approx \boxed{1.3}$.

From (24) of Question 7.13, $\boxed{v_\phi} = \dfrac{c}{\sqrt{1 - 5^2/8^2}} \simeq \boxed{1.3c}$ and $\boxed{v_g} = \dfrac{c^2}{1.3\,c} \simeq \boxed{0.77c}$.

(c) With $m = 1$ and $n = 0$ the electric-field amplitude given by $(19)_2$ of Question 7.13 is waB_0/π. Equating this to the dielectric breakdown of air and solving for B_0 gives

$$B_0 = \frac{\pi \times 30 \times 10^5}{2\pi(8 \times 10^9) \times (3 \times 10^{-2})} = 6.25 \times 10^{-3}\,\text{T}.$$

For the TE_{10} mode, the power formula (3) derived in Question 7.13 simplifies to

$$P = \frac{\omega k\, a^3 b\, B_0^2}{4\pi^2 \mu_0}. \tag{1}$$

Inserting the numerical values in (1) yields

$$\mathcal{P}_{\text{max}} = \frac{2\pi(8 \times 10^9) \times (130.8) \times (0.03)^3 \times (0.01) \times 0.00625^2}{4\pi^2 \times (4\pi \times 10^{-7})} \approx \boxed{1.4 \times 10^6\,\text{W}}.$$

(d) From (4) and (17) of Question 7.13 we obtain $q = 2$ (since $m = 1$; $n = 0$) and therefore $\gamma = \pi/a = 104.7\,\text{m}^{-1}$. Substituting these results in (3) of Question 7.14 gives

$$d_{10} = \frac{2}{\delta} \frac{\sqrt{\omega^2/c^2 - \pi^2/a^2}}{(\pi^2/a^3 + \omega^2/c^2 b)}. \tag{2}$$

At this frequency

$$\delta = \sqrt{\frac{2}{4\pi \times 10^{-7} \times 6.5 \times 10^7 \times (2\pi \times 8 \times 10^9)}} = 6.98 \times 10^{-7}\,\text{m},$$

and so

$$d_{10} = \frac{2}{6.98 \times 10^{-7}} \left(\frac{\sqrt{167.6^2 - 104.7^2}}{\pi^2/0.03^3 + 167.6^2/0.01} \right) \approx \boxed{118\,\text{m}}.$$

(e) The field distribution is sketched below:

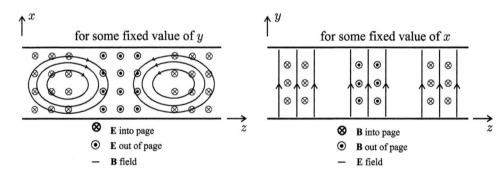

for some fixed value of y	for some fixed value of x
⊗ E into page	⊗ B into page
⊙ E out of page	⊙ B out of page
— B field	— E field

Comments

(i) The cut-off frequency for the TE_{10} mode corresponds to a wavelength (in free space) of $\lambda_0 = 2\pi c/\omega_{10} = 2a$. So the cut-off frequency is that for which the wider dimension is half a free-space wavelength.

(ii) Square wave guides for which $a = b$ exhibit degeneracy: the modes TE_{mn} and TE_{nm} differ by a rotation of $90°$. The cut-off frequencies for these modes are identical, and there is no dominant mode. Any symmetry in cross-section usually gives rise to degeneracy, as we shall see again in the Question 7.16.

Question 7.16 **

Suppose the wave guide of Question 7.12 has a circular cross-section of radius a. Consider the propagation of electromagnetic waves in the guide of the form

$$\psi(r, \theta, z) = R(r)\,\Theta(\theta)\,e^{i(kz - \omega t)}, \tag{1}$$

where R and Θ are functions of a single variable and ψ is either E_z or B_z.

(a) Use the method of separation of variables to show that

$$\psi = \psi_0\, J_m(\gamma r)\,\cos m\theta\, e^{i(kz - \omega t)}, \tag{2}$$

where m is an integer having the values $0, 1, 2, \ldots$, $J_m(\gamma r)$ is a Bessel function of the first kind of order m, and $\gamma^2 = \omega^2/c^2 - k^2$.

(b) Use (6) of Question 7.12 to determine the dispersion relation satisfied by TE modes in this guide. Repeat for TM modes.

Hint: See (III) and (IV) of Appendix G and the related discussion on the roots of $J_m(x)$ and its derivatives.

(c) Use *Mathematica* to calculate the values of s_{mn} and t_{mn} (as defined in Appendix G), then tabulate the cut-off frequencies for TE modes taking $m = 0, 1, 2$ and $n = 1, 2$. Repeat for TM modes.

(d) Use (4) of Question 7.12 to calculate the field components for TE modes. Repeat for TM modes.

Solution

(a) Using ∇^2 for cylindrical polar coordinates $\big($see $(VIII)_4$ of Appendix D$\big)$ gives the wave equation

$$\left[\frac{1}{r}\frac{\partial}{\partial r}\left(r\frac{\partial}{\partial r}\right) + \frac{1}{r^2}\frac{\partial^2}{\partial \theta^2} + \frac{\partial^2}{\partial z^2} - \frac{1}{c^2}\frac{\partial^2}{\partial t^2}\right]\psi = 0,$$

and with solutions of the form (1) we may separate the variables in the usual way (see Question 7.13) to obtain

$$\frac{r}{R}\frac{d}{dr}\left(r\frac{dR}{dr}\right) + \frac{1}{\Theta}\frac{d^2\Theta}{d\theta^2} + \gamma^2 r^2 = 0. \tag{3}$$

Since the middle term in (3) is a function of θ only and the other two terms are independent of θ, we let

$$\frac{1}{\Theta}\frac{d^2\Theta}{d\theta^2} = -m^2. \tag{4}$$

The separation constant in (4) must have the following properties:

1. it must be negative to produce periodic functions, and
2. it must be an integer since $\psi(r,\theta)$ must obviously equal $\psi(r, \theta + 2m\pi)$.

The general solution of (4) is thus

$$\Theta(\theta) = C_1 \sin m\theta + C_2 \cos m\theta, \tag{5}$$

where the C_i are arbitrary constants and $m = 0, 1, 2, \ldots$ Substituting (5) in (3) gives

$$r\frac{d}{dr}\left(r\frac{dR}{dr}\right) + (\gamma^2 r^2 - m^2)R = 0,$$

which is a Bessel equation having the general solution

$$R(r) = C_3 J_m(\gamma r) + C_4 N_m(\gamma r), \tag{6}$$

discussed in Appendix G. Here $J_m(\gamma r)$ and $N_m(\gamma r)$ are mth order Bessel functions of the first and second kind respectively and C_3 and C_4 are arbitrary constants. The general solution of (3) which follows from (1), (5) and (6) is thus

$$\psi = (C_1 \sin m\theta + C_2 \cos m\theta)(C_3 J_m(\gamma r) + C_4 N_m(\gamma r)) e^{i(kz - \omega t)}. \tag{7}$$

The values of the C_i are determined as follows: since the choice of the origin of θ is arbitrary, we make it such that $C_1 = 0$. Now $N_m(\gamma r) \to -\infty$ as $r \to 0$ and, because all field components must be finite everywhere inside the wave guide, this requires that $C_4 = 0$. Hence (2), where the constant $\psi_0 = C_2 C_3$ is determined by the boundary conditions.

(b) TE modes

ψ in (2) represents B_z, and the boundary condition $\partial B_z/\partial r = 0$ at $r = a$ requires $J'_m(\gamma a) = 0$ or $\gamma = t_{mn}/a$ (recall from Appendix G that t_{mn} are the roots of $J'_m(\gamma r)$). But $\gamma^2 = \omega^2/c^2 - k^2$, and so

$$\omega^2 = c^2\left[\frac{t_{mn}^2}{a^2} + k^2\right]. \tag{8}$$

Associated with each value of t_{mn} is the corresponding TE_{mn} mode.

TM modes

ψ in (2) represents E_z, and the boundary condition $E_z = 0$ at $r = a$ requires $J_m(\gamma a) = 0$ or $\gamma = s_{mn}/a$ (recall from Appendix G that s_{mn} are the roots of $J_m(\gamma r)$). Now

$$\omega^2 = c^2 \left[\frac{s_{mn}^2}{a^2} + k^2 \right], \tag{9}$$

and associated with each value of s_{mn} is the corresponding TM_{mn} mode.

(c) The roots of $J_m(u)$ and $J'_m(u)$ are conveniently calculated using *Mathematica*'s `BesselJZero` function.[‡] The first few are tabulated below.

	s_{mn}			t_{mn}	
$m\backslash n$	1	2	$m\backslash n$	1	2
0	2.405	5.520	0	3.832	7.016
1	3.832	7.016	1	1.841	5.331
2	5.136	8.417	2	3.054	6.706

Substituting the values for s_{mn} and t_{mn} in (8) and (9) and putting $k = 0$ give the following cut-off frequencies:

mode	$f_{mn} \times a$ (GHz cm)	mode	$f_{mn} \times a$ (GHz cm)
TE_{11}	8.79	TE_{12}	25.45
TM_{01}	11.48	TM_{02}	26.36
TE_{21}	14.58	TE_{22}	32.02
$\left.\begin{array}{c}\text{TE}_{01}\\\text{TM}_{11}\end{array}\right\}$	18.30	$\left.\begin{array}{c}\text{TE}_{02}\\\text{TM}_{12}\end{array}\right\}$	33.50
TM_{21}	24.52	TM_{22}	40.19

(d) ## TE modes

With $B_z = B_0 \, J_m(\gamma r) \cos m\theta \, e^{i(kz - \omega t)}$ and $\nabla_t = \frac{\partial}{\partial r} \hat{r} + \frac{1}{r} \frac{\partial}{\partial \theta} \hat{\theta}$ we obtain

$$\nabla_t B_z = B_0 \left(\gamma J'_m(\gamma r) \cos m\theta \, \hat{r} - \frac{m}{r} J_m(\gamma r) \sin m\theta \, \hat{\theta} \right) e^{i(kz - \omega t)}, \tag{10}$$

where

$$J'_m(\gamma r) \text{ means } \left[\frac{d}{du} J_m(u) \right]_{u = \gamma r}.$$

[‡]For example, `N[BesselJZero[2, Range[3]],6]` returns numerical values for the first three zeros of $J_2(u)$ to six-digit precision. The roots of $J'_m(u)$ may be calculated in different ways such as: `FindRoot[D[BesselJ[2, u], u], {u, 2}]`.

Substituting (10) in $\mathbf{E_t} = -\dfrac{i\omega}{\gamma^2} \hat{\mathbf{z}} \times \boldsymbol{\nabla}_t B_z$ and taking the real part give

$$
\left.
\begin{aligned}
E_r &= \frac{\omega B_0}{\gamma^2} \frac{m}{r} J_m(\gamma r) \sin m\theta \, \sin(kz - \omega t) \\[2mm]
E_\theta &= \frac{\omega B_0}{\gamma} J_m'(\gamma r) \cos m\theta \, \sin(kz - \omega t) \\[2mm]
B_r &= -\frac{kB_0}{\gamma} J_m'(\gamma r) \cos m\theta \, \sin(kz - \omega t) \\[2mm]
B_\theta &= \frac{kB_0}{\gamma^2} \frac{m}{r} J_m(\gamma r) \sin m\theta \, \sin(kz - \omega t)
\end{aligned}
\right\},
$$

where $\gamma = t_{mn}/a$ and $k = (\omega^2/c^2 - \gamma^2)^{1/2}$.

<mark>TM modes</mark>

Here $E_z = E_0 \, J_m(\gamma r) \cos m\theta \, e^{i(kz - \omega t)}$, and now

$$
\boldsymbol{\nabla}_t E_z = E_0 \left(\gamma J_m'(\gamma r) \cos m\theta \, \hat{\mathbf{r}} - \frac{m}{r} J_m(\gamma r) \sin m\theta \, \hat{\boldsymbol{\theta}} \right) e^{i(kz - \omega t)}. \qquad (11)
$$

Substituting (11) into $\mathbf{E_t} = \dfrac{ik}{\gamma^2} \boldsymbol{\nabla}_t E_z$ and taking the real part give

$$
\left.
\begin{aligned}
E_r &= -\frac{kE_0}{\gamma^2} J_m'(\gamma r) \cos m\theta \, \sin(kz - \omega t) \\[2mm]
E_\theta &= \frac{kE_0}{\gamma^2} \frac{m}{r} J_m(\gamma r) \sin m\theta \, \sin(kz - \omega t) \\[2mm]
B_r &= -\frac{k^2 E_0}{\gamma^2 c^2 \omega} \frac{m}{r} J_m(\gamma r) \sin m\theta \, \sin(kz - \omega t) \\[2mm]
B_\theta &= -\frac{k^2 E_0}{\gamma^2 c^2 \omega} J_m'(\gamma r) \cos m\theta \, \sin(kz - \omega t)
\end{aligned}
\right\},
$$

where $\gamma = s_{mn}/a$ and other symbols have their previous meaning.

Comments

(i) Unlike a rectangular wave guide, the dominant mode here is TE_{11} not TE_{10}. It is easily shown that the free-space wavelength corresponding to the cut-off frequency is approximately half the circumference of this guide.

(ii) As was the case for a rectangular wave guide, the symmetrical cross-section of this guide leads to degenerate modes, such as TE_{01} and TM_{11}, TE_{02} and TM_{12}, etc.

8

The electromagnetic potentials

In earlier chapters of this book, we derived the static scalar and vector potentials and then used them to calculate, for example, time-independent **E**- and **B**-fields. Now we extend the discussion and relate the time-dependent potentials $\Phi(\mathbf{r}, t)$ and $\mathbf{A}(\mathbf{r}, t)$ to the time-dependent fields $\mathbf{E}(\mathbf{r}, t)$ and $\mathbf{B}(\mathbf{r}, t)$. In doing this, we will discover that the electromagnetic potentials are defined up to a gauge transformation only. See Question 8.1 for further discussion of this concept, and the associated notion of gauge invariance.

Classical electrodynamics describes **E**- and **B**-fields in free space and in matter, where the potentials are treated as convenient mathematical devices. However, at the microscopic level $\Phi(\mathbf{r}, t)$ and $\mathbf{A}(\mathbf{r}, t)$ are not simply conveniences: they are crucial for further development of the theory. For example, the potentials play an indispensable role in the quantization of the electromagnetic field which leads to the photon, and in the theory of fundamental interactions between particles and fields. Readers might obtain a glimpse of this in Questions 8.16–8.19 which provide a stepping stone from the classical to the semi-classical theory.

The retarded potentials are important results and they appear first in Question 8.2. Also considered in this chapter are the Liénard–Wiechert potentials for an arbitrarily moving point charge, together with its electromagnetic field. We conclude by deriving multipole expansions for $\Phi(\mathbf{r}, t)$ and $\mathbf{A}(\mathbf{r}, t)$. These will form the starting point for our treatment of electromagnetic radiation later on in Chapter 11.

Question 8.1

(a) Use Maxwell's equations to show that in electrodynamics

$$\mathbf{E}(\mathbf{r}, t) \;=\; -\nabla\Phi - \frac{\partial \mathbf{A}}{\partial t} \qquad \text{and} \qquad \mathbf{B}(\mathbf{r}, t) \;=\; \nabla \times \mathbf{A}, \qquad (1)$$

where $\Phi(\mathbf{r}, t)$ and $\mathbf{A}(\mathbf{r}, t)$ are the electric scalar potential and magnetic vector potential respectively.

Hint: Use the vector identity $\nabla \cdot (\nabla \times \mathbf{F}) = 0$.

(b) Prove that the potentials leading to the electric and magnetic fields in (1) are not uniquely defined.

Hint: Use the vector identity $\nabla \times \nabla f = 0$.

Solved Problems in Classical Electromagnetism. J. Pierrus, Oxford University Press (2018).
© J. Pierrus. DOI: 10.1093/oso/9780198821915.001.0001

Solution

(a) Because of (2) of Question 1.15, the Maxwell equation $\nabla \cdot \mathbf{B} = 0$ can always be written as $\mathbf{B} = \nabla \times \mathbf{A}$. Substituting this in $\nabla \times \mathbf{E} = -\partial \mathbf{B}/\partial t$ gives

$$\nabla \times \mathbf{E} = -\frac{\partial}{\partial t}(\nabla \times \mathbf{A}) = -\nabla \times \left(\frac{\partial \mathbf{A}}{\partial t}\right)$$

or

$$\nabla \times \left(\mathbf{E} + \frac{\partial \mathbf{A}}{\partial t}\right) = 0. \tag{2}$$

The quantity in brackets in (2) is irrotational, and can therefore be expressed as the gradient of a scalar field (see (2) of Question 1.14). Hence $\mathbf{E} + \partial \mathbf{A}/\partial t = -\nabla \Phi$, which is $(1)_1$. (From a mathematical point of view, it is immaterial whether the gradient is taken with a positive or negative sign. The latter choice is due to a sign convention of physics.)

(b) Motivated by the hint, we take an arbitrary yet suitably differentiable scalar field $\chi = \chi(\mathbf{r}, t)$, and make the transformation

$$\mathbf{A} \rightarrow \mathbf{A}' = \mathbf{A} + \nabla \chi. \tag{3}$$

Since $\nabla \times \nabla \chi = 0$, the curl of $\mathbf{A}'$ gives the same magnetic field $\mathbf{B}(\mathbf{r}, t)$ as the curl of $\mathbf{A}$. However, the effect of the transformation (3) changes $\mathbf{E}(\mathbf{r}, t)$, as follows:

$$\mathbf{E} = -\nabla \Phi - \frac{\partial \mathbf{A}'}{\partial t} = -\nabla \Phi - \frac{\partial}{\partial t}(\mathbf{A} + \nabla \chi) = -\nabla \left(\Phi + \frac{\partial \chi}{\partial t}\right) - \frac{\partial \mathbf{A}}{\partial t}. \tag{4}$$

Because the electric field must also remain unchanged when $\nabla \chi$ is added to $\mathbf{A}$, (4) shows that (3) must be accompanied by the simultaneous transformation

$$\Phi \rightarrow \Phi' = \Phi - \frac{\partial \chi}{\partial t}. \tag{5}$$

Clearly, (3) and (5) imply that Φ and $\mathbf{A}$ are not unique: there are infinitely many ways of choosing χ (and hence the potentials) which will produce the required $\mathbf{E}$- and $\mathbf{B}$-fields.

Comments

(i) The transformation

$$\left. \begin{array}{l} \Phi \rightarrow \Phi' = \Phi - \dfrac{\partial \chi}{\partial t} \\[2mm] \mathbf{A} \rightarrow \mathbf{A}' = \mathbf{A} + \nabla \chi \end{array} \right\} \tag{6}$$

is known as a *gauge transformation* and $\chi(\mathbf{r}, t)$ is called the *gauge function*. The invariance of the fields $\mathbf{E}$ and $\mathbf{B}$ (and hence Maxwell's equations) under the gauge transformation (6) is known as *gauge invariance*.

(ii) The non-uniqueness in the definitions of Φ and $\mathbf{A}$ is often exploited in order 'to impose conditions on the potentials without affecting the results of measurements made on the system being studied. Such choices are commonly called gauges, and perhaps the most common of these are the Lorentz [*sic*] gauge and the Coulomb gauge'.[1] We discuss both of these gauges further in Question 8.2.

(iii) Notice that for time-independent fields, $(1)_1$ reduces to $\mathbf{E}(\mathbf{r}) = -\nabla\Phi(\mathbf{r})$ which is the usual electrostatic result. Electrostatic fields possess an obvious invariance: they are unaffected by the transformation $\Phi \to \Phi' = \Phi + \Phi_0$, where Φ_0 is a constant. Because Φ changes by the *same* amount everywhere in space, this invariance is known as a 'global' gauge invariance. However, the invariance discussed in Comment (i) above is of a more general kind, because here Φ and $\mathbf{A}$ can be changed by *different* amounts at each point in space-time. This invariance, known as 'local' gauge invariance, is an important concept in understanding fundamental interactions.[2,3]

(iv) We conclude these comments with the following remarks:

classically, gauge invariance does not seem to be very profound. It merely represents the freedom in the choice of the electromagnetic potential for the description of electromagnetic phenomena. And we can describe the classical system completely in terms of the physically measurable $\mathbf{E}$ and $\mathbf{B}$ fields without any arbitrariness. However, when we go from classical to quantum systems, the situation changes. The quantum mechanics of electromagnetic interactions are most simply formulated in terms of the electromagnetic potentials Φ and $\mathbf{A}$ rather than the $\mathbf{E}$ and $\mathbf{B}$ fields. Thus gauge invariance plays an essential role in the quantum description of electromagnetism. ... The term *eichinvarianz* (gauge invariance), meaning an invariance under the change of the scale (the gauge), was coined by Weyl in 1919 in the framework of his attempt to geometrize the electromagnetic interaction and to construct in this way a unified geometrical theory of gravity and electromagnetism.[4]

Question 8.2

Show that Maxwell's four first-order equations involving the fields can be expressed as the following two second-order equations involving the potentials:

$$\left.\begin{array}{c} \nabla^2\Phi + \dfrac{\partial}{\partial t}\left(\nabla \cdot \mathbf{A}\right) = -\dfrac{\rho}{\epsilon_0} \\[2ex] \nabla^2\mathbf{A} - \epsilon_0\mu_0\dfrac{\partial^2 \mathbf{A}}{\partial t^2} = -\mu_0\mathbf{J} + \nabla\left[\nabla \cdot \mathbf{A} + \epsilon_0\mu_0\dfrac{\partial\Phi}{\partial t}\right] \end{array}\right\}. \tag{1}$$

[1]   O. L. Brill and B. Goodman, 'Causality in the Coulomb gauge', *American Journal of Physics*, vol. 35, p. 832, 1967.

[2]   See, for example, G. 't Hooft, 'Gauge theories of the forces between elementary particles', *Scientific American*, vol. 242, pp. 90–116, June 1980.

[3]   L. O'Raifeartaigh, *The dawning of gauge theory*. Princeton: Princeton University Press, 1997.

[4]   T. P. Cheng and L. Li, 'Resource letter: G1-1 gauge invariance', *American Journal of Physics*, vol. 56, pp. 586–600, July 1988.

Solution

In (a) of Question 8.1 we showed that Maxwell's homogeneous equations give

$$\left.\begin{array}{l} \mathbf{E} = -\nabla\Phi - \dfrac{\partial\mathbf{A}}{\partial t} \\[2mm] \mathbf{B} = \nabla\times\mathbf{A} \end{array}\right\}. \tag{2}$$

Combining (2) with Maxwell's inhomogeneous equations leads to (1), as we now show.

Maxwell–Gauss

$$\nabla\cdot\mathbf{E} = -\nabla\cdot\left(\nabla\Phi + \frac{\partial\mathbf{A}}{\partial t}\right) = \rho/\epsilon_0 \text{ which is } (1)_1 \text{ because } \nabla\cdot\frac{\partial\mathbf{A}}{\partial t} = \frac{\partial}{\partial t}(\nabla\cdot\mathbf{A}).$$

Maxwell–Ampère

$$\nabla\times\mathbf{B} = \mu_0\left(\mathbf{J} + \epsilon_0\frac{\partial\mathbf{E}}{\partial t}\right)$$

$$\nabla\times(\nabla\times\mathbf{A}) = \mu_0\left[\mathbf{J} - \epsilon_0\frac{\partial}{\partial t}\left(\nabla\phi + \frac{\partial\mathbf{A}}{\partial t}\right)\right]$$

$$-\nabla^2\mathbf{A} + \nabla(\nabla\cdot\mathbf{A}) = \mu_0\left[\mathbf{J} - \epsilon_0\frac{\partial}{\partial t}(\nabla\phi) - \epsilon_0\frac{\partial^2\mathbf{A}}{\partial t^2}\right]$$

because of the vector identity (11) of Question 1.8. Interchanging the order of differentiation and rearranging terms give $(1)_2$.

Comments

(i) It is often convenient to exploit the latitude in the definitions of the electromagnetic potentials by specifying an appropriate form for $\mathbf{A}$ (or more usually the divergence of $\mathbf{A}$) to decouple the pair of equations in (1) above. Two commonly encountered gauges are:

☞ $\nabla\cdot\mathbf{A} = 0,$ (3)

known variously as the Coulomb, radiation or transverse gauge.

☞ $\nabla\cdot\mathbf{A} = -\epsilon_0\mu_0\dfrac{\partial\Phi}{\partial t},$ (4)

known as the Lorenz gauge.

We now consider these gauges in turn.

☞ Coulomb gauge

With $\nabla \cdot \mathbf{A} = 0$ the two coupled equations in (1) become

$$\left.\begin{array}{rcl} \nabla^2 \Phi & = & -\rho/\epsilon_0 \\[2mm] \nabla^2 \mathbf{A} - \epsilon_0\mu_0\dfrac{\partial^2 \mathbf{A}}{\partial t^2} & = & -\mu_0\left[\mathbf{J} - \epsilon_0\nabla\left(\dfrac{\partial \Phi}{\partial t}\right)\right] \end{array}\right\} . \tag{5}$$

Equation $(5)_1$ is Poisson's equation. By analogy with electrostatics, and recalling that here both ρ and Φ are also functions of time, we write down the solution

$$\Phi(\mathbf{r}, t) = \frac{1}{4\pi\epsilon_0} \int_v \frac{\rho(\mathbf{r}', t)}{|\mathbf{r} - \mathbf{r}'|}\, dv' \tag{6}$$

with $\Phi = 0$ at infinity. Thus, $\Phi(\mathbf{r}, t)$ is the instantaneous Coulomb potential (hence the name of this gauge) and the contribution of the scalar potential to $\mathbf{E}$ via $-\nabla\Phi$ takes the form of a superposition of $1/r^2$ Coulomb-like fields. The vector potential which satisfies $(5)_2$, on the other hand, does not possess a simple solution. In the absence of sources ($\rho = 0$; $\mathbf{J} = 0$) the scalar potential $\Phi = 0$. Then the vector potential satisfies the homogeneous wave equation

$$\nabla^2 \mathbf{A} - \epsilon_0\mu_0\frac{\partial^2 \mathbf{A}}{\partial t^2} = 0. \tag{7}$$

The field is then given by $\mathbf{A}$ alone:

$$\mathbf{E}(\mathbf{r}, t) = -\frac{\partial \mathbf{A}}{\partial t} \qquad \text{and} \qquad \mathbf{B}(\mathbf{r}, t) = \nabla \times \mathbf{A}. \tag{8}$$

Equation (8) can be used, for example, to derive radiation fields that are known to vary as $1/r$ (see, for example, Question 11.6). This explains why $\nabla \cdot \mathbf{A} = 0$ is also known as the 'radiation' gauge. The origin of the alternative name (transverse gauge) will become apparent in Question 8.11.

☞ Lorenz gauge

Substituting $\nabla \cdot \mathbf{A} = -\epsilon_0\mu_0\dfrac{\partial \Phi}{\partial t}$ in (1) gives the inhomogeneous wave equations

$$\left.\begin{array}{rcl} \nabla^2 \Phi - \epsilon_0\mu_0\dfrac{\partial^2 \Phi}{\partial t^2} & = & -\dfrac{\rho}{\epsilon_0} \\[3mm] \nabla^2 \mathbf{A} - \epsilon_0\mu_0\dfrac{\partial^2 \mathbf{A}}{\partial t^2} & = & -\mu_0\mathbf{J} \end{array}\right\} , \tag{9}$$

whose solutions are the retarded potentials

$$\left.\begin{aligned}\Phi(\mathbf{r},t) &= \frac{1}{4\pi\epsilon_0}\int_v \frac{\rho(\mathbf{r}',t-|\mathbf{r}-\mathbf{r}'|/c)}{|\mathbf{r}-\mathbf{r}'|}\,dv'\\[2mm]\mathbf{A}(\mathbf{r},t) &= \frac{\mu_0}{4\pi}\int_v \frac{\mathbf{J}(\mathbf{r}',t-|\mathbf{r}-\mathbf{r}'|/c)}{|\mathbf{r}-\mathbf{r}'|}\,dv'\end{aligned}\right\}. \tag{10}$$

The Lorenz gauge is one of the most important and commonly used gauges 'first because it leads to wave equations (9) which treat Φ and $\mathbf{A}$ on equivalent footings, and second because it is a concept independent of the coordinate system chosen and so fits naturally into the considerations of special relativity'.[5] For this reason it is sometimes useful to introduce the four-dimensional operator $\Box = \left(\dfrac{\partial}{\partial x}, \dfrac{\partial}{\partial y}, \dfrac{\partial}{\partial z}, \dfrac{1}{ic}\dfrac{\partial}{\partial t}\right)$. The scalar product of $\Box$ with itself is

$$\Box^2 = \frac{\partial^2}{\partial x^2} + \frac{\partial^2}{\partial y^2} + \frac{\partial^2}{\partial z^2} - \frac{1}{c^2}\frac{\partial^2}{\partial t^2} = \nabla^2 - \frac{1}{c^2}\frac{\partial^2}{\partial t^2}, \tag{11}$$

known as the d'Alembertian. In this notation, the inhomogeneous wave equations can be expressed in the compact form

$$\left.\begin{aligned}\Box^2\Phi &- -\frac{\rho}{\epsilon_0}\\[2mm]\Box^2\mathbf{A} &= -\mu_0\mathbf{J}\end{aligned}\right\}. \tag{12}$$

(ii) The Lorenz condition (4) is often written in the equivalent form

$$\nabla\cdot\mathbf{A} + \frac{1}{c^2}\frac{\partial\Phi}{\partial t} = 0. \tag{13}$$

Some textbooks (especially the older editions) incorrectly attribute the name of this gauge to the Dutch physicist H. A. Lorentz instead of the Danish physicist L. V. Lorenz. This point is clarified in a note of explanation in Ref. [6].

(iii) The inhomogeneous wave equations (9) have both a retarded time solution, given by (10), and an advanced time solution.‡ The latter—which has some obvious implications relating to causality—is not considered further here.

(iv) The term $\epsilon_0\mu_0\nabla(\partial\Phi/\partial t)$ in (5) is clearly irrotational and can, in principle, be cancelled by a corresponding part of the current density. See Question 8.11.

---

‡This would require that a signal observed at a field point at $\mathbf{r}$ at time t originate from a source located at $\mathbf{r}'$ and emitted at a time $t+|\mathbf{r}-\mathbf{r}'|/c$ in the *future*.

---

[5]   J. D. Jackson, *Classical electrodynamics*, Chap. 6, p. 241. New York: Wiley, 3rd edn, 1998.
[6]   J. D. Jackson, *Classical electrodynamics*, Chap. 6, p. 294. New York: Wiley, 3rd edn, 1998.

(v) Although the scalar potential in the Coulomb gauge is unretarded and thus at variance with the finite speed of propagation of signals, this has no effect on observable quantities like the **E**- and **B**-fields (Φ is not a measurable quantity).[1] Ref. [7] explains that 'somehow it is built into the vector potential, that whereas Φ instantaneously reflects all changes in ρ, the combination $-\nabla\Phi - \partial\mathbf{A}/\partial t$ does *not*. **E** will change only after sufficient time has elapsed for the "news" to arrive'.[7]

Question 8.3

Consider a point charge q at rest at the origin. An obvious choice of potentials is

$$\left.\begin{array}{l} \Phi = \dfrac{1}{4\pi\epsilon_0}\dfrac{q}{r} \\[2mm] \mathbf{A} = 0 \end{array}\right\}. \tag{1}$$

(a) Show that an equivalent (although less obvious) set of potentials is

$$\left.\begin{array}{l} \Phi' = 0 \\[2mm] \mathbf{A}' = -\dfrac{1}{4\pi\epsilon_0}\dfrac{qt\mathbf{r}}{r^3} \end{array}\right\}. \tag{2}$$

(b) Are the potentials given by (1) in the ☞ Coulomb gauge, ☞ Lorenz gauge?

(c) Repeat (b) for the potentials given by (2).

Solution

(a) Putting $\Phi' = 0$ in $(6)_1$ of Question 8.1 gives $\partial\chi/\partial t = \Phi = \dfrac{1}{4\pi\epsilon_0}\dfrac{q}{r}$. It then follows that $\chi = \dfrac{1}{4\pi\epsilon_0}\dfrac{qt}{r}$ (where, because of the arbitrary nature of the gauge function, the constant of integration is chosen equal to zero). Transforming the vector potential using $(6)_2$ of Question 8.1 yields $(2)_2$ since $\nabla r^{-1} = -\mathbf{r}/r^3$.

(b) Clearly, $\nabla \cdot \mathbf{A} = 0$ and $\partial\Phi/\partial t = 0$. So the potentials satisfy both the Coulomb and Lorenz gauge conditions (see (3) and (4) of Question 8.2 respectively).

(c) Here $\nabla \cdot \mathbf{A}' = (4\pi\epsilon_0)^{-1}qt\nabla^2 r^{-1} = -\epsilon_0^{-1}qt\delta(\mathbf{r})$ because of (1) of Question 1.21. Also, $\partial\Phi'/\partial t = 0$. The gauge conditions $\nabla \cdot \mathbf{A}' = 0$ and $\nabla \cdot \mathbf{A}' = -\partial\Phi'/\partial t$ are both satisfied everywhere *except* at the origin. So, for this reason, Φ' and $\mathbf{A}'$ are in neither the Coulomb nor the Lorenz gauge.

[7] D. J. Griffiths, *Introduction to electrodynamics*, Chap. 10, p. 421. New York: Prentice Hall, 3 edn, 1999.

Comment

This simple example illustrates two important points:

☞ these gauges are not mutually exclusive. That is, the potentials can satisfy two (or more) gauges simultaneously (see (b) above).

☞ Although the Coulomb and Lorenz gauges are often used, the electromagnetic potentials are not *required* to satisfy either of these gauges per se (see (c) above). Indeed, one sometimes encounter other gauges in the literature, e.g., the Barron–Gray gauge.

Question 8.4

(a) A vector potential for a uniform magnetic field **B** is:

$$\mathbf{A}(\mathbf{r}) \;=\; -\tfrac{1}{2}(\mathbf{r} \times \mathbf{B}) \tag{1}$$

(see (b) of Question 1.25). Show that (1) satisfies the Coulomb gauge condition given by (3) of Question 8.2.

(b) Suppose $\mathbf{B} = (0, 0, 1)B$, $\mathbf{A} = (0, x, 0)B$ and $\mathbf{A}' = (-y, 0, 0)B$. Determine a gauge function $\chi(\mathbf{r})$ for the transformation $\mathbf{A} \to \mathbf{A}'$.

Solution

(a) Clearly, $\nabla \cdot \mathbf{A} = -\tfrac{1}{2}\varepsilon_{ijk}B_k\nabla_i r_j = -\tfrac{1}{2}\varepsilon_{ijk}B_k\delta_{ij} = -\tfrac{1}{2}\varepsilon_{iik}B_k$. Now because $\varepsilon_{iik} = 0$ we obtain $\nabla \cdot \mathbf{A} = 0$ which is the Coulomb gauge condition.

(b) By definition, $\mathbf{A}' = \mathbf{A} + \nabla\chi$ and so $\nabla\chi = \mathbf{A}' - \mathbf{A} = \left(\dfrac{\partial\chi}{\partial x}, \dfrac{\partial\chi}{\partial y}, \dfrac{\partial\chi}{\partial z} \right) = (-y, -x, 0)$.

Integration gives

$$\left.
\begin{aligned}
\chi(\mathbf{r}) &= -y(x - x_0) + f(y, z) \\
\chi(\mathbf{r}) &= -x(y - y_0) + g(x, z) \\
\chi(\mathbf{r}) &= z_0 + h(x, y)
\end{aligned}
\right\} , \tag{2}$$

where x_0, y_0 and z_0 are constants and f, g and h are arbitrary functions. The simplest solution of (2) is

$$\left.
\begin{aligned}
x_0 = y_0 &= z_0 = 0 \\
f(y, z) = g(x, z) &= 0 \\
h(x, y) &= -xy
\end{aligned}
\right\} ,$$

and a suitable gauge function is therefore

$$\chi(\mathbf{r}) \;=\; -xy.$$

Question 8.5

Deduce the condition that must be imposed on a gauge function $\chi(\mathbf{r}, t)$ for the magnetic flux ϕ through an arbitrary surface s to be gauge-invariant.

Solution

By definition $\phi = \displaystyle\int_s \mathbf{B} \cdot d\mathbf{a} = \int_s (\nabla \times \mathbf{A}) \cdot d\mathbf{a} = \oint_c \mathbf{A} \cdot d\mathbf{l}$, because of Stokes's theorem.

Under the gauge transformation (6) of Question 8.1 we have

$$\phi' = \oint_c \mathbf{A}' \cdot d\mathbf{l} = \oint_c (\mathbf{A} + \nabla\chi) \cdot d\mathbf{l}$$

$$= \phi + \oint_c \nabla\chi \cdot d\mathbf{l}$$

$$= \phi + \oint_c d\chi. \tag{1}$$

Now $\phi' = \phi$ requires $\displaystyle\oint_c d\chi = 0$, and the restriction we seek is that χ must be a *single-valued* function.

Question 8.6

(a) Find the condition which a gauge function $\chi(\mathbf{r})$ must satisfy in order to preserve the Coulomb gauge condition (see (3) of Question 8.2).

(b) In a certain gauge, the vector potential $\mathbf{A}(\mathbf{r})$ satisfies $\nabla \cdot \mathbf{A} = f(\mathbf{r}) \neq 0$. Determine the condition which a gauge function $\chi(\mathbf{r})$ must satisfy in order to give $\nabla \cdot \mathbf{A}' = 0$. Express your answer in terms of $f(\mathbf{r})$.

Solution

(a) $\nabla \cdot \mathbf{A}' = \nabla \cdot (\mathbf{A} + \nabla\chi) = \nabla \cdot \mathbf{A} + \nabla^2\chi = 0$. Since $\nabla \cdot \mathbf{A} = 0$ ($\mathbf{A}$ satisfies the Coulomb gauge) we obtain $\nabla^2\chi = 0$. Thus $\chi(\mathbf{r})$ must be a solution of Laplace's equation.

(b) Now $\nabla^2\chi = -f(\mathbf{r})$, where $\chi(\mathbf{r})$ must satisfy Poisson's equation. Then, making an analogy with the solution of the electrostatic equation $\nabla^2\Phi = -\epsilon_0^{-1}\rho(\mathbf{r})$ gives

$$\chi(\mathbf{r}) = \frac{1}{4\pi} \int_v \frac{f(\mathbf{r}')}{|\mathbf{r} - \mathbf{r}'|} dv'.$$

Question 8.7

(a) Find the condition which a gauge function $\chi(\mathbf{r})$ must satisfy in order to preserve the condition for the Lorenz gauge (see (4) of Question 8.2).

(b) In a certain gauge $\mathbf{A}$ and Φ satisfy

$$\nabla \cdot \mathbf{A} + \epsilon_0 \mu_0 \frac{\partial \Phi}{\partial t} = f(\mathbf{r}, t) \neq 0. \tag{1}$$

Determine the condition on the gauge function $\chi(\mathbf{r}, t)$, given that the transformed potentials $\mathbf{A}'$ and Φ' must remain in Lorenz gauge.

Solution

(a) Substituting (6) of Question 8.1 in $\nabla \cdot \mathbf{A}' + \epsilon_0 \mu_0 \frac{\partial \Phi'}{\partial t} = 0$ yields

$$\nabla \cdot \left(\mathbf{A} + \nabla \chi \right) + \epsilon_0 \mu_0 \frac{\partial}{\partial t} \left(\Phi - \frac{\partial \chi}{\partial t} \right) = \left(\nabla \cdot \mathbf{A} + \epsilon_0 \mu_0 \frac{\partial \Phi}{\partial t} \right) + \left(\nabla^2 \chi - \epsilon_0 \mu_0 \frac{\partial^2 \chi}{\partial t^2} \right) = 0.$$

Now $\nabla \cdot \mathbf{A} + \epsilon_0 \mu_0 \partial \Phi / \partial t = 0$ because $\mathbf{A}$ and Φ satisfy the Lorenz gauge condition. Hence the restriction on $\chi(\mathbf{r}, t)$ is that it should satisfy the homogeneous wave equation

$$\nabla^2 \chi - \epsilon_0 \mu_0 \frac{\partial^2 \chi}{\partial t^2} = 0. \tag{2}$$

(b) Here $\nabla^2 \chi - \epsilon_0 \mu_0 \frac{\partial^2 \chi}{\partial t^2} = -f(\mathbf{r}, t)$, which is the inhomogeneous wave equation whose solutions

$$\chi(\mathbf{r}, t) = \frac{1}{4\pi} \int_v \frac{f(\mathbf{r}', t \mp |\mathbf{r} - \mathbf{r}'|/c))}{|\mathbf{r} - \mathbf{r}'|} dv' \tag{3}$$

are discussed in Comments (i) and (iii) of Question 8.2. Equation (3) is the condition we seek.

Comment

The transformation

$$\left. \begin{array}{l} \Phi' = \Phi - \partial \chi / \partial t \\ \mathbf{A}' = \mathbf{A} + \nabla \chi \end{array} \right\}, \tag{4}$$

where the gauge function $\chi(\mathbf{r}, t)$ is constrained to satisfy some condition—like (2) or (3)—in order to preserve the original gauge, is known as a *restricted* gauge transformation.

Question 8.8**

The retarded potentials discussed in Question 8.2 are

$$
\left.
\begin{array}{l}
\Phi(\mathbf{r}, t) \;=\; \dfrac{1}{4\pi\epsilon_0} \displaystyle\int_v \dfrac{\rho(\mathbf{r}', t - R/c)}{R}\, dv' \\[3ex]
\mathbf{A}(\mathbf{r}, t) \;=\; \dfrac{\mu_0}{4\pi} \displaystyle\int_v \dfrac{\mathbf{J}(\mathbf{r}', t - R/c)}{R}\, dv'
\end{array}
\right\},
\tag{1}
$$

where $R = |\mathbf{r} - \mathbf{r}'|$.

(a) Show that for the retarded source densities, the familiar continuity equation ((1) of Question 7.1) is modified to read

$$
(\boldsymbol{\nabla}' + \boldsymbol{\nabla}) \cdot \mathbf{J}(\mathbf{r}', t - R/c) \;+\; \frac{\partial}{\partial t}\rho(\mathbf{r}', t - R/c) \;=\; 0.
\tag{2}
$$

(b) Hence prove that the potentials $(1)_1$ and $(1)_2$ satisfy the Lorenz condition

$$
\boldsymbol{\nabla} \cdot \mathbf{A} \;+\; \epsilon_0\mu_0 \frac{\partial \Phi}{\partial t} \;=\; 0.
\tag{3}
$$

Solution

(a) The continuity equation is

$$
\boldsymbol{\nabla}'_{\mathrm{e}} \cdot \mathbf{J}(\mathbf{r}', t - R/c) \;+\; \frac{\partial}{\partial t}\rho(\mathbf{r}', t - R/c) \;=\; 0,
\tag{4}
$$

where the subscript e denotes differentiation with respect to the 'explicit' dependence of $\mathbf{J}$ on $\mathbf{r}'$. It excludes differentiation $\boldsymbol{\nabla}'_{\mathrm{i}}$ with respect to the 'implicit' dependence of $\mathbf{J}$ on $\mathbf{r}'$ contained in the retarded time (i.e. in R). Now $\boldsymbol{\nabla}'$ involves both of these

$$
\boldsymbol{\nabla}' \cdot \mathbf{J} \;=\; (\boldsymbol{\nabla}'_{\mathrm{e}} + \boldsymbol{\nabla}'_{\mathrm{i}}) \cdot \mathbf{J}.
\tag{5}
$$

Also, $\partial R/\partial \mathbf{r}' = -\partial R/\partial \mathbf{r}$, and therefore

$$
\boldsymbol{\nabla}'_{\mathrm{i}} \cdot \mathbf{J} \;=\; -\boldsymbol{\nabla} \cdot \mathbf{J}.
\tag{6}
$$

Equations (4)–(6) yield (2).

(b) First, differentiate $(1)_1$ with respect to t and use (2). Then

$$\epsilon_0\mu_0\frac{\partial\Phi}{\partial t} = -\frac{\mu_0}{4\pi}\int_v\left(\frac{\nabla'\cdot\mathbf{J}}{R} + \frac{\nabla\cdot\mathbf{J}}{R}\right)dv'$$

$$= -\frac{\mu_0}{4\pi}\left\{\int_v\nabla'\cdot\left(\frac{\mathbf{J}}{R}\right)dv' - \int_v\mathbf{J}\cdot\nabla'\left(\frac{1}{R}\right)dv' + \int_v\frac{\nabla\cdot\mathbf{J}}{R}dv'\right\}. \quad (7)$$

Applying Gauss's theorem and the result $\nabla'R^{-1} = -\nabla R^{-1}$ enable us to rewrite (7) as

$$\epsilon_0\mu_0\frac{\partial\Phi}{\partial t} = -\frac{\mu_0}{4\pi}\left\{\oint_s\frac{\mathbf{J}\cdot d\mathbf{a}'}{R} + \int_v\mathbf{J}\cdot\nabla\left(\frac{1}{R}\right)dv' + \int_v\frac{\nabla\cdot\mathbf{J}}{R}dv'\right\}$$

$$= -\frac{\mu_0}{4\pi}\left\{\int_v\mathbf{J}\cdot\nabla\left(\frac{1}{R}\right)dv' + \int_v\frac{\nabla\cdot\mathbf{J}}{R}dv'\right\}, \quad (8)$$

because the surface integral is zero for a bounded distribution. Next, take the divergence of $(1)_2$:

$$\nabla\cdot\mathbf{A} = \frac{\mu_0}{4\pi}\int_v\nabla\cdot\left(\frac{\mathbf{J}}{R}\right)dv'$$

$$= \frac{\mu_0}{4\pi}\left\{\int_v\frac{\nabla\cdot\mathbf{J}}{R}dv' + \int_v\mathbf{J}\cdot\nabla\left(\frac{1}{R}\right)dv'\right\}. \quad (9)$$

The sum of (8) and (9) is (3).

Comment

In the limit $c\to\infty$, the continuity equation (2) reduces to the familiar form

$$\nabla'\cdot\mathbf{J}(\mathbf{r}',t) + \frac{\partial}{\partial t}\rho(\mathbf{r}',t) = 0,$$

since the contribution arising from the implicit differentiation of R vanishes.

Question 8.9

Suppose that $\Phi(\mathbf{r},t)$ and $\mathbf{A}(\mathbf{r},t)$ are time-harmonic potentials in Lorenz gauge.

(a) Show that

$$\Phi(\mathbf{r},t) = -\frac{ic^2}{\omega}(\nabla\cdot\mathbf{A}). \quad (1)$$

(b) Use (1) to express the electric field $\mathbf{E}(\mathbf{r},t)$ in terms of $\mathbf{A}$ only.

Solution

(a) The time-harmonic potentials Φ and $\mathbf{A}$ must vary as $e^{-i\omega t}$. So $\partial \Phi / \partial t = -i\omega \Phi$. Substituting this in $\nabla \cdot \mathbf{A} + c^{-2}\partial\Phi/\partial t = 0$ (see (13) of Question 8.2) gives (1).

(b) The electric field, given by $(1)_1$ of Question 8.1, is $\mathbf{E} = -\nabla\Phi - \partial\mathbf{A}/\partial t$. Now since $\mathbf{A}$ is time-harmonic, $\mathbf{E} = -\nabla\Phi + i\omega\mathbf{A}$. Then using (1) to eliminate Φ from $\mathbf{E}$ yields

$$\mathbf{E}(\mathbf{r}, t) = i\omega \left[\frac{c^2}{\omega^2} \nabla(\nabla \cdot \mathbf{A}) + \mathbf{A} \right]. \tag{2}$$

Comments

(i) Since $\mathbf{B}$ is always equal to the curl of $\mathbf{A}$, we see from (2) that the electromagnetic field in the Lorenz gauge can be calculated from the vector potential alone.

(ii) Using identity (11) of Question 1.8 gives $\nabla(\nabla \cdot \mathbf{A}) = \nabla^2\mathbf{A} + \nabla \times (\nabla \times \mathbf{A})$, and so (2) becomes

$$\mathbf{E}(\mathbf{r}, t) = \frac{ic^2}{\omega} \left[\nabla^2\mathbf{A} + \nabla \times (\nabla \times \mathbf{A}) \right] + i\omega\mathbf{A}.$$

Now for plane waves $\nabla^2 = (i\mathbf{k}) \cdot (i\mathbf{k}) = -k^2 = -\omega^2/c^2$ (see (9) of Question 1.1 and (7) of Question 7.4), and so

$$\mathbf{E}(\mathbf{r}, t) = \frac{ic^2}{\omega} \left[\nabla \times (\nabla \times \mathbf{A}) \right]. \tag{3}$$

(iii) It is, of course, easily verified that (3) also follows directly from the Maxwell–Ampère equation for a source-free vacuum.

Question 8.10

(a) Consider a bounded time-independent current distribution having density $\mathbf{J}(\mathbf{r})$. Express the Maxwell equation $\nabla \times \mathbf{B} = \mu_0 \mathbf{J}$ in terms of the vector potential, then deduce the condition for which

$$\mathbf{A}(\mathbf{r}) = \frac{\mu_0}{4\pi} \int_v \frac{\mathbf{J}(\mathbf{r}')}{|\mathbf{r} - \mathbf{r}'|} \, dv'. \tag{1}$$

(b) Show, by differentiating $\mathbf{A}(\mathbf{r})$ explicitly, that $\nabla \cdot \mathbf{A} = 0$.

Solution

(a) With $\mathbf{B} = \nabla \times \mathbf{A}$ and the vector identity (11) of Question 1.8 we obtain

$$\nabla \times (\nabla \times \mathbf{A}) = \nabla(\nabla \cdot \mathbf{A}) - \nabla^2 \mathbf{A} = \mu_0 \mathbf{J}. \tag{2}$$

If we choose $\nabla \cdot \mathbf{A} = 0$, the solution of (2) is (1).‡

(b) Taking the divergence of (1) gives

$$\nabla \cdot \mathbf{A} = \frac{\mu_0}{4\pi} \nabla \cdot \int_v \frac{\mathbf{J}(\mathbf{r}')}{|\mathbf{r} - \mathbf{r}'|} dv' = \int_v \mathbf{J}(\mathbf{r}') \cdot \nabla \left[\frac{1}{|\mathbf{r} - \mathbf{r}'|} \right] dv', \tag{3}$$

since $\nabla \cdot \mathbf{J}(\mathbf{r}') = 0$ (recall that $\mathbf{J}(\mathbf{r}')$ is a function of primed coordinates only). Now $\nabla(|\mathbf{r} - \mathbf{r}'|)^{-1} = -\nabla'(|\mathbf{r} - \mathbf{r}'|)^{-1}$ and (3) becomes

$$\nabla \cdot \mathbf{A} = -\int_v \mathbf{J}(\mathbf{r}') \cdot \nabla' \left[\frac{1}{|\mathbf{r} - \mathbf{r}'|} \right] dv' = -\int_v \nabla' \cdot \left[\frac{\mathbf{J}(\mathbf{r}')}{|\mathbf{r} - \mathbf{r}'|} \right] dv', \tag{4}$$

because $\nabla' \cdot \mathbf{J}(\mathbf{r}') = 0$ for stationary currents. Using Gauss's theorem to transform the right-hand side of (4) yields

$$\nabla \cdot \mathbf{A} = -\oint_s \left[\frac{\mathbf{J}(\mathbf{r}')}{|\mathbf{r} - \mathbf{r}'|} \right] \cdot d\mathbf{a}' = 0, \tag{5}$$

where the last step follows because $\mathbf{J}(\mathbf{r}') \cdot d\mathbf{a}' = 0$ everywhere on s for a bounded distribution.

Comment

Equation (5) confirms that the magnetostatic vector potential given by (1) satisfies the Coulomb gauge condition.

Question 8.11*

Suppose the current density $\mathbf{J}(\mathbf{r}, t)$ is expressed as the sum of irrotational and solenoidal parts

$$\mathbf{J}(\mathbf{r}, t) = \mathbf{J}_l + \mathbf{J}_t, \tag{1}$$

where $\mathbf{J}_l$ is the longitudinal current density having zero curl ($\nabla \times \mathbf{J}_l = 0$) and $\mathbf{J}_t$ is the transverse current density having zero divergence ($\nabla \cdot \mathbf{J}_t = 0$). Show that the vector potential in the Coulomb gauge satisfies

‡Note that $\nabla \cdot \mathbf{A} = 0$ leads to the vector form of Poisson's equation $\nabla^2 \mathbf{A} = -\mu_0 \mathbf{J}$, and so we make the same analogy used in Question 8.6.

$$\nabla^2 \mathbf{A} - \epsilon_0 \mu_0 \frac{\partial^2 \mathbf{A}}{\partial t^2} = -\mu_0 \mathbf{J}_t. \tag{2}$$

Hint: Start with Helmholtz's theorem (see Comment (i) of Question 1.24) and make use of the continuity equation.

Solution

Because of the hint, (1) can be written in the form

$$\mathbf{J}(\mathbf{r}, t) = -\frac{1}{4\pi} \nabla \int_v \frac{\nabla' \cdot \mathbf{J}}{|\mathbf{r} - \mathbf{r}'|} \, dv' + \frac{1}{4\pi} \nabla \times \int_v \frac{\nabla' \times \mathbf{J}}{|\mathbf{r} - \mathbf{r}'|} \, dv', \tag{3}$$

where

$$\mathbf{J}_1 = -\frac{1}{4\pi} \nabla \int_v \frac{\nabla' \cdot \mathbf{J}}{|\mathbf{r} - \mathbf{r}'|} \, dv', \tag{4}$$

and

$$\mathbf{J}_t = \frac{1}{4\pi} \nabla \times \int_v \frac{\nabla' \times \mathbf{J}}{|\mathbf{r} - \mathbf{r}'|} \, dv'. \tag{5}$$

The continuity equation and (4) then give

$$\mathbf{J}_1 = \frac{1}{4\pi} \nabla \int_v \frac{\partial \rho / \partial t}{|\mathbf{r} - \mathbf{r}'|} \, dv' = \epsilon_0 \nabla \frac{\partial \Phi}{\partial t}, \tag{6}$$

where in the last step we use (6) of Question 8.2. Substituting (1) and (6) in (5)$_2$ of Question 8.2 yields (2).

Comment

The source term in the inhomogeneous wave equation (2) is a solenoidal current given by (5). Hence the alternative name 'transverse' gauge for the condition $\nabla \cdot \mathbf{A} = 0$.

Question 8.12**

Consider a point charge q moving arbitrarily along a trajectory described by the vector function of time $\mathbf{r}_q(t)$. The velocity of the charge is thus $\mathbf{v}(t) = d\mathbf{r}_q/dt$. Suppose Q and Q' represent points on the trajectory where the charge *is* at time t and *was* at the earlier time t'. Let $\mathbf{R}(t) = \mathbf{r} - \mathbf{r}_q(t)$ be the vector from the charge at time t to the fixed field point P, as shown in the figure on p. 388.

(a) Prove that the velocity of the charge is given by

$$\mathbf{v}(t) = -d\mathbf{R}/dt. \tag{1}$$

(b) Show that the retarded time satisfies the equation

$$t' = t - R(t')/c. \tag{2}$$

(c) Use the retarded potentials (10) of Question 8.2 to show that

$$
\left.
\begin{aligned}
\Phi(\mathbf{r},t) &= \frac{1}{4\pi\epsilon_0} \frac{q}{[R - \mathbf{R}\cdot\mathbf{v}/c]} \\[2mm]
\mathbf{A}(\mathbf{r},t) &= \frac{\mu_0}{4\pi} \frac{q[\mathbf{v}]}{[R - \mathbf{R}\cdot\mathbf{v}/c]}
\end{aligned}
\right\}, \tag{3}
$$

where the quantities in square brackets are to be evaluated at the retarded time t'.

Hint: Start by expressing the charge and current densities in terms of a delta function.

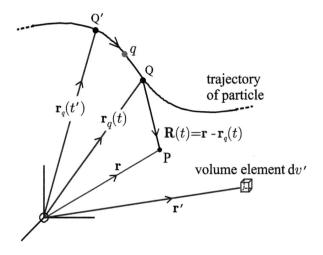

Solution

(a) Recall that $\mathbf{r}$ is the fixed point of observation. Differentiation of $\mathbf{R}(t) = \mathbf{r} - \mathbf{r}_q(t)$ with respect to t therefore gives (1).

(b) Suppose the charge, at the retarded position, emits an electromagnetic 'signal' (travelling at speed c) which is received at the field point P at time t. The travel time of this signal is $t - t'$. Now obviously $t - t' = R(t')/c$, which is (2).

(c) We begin with $\Phi(\mathbf{r}, t)$ and substitute

$$\rho(\mathbf{r}, t) = q\delta(\mathbf{r} - \mathbf{r}_q(t)) \tag{4}$$

in the retarded scalar potential, which gives

$$\Phi(\mathbf{r}, t) = \frac{q}{4\pi\epsilon_0} \int_v \frac{\delta(\mathbf{r}' - \mathbf{r}_q(t - |\mathbf{r} - \mathbf{r}'|/c))}{|\mathbf{r} - \mathbf{r}'|} dv'.$$

Introducing a dummy variable t' yields

$$\Phi(\mathbf{r}, t) = \frac{q}{4\pi\epsilon_0} \int_{-\infty}^{\infty} \int_v \frac{\delta(\mathbf{r}' - \mathbf{r}_q(t')) \delta(t' - t + |\mathbf{r} - \mathbf{r}'|/c)}{|\mathbf{r} - \mathbf{r}'|} dv' dt'$$

$$= \frac{q}{4\pi\epsilon_0} \int_{-\infty}^{\infty} \frac{\delta(t' - t + |\mathbf{r} - \mathbf{r}_q(t')|/c)}{|\mathbf{r} - \mathbf{r}_q(t')|} dt'$$

$$= \frac{q}{4\pi\epsilon_0} \int_{-\infty}^{\infty} \frac{\delta(t' - t + R(t')/c)}{R(t')} dt'. \tag{5}$$

In order to evaluate (5) we define another dummy variable

$$t'' = t' - t + R(t')/c, \tag{6}$$

which will ensure that the argument of the delta function in (5) is the same as the variable of integration. Differentiating (6) and noting that $dt = 0$ (since t is the fixed time of observation) give

$$dt'' = dt' \left\{ 1 + \frac{1}{c} \frac{d}{dt'} R(t') \right\}. \tag{7}$$

Clearly $R(t') = \sqrt{\mathbf{R}(t') \cdot \mathbf{R}(t')}$, and so $\dfrac{1}{c} \dfrac{dR(t')}{dt'} = \dfrac{\dfrac{d\mathbf{R}(t')}{dt'} \cdot \mathbf{R}(t')}{cR(t')} = -\dfrac{\mathbf{v}(t') \cdot \mathbf{R}(t')}{cR(t')}$ because of (1). Then from (7)

$$dt' = \frac{R(t')}{R(t') - \mathbf{R}(t') \cdot \mathbf{v}(t')/c} dt''. \tag{8}$$

Substituting (8) in (5) yields

$$\Phi(\mathbf{r},t) = \frac{q}{4\pi\epsilon_0}\int_{-\infty}^{\infty} \frac{\delta(t'')}{R(t') - \mathbf{R}(t')\cdot\mathbf{v}(t')/c}\, dt''$$

$$= \frac{q}{4\pi\epsilon_0}\frac{1}{R(t') - \mathbf{R}(t')\cdot\mathbf{v}(t')/c}\bigg|_{t''=0}.$$

Now it follows from (6) that $t'' = 0$ corresponds to $t' = t - R/c$. But this is the retarded time (2). Hence $(3)_1$. The vector potential $(3)_2$ can be found in a similar way by expressing the current density as $\mathbf{J}(\mathbf{r},t) = \rho(\mathbf{r},t)\mathbf{v}(t) = q\mathbf{v}(t)\,\delta(\mathbf{r}-\mathbf{r}_q(t))$.

Comments

(i) Equations $(1)_1$ and $(1)_2$ explicitly exhibit the dependence of the potentials on the velocity of the particle. They are the famous Liénard–Wiechert potentials for a moving point charge and are correct relativistically, even though their discovery (c. 1890s) predates Einstein's formulation of the theory of special relativity in 1905. However, the interpretation that $\mathbf{v}$ could be regarded simply as the *relative* velocity between the particle and the observer became apparent only after 1905.

(ii) Ref. [8] gives three alternative derivations of the Liénard–Wiechert potentials. In addition to the delta-function approach used above, a heuristic and a covariant derivation are also presented. These different methods each have their individual merits; lecturers will no doubt choose an approach which best suits its intended purpose.

(iii) We mention in passing that at most only *one* retarded point Q' on the past trajectory of the particle contributes to the field at P$(\mathbf{r},t)$.[9] A weaker version of this statement can easily be proved by contradiction. Suppose there are two such points having retarded times t_1' and t_2'. Then $R_1(t_1') = c(t-t_1')$ and $R_2(t_2') = c(t-t_2')$. So $R_2 - R_1 = c(t_1' - t_2')$, which shows that the average velocity $(R_2 - R_1)/(t_1' - t_2')$ of the particle in the direction of P is c. (Note: this excludes any velocity the particle may have in the perpendicular direction.) Since no particle can travel at (or faster than) the speed of light, there cannot be more than one point on the trajectory 'in communication' with P at the time t.

(iv) The Liénard–Wiechert potentials are related in a simple way according to:

$$\mathbf{A}(\mathbf{r},t) = \Phi(\mathbf{r},t)\frac{[\mathbf{v}]}{c^2}. \tag{9}$$

(v) The electric and magnetic fields of this arbitrarily moving point charge can be found from the potentials: $\mathbf{E} = -\nabla\Phi - \partial\mathbf{A}/\partial t$ and $\mathbf{B} = \nabla\times\mathbf{A}$. But because the retarded time is determined implicitly by (2), the differentiations leading to the fields are cumbersome. We attempt this task in Question 8.14.

[8] A. Zangwill, *Modern electrodynamics*, Chap. 23, pp. 870–4. Cambridge: Cambridge University Press, 2013.

[9] D. J. Griffiths, *Introduction to electrodynamics*, Chap. 10, p. 430. New York: Prentice Hall, 3 edn, 1999.

Question 8.13*

Consider a point charge q moving with constant velocity $\mathbf{v}$. Suppose that q is at the origin O at time $t = 0$.

(a) Use the Liénard–Wiechert potentials of Question 8.12 to show that

$$\left.\begin{aligned}\Phi(\mathbf{r}, t) &= \frac{1}{4\pi\epsilon_0} \frac{qc}{\sqrt{(c^2 - v^2)(r^2 - c^2t^2) + (c^2t - \mathbf{r}\cdot\mathbf{v})^2}} \\[2ex] \mathbf{A}(\mathbf{r}, t) &= \frac{\mu_0}{4\pi} \frac{qc\mathbf{v}}{\sqrt{(c^2 - v^2)(r^2 - c^2t^2) + (c^2t - \mathbf{r}\cdot\mathbf{v})^2}}\end{aligned}\right\}. \tag{1}$$

(b) Hence determine the electromagnetic field $\mathbf{E}(\mathbf{r}, t)$ and $\mathbf{B}(\mathbf{r}, t)$.

Solution

(a) Since $\mathbf{v}$ is constant, the trajectory of the charge is a straight line (see the figure below) and is given by $\mathbf{r}_q(t) = \mathbf{v}t$. Thus the retarded scalar potential $(3)_1$ of Question 8.12 becomes

$$\Phi(\mathbf{r}, t) = \frac{1}{4\pi\epsilon_0} \frac{q}{R' - \mathbf{R}'\cdot\boldsymbol{\beta}}, \tag{2}$$

where $\boldsymbol{\beta} = \mathbf{v}/c$, $\mathbf{R}' = [\mathbf{R}]$ and $R' = [R]$.

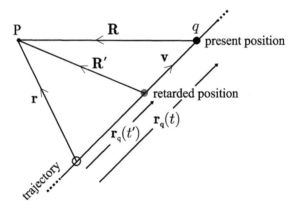

Now

$$\mathbf{R}(t) = \mathbf{r} - \mathbf{r}_q(t) = \mathbf{r} - \mathbf{v}t \quad \text{and} \quad \mathbf{R}(t') = \mathbf{R}' = \mathbf{r} - \mathbf{v}t', \tag{3}$$

where $t' = t - R'/c$ is the retarded time $\big($see (2) of Question 8.12$\big)$. So

$$R' = |\mathbf{r} - \mathbf{v}t'| = c(t - t').$$

(4)

Squaring (4) leads to the quadratic equation

$$(c^2 - v^2)t'^2 - 2(c^2 t - \mathbf{r} \cdot \mathbf{v})t' + (c^2 t^2 - r^2) = 0,$$

(5)

whose roots are

$$t' = \frac{ct - \boldsymbol{\beta} \cdot \mathbf{r} \pm \sqrt{(\boldsymbol{\beta} \cdot \mathbf{r})^2 + r^2(1 - \beta^2) + v^2 t^2 - 2(\mathbf{v} \cdot \mathbf{r})t}}{c(1 - \beta^2)}.$$

(6)

In the limit $\beta \to 0$, $t' \to t \pm r/c$. Comparing this result with (2) of Question 8.12 shows that we require the negative root in (6), and so

$$t' = \frac{ct - \boldsymbol{\beta} \cdot \mathbf{r} - \sqrt{(\boldsymbol{\beta} \cdot \mathbf{r})^2 + r^2(1 - \beta^2) + v^2 t^2 - 2(\mathbf{v} \cdot \mathbf{r})t}}{c(1 - \beta^2)}.$$

(7)

Next we use (3) and (4) to evaluate the denominator in (2):

$$
\begin{aligned}
R' - \boldsymbol{\beta} \cdot \mathbf{R}' &= c(t - t') - \boldsymbol{\beta} \cdot (\mathbf{r} - \mathbf{v}t') = c(t - t') - \boldsymbol{\beta} \cdot \mathbf{r} + v^2 t'/c \\
&= ct'(v^2/c^2 - 1) + ct - \boldsymbol{\beta} \cdot \mathbf{r} \\
&= ct - (1 - \beta^2)ct' - \boldsymbol{\beta} \cdot \mathbf{r} \\
&= \sqrt{r^2 + v^2 t^2 - 2\mathbf{r} \cdot \mathbf{v}t + (\boldsymbol{\beta} \cdot \mathbf{r})^2 - r^2 \beta^2} \qquad \text{(using (7))} \\
&= c^{-1}\sqrt{r^2 c^2 + v^2 c^2 t^2 - 2\mathbf{r} \cdot \mathbf{v}c^2 t + (\mathbf{r} \cdot \mathbf{v})^2 - r^2 v^2} \qquad \text{(eliminating } \boldsymbol{\beta}) \\
&= c^{-1}\sqrt{(c^2 - v^2)(r^2 - c^2 t^2) + (c^2 t - \mathbf{r} \cdot \mathbf{v})^2}.
\end{aligned}
$$

(8)

Substituting (8) in (2) gives $(1)_1$. The vector potential $(1)_2$ then follows immediately from (9) of Question 8.12.

(b) electric field

Substitute (9) of Question 8.12 in $(1)_1$ of Question 8.1. Then

$$\mathbf{E}(\mathbf{r}, t) = -\boldsymbol{\nabla}\Phi - \frac{\partial \mathbf{A}}{\partial t} = -\boldsymbol{\nabla}\Phi - \frac{\mathbf{v}}{c^2}\frac{\partial \Phi}{\partial t}$$

(9)

(because $\mathbf{v}$ is a constant) and so $E_i = -\nabla_i \Phi - \dfrac{v_i}{c^2}\dfrac{\partial \Phi}{\partial t}$. We evaluate these two derivatives below:

$$-\nabla_i \Phi = \frac{qc}{4\pi\epsilon_0} \frac{(c^2 - v^2)2r\partial r/\partial r_i + 2(c^2t - \mathbf{r}\cdot\mathbf{v})(-\delta_{ij}v_j)}{2\left[(c^2-v^2)(r^2-c^2t^2) + (c^2t-\mathbf{r}\cdot\mathbf{v})^2\right]^{3/2}}$$

$$= \frac{qc}{4\pi\epsilon_0} \frac{\left[r_i(c^2-v^2) - v_i(c^2t - \mathbf{r}\cdot\mathbf{v})\right]}{\left[(c^2-v^2)(r^2-c^2t^2) + (c^2t-\mathbf{r}\cdot\mathbf{v})^2\right]^{3/2}}. \tag{10}$$

Similarly,

$$-\frac{v_i}{c^2}\frac{\partial\Phi}{\partial t} = \frac{qc}{4\pi\epsilon_0}\frac{v_i}{2c^2}\frac{\left[(c^2-v^2)(-2c^2t) + 2c^2(c^2t-\mathbf{r}\cdot\mathbf{v})\right]}{\left[(c^2-v^2)(r^2-c^2t^2) + (c^2t-\mathbf{r}\cdot\mathbf{v})^2\right]^{3/2}}$$

$$= \frac{qc}{4\pi\epsilon_0}\frac{\left[v^2v_it - v_i(\mathbf{r}\cdot\mathbf{v})\right]}{\left[(c^2-v^2)(r^2-c^2t^2) + (c^2t-\mathbf{r}\cdot\mathbf{v})^2\right]^{3/2}}. \tag{11}$$

Adding (10) and (11) gives $E_i = \dfrac{qc}{4\pi\epsilon_0}\dfrac{(c^2-v^2)(r_i - v_it)}{\left[(c^2-v^2)(r^2-c^2t^2) + (c^2t-\mathbf{r}\cdot\mathbf{v})^2\right]^{3/2}}.$

Hence

$$\mathbf{E}(\mathbf{r},t) = \frac{qc}{4\pi\epsilon_0}\frac{(c^2-v^2)(\mathbf{r}-\mathbf{v}t)}{\left[(c^2-v^2)(r^2-c^2t^2) + (c^2t-\mathbf{r}\cdot\mathbf{v})^2\right]^{3/2}}. \tag{12}$$

magnetic field

Now $\mathbf{B}(\mathbf{r},t) = \nabla\times\mathbf{A}$ gives

$$B_i = \varepsilon_{ijk}\nabla_j A_k$$

$$= \varepsilon_{ijk}\frac{\mu_0}{4\pi}qcv_k\nabla_j\left[(c^2-v^2)(r^2-c^2t^2) + (c^2t-\mathbf{r}\cdot\mathbf{v})^2\right]^{-1/2}$$

$$= -\varepsilon_{ijk}\frac{\mu_0 q cv_k\left[(c^2-v^2)(2r\partial r/\partial r_j) + 2(c^2t-\mathbf{r}\cdot\mathbf{v})(-\delta_{jl}v_l)\right]}{8\pi\left[(c^2-v^2)(r^2-c^2t^2) + (c^2t-\mathbf{r}\cdot\mathbf{v})^2\right]^{3/2}}$$

$$= \varepsilon_{ikj}\frac{\mu_0 q cv_k\left[r_j(c^2-v^2) - v_j(c^2t-\mathbf{r}\cdot\mathbf{v})\right]}{4\pi\left[(c^2-v^2)(r^2-c^2t^2) + (c^2t-\mathbf{r}\cdot\mathbf{v})^2\right]^{3/2}}$$

$$= \frac{\mu_0}{4\pi}\frac{qc(c^2-v^2)(\mathbf{v}\times\mathbf{r})_i}{\left[(c^2-v^2)(r^2-c^2t^2) + (c^2t-\mathbf{r}\cdot\mathbf{v})^2\right]^{3/2}},$$

since $\varepsilon_{ikj}v_kr_j = (\mathbf{v}\times\mathbf{r})_i$ and $\varepsilon_{ikj}v_kv_j = (\mathbf{v}\times\mathbf{v})_i = 0$. Therefore,

$$\mathbf{B}(\mathbf{r}, t) = \frac{\mu_0}{4\pi} \frac{qc(c^2 - v^2)(\mathbf{v} \times \mathbf{r})}{\left[(c^2 - v^2)(r^2 - c^2t^2) + (c^2t - \mathbf{r} \cdot \mathbf{v})^2\right]^{3/2}}. \tag{13}$$

Comment

Equations (12) and (13) show how the **E**- and **B**-fields of a point charge, moving with constant velocity, depend explicitly on time. We will return to this important problem in Question 8.15 and again in Question 12.14, where these fields are derived more elegantly by making a Lorentz transformation between relatively moving reference frames.

Question 8.14**

Consider a point charge q moving arbitrarily with velocity $\mathbf{v} = c\boldsymbol{\beta}(t)$ and acceleration $c\dot{\boldsymbol{\beta}}(t)$. Suppose $\mathbf{r}$ is the vector from an origin O to the field point P and $\mathbf{r}'(t)$ is the vector from O to the position of q at time t. Then $\mathbf{R}(t) = \mathbf{r} - \mathbf{r}'(t)$ is the position of q relative to P at time t. As before, $[\mathbf{R}] = \mathbf{r} - [\mathbf{r}']$ is the retarded position and $t' = t - R(t')/c = t - [R]/c$ is the retarded time.

(a) Prove the following identities:

☞ $\qquad c\boldsymbol{\nabla}t' = -\boldsymbol{\nabla}[R],$ \hfill (1)

☞ $\qquad \nabla_i[R_j] = \delta_{ij} + [\beta_j]\nabla_i[R],$ \hfill (2)

☞ $\qquad \boldsymbol{\nabla}[R] = \dfrac{[\hat{\mathbf{R}}]}{1 - [\boldsymbol{\beta}] \cdot [\hat{\mathbf{R}}]},$ \hfill (3)

☞ $\qquad c\nabla_i[\beta_j] = -[\dot{\beta}_j]\nabla_i[R],$ \hfill (4)

☞ $\qquad \dfrac{\partial[\mathbf{R}]}{\partial t} = -c[\boldsymbol{\beta}]\dfrac{\partial t'}{\partial t},$ \hfill (5)

☞ $\qquad \dfrac{\partial[R]}{\partial t} = -c[\boldsymbol{\beta}] \cdot [\hat{\mathbf{R}}]\dfrac{\partial t'}{\partial t},$ \hfill (6)

☞ $\qquad \dfrac{\partial t'}{\partial t} = \dfrac{1}{1 - [\boldsymbol{\beta}] \cdot [\hat{\mathbf{R}}]},$ \hfill (7)

☞ $\qquad c\boldsymbol{\nabla} \times [\boldsymbol{\beta}] = \dfrac{[\dot{\boldsymbol{\beta}}] \times [\hat{\mathbf{R}}]}{1 - [\boldsymbol{\beta}] \cdot [\hat{\mathbf{R}}]}.$ \hfill (8)

(b) Using the results (1)–(8), differentiate the Liénard–Wiechert potentials ((3) of Question 8.12) and show that the fields $\mathbf{E}(\mathbf{r}, t)$ and $\mathbf{B}(\mathbf{r}, t)$ of the charge are given by

$$\mathbf{E} = \frac{1}{4\pi\epsilon_0} \frac{q}{R^2(1 - \boldsymbol{\beta} \cdot \hat{\mathbf{R}})^3} \left\{ (1 - \beta^2)(\hat{\mathbf{R}} - \boldsymbol{\beta}) + \mathbf{R} \times \left((\hat{\mathbf{R}} - \boldsymbol{\beta}) \times \dot{\boldsymbol{\beta}}/c \right) \right\}, \qquad (9)$$

$$\mathbf{B} = \frac{\mu_0}{4\pi} \frac{q}{R^2(1 - \boldsymbol{\beta} \cdot \hat{\mathbf{R}})^3} \left\{ (1 - \beta^2)(c\boldsymbol{\beta} \times \hat{\mathbf{R}}) + \hat{\mathbf{R}} \times \left(\mathbf{R} \times \left((\hat{\mathbf{R}} - \boldsymbol{\beta}) \times \dot{\boldsymbol{\beta}} \right) \right) \right\}. \qquad (10)$$

In (9) and (10) we have, for reasons of clarity, suppressed the square bracket notation. It is understood that all quantities in these equations involving $\boldsymbol{\beta}$ and $\mathbf{R}$ are to be evaluated at the retarded time.

Solution

(a) ☞   Clearly $R(t') = [R] = c(t - t')$. Hence (1).

☞   Because $[R_j] = r_j - [r_j']$, we have $\nabla_i[R_j] = \dfrac{\partial[R_j]}{\partial r_i} = \dfrac{\partial r_j}{\partial r_i} - \dfrac{\partial[r_j']}{\partial t'}\dfrac{\partial t'}{\partial r_i} =$

$\delta_{ij} - \dfrac{\partial[r_j']}{\partial t'}\nabla_i t'$. Now $\dfrac{\partial[r_j']}{\partial t'} = [v_j]$, and so (2) follows from (1).

☞   From $[\mathbf{R}] = \mathbf{r} - [\mathbf{r}']$ we obtain $[R]^2 = r^2 + [r']^2 - 2r_j[r_j']$, and so

$$\frac{\partial[R]^2}{\partial[R]}\frac{\partial[R]}{\partial r_i} = 2[R]\nabla_i[R] = 2\left(r_j\frac{\partial r_j}{\partial r_i} + [r_j']\frac{\partial[r_j']}{\partial r_i} - r_j\frac{\partial[r_j']}{\partial r_i} - [r_j']\frac{\partial r_j}{\partial r_i} \right)$$

$$[R]\nabla_i[R] = \left(r_i - [r_i'] \right) + [r_j']\frac{\partial[r_j']}{\partial t'}\frac{\partial t'}{\partial r_i} - r_j\frac{\partial[r_j']}{\partial t'}\frac{\partial t'}{\partial r_i}$$

$$= \left(r_i - [r_i'] \right) + [r_j'][v_j]\nabla_i t' - r_j[v_j]\nabla_i t'$$

$$= \left(r_i - [r_i'] \right) - [r_j'][\beta_j]\nabla_i[R] + r_j[\beta_j]\nabla_i[R]$$

$$= \left(r_i - [r_i'] \right) + [\beta_j](r_j - [r_j'])\nabla_i[R].$$

Rearranging this equation gives (3).

☞ $c\nabla_i[\beta_j] = \dfrac{\partial[v_j]}{\partial r_i} = \dfrac{\partial[v_j]}{\partial t'}\dfrac{\partial t'}{\partial r_i} = c[\dot{\beta}_j]\nabla_i t' = -[\dot{\beta}_j]\nabla_i[R].$

☞ $\dfrac{\partial[\mathbf{R}]}{\partial t} = -\dfrac{\partial[\mathbf{r}']}{\partial t'}\dfrac{\partial t'}{\partial t} = -c[\boldsymbol{\beta}]\dfrac{\partial t'}{\partial t}.$

☞ As before $[R]^2 = r^2 + [r']^2 - 2r_j[r'_j]$. Therefore

$$\frac{\partial[R]^2}{\partial[R]}\frac{\partial[R]}{\partial t} = 2[R]\frac{\partial[R]}{\partial t} = 2[r'_j]\frac{\partial[r'_j]}{\partial t'}\frac{\partial t'}{\partial t} - 2r_j\frac{\partial[r'_j]}{\partial t'}\frac{\partial t'}{\partial t}$$

or

$$[R]\frac{\partial[R]}{\partial t} = ([r'_j] - r_j)\frac{\partial[r'_j]}{\partial t'}\frac{\partial t'}{\partial t} = -[R_j][v_j]\frac{\partial t'}{\partial t} = -c[\boldsymbol{\beta}]\cdot[\mathbf{R}]\frac{\partial t'}{\partial t}$$

which is (6).

☞ Since $t' = t - [R]/c$, it follows that

$$\frac{\partial t'}{\partial t} = 1 - \frac{1}{c}\frac{\partial[R]}{\partial t} = 1 + \frac{[\boldsymbol{\beta}]\cdot[\mathbf{R}]}{[R]}\frac{\partial t'}{\partial t}.$$

Rearranging this equation yields (7).

☞ $$c\left(\boldsymbol{\nabla}\times[\boldsymbol{\beta}]\right)_i = c\,\varepsilon_{ijk}\nabla_j[\beta_k] = c\,\varepsilon_{ijk}\frac{\partial[\beta_k]}{\partial r_j} = c\,\varepsilon_{ijk}\frac{\partial[\beta_k]}{\partial t'}\frac{\partial t'}{\partial r_j}$$

$$= c\,\varepsilon_{ijk}[\dot{\beta}_k]\nabla_j t'$$

$$= \varepsilon_{ikj}[\dot{\beta}_k]\nabla_j[R]$$

$$= \left([\dot{\boldsymbol{\beta}}]\times\boldsymbol{\nabla}[R]\right)_i.$$

Substituting (2) in this last equation gives (8).

(b) With the square bracket suppressed as before, we obtain:

electric field

$$\mathbf{E}(\mathbf{r},t) = -\boldsymbol{\nabla}\Phi - \frac{\partial\mathbf{A}}{\partial t}$$

$$= -\boldsymbol{\nabla}\Phi - \frac{1}{c}\frac{\partial(\boldsymbol{\beta}\Phi)}{\partial t} \qquad \text{(using (9) of Question 8.12)}$$

$$= -\frac{q/c}{4\pi\epsilon_0}\left\{c\boldsymbol{\nabla}\left(\frac{1}{R - \boldsymbol{\beta}\cdot\mathbf{R}}\right) + \frac{1}{R - \boldsymbol{\beta}\cdot\mathbf{R}}\frac{\partial\boldsymbol{\beta}}{\partial t} + \boldsymbol{\beta}\frac{\partial}{\partial t}\left(\frac{1}{R - \boldsymbol{\beta}\cdot\mathbf{R}}\right)\right\}. \quad (11)$$

Substituting (1)–(7) in (11) gives (9), after some tedious algebra.

magnetic field

$\mathbf{B}(\mathbf{r},t) = \boldsymbol{\nabla}\times\mathbf{A}$ with $\mathbf{A}$ given by (9) of Question 8.12 yields

$$\mathbf{B}(\mathbf{r},t) = \frac{\boldsymbol{\nabla}\times\boldsymbol{\beta}\Phi}{c} = \frac{\boldsymbol{\nabla}\Phi\times\boldsymbol{\beta}}{c} + \frac{\Phi\boldsymbol{\nabla}\times\boldsymbol{\beta}}{c}$$

$$= \frac{\mu_0}{4\pi}qc\left\{\boldsymbol{\nabla}\left(\frac{1}{R - \boldsymbol{\beta}\cdot\mathbf{R}}\right)\times\boldsymbol{\beta} + \frac{\boldsymbol{\nabla}\times\boldsymbol{\beta}}{R - \boldsymbol{\beta}\cdot\mathbf{R}}\right\}. \quad (12)$$

Substituting (1)–(4) and (8) in (12) gives (10), again after some algebra.

Comments

(i) We can express the fields (9) and (10) in the form

$$\left.\begin{array}{l} \mathbf{E}(\mathbf{r}, t) = \mathbf{E_v} + \mathbf{E_a} \\[2mm] \text{and} \\[2mm] \mathbf{B}(\mathbf{r}, t) = \mathbf{B_v} + \mathbf{B_a} \end{array}\right\},$$

where

$$\mathbf{E_v} = \frac{1}{4\pi\epsilon_0} q \left[\frac{(1 - \beta^2)(\hat{\mathbf{R}} - \boldsymbol{\beta})}{R^2(1 - \boldsymbol{\beta}\cdot\hat{\mathbf{R}})^3} \right], \tag{13}$$

$$\mathbf{E_a} = \frac{1}{4\pi\epsilon_0} \frac{q}{c} \left[\frac{\hat{\mathbf{R}} \times (\hat{\mathbf{R}} - \boldsymbol{\beta}) \times \dot{\boldsymbol{\beta}}}{R(1 - \boldsymbol{\beta}\cdot\hat{\mathbf{R}})^3} \right], \tag{14}$$

$$\mathbf{B_v} = \frac{\mu_0}{4\pi} qc \left[\frac{(1 - \beta^2)(\boldsymbol{\beta} \times \hat{\mathbf{R}})}{R^2(1 - \boldsymbol{\beta}\cdot\hat{\mathbf{R}})^3} \right], \tag{15}$$

$$\mathbf{B_a} = \frac{\mu_0}{4\pi} q \left[\frac{\hat{\mathbf{R}} \times \left(\hat{\mathbf{R}} \times ((\hat{\mathbf{R}} - \boldsymbol{\beta}) \times \dot{\boldsymbol{\beta}}) \right)}{R(1 - \boldsymbol{\beta}\cdot\hat{\mathbf{R}})^3} \right]. \tag{16}$$

Quantities enclosed in square brackets are to be evaluated at the retarded time. Equations (13) and (15) are known as the velocity or the generalized Coulomb fields and they vary as $\mathbf{R}^{-2}$. Equations (14) and (16) are known as the acceleration or radiation fields and they vary as $\mathbf{R}^{-1}$. In Chapter 11, we will use (14) and (16) to study the radiation produced by an accelerated charge.

(ii) It is straightforward to show that the fields given by (9) and (10) are related by

$$\mathbf{B} = \frac{\hat{\mathbf{R}} \times \mathbf{E}}{c}, \tag{17}$$

where $\mathbf{R}$ is the vector from the retarded position of q. That is, the B-field is perpendicular to $\mathbf{E}$ and $\mathbf{R}$. Furthermore, $\mathbf{E_v}$ is perpendicular to $\mathbf{B_v}$ and $\mathbf{E_a}$ is perpendicular to $\mathbf{B_a}$.

(iii) Equations (13)–(16) reduce to the correct electrostatic field for a stationary point charge if we let $\boldsymbol{\beta} = 0$ and $\dot{\boldsymbol{\beta}} = 0$. They also give the correct electromagnetic field when $\beta \to 1$. As we remarked earlier, the Liénard–Wiechert potentials are correct relativistically, even though they predate Einstein's theory of special relativity.

Question 8.15*

Consider a point charge q moving with *constant* velocity $c\boldsymbol{\beta}$. Use the results of Question 8.14 to show that the electromagnetic field at P can be expressed as

$$\mathbf{E} = \frac{1}{4\pi\epsilon_0}\frac{q}{R^2}\frac{(1-\beta^2)\,\hat{\mathbf{R}}}{(1-\beta^2\sin^2\theta)^{3/2}} \qquad \text{and} \qquad \mathbf{B} = \frac{\mu_0}{4\pi}\frac{qc}{R^2}\frac{(\boldsymbol{\beta}\times\hat{\mathbf{R}})(1-\beta^2)}{(1-\beta^2\sin^2\theta)^{3/2}}, \qquad (1)$$

where $\mathbf{R}(t)$ is the vector from the present position of q to P and $\theta(t)$ is the angle between $\mathbf{R}(t)$ and $\boldsymbol{\beta}$.

Solution

Substituting $\dot{\boldsymbol{\beta}} = 0$ in (13) of Question 8.14 gives

$$\mathbf{E} = \mathbf{E}_v = \frac{q}{4\pi\epsilon_0}\left[\frac{(1-\beta^2)(\hat{\mathbf{R}}-\boldsymbol{\beta})}{R^2(1-\boldsymbol{\beta}\cdot\hat{\mathbf{R}})^3}\right] = \frac{q}{4\pi\epsilon_0}\left[\frac{(1-\beta^2)(\mathbf{R}-\beta R)}{(R-\boldsymbol{\beta}\cdot\mathbf{R})^3}\right]$$

$$= \frac{q}{4\pi\epsilon_0}\frac{(1-\beta^2)(\mathbf{R}'-\beta R')}{(R'-\boldsymbol{\beta}\cdot\mathbf{R}')^3}. \qquad (2)$$

In the figure below, let Q represent the position of q at time t and Q′ the position of q at the retarded time t'. As before, $(t-t') = R'/c$ and $\mathbf{R}' - \mathbf{R} = c\boldsymbol{\beta}\times(t-t')$. So

$$\mathbf{R}' = \mathbf{R} + c\boldsymbol{\beta}(t-t') = \mathbf{R} + \beta R'. \qquad (3)$$

Squaring (3) yields the quadratic equation

$$R'^2(1-\beta^2) - (2R\beta\cos\theta)R' - R^2 = 0,$$

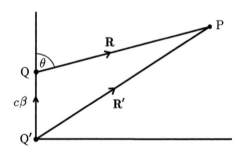

which has the solution[‡]

$$R' = \frac{R\beta\cos\theta + R\sqrt{1-\beta^2\sin^2\theta}}{1-\beta^2}. \qquad (4)$$

[‡]We reject the negative root. To see why, consider what happens in the limit $\beta \to 0$.

From (3), $\boldsymbol{\beta} \cdot \mathbf{R}' = \boldsymbol{\beta} \cdot \mathbf{R} + \beta^2 R' = R\beta \cos\theta + \beta^2 R'$, and so

$$R' - \boldsymbol{\beta} \cdot \mathbf{R}' = R' - R\beta\cos\theta - \beta^2 R'$$
$$= R'(1 - \beta^2) - R\beta\cos\theta$$
$$= R\sqrt{1 - \beta^2 \sin^2\theta}, \tag{5}$$

because of (4). Substituting (3) and (5) in (2) yields $(1)_1$. Equation $(1)_2$ follows immediately from $(1)_1$ and (17) of Question 8.14.

Comments

(i) Since q moves with constant velocity, the acceleration fields $\mathbf{E}_a$ and $\mathbf{B}_a$ (see (14) and (16) of Question 8.14) are both zero.

(ii) It is worth emphasizing again that the vector $\mathbf{R}$ in (1) points from the *present* position of charge q to the field point P. In the words of Ref. [10], this is 'an *extraordinary* coincidence, since the "message" came from the *retarded* position'. See also Comment (ii) of Question 12.14.

Question 8.16

Consider a particle having mass m and charge q moving non-relativistically with velocity $\dot{\mathbf{r}}$ in an electromagnetic field.

(a) Use the Lorentz force to express the equation of motion of the particle in the form

$$\frac{d}{dt}(m\dot{r}_i + qA_i) = \frac{\partial}{\partial r_i}\left[q(\dot{r}_j A_j - \Phi)\right]. \tag{1}$$

(b) Hence deduce that the Lagrangian of the particle is

$$\mathsf{L} = \tfrac{1}{2}m\dot{\mathbf{r}}^2 - q(\Phi - \dot{\mathbf{r}}\cdot\mathbf{A}). \tag{2}$$

Hint: Use the Euler–Lagrange equations

$$\frac{d}{dt}\left(\frac{\partial\mathsf{L}}{\partial\dot{r}_i}\right) = \left(\frac{\partial\mathsf{L}}{\partial r_i}\right), \tag{3}$$

where $i = x, y, z$.

(c) Prove that (2) is not invariant under a gauge transformation.

[10] D. J. Griffiths, *Introduction to electrodynamics*, Chap. 10, p. 439. New York: Prentice Hall, 3 edn, 1999.

Solution

(a) The Lorentz force acting on the particle is $\mathbf{F} = m\ddot{\mathbf{r}} = q[\mathbf{E} + \dot{\mathbf{r}} \times \mathbf{B}]$, where $\mathbf{E} = -\nabla\Phi - \partial\mathbf{A}/\partial t$ and $\mathbf{B} = \nabla \times \mathbf{A}$. So $m\ddot{\mathbf{r}} = q[-\nabla\Phi - \partial\mathbf{A}/\partial t + \dot{\mathbf{r}} \times (\nabla \times \mathbf{A})]$,

or

$$m\ddot{r}_i = q\left[-\nabla_i\Phi - \partial A_i/\partial t + \varepsilon_{ijk}\varepsilon_{klm}\dot{r}_j\nabla_l A_m \right]$$

$$= q\left[-\nabla_i\Phi - \partial A_i/\partial t + (\delta_{il}\delta_{jm} - \delta_{im}\delta_{jl})\dot{r}_j\nabla_l A_m \right]$$

$$= q\left[-\nabla_i\Phi - \partial A_i/\partial t + \dot{r}_j(\nabla_i A_j - \nabla_j A_i) \right].$$

Rearranging yields

$$m\ddot{r}_i + q\partial A_i/\partial t + q\dot{r}_j\nabla_j A_i = q\dot{r}_j\nabla_i A_j - q\nabla_i\Phi$$

$$m\ddot{r}_i + q[\partial/\partial t + \dot{\mathbf{r}} \cdot \nabla]A_i = \nabla_i q(\dot{r}_j A_j - \Phi), \qquad (4)$$

where in the last step we recognize that the velocity and position coordinates are independent of each other $(\partial\dot{r}_j/\partial r_i = 0)$. Now the term in square brackets in (4) is the convective derivative (see $(1)_2$ of Question 1.9), and so

$$m\ddot{r}_i + q\frac{dA_i}{dt} = \frac{d}{dt}(m\dot{r}_i + qA_i)$$

$$= \frac{\partial}{\partial r_i}\left[q(\dot{r}_j A_j - \Phi) \right],$$

which is (1).

(b) Comparing (1) and (3) shows that

$$\left.\begin{aligned} \frac{\partial L}{\partial \dot{r}_i} &= m\dot{r}_i + qA_i \\[2mm] \frac{\partial L}{\partial r_i} &= \frac{\partial}{\partial r_i}\left[q(\dot{r}_j A_j - \Phi) \right] \end{aligned}\right\}, \qquad (5)$$

which can be integrated to give

$$\left.\begin{aligned} L &= \tfrac{1}{2}m\dot{r}_i^2 + q\dot{r}_i A_i + g(\mathbf{r}) \\[2mm] L &= q(\dot{r}_i A_i - \Phi) + h(\dot{\mathbf{r}}) \end{aligned}\right\}. \qquad (6)$$

Here $g(\mathbf{r})$ and $h(\dot{\mathbf{r}})$ are arbitrary functions of the position and velocity coordinates respectively. Subtracting $(6)_1$ and $(6)_2$ yields

$$\{g(\mathbf{r}) + q\Phi(\mathbf{r})\} = \{h(\dot{\mathbf{r}}) - \tfrac{1}{2}m\dot{\mathbf{r}}^2\}. \tag{7}$$

Since $\mathbf{r}$ and $\dot{\mathbf{r}}$ are independent variables, (7) can only be satisfied if each of the terms in curly brackets equals a constant; zero, say. Taking $g(\mathbf{r}) = -q\Phi(\mathbf{r})$ and $h(\dot{\mathbf{r}}) = \tfrac{1}{2}m\dot{\mathbf{r}}^2$ in (6) produces the desired Lagrangian.

(c) Substituting (6) of Question 8.1 in (2) gives

$$
\begin{aligned}
\mathsf{L}' = \tfrac{1}{2}m\dot{\mathbf{r}}^2 - q\Big[\Phi' - \dot{\mathbf{r}}\cdot\mathbf{A}'\Big] &= \tfrac{1}{2}m\dot{\mathbf{r}}^2 - q\left[\left(\Phi - \frac{\partial\chi}{\partial t}\right) - \dot{\mathbf{r}}\cdot(\mathbf{A} + \nabla\chi)\right] \\
&= \tfrac{1}{2}m\dot{\mathbf{r}}^2 - q\left[\Phi - \dot{\mathbf{r}}\cdot\mathbf{A}\right] + q\left[\frac{\partial\chi}{\partial t} + \dot{\mathbf{r}}\cdot\nabla\chi\right] \\
&= \mathsf{L} + q\frac{d\chi}{dt}, \tag{8}
\end{aligned}
$$

where $\dfrac{d}{dt} = \dfrac{\partial}{\partial t} + (\dot{\mathbf{r}}\cdot\nabla)$ is the convective derivative. Now χ is an arbitrary gauge function, and so $\mathsf{L} \neq \mathsf{L}'$ in general.

Comments

(i) The term $\tfrac{1}{2}m\dot{\mathbf{r}}^2$ in (2) represents the kinetic energy T of the particle and the term $q(\Phi - \dot{\mathbf{r}}\cdot\mathbf{A})$ is its generalized potential energy V. Therefore $\mathsf{L} = T - V$, as one would expect.

(ii) In classical mechanics $p_i = \partial\mathsf{L}/\partial\dot{r}_i$ is the canonical momentum. It follows from (2) that $p_i = m\dot{r}_i + qA_i$ because $\Phi(\mathbf{r})$ is a velocity-independent potential. Thus

$$\mathbf{p} = m\dot{\mathbf{r}} + q\mathbf{A}, \tag{9}$$

where $m\dot{\mathbf{r}}$ is the kinetic (or mechanical) momentum of the particle. It is obvious, on dimensional grounds, that the units of $\mathbf{A}$ are those of momentum per unit charge. For further details on this interpretation the reader is referred to Ref. [11], from which we quote:

> A second obstacle to our students' understanding of the vector potential is the still prevalent view that $\mathbf{A}$ is merely a mathematical fiction whose only role is to express $\mathbf{B}$ as $\nabla\times\mathbf{A}$. Curiously, the founder of our subject, Maxwell himself, advocated in 1865 a quite opposite view, which we shall echo, that the vector potential can be seen as a stored momentum per unit charge. Indeed, one of Maxwell's several names for the vector potential was 'electromagnetic momentum'.

[11] M. D. Semon and J. R. Taylor, 'Thoughts on the magnetic vector potential', *American Journal of Physics*, vol. 64, no. 11, pp. 1361–9, 1996.

Question 8.17

(a) Consider a particle having mass m and charge q moving in a potential $V(\mathbf{r})$. Show that the Hamiltonian, in the presence of an electromagnetic field, is

$$\mathsf{H} = \frac{1}{2m}(\mathbf{p} - q\mathbf{A})^2 + V + q\Phi, \tag{1}$$

where Φ and $\mathbf{A}$ have their usual meaning.

Hint: Start with the definition of the classical Hamiltonian

$$\mathsf{H} = \mathbf{p}\cdot\dot{\mathbf{r}} - \mathsf{L}, \tag{2}$$

then use (2) and (9) of Question 8.16.

(b) For the choice of gauge

$$\left.\begin{array}{c}\Phi = 0 \\ \nabla\cdot\mathbf{A} = 0\end{array}\right\}, \tag{3}$$

show that (1) can be expressed as

$$\mathsf{H} = \mathsf{H}^{(0)} + \mathsf{H}', \tag{4}$$

where

$$\left.\begin{array}{l}\mathsf{H}^{(0)} = \dfrac{p^2}{2m} + V \\[2ex] \mathsf{H}' = -\dfrac{q}{m}\mathbf{A}\cdot\mathbf{p} + \dfrac{q^2 A^2}{2m}\end{array}\right\}. \tag{5}$$

Hint: Recall from quantum mechanics that:

☞ The commutator of two operators A and B is $[A, B] = AB - BA$, and

☞ $[p_i, A_i] = -i\hbar\dfrac{\partial A_i}{\partial r_i}$.

Solution

(a) Substituting $\mathsf{L} = \frac{1}{2}m\dot{\mathbf{r}}^2 - q(\Phi - \dot{\mathbf{r}}\cdot\mathbf{A})$ and $\mathbf{p} = m\dot{\mathbf{r}} + q\mathbf{A}$ (see Question 8.16) in (2) yields

$$\mathsf{H} = (m\dot{\mathbf{r}} + q\mathbf{A})\cdot\dot{\mathbf{r}} - \tfrac{1}{2}m\dot{\mathbf{r}}^2 + q\Phi - q\dot{\mathbf{r}}\cdot\mathbf{A} = \tfrac{1}{2}m\dot{\mathbf{r}}^2 + q\Phi. \tag{6}$$

Now in quantum mechanics the Hamiltonian must be expressed in terms of the conjugate momentum, and so we eliminate $\dot{\mathbf{r}}$ from (6) using $\mathbf{p} = m\dot{\mathbf{r}} + q\mathbf{A}$ which gives

$$\mathsf{H} \; = \; \frac{1}{2m}(\mathbf{p} - q\mathbf{A})^2 + q\Phi. \tag{7}$$

If the particle in the absence of the electromagnetic field is exposed to the potential V, then $\mathsf{H} = \mathbf{p}^2/2m + V$. In the presence of the field, the Hamiltonian is therefore given by (7) with the term V added to the right-hand side. Hence (1).

(b) Expanding (1) and allowing for the possibility that $\mathbf{p}$ and $\mathbf{A}$ do not necessarily commute yield

$$
\begin{aligned}
\mathsf{H} \; &= \; \frac{1}{2m}\Big[p^2 - q(\mathbf{p}\cdot\mathbf{A} + \mathbf{A}\cdot\mathbf{p}) + q^2 A^2\Big] + V + q\Phi \\[4pt]
&= \; \frac{p^2}{2m} + V - \frac{q}{2m}\big(p_i A_i + A_i p_i\big) + \frac{q^2 A^2}{2m} + q\Phi \\[4pt]
&= \; \frac{p^2}{2m} + V - \frac{q}{2m}\big(A_i p_i + [p_i, A_i]\big) - \frac{q}{2m} A_i p_i + \frac{q^2 A^2}{2m} + q\Phi \\[4pt]
&= \; \frac{p^2}{2m} + V - \frac{q}{2m}[p_i, A_i] - \frac{q}{m} A_i p_i + \frac{q^2 A^2}{2m} + q\Phi \\[4pt]
&= \; \frac{p^2}{2m} + V + \frac{i\hbar q}{2m}\,\boldsymbol{\nabla}\cdot\mathbf{A} - \frac{q}{m}\mathbf{A}\cdot\mathbf{p} + \frac{q^2 A^2}{2m} + q\Phi \qquad \text{(see the hint)} \\[4pt]
&= \; \frac{p^2}{2m} + V - \frac{q}{m}\mathbf{A}\cdot\mathbf{p} + \frac{q^2 A^2}{2m} \qquad\qquad\quad \text{(use the gauge choice (3))},
\end{aligned}
$$

which is the result we seek.

Comments

(i) The Hamiltonian in (1), like the Lagrangian of Question 8.16, inherits the gauge dependence of the potentials $\mathbf{A}$ and Φ.

(ii) Note that the potentials in (3) satisfy the Lorenz gauge condition (see (4) of Question 8.2).

(iii) It is sometimes possible to treat $\mathbf{A}$ and Φ classically (in, for example, an atomic or molecular system) rather than as quantum-mechanical operators.[‡] In these circumstances H is a semi-classical (rather than a fully-quantized) Hamiltonian.

[‡] The momentum operator remains $\mathbf{p} = -i\hbar\boldsymbol{\nabla}$.

Question 8.18

Consider a particle having mass m and charge q moving in a uniform magnetic field **B**. Show that the perturbation Hamiltonian of Question 8.17 is

$$H' = -\frac{q}{2m}\mathbf{B}\cdot\mathbf{L} + \frac{q^2}{8m}\left(r^2B^2 - (\mathbf{r}\cdot\mathbf{B})^2\right), \tag{1}$$

where **L** is the angular momentum operator $\mathbf{r}\times\mathbf{p}$.

Solution

A possible vector potential for a uniform magnetic field (see (b) of Question 1.25) is $\mathbf{A} = -\frac{1}{2}(\mathbf{r}\times\mathbf{B})$. Substituting this in $(5)_2$ of Question 8.17 and using Cartesian tensors gives

$$
\begin{aligned}
H' &= -\frac{q}{m}\mathbf{A}\cdot\mathbf{p} + \frac{q^2A^2}{2m}\\[2mm]
&= \frac{q}{2m}(\mathbf{r}\times\mathbf{B})\cdot\mathbf{p} + \frac{q^2}{8m}(\mathbf{r}\times\mathbf{B})\cdot(\mathbf{r}\times\mathbf{B})\\[2mm]
&= \frac{q}{2m}(\mathbf{r}\times\mathbf{B})_i p_i + \frac{q^2}{8m}(\mathbf{r}\times\mathbf{B})_i(\mathbf{r}\times\mathbf{B})_i\\[2mm]
&= \frac{q}{2m}\varepsilon_{ijk}r_j B_k p_i + \frac{q^2}{8m}\varepsilon_{ijk}\varepsilon_{ilm}r_j B_k r_l B_m\\[2mm]
&= -\frac{q}{2m}\varepsilon_{kji}B_k r_j p_i + \frac{q^2}{8m}(\delta_{jl}\delta_{km}-\delta_{jm}\delta_{kl})r_j r_l B_k B_m \qquad \text{(properties of } \varepsilon_{ijk})\\[2mm]
&= -\frac{q}{2m}B_i(\mathbf{r}\times\mathbf{B})_i + \frac{q^2}{8m}(r^2 B_i B_i - r_i r_j B_i B_j) \qquad \text{(subscripts are arbitrary)}\\[2mm]
&= -\frac{q}{2m}\mathbf{B}\cdot(\mathbf{r}\times\mathbf{B}) + \frac{q^2}{8m}\left(r^2 B^2 - (r_i B_i)(r_i B_j)\right),
\end{aligned}
$$

which is (1).

Comment

In the spectral splitting of atoms by a magnetic field, the term involving $\mathbf{B}\cdot\mathbf{L}$ in (1) is responsible for the normal Zeeman effect.

Question 8.19*

Prove that Schrödinger's equation

$$\frac{1}{2m}(\mathbf{p} - q\mathbf{A})^2 \psi = \left(i\hbar\frac{\partial}{\partial t} - q\Phi\right)\psi \tag{1}$$

is invariant under the gauge transformation

$$\left.\begin{aligned}\Phi \to \Phi' &= \Phi - \frac{\partial\chi}{\partial t}\\[4pt]\mathbf{A} \to \mathbf{A}' &= \mathbf{A} + \nabla\chi\\[4pt]\psi \to \psi' &= e^{iq\chi/\hbar}\,\psi\end{aligned}\right\}. \tag{2}$$

In the above, $\psi = \psi(\mathbf{r}, t)$ and $\chi = \chi(\mathbf{r}, t)$ are arbitrary wave and gauge functions respectively.

Hint: Use the momentum operator $\mathbf{p} = -i\hbar\nabla$.

Solution

We need to prove that (1) is restored following a gauge transformation. That is,

$$\frac{1}{2m}(\mathbf{p} - q\mathbf{A}')^2 \psi' = \left(i\hbar\frac{\partial}{\partial t} - q\Phi'\right)\psi'. \tag{3}$$

Substituting (2) in (3) gives

$$\begin{aligned}\frac{1}{2m}(\mathbf{p} - q\mathbf{A} - q\nabla\chi)^2 \psi' &= \left(i\hbar\frac{\partial}{\partial t} - q\Phi + q\frac{\partial\chi}{\partial t}\right)\psi'\\[6pt]&= e^{iq\chi/\hbar}\left(i\hbar\frac{\partial\psi}{\partial t} - q\frac{\partial\chi}{\partial t} - q\Phi + q\frac{\partial\chi}{\partial t}\right)\psi\\[6pt]&= e^{iq\chi/\hbar}\left(i\hbar\frac{\partial\psi}{\partial t} - q\Phi\right)\psi. \tag{4}\end{aligned}$$

Now $\mathbf{p}\psi' = -i\hbar\nabla\left(e^{iq\chi/\hbar}\psi\right) = q\nabla\chi\psi' + e^{iq\chi/\hbar}\,\mathbf{p}\psi$ or $(\mathbf{p} - q\nabla\chi)\psi' = e^{iq\chi/\hbar}\,\mathbf{p}\psi$ and so $(\mathbf{p} - q\mathbf{A} - q\nabla\chi)\psi' = e^{iq\chi/\hbar}\,(\mathbf{p} - q\mathbf{A})\psi$. Therefore

$$\frac{1}{2m}(\mathbf{p}-q\mathbf{A}-q\boldsymbol{\nabla}\chi)^2\psi' = \frac{1}{2m}(\mathbf{p}-q\mathbf{A}-q\boldsymbol{\nabla}\chi)\cdot\left\{e^{iq\chi/\hbar}(\mathbf{p}-q\mathbf{A})\psi\right\}$$

$$= \frac{e^{iq\chi/\hbar}}{2m}\left\{q\boldsymbol{\nabla}\chi\cdot(\mathbf{p}-q\mathbf{A})+\mathbf{p}\cdot(\mathbf{p}-q\mathbf{A})-\right.$$
$$\left. q(\mathbf{A}+\boldsymbol{\nabla}\chi)\cdot(\mathbf{p}-q\mathbf{A})\right\}\psi$$

$$= \frac{e^{iq\chi/\hbar}}{2m}(\mathbf{p}-q\mathbf{A})^2\psi. \tag{5}$$

Equations (4) and (5) yield

$$\frac{e^{iq\chi/\hbar}}{2m}(\mathbf{p}-q\mathbf{A})^2\psi = e^{iq\chi/\hbar}\left(i\hbar\frac{\partial\psi}{\partial t}-q\Phi\right)\psi, \tag{6}$$

which is (1).

Comment

In both classical and quantum mechanics all physical observables remain unchanged by a gauge transformation.

Question 8.20**

(This question and its solution adopt the approach of Ref. [12].)

Consider an arbitrary distribution of electric currents contained within a bounded region having volume v'. Show that the retarded vector potential

$$\mathbf{A}(\mathbf{r},t) = \frac{\mu_0}{4\pi}\int_v \frac{\mathbf{J}(\mathbf{r}',t-|\mathbf{r}-\mathbf{r}'|/c)}{|\mathbf{r}-\mathbf{r}'|}dv', \tag{1}$$

can be expanded as the infinite series

$$A_i(\mathbf{r},t) = \frac{\mu_0}{4\pi}\left[\frac{1}{r}\int_v J_i\,dv' + \frac{r_j}{r^3}\left(\int_v r'_j J_i\,dv' + \frac{r}{c}\int_v r'_j \dot{J}_i\,dv'\right)+\right.$$

$$\left.\frac{(3r_j r_k - r^2\delta_{jk})}{2r^5}\left(\int_v r'_j r'_k J_i\,dv' + \frac{r}{c}\int_v r'_j r'_k \dot{J}_i\,dv'\right)+\frac{r_j r_k}{2c^2 r^3}\int_v r'_j r'_k \ddot{J}_i\,dv' + \cdots\right]. \tag{2}$$

Hint: Use (1) and (2) of Question 1.29.

[12] R. E. Raab and O. L. de Lange, *Multipole theory in electromagnetism*, Chap. 1, pp. 15–17. Oxford: Clarendon Press, 2005.

Solution

Equation (1) of Question 1.29 is

$$|\mathbf{r} - \mathbf{r}'| = r - \frac{r_i}{r} r_i' - \frac{r_i r_j - r^2 \delta_{ij}}{2r^3} r_i' r_j' - \cdots,$$

which is used to expand the retarded time $t - |\mathbf{r} - \mathbf{r}'|/c$ as follows:

$$t - |\mathbf{r} - \mathbf{r}'|/c = (t - r/c) + \left\{ \frac{r_i}{cr} r_i' + \frac{r_i r_j - r^2 \delta_{ij}}{2cr^3} r_i' r_j' + \cdots \right\} = t' + \Delta t, \qquad (3)$$

where

$$t' = t - r/c, \qquad (4)$$

is the retarded time at the origin O, and

$$\Delta t = \frac{r_i}{cr} r_i' + \frac{r_i r_j - r^2 \delta_{ij}}{2cr^3} r_i' r_j' + \cdots. \qquad (5)$$

Expanding $J_i(\mathbf{r}', t' + \Delta t)$ about t' using Taylor's theorem yields

$$J_i(\mathbf{r}', t' + \Delta t) = J_i(\mathbf{r}', t') + \frac{\partial J_i(\mathbf{r}', t')}{\partial t'} \Delta t + \frac{1}{2!} \frac{\partial^2 J_i(\mathbf{r}', t')}{\partial t'^2} (\Delta t)^2 + \cdots$$

$$= J_i + \dot{J}_i \Delta t + \tfrac{1}{2} \ddot{J}_i (\Delta t)^2 + \cdots, \qquad (6)$$

where $J_i = J_i(\mathbf{r}', t')$, $\dot{J}_i = \dfrac{\partial J_i(\mathbf{r}', t')}{\partial t'}$, and so on. Substituting (5) in (6) gives

$$J_i(\mathbf{r}', t' + \Delta t) = J_i + \frac{r_j}{cr} r_j' \dot{J}_i + \frac{(r_j r_k - r^2 \delta_{jk})}{2cr^3} r_j' r_k' \dot{J}_i + \frac{r_j r_k}{2c^2 r^2} r_j' r_k' \ddot{J}_i + \cdots. \qquad (7)$$

Next we use (2) of Question 1.29 and (7) to expand the integrand in (1) which yields

$$\frac{J_i(\mathbf{r}', t')}{|\mathbf{r} - \mathbf{r}'|} = \left\{ J_i + \frac{r_j}{cr} r_j' \dot{J}_i + \frac{(r_j r_k - r^2 \delta_{jk})}{2cr^3} r_j' r_k' \dot{J}_i + \frac{r_j r_k}{2c^2 r^2} r_j' r_k' \ddot{J}_i + \cdots \right\} \times$$

$$\left\{ \frac{1}{r} + \frac{r_l}{r^3} r_l' + \frac{(3r_l r_m - r^2 \delta_{lm})}{2r^5} r_l' r_m' + \cdots \right\}.$$

Multiplying the two terms in curly brackets and omitting terms of order $(r'/r)^3$ and higher give

$$\frac{J_i(\mathbf{r}', t')}{|\mathbf{r} - \mathbf{r}'|} = \frac{J_i}{r} + \frac{r_j}{r^3} \left(J_i + \frac{r}{c} \dot{J}_i \right) r_j' + \left[\frac{(3 r_j r_k - r^2 \delta_{jk})}{2r^5} \left(J_i + \frac{r}{c} \dot{J}_i \right) + \frac{r_j r_k}{2c^2 r^3} \ddot{J}_i \right] r_j' r_k' + \cdots,$$

which is (2).

Comments

(i) The retarded scalar potential can be expanded in a similar way. The result is

$$\Phi(\mathbf{r},t) \;=\; \frac{1}{4\pi\epsilon_0}\left[\frac{1}{r}\int_v \rho\,dv' \;+\; \frac{r_i}{r^3}\int_v r'_i\Big(\rho + \frac{r}{c}\dot{\rho}\Big)dv' \;+\right.$$

$$\left.\frac{(3r_i r_j - r^2\delta_{ij})}{2r^5}\int_v r'_i r'_j\Big(\rho + \frac{r}{c}\dot{\rho}\Big)dv' \;+\; \frac{r_i r_j}{2c^2 r^3}\int_v r'_i r'_j \ddot{\rho}\,dv' \;+\;\cdots\right]. \tag{8}$$

(ii) Equation (2) is a useful result. We will use it in Chapter 11 as the starting point of our treatment of electromagnetic radiation. Equation (8), as we have implied earlier, is essentially redundant in the Lorenz gauge.[‡]

(iii) A necessary condition for the convergence of the series expansions (2) and (8) is that $r' \ll r$ for all $r' \in v'$. If we let d represent a typical length scale of the distribution (i.e. $d \sim \sqrt[3]{v'}$), then this condition can be expressed as

$$d \ll r. \tag{9}$$

Equation (9) is used in Question 11.2 when we consider the radiation from a time-dependent distribution of electric charges in arbitrary (but non-relativistic) motion.

[‡]See, for example, (1) of Question 8.9.

9

Static electric and magnetic fields in matter

Under high resolution at the atomic level, matter can be regarded as a distribution of electric charges (electrons and nuclei) in vacuum. At each mathematical point in this vacuum, we denote the microscopic electromagnetic fields produced by these charges as $\mathbf{e}(\mathbf{r}, t)$ and $\mathbf{b}(\mathbf{r}, t)$. Because the fields vary extremely rapidly in both space and time, it is not possible to measure them. The only meaningful way of specifying a field inside matter is in terms of its macroscopic average: $\mathbf{E}(\mathbf{r}, t)_{\text{macro}} = \langle \mathbf{e}(\mathbf{r}, t) \rangle$ and $\mathbf{B}(\mathbf{r}, t)_{\text{macro}} = \langle \mathbf{b}(\mathbf{r}, t) \rangle$. This is done as the volume average of the microscopic field over an infinitesimally small macroscopic volume element[‡] Δv, where

$$\mathbf{E}_{\text{macro}} = \frac{1}{\Delta v} \int_{\Delta v} \mathbf{e}(\mathbf{r}, t) \, dv_{\text{micro}} \quad \text{and} \quad \mathbf{B}_{\text{macro}} = \frac{1}{\Delta v} \int_{\Delta v} \mathbf{b}(\mathbf{r}, t) \, dv_{\text{micro}}. \tag{I}$$

This averaging process is not trivial, and the reader is referred to Ref. [1] for further information on this topic. Henceforth, we will write $\mathbf{E}_{\text{macro}}$ and $\mathbf{B}_{\text{macro}}$ simply as $\mathbf{E}$ and $\mathbf{B}$, without subscripts. Of course, the understanding that we are dealing with macroscopic fields remains.

Part 1: Electric fields

Suppose we consider a non-conducting material (an electrical insulator) that is placed in an external electrostatic field. The electrons of the individual atoms respond by moving a small amount (some fraction of an atomic radius), although they cannot detach from their parent atoms and rearrange themselves like delocalized electrons. Because of this, the electric field inside these materials (we call them dielectrics) is only partially cancelled out, leading to a finite electric field in the interior. Of course, in a conducting material the cancellation is complete, as we saw in Chapter 3 (at least for static fields). This effect—where the electrons in a dielectric are displaced from their equilibrium positions (relative to their nuclei)—is called *polarization*. It has some interesting consequences. For example, we will discover that *bound charges*

[‡]That is, a volume element which is small enough to be treated as a mathematical point, but large enough to contain sufficiently many molecules so that molecular fluctuations average out to a constant value in a static external field.

---

[1]   A. Zangwill, *Modern electrodynamics*, Chap. 2, pp. 38–42. Cambridge: Cambridge University Press, 2013.

Solved Problems in Classical Electromagnetism. J. Pierrus, Oxford University Press (2018).
© J. Pierrus. DOI: 10.1093/oso/9780198821915.001.0001

can be induced on the surface and inside a dielectric. These bound charges are just as important as free charges for the purpose of determining the electric fields in and around the material. In Chapter 7 we discussed Maxwell's equations for a vacuum and the relationship of the fields to their free sources. Now, in this chapter and the next, we will investigate the role of bound charges (and later bound currents) and how these sources modify Maxwell's equations in the presence of matter. See Question 9.2 (and also Question 9.13).

In Question 2.20 we learnt that for any charge distribution the electric field and potential at a distant point are those of the leading non-vanishing multipole. As this cannot be charge for a bulk sample of neutral matter, the next possibility is the electric dipole moment. Even for a bulk sample comprising non-polar molecules, an external field in which the sample is placed will most likely induce an electric dipole in the molecules of the medium, as will interactions between neighbouring molecules. The neglect of contributions to the field from all the higher multipoles is known as the electric dipole approximation. Textbooks often assume implicitly that the electric dipole approximation holds, although there are instances where it fails. Then it must be replaced by a suitable higher-order multipole approximation; see also Question 10.4. In this chapter we consider no multipole contributions beyond the dipole.

Consider a polarized dielectric, and let $n(\mathbf{r})$ represent the number density of molecules at a macroscopic volume element at $\mathbf{r}$. The total electric dipole moment of the volume element is $n\bar{\mathbf{p}}\Delta v$, where $\bar{\mathbf{p}}$ is the average dipole moment of a molecule inside Δv. We now introduce a new quantity called the polarization density $\mathbf{P}(\mathbf{r})$, defined as the average electric dipole moment per unit macroscopic volume. Then

$$\mathbf{P} = n\bar{\mathbf{p}}. \tag{II}$$

Certain dielectric materials which are isotropic on the macroscopic scale have the property that a dipole induced in them by a weak electric field $\mathbf{E}$ is parallel to $\mathbf{E}$. So

$$\mathbf{P} = \epsilon_0 \chi_e \mathbf{E}, \tag{III}$$

where χ_e is a dimensionless macroscopic property called the electric susceptibility (ϵ_0 in the constant of proportionality in (III) ensures that χ_e is dimensionless). Materials for which (III) holds are called linear, isotropic and homogeneous[#] dielectrics (lih for short). See Questions 9.1–9.9 for some applications of the above. Many materials are of course anisotropic, and, as we discussed earlier for an anisotropic molecule,[†] it is possible that an induced dipole is not parallel to the field. This would imply that $\mathbf{P}$ is not parallel to $\mathbf{E}$. Then for a linear response, (III) is replaced by

$$P_i = \epsilon_0 \chi_{ij} E_j,$$

where the second-rank electric susceptibility tensor χ_{ij} plays, for anisotropic matter, an analogous role to the polarizability tensor α_{ij} for a single molecule.

[#] A homogeneous dielectric has the property that if all its macroscopic points experience the same $\mathbf{E}$, then $\mathbf{P}(\mathbf{r})$ is independent of $\mathbf{r}$.

[†] See Comments (v)–(viii) of Question 2.26.

Changes occur when a sample of dielectric material is placed in the vicinity of an external electric field. For example, the field in the vacuum outside the sample is affected, or the capacitance of a capacitor filled with the material increases. This latter effect enables us to define another bulk (or macroscopic) property, namely, the dielectric constant (or relative permittivity ϵ_r) of the material

$$\epsilon_r = C/C_0, \tag{IV}$$

where C (C_0) is the capacitance of the capacitor in the presence (absence) of the material. Now $C_0 = \epsilon_0 A/d$ (the symbols have their usual meaning), and so, from (IV), $C = \epsilon_r \epsilon_0 A/d = \epsilon A/d$ where the macroscopic property

$$\epsilon = \epsilon_0 \epsilon_r, \tag{V}$$

is the permittivity of the material.[b] In Question 9.2 we show that the polarization of matter leads us to a new macroscopic field $\mathbf{D}$ called the electric displacement, where in the electric dipole approximation

$$\mathbf{D} = \epsilon_0 \mathbf{E} + \mathbf{P}. \tag{VI}$$

The energy of an electrostatic field in vacuum discussed in Question 2.17,

$$U' = \tfrac{1}{2}\epsilon_0 \int_v \mathbf{E} \cdot \mathbf{E} \, dv, \tag{VII}$$

also holds inside a dielectric if we take $\mathbf{E}$ and dv to be macroscopic quantities (the reason for the prime in (VII) will become apparent below). However, in addition to the energy required to establish the field there is also in a dielectric the energy U'' required to distort the molecules (the energy of polarization). For a given electric field U' is the same whether $\mathbf{E}$ permeates a vacuum or a dielectric. Of course, the energy associated with the polarization exists only in the presence of matter, and its contribution must also be included in the total energy U. We attempt this task in Questions 9.9 and 9.10.

Part 2: Magnetic fields

All atoms and molecules experience a force in a non-uniform magnetic field $\mathbf{B}$ (i.e. where the field *gradient* $\nabla B \neq 0$), and so does bulk matter. But depending on the type of material used, very different forces are observed. Experiments show that these forces can be divided into three different categories. The important facts are:

☞ Some substances are expelled from the field with a weak force, others are attracted into the field with an equally weak force and there is another class which is pulled into regions of higher $\mathbf{B}$ with a very strong force.[‡] These materials are said to be diamagnetic, paramagnetic and ferromagnetic respectively.[#]

---

[b] Although shown here for the special case of a parallel-plate capacitor, (V) is a general result.

[‡] It is important to add that these observations remain the same if the direction of the magnetic field is *reversed*.

[#] We mention in passing that all atoms display diamagnetic behaviour, although this feature is completely masked when either para- or ferromagnetism is present.

☞ When the field is turned off, ferromagnetic materials often become permanently magnetized, whereas dia- and paramagnetic substances do not.

☞ If $\mathbf{B}$ and ∇B are reduced to half their original value (by halving the current in a solenoid, say) it is found that the force on dia- and paramagnetic substances decreases by a factor of four, while that for a ferromagnetic material drops by a factor of about two.

We must therefore conclude that the 'magnetism' in diamagnetic and paramagnetic materials is *induced* by the field, and that this magnetism disappears when $B = 0$. The quadratic dependence described above is consistent with a magnetism which is proportional to the inducing $\mathbf{B}$-field, and which interacts linearly with the magnetic field gradient ∇B.

The orbital motion of the electrons in a molecule (or atom), and the intrinsic magnetic moments of the electrons themselves, may result in the molecule having a magnetic dipole moment. Sometimes the electronic states of the molecule are filled in pairs in such a way that, in the absence of an external magnetic field, the net magnetic moment of the molecule is zero. However, other molecules may possess an odd number of electrons, while those with an even number can still have unpaired electrons and a permanent magnetic moment. Ordinarily, these molecular moments cancel out in a macroscopic volume element because their orientations are randomized by thermal agitation. But in an external magnetic field it is possible for a net alignment of these magnetic dipoles to occur, which results in the medium becoming magnetized. In all materials, we will assume that a smooth macroscopic property $\mathbf{M}(\mathbf{r})$ can be defined such that the total magnetic dipole moment of a macroscopic volume v is given by

$$\mathbf{m} = \int_v \mathbf{M}(\mathbf{r}') \, dv'. \tag{VIII}$$

Here $\mathbf{M}(\mathbf{r})$ is the magnetic dipole moment per unit macroscopic volume, analogous to the electric polarization $\mathbf{P}(\mathbf{r})$. Usually, we will not concern ourselves as to the cause of this magnetization; it could be dia-, para- or even ferro-magnetism. The magnetization of matter produces bound surface and volume currents (see Question 9.13). Like bound charges, these bound currents must also feature in Maxwell's equations where matter is present. In Questions 9.12–9.16 we use the notion of bound currents to find the fields of various magnetized objects with a simple geometry.

Our study of magnetized matter proceeds, at least initially, in a parallel way with electrical polarization. Firstly, we will assume that the materials used in the questions of this chapter are lih (unless they are ferromagnetic). Secondly, we introduce the permeability μ and the relative permeability μ_{r}, where

$$\mu = \mu_0 \mu_{\mathrm{r}}. \tag{IX}$$

(For diamagnetic (paramagnetic) materials μ_{r} is less (greater) than unity by a few parts in 10^5. Ferromagnetic materials, on the other hand, usually have relative permeabilities which depend on the applied field and range typically from 10^3 to 10^6.) Thirdly, we introduce a macroscopic field known as the **H**-field[‡] (which is the magnetic counterpart of **D** for polarized matter). To magnetic dipole order this field is given by

$$\mathbf{H} = \mu_0^{-1}\mathbf{B} - \mathbf{M}, \tag{X}$$

and is discussed elsewhere in this chapter (see, for example, Question 9.13). There are some important advantages associated with the use of **H**, which we mention in Comment (ii) of Question 10.3. Unfortunately, the parallel treatment of the electric and magnetic cases does not continue indefinitely. For historical reasons which have been the source of some debate in the past, **M** is not expressed in terms of **B**, as one might anticipate, but in terms of **H**. That is,

$$\mathbf{M} = \chi_m\,\mathbf{H}. \tag{XI}$$

Here χ_m is the magnetic susceptibility. It is a dimensionless quantity and is negative (positive) for diamagnetic (paramagnetic) materials. Particular examples involving cylinders, discs, rings and needles are considered in some of the questions below.

Before concluding this introduction, we mention the phenomenon of hysteresis. For ferromagnetic materials, a non-linear relationship exists between **B** and **H**. Because **H** depends on how the material was treated in the past (i.e. on its history), it cannot be expressed as a single-valued function of **B**. This is illustrated schematically below. Most ferromagnetic materials reach a state called the saturation magnetization. Increasing H beyond the point P shown in the figure does not produce any further increase in M, and the magnetic field more-or-less approaches a maximum value asymptotically. Hysteresis is an important topic with regard to the design and in particular functioning of devices that operate through their magnetic fields (e.g. transformers); see Questions 9.18 and 9.19.

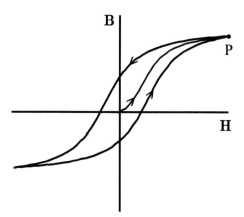

Question 9.1

Two ideal parallel-plate capacitors having the same size and shape are connected to a battery (emf $\mathcal{E}$) as shown in Figs (I) and (II) below. The capacitor labelled C_0 is a vacuum device; the other labelled C has an lih dielectric completely filling the space between its plates.

(a) For the parallel combination (I), express the charge on C in terms of the charge on C_0.

(b) For the series combination (II), express the voltage across C in terms of the voltage across C_0.

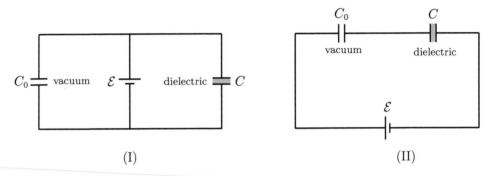

(I)   (II)

Solution

(a) In this configuration, the voltage across each capacitor is $\mathcal{E}$. So $q_0 = C_0\mathcal{E}$ and $q = C\mathcal{E}$. It follows from (IV) of the introduction that $q = \epsilon_r C_0 \mathcal{E}$. Thus

$$q = \epsilon_r q_0. \tag{1}$$

(b) Here the charge on each capacitor is the same, and $V_0 = \dfrac{q}{C_0}$. Then $V = \dfrac{q}{C} = \dfrac{q}{\epsilon_r C_0}$, and

$$V = V_0/\epsilon_r. \tag{2}$$

Comments

(i) The dielectric constants of some common materials can easily be found (internet, handbooks, etc.). For gases at standard temperature and pressure, ϵ_r differs from unity in typically the fourth decimal place (so in this respect, air behaves like a vacuum). Some condensed substances comprising polar molecules may have ϵ_r values much greater than 1. For instance, the relative permittivity of water at room temperature and pressure is ~ 80.

(ii) It follows from the above results that a capacitor containing a dielectric can store more electric charge than its vacuum equivalent for the same applied voltage.

Question 9.2

Consider a polarized dielectric body of arbitrary shape having surface s and volume v as shown. Suppose $\mathbf{P} = \mathbf{P}(\mathbf{r'})$ is a time-independent polarization density and that the body is electrically neutral.

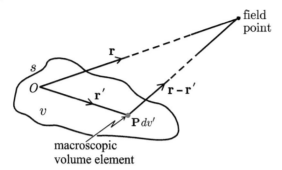

(a) Show that, within the electric dipole approximation, the scalar potential at an arbitrary field point located at $\mathbf{r}$ (relative to origin O) is

$$\Phi(\mathbf{r}) \;=\; \frac{1}{4\pi\epsilon_0}\int_v \frac{\mathbf{P}(\mathbf{r'})\cdot(\mathbf{r}-\mathbf{r'})}{|\mathbf{r}-\mathbf{r'}|^3}\, dv'. \tag{1}$$

Hint: Start with (1) of Question 2.11.

(b) Hence prove, by transforming the integral in (1), that

$$\Phi(\mathbf{r}) \;=\; \frac{1}{4\pi\epsilon_0}\oint_s \frac{(\mathbf{P}\cdot\hat{\mathbf{n}})}{|\mathbf{r}-\mathbf{r'}|}\, da' + \frac{1}{4\pi\epsilon_0}\int_v \frac{(-\nabla'\cdot\mathbf{P})}{|\mathbf{r}-\mathbf{r'}|}\, dv'. \tag{2}$$

Hint: Use the result $\nabla'\left(\dfrac{1}{|\mathbf{r}-\mathbf{r'}|}\right) = \dfrac{(\mathbf{r}-\mathbf{r'})}{|\mathbf{r}-\mathbf{r'}|^3}$ derived in Question 1.6.

Solution

(a) Consider a macroscopic volume element dv' located at $\mathbf{r'}$. The electric dipole moment of this element is $d\mathbf{p} = \mathbf{P}(\mathbf{r'})\,dv'$. The dipole potential $\Phi(\mathbf{r}) \sim \mathbf{p}\cdot\mathbf{r}/r^3$ gives the contribution of dv' to the scalar potential at $\mathbf{r}$:

$$d\Phi(\mathbf{r}) \;=\; \frac{1}{4\pi\epsilon_0}\frac{d\mathbf{p}(\mathbf{r'})\cdot(\mathbf{r}-\mathbf{r'})}{|\mathbf{r}-\mathbf{r'}|^3} \;=\; \frac{1}{4\pi\epsilon_0}\frac{\mathbf{P}(\mathbf{r'})\cdot(\mathbf{r}-\mathbf{r'})}{|\mathbf{r}-\mathbf{r'}|^3}\, dv', \tag{3}$$

and then $\Phi(\mathbf{r})$ in (1) follows by integrating (3).

(b) Because of the hint,

$$\Phi(\mathbf{r}) = \frac{1}{4\pi\epsilon_0}\int_v \mathbf{P}(\mathbf{r'}) \cdot \nabla'\left(\frac{1}{|\mathbf{r}-\mathbf{r'}|}\right)dv'. \tag{4}$$

Using the identity (5) of Question 1.8 yields

$$\mathbf{P}(\mathbf{r'}) \cdot \nabla'\left(\frac{1}{|\mathbf{r}-\mathbf{r'}|}\right) = \nabla'\cdot\left(\frac{\mathbf{P}(\mathbf{r'})}{|\mathbf{r}-\mathbf{r'}|}\right) - \frac{\nabla'\cdot\mathbf{P}(\mathbf{r'})}{|\mathbf{r}-\mathbf{r'}|},$$

and then substituting this result in (4) gives

$$\Phi(\mathbf{r}) = \frac{1}{4\pi\epsilon_0}\int_v \nabla'\cdot\left(\frac{\mathbf{P}(\mathbf{r'})}{|\mathbf{r}-\mathbf{r'}|}\right)dv' + \frac{1}{4\pi\epsilon_0}\int_v \frac{(-\nabla'\cdot\mathbf{P}(\mathbf{r'}))}{|\mathbf{r}-\mathbf{r'}|}dv'.$$

Now the first integral on the right-hand side above can be transformed using Gauss's theorem, and so

$$\Phi(\mathbf{r}) = \frac{1}{4\pi\epsilon_0}\oint_s \frac{\mathbf{P}(\mathbf{r'})\cdot d\mathbf{a'}}{|\mathbf{r}-\mathbf{r'}|} + \frac{1}{4\pi\epsilon_0}\int_v \frac{(-\nabla'\cdot\mathbf{P}(\mathbf{r'}))}{|\mathbf{r}-\mathbf{r'}|}dv', \tag{5}$$

which is (2) because $d\mathbf{a'} = \hat{\mathbf{n}}\,da'$.

Comments

(i) The bracketed terms $\mathbf{P}\cdot\hat{\mathbf{n}}$ and $-\nabla\cdot\mathbf{P}$ (for convenience we omit the primes) on the right-hand side of (2) have the dimensions of charge per unit area and charge per unit volume respectively. This suggests that we define a bound surface-charge density σ_b and a bound volume-charge density ρ_b, as follows:

$$\sigma_b = \mathbf{P}\cdot\hat{\mathbf{n}} \quad\text{and}\quad \rho_b = -\nabla\cdot\mathbf{P}. \tag{6}$$

(These results are, of course, only correct within the electric dipole approximation.)

(ii) Two obvious conclusions which follow from (6) are:

☞ Bound surface charge arises only from the *normal* component of $\mathbf{P}$, unlike the tangential component $\mathbf{P}$ which effectively displaces charge *parallel* to the surface.

☞ Bound charge appears at a macroscopic point inside a dielectric only where $\mathbf{P}$ has a non-zero divergence.

(iii) We ought to check that the net bound charge q_b carried by the dielectric body is zero. To do this, we begin with the definition of the charge densities and make use of (6). Then

$$q_b = \oint_s \sigma_b \, da + \int_v \rho_b \, dv = \oint_s \mathbf{P} \cdot da + \int_v (-\boldsymbol{\nabla} \cdot \mathbf{P}) \, dv.$$

Applying Gauss's theorem to the surface integral gives

$$\int_v (\boldsymbol{\nabla} \cdot \mathbf{P}) \, dv + \int_v (-\boldsymbol{\nabla} \cdot \mathbf{P}) \, dv = 0,$$

as required.

(iv) A polarized dielectric, as we have shown above, carries bound charge distributed both on its surface s and also throughout its volume v. This being so, it should be possible to remove the polarized dielectric altogether and replace it with an equivalent distribution of *free charge in vacuum*. In this alternative representation, the surface of the dielectric would be replaced by an identical surface s in vacuum, carrying free charge and having surface-charge density $\sigma_f = \mathbf{P} \cdot \hat{\mathbf{n}}$. At the same time, we would imitate the interior of the dielectric with an identical volume v also in vacuum carrying free charge and having volume-charge density $\rho_f = -\boldsymbol{\nabla} \cdot \mathbf{P}$. The two representations (polarized dielectric on the one hand, vacuum distribution of free charges on the other) are equivalent for the purpose of determining macroscopic fields. See, for example, Question 9.4.

(v) As a result of the averaging described by (I) of the introduction, Gauss's law remains valid on the macroscopic scale:

$$\oint_s \mathbf{E} \cdot da = \frac{1}{\epsilon_0} \int_v \rho_{\text{total}} \, dv,$$

where $\mathbf{E}$ is understood to be the macroscopic field on a macroscopic area element da at the surface s, and ρ_{total} means the density of all the charge, free and bound. Suppose we consider the case where the surface s lies entirely *within* the polarized dielectric. Then

$$\oint_s \mathbf{E} \cdot da = \frac{1}{\epsilon_0} \int_v (\rho_f + \rho_b) \, dv. \tag{7}$$

Using Gauss's theorem to transform the surface integral yields

$$\int_v \boldsymbol{\nabla} \cdot \mathbf{E} \, dv = \frac{1}{\epsilon_0} \int_v (\rho_f + \rho_b) \, dv, \tag{8}$$

and because v is arbitrary we obtain the differential form

$$\boldsymbol{\nabla} \cdot \mathbf{E} = (\rho_f + \rho_b)/\epsilon_0, \tag{9}$$

which holds at each macroscopic point.

(vi) In the electric dipole approximation $\rho_b = -\nabla \cdot \mathbf{P}$. Substituting this result in (7) and (9) and rearranging give

$$\oint_s (\epsilon_0 \mathbf{E} + \mathbf{P}) \cdot d\mathbf{a} = q_f \quad \text{and} \quad \nabla \cdot (\epsilon_0 \mathbf{E} + \mathbf{P}) = \rho_f, \quad (10)$$

respectively (to obtain $(10)_1$ apply Gauss's theorem). Motivated by (10), we define—for the electric dipole approximation—a new vector, $\mathbf{D}$, at a macroscopic point

$$\mathbf{D} = \epsilon_0 \mathbf{E} + \mathbf{P}, \quad (11)$$

called the electric displacement, and then

$$\int_v \nabla \cdot \mathbf{D} \, dv = q_f \quad \text{and} \quad \nabla \cdot \mathbf{D} = \rho_f. \quad (12)$$

Notice from (12) that only the *free* charges contribute to the electric displacement. The contributions from the *bound* charges have not disappeared, but have been absorbed into this new vector $\mathbf{D}$.

(vii) For an lih dielectric with $\mathbf{P} = \epsilon_0 \chi_e \mathbf{E}$ (see (III) in the introduction) we obtain the following results:

☞ From (11), $\mathbf{D} = \epsilon_0 \mathbf{E} + \mathbf{P} = \epsilon_0 (1 + \chi_e) \mathbf{E}$. Now $(1 + \chi_e) = \epsilon_r$ as we will show in Comment (iii) of Question 9.4. So

$$\mathbf{D} = \epsilon \mathbf{E}, \quad (13)$$

which holds within the electric dipole approximation.

☞ From $(6)_2$ we have

$$\rho_b = -\nabla \cdot \mathbf{P} = -\epsilon_0 \chi_e \nabla \cdot \mathbf{E} = -\chi_e \rho_f, \quad (14)$$

because of Gauss's law. This last equation shows that the bound volume-charge density is zero in all cases except where free charge is present inside the material.

Question 9.3

An lih dielectric ($\epsilon_r = 3$; $\chi_e = 2$) in the shape of a circular cylinder has length ℓ and radius r. It is placed with its axis parallel to a uniform (vacuum) electric field $E_0 = 1.5 \times 10^6 \text{ V m}^{-1}$. Calculate E, D, P, ρ_b and σ_b inside the dielectric when:

(a) $r \gg \ell$ (here the cylinder is a disc).

(b) $r \ll \ell$ (here the cylinder is a needle or whisker).

Hint: You may assume that the normal components of $\mathbf{D}$ and tangential components of $\mathbf{E}$ are continuous at all vacuum–dielectric boundaries. (These 'matching conditions' are derived in Question 10.6.)

Solution

(a) The electric-field lines are essentially perpendicular to the flat faces, and so the matching condition for **D** requires that $D_{\text{in}} = D_0$. Now $D_0 = \epsilon_0 E_0$ and

$$D_{\text{in}} = \epsilon_0 E_0 = \frac{1}{4\pi} \times 4\pi\epsilon_0 \times E_0 = \frac{1}{4\pi \times 9.0 \times 10^9} \times 1.5 \times 10^6 = \boxed{1.33 \times 10^{-5}\,\mathrm{C\,m^{-2}}},$$

and then $\boxed{E_{\text{in}}} = \frac{D_{\text{in}}}{\epsilon} = \frac{E_0}{\epsilon_r} = \frac{1.5 \times 10^6}{3} = \boxed{5.00 \times 10^5\,\mathrm{V\,m^{-1}}},$

$$\boxed{P_{\text{in}}} = \epsilon_0 \chi_e E_{\text{in}} = \frac{4\pi\epsilon_0}{4\pi}(\epsilon_r - 1) E_{\text{in}} = \frac{2 \times 5.0 \times 10^5}{4\pi \times 9.0 \times 10^9} = \boxed{8.84 \times 10^{-6}\,\mathrm{C\,m^{-2}}},$$

$$\boxed{\rho_b} = -\chi_e \rho_f = \boxed{0},$$

$$\boxed{\sigma_b} = \mathbf{P}\cdot\hat{\mathbf{n}} = \pm P = \boxed{\pm 8.84 \times 10^{-6}\,\mathrm{C\,m^{-2}}} \text{ on the top/bottom faces; 0 elsewhere.}$$

(b) Here the electric-field lines are parallel to the curved face, and the matching condition for **E** requires that $E_{\text{in}} = E_0$. So

$$\boxed{E_{\text{in}}} = 1.5 \times 10^6\,\mathrm{V\,m^{-1}},$$

and then $\boxed{D_{\text{in}}} = \epsilon_0 \epsilon_r E_{\text{in}} = \frac{3 \times 1.5 \times 10^6}{4\pi \times 9.0 \times 10^9} = \boxed{3.98 \times 10^{-5}\,\mathrm{C\,m^{-2}}},$

$$\boxed{P_{\text{in}}} = \epsilon_0 \chi_e E_{\text{in}} = \frac{4\pi\epsilon_0}{4\pi}(\epsilon_r - 1) E_{\text{in}} = \frac{2 \times 1.5 \times 10^6}{4\pi \times 9.0 \times 10^9} = \boxed{2.65 \times 10^{-5}\,\mathrm{C\,m^{-2}}},$$

$$\boxed{\rho_b} = -\chi_e \rho_f = \boxed{0},$$

$$\boxed{\sigma_b} = \mathbf{P}\cdot\hat{\mathbf{n}} = \pm P = \boxed{\pm 8.84 \times 10^{-6}\,\mathrm{C\,m^{-2}}} \text{ on the top/bottom faces; 0 elsewhere.}$$

Question 9.4

The figure below illustrates a large slab of lih dielectric which has acquired a uniform polarization **P** perpendicular to its end faces.

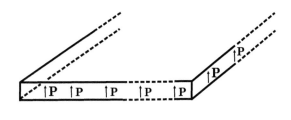

(a) Show that the electric field $\mathbf{E}_p$ arising from the *polarization* is

$$\mathbf{E}_p = \begin{cases} -\mathbf{P}/\epsilon_0 & \text{(inside the slab)} \\ 0 & \text{(outside the slab).} \end{cases} \tag{1}$$

Hint: Find the vacuum equivalent[‡] of the polarized slab.

(b) Let $\mathbf{E}_0$ be the applied electric field which produces the polarization.[♯] Use (1) to show that the (macroscopic) electric field $\mathbf{E}$ inside the polarized dielectric is

$$\mathbf{E} = \frac{\mathbf{E}_0}{1 + \chi_e}. \tag{2}$$

Solution

(a) We begin by evaluating σ_b and ρ_b using (6) of Question 9.2, and then arrange for charges in vacuum to have the same distribution. Since $\mathbf{P}$ is uniform $\nabla \cdot \mathbf{P} = 0$, and no bound charge appears within the dielectric. On the top flat face of the slab $\sigma_b = \mathbf{P} \cdot \hat{\mathbf{n}} = +P$, whilst on the bottom flat face $\sigma_b = \mathbf{P} \cdot \hat{\mathbf{n}} = -P$. All four side faces carry no bound charge, because for them $\mathbf{P} \cdot \hat{\mathbf{n}} = 0$. The vacuum equivalent of the polarized dielectric thus comprises two parallel layers of free charge having uniform density $\pm P$ as shown.

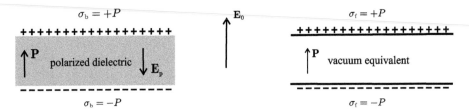

The polarization field is given by (4) of Question 2.5 and the principle of superposition. So

$$\mathbf{E}_p = \begin{cases} E_{+p} + E_{-p} = P/2\epsilon_0 + P/2\epsilon_0 = P/\epsilon_0 & \text{(inside the slab)} \\ E_{+p} + E_{-p} = -P/2\epsilon_0 + P/2\epsilon_0 = 0 & \text{(outside the slab).} \end{cases}$$

Equation (1) then follows because $\mathbf{E}_p$ is clearly in the opposite direction to $\mathbf{P}$.

(b) The total macroscopic electric field at any point inside the uniformly polarized slab (and also outside for that matter) is the superposition of the two fields: the external field $\mathbf{E}_0$ inducing the polarization and the polarization field $\mathbf{E}_p$. Both of these are uniform. So

$$\mathbf{E} = \mathbf{E}_0 + \mathbf{E}_p = \mathbf{E}_0 - \epsilon_0^{-1}\mathbf{P}, \tag{3}$$

where for an lih material $\mathbf{P} = \epsilon_0 \chi_e \mathbf{E}$. Then $\mathbf{E} = \mathbf{E}_0 - \chi_e \mathbf{E}$. Hence (2).

[‡]See Comment (iv) of Question 9.2.
[♯]See Comment (i) on p. 421.

Comments

(i) Since **P** is a uniform polarization, and because the dielectric is *homogeneous*, we must conclude that the field $\mathbf{E}_0$ is itself uniform; and so a parallel-plate capacitor would serve as a convenient device for this purpose. However, the sources of this field (the charges on each plate) would need to be sufficiently distant from the slab so that they remain essentially fixed in position as the dielectric is inserted.

(ii) Notice from (2) that $0 < E < E_0$ implies that the induced surface charges succeed in only *partially* cancelling the field $\mathbf{E}_0$ inside the dielectric, and that $E \to 0$ as $\chi_e \to \infty$, which mimics the behaviour of a good conductor.

(iii) The constant $(1 + \chi_e)$ in (2) is simply the relative permittivity, as the following argument shows. Consider a parallel-plate capacitor in vacuum carrying charge q which has been isolated from its power supply. By definition, $C_0 = q/(E_0 d)$ where d is the distance between the plates. Now suppose a dielectric slab is inserted between the plates, completely filling the air gap. Since q cannot change (the power supply was disconnected), the new capacitance is $C = \dfrac{q}{Ed} = \dfrac{q(1 + \chi_e)}{E_0 d}$ because of (2). Then $C = (1 + \chi_e)C_0$ with $C/C_0 = \epsilon_r$ from (IV). Hence the result

$$\epsilon_r = (1 + \chi_e). \tag{4}$$

Although derived here for a rectangular geometry, (4) is a general result which holds for an lih material of arbitrary shape.

(iv) The polarization density of the slab can be expressed in terms of $\mathbf{E}_0$ by substituting (2) and (4) in $\mathbf{P} = \epsilon_0 \chi_e \mathbf{E}$. Then

$$\mathbf{P} = \frac{\epsilon_0 (\epsilon_r - 1) \mathbf{E}_0}{\epsilon_r}. \tag{5}$$

Question 9.5

Consider a gas comprising n polar molecules per unit volume, each having permanent electric dipole moment **p**. Suppose the gas is in thermal equilibrium at absolute temperature T and is permeated by a uniform electric field $\mathbf{E} = E\hat{\mathbf{z}}$.

(a) Show that the polarization $\mathbf{P}_d$ arising from the permanent dipoles is

$$\mathbf{P}_d = -n \langle U \rangle \mathbf{E}/E^2, \tag{1}$$

where $\langle U \rangle = -\langle \mathbf{p} \cdot \mathbf{E} \rangle$ is the average potential energy of a molecule in the field.

(b) The probability that a molecule has potential energy in the interval dU at U is proportional to $e^{-U/k_B T}$, where k_B is Boltzmann's constant. Since U can assume all values between $-pE$ and pE (see Comment (v) of Question 2.28), we have

$$\langle U \rangle = \frac{\int_{-pE}^{pE} U e^{-U/k_B T} dU}{\int_{-pE}^{pE} e^{-U/k_B T} dU}.$$ (2)

Recalling that $pE \ll \frac{3}{2} k_B T$ over a wide range of temperatures (see Comment (vi) of Question 2.28), use (1) and (2) to show that the electric susceptibility χ_e of the gas is given by

$$\chi_e = n \left[\alpha + \frac{p^2}{3 \epsilon_0 k_B T} \right],$$ (3)

where α is an isotropic polarizability averaged over all orientations of the molecule.[‡] It results in an additional *induced* polarization $\mathbf{P}_i = n \alpha \epsilon_0 \mathbf{E}$.

Solution

(a) By symmetry it is clear that the components of $\mathbf{P}_d$ in the x and y directions are zero, and so

$$\mathbf{P}_d = n \langle p \cos \theta \rangle \hat{\mathbf{z}} = \frac{n \langle p E \cos \theta \rangle}{E} \hat{\mathbf{z}},$$ (4)

where θ is the angle which a dipole makes with the z-axis. Substituting $\langle p E \cos \theta \rangle = \langle \mathbf{p} \cdot \mathbf{E} \rangle = -\langle U \rangle$ in (4) gives (1) since $\hat{\mathbf{z}} = \mathbf{E}/E$.

(b) Because $U/k_B T \ll 1$, the Boltzmann factor in (2) can be expanded in powers of $U/k_B T$. Retaining only the first-order term gives

$$\langle U \rangle = \frac{\int_{-pE}^{pE} U(1 - U/k_B T) dU}{\int_{-pE}^{pE} (1 - U/k_B T) dU} = \frac{-\frac{2}{3} \frac{p^3 E^3}{k_B T}}{2pE} = -\frac{1}{3} \frac{p^2 E^2}{k_B T},$$

and then from (1)

$$\mathbf{P}_d = \frac{n}{3} \frac{p^2}{k_B T} \mathbf{E}.$$

Now the total polarization is the sum of $\mathbf{P}_i$ and $\mathbf{P}_d$, and so

$$\mathbf{P} = \mathbf{P}_i + \mathbf{P}_d = n \alpha \epsilon_0 \mathbf{E} + \frac{n}{3} \frac{p^2}{k_B T} \mathbf{E} = \epsilon_0 n \left[\alpha + \frac{p^2}{3 \epsilon_0 k_B T} \right] \mathbf{E}.$$ (5)

Comparing (5) with $\mathbf{P} = \epsilon_0 \chi_e \mathbf{E}$ gives (3).

[‡]Suppose α_{11}, α_{22} and α_{33} are the principal components of the polarizability of a molecule (see Comment (vi) of Question 2.26). Then $\alpha = \langle \alpha_{ij} \rangle = \frac{1}{3}(\alpha_{11} + \alpha_{22} + \alpha_{33})$.

Comments

(i) In the above derivation we have assumed that the electric field acting on a molecule is the applied external field **E** only. This neglect of molecular interactions is a good approximation at low pressures and/or high temperatures.

(ii) Equation (3) suggests an experimental technique for measuring α and p. If χ_e is determined over a range of different temperatures, then a plot of χ_e vs T^{-1} is a straight line having intercept $n\alpha$ and slope $np^2/3\epsilon_0 k_B$ (here $n =$ Avogadro's number divided by the molar volume).

(iii) This classical approach using a Boltzmann weighting to derive the average of a permanent dipole moment in the field direction was first considered by Langevin, and the Langevin relation follows from the exact integration of (2):

$$\mathbf{P}_d = np\left[\coth\left(\frac{pE}{k_BT}\right) - \left(\frac{k_BT}{pE}\right)\right]\hat{\mathbf{z}}.$$

Question 9.6

Consider a sphere of radius a made from an lih dielectric having relative permittivity ϵ_r. The sphere is placed in a uniform electric field[‡] $\mathbf{E} = E_0\hat{\mathbf{z}}$ and acquires a uniform polarization $\mathbf{P} = P_0\hat{\mathbf{z}}$.

(a) Show that the scalar potential, in spherical polar coordinates, is

$$\Phi(r,\theta) = \begin{cases} \dfrac{P_0 r \cos\theta}{3\epsilon_0} & \text{for } r \leq a, \\[4mm] \dfrac{P_0 \cos\theta}{3\epsilon_0}\left(\dfrac{a^3}{r^2}\right) & \text{for } r \geq a. \end{cases} \tag{1}$$

Hint: Use (5) of Question 9.2.

(b) Hence show that the electric field $\mathbf{E}_p$ arising from the *polarization* is

$$\mathbf{E}_p = \begin{cases} -\dfrac{\mathbf{P}}{3\epsilon_0} & \text{for } r \leq a, \\[4mm] \dfrac{P}{3\epsilon_0}\left(\dfrac{a}{r}\right)^3(2\cos\theta\,\hat{\mathbf{r}} + \sin\theta\,\hat{\boldsymbol{\theta}}) & \text{for } r \geq a. \end{cases} \tag{2}$$

(c) Show that the macroscopic fields **E** and **D** inside the polarized dielectric are

$$\mathbf{E}_{in} = \frac{3\mathbf{E}_0}{2 + \epsilon_r} \quad \text{and} \quad \mathbf{D}_{in} = \frac{3\epsilon_0\epsilon_r\mathbf{E}_0}{2 + \epsilon_r}. \tag{3}$$

[‡]The external electric field is uniform before the sphere is inserted; afterwards, its uniformity is perturbed locally by the presence of the sphere.

Solution

(a) Because $\mathbf{P}$ is uniform, $\nabla \cdot \mathbf{P} = 0$ inside the sphere. Then

$$\Phi(\mathbf{r}) = \frac{1}{4\pi\epsilon_0} \oint_s \frac{(\mathbf{P} \cdot \hat{\mathbf{n}})}{|\mathbf{r} - \mathbf{r}'|} \, da' = \frac{1}{4\pi\epsilon_0} \mathbf{P} \cdot \oint_s \frac{da'}{|\mathbf{r} - \mathbf{r}'|}. \qquad (4)$$

Now the integral on the right-hand side of (4) is given by (3) of Question 1.26. Hence (1).

(b) Using $\mathbf{E}_p = -\nabla\Phi$ (with the gradient operator for spherical polar coordinates given by $(XI)_1$ of Appendix C) yields (2).

(c) Proceeding as in (b) of Question 9.4 gives

$$\mathbf{E} = \mathbf{E}_0 + \mathbf{E}_p = \mathbf{E}_0 - \frac{\mathbf{P}}{3\epsilon_0}, \qquad (5)$$

with $\mathbf{P} = \epsilon_0 \chi_e \mathbf{E} = \epsilon_0(\epsilon_r - 1)\mathbf{E}$. Then $\mathbf{E} = \mathbf{E}_0 - \frac{1}{3}(\epsilon_r - 1)\mathbf{E}$. Hence $(3)_1$, and taking $\mathbf{D} = \epsilon\mathbf{E} = \epsilon_0\epsilon_r\mathbf{E}$ gives $(3)_2$.

Comments

(i) The polarization field $\mathbf{E}_p$, given by $(2)_1$, points in the *opposite direction* to both $\mathbf{P}$ and $\mathbf{E}$.

(ii) Because the electric dipole moment of the sphere is $\mathbf{p} = \frac{4}{3}\pi a^3 \mathbf{P}$, we can express (2) in the alternative form

$$\mathbf{E}_p = \begin{cases} -\dfrac{1}{4\pi\epsilon_0} \dfrac{\mathbf{p}}{a^3} & \text{for } r \leq a, \\[3mm] \dfrac{1}{4\pi\epsilon_0} \dfrac{3(\mathbf{p} \cdot \hat{\mathbf{r}})\hat{\mathbf{r}} - \mathbf{p}}{r^3} & \text{for } r \geq a. \end{cases} \qquad (6)$$

It is clear from $(6)_1$ that the electric field inside the sphere is constant, whilst outside, $(6)_2$ shows that $\mathbf{E}$ has an exact dipole nature, not only asymptotically but in fact for all $r \geq a$; the higher electric multipoles make no contributions whatsoever.

(iii) It follows from (3) that $\mathbf{E}$ and $\mathbf{D}$ are proportional to $\mathbf{E}_0$. They vanish, as one would expect, when $\mathbf{E}_0 = 0$.

(iv) Substituting (3) in $\mathbf{P} = \epsilon_0 \chi_e \mathbf{E}$ gives $\mathbf{P} = 3\epsilon_0 \dfrac{(\epsilon_r - 1)}{(\epsilon_r + 2)} \mathbf{E}_0$. Since $\mathbf{P} = n\bar{\mathbf{p}} = \dfrac{\mathcal{N}}{V_m}\bar{\mathbf{p}}$, where $\mathcal{N}$ is Avogadro's number and V_m the molar volume, we have

$$\frac{(\epsilon_r - 1)}{(\epsilon_r + 2)}V_m = \frac{\mathcal{N}}{3\epsilon_0}\frac{\bar{p}}{E_0}. \qquad (7)$$

Equation (7) may be used to interpret ϵ_r for a gas in terms of molecular properties.

Question 9.7

A spherical conductor of radius R carrying charge q
is surrounded to a depth R by an lih dielectric shell
as indicated in the figure alongside.

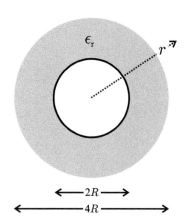

(a) Show that the electric field is given by

$$
\mathbf{E}(\mathbf{r}) =
\begin{cases}
\dfrac{1}{4\pi\epsilon_0}\dfrac{q}{r^2}\hat{\mathbf{r}} & \text{(for } r > 2R\text{)}, \\[2mm]
\dfrac{1}{4\pi\epsilon_0\epsilon_\mathrm{r}}\dfrac{q}{r^2}\hat{\mathbf{r}} & \text{(for } R < r < 2R\text{)}, \\[2mm]
0 & \text{(for } r < R\text{)}.
\end{cases}
\tag{1}
$$

(b) Hence determine the capacitance of the conductor.

Solution

(a) We use Gauss's law in the form of $(12)_1$ of Question 9.2. Integrating the flux of
$\mathbf{D}$ over a spherical Gaussian surface of radius r centred on the conductor gives
$D \times 4\pi r^2 = q_\mathrm{f}$, or $\mathbf{D} = q/(4\pi r^2)\hat{\mathbf{r}}$ (this vector is necessarily radial because of
the spherical symmetry). Now $\mathbf{E} = \mathbf{D}/\epsilon$ (see (13) of Question 9.2). So in the
vacuum, where $r > 2R$, the permittivity is ϵ_0; whereas inside the dielectric (here
$R < r < 2R$) the permittivity is $\epsilon_0\,\epsilon_\mathrm{r}$. In the region $r < R$, the free charge $q_\mathrm{f} = 0$
(recall the properties of conductors in electrostatic equilibrium). Hence (1).

(b) The potential of the conductor relative to a point at infinity $\big(\text{where } \Phi(\infty) = 0\big)$
is

$$
V = \Phi(R) - \Phi(\infty) = -\int_\infty^{2R} \mathbf{E}_\mathrm{vac}\cdot d\mathbf{r} - \int_{2R}^{R} \mathbf{E}_\mathrm{diel}\cdot d\mathbf{r}
$$

$$
= -\frac{q}{4\pi\epsilon_0}\left[\int_\infty^{2R}\frac{dr}{r^2} + \frac{1}{\epsilon_\mathrm{r}}\int_{2R}^{R}\frac{dr}{r^2}\right]
$$

$$
= \frac{q}{4\pi\epsilon_0}\left[\left(\frac{1}{2R}\right) + \frac{1}{\epsilon_\mathrm{r}}\left(\frac{2}{2R} - \frac{1}{2R}\right)\right].
$$

Simplifying this last equation gives

$$
V = \frac{q}{8\pi\epsilon_0 R}\left(\frac{\epsilon_\mathrm{r} + 1}{\epsilon_\mathrm{r}}\right),
$$

and hence the capacitance

$$
C = \frac{q}{V} = 8\pi\epsilon_0\left(\frac{\epsilon_\mathrm{r}}{1 + \epsilon_\mathrm{r}}\right)R.
\tag{2}
$$

Comments

(i) Notice the following limits:

☞ As $\epsilon_r \to 1$, $C \to 4\pi\epsilon_0 R$, which is the expected result.

☞ As $\epsilon_r \to \infty$, $C \to 8\pi\epsilon_0 R$.

(ii) The bound surface-charge and volume-charge densities σ_b and ρ_b induced on/in the dielectric are of interest. These quantities can be calculated from the polarization density $\mathbf{P} = \epsilon_0(\epsilon_r - 1)\mathbf{E}$. With $\mathbf{E}$ given by $(1)_2$, we obtain

$$\mathbf{P} = \frac{1}{4\pi}\frac{(\epsilon_r - 1)}{\epsilon_r}\frac{q}{r^2}\hat{\mathbf{r}}. \tag{3}$$

On the outer surface at $r = 2R$ we obtain: $\sigma_b = \mathbf{P}\cdot\hat{\mathbf{r}} = \frac{1}{4\pi}\frac{(\epsilon_r-1)}{\epsilon_r}\frac{q}{(2R)^2}$, and

on the inner surface at $r = R$: $\sigma_b = \mathbf{P}\cdot\hat{\mathbf{r}} = -\frac{1}{4\pi}\frac{(\epsilon_r-1)}{\epsilon_r}\frac{q}{R^2}$. The bound volume-charge density $\rho_b = 0$ because of (14) of Question 9.2.

Question 9.8

A parallel-plate capacitor (area A and separation d) initially contains air for which $\epsilon_r = 1$. Its plates are connected, via a closed switch S, to a battery which maintains a potential difference V_0 across them. In all subsequent calculations, assume that fringing effects are negligible.

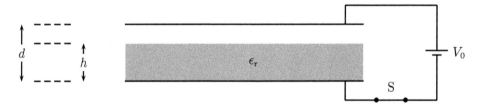

(a) Suppose the space between the plates is *partially* filled with a slab of lih dielectric material (area A and thickness h) as shown above. Calculate the following:

☞ The electric field in the air gap,

☞ the electric field in the dielectric,

☞ the capacitance C (expressed in terms of the vacuum capacitance C_0).

(b) The dielectric slab is now removed, S is opened and the slab returned to its original position between the plates (note the sequence in which these changes are made). Repeat the calculations for (a) above.

Solution

(a) Consider a cylindrical Gaussian cylinder G (infinitesimal area δa) straddling the top plate of the capacitor. One end cap of G is located inside the metal plate, the other lies either in the air gap or in the dielectric. Assume that no flux emerges through the sides of G. Then Gauss's law in the form $\oint_s \mathbf{D} \cdot d\mathbf{a} = q_f$ gives: $D \times \delta a = \sigma_f \times \delta a \Rightarrow D = \sigma_f$. In the air gap $E_{air} = D/\epsilon_0 = \sigma_f/\epsilon_0$ and in the dielectric $E_{diel} = \sigma_f/\epsilon = E_{air}/\epsilon_r$. For a uniform field the potential difference between two points is obtained from field $\times$ distance, and so

$$V_0 = E_{air} \times (d - h) + E_{diel} \times h = (d - h)E_{air} + hE_{air}/\epsilon_r. \tag{1}$$

Solving (1) yields

$$E_{air} = \frac{\epsilon_r V_0}{\epsilon_r(d - h) + h} \quad \text{and therefore} \quad E_{diel} = \frac{E_{air}}{\epsilon_r} = \frac{V_0}{\epsilon_r(d - h) + h}. \tag{2}$$

Now $C = q/V_0$ where the charge $q = \sigma_f A = \epsilon_0 E_{air} A = \dfrac{\epsilon_0 \epsilon_r A V_0}{\epsilon_r(d - h) + h}$. Thus

$$C = \frac{\epsilon_r d}{\epsilon_r(d - h) + h} C_0, \tag{3}$$

where $C_0 = \epsilon_0 A/d$.

(b) The electric field between the plates reverts to the vacuum (or air) value E_0 after the dielectric is removed. Hence $E_0 = V_0/d = \sigma_f/\epsilon_0 \Rightarrow \sigma_f = \epsilon_0 V_0/d$. Now when S is opened, the charge on each plate (and therefore σ_f) cannot change. Then, as in (a), the field in the air gap is $\boxed{E_0 = \sigma_f/\epsilon_0 = V_0/d}$ and in the dielectric $\boxed{E = \sigma_f/\epsilon = E_0/\epsilon_r}$. Using these fields to calculate the potential V between the plates gives

$$V = E_0 \times (d - h) + E \times h = (d - h)\frac{V_0}{d} + h\frac{V_0}{\epsilon_r d}$$

$$= \frac{V_0}{d}\frac{\epsilon_r(d - h) + h}{\epsilon_r}. \tag{4}$$

From $C = q/V = C_0 V_0/V$ we again obtain (3) using (4).

Comments

(i) In (a) the voltage across the capacitor is maintained constant by the battery. When the dielectric is inserted between the plates, the charge on each increases by Δq (obviously the battery provides this additional charge). It is a simple matter to show that

$$\Delta q = \frac{(\epsilon_r - 1)h}{\epsilon_r(d - h) + h}\left(\frac{C_0}{V_0}\right). \tag{5}$$

(ii) The charge on the capacitor in (b), however, cannot change when S remains open. Now the potential difference between the plates must decrease when the dielectric is inserted. This is easily seen to be true if (4) is expressed in the alternative form

$$V = V_0\left[1 - \frac{h}{d}\left(\frac{\epsilon_r - 1}{\epsilon_r}\right)\right].\tag{6}$$

Question 9.9

Consider a plane boundary at $z = 0$ between two lih dielectrics (permittivities ϵ_1 and ϵ_2), which lie in a uniform electric field. Let $\hat{n}$ be the unit vector normal to the interface, as shown alongside. Suppose that no free charges are present on the boundary surface.

(a) Prove the result

$$\epsilon_1 \cot\theta_1 = \epsilon_2 \cot\theta_2,\tag{1}$$

where θ_1 and θ_2 are the angles which $\hat{n}$ makes with the **E**-field vector on either side of the boundary.

Hint: Use the matching conditions

$$(\mathbf{D}_2 - \mathbf{D}_1)\cdot\hat{n} = \sigma_f \quad\text{and}\quad (\mathbf{E}_2 - \mathbf{E}_1)\times\hat{n} = 0,\tag{2}$$

derived in Question 10.6.

(b) Consider the cases $\epsilon_1 < \epsilon_2$ and $\epsilon_1 > \epsilon_2$ separately. For each, sketch the 'refraction' of an electric-field line through the boundary.

Solution

(a) In the absence of free charge $\sigma_f = 0$. Then the matching condition $(2)_1$ gives $\mathbf{D}_1\cdot\hat{n} = \mathbf{D}_2\cdot\hat{n}$, or $D_1\cos\theta_1 = D_2\cos\theta_2$. So

$$\epsilon_1 E_1\cos\theta_1 = \epsilon_2 E_2\cos\theta_2.\tag{3}$$

Similarly, from the matching condition $(2)_2$ we obtain

$$E_1\sin\theta_1 = E_2\sin\theta_2.\tag{4}$$

Dividing (3) by (4) gives (1).

(b) We obtain the following 'field-line' diagrams:

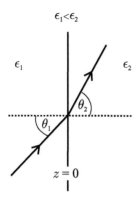

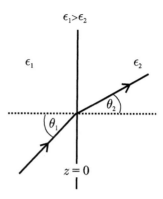

Comments

(i) The bending of a line of electric field at an interface between two dielectrics is reminiscent of Snell's law, and the behaviour of a light ray as it travels between two media having different refractive indices. This analogy is perhaps less surprising when one recalls that the laws of geometrical optics can also be derived from the matching conditions for the electromagnetic field. See Questions 10.14 and 10.15.

(ii) One expects that the magnetic-field lines obey a similar equation at the boundary between two media having different relative permeabilities. It is possible to show that they do, where the appropriate matching conditions are now $(\mathbf{B}_2 - \mathbf{B}_1) \cdot \hat{\mathbf{n}} = 0$; $(\mathbf{H}_2 - \mathbf{H}_1) \times \hat{\mathbf{n}} = 0$ (assuming $\mathbf{K}_f = 0$) with $\mathbf{B} = \mu \mathbf{H}$ (see $(13)_2$ of Question 9.13).

Question 9.10*

The plane interface between two semi-infinite dielectric slabs (having permittivities ϵ_1 and ϵ_2) lies in the xy-plane at $z = 0$, as shown alongside. A point charge q, embedded in medium 1, is located at $(0, 0, d)$. We seek the solution to the equations

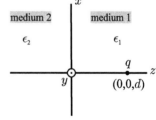

$$\left. \begin{array}{ll} \nabla^2\Phi = -\rho/\epsilon_1 & z > 0 \\ \nabla^2\Phi = 0 & z < 0 \end{array} \right\}, \qquad (1)$$

subject to the boundary conditions at $z = 0$:

$$\lim_{z \to 0_+} \left\{ \begin{array}{c} E_{\mathrm{t}} \\ \epsilon_1 E_z \end{array} \right\} = \lim_{z \to 0_-} \left\{ \begin{array}{c} E_{\mathrm{t}} \\ \epsilon_2 E_z \end{array} \right\}. \qquad (2)$$

These boundary conditions are just the matching conditions derived in Question 10.6, and E_{t} and E_z are the electric-field components tangential and normal to the boundary respectively.

(a) Attempt a solution based on the method of images[‡] using the image charges q' and q'' shown in the figure below. Write down expressions for Φ in each dielectric, and then prove that

$$q' = \left(\frac{\epsilon_1 - \epsilon_2}{\epsilon_1 + \epsilon_2}\right) q \quad \text{and} \quad q'' = \left(\frac{2\epsilon_2}{\epsilon_1 + \epsilon_2}\right) q. \tag{3}$$

the region $z>0$	the region $z<0$
Remove slab 2. Extend slab 1 throughout all space. Position image charge q' at $(0, 0, -d)$.	Remove slab 1. Extend slab 2 throughout all space. Replace q with image charge q'' at $(0, 0, d)$.

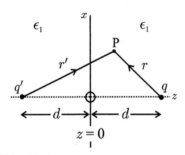

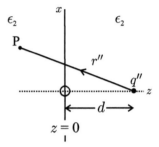

(b) Hence show that the total bound charge induced at the interface is

$$q_b = \frac{\epsilon_0(\epsilon_1 - \epsilon_2)}{\epsilon_1(\epsilon_1 + \epsilon_2)} q. \tag{4}$$

(c) Calculate the resultant force on the charge q.

Solution

(a) Applying the principle of superposition gives

$$\Phi_P(x, y, z) = \begin{cases} \dfrac{1}{4\pi\epsilon_1}\left(\dfrac{q}{r} + \dfrac{q'}{r'}\right) & z > 0 \\[3mm] \dfrac{1}{4\pi\epsilon_2}\dfrac{q''}{r''} & z < 0, \end{cases} \tag{5}$$

where $r' = \sqrt{R^2 + (z + d)^2}$, $r = \sqrt{R^2 + (z - d)^2}$, $r'' = \sqrt{R^2 + (z - d)^2}$ and $R^2 = x^2 + y^2$. The field components follow from $\mathbf{E} = -\nabla\Phi$:

[‡]If necessary, refer to Chapter 3 where this method is applied to point charges near conducting surfaces. The procedure with dielectrics is similar: attempt to mimic the boundary conditions of the actual problem using image charges which must be located 'outside the region of interest', so that Poisson's equation remains unchanged there.

$$E_t = -\frac{\partial \Phi}{\partial R} = \begin{cases} \dfrac{1}{4\pi\epsilon_1}\left(\dfrac{Rq}{r^3} + \dfrac{Rq'}{r'^3}\right) & z > 0, \\[4mm] \dfrac{1}{4\pi\epsilon_2}\dfrac{Rq''}{r''^3} & z < 0 \end{cases} \tag{6}$$

and

$$E_z = -\frac{\partial \Phi}{\partial z} = \begin{cases} \dfrac{1}{4\pi\epsilon_1}\left(\dfrac{(z-d)q}{r^3} + \dfrac{(z+d)q'}{r'^3}\right) & z > 0, \\[4mm] \dfrac{1}{4\pi\epsilon_2}\dfrac{(z-d)q''}{r''^3} & z < 0. \end{cases} \tag{7}$$

Then (2), (6) and (7) yield $\epsilon_1 q'' = \epsilon_2 q + \epsilon_2 q'$ and $q - q' = q''$. Solving these equations simultaneously gives (3).

(b) The bound surface-charge density (see $(6)_1$ of Question 9.2) is $\sigma_b = P_1 \cdot (-\hat{z}) + P_2 \cdot \hat{z} = -P_{1z} + P_{2z}$ where $P_{iz} = (\epsilon_i - \epsilon_0)E_{iz}$ with $i = 1, 2$, and so

$$\sigma_b = -(\epsilon_1 - \epsilon_0)E_{1z} + (\epsilon_2 - \epsilon_0)E_{2z}. \tag{8}$$

The field components E_{iz} at $z = 0$ follow from (7), and are:

$$\left. \begin{aligned} E_{1z} &= \frac{-1}{4\pi\epsilon_1}\left(\frac{qd}{(R^2+d^2)^{\frac{3}{2}}} - \frac{q'd}{(R^2+d^2)^{\frac{3}{2}}}\right) = \frac{-\epsilon_2}{4\pi(\epsilon_1+\epsilon_2)\epsilon_1}\frac{2qd}{(R^2+d^2)^{\frac{3}{2}}} \\[3mm] E_{2z} &= -\frac{1}{4\pi\epsilon_1}\frac{q''d}{(R^2+d^2)^{\frac{3}{2}}} = -\frac{1}{4\pi(\epsilon_1+\epsilon_2)}\frac{2qd}{(R^2+d^2)^{\frac{3}{2}}} = \frac{\epsilon_1}{\epsilon_2}E_{1z} \end{aligned} \right\}, \tag{9}$$

where we make use of (3) to eliminate q' and q'' in favour of q. Substituting (9) in (8) yields

$$\sigma_b = \frac{1}{2\pi}\frac{qd}{(R^2+d^2)^{3/2}}\frac{\epsilon_0(\epsilon_1-\epsilon_2)}{\epsilon_1(\epsilon_1+\epsilon_2)}. \tag{10}$$

Now $q_b = \displaystyle\int_s \sigma_b \, da$ where s represents the infinite plane interface. To evaluate this integral we consider two circles having radii R and $R+dR$, centred on the origin and lying in the interface. The bound charge contained in the infinitesimal area between these circles is then $dq_b = 2\pi R dR \sigma_b$. Substituting (10) in this last result gives

$$q_b = \frac{qd}{2\pi}\frac{\epsilon_0(\epsilon_1-\epsilon_2)}{\epsilon_1(\epsilon_1+\epsilon_2)}\int_0^\infty \frac{2\pi R dR}{(R^2+d^2)^{3/2}}. \tag{11}$$

The change of variable $r^2 = R^2 + d^2$ makes the integration of (11) trivial. Hence (4).

(c) Because the field for the image problem is equivalent to that for the real problem with the two dielectric slabs, the charge q cannot distinguish between the force exerted on it by q' and the force exerted on it by q_b distributed over the plane interface. Exploiting this equivalence, we obtain from Coulomb's law

$$\mathbf{F} = \frac{1}{4\pi\epsilon_1} \frac{qq'}{(2d)^2} \hat{\mathbf{z}}$$

$$= \frac{1}{4\pi\epsilon_1} \left(\frac{\epsilon_1 - \epsilon_2}{\epsilon_1 + \epsilon_2} \right) \frac{q^2}{4d^2} \hat{\mathbf{z}}, \tag{12}$$

where, in the last step, we make use of $(3)_1$.

Comments

(i) Notice that the potential given by (5) has the properties that $\Phi \to 0$ as $r \to \infty$, it satisfies the boundary conditions for arbitrary x and y on the $z = 0$ interface and it does not change Poisson's equation anywhere. The image solution is therefore the only possible solution of the actual problem with the dielectric slabs. (The reader should recall that it is the uniqueness theorem which validates the method of images. Of course, as is always the case with this method, there is no guarantee *in advance* that it will succeed in solving the actual physical problem.)

(ii) We see from (4) that the sign of the induced charge can be the same as that of q if $\epsilon_1 > \epsilon_2$, in which case $\mathbf{F}$, given by (12), will be a repulsive force.

Question 9.11

(a) At a macroscopic point inside a dielectric let the number density of molecules be n and the electric field $\mathbf{E}$. Show that the total energy per unit volume to polarize the molecules in the dielectric is

$$\frac{dU''}{dv} = \int_0^{\mathbf{E}} \mathbf{E} \cdot d\mathbf{P}. \tag{1}$$

Hint: Calculate the work done in a process where the charge q in each molecule undergoes an average displacement $d\bar{\mathbf{r}}$.

(b) Hence show that the total energy per unit volume of a polarized dielectric is

$$\frac{dU}{dv} = \int_0^{\mathbf{E}} \mathbf{E} \cdot d\mathbf{D}. \tag{2}$$

Solution

(a) Suppose that the macroscopic electric field increases by an infinitesimal amount $d\mathbf{E}$. A further polarization of a molecule occurs, producing an average increase $d\bar{\mathbf{p}}$ in its electric dipole moment. The work expended in this process is $dW = \mathbf{F} \cdot d\bar{\mathbf{r}}$, where $\mathbf{F} = q\mathbf{E}$ is the force on a charge q. So $dW = q\mathbf{E} \cdot d\bar{\mathbf{r}} = \mathbf{E} \cdot d\bar{\mathbf{p}}$, since $d\bar{\mathbf{p}} = q\,d\bar{\mathbf{r}}$. Now the work per unit volume to polarize the dielectric is $n\mathbf{E} \cdot d\bar{\mathbf{p}}$. But $\mathbf{P} = n\bar{\mathbf{p}}$, and so $dU''/dv = \mathbf{E} \cdot d\mathbf{P}$. Integrating from $\mathbf{E}_{\text{initial}} = 0$ to $\mathbf{E}_{\text{final}} = \mathbf{E}$ gives (1).

(b) Using (VII), we obtain the energy density stored in the field

$$\frac{dU'}{dv} = \tfrac{1}{2}\,\epsilon_0 \int_0^{\mathbf{E}} d(\mathbf{E} \cdot \mathbf{E}) = \tfrac{1}{2}\,\epsilon_0 \int_0^{\mathbf{E}} \frac{d(\mathbf{E} \cdot \mathbf{E})}{d\mathbf{E}} \frac{d\mathbf{E}}{dv}\,dv$$

$$= \epsilon_0 \int_0^{\mathbf{E}} \mathbf{E} \cdot d\mathbf{E}. \tag{3}$$

Adding this to (1) gives the total energy density

$$\frac{dU}{dv} = \frac{dU'}{dv} + \frac{dU''}{dv} = \epsilon_0 \int_0^{\mathbf{E}} \mathbf{E} \cdot d\mathbf{E} + \int_0^{\mathbf{E}} \mathbf{E} \cdot d\mathbf{P}$$

$$= \int_0^{\mathbf{E}} \mathbf{E} \cdot d(\epsilon_0 \mathbf{E} + \mathbf{P}),$$

which is (2) since $\mathbf{D} = \epsilon_0 \mathbf{E} + \mathbf{P}$.

Comments

(i) Notice that in deriving (2) we have assumed nothing about the nature of the dielectric. In the special case of an lih dielectric $\mathbf{D} = \epsilon\mathbf{E}$, and assuming ϵ is independent of the field, we have

$$\frac{dU}{dv} = \int_0^{\mathbf{E}} \mathbf{E} \cdot d(\epsilon\mathbf{E}) = \epsilon \int_0^{\mathbf{E}} \mathbf{E} \cdot d\mathbf{E} = \tfrac{1}{2}\epsilon E^2,$$

or

$$U = \tfrac{1}{2} \int_v \mathbf{E} \cdot \mathbf{D}\,dv. \tag{4}$$

Equation (4) shows that

☞ the energy density in a polarized lih dielectric is

$$u = \tfrac{1}{2}\mathbf{E} \cdot \mathbf{D}. \tag{5}$$

☞ the energy U is greater by the factor ϵ_r than the energy required to establish the same field in vacuum.

(ii) In Question 9.12, we show more generally that U is given by (4), whether the dielectric is lih or not.

Question 9.12

The electrostatic energy U in a region of space v where there is a free-charge density $\rho_f(\mathbf{r})$ is given by (4) of Question 2.17:

$$U = \tfrac{1}{2}\int_v \rho_f \Phi \, dv, \tag{1}$$

where $\Phi(\mathbf{r})$ is the scalar potential. Show that U can be expressed as

$$U = \tfrac{1}{2}\int_v \mathbf{E} \cdot \mathbf{D} \, dv. \tag{2}$$

Solution

Eliminating the free-charge density from (1) using $\nabla \cdot \mathbf{D} = \rho_f$ gives

$$U = \tfrac{1}{2}\int_v \Phi(\nabla \cdot \mathbf{D}) \, dv,$$

and then applying the vector identity (5) of Question 1.8 yields

$$U = \tfrac{1}{2}\int_v \nabla \cdot (\Phi\mathbf{D}) \, dv - \tfrac{1}{2}\int_v \mathbf{D} \cdot \nabla\Phi \, dv.$$

Converting the first integral above using Gauss's theorem and putting $\mathbf{E} = -\nabla\Phi$ give

$$U = \tfrac{1}{2}\oint_s \Phi\mathbf{D} \cdot d\mathbf{a} + \tfrac{1}{2}\int_v \mathbf{E} \cdot \mathbf{D} \, dv. \tag{3}$$

The **D**-field (like **E**) varies at least as rapidly as r^{-2}, whilst Φ behaves as r^{-1} (if in doubt, recall the field and potential of a point charge). Then $\Phi\mathbf{D} \cdot d\mathbf{a}$ in (3) scales as $r^2/r^3 \sim r^{-1}$. Now since s is any surface which encloses all the charges, we may choose a sphere having an arbitrarily large radius R. In the limit $R \to \infty$, the surface integral in (3) tends to zero. Hence (2).

Question 9.13

Consider a magnetized material of arbitrary shape having surface s and volume v as shown on p. 435. Suppose $\mathbf{M} = \mathbf{M}(\mathbf{r}')$ is a time-independent magnetization density.

(a) Show that the magnetic vector potential at an arbitrary field point located at $\mathbf{r}$ (relative to origin O) is

$$\mathbf{A}(\mathbf{r}) = \frac{\mu_0}{4\pi}\int_v \frac{\mathbf{M}(\mathbf{r}') \times (\mathbf{r} - \mathbf{r}')}{|\mathbf{r} - \mathbf{r}'|^3} \, dv'. \tag{1}$$

Hint: Start with (1) of Question 4.10.

(b) Hence prove, by transforming the integral in (1), that

$$\mathbf{A}(\mathbf{r}) = \frac{\mu_0}{4\pi} \oint_s \frac{(\mathbf{M} \times \hat{\mathbf{n}})}{|\mathbf{r} - \mathbf{r}'|} da' + \frac{\mu_0}{4\pi} \int_v \frac{(\nabla' \times \mathbf{M})}{|\mathbf{r} - \mathbf{r}'|} dv'. \tag{2}$$

Hint: Use the hint given previously in Question 9.2 and also (2) of Question 1.22.

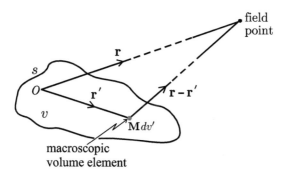

field
point

macroscopic
volume element

Solution

(a) Consider a macroscopic volume element dv' located at $\mathbf{r}'$. The magnetic dipole moment of this element is $d\mathbf{m} = \mathbf{M}(\mathbf{r}')dv'$, and the contribution of dv' to the vector potential at $\mathbf{r}$ follows from the dipole potential $\mathbf{A}(\mathbf{r}) = \frac{\mu_0}{4\pi} \frac{\mathbf{m} \times \mathbf{r}}{r^3}$. So

$$d\mathbf{A}(\mathbf{r}) = \frac{\mu_0}{4\pi} \frac{d\mathbf{m}(\mathbf{r}') \times (\mathbf{r} - \mathbf{r}')}{|\mathbf{r} - \mathbf{r}'|^3} = \frac{\mu_0}{4\pi} \frac{\mathbf{M}(\mathbf{r}') \times (\mathbf{r} - \mathbf{r}')}{|\mathbf{r} - \mathbf{r}'|^3} dv'. \tag{3}$$

Integrating (3) yields (1).

(b) We begin by expressing (1) in the form[‡]

$$\mathbf{A}(\mathbf{r}) = \frac{\mu_0}{4\pi} \int_v \mathbf{M}(\mathbf{r}') \times \nabla' \left(\frac{1}{|\mathbf{r} - \mathbf{r}'|} \right) dv'. \tag{4}$$

Then, using the identity (6) of Question 1.8, we obtain

$$\nabla' \times \left(\frac{\mathbf{M}(\mathbf{r}')}{|\mathbf{r} - \mathbf{r}'|} \right) = \nabla' \left(\frac{1}{|\mathbf{r} - \mathbf{r}'|} \right) \times \mathbf{M}(\mathbf{r}') + \frac{\nabla' \times \mathbf{M}(\mathbf{r}')}{|\mathbf{r} - \mathbf{r}'|},$$

which when substituted in (4) gives

$$\mathbf{A}(\mathbf{r}) = -\frac{\mu_0}{4\pi} \int_v \nabla' \times \left(\frac{\mathbf{M}(\mathbf{r}')}{|\mathbf{r} - \mathbf{r}'|} \right) dv' + \frac{\mu_0}{4\pi} \int_v \frac{\nabla' \times \mathbf{M}(\mathbf{r}')}{|\mathbf{r} - \mathbf{r}'|} dv'.$$

[‡]See the hint for Question 9.2.

The first integral on the right-hand side above can be transformed using Gauss's theorem in the form given by (2) of Question 1.22. So

$$\mathbf{A}(\mathbf{r}) = \frac{\mu_0}{4\pi} \oint_s \frac{\mathbf{M}(\mathbf{r}') \times d\mathbf{a}'}{|\mathbf{r} - \mathbf{r}'|} + \frac{\mu_0}{4\pi} \int_v \frac{\nabla' \times \mathbf{M}(\mathbf{r}')}{|\mathbf{r} - \mathbf{r}'|} \, dv', \tag{5}$$

which is (2) because $d\mathbf{a}' = \hat{\mathbf{n}} \, da'$.

Comments

(i) The bracketed terms $\mathbf{M} \times \hat{\mathbf{n}}$ and $\nabla \times \mathbf{M}$ on the right-hand side of (2) have the dimensions of current per unit length and current per unit area respectively (as before, we omit the primes for convenience). This suggests that we define a bound surface-current density $\mathbf{K}_b$ and a bound volume-current density $\mathbf{J}_b$ as follows:

$$\mathbf{K}_b = \mathbf{M} \times \hat{\mathbf{n}} \quad \text{and} \quad \mathbf{J}_b = \nabla \times \mathbf{M}. \tag{6}$$

(ii) Two obvious conclusions which follow from (6) are:

  ☞ Bound surface current arises only from the *parallel* component of $\mathbf{M}$. Contrast this behaviour with a polarized dielectric, where bound surface charges arise from the *normal* component of $\mathbf{P}$ (see Comment (ii) of Question 9.2).

  ☞ Bound currents appear inside a magnetized body only where $\mathbf{M}$ has a non-zero curl. (Recall that in a dielectric, volume charges arise from $-\nabla \cdot \mathbf{P}$. The difference in sign between ρ_b and $\mathbf{J}_b$ does have implications. Compare, for example, the definitions of $\mathbf{D}$ and $\mathbf{H}$.)

(iii) We have already seen that a polarized dielectric can be replaced by an equivalent distribution of free charge in vacuum. It therefore seems reasonable to suppose that a similar situation exists for magnetized matter. It does. The field of a magnetized object can be calculated by replacing the body with an equivalent distribution of free currents in vacuum, having surface-current density $\mathbf{K}_f = \mathbf{M} \times \hat{\mathbf{n}}$ over a surface s identical to that of the object, and volume-current density $\mathbf{J}_f = \nabla \times \mathbf{M}$ over the enclosed volume. As in the dielectric case, these two representations (magnetized body on the one hand, vacuum distribution of free currents on the other) are equivalent for the purpose of determining macroscopic fields. See Question 9.15 for an example.

(iv) As a result of the averaging described by (I) of the introduction, Ampère's law is valid on the macroscopic scale:

$$\oint_c \mathbf{B} \cdot d\mathbf{l} = \mu_0 \int_s \mathbf{J}_{total} \cdot d\mathbf{a},$$

where $\mathbf{B}$ is the macroscopic field at a macroscopic element $d\mathbf{l}$ on the contour, and $\mathbf{J}_{total}$ is the density of all the current, free and bound. Suppose we consider the case where the contour c lies entirely *within* the magnetized material. Then

$$\oint_s \mathbf{B} \cdot dl = \mu_0 \int_s (\mathbf{J}_\mathrm{f} + \mathbf{J}_\mathrm{b}) \cdot d\mathbf{a}. \tag{7}$$

Using Stokes's theorem to transform the contour integral gives

$$\int_s (\nabla \times \mathbf{B}) \cdot d\mathbf{a} = \mu_0 \int_s (\mathbf{J}_\mathrm{f} + \mathbf{J}_\mathrm{b}) \cdot d\mathbf{a}, \tag{8}$$

and because s is arbitrary we obtain the differential form

$$\nabla \times \mathbf{B} = \mu_0 (\mathbf{J}_\mathrm{f} + \mathbf{J}_\mathrm{b}), \tag{9}$$

which holds at each macroscopic point.

(v) Substituting $\mathbf{J}_\mathrm{b} = \nabla \times \mathbf{M}$ in (7) and (9) and rearranging give

$$\oint_c (\mu_0{}^{-1}\mathbf{B} - \mathbf{M}) \cdot dl = I_\mathrm{f} \qquad \text{and} \qquad \nabla \times (\mu_0{}^{-1}\mathbf{B} - \mathbf{M}) = \mathbf{J}_\mathrm{f}, \tag{10}$$

respectively $\big($to obtain $(10)_1$ apply Stokes's theorem$\big)$. Motivated by (10), we define a new vector, $\mathbf{H},^{\sharp}$ at a macroscopic point

$$\mathbf{H} = \mu_0{}^{-1}\mathbf{B} - \mathbf{M}, \tag{11}$$

and then

$$\oint_c \mathbf{H} \cdot dl = I_\mathrm{f} \qquad \text{and} \qquad \nabla \times \mathbf{H} = \mathbf{J}_\mathrm{f}. \tag{12}$$

Equation (12) shows that only the *free* currents contribute to the **H**-field. The contributions from the *bound* currents have not disappeared, but have been absorbed into the vector **H**.

(vi) For an lih material $\mathbf{M} = \chi_\mathrm{m}\,\mathbf{H}$ $\big($see (XI) of the introduction$\big)$, and we obtain the following results.

☞ From (11):

$$(1 + \chi_\mathrm{m})\mathbf{H} = \mu_0{}^{-1}\mathbf{B} \qquad \text{or} \qquad \mathbf{H} = \mu^{-1}\mathbf{B}, \tag{13}$$

where $\mu = \mu_\mathrm{r}\mu_0 = (1 + \chi_\mathrm{m})\mu_0$. Contrast $(13)_2$ with $\mathbf{D} = \epsilon\mathbf{E}$.

☞ From $(6)_2$:

$$\mathbf{J}_\mathrm{b} = \nabla \times \mathbf{M} = \chi_\mathrm{m}\nabla \times \mathbf{H} = \chi_\mathrm{m}\mathbf{J}_\mathrm{f}, \tag{14}$$

which shows that the bound volume-current density is zero except if $\mathbf{J}_\mathrm{f} \neq 0$ inside the material.

$^{\sharp}$We refer to this field as 'the **H**-field'. See the footnote on p. 413.

(vii) In cases where there are no free currents, $\nabla \times \mathbf{H} = 0$ implies that

$$\mathbf{H} = -\nabla \Phi_{\mathrm{m}}, \tag{15}$$

where Φ_{m} is a magnetic scalar potential (see Question 4.18). Now $\mathbf{B} = \mu\mathbf{H}$ and the Maxwell equation $\nabla \cdot \mathbf{B} = 0$ give $\nabla \cdot \mathbf{H} = 0$. Substituting (15) in this last result shows that Φ_{m} satisfies Laplace's equation

$$\nabla^2 \Phi_{\mathrm{m}} = 0. \tag{16}$$

In Question 9.21 we use the magnetic scalar potential to analyse the problem of a ferromagnetic shell placed in an external magnetic field.

Question 9.14

A ferromagnetic material in the shape of a circular cylinder has length ℓ and radius r. It is placed with its axis parallel to a uniform (vacuum) magnetic field $B_0 = 6.00 \times 10^{-4}$ T. For this value of B_0 assume that the effective relative permeability is $\mu_{\mathrm{r}} = 1000$, and calculate the following quantities: B, H, M, J_{b} and K_{b} inside the medium when

(a) $r \gg \ell$ (here the cylinder is a disc),

(b) $r \ll \ell$ (here the cylinder is a needle or whisker).

(c) Now suppose the needle of (b) is placed in a much stronger magnetic field. Take $B_0 = 1.50$ T and repeat the above calculations. Assume that the material has reached its saturation magnetization (where $M = 8.00 \times 10^5$ A m^{-1}) and note that the value of $\mu_{\mathrm{r}} \neq 1000$ under these conditions.

Hint: Assume that the normal components of **B** and tangential components of **H** are continuous at all vacuum–ferromagnet boundaries. (These 'matching conditions' are derived in Question 10.6.)

Solution

(a) The magnetic-field lines are essentially perpendicular to the flat faces, and so the matching condition for **B** requires that $B_{\mathrm{in}} = B_0$. Then

$B_{\mathrm{in}} = 6.0 \times 10^{-4}$ T,

and so $\boxed{H_{\mathrm{in}}} = \dfrac{B_{\mathrm{in}}}{\mu} = \dfrac{B_{\mathrm{in}}}{\mu_0\mu_{\mathrm{r}}} = \dfrac{6.00 \times 10^{-4}}{4\pi \times 10^{-7} \times 1000} = \boxed{4.77 \times 10^{-1} \text{ A m}^{-1}}$,

$M_{\mathrm{in}} = \chi_{\mathrm{m}}H_{\mathrm{in}} = (\mu_{\mathrm{r}}-1)H_{\mathrm{in}} = 999 \times 4.77 \times 10^{-1} = \boxed{4.77 \times 10^2 \text{ A m}^{-1}}$,

$J_{\mathrm{b}} = \chi_{\mathrm{m}}J_{\mathrm{f}} = \boxed{0}$,

$K_{\mathrm{b}} = M_{\mathrm{in}} = \boxed{4.77 \times 10^2 \text{ A m}^{-1}}$ on the curved surface; 0 elsewhere.

(b) Here magnetic-field lines are parallel to the curved surface, and so the matching condition for **H** requires that $H_{\text{in}} = H_0$. Then

$$H_{\text{in}} = \frac{B_0}{\mu_0} = \frac{6.0 \times 10^{-4}}{4\pi \times 10^{-7}} = \boxed{4.77 \times 10^2 \, \text{A m}^{-1}},$$

and so $\boxed{B_{\text{in}}} = \mu_0 \mu_r H_{\text{in}} = 4\pi \times 10^{-7} \times 1000 \times 4.77 \times 10^2 = \boxed{0.60 \, \text{T}},$

$$\boxed{M_{\text{in}}} = \chi_m H_{\text{in}} = (\mu_r - 1) H_{\text{in}} = 999 \times 4.77 \times 10^2 = \boxed{4.77 \times 10^5 \, \text{A m}^{-1}},$$

$$\boxed{J_b} = \chi_m J_f = \boxed{0},$$

$$\boxed{K_b} = M_{\text{in}} = \boxed{4.77 \times 10^5 \, \text{A m}^{-1}} \text{ on the curved surface; 0 elsewhere.}$$

(c) With $\boxed{M_{\text{in}} = M_{\text{sat}} = 8.00 \times 10^5 \, \text{A m}^{-1}}$, the boundary condition requires as before that $H_{\text{in}} = H_0$. So

$$H_{\text{in}} = \frac{B_0}{\mu_0} = \frac{1.5}{4\pi \times 10^{-7}} = \boxed{1.19 \times 10^6 \, \text{A m}^{-1}},$$

then $\boxed{B_{\text{in}}} = \mu_0 (H_{\text{in}} + M_{\text{in}}) = 4\pi \times 10^{-7} \times (1.19 \times 10^6 + 8.00 \times 10^5) = \boxed{2.50 \, \text{T}},$

$$\boxed{J_b} = \chi_m J_f = \boxed{0},$$

$$\boxed{K_b} = M_{\text{in}} = \boxed{8.00 \times 10^5 \, \text{A m}^{-1}} \text{ on the curved surface; 0 elsewhere.}$$

Question 9.15

Consider a long (assume infinite) circular cylinder of lih material (either paramagnetic or diamagnetic) which has a relative permeability μ_r. Suppose the material has a uniform magnetization‡ **M** parallel to the axis. See Fig. (I) below.

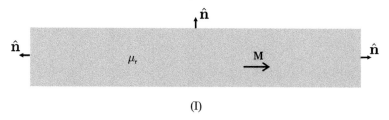

(I)

‡Induced by an externally applied magnetic field.

(a) Construct the vacuum equivalent for the magnetized cylinder.

(b) Hence show that the macroscopic magnetization field of the cylinder is

$$
\left.
\begin{aligned}
\mathbf{B}_{\mathrm{m}} &= 0 && \text{outside} \\
\mathbf{B}_{\mathrm{m}} &= \mu_0 \mathbf{M} && \text{inside}
\end{aligned}
\right\}. \tag{1}
$$

(c) Suppose the magnetization is induced by a uniform magnetic field $\mathbf{B}_0$. Express the magnetic field $\mathbf{B}$ (both inside and outside the cylinder) in terms of $\mathbf{B}_0$.

Solution

(a) The vacuum equivalent of this cylinder has the same surface with the same current distribution, except that it is free current. This is illustrated in Fig. (II) below.

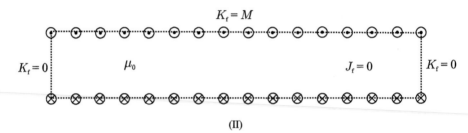

$$K_{\mathrm{f}} = M$$

$$K_{\mathrm{f}} = 0 \qquad \mu_0 \qquad J_{\mathrm{f}} = 0 \qquad K_{\mathrm{f}} = 0$$

(II)

Since $\mathbf{M}$ is uniform, $\mathbf{J}_{\mathrm{b}} = \nabla \times \mathbf{M} = 0$. The only bound current that exists is on the surface, having density $\mathbf{K}_{\mathrm{b}} = \mathbf{M} \times \hat{\mathbf{n}}$. This vanishes on the end faces to which $\mathbf{M}$ is parallel (or anti-parallel) but exists elsewhere over the curved surface, where it encircles the axis in the direction shown. Its magnitude is $K_{\mathrm{b}} = M$. So the vacuum equivalent is an infinite solenoid for which

$$
K_{\mathrm{f}} = M \qquad \text{and} \qquad J_{\mathrm{f}} = 0. \tag{2}
$$

(b) The field of an infinite solenoid is an elementary result: it is zero everywhere outside, and inside it is uniform and parallel to the axis. So $B = \mu_0 n I = \mu_0 K$, where n is the number of turns per unit length. The macroscopic magnetization field of the cylinder is thus

$$
\left.
\begin{aligned}
B_{\mathrm{m}} &= 0 && \text{outside} \\
B_{\mathrm{m}} &= \mu_0 K = \mu_0 M && \text{inside}
\end{aligned}
\right\},
$$

because of (2). Now since the current determines by the right-hand rule the direction of both $\mathbf{M}$ and $\mathbf{B}_{\mathrm{m}}$, these two vectors are parallel. Hence (1).

(c) Superimposing the applied field $\mathbf{B}_0$ gives the total field in the presence of the uniformly magnetized cylinder

$$\left.\begin{aligned}\mathbf{B} &= \mathbf{B}_0 + \mathbf{B}_\mathrm{m} = \mathbf{B}_0 + 0 = \mathbf{B}_0 \qquad &\text{outside}\\ \mathbf{B} &= \mathbf{B}_0 + \mathbf{B}_\mathrm{m} = \mathbf{B}_0 + \mu_0 \mathbf{M} \qquad &\text{inside}\end{aligned}\right\}, \qquad (3)$$

where $\mathbf{M} = \chi_\mathrm{m} \mathbf{H} = \chi_\mathrm{m} \mathbf{B}/\mu$. Substituting this in (3) and rearranging give

$$\left.\begin{aligned}\mathbf{B} &= \mathbf{B}_0 \qquad &\text{outside}\\ \mathbf{B} &= \mu_\mathrm{r} \mathbf{B}_0 \qquad &\text{inside}\end{aligned}\right\}, \qquad (4)$$

because $\mu = \mu_\mathrm{r}\mu_0$ and $\mu_\mathrm{r} = 1 + \chi_\mathrm{m}$.

Question 9.16

Consider a long (assume infinite) cylindrical rod of radius R that is made from a hard ferromagnetic material,[‡] which has a uniform magnetization $\mathbf{M}$ parallel to its axis. Suppose that a transverse disc of material of width $w \ll R$ is excised from the central portion of the rod (see the figure below) without disturbing the magnetization in the remainder. Show that the magnetic field B at the centre of the air gap is given by

$$\mathbf{B} = \mu_0\left(1 - \frac{w}{2R}\right)\mathbf{M}. \qquad (1)$$

Hint: Use the results of Question 9.15 and then apply the principle of superposition.

Solution

As we showed in Question 9.15, the vacuum equivalent of a magnetized rod has the same surface and volume-current densities, $\mathbf{K}$ and $\mathbf{J}$ respectively, in a vacuum, as does the magnetized material. That is,

$$\mathbf{J}_\mathrm{b} = \nabla \times \mathbf{M} \qquad \text{and} \qquad \mathbf{K}_\mathrm{b} = \mathbf{M} \times \hat{\mathbf{n}}.$$

For uniform $\mathbf{M}$ parallel to the axis, $\nabla \times \mathbf{M} = 0$. Then for the vacuum equivalent

[‡]Typical examples are alloys of nickel, cobalt, iron and—more recently—rare-earth metals.

$$J_{\mathrm{f}} = 0 \quad \text{and} \quad K_{\mathrm{f}} = M \text{ over the curved surfaces.}$$

Now the excised disc can be treated like a ring of current $I = wK = wM$, where the field at the centre of a current ring is an elementary result.[#] So

$$B_{\mathrm{ring}} = \frac{\mu_0}{2} \frac{I}{R} = \frac{\mu_0}{2} \frac{wM}{R}. \tag{2}$$

Furthermore, the field inside the original rod is that inside a long solenoid with

$$B_{\mathrm{rod}} = \mu_0 nI = \mu_0 K = \mu_0 M. \tag{3}$$

Applying the principle of superposition gives $\mathbf{B}_{\mathrm{gap}} = \mathbf{B}_{\mathrm{rod}} - \mathbf{B}_{\mathrm{ring}}$. Then from (2) and (3) we obtain $\mathbf{B}_{\mathrm{gap}} = \mu_0 \mathbf{M} - \dfrac{\mu_0 wM}{2R}$. Hence (1).

Question 9.17

An iron ring of mean radius R and cross-sectional area πb^2 contains an air gap of width w. Suppose $w \ll b \ll R$ and that there are N turns of wire carrying current I wound around the ring in a single layer, as illustrated in the figure below. Show that the magnetic field near the centre of the air gap, away from the edges, is given by

$$B_{\mathrm{air}} = \frac{\mu_0 \mu_{\mathrm{r}} NI}{2\pi R + (\mu_{\mathrm{r}} - 1)w}. \tag{1}$$

Hint: Although the relative permittivity μ_{r} of a ferromagnetic material depends on the applied field, assume nevertheless that the result $\mathbf{B} = \mu_0 \mu_{\mathrm{r}} \mathbf{H}$ still applies.

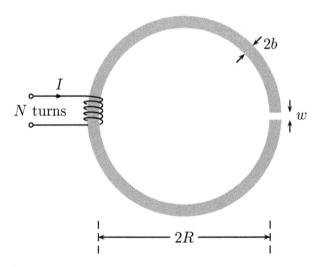

[#]See Comment (i) of Question 4.8.

Solution

The circulation of $\mathbf{H}$ around a closed contour c equals the total free current NI (recall $(12)_1$ of Question 9.13). Now because of the symmetry, we choose for c a circle of radius R centred on the ring, and so

$$(2\pi R - w)H_{\text{iron}} + wH_{\text{air}} = NI. \tag{2}$$

Then because of the hint,

$$\frac{(2\pi R - w)B_{\text{iron}}}{\mu_r \mu_0} + \frac{wB_{\text{air}}}{\mu_0} = NI. \tag{3}$$

Furthermore, at the iron–air interface the normal component of $\mathbf{B}$ is continuous (see the matching condition $(2)_2$ of Question 10.6), and since $\mathbf{B}$ is parallel to the unit normal $\hat{\mathbf{n}}$, it follows that

$$B_{\text{iron}} = B_{\text{air}}. \tag{4}$$

Solving (3) and (4) simultaneously gives (1).

Comments

(i) Because the μ_r values of iron are of the order 1000 or more, it is reasonable to suppose that $w(\mu_r - 1) \gg 2\pi R$ (except in the case of a very small gap when $w \to 0$). Then (1) shows that

$$B_{\text{air}} \simeq \frac{\mu_0 NI}{w}. \tag{5}$$

Clearly (4) is much larger than the magnetic field $B_{\text{air}} = \mu_0 NI/2\pi R$ in a non-ferromagnetic material for which $\mu_r \approx 1$. Obviously it is the presence of the iron core, and its large magnetizability, which account for this increased field.

(ii) The fields inside the iron core must satisfy:

☞ the linear equation

$$B = \frac{\mu_0 NI}{w} - \frac{\mu_0(2\pi R - w)H}{w}$$

(which follows from straightforward manipulation of (3) and (4)), and

☞ the $B(H)$ relation for iron.

The intersection of the straight line for a particular value of I with the $B(H)$ curve gives B and hence B_{air} because of (4). See Question 9.18 for a numerical example.

Question 9.18

Consider again the iron ring of Question 9.17. Suppose $N = 400$, $R = 100\,\text{mm}$ and $w = 10\,\text{mm}$. Assume that the following B–H data for iron applies.

B/T	0	0.20	0.40	0.80	1.13	1.33	1.60	1.87	1.95	2.00
$H/\text{A m}^{-1}$	0	40	80	160	240	320	480	800	1200	1600

(a) Use *Mathematica* to fit a fourth-order polynomial to these data, and show that

$$B = (-2.67 + 6.37H - 8.15 \times 10^{-3}H^2 + 4.75 \times 10^{-6}H^3 - 1.03 \times 10^{-9}H^4) \times 10^{-3}\text{T}. \quad (1)$$

Plot (1) together with these data points on the same set of axes.

(b) Suppose that the iron ring is not magnetized initially, and that its $B(H)$ curve may be represented by (1). Extend your *Mathematica* notebook to calculate the intersection of (1) with the straight line $B = \mu_0 NI/w - \mu_0(2\pi R - w)H/w^{\ddagger}$ for different values of the current I in the interval $0 < I \leq 4\,\text{A}$. Starting at $I = 0.05\,\text{A}$ and using a step size $\delta I = 0.1\,\text{A}$, calculate the effective μ_r at each value of I. Then plot a graph of μ_r vs H.

Solution

(a) See cell 1 of the notebook below.

(b) The point (H_1, B_1) shown in the first graph on p. 445 simultaneously satisfies both the straight-line equation and the $B(H)$ curve. Substituting B_1 in (1) of Question 9.17 and solving for μ_r give the effective relative permeability of iron. Repeating this procedure for different values of I leads to the second graph on p. 445.

Comment

If a complete $B(H)$ cycle (hysteresis curve) is used, there are two points of intersection between this curve and the straight line, each generating a separate value for μ_r. Plotting these values vs H generates a hysteresis loop of its own.

```
In[1]:= BHdata = {{0, 0}, {40, .2}, {80, .4}, {160, .8}, {240, 1.13}, {320, 1.33},
           {480, 1.6}, {800, 1.87}, {1200, 1.95}, {1600, 2}};
        fun = a1 + a2 H + a3 H² + a4 H³ + a5 H⁴;
        fitt = NonlinearModelFit[BHdata, fun, {a1, a2, a3, a4, a5}, H];
        B1[H_] := Normal[fitt]
        Show[ListPlot[BHdata], Plot[B1[H], {H, 0, 1600}], AspectRatio → 0.55]
```

‡See Comment (ii) of Question 9.17.

```
In[2]:= μ0 = 4 π×10⁻⁷; No = 4000; R = 100/1000; w = 10/1000; data1 = {};

       B2[I_, H_] := μ0 No/w I - μ0 (2 π R - w)/w H

       Do[intercept = Solve[B1[H] - B2[current, H] == 0, H, Reals];
         h = H /. First[intercept];

         μr = (2 π R - w) B2[current, h]/(μ0 No × current - B2[current, h] × w);

         AppendTo[data1, {h, μr}], {current, 0.05, 4, 0.1}]
       ListLinePlot[data1]
```

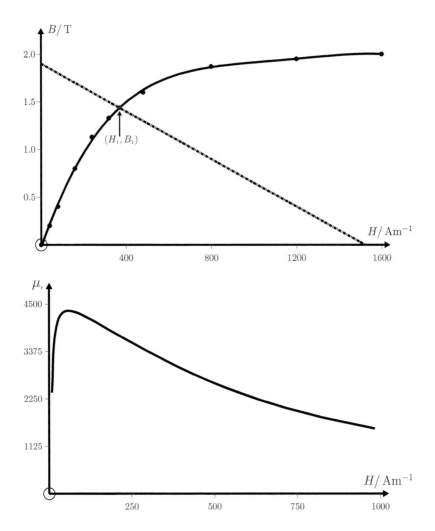

Question 9.19

Consider a sphere of radius a made from a hard ferromagnetic material that carries a uniform magnetization $\mathbf{M} = M_0\hat{\mathbf{z}}$ throughout.

(a) Show that the vector potential in spherical polar coordinates is given by

$$\mathbf{A}(r,\theta) = \begin{cases} \dfrac{\mu_0}{3} M_0 r \sin\theta\, \hat{\boldsymbol{\phi}} & \text{for } r \leq a, \\[2ex] \dfrac{\mu_0}{3} \dfrac{M_0 a^3 \sin\theta}{r^2}\, \hat{\boldsymbol{\phi}} & \text{for } r \geq a. \end{cases} \tag{1}$$

Hint: Use (5) of Question 9.13.

(b) Hence show that the magnetic field arising from the *magnetization* is

$$\mathbf{B}_{\mathrm{m}} = \begin{cases} \dfrac{2\mu_0 \mathbf{M}}{3} & \text{for } r \leq a, \\[2ex] \dfrac{\mu_0}{3} \dfrac{M_0 a^3}{r^3} (2\cos\theta\,\hat{\mathbf{r}} + \sin\theta\,\hat{\boldsymbol{\theta}}) & \text{for } r \geq a. \end{cases} \tag{2}$$

Solution

(a) Because $\mathbf{M}$ is uniform, $\nabla \times \mathbf{M} = 0$ inside the sphere. Then

$$\mathbf{A}(\mathbf{r}) = \frac{\mu_0}{4\pi} \oint_s \frac{(\mathbf{M} \times \hat{\mathbf{n}})}{|\mathbf{r} - \mathbf{r}'|}\, da' = \frac{\mu_0}{4\pi} \mathbf{M} \times \oint_s \frac{da'}{|\mathbf{r} - \mathbf{r}'|}. \tag{3}$$

Equation (1) follows directly from (3) of Question 1.26.

(b) Using $\mathbf{B} = \nabla \times \mathbf{A}$ and the curl operator for spherical polar coordinates $\big($see $(\mathrm{XI})_3$ of Appendix C$\big)$ yields (2).

Comments

(i) The $\mathbf{H}$-field inside the sphere is given by $\mathbf{H} = \mu_0^{-1}\mathbf{B}_{\mathrm{m}} - \mathbf{M} = -\mathbf{M}/3$ from $(2)_1$, and it points in the *opposite direction* to $\mathbf{M}$ and $\mathbf{B}_{\mathrm{m}}$.

(ii) Because the dipole moment of the sphere is $\mathbf{m} = \frac{4}{3}\pi a^3 \mathbf{M}$, we can express (2) as

$$\mathbf{B}_{\mathrm{m}} = \begin{cases} \dfrac{\mu_0}{2\pi} \dfrac{\mathbf{m}}{a^3} & \text{for } r \leq a, \\[2ex] \dfrac{\mu_0}{4\pi} \left[\dfrac{3(\mathbf{m}\cdot\hat{\mathbf{r}})\hat{\mathbf{r}} - \mathbf{m}}{r^3} \right] & \text{for } r \geq a. \end{cases} \tag{4}$$

Notice that inside the sphere the magnetic field is constant $\bigl(\text{see } (4)_1\bigr)$, whilst outside the field has an exact dipole nature, not only asymptotically but for all $r \geq a$ $\bigl(\text{see } (4)_2\bigr)$; the higher magnetic multipoles make no contribution at all.

Question 9.20

Consider a sphere of radius a made from a non-ferromagnetic lih material which carries a uniform magnetization $\mathbf{M} = M_0\hat{\mathbf{z}}$, that is induced by a previously uniform magnetic field $\mathbf{B}_0$. Use the results of Question 9.19 to express the macroscopic fields $\mathbf{B}$ and $\mathbf{H}$ in terms of $\mathbf{B}_0$ in the region $r \leq a$.

Solution

Since the field equations are linear, we can superimpose a uniform magnetic field $\mathbf{B}_0$ everywhere in space. Then the magnetic field $\mathbf{B}_{\text{in}}$ inside the sphere is given by

$$\mathbf{B}_{\text{in}} = \mathbf{B}_0 + \frac{2\mu_0 \mathbf{M}}{3}, \tag{1}$$

because of $(2)_1$ of Question 9.19. Now $\mathbf{H}_{\text{in}} = \mu_0^{-1}\mathbf{B}_{\text{in}} - \mathbf{M} = \mu_0^{-1}\mathbf{B}_0 - \frac{1}{3}\mathbf{M}$. But $\mathbf{B}_{\text{in}} = \mu\mathbf{H}_{\text{in}} = \mu_0\mu_r\mathbf{H}_{\text{in}}$, and so from (1)

$$\mathbf{B}_0 + \frac{2\mu_0 \mathbf{M}}{3} = \mu\left(\frac{\mathbf{B}_0}{\mu_0} - \frac{\mathbf{M}}{3}\right)$$

$$= \mu_r\mathbf{B}_0 - \mu_0\mu_r\frac{\mathbf{M}}{3}.$$

Solving this equation for $\mathbf{M}$ gives

$$\mathbf{M} = \frac{3}{\mu_0}\frac{(\mu_r - 1)}{(\mu_r + 2)}\mathbf{B}_0. \tag{2}$$

Substituting (2) in (1) yields

$$\mathbf{B}_{\text{in}} = \frac{3\mu_r\mathbf{B}_0}{2 + \mu_r}, \tag{3}$$

and so

$$\mathbf{H}_{\text{in}} = \frac{\mathbf{B}_{\text{in}}}{\mu_0\mu_r} = \frac{3\mathbf{B}_0}{\mu_0(2 + \mu_r)}. \tag{4}$$

Comments

(i) We see from (2)–(4) that $\mathbf{M}$, $\mathbf{B}$ and $\mathbf{H}$ are proportional to $\mathbf{B}_0$. They vanish, as one would expect, when $\mathbf{B}_0 = 0$.

(ii) It is worth emphasizing the equivalence between the uniformly polarized and magnetized spheres (compare, for example, (3) of Question 9.6 with (3) and (4) above). This similarity is perhaps less surprising than one might think, because the fields $\mathbf{E}$ and $\mathbf{D}$ obey the same mathematical equations as the fields $\mathbf{H}$ and $\mathbf{B}$:

$$\nabla \times \mathbf{E} = 0, \quad \nabla \cdot \mathbf{D} = 0 \quad \text{and} \quad \nabla \times \mathbf{H} = 0, \quad \nabla \cdot \mathbf{B} = 0.$$

Either problem can be converted into the other by means of the transformation

$$\mathbf{E} \leftrightarrow \mathbf{H}, \quad \mathbf{D} \leftrightarrow \mathbf{B}, \quad \mathbf{P} \leftrightarrow \mu_0 \mathbf{M}, \quad \epsilon \leftrightarrow \mu.$$

Question 9.21**

Consider a spherical shell having inner radius a and outer radius b that is made from a material whose permeability is μ. Suppose the shell is placed in a uniform magnetic field $\mathbf{B}_0 = B_0 \hat{\mathbf{z}}$. Because $\nabla \times \mathbf{H} = 0$ everywhere, we can express $\mathbf{H}$ in terms of a magnetic scalar potential Φ_m where $\mathbf{H} = -\nabla \Phi_m$ (see (15) of Question 9.13).

(a) Show that a suitable trial solution of Laplace's equation is given by

$$
\left.
\begin{aligned}
\Phi_1(r,\theta) &= \sum_{n=0}^{\infty} \alpha_n r^n P_n(\cos\theta) & r \leq a \\[2mm]
\Phi_2(r,\theta) &= \sum_{n=0}^{\infty} \left[\beta_n r^n + \lambda_n r^{-(n+1)}\right] P_n(\cos\theta) & a \leq r \leq b \\[2mm]
\Phi_3(r,\theta) &= -\mu_0^{-1} B_0 r \cos\theta + \sum_{n=0}^{\infty} \gamma_n r^{-(n+1)} P_n(\cos\theta) & r \geq b
\end{aligned}
\right\}, \quad (1)
$$

where α_n, β_n, λ_n and γ_n are constant coefficients and $P_n(\cos\theta)$ is the nth order Legendre polynomial in $\cos\theta$ (see Appendix F).

Hint: Start with (3) of Question 1.19.

(b) Using the hint below (and *Mathematica* where necessary), prove that all coefficients for $n \neq 1$ are zero. Prove also that the first-order coefficients satisfy

$$
\left.
\begin{aligned}
\mu_0 a^3 \alpha_1 - \mu a^3 \beta_1 + 2\mu\lambda_1 &= 0 \\
\mu b^3 \beta_1 - 2\mu\lambda_1 + 2\mu_0\gamma_1 + B_0 b^3 &= 0 \\
a^3 \alpha_1 - a^3 \beta_1 - \lambda_1 &= 0 \\
\mu_0 b^3 \beta_1 + \mu_0\lambda_1 - \mu_0\gamma_1 + B_0 b^3 &= 0
\end{aligned}
\right\}. \quad (2)
$$

Hint: Assume the following:

☞ The function Φ_m defined by (1) is piecewise continuous.

☞ At both material–vacuum boundaries the normal components of the magnetic field **B** and the tangential components of **H** are continuous. (These 'matching conditions' are derived in Question 10.6.)

(c) Solve (2) for the first-order coefficients with the help of *Mathematica*, and show that

$$
\left.
\begin{aligned}
\alpha_1 &= \frac{-9\mu B_0}{(2\mu + \mu_0)(\mu + 2\mu_0) - 2(a^3/b^3)(\mu - \mu_0)^2} \\[2mm]
\beta_1 &= \frac{-3(2\mu + \mu_0)B_0}{(2\mu + \mu_0)(\mu + 2\mu_0) - 2(a^3/b^3)(\mu - \mu_0)^2} \\[2mm]
\lambda_1 &= \frac{-3a^3(\mu - \mu_0)B_0}{(2\mu + \mu_0)(\mu + 2\mu_0) - 2(a^3/b^3)(\mu - \mu_0)^2} \\[2mm]
\gamma_1 &= \frac{1}{\mu_0}\frac{(\mu - \mu_0)(2\mu + \mu_0)(b^3 - a^3)B_0}{(2\mu + \mu_0)(\mu + 2\mu_0) - 2(a^3/b^3)(\mu - \mu_0)^2}
\end{aligned}
\right\}.
\tag{3}
$$

(d) Calculate the magnetic field **B** in the regions $r \le a$ and $r \ge b$.

Solution

(a) Because Φ_1 must remain finite as $r \to 0$, we require that all the B_n coefficients in (3) of Question 1.19 are zero. Hence $(1)_1$. It is reasonable to suppose that the perturbation produced by the magnetized shell is negligible as $r \to \infty$. Then $-\mu_0{}^{-1}B_0 z = -\mu_0{}^{-1}B_0 r \cos\theta$ gives the correct (uniform) magnetic field at infinity. This requires that all the A_n coefficients in (3) of Question 1.19 are zero except that for $n = 1$. Hence $(1)_3$.

(b) We apply the hint and proceed order by order.

coefficients for $n = 0$

$$
\left.
\begin{aligned}
(\Phi_2 - \Phi_1)_{r=a} &= 0 \\
(\Phi_2 - \Phi_3)_{r=b} &= 0
\end{aligned}
\right\}
\quad\Rightarrow\quad
\left.
\begin{aligned}
B_0 + \lambda_0/a &= \alpha_0 \\
B_0 + \lambda_0/b &= \gamma_0/b
\end{aligned}
\right\},
\tag{4}
$$

$$
\left.
\begin{aligned}
\frac{\partial}{\partial r}(\mu\Phi_2 - \mu_0\Phi_1)\Big|_{r=a} &= 0 \\
\frac{\partial}{\partial r}(\mu_0\Phi_3 - \mu\Phi_2)\Big|_{r=b} &= 0
\end{aligned}
\right\}
\quad\Rightarrow\quad
\left.
\begin{aligned}
\lambda_0 &= 0 \\
\mu_0\gamma_0 &= \mu\lambda_0
\end{aligned}
\right\}.
\tag{5}
$$

Solving (4) and (5) simultaneously gives

$$
\alpha_0 = \beta_0 = 0 \quad \text{and} \quad \lambda_0 = \gamma_0 = 0.
\tag{6}
$$

coefficients for $n = 1$

$$\left.\frac{\partial}{\partial r}(\mu\Phi_2 - \mu_0\Phi_1)\right|_{r=a} = 0 \atop \left.\frac{\partial}{\partial r}(\mu_0\Phi_3 - \mu\Phi_2)\right|_{r=b} = 0 \right\} \Rightarrow \left. \begin{array}{rcl} \mu_0\alpha_1 &=& \mu(\beta_1 - 2\lambda_1/a^3) \\ -\mu_0^{-1}B_0 - 2\gamma_1/b^3 &=& \mu(\beta_1 - 2\lambda_1/b^3) \end{array} \right\}, \quad (7)$$

$$\left.\frac{\partial}{\partial\theta}(\Phi_2 - \Phi_1)\right|_{r=a} = 0 \atop \left.\frac{\partial}{\partial\theta}(\Phi_3 - \Phi_2)\right|_{r=b} = 0 \right\} \Rightarrow \left. \begin{array}{rcl} \alpha_1 a &=& \beta_1 a + \lambda_1/a^2 \\ -\mu_0^{-1}B_0 b + \gamma_1/b^2 &=& \beta_1 b + \lambda_1/b^2 \end{array} \right\}. \quad (8)$$

Rearranging (7) and (8) gives (2).

coefficients for $n \geq 2$

The continuity of the Φ_i and B_{ri} (with $i = 1, 2, 3$) at $r = a$ and $r = b$ leads immediately to the following equations

$$\left. \begin{array}{rcl} a^{2n+1}\alpha_n - a^{2n+1}\beta_n - \lambda_n &=& 0 \\[2mm] b^{2n+1}\beta_n + \lambda_n - \gamma_n &=& 0 \\[2mm] na^{2n+1}\mu_0\alpha_n - na^{2n+1}\mu\beta_n + (n+1)\mu\lambda_n &=& 0 \\[2mm] nb^{2n+1}\mu\beta_n - (n+1)\mu\lambda_n + (n+1)\mu_0\gamma_n &=& 0 \end{array} \right\}. \quad (9)$$

The determinant of the coefficients in (9) is most easily calculated using *Mathematica* (see cell 1 in the notebook on p. 452). This gives

$$\text{Det coeff} = n(n+1)\left[(\mu^2 + \mu_0^2)b^{1+2n} - (\mu^2 - \mu_0^2)a^{1+2n}\right] + (2n^2 + 2n + 1)\mu\mu_0 b^{1+2n},$$

which is obviously greater than zero, and so only trivial solutions of (9) exist. Hence

$$\left. \begin{array}{rcl} \alpha_2 = \alpha_3 = \cdots = \beta_2 = \beta_3 = \cdots = 0 \\[2mm] \lambda_2 = \lambda_3 = \cdots = \gamma_2 = \gamma_3 = \cdots = 0 \end{array} \right\}. \quad (10)$$

(c) Solving (2) using *Mathematica*'s Solve function gives (3). See cell 2 of the notebook on p. 452.

(d) From $\mathbf{H} = -\nabla\Phi_m$ where $\mathbf{H} = \mu_0^{-1}\mathbf{B}$, we have $\mathbf{B} = -\mu_0\nabla\Phi_m$. Then for:

$r \leq a$

and with $\Phi_1 = \alpha_1 r \cos\theta$ (all the other α_i are zero) we obtain

$$\mathbf{B}_1 = -\mu_0 \left(\frac{\partial \Phi_1}{\partial r} \hat{\mathbf{r}} + \frac{1}{r} \frac{\partial \Phi_1}{\partial \theta} \hat{\boldsymbol{\theta}} \right) = -\mu_0 (\hat{\mathbf{r}} \cos \theta - \hat{\boldsymbol{\theta}} \sin \theta) \alpha_1, \qquad (11)$$

with α_1 given by $(3)_1$.

$\boxed{r \geq b}$

and with $\Phi_3 = -\mu_0^{-1} B_0 r \cos \theta + \gamma_1 \cos \theta / r^2$ (all the other γ_i are zero) we obtain

$$\mathbf{B}_3 = -\mu_0 \left(\frac{\partial \Phi_3}{\partial r} \hat{\mathbf{r}} + \frac{1}{r} \frac{\partial \Phi_3}{\partial \theta} \hat{\boldsymbol{\theta}} \right)$$

$$= B_0 (\hat{\mathbf{r}} \cos \theta - \hat{\boldsymbol{\theta}} \sin \theta) + \mu_0 \gamma_1 (2 \cos \theta \hat{\mathbf{r}} + \sin \theta \hat{\boldsymbol{\theta}}) / r^3, \qquad (12)$$

with γ_1 given by $(3)_4$.

Comments

(i) From $(3)_1$ and because $\hat{\mathbf{z}} = \hat{\mathbf{r}} \cos \theta - \hat{\boldsymbol{\theta}} \sin \theta$, we can express (11) in the form

$$\mathbf{B}_1 = \frac{9 \mu_0 \mu B_0 \hat{\mathbf{z}}}{(2\mu + \mu_0)(\mu + 2\mu_0) - 2(a^3/b^3)(\mu - \mu_0)^2}. \qquad (13)$$

Evidently the magnetic field inside the shell $r \leq a$ is constant and in the same direction as $\mathbf{B}_0$. For the special case where the material is ferromagnetic and $\mu \gg \mu_0$, (13) becomes

$$\mathbf{B}_1 \simeq \frac{9 \mu_0 B_0 \hat{\mathbf{z}}}{2\mu (1 - a^3/b^3)}. \qquad (14)$$

Equation (14) shows that the internal field can be very much less than B_0 (even for a relatively thin shell when μ/μ_0 lies in the range 10^3–10^6). This reduction in the field due to the presence of the permeable material is an example of magnetic shielding. It has important practical applications where it may be desirable to screen unwanted or stray magnetic fields from certain regions (e.g. a laboratory containing sensitive apparatus).

(ii) It is also useful to express (12) in an alternative form. As before, $\hat{\mathbf{z}} = \hat{\mathbf{r}} \cos \theta - \hat{\boldsymbol{\theta}} \sin \theta$, and so

$$\mathbf{B}_3 = \mathbf{B}_0 + \frac{\mu_0}{4\pi} \frac{3 (\mathbf{m} \cdot \hat{\mathbf{r}}) \hat{\mathbf{r}} - \mathbf{m}}{r^3}, \qquad (15)$$

where

$$\mathbf{m} = 4\pi \gamma_1 \hat{\mathbf{z}} = \frac{4\pi}{\mu_0} \frac{(\mu - \mu_0)(2\mu + \mu_0)(b^3 - a^3) \mathbf{B}_0}{(2\mu + \mu_0)(\mu + 2\mu_0) - 2(a^3/b^3)(\mu - \mu_0)^2}. \qquad (16)$$

We see from (15) that the magnetic field outside the shell $r \geq b$ is the superposition of a uniform field and a dipole field, with **m** a point magnetic dipole located at O whose moment is given by (16).

(iii) The plot shown below (generated using `cell 3` of the notebook on p. 453) illustrates the magnetic field in the yz-plane for $\mu = 1000\,\mu_0$, $a = 4\,\text{cm}$ and $b = 5\,\text{cm}$.

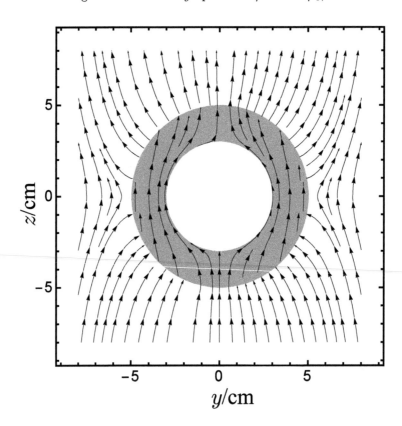

```
In[1]:=  coeffs = {{a^(2n+1), a^(2n+1), -1, 0}, {0, b^(2n+1), 1, -1},
                   {n μ0 a^(2n+1), -n μ a^(2n+1), (n+1) μ, 0},
                   {0, n μ b^(2n+1), -(n+1) μ, (n+1) μ0 }};
         FullSimplify[Det[coeffs]];

In[2]:=  eqn1 = μ0 a^3 α1 - μ a^3 β1 + 2 μ λ1 == 0;
         eqn2 = μ b^3 β1 - 2 μ λ1 + 2 μ0 γ1 + B0 b^3 == 0;
         eqn3 = α1 a^3 - β1 a^3 - λ1 == 0;   eqn4 = μ0 β1 b^3 + μ0 λ1 - μ0 γ1 + B0 b^3 == 0;
         sol = Solve[eqn1 && eqn2 && eqn3 && eqn4, {α1, β1, γ1, λ1}];
         α1 = First[α1 /. sol];   β1 = First[β1 /. sol];
         λ1 = First[λ1 /. sol];   γ1 = First[γ1 /. sol];
```

```
In[3]:=  μ = 1000 μ0 ;  B0 = 1;  b = 5;  a = 4;  range = 8;

Φ1[r_, θ_] = α1 r Cos[θ] ;  Φ2[r_, θ_] = (β1 r + λ1/r²) Cos[θ] ;

Φ3[r_, θ_] = (- B0/μ0 r + γ1/r²) Cos[θ] ;

Φ[r_, θ_] = Piecewise[{{Φ1[r, θ], r ≤ a},
                       {Φ2[r, θ], a ≤ r ≤ b}, {Φ3[r, θ], r ≥ b}}];
B1r[r_, θ_] = -μ0 D[Φ1[r, θ], r];    B1θ[r_, θ_] = -μ0 r⁻¹ D[Φ1[r, θ], θ];
B2r[r_, θ_] = -μ D[Φ2[r, θ], r];   B2θ[r_, θ_] = -μ r⁻¹ D[Φ2[r, θ], θ];
B3r[r_, θ_] = -μ0 D[Φ3[r, θ], r];    B3θ[r_, θ_] = -μ0 r⁻¹ D[Φ3[r, θ], θ];
Br[r_, θ_] = Piecewise[{{B1r[r, θ], r ≤ a}, {B2r[r, θ], a ≤ r ≤ b},
                       {B3r[r, θ], r ≥ b}}];
Bθ[r_, θ_] = Piecewise[{{B1θ[r, θ], r ≤ a}, {B2θ[r, θ], a ≤ r ≤ b},
                       {B3θ[r, θ], r ≥ b}}];

listt = {Cos[θ] → z/√(y² + z²), Sin[θ] → y/√(y² + z²), r → √(y² + z²)};

con1 = (√(y² + z²) ≤ a); con2 = (a ≤ √(y² + z²) ≤ b); con3 = (√(y² + z²) ≥ b);

B1y[y_, z_] = (Sin[θ] B1r[r, θ] + Cos[θ] B1θ[r, θ]) /. listt;
B1z[y_, z_] = (Cos[θ] B1r[r, θ] - Sin[θ] B1θ[r, θ]) /. listt;
B2y[y_, z_] = (Sin[θ] B2r[r, θ] + Cos[θ] B2θ[r, θ]) /. listt;
B2z[y_, z_] = (Cos[θ] B2r[r, θ] - Sin[θ] B2θ[r, θ]) /. listt;
B3y[y_, z_] = (Sin[θ] B3r[r, θ] + Cos[θ] B3θ[r, θ]) /. listt;
B3z[y_, z_] = (Cos[θ] B3r[r, θ] - Sin[θ] B3θ[r, θ]) /. listt;
By[y_, z_] = Piecewise[{{B1y[y, z], con1}, {B2y[y, z], con2},
                       {B3y[y, z], con3}}];
Bz[y_, z_] = Piecewise[{{B1z[y, z], con1}, {B2z[y, z], con2},
                       {B3z[y, z], con3}}];
gr1 = StreamPlot[{By[y, z], Bz[y, z]}, {y, -range, range}, {z, -range, range},
    RegionFunction → Function[{y, z}, y² + z² ≥ a²], AspectRatio → 1,
    StreamStyle → {Black, Thickness[0.015]},
    FrameStyle → Directive[Black, Thickness[0.005]], AxesLabel → {y, z},
    LabelStyle → Directive[FontSize → 15]];
gr2 = Graphics[{Gray, Disk[{0, 0}, b]}];
gr3 = Graphics[{White, Disk[{0, 0}, a]}];      Show[gr1, gr2, gr3]
```

Question 9.22

Consider a long solenoid (assume it to be ideal) having length L, cross-sectional area A and n turns per unit length connected to a power supply which maintains a steady current I.

(a) Briefly describe what happens when the space inside the solenoid is completely filled with an lih material having relative permeability μ_r.

(b) Calculate the **H**-field inside the solenoid.

(c) Suppose the current changes to $I + \delta I$ (and the magnetic field to $B + \delta B$) in a time δt. Show that there is an induced back emf

$$\mathcal{E}_{ind} = -nLA\frac{\delta B}{\delta t}. \tag{1}$$

Solution

(a) Insertion of the material will be accompanied by a brief surge in current, where the extra energy delivered by the power supply is used to magnetize the material. This energy is stored in the magnetic field. Once this has occurred, the current reverts to its original value I.

(b) We apply Ampère's law $\oint_c \mathbf{H} \cdot d\mathbf{l} = \int_s \mathbf{J}_f \cdot d\mathbf{a}$ to the rectangular path abcd shown below.

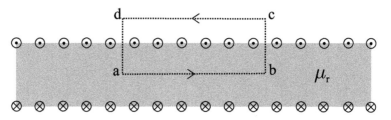

Because of the assumed ideal behaviour, **H** inside the solenoid is parallel to its axis and is zero everywhere outside. Hence $\int_c^d \mathbf{H}\cdot d\mathbf{l} = 0$. Also, $\int_b^c \mathbf{H}\cdot d\mathbf{l} = \int_d^a \mathbf{H}\cdot d\mathbf{l} = 0$ because **H** is perpendicular to $d\mathbf{l}$ here. Then

$$\oint_c \mathbf{H} \cdot d\mathbf{l} = \int_a^b \mathbf{H} \cdot d\mathbf{l} = H\ell = n\ell I,$$

and so

$$H = nI. \tag{2}$$

(c) Accompanying the change in current is a change of flux in the solenoid, given by $\delta\phi = nL \times \delta B \times A$. Application of Lenz's law $\mathcal{E}_{ind} = -\delta\phi/\delta t$ leads immediately to (1).

Comments

(i) The work done by the power supply against the back emf is $dW = -I \times \mathcal{E}_{\text{ind}} \times \delta t = nLAI\,\delta B$, giving an energy density δu (energy per unit volume) of

$$\delta u = nI\delta B = H\,\delta B,$$

because of (2).

(ii) The change in energy δU which occurs when a magnetic field $\mathbf{B}$ changes to $\mathbf{B}+\delta\mathbf{B}$ is

$$\delta U = \int_v \mathbf{H} \cdot \delta\mathbf{B}\,dv. \tag{3}$$

In the particular case of an lih material for which $\mathbf{B} = \mu\mathbf{H}$, this last equation may be integrated with respect to the field to give

$$U = \tfrac{1}{2}\int_v \mathbf{H} \cdot \mathbf{B}\,dv. \tag{4}$$

In Question 9.23 we show more generally that U is given by (4), whether the material is lih or not.

Question 9.23

The magnetic energy U in a region of space v, where there is a free-current density $\mathbf{J}_{\text{f}}(\mathbf{r})$, is given by (4) of Question 5.15:

$$U = \tfrac{1}{2}\int_v \mathbf{J}_{\text{f}} \cdot \mathbf{A}\,dv, \tag{1}$$

where $\mathbf{A}(\mathbf{r})$ is the vector potential. Show that U can be expressed as

$$U = \tfrac{1}{2}\int_v \mathbf{B} \cdot \mathbf{H}\,dv. \tag{2}$$

Solution

Eliminating the free-current density from (1) using $\nabla \times \mathbf{H} = \mathbf{J}_{\text{f}}$ gives

$$U = \tfrac{1}{2}\int_v (\nabla \times \mathbf{H}) \cdot \mathbf{A}\,dv,$$

and then applying the vector identity (7) of Question 1.8 yields

$$U = \tfrac{1}{2}\int_v \nabla \cdot (\mathbf{H} \times \mathbf{A})\, dv + \tfrac{1}{2}\int_v (\nabla \times \mathbf{A}) \cdot \mathbf{H}\, dv.$$

Converting the first integral above using Gauss's theorem and putting $\mathbf{B} = \nabla \times \mathbf{A}$ give

$$U = \tfrac{1}{2}\oint_s (\mathbf{H} \times \mathbf{A}) \cdot d\mathbf{a} + \tfrac{1}{2}\int_v \mathbf{B} \cdot \mathbf{H}\, dv. \qquad (3)$$

The $\mathbf{H}$-field (like $\mathbf{B}$) behaves asymptotically as r^{-3}, whilst $\mathbf{A}$ behaves as r^{-2} (if in doubt, recall the field and vector potential of a point magnetic dipole). Then $(\mathbf{H} \times \mathbf{A}) \cdot d\mathbf{a}$ in (3) scales as $r^2/r^5 \sim r^{-3}$. Now since s is any surface which encloses all the currents, we may choose a sphere having an arbitrarily large radius R. In the limit $R \to \infty$, the surface integral in (3) tends to zero. Hence (2).

Comment

From (2) we can deduce the following:

☞ The quantity $\tfrac{1}{2}\mathbf{B} \cdot \mathbf{H}$ is the energy density of the magnetic field. Note its similarity to the energy density $\tfrac{1}{2}\mathbf{E} \cdot \mathbf{D}$ in the electrostatic field (see (5) of Question 9.11).

☞ The energy U is greater by the factor μ_r than the energy required to establish the same magnetic field in vacuum.

10

Some applications of Maxwell's equations in matter

This chapter comprises a range of questions of a miscellaneous nature. They mostly have little in common except that all processes are time-dependent, and occur within a material medium. We begin with some important preliminaries, and then show how Maxwell's equations for a vacuum must be modified in the presence of matter; a task we perform in Question 10.2. These equations are completely general, apply in any type of material medium (whether lih or not), and are the foundation upon which the theory of macroscopic electromagnetism is based. The behaviour of the electromagnetic field at the boundary (or interface) between two media having different properties (e.g. permittivity, permeability, conductivity) is an important topic, and the matching conditions (as we refer to them) are discussed in Questions 10.6 and 10.7. Certain specific examples then follow. In particular, we will consider some simple applications involving conductors (see Questions 10.10–10.13), and dielectrics where we use Maxwell's equations to derive the laws of geometrical optics (see Questions 10.14–10.17). The chapter ends with Questions 10.18–10.20 which deal with electromagnetic waves propagating in a tenuous electronic plasma.

Question 10.1

Suppose $\mathbf{P}(\mathbf{r}, t)$ is a time-dependent polarization density. Show that there is a bound-charge current density

$$\mathbf{J}_{\mathrm{p}} \;=\; \frac{\partial \mathbf{P}}{\partial t}. \tag{1}$$

Hint: Start with (II) from the introduction to Chapter 9.

Solution

At an arbitrary macroscopic point inside the medium we have $\mathbf{P}(\mathbf{r}, t) = n(\mathbf{r})\bar{\mathbf{p}}(t)$, where $n(\mathbf{r})$ is the number of dipoles per unit volume and $\bar{\mathbf{p}}(t)$ is the average electric dipole moment at the macroscopic point at time t. In an infinitesimal interval dt the change in polarization density, $d\mathbf{P} = n\,d\bar{\mathbf{p}}$, arises from a distortion of the electronic charge distribution in the dielectric at the macroscopic point, and so $d\mathbf{P}/dt = n\,d\bar{\mathbf{p}}/dt$.

Solved Problems in Classical Electromagnetism. J. Pierrus, Oxford University Press (2018).
© J. Pierrus. DOI: 10.1093/oso/9780198821915.001.0001

This quantity has the dimensions of a current density, and because it arises from a time-dependent polarization we write $\mathbf{J}_p = d\mathbf{P}/dt$. However, at the fixed macroscopic point under consideration the only change in $\mathbf{P}$ is due to an *explicit* time variation. Hence (1).

Comments

(i) It seems reasonable to assume that the result for the bound-charge density,

$$\rho_b(\mathbf{r}, t) = -\nabla \cdot \mathbf{P}, \qquad (2)$$

remains valid even for a time-dependent polarization. It does.[1]

(ii) The bound-charge current density (1) exists quite independently of $\nabla \times \mathbf{M}$ (recall $(6)_2$ of Question 9.13), which may or may not be zero. Suppose that, in addition to $\mathbf{P}(\mathbf{r}, t)$, there is also a time-dependent magnetization $\mathbf{M}(\mathbf{r}, t)$. Then there are two separate contributions to $\mathbf{J}_b$, giving in general

$$\mathbf{J}_b(\mathbf{r}, t) = \mathbf{J}_p + \nabla \times \mathbf{M} = \frac{\partial \mathbf{P}}{\partial t} + \nabla \times \mathbf{M}. \qquad (3)$$

Question 10.2

Show that $\rho_b(\mathbf{r}, t)$ and $\mathbf{J}_b(\mathbf{r}, t)$, given by (2) and (3) of Question 10.1, satisfy the bound-charge continuity equation

$$\nabla \cdot \mathbf{J}_b + \frac{\partial \rho_b}{\partial t} = 0. \qquad (1)$$

Solution

Substituting $\mathbf{J}_b = \partial \mathbf{P}/\partial t + \nabla \times \mathbf{M}$ in the left-hand side of (1) yields

$$\nabla \cdot \mathbf{J}_b + \frac{\partial \rho_b}{\partial t} = \nabla \cdot \left(\frac{\partial \mathbf{P}}{\partial t} + \nabla \times \mathbf{M} \right) + \frac{\partial \rho_b}{\partial t} = \frac{\partial}{\partial t}(\nabla \cdot \mathbf{P} + \rho_b) \qquad (2)$$

because the divergence of any curl is identically zero. Then $\rho_b = -\nabla \cdot \mathbf{P}$ in (2) gives zero as required.

Comment

We might expect that bound charge (like free charge) is also conserved. Equation (1) shows that it is.

---

[1] R. E. Raab and O. L. de Lange, *Multipole theory in electromagnetism*, Chap. 1, p. 19. Oxford: Clarendon Press, 2005.

Question 10.3*

The macroscopic fields $\mathbf{E}(\mathbf{r}, t)$ and $\mathbf{B}(\mathbf{r}, t)$ are defined in terms of the corresponding microscopic fields $\mathbf{e}(\mathbf{r}, t)$ and $\mathbf{b}(\mathbf{r}, t)$ by $\mathbf{E} = \langle \mathbf{e} \rangle$ and $\mathbf{B} = \langle \mathbf{b} \rangle$,[‡] where $\mathbf{e}$ and $\mathbf{b}$ satisfy the microscopic Maxwell equations

$$\left.\begin{aligned}
\nabla \cdot \mathbf{e} &= \epsilon_0^{-1} \eta(\mathbf{r}, t) \\
\nabla \times \mathbf{e} &= -\dot{\mathbf{b}} \\
\nabla \cdot \mathbf{b} &= 0 \\
\nabla \times \mathbf{b} &= \mu_0 \mathbf{j}(\mathbf{r}, t) + \epsilon_0 \mu_0 \dot{\mathbf{e}}
\end{aligned}\right\}. \tag{1}$$

Here $\eta(\mathbf{r}, t)$ and $\mathbf{j}(\mathbf{r}, t)$ are the microscopic charge and current densities. They can be decomposed into *free* and *bound* parts:

$$\left.\begin{aligned}
\eta &= \eta(\mathbf{r}, t) = \eta_{\mathrm{f}} + \eta_{\mathrm{b}} \\
\mathbf{j} &= \mathbf{j}(\mathbf{r}, t) = \mathbf{j}_{\mathrm{f}} + \mathbf{j}_{\mathrm{b}}
\end{aligned}\right\}, \tag{2}$$

where

$$\langle \eta_{\mathrm{f}} \rangle = \rho_{\mathrm{f}}, \qquad \langle \eta_{\mathrm{b}} \rangle = \rho_{\mathrm{b}}, \qquad \langle \mathbf{j}_{\mathrm{f}} \rangle = \mathbf{J}_{\mathrm{f}}, \qquad \langle \mathbf{j}_{\mathrm{b}} \rangle = \mathbf{J}_{\mathrm{b}}. \tag{3}$$

(The quantities on the right-hand sides of (3) represent macroscopic averages. A formal derivation of $(3)_2$ and $(3)_4$ can be found in Ref. [2].)

(a) Use (1)–(3) and the definitions of $\mathbf{E}(\mathbf{r}, t)$ and $\mathbf{B}(\mathbf{r}, t)$ to derive Maxwell's macroscopic equations.

(b) Now eliminate ρ_{b} and $\mathbf{J}_{\mathrm{b}}$ from the two inhomogeneous equations derived in (a) above. *Hint:* Refer to (2) and (3) of Question 10.1.

Solution

(a) We consider each equation in turn:

Maxwell–Gauss

From $(1)_1$ and $(2)_1$ we obtain

$$\epsilon_0 \langle \nabla \cdot \mathbf{e} \rangle = \langle \eta(\mathbf{r}, t) \rangle = \langle \eta_{\mathrm{f}} \rangle + \langle \eta_{\mathrm{b}} \rangle \quad \Rightarrow \quad \boxed{\epsilon_0 \nabla \cdot \mathbf{E} = \rho_{\mathrm{f}} + \rho_{\mathrm{b}}}, \tag{4}$$

because of $(3)_1$ and $(3)_2$.

[‡] The remarks about averaging, given in the introduction to Chapter 9, for static **E**- and **B**-fields are also relevant here.

[2] J. D. Jackson, *Classical electrodynamics*, Chap. 6, pp. 248–50. New York: Wiley, 3 edn, 1998.

Maxwell–Faraday

Now from $(2)_1$

$$\langle \boldsymbol{\nabla} \times \mathbf{e} + \dot{\mathbf{b}} \rangle \;=\; \langle \boldsymbol{\nabla} \times \mathbf{e} \rangle + \langle \dot{\mathbf{b}} \rangle \;=\; 0 \;\Rightarrow\; \boxed{\boldsymbol{\nabla} \times \mathbf{E} + \frac{\partial \mathbf{B}}{\partial t} = 0} \,. \tag{5}$$

Maxwell–Gauss

Similarly from $(2)_3$

$$\langle \boldsymbol{\nabla} \cdot \mathbf{b} \rangle = 0 \;\Rightarrow\; \boxed{\boldsymbol{\nabla} \cdot \mathbf{B} = 0} \,. \tag{6}$$

Maxwell–Ampère

Equations $(1)_4$ and $(2)_4$ yield

$$\langle \boldsymbol{\nabla} \times \mathbf{b} - \epsilon_0 \mu_0 \dot{\mathbf{e}} \rangle \;=\; \langle \boldsymbol{\nabla} \times \mathbf{b} \rangle - \epsilon_0 \mu_0 \langle \dot{\mathbf{e}} \rangle \;=\; \mu_0 \langle \mathbf{j}(\mathbf{r}, t) \rangle \;=\; \mu_0 \langle \mathbf{j}_f \rangle + \mu_0 \langle \mathbf{j}_b \rangle$$

$$\Rightarrow\; \boxed{\boldsymbol{\nabla} \times \mathbf{B} = \mu_0 \mathbf{J}_f + \mu_0 \mathbf{J}_b + \epsilon_0 \mu_0 \frac{\partial \mathbf{E}}{\partial t}} \,, \tag{7}$$

where in the last step we use $(3)_3$ and $(3)_4$.

(b) Taking the inhomogeneous equations (4) and (7) in turn gives:

Maxwell–Gauss

Substituting $\rho_b = -\boldsymbol{\nabla} \cdot \mathbf{P}$ in (4) gives:

$$\epsilon_0 \boldsymbol{\nabla} \cdot \mathbf{E} \;=\; \rho_f + \rho_b \;=\; \rho_f - \boldsymbol{\nabla} \cdot \mathbf{P}$$

$$\boldsymbol{\nabla} \cdot \left(\epsilon_0 \mathbf{E} + \mathbf{P} \right) \;=\; \rho_f .$$

Defining the term in brackets as the time-dependent field $\mathbf{D}(\mathbf{r}, t)$ yields

$$\mathbf{D} \;=\; \left(\epsilon_0 \mathbf{E} + \mathbf{P} \right), \qquad \text{and so} \qquad \boxed{\boldsymbol{\nabla} \cdot \mathbf{D} = \rho_f} \,. \tag{8}$$

Maxwell–Ampère

Substituting $\mathbf{J}_b = \partial \mathbf{P} / \partial t + \boldsymbol{\nabla} \times \mathbf{M}$ in (7) gives:

$$\mu_0^{-1} \boldsymbol{\nabla} \times \mathbf{B} \;=\; \mathbf{J}_f + \mathbf{J}_b + \epsilon_0 \frac{\partial \mathbf{E}}{\partial t}$$

$$\mu_0^{-1} \boldsymbol{\nabla} \times \mathbf{B} \;=\; \mathbf{J}_f + \frac{\partial \mathbf{P}}{\partial t} + \boldsymbol{\nabla} \times \mathbf{M} + \epsilon_0 \frac{\partial \mathbf{E}}{\partial t}$$

$$\boldsymbol{\nabla} \times \left(\mu_0^{-1} \mathbf{B} - \mathbf{M} \right) \;=\; \mathbf{J}_f + \frac{\partial}{\partial t} \left(\epsilon_0 \mathbf{E} + \mathbf{P} \right).$$

Defining the term in brackets on the left-hand side as the time-dependent field $\mathbf{H}(\mathbf{r}, t)$ and using the definition $(8)_1$ yield

$$\mathbf{H} = \left(\mu_0^{-1}\mathbf{B} - \mathbf{M}\right), \qquad \text{and so} \qquad \boxed{\nabla \times \mathbf{H} = \mathbf{J}_{\mathrm{f}} + \frac{\partial \mathbf{D}}{\partial t}}. \tag{9}$$

Comments

(i) The integral forms of the differential equations $(8)_2$ and $(9)_2$ follow in the usual way from Gauss's theorem and Stokes's theorem, and are:

$$\left.\begin{array}{l} \oint_s \mathbf{D} \cdot d\mathbf{a} = \displaystyle\int_v \rho_{\mathrm{f}}\, dv = q_{\mathrm{f}} \\[3mm] \oint_c \mathbf{H} \cdot d\mathbf{l} = \displaystyle\int_s \left(\mathbf{J}_{\mathrm{f}} + \frac{\partial \mathbf{D}}{\partial t}\right) \cdot d\mathbf{a} = I_{\mathrm{f}} + \int_s \dot{\mathbf{D}} \cdot d\mathbf{a} \end{array}\right\}. \tag{10}$$

(ii) Often $\dot{\mathbf{D}} = 0$, and $(10)_2$ shows that the $\mathbf{H}$-field is directly related to the free current I_{f}. Because I_{f} is a quantity that is readily controlled and easily measured in the laboratory, $\mathbf{H}$ is a very useful field in practice. By contrast, the free charge q_{f} carried by a body is harder to manipulate and usually more difficult to measure: 'charge meters', unlike ammeters, are not off-the-shelf items of equipment.

(iii) $\mathbf{D}$ and $\mathbf{H}$ in the above equations are called 'response fields'; the reason for this name is explained in Comment (i) of Question 10.4.

Question 10.4 **

Suppose we relax the electric dipole approximation (discussed in the introduction to Chapter 9) and allow contributions to $\mathbf{P}$ and $\mathbf{M}$ from the higher electric and magnetic multipole moment densities Q_{ij}, $Q_{ijk} \ldots$ and $M_{ij} \ldots$ respectively. By making suitable expansions of the retarded scalar and vector potentials, it is possible to derive multipole expressions for ρ_{b} and $J_{\mathrm{b}i}$ in terms of these densities.[‡] In particular

$$\left.\begin{array}{l} \rho_{\mathrm{b}} = -\nabla_i P_i + \frac{1}{2}\nabla_i\nabla_j Q_{ij} - \cdots \\[3mm] J_{\mathrm{b}i} = \dot{P}_i - \nabla_j\left(\frac{1}{2}\dot{Q}_{ij} - \varepsilon_{ijk}M_k\right) - \cdots \end{array}\right\}, \tag{1}$$

where the dot notation denotes partial differentiation with respect to time. Use (1) to show that the response fields to electric quadrupole–magnetic dipole order are given by

$$\left.\begin{array}{l} D_i = \epsilon_0 E_i + P_i - \frac{1}{2}\nabla_j Q_{ij} \\[3mm] H_i = \mu_0^{-1}B_i - M_i \end{array}\right\}. \tag{2}$$

[‡]Readers who are interested in the details should consult Ref. [3].

[3]   R. Raab and O. de Lange, *Multipole theory in electromagnetism*. Oxford: Clarendon Press, 2005.

Solution

☞ Substituting $(1)_1$ in $\epsilon_0 \nabla \cdot \mathbf{E} = \rho_f + \rho_b$ and proceeding as for Question 10.3 give:

$$\epsilon_0 \nabla_i E_i = \rho_f + \rho_b = \rho_f - \nabla_i P_i + \tfrac{1}{2} \nabla_i \nabla_j Q_{ij}$$

$$\nabla_i \left(\epsilon_0 E_i + P_i - \tfrac{1}{2} \nabla_j Q_{ij} \right) = \rho_f.$$

Then because $\nabla \cdot \mathbf{D} = \rho_f$ we obtain $(2)_1$.

☞ Substituting $(1)_2$ in $\mu_0^{-1} \nabla \times \mathbf{B} = \mathbf{J}_f + \mathbf{J}_b + \epsilon_0 \dfrac{\partial \mathbf{E}}{\partial t}$ and proceeding as above give:

$$\mu_0^{-1} \nabla \times \mathbf{B} = \mathbf{J}_f + \mathbf{J}_b + \epsilon_0 \frac{\partial \mathbf{E}}{\partial t}$$

$$\mu_0^{-1} \varepsilon_{ijk} \nabla_j B_k = J_{if} + \left\{ \dot{P}_i - \nabla_j \left(\tfrac{1}{2} \dot{Q}_{ij} - \varepsilon_{ijk} M_k \right) \right\} + \epsilon_0 \dot{E}_i$$

$$\varepsilon_{ijk} \nabla_j \left(\mu_0^{-1} B_k - M_k \right) = J_{if} + \left(\epsilon_0 \dot{E}_i + \dot{P}_i - \tfrac{1}{2} \nabla_j \dot{Q}_{ij} \right).$$

Equation $(2)_2$ then follows from $\nabla \times \mathbf{H} = \mathbf{J}_f + \dfrac{\partial \mathbf{D}}{\partial t}$ with $\mathbf{D}$ given by $(2)_1$.

Comments

(i) The moment densities $P_i, Q_{ij}, \ldots$ and $M_i, \ldots$ used in the definitions of $\mathbf{D}$ and $\mathbf{H}$ (see (2)) include induced contributions which arise from the response of matter to externally applied fields. Hence their name.

(ii) Like the static equations for $\mathbf{D}$ and $\mathbf{H}$ discussed in Chapter 9, both $\nabla \cdot \mathbf{D} = \rho_f$ and $\nabla \times \mathbf{H} = \mathbf{J}_f + \partial \mathbf{D}/\partial t$, written in terms of the response fields, are convenient ways of ignoring what happens inside the medium. All the sources of $\mathbf{E}$ and $\mathbf{B}$—except the free sources—are effectively 'buried' within the definitions of $\mathbf{D}$ and $\mathbf{H}$. The response-field equations, although convenient to use, do not introduce any new physics as such.

(iii) A well-known and important feature of Maxwell's two inhomogeneous equations is that the response fields in them are not uniquely defined.[3] This fact can be most easily understood by recalling Helmholtz's theorem, which requires that $\mathbf{F}(\mathbf{r}, t)$ is a uniquely defined vector field provided both $\nabla \cdot \mathbf{F}$ *and* $\nabla \times \mathbf{F}$ are specified, together with an appropriate boundary condition. In the case of the response fields, there is no equation specifying either $\nabla \times \mathbf{D}$ or $\nabla \cdot \mathbf{H}$.

(iv) In Chapter 9 we expressed the static response fields in the form

$$\left. \begin{array}{l} \mathbf{D} = \epsilon \mathbf{E} \\ \mathbf{H} = \mu^{-1} \mathbf{B} \end{array} \right\}, \tag{3}$$

where ϵ and μ are the permittivity and permeability of the medium. Equations like (3) and $\mathbf{J} = \sigma \mathbf{E}$ can be used for dynamic fields as well. Because they relate the response of the constitutive charges of a medium to an **E**- or **B**-field, they are known as *constitutive relations*. In general, the permittivity, permeability and conductivity depend on the frequency of the oscillation and can be complex. (Recall that for static fields, these three quantities are real constants.)

(v) To electric dipole order, the correct dynamic response fields are

$$\left.\begin{array}{c} \mathbf{D} = \epsilon_0 \mathbf{E} + \mathbf{P} \\ \mathbf{H} = \mu_0^{-1}\mathbf{B} \end{array}\right\} \quad \underline{\text{and not}} \quad \left.\begin{array}{c} \mathbf{D} = \epsilon_0 \mathbf{E} + \mathbf{P} \\ \mathbf{H} = \mu_0^{-1}\mathbf{B} - \mathbf{M} \end{array}\right\}. \tag{4}$$

Some textbooks retain the magnetization term in $(2)_2$, omit the quadrupole term in $(2)_1$ and report the set of equations on the right-hand side of (4) above. This practice should be strongly discouraged, because M_i and $\frac{1}{2}\nabla_j Q_{ij}$ are of the same multipole order in an electromagnetic response. That said, it is often true that $\nabla_j \dot{Q}_{ij} \ll (\nabla \times \mathbf{M})_i$, and then the quadrupole contribution is justifiably neglected. However, this is not always the case and the reader is urged to be cautious. Certain physical effects can only be properly explained by including the quadrupole contribution: optical activity in a quartz crystal being an example.[3] See also Comment (ii) of Question 10.5, where inconsistent multipole contributions result in a set of origin-dependent Maxwell equations.

Question 10.5*

Consider an arbitrary distribution of electric charges and currents, that is both bounded and electrically neutral. Using the definitions of the dipole moments (see the footnote on p. 507), prove that the:

(a) electric dipole moment $\mathbf{p}(t)$ of the distribution is origin-*independent*.

(b) magnetic dipole moment $\mathbf{m}(t)$ of the distribution is, in general, origin-*dependent*.

Solution

(a) In the definition $\mathbf{p}(t) = \displaystyle\int_v \mathbf{r}' \rho(\mathbf{r}', t)\, dv'$ we let $\bar{\mathbf{r}}' = \mathbf{r}' - \mathbf{d}$, where $\mathbf{d}$ is the displacement of origin $\bar{O}$ relative to O. Then

$$\mathbf{p}_{\bar{\sigma}} = \mathbf{p}_o - \mathbf{d}\int_v \rho(\mathbf{r}', t)\, dv', \tag{1}$$

because $\mathbf{d}$ is a constant. Now $\displaystyle\int_v \rho(\mathbf{r}', t)\, dv' = 0$ for a neutral distribution, and so $\mathbf{p}_{\bar{\sigma}} = \mathbf{p}_o$ as required.

(b) Here $\mathbf{m}(t) = \frac{1}{2} \int_v \mathbf{r}' \times \mathbf{J}(\mathbf{r}', t) \, dv'$, and

$$\mathbf{m}_{\overline{o}} = \mathbf{m}_o - \mathbf{d} \times \int_v \mathbf{J}(\mathbf{r}', t) \, dv', \tag{2}$$

since $\mathbf{d}$ is a constant as before. Now it follows from identity (5) of Question 1.8 that $J_i = \mathbf{\nabla}' \cdot (r_i' \mathbf{J}) - r_i' \mathbf{\nabla}' \cdot \mathbf{J}$. Then by Gauss's theorem

$$\int_v J_i \, dv' = \oint_s r_i' (\mathbf{J} \cdot \hat{\mathbf{n}}) \, da' - \int_v r_i' (\mathbf{\nabla}' \cdot \mathbf{J}) \, dv'. \tag{3}$$

Because the distribution is bounded $\mathbf{J} \cdot \hat{\mathbf{n}} = 0$, and so

$$\int_v \mathbf{J} \, dv' = -\int_v \mathbf{r}' (\mathbf{\nabla}' \cdot \mathbf{J}) \, dv' = \int_v \mathbf{r}' \frac{\partial \rho}{\partial t} \, dv' = \frac{d}{dt} \int_v \mathbf{r}' \rho \, dv' = \frac{d\mathbf{p}}{dt}, \tag{4}$$

where we have used the continuity equation $\mathbf{\nabla}' \cdot \mathbf{J} = -\dfrac{\partial \rho}{\partial t}$. Substituting (4) in (2) and since $\mathbf{d}$ is a constant displacement, we obtain

$$\mathbf{m}_{\overline{o}} = \mathbf{m}_o - \frac{d}{dt} (\mathbf{d} \times \mathbf{p}). \tag{5}$$

Now the cross-product in (5) is not always zero; hence the result.

Comments

(i) It is interesting to contrast the origin dependences of the static dipole moments with their dynamic counterparts. For a *neutral* distribution of charge, whether time-dependent or time-independent, $\mathbf{p}$ is always independent of origin; whereas only the magnetostatic dipole moment is origin-independent, evident by putting the last term in (5) equal to zero (see also Comment (iv) of Question 4.24).

(ii) The volume element $dv' = dx' \, dy' \, dz'$ is obviously origin-independent, and so the dipole moment densities $\mathbf{P} = d\mathbf{p}/dv$ and $\mathbf{M} = d\mathbf{m}/dv$ inherit the traits of $\mathbf{p}$ and $\mathbf{m}$ respectively. Thus for a neutral distribution $\mathbf{P}$ is origin-independent and therefore $\mathbf{D} = \epsilon_o \mathbf{E} + \mathbf{P}$ is also origin-independent. By contrast, $\mathbf{M}$ and $\mathbf{H}$ (given by $\mathbf{H} = (\mu_0^{-1} \mathbf{B} - \mathbf{M})$) are both origin-dependent fields. Now $\mathbf{D}$ and $\mathbf{H}$ with their different origin dependences are related by the Maxwell–Ampère equation, and so this equation becomes origin-dependent[‡] too. The difficulty is resolved by either excluding $\mathbf{M}$ from $(9)_1$ of Question 10.3, which then with $(8)_1$ is correct to electric dipole order; or retaining $\mathbf{M}$ in $(2)_2$ of Question 10.4 and including the electric quadrupole contribution (an origin-dependent term) in $(2)_1$. The latter procedure ensures that the Maxwell–Ampère equation, now to the order of electric quadrupole-magnetic dipole, is origin-independent as required.

[‡]This is obviously unacceptable. A necessary requirement which any physical law must satisfy is that it does not depend on an arbitrary choice of origin.

Question 10.6

Two semi-infinite homogeneous media 1 and 2 having different electromagnetic properties occupy the half-spaces $z < 0$ and $z > 0$ respectively, with the interface lying in the
xy-plane (see the figure below). For the sake of generality, suppose that free charges
and currents are present at the interface. Relations exist which connect the **E**, **B**, **D**
and **H** fields on either side of the interface at $z = 0$. In this question, we determine
these relations by applying Maxwell's *integral* equations to some simple geometrical
shape (e.g. a cylindrical surface or rectangular contour) which spans a representative
section of the interface. In Question 10.7, we repeat the derivation using Maxwell's
differential equations applied at a point in the interface, and comment on important
differences between the two approaches.

(a) Choose an infinitesimal Gaussian cylinder s oriented such that its end caps are
parallel to the xy-plane as shown. Suppose the cylinder has cross-sectional area
δa and is several atomic lengths in depth. Stating any assumptions which you
make, use the flux equations

$$\oint_s \mathbf{D} \cdot d\mathbf{a} = \int_v \rho_f \, dv \qquad \text{and} \qquad \oint_s \mathbf{B} \cdot d\mathbf{a} = 0 \qquad (1)$$

to prove that

$$(\mathbf{D}_2 - \mathbf{D}_1) \cdot \hat{\mathbf{n}} = \sigma_f \qquad \text{and} \qquad (\mathbf{B}_2 - \mathbf{B}_1) \cdot \hat{\mathbf{n}} = 0, \qquad (2)$$

where $\hat{\mathbf{n}} = \hat{\mathbf{z}}$ and σ_f is the free surface-charge density.

(b) Choose an infinitesimal Stokesian contour c oriented such that the normal $\hat{\mathbf{t}}$ to its
surface is tangential to the interface (i.e. $\hat{\mathbf{t}}$ lies in the xy-plane) as shown. Suppose
the long sides of c have length $\delta l = \pm(\hat{\mathbf{n}} \times \hat{\mathbf{t}}) \, \delta l$ and the short sides are several
atomic lengths. Stating any assumptions which you make, use the circulation
equations

$$\oint_c \mathbf{E} \cdot d\mathbf{l} + \int_s \frac{\partial \mathbf{B}}{\partial t} \cdot d\mathbf{a} = 0 \qquad \text{and} \qquad \oint_c \mathbf{H} \cdot d\mathbf{l} = \int_s \left(\mathbf{J}_f + \frac{\partial \mathbf{D}}{\partial t} \right) \cdot d\mathbf{a} \qquad (3)$$

to prove that

$$(\mathbf{E}_2 - \mathbf{E}_1) \times \hat{\mathbf{n}} = 0 \qquad \text{and} \qquad (\mathbf{H}_2 - \mathbf{H}_1) \times \hat{\mathbf{n}} = -\mathbf{K}_f, \qquad (4)$$

where $\mathbf{K}_f$ is the free surface-current density.

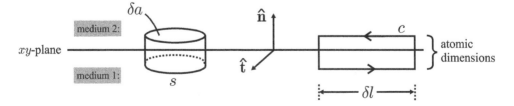

Solution

(a) The net outward flux F of $\mathbf{D}$ through the cylindrical Gaussian surface may be expressed as the sum of two terms: $F_{\text{end caps}} + F_{\text{curved}}$, each representing flux through the corresponding part of s. The Maxwell equation $(1)_1$ gives

$$(\mathbf{D}_2 - \mathbf{D}_1) \cdot \delta \mathbf{a} + F_{\text{curved}} = (\mathbf{D}_2 - \mathbf{D}_1) \cdot \hat{\mathbf{n}} \, \delta a + F_{\text{curved}}$$

$$= \sigma_{\mathrm{f}} \, \delta a. \tag{5}$$

Equation $(2)_1$ follows directly from (5) assuming that $F_{\text{curved}} \to 0$ for an infinitesimal cylinder. Similar reasoning applied to $(1)_2$ yields $(2)_2$.

(b) The circulation C of $\mathbf{E}$ around the Stokesian contour c may also be expressed as the sum of two terms: $C_{\text{long sides}} + C_{\text{short sides}}$, each representing the line integral of the electric field along the corresponding part of c. The Maxwell equation $(3)_1$ gives

$$(\mathbf{E}_1 - \mathbf{E}_2) \cdot (\hat{\mathbf{n}} \times \hat{\mathbf{t}}) \, \delta l + C_{\text{short sides}} = -\int_s \frac{\partial \mathbf{B}}{\partial t} \cdot d\mathbf{a}. \tag{6}$$

In (6), the quantity $C_{\text{short sides}}$ and the area of the contour include, as a factor, the length of the short sides of c. Because of this, and assuming that $\partial \mathbf{B}/\partial t$ remains finite in the interface, both the second and third term in (6) tend to zero for an infinitesimal contour. Thus $(\mathbf{E}_1 - \mathbf{E}_2) \cdot (\hat{\mathbf{n}} \times \hat{\mathbf{t}}) \, \delta l = 0$ which can be written as $[(\mathbf{E}_2 - \mathbf{E}_1) \times \hat{\mathbf{n}}] \cdot \hat{\mathbf{t}} \, \delta l = 0$ (see of (1) of Question 1.8). Equation $(4)_1$ follows immediately because c can have any orientation in the surface. Similar reasoning applied to $(4)_2$ gives

$$\left[(\mathbf{H}_1 - \mathbf{H}_2) \times \hat{\mathbf{n}} \right] \cdot \hat{\mathbf{t}} \, \delta l = I_{\mathrm{f}}, \tag{7}$$

where $I_{\mathrm{f}} = \mathbf{K} \cdot \hat{\mathbf{t}} \, \delta l$ is the free current enclosed by the path of integration. Substituting this in (7) yields $(4)_2$.

Comments

(i) The sets of equations (2) and (4) are often used to connect the solutions of Maxwell's equations on either side of an interface to obtain a solution throughout all space. They are often referred to as 'the boundary conditions for the fields at an interface' which is a misnomer because the term *boundary conditions* has an entirely separate meaning. Alternative names, which are in use, include 'matching conditions'[4] and 'discontinuity equations'.[5] In this book, we will refer to them as the *matching conditions*.

[4] D. G. Hall, 'A few remarks on the matching conditions at interfaces in electromagnetic theory', *American Journal of Physics*, vol. 63, pp. 508–12, 1995.

[5] J. D. Jackson, *Classical electrodynamics*, Chap. I, p. 18. New York: Wiley, 3 edn, 1998.

(ii) Often $\sigma_f = 0$ and $\mathbf{K}_f = 0$ (for example, at the interface between two uncharged dielectric media). Then

$$
\left.
\begin{aligned}
(\mathbf{D}_2 - \mathbf{D}_1) \cdot \hat{\mathbf{n}} &= 0 &&\Rightarrow& D_{2_\perp} &= D_{1_\perp} \\
(\mathbf{B}_2 - \mathbf{B}_1) \cdot \hat{\mathbf{n}} &= 0 &&\Rightarrow& B_{2_\perp} &= B_{1_\perp} \\
(\mathbf{E}_2 - \mathbf{E}_1) \times \hat{\mathbf{n}} &= 0 &&\Rightarrow& E_{2_\parallel} &= E_{1_\parallel} \\
(\mathbf{H}_2 - \mathbf{H}_1) \times \hat{\mathbf{n}} &= 0 &&\Rightarrow& H_{2_\parallel} &= H_{1_\parallel}
\end{aligned}
\right\}, \tag{8}
$$

where the subscripts $\perp$ and $\parallel$ refer to field components perpendicular and parallel to the interface respectively.

(iii) Suppose medium 1 is a non-conductor and medium 2 a perfect conductor. Inside the latter $\mathbf{E} = 0$,[‡] and from the Maxwell equation $\nabla \times \mathbf{E}_2 = -\dot{\mathbf{B}}_2$ we conclude that there can be no time-dependent magnetic field either. It then follows from $(2)_2$ and $(4)_1$ that just above the conducting surface

$$
\mathbf{B}_\perp = 0 \quad \text{and} \quad \mathbf{E}_\parallel = 0. \tag{9}
$$

The matching conditions $(2)_1$ and $(4)_2$ place no further restrictions on the field. Instead, they require that the time-dependent σ_f and $\mathbf{K}_f$, given by

$$
\sigma_f = -\mathbf{D} \cdot \hat{\mathbf{n}} \quad \text{and} \quad \mathbf{K}_f = \mathbf{H} \times \hat{\mathbf{n}}, \tag{10}
$$

adjust themselves as rapidly as is necessary to ensure that the field remains zero everywhere inside the conductor at all times.

(iv) For a good, but less-than-perfect, conductor (i.e. finite conductivity) a surface layer of current is an idealization. Instead, the current is confined to a thin skin at the surface; the width of which is of order δ.[#] The skin width can be regarded as a transition zone where the fields decay exponentially to zero. It is now possible to have $E_{2\parallel} \neq 0$ and $B_{2\perp} \neq 0$ at the surface and, in a first approximation, the following field components are continuous at the interface:

$$
D_\perp, \quad B_\perp, \quad E_\parallel \quad \text{and} \quad H_\parallel.^{[6]} \tag{11}
$$

(v) The derivation of the matching conditions used in this question hinges on two crucial assumptions:

$$
\left.
\begin{aligned}
F_{\text{curved}} &\to 0 && \text{for an infinitesimal Gaussian cylinder} \\
C_{\text{short sides}} &\to 0 && \text{for an infinitesimal Stokesian loop}
\end{aligned}
\right\}. \tag{12}
$$

Many textbooks fail to comment on the scope and validity of (12). As a result, readers are left with the mistaken impression that matching conditions like (8)

[‡] This is a necessary consequence of Ohm's law. Non-zero $\mathbf{E}$ implies $\mathbf{J} \to \infty$, which is unphysical because of the infinite energy that would be associated with these currents.

[#] The skin depth δ is defined in Question 10.10.

---

[6] J. D. Jackson, *Classical electrodynamics*, Chap. 8, pp. 353–6. New York: Wiley, 3 edn, 1998.

are 'exact', whereas the converse is true. To illustrate this important point we mention two common cases:

a dielectric–dielectric interface

At the surface of a dielectric, singular terms in $\mathbf{E}$, $\mathbf{B}$, $\mathbf{D}$, $\mathbf{H}$, $\mathbf{J}$ and ρ veto the use of (12). Matching conditions such as (8) are correct only in a first approximation: namely at electric dipole order. An alternative approach, based on a differential rather than an integral formulation is required. See Question 10.7.

a vacuum–conductor interface

Non-singular fields may also occur at sharp corners and at points on a conducting surface, and the task of finding suitable matching conditions must be considered with care. See, for example, Ref. [7].

(vi) Notwithstanding our remarks in the last two comments, the matching conditions derived in this question are in widespread use. However, the reader should be aware that certain physical effects (e.g. reflection from the surface of an anti-ferromagnetic medium) cannot be adequately explained in terms of them. In such cases, matching conditions derived to a higher multipole order must be used. See also Question 10.7 and the references therein.

Question 10.7 **

In this question the *differential* forms of Maxwell's equations are used to determine the matching conditions for the fields at a vacuum–dielectric interface. We choose axes such that the vacuum and dielectric occupy the half-spaces $z < 0$ and $z > 0$ respectively, with the interface lying in the xy-plane. The approach is based on that of Ref. [8] and we illustrate the application of this technique at electric dipole order. To this multipole order, the bound source densities for an infinite dielectric medium are given by

$$\rho^{(\infty)} = -\boldsymbol{\nabla} \cdot \mathbf{P} \qquad \text{and} \qquad \mathbf{J}^{(\infty)} = \frac{\partial \mathbf{P}}{\partial t} . \tag{1}$$

(a) The source densities for the 'compound' medium are obtained by multiplying $\mathbf{P}$ in (1) with the step function $u(z)$, which is defined by

$$u(z) = \begin{cases} 1 & z > 0 \\ 0 & z < 0. \end{cases} \tag{2}$$

Prove that

$$\rho = u(z)\rho^{(\infty)} - \delta(z)P_z \qquad \text{and} \qquad \mathbf{J} = u(z)\mathbf{J}^{(\infty)}, \tag{3}$$

where $\delta(z)$ is the Dirac delta function, and is related to $u(z)$ by

$$\nabla_z u(\pm z) = \pm \delta(z). \tag{4}$$

[7]   D. S. Jones, *The theory of electromagnetism*. New York: Macmillan, 1964.
[8]   O. L. de Lange and R. E. Raab, 'Electromagnetic boundary conditions in multipole theory', *Journal of Mathematical Physics*, vol. 54, pp. 093513–11, 2013.

(b) With the source densities given by (3), attempt a solution of Maxwell's macro-scopic equations of the form

$$\left.\begin{array}{l} \mathbf{E} = u(-z)\mathbf{E}_1 + u(z)\mathbf{E}_2 \\ \mathbf{B} = u(-z)\mathbf{B}_1 + u(z)\mathbf{B}_2 \end{array}\right\}, \tag{5}$$

and show that

$$\left.\begin{array}{l} \nabla \cdot \mathbf{E} - \epsilon_0^{-1}\rho = \delta(z)(-E_{1z} + E_{2z} + \epsilon_0^{-1}P_z) \\[4pt] \nabla \times \mathbf{E} - \partial\mathbf{B}/\partial t = \delta(z)(E_{1y} - E_{2y},\ -E_{1x} + E_{2x},\ 0) \\[4pt] \nabla \cdot \mathbf{B} = \delta(z)(-B_{1z} + B_{2z}) \\[4pt] \nabla \times \mathbf{B} - \epsilon_0\mu_0\partial\mathbf{E}/\partial t - \mu_0\mathbf{J} = \delta(z)(B_{1y} - B_{2y},\ -B_{1x} + B_{2x},\ 0) \end{array}\right\}. \tag{6}$$

Here the subscripts 1 and 2 indicate fields in an infinite vacuum and dielectric respectively.

Hint: The fields $\mathbf{E}_1$, $\mathbf{B}_1$ satisfy Maxwell's vacuum equations $\nabla \cdot \mathbf{E}_1 = 0$, etc., and the fields $\mathbf{E}_2$, $\mathbf{B}_2$ satisfy $\nabla \cdot \mathbf{E}_2 = \epsilon_0^{-1}\rho^{(\infty)}$, etc.

(c) Hence deduce the requirements for the trial solution (5) to satisfy Maxwell's equations.

Solution

(a) By construction, $\rho = -\nabla \cdot (u\mathbf{P}) = -u\nabla_i P_i - P_i\nabla_i u = -u\nabla_i P_i - P_z\nabla_z u$, since u is a function of z only. Substituting $(1)_1$ and (4) in this equation for ρ gives $(3)_1$.

Similarly, $\mathbf{J} = \dfrac{\partial}{\partial t}(u\mathbf{P}) = u\dfrac{\partial \mathbf{P}}{\partial t}$, which is $(3)_2$ because of $(1)_2$.

(b) Taking the divergence of $(5)_1$, applying the hint and using $(3)_1$ give:

$$\begin{aligned} \nabla \cdot \mathbf{E} &= \nabla u(-z) \cdot \mathbf{E}_1 + u(-z)\nabla \cdot \mathbf{E}_1 + \nabla u(z) \cdot \mathbf{E}_2 + u(z)\nabla \cdot \mathbf{E}_2 \\ &= -\delta(z)E_{1z} + \delta(z)E_{2z} + \epsilon_0^{-1}u(z)\rho^{(\infty)} \\ &= \big(-E_{1z} + E_{2z}\big)\delta(z) + \epsilon_0^{-1}\big[\rho + \delta(z)P(z)\big], \end{aligned}$$

which can be rearranged to yield $(6)_1$. Equation $(6)_3$ follows in the same way. Similarly, taking the curl of $(5)_2$, applying the hint and using $(3)_2$ give

$$\begin{aligned} \nabla \times \mathbf{B} &= \nabla \times \{u(-z)\mathbf{B}_1 + u(z)\mathbf{B}_2\} \\ &= \nabla u(-z) \times \mathbf{B}_1 + u(-z)\nabla \times \mathbf{B}_1 + \nabla u(z) \times \mathbf{B}_2 + u(z)\nabla \times \mathbf{B}_2 \\ &= \delta(z)\hat{\mathbf{z}} \times (\mathbf{B}_2 - \mathbf{B}_1) + \epsilon_0\mu_0\big(u(-z)\dot{\mathbf{E}}_1 + u(z)\dot{\mathbf{E}}_2\big) + u(z)\mu_0\mathbf{J}^{(\infty)} \\ &= \delta(z)\hat{\mathbf{z}} \times \big((B_{2x} - B_{1x}),\ (B_{2y} - B_{1y}),\ 0\big) + \epsilon_0\mu_0\dot{\mathbf{E}} + \mu_0\mathbf{J} \\ &= \delta(z)\Big((B_{1y} - B_{2y}),\ (B_{2x} - B_{1x}),\ 0\Big) + \epsilon_0\mu_0\dot{\mathbf{E}} + \mu_0\mathbf{J}, \end{aligned}$$

which can be rearranged to yield $(6)_4$. Equation $(6)_2$ follows similarly.

(c) The trial solutions can be made to satisfy Maxwell's equations by requiring that the right-hand sides of (6) are zero everywhere. Since $\delta(z)$ is non-zero only in the interface, we let

$$
\left.
\begin{aligned}
-E_{1z} + E_{2z} + \epsilon_0^{-1} P_z &= 0 &&\Rightarrow& E_{1z} - E_{2z} &= \epsilon_0^{-1} P_z \\
(E_{1y} - E_{2y}, \; -E_{1x} + E_{2x}, \; 0) &= 0 &&\Rightarrow& E_{1x} &= E_{2x} \text{ and } E_{1y} = E_{2y} \\
-B_{1z} + B_{2z} &= 0 &&\Rightarrow& B_{1z} &= B_{2z} \\
(B_{1y} - B_{2y}, \; -B_{1x} + B_{2x}, \; 0) &= 0 &&\Rightarrow& B_{1x} &= B_{2x} \text{ and } B_{1y} = B_{2y}
\end{aligned}
\right\}.
$$

Expressed in coordinate-free form, these matching conditions are:

$$
\left.
\begin{aligned}
(\mathbf{E}_1 - \mathbf{E}_2) \cdot \hat{\mathbf{n}} &= \sigma_b/\epsilon_0 \\
(\mathbf{E}_1 - \mathbf{E}_2) \times \hat{\mathbf{n}} &= 0 \\
(\mathbf{B}_1 - \mathbf{B}_2) \cdot \hat{\mathbf{n}} &= 0 \\
(\mathbf{B}_1 - \mathbf{B}_2) \times \hat{\mathbf{n}} &= 0
\end{aligned}
\right\},
\tag{7}
$$

where the surface normal $\hat{\mathbf{n}}$ is directed from the vacuum side of the interface towards the dielectric medium.

Comments

(i) With a dielectric in the half-space $z < 0$ (region 1), we can readily obtain matching conditions for a dielectric–dielectric interface by replacing ρ in (3) with $-\nabla \cdot (u\mathbf{P}_1 + u\mathbf{P}_2)$. In the manner outlined above, this then leads to $(\mathbf{E}_1 - \mathbf{E}_2) \cdot \hat{\mathbf{n}} = \epsilon_0^{-1}(\mathbf{P}_2 - \mathbf{P}_1) \cdot \hat{\mathbf{n}}$ or $(\mathbf{D}_1 - \mathbf{D}_2) \cdot \hat{\mathbf{n}} = 0$, which together with $(7)_2$–$(7)_4$ are identical to the set of equations (8) of Question 10.6.

(ii) At the interface between vacuum and a good conductor, $(3)_1$ and $(3)_2$ become

$$
\rho = u(z)\rho^{(\infty)} + \delta(z)\sigma_f \qquad \text{and} \qquad \mathbf{J} = u(z)\mathbf{J}^{(\infty)} + \delta(z)\mathbf{K}_f,
\tag{8}
$$

where σ_f and $\mathbf{K}_f$ are the free surface charge and current densities respectively. Proceeding as before leads to the set of matching conditions given by (9) of Question 10.6.

(iii) The differential method for deriving matching conditions outlined above is based upon multipole expansions for the macroscopic source densities. In principle, this technique can be extended to any multipole order. The calculation performed here, to electric dipole order, shows that all electric- and magnetic-field components are continuous at a dielectric–dielectric interface except the normal component of $\mathbf{E}$. A calculation at higher multipole order reveals that all electric- and magnetic-field components are discontinuous *in general* and that the matching conditions contain these multipole moment densities and their gradients.[8]

(iv) If one attempts to derive the general matching conditions beyond electric dipole order using Maxwell's integral equations, the appearance of singular terms in the fields prevents the standard assumptions $(F_{\text{curved}} \to 0$ and $C_{\text{short sides}} \to 0$ in the limit of infinitesimal cylinders and rectangles (see Question 10.6)) being made. Instead, 'the integral theorems have to be supplemented by the differential approach, and in the process they become superfluous'.[8]

Question 10.8

At the turn of the twentieth century, Drude proposed a model for the transport properties of a metal by applying kinetic theory to its 'gas' of conduction electrons. Keep in mind the following assumptions of this model, when answering the questions below.

☞ Between collisions the conduction electrons interact neither with each other nor with the ionic cores.

☞ Abrupt changes in the velocity of a conduction electron occur as a result of interactions with the ionic core (a 'collision', or scattering).

☞ This scattering occurs with a probability per unit time of τ_c^{-1}, where τ_c is the mean time between collisions (sometimes called the collision time[‡]).

(a) Suppose a static electric field $\mathbf{E}$ is applied to the metal. Use Ohm's law for an isotropic medium $\mathbf{J} = \sigma\mathbf{E}$ and a classical equation of motion to show that the dc conductivity is

$$\sigma = \frac{Ne^2\tau_c}{m_e}. \tag{1}$$

Here N is the density[♯] of conduction electrons, each having mass m_e and charge e.

(b) Consider a time-harmonic electric field $\mathbf{E} = \mathbf{E}_0 e^{-i\omega t}$ and assume that the spatial variation of $\mathbf{E}$ is negligible over the mean free path of an electron. Now suppose that the electrons experience a velocity-dependent, damping force[†] of the form $-m_e\gamma\bar{\mathbf{v}}$, where γ is a phenomenological constant and $\bar{\mathbf{v}}$ is the average electron velocity. Include this frictional force in the above equation of motion, and then show that the ac conductivity is

$$\sigma(\omega) = \frac{Ne^2}{m_e(\gamma - i\omega)}. \tag{2}$$

(c) Hence show that

$$\sigma(\omega) = \frac{\sigma_0}{1 - i\omega\tau_c}, \quad \text{where} \quad \sigma_0 = \frac{Ne^2\tau_c}{m_e} \tag{3}$$

is the dc conductivity.

[‡]Accurate quantitative treatments of τ_c are not trivial and remain a weak link in modern theories of metallic conductivity.
[♯]For the remainder of this chapter we reserve the symbol n for refractive index, and use N for a number density.
[†]This damping force arises from 'collisions', radiation of energy, etc. See also the footnote on p. 493.

(d) Use the information given below to estimate τ_c and γ for copper (a typical mono-valent metal):

$$\sigma_0 = 6.5 \times 10^7 \, \mathrm{S \, m^{-1}}, \quad \rho = 8.94 \, \mathrm{g \, cm^{-3}} \text{ and an atomic mass of } 63.5 \, \mathrm{g \, mol^{-1}}.$$

(*Note:* The SI unit of conductivity is siemens per metre, where $1\,\mathrm{S} = 1\,\Omega^{-1}$.)

Solution

(a) Consider a typical electron at any time t. It is convenient to choose $t = 0$. Suppose this electron experienced its last collision at $t = -\tau$ and that its velocity was $\mathbf{v}_0$ immediately after that collision. Integrating the equation of motion $m_e \, d\mathbf{v}/dt = -e\mathbf{E}$ then yields

$$\int_{\mathbf{v}_0}^{\mathbf{v}} d\mathbf{v} = -\frac{e\mathbf{E}}{m_e} \int_{-\tau}^{0} dt \qquad \text{or} \qquad \mathbf{v} = \mathbf{v}_0 - \frac{e\mathbf{E}}{m_e}\tau. \tag{4}$$

Averaging (4) over all electron velocities and recognizing that there will be no contribution[b] from $\mathbf{v}_0$ to $\bar{\mathbf{v}}$ give

$$\bar{\mathbf{v}} = -\frac{e\mathbf{E}}{m_e}\langle\tau\rangle = -\frac{e\mathbf{E}}{m_e}\tau_c, \tag{5}$$

since $\langle\tau\rangle = \tau_c$. From (5) and (4) of Question 4.3 we have $\mathbf{J} = N(-e)\bar{\mathbf{v}} = \dfrac{Ne^2\tau_c}{m_e}\mathbf{E}$.

Comparing this with $\mathbf{J} = \sigma\mathbf{E}$ yields (1).

(b) The equation of motion is now $-m_e\gamma\bar{\mathbf{v}} - e\mathbf{E} = m_e\,d\bar{\mathbf{v}}/dt$. Assuming a trial solution of the form $\bar{\mathbf{v}} = \bar{\mathbf{v}}_0 e^{-i\omega t}$ gives $\bar{\mathbf{v}}_0 = \dfrac{-e\mathbf{E}_0}{m_e(\gamma - i\omega)}$ or $\bar{\mathbf{v}} = \dfrac{-e\mathbf{E}}{m_e(\gamma - i\omega)}$.

Writing $\mathbf{J}(\omega) = \sigma(\omega)\mathbf{E}$ and proceeding as in (a) yield (2).

(c) Comparing (1) and (2) at $\omega = 0$ gives

$$\gamma = \tau_c^{-1}. \tag{6}$$

Substituting (6) in (2) yields (3).

(d) In one mole of copper the conduction-electron density is

$$N = \frac{6.023 \times 10^{23} \, \text{atoms} \, \times \, 1\,\text{electron}/\text{atom}}{\frac{63.5}{8.94} \times 10^{-6} \, \mathrm{m^3}} = 8.5 \times 10^{28} \, \mathrm{m^{-3}}. \tag{7}$$

Substituting (7) in (3)$_2$ gives

$$\tau_c = \frac{\sigma_0 m_e}{Ne^2} = \frac{6.5 \times 10^7 \times 9.11 \times 10^{-31}}{8.5 \times 10^{28} \times (1.6 \times 10^{-19})^2} \sim 3 \times 10^{-14} \, \mathrm{s}, \tag{8}$$

and from (6)

$$\gamma = \tau_c^{-1} \sim 3 \times 10^{13} \, \mathrm{s^{-1}}. \tag{9}$$

[b] We assume that, after a collision, electrons emerge in all directions with equal probability.

Comments

(i) Equation (2) shows that, provided $\gamma \gg \omega$, the conductivity $\sigma(\omega)$ of a typical metal is real, independent of frequency and equal to σ_0. Since $\gamma \sim 10^{13}\,\text{s}^{-1}$, the condition is valid over a very wide range, from dc up into the microwave region ($f \sim 10^{11}\,\text{Hz}$) of the electromagnetic spectrum.

(ii) At higher frequencies and beyond the infrared, the classical kinetic-theory approach fails completely. A proper description of electrical conductivity requires quantum mechanics, since Pauli's exclusion principle has a crucial role. Yet in spite of this:

> the successes of the Drude model were considerable, and it is still used today as a quick practical way to form simple pictures and rough estimates of properties whose more precise comprehension may require analysis of considerable complexity. The failures of the Drude model to account for some experiments, and the conceptual puzzles it raised, defined the problems with which the theory of metals was to grapple over the next quarter century. These found their resolution only in the rich and subtle structure of the quantum theory of solids.[9]

(iii) A comprehensive review of the basic assumptions of Drude's free electron model (including those stated above) together with its failures can be found in Chapter 3 of Ref. [9].

(iv) An example of a conducting medium for which $\gamma \ll \omega$ is considered in Question 10.12.

Question 10.9

Excess electric charge is introduced[‡] into a material which is a poor conductor. Assume that the conductivity σ and permittivity ϵ of the medium are independent of position and that Ohm's law $\mathbf{J} = \sigma \mathbf{E}$ holds. Use the continuity equation to show that the charge density $\rho(\mathbf{r}, t)$ inside the medium decays exponentially with time.

Solution

From $\nabla \cdot \mathbf{J} + \dfrac{\partial \rho_f}{\partial t} = 0$ we have $\dfrac{\partial \rho_f}{\partial t} = -\nabla \cdot \mathbf{J} = -\sigma \nabla \cdot \mathbf{E}$. Gauss's law, $\nabla \cdot \mathbf{D} = \rho_f$ where $\mathbf{D} = \epsilon \mathbf{E}$, then gives

$$\frac{\partial \rho_f}{\partial t} = -\frac{\sigma}{\epsilon} \nabla \cdot \mathbf{D} = -\frac{\rho_f}{\tau_r} \tag{1}$$

[‡]For example, by bombarding the material with a pulse of high-energy electrons which come to rest in the interior.

[9] N. W. Ashcroft and N. D. Mermin, *Solid state physics*, Chap. 1, pp. 2–6. New York: Holt, Rinehart and Winston, 1976.

where

$$\tau_{\rm r} = \epsilon/\sigma. \tag{2}$$

Integrating (1) with respect to time yields

$$\rho_{\rm f}(\mathbf{r},t) = \rho_{\rm f}(\mathbf{r},0)\,e^{-t/\tau_{\rm r}}. \tag{3}$$

Comments

(i) The quantity $\tau_{\rm r}$, known as the relaxation time, is the time for $\rho_{\rm f}$ to decay by a factor e^{-1}, and it sets the time scale for the attainment of equilibrium ($\rho_{\rm f} = 0$ everywhere inside the dielectric). At equilibrium ($t \gg \tau_{\rm r}$), the excess charge resides on the outer surface of the material.

(ii) It follows from $\boldsymbol{\nabla} \cdot \mathbf{D} = \rho_{\rm f}$ that the electric field inside the dielectric will have the time dependence $e^{-t/\tau_{\rm r}}$ and the total current $\mathbf{J}_{\rm f} + \epsilon_0 \dfrac{\partial \mathbf{E}}{\partial t} = \sigma \mathbf{E} - \epsilon \dfrac{\mathbf{E}}{\tau_{\rm r}} = 0$ because of (2). The approach to equilibrium therefore involves no magnetic fields nor any induced electric fields. This contrasts with the physics for a good conductor as we see in (iv) below.

(iii) The analysis in (ii) above is not valid for good conductors because of a 'failure' of Ohm's law. To make sense of this last statement, we begin by considering the numerical value of ϵ_0/σ for copper (a typical metal). Taking $\sigma \sim 6.4 \times 10^7\,{\rm S\,m^{-1}}$ gives

$$\frac{\epsilon_0}{\sigma} = \frac{4\pi\epsilon_0}{4\pi\sigma} = \frac{1}{4\pi \times 9 \times 10^9 \times 6.5 \times 10^7} \sim 10^{-19}\,{\rm s}\,.$$

The physical assumptions underlying Ohm's law require the creation of an average balance between the momentum gained by a conduction electron from the electric field and the momentum lost in collisions, which for copper is $\tau_{\rm c} \sim 10^{-14}\,{\rm s}$ (see Question 10.8). It is unreasonable to suppose that the requisite momentum gain could occur on a time scale some five orders of magnitude smaller than this. In fact, we would expect that a time *at least* as long as $\tau_{\rm c}$ would be necessary to achieve the steady-state conditions in the momentum balance.[10]

(iv) Unlike a dielectric, the transient and displacement currents in a good conductor generate a time-dependent magnetic field which induces a time-dependent electric field and so forth. The approach to equilibrium in a good conductor should be thought of as a three-step process as Ref. [10] explains:

> first, the free charges are expelled from the volume; second, the currents and the dynamic electric and magnetic fields are expelled from the volume; and third, the surface currents and wave fields are damped. Of course, these three relaxation processes overlap to some extent; but for a rough estimate of the overall relaxation time we may take the sum of the individual relaxation times.

[10] H. C. Ohanian, 'On the approach to electro- and magneto-static equilibrium', *American Journal of Physics*, vol. 51, pp. 1020–2, 1983.

Since the first step is very quick, the time scale for the approach to equilibrium is dominated by steps two and three.

In this same paper, the author shows that steps two and three themselves depend on the geometry of the conductor and the initial charge distribution.

(v) It is clear from the above that no simple theory for the relaxation time of a good conductor exists. Some textbooks use $\tau_r = \epsilon_0/\sigma$, whilst others claim $\tau_r \sim \tau_c$. Both are wrong and 'the approach to electrostatic and magnetostatic equilibrium takes much longer than many textbooks would have us believe'.[10]

Question 10.10

An electromagnetic wave propagates in a charge-free ($\rho_f = 0$), isotropic medium in which the constitutive relations are

$$\mathbf{D} = \epsilon\mathbf{E}, \qquad \mathbf{H} = \mu^{-1}\mathbf{B} \qquad \text{and} \qquad \mathbf{J}_f = \sigma\mathbf{E}. \qquad (1)$$

Assume that $\epsilon(\omega)$, $\mu(\omega)$ and $\sigma(\omega)$ are independent of space and time coordinates.

(a) Show that all components of the electromagnetic field satisfy the wave equation

$$\nabla^2\psi - \mu\sigma\frac{\partial\psi}{\partial t} - \epsilon\mu\frac{\partial^2\psi}{\partial t^2} = 0, \qquad (2)$$

where ψ represents any component of $\mathbf{E}$ or $\mathbf{B}$.

(b) Attempt plane wave solutions to (2) of the form $\psi = \psi_0 e^{i(\mathbf{k}\cdot\mathbf{r}-\omega t)}$, and show that

$$k^2 = \epsilon\mu\omega^2\left(1 + \frac{i\sigma}{\epsilon\omega}\right). \qquad (3)$$

(c) Express $\mathbf{k}$ as[‡]

$$\mathbf{k} = \mathbf{k}_+ + i\mathbf{k}_-, \qquad (4)$$

where $\mathbf{k}_+$ and $\mathbf{k}_-$ are real and colinear. Then show that

$$k_+ = \sqrt{\frac{\epsilon\mu\omega^2}{2}}\sqrt{\sqrt{1+\frac{1}{Q^2}}+1} \quad \text{and} \quad k_- = \sqrt{\frac{\epsilon\mu\omega^2}{2}}\sqrt{\sqrt{1+\frac{1}{Q^2}}-1}, \qquad (5)$$

where

$$Q = \frac{\epsilon\omega}{\sigma}$$

is a dimensionless quantity analogous to the quality factor of a mechanical or electrical oscillator (see Comment (iii) on p. 477).

[‡]The physical implication of a complex $\mathbf{k}$ is that the wave is attenuated. See Comment (ii) on p. 476.

Solution

(a) The proof is identical to the corresponding vacuum derivation (see the steps leading to (1) of Question 7.4), except we use the following Maxwell equations

$$\left.\begin{array}{ll} \nabla \cdot \mathbf{D} = 0 & \nabla \cdot \mathbf{B} = 0 \\ \nabla \times \mathbf{E} = -\dfrac{\partial \mathbf{B}}{\partial t} & \nabla \times \mathbf{H} = \mathbf{J}_\mathrm{f} + \epsilon \dfrac{\partial \mathbf{E}}{\partial t} \end{array}\right\}, \tag{6}$$

where $\mathbf{J}_\mathrm{f}$ is given by $(1)_3$.

(b) Substituting the trial solution in (2) gives $(-k^2 + \mu\sigma i\omega + \epsilon\mu\omega^2)\psi = 0$. Since this must be true for all values of ψ, we have $k^2 - \mu\sigma i\omega - \epsilon\mu\omega^2 = 0$ which is (3).

(c) Substituting (4) in (3) and rearranging give $(k_+^2 - k_-^2 - \epsilon\mu\omega^2) + i(2k_+k_- - \mu\sigma\omega) = 0$. Each term in brackets is real, so

$$k_+^2 - k_-^2 - \epsilon\mu\omega^2 = 0 \quad \text{and} \quad 2k_+k_- - \mu\sigma\omega = 0. \tag{7}$$

Eliminating k_- from (7) yields the quadratic equation $k_+^4 - \epsilon\mu\omega^2 k_+^2 - \dfrac{\epsilon^2\mu^2\omega^4}{4Q^2} = 0$ whose roots are $(5)_1$, and similarly for the derivation of $(5)_2$.

Comments

(i) It is sometimes useful to express (4) in polar form. Then $k = \sqrt{k_+^2 + k_-^2}\, e^{i\varphi}$ where $\varphi = \tan^{-1}(k_-/k_+)$, or

$$k = \sqrt{\epsilon\mu\omega^2}\left(1 + \dfrac{1}{Q^2}\right)^{\frac{1}{4}} e^{i\varphi}, \quad \varphi = \tan^{-1}\sqrt{\dfrac{\sqrt{1 + \frac{1}{Q^2}} - 1}{\sqrt{1 + \frac{1}{Q^2}} + 1}} \tag{8}$$

because of (5).

(ii) In a medium of finite conductivity, the general solution of (2) is

$$\psi = \psi_0 e^{-\mathbf{k}_- \cdot \mathbf{r}} e^{i(\mathbf{k}_+ \cdot \mathbf{r} - \omega t)}. \tag{9}$$

For the specific case of waves travelling along the z-axis

$$\psi = \psi_0 e^{-k_- z} e^{i(k_+ z - \omega t)}, \tag{10}$$

and (10) shows that the wave amplitude decreases exponentially along its path. The quantity k_-^{-1} is the distance required for ψ to decay by e^{-1}. It is called the 'penetration' or 'skin' depth δ and is defined as

$$\delta = \dfrac{1}{k_-} = \sqrt{\dfrac{2}{\epsilon\mu\omega^2}}\left(\sqrt{1 + \dfrac{1}{Q^2}} - 1\right)^{-1/2}. \tag{11}$$

(iii) The ratio $|\mathbf{J}_d|/|\mathbf{J}_f|$ for a time-harmonic field is

$$\frac{\left|\epsilon\dfrac{\partial \mathbf{E}}{\partial t}\right|}{|\mathbf{J}_f|} = \frac{\epsilon|i\,\omega\,\mathbf{E}|}{\sigma|\mathbf{E}|} = \frac{\epsilon\omega}{\sigma},$$

which is the quality factor Q. Thus, in effect, Q is a comparison of the relative magnitudes of the displacement and conduction current densities. For good conductors we would expect $\mathbf{J}_d \ll \mathbf{J}_f$, and vice versa for poor conductors. Hence we adopt the criteria:

$$\left.\begin{array}{l}\text{if } Q \ll 1 \text{ the medium behaves like a good conductor, and}\\[4pt]\text{if } Q \gg 1 \text{ the medium behaves like a poor conductor}\end{array}\right\}. \tag{12}$$

Q is clearly a frequency-dependent quantity, not only because it varies explicitly with ω but also through the implicit dependences $\sigma(\omega)$ (see (2) Question 10.8) and $\epsilon(\omega)$.

(iv) Even though $\mathbf{k}$ is complex, two properties of plane waves established in Question 7.5 also apply here. They are:

☞ the field is a transverse wave,

☞ $\mathbf{E}$ and $\mathbf{B}$ are mutually perpendicular and satisfy $\mathbf{B} = \dfrac{\mathbf{k} \times \mathbf{E}}{\omega}$. So

$$B = \frac{kE}{\omega} = \sqrt{\epsilon\mu}\left(1 + \frac{1}{Q^2}\right)^{\frac{1}{4}} E_0\, e^{-\mathbf{k}_- \cdot \mathbf{r}} e^{i(\mathbf{k}_+ \cdot \mathbf{r} - \omega t + \varphi)}, \tag{13}$$

because of (4) and $(8)_1$. Equation (13) shows that B lags E in phase by φ given by $(8)_2$.

(v) The phase velocity of the wave is

$$v_\phi = \frac{\omega}{k_+} = \frac{1}{\sqrt{\epsilon\mu}}\frac{\sqrt{2}}{\sqrt{\sqrt{1+\frac{1}{Q^2}}+1}} = \frac{c}{\sqrt{\epsilon_r\mu_r}}\frac{\sqrt{2}}{\sqrt{\sqrt{1+\frac{1}{Q^2}}+1}}. \tag{14}$$

We see that

☞ the phase velocity of a wave travelling in a conducting medium is less than the phase velocity for a non-conducting medium, and

☞ the medium is dispersive because v_ϕ is a function of ω.

(vi) By analogy with (2) of Question 7.6 we write the energy density in the electromagnetic field as

$$u = u_{\text{el}} + u_{\text{mag}}$$

$$= \tfrac{1}{2}\epsilon E^2 + \tfrac{1}{2}\mu^{-1}B^2$$

$$= \tfrac{1}{2}\epsilon E_0^2\, e^{-2k_- z}\left(\cos^2(k_+z - \omega t) + \sqrt{1+\frac{1}{Q^2}}\cos^2(k_+z - \omega t + \varphi)\right), \tag{15}$$

where in the last step we use (10) and (13) for waves travelling along the z-axis. The time-average of u is thus

$$\langle u \rangle = \langle u_{\text{el}} \rangle + \langle u_{\text{mag}} \rangle$$

$$= \tfrac{1}{2}\epsilon E_0^2 e^{-2k_- z}\left(\langle \cos^2(k_+ z - \omega t)\rangle + \sqrt{1 + \frac{1}{Q^2}}\,\langle \cos^2(k_+ z - \omega t + \varphi)\rangle\right)$$

$$= \tfrac{1}{4}\epsilon E_0^2 e^{-2k_- z}\left(1 + \sqrt{1 + \frac{1}{Q^2}}\right), \tag{16}$$

because $\langle \cos^2(k_+ z - \omega t)\rangle = \langle \cos^2(k_+ z - \omega t + \varphi)\rangle = \tfrac{1}{2}$.

(vii) In a similar way we calculate the time-average Poynting vector $\langle \mathbf{S} \rangle$. Consider waves travelling in the $\hat{\mathbf{z}}$ direction and polarized along $\hat{\mathbf{x}}$. Then

$$\mathbf{E} = E_0 e^{-k_- z}\cos(k_+ z - \omega t)\,\hat{\mathbf{x}}, \qquad \mathbf{B} = \frac{k E_0}{\omega}e^{-k_- z}\cos(k_+ z - \omega t + \varphi)\,\hat{\mathbf{y}}$$

and $\mathbf{S} = \mathbf{E} \times \dfrac{\mathbf{B}}{\mu} = \dfrac{k E_0^2}{\mu\omega}e^{-2k_- z}\cos(k_+ z - \omega t)\cos(k_+ z - \omega t + \varphi)\,\hat{\mathbf{z}}$. Thus

$$\langle \mathbf{S} \rangle = \frac{k E_0^2}{\mu\omega}e^{-2k_- z}\big\langle \cos(k_+ z - \omega t)\cos(k_+ z - \omega t + \varphi)\big\rangle\,\hat{\mathbf{z}}$$

$$= \frac{k E_0^2}{\mu\omega}e^{-2k_- z}\big\langle \cos^2(k_+ z - \omega t)\cos\varphi\big\rangle - \big\langle \tfrac{1}{2}\sin 2(k_+ z - \omega t)\sin\varphi\big\rangle\,\hat{\mathbf{z}}$$

$$= \frac{k E_0^2}{2\mu\omega}e^{-2k_- z}\cos\varphi\,\hat{\mathbf{z}}, \tag{17}$$

because $\langle \cos^2(k_+ z - \omega t)\rangle = \tfrac{1}{2}$ and $\langle \sin 2(k_+ z - \omega t)\rangle = 0$. Now $k\cos\varphi = k_+$, and so

$$\langle \mathbf{S} \rangle = \frac{k_+ E_0^2}{2\mu\omega}e^{-2k_- z}\,\hat{\mathbf{z}} = \left(\frac{1}{2}\sqrt{\frac{\epsilon}{\mu}}\,E_0^2\right)\left(\sqrt{\frac{\sqrt{1 + \frac{1}{Q^2}} + 1}{2}}\,e^{-2k_- z}\right)\hat{\mathbf{z}}, \tag{18}$$

where the first term in brackets in (18) is Poynting's vector with $\sigma = 0$ $\big($see (10) of Question 7.5$\big)$, whilst the second bracketed term is the modification introduced by a finite conductivity.

(viii) For a 'good' conductor $(\omega\tau \ll 1)$ the above results simplify:

☞ $\quad k_+ = k_- = \delta^{-1} = \sqrt{\tfrac{1}{2}\mu\sigma\omega}\,,$ (19)

☞ $\quad \varphi = \tfrac{1}{4}\pi,$ (20)

☞ $\quad B_0 = \sqrt{\dfrac{\mu\sigma}{\omega}}\,E_0,$ (21)

☞ $\quad v_\phi = \sqrt{\dfrac{2\omega}{\mu\sigma}}\,,$ (22)

☞ $\langle u \rangle = \left(\dfrac{\sigma}{4\omega} \right) E_0^2 e^{-2z/\delta}$,   (23)

☞ $\langle \mathbf{S} \rangle = \sqrt{\dfrac{\sigma}{8\mu\omega}}\, E_0^2\, e^{-2z/\delta}\, \hat{\mathbf{z}} = \left(\dfrac{1}{2}\sqrt{\dfrac{\epsilon}{\mu}}\, E_0^2 \right)\left(\dfrac{1}{\sqrt{2Q}}\, e^{-2z/\delta} \right)\hat{\mathbf{z}}.$ (24)

Question 10.11

Use the results of Question 10.10 to show that for copper (a typical metal):

(a) $Q \sim 10^{-18} f$,   (b) $\delta \sim 0.06\, f^{-1/2}$ and   (c) $\langle u_{\text{mag}} \rangle / \langle u_{\text{el}} \rangle \sim 10^{18} f^{-1}$,

where f is the frequency in hertz. Then evaluate these quantities at the microwave frequency $f = 1 \times 10^9$ Hz. Take $\sigma = 6.5 \times 10^7\,\text{S}\,\text{m}^{-1}$, $\epsilon = \epsilon_0$, and $\mu = \mu_0$.

Solution

(a) $Q = \dfrac{\epsilon_0 \omega}{\sigma} = \dfrac{4\pi\epsilon_0 \omega}{4\pi\sigma} = \dfrac{1}{4\pi} \times \dfrac{1}{9 \times 10^9} \times \dfrac{2\pi f}{6.5 \times 10^7} \sim 10^{-18} f$, and at 1 GHz

$Q \sim 9 \times 10^{-10}$.

(b) From (19) of Question 10.10

$$\delta = \sqrt{\dfrac{2}{\mu_0 \sigma \omega}} = \sqrt{\dfrac{2}{4\pi \times 10^{-7} \times 6.5 \times 10^7 \times 2\pi \times f}} \sim 0.06\, f^{-1/2}, \text{ and at 1 GHz}$$

$\delta \sim 2\,\mu\text{m}$.

(c) From (16) of Question 10.10

$$\dfrac{\langle u_{\text{mag}} \rangle}{\langle u_{\text{el}} \rangle} = \sqrt{1 + \dfrac{1}{Q^2}} \simeq \dfrac{1}{Q} \sim \dfrac{10^{18}}{f}, \text{ and at 1GHz } \dfrac{\langle u_{\text{mag}} \rangle}{\langle u_{\text{el}} \rangle} \sim 10^9.$$

Comment

The solution above illustrates the following:

☞   Copper and indeed other metals (such as aluminium and silver) remain good conductors ($Q \ll 1$) over a wide range of frequencies which extend beyond the microwave part of the electromagnetic spectrum. Near the visible region the frequency dependence of σ becomes significant. See Comment (i) of Question 10.8.

☞   At microwave frequencies in metals the wave penetrates just beneath the surface of the conductor,[‡] hence the name 'skin depth'. Many practical devices, designed

[‡]Compare $\delta \sim 2\,\mu\text{m}$ at 1 GHz to $\delta \sim 1\,\text{cm}$ at 50 Hz, which is a typical mains supply frequency.

to operate at high frequencies, are made by coating an insulating material with a suitably thin layer of silver which is a better conductor than copper (although more expensive).

☞ Induced currents near the surface of a good conductor result in magnetic field energies that are orders of magnitude larger than those stored in the electric field.

Question 10.12

Consider a conducting medium for which the damping term $\gamma \ll \omega$ (here γ is the phenomenological constant mentioned in (b) of Question 10.8).

(a) Show that the conductivity is

$$\sigma(\omega) = i\epsilon_0 \frac{\omega_p^2}{\omega}, \tag{1}$$

where

$$\omega_p^2 = \frac{Ne^2}{\epsilon_0 m_e} \tag{2}$$

is a constant (recall that N is the number density of conduction electrons).

(b) Hence show that the dispersion relation is

$$\frac{k^2}{\omega^2} = \epsilon_0 \mu_0 \left(1 - \frac{\omega_p^2}{\omega^2}\right). \tag{3}$$

Solution

(a) Putting $\gamma = 0$ in (2) of Question 10.8 gives (1).

(b) The result follows from (3) of Question 10.10 and (1) with $\epsilon = \epsilon_0$ and $\mu = \mu_0$.

Comments

(i) The constant ω_p in (2) is known as the plasma frequency. An example of such a conducting medium is a dilute or tenuous electronic plasma in which collisions are infrequent events ($\tau_c \gg \omega^{-1}$). The term $(1 - \omega_p^2/\omega^2)$ in (3) can be interpreted as the relative permittivity $\epsilon_r(\omega)$. Since $\mu_r = 1$ we define the refractive index n as

$$n = \sqrt{\epsilon_r} = \sqrt{1 - \frac{\omega_p^2}{\omega^2}}. \tag{4}$$

For $\omega > \omega_p$, a travelling wave will propagate in the plasma with phase velocity $v_\phi = c/n$. For $\omega < \omega_p$, the refractive index is complex, the wave is attenuated and has a penetration depth

$$\delta_{\text{plasma}} = \frac{c}{\sqrt{\omega_p^2 - \omega^2}} . \tag{5}$$

(ii) Equation (1) shows that since $\mathbf{J}_f = \sigma \mathbf{E}$ the current lags the driving electric field in phase by $90°$ and there is no dissipation of energy.[‡] (This is not a surprising result because the plasma is collisionless.)

(iii) An interesting consequence of the complex conductivity (1) is the appearance of charge-density oscillations inside the plasma at $\omega = \omega_p$. To see this, suppose that the charge density ρ has a time-harmonic dependence $e^{-i\omega t}$. The continuity equation then gives $\nabla \cdot \mathbf{J} + \partial \rho / \partial t = \nabla \cdot \mathbf{J} - i\omega \rho = 0$. But Ohm's law and Gauss's law require that $\nabla \cdot \mathbf{J} = \sigma \nabla \cdot \mathbf{E} = \sigma \rho / \epsilon_0$. Thus $(\sigma/\epsilon_0 - i\omega)\rho = 0$ and

$$i\omega \left(\frac{\omega_p^2}{\omega^2} - 1 \right) \rho = 0 \tag{6}$$

because of (1). Clearly, (6) is satisfied for $\omega \neq 0$, provided $\omega = \omega_p$.

(iv) In the ionosphere,[♯] ionization of air molecules by solar radiation produces a dilute plasma having a free-electron density[†] $N \sim 10^{11} \, \text{m}^{-3}$ with a plasma frequency calculated from (2) of $\sim 10^6 \, \text{Hz}$. The presence of this plasma makes long-distance radio communication possible at the Earth's surface. Radio waves whose frequency $f \lesssim 1 \, \text{MHz}$ $(\omega < \omega_p)$ can be broadcast around the globe by bouncing them between the ionosphere and suitably placed stations on the ground. At radio frequencies higher than ω_p the receiver and transmitter must be line-of-sight. Note that things are not quite this simple in practice, because the Earth's magnetic field has an important effect on the propagation of these waves (for example, the plasma becomes birefringent;[♭] see Question 10.18).

(v) Recall from Comment (i) of Question 10.8 that the conductivity of a typical metal (e.g. copper) is constant and independent of frequency from dc up to about $10^{11} \, \text{Hz}$. At much higher frequencies the conductivity of a metal is given by (1), where the conduction-electron gas can be treated as a dilute plasma $(\tau_c \gg \omega^{-1})$. From (2) and (7) of Question 10.8 we calculate ω_p for copper

$$\omega_p = \sqrt{\frac{4\pi \times 9 \times 10^9 \times 8.5 \times 10^{28} \times (1.6 \times 10^{-19})^2}{9.11 \times 10^{-31}}}$$

$$\simeq 1.6 \times 10^{16} \, \text{s}^{-1} ,$$

which lies at the short-wavelength end of the visible spectrum. That metals will suddenly transmit radiation at these frequencies is a phenomenon known as 'ultraviolet transparency'.

[‡] Compare this with a purely reactive circuit where i and V are also $90°$ out of phase and the power factor is zero.

[♯] That part of the atmosphere between about 10^2–10^3 km above the Earth's surface.

[†] This is a simplistic model. It neglects, for example, diurnal and altitudinal variations.

[♭] That is, different polarization states of the travelling wave have different refractive indices.

Question 10.13*

The xy-plane at $z = 0$ of Cartesian coordinates forms the boundary between two media 1 and 2, whose permittivity, permeability and conductivity are $(\epsilon_0, \mu_0, 0)$ and $(\epsilon_0, \mu_0, \sigma)$ respectively. A plane monochromatic electromagnetic wave polarized along $\hat{\mathbf{x}}$ and travelling along $\hat{\mathbf{z}}$ in medium 1 towards medium 2 is partially reflected and partially transmitted at the boundary. The figure below shows the **E**- and **H**-fields for the incident, reflected and transmitted waves.

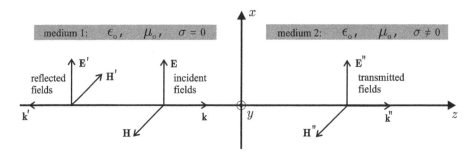

(a) Show that at the boundary the fields satisfy the equations

$$(E_0 + E_0') - E_0'' \quad \text{and} \quad (F_0 - F_0') = \left(\sqrt{\frac{\sqrt{\alpha} + 1}{2}} + i \sqrt{\frac{\sqrt{\alpha} - 1}{2}} \right) E_0'', \quad (1)$$

where

$$\alpha = 1 + \frac{\sigma^2}{\omega^2 \epsilon_0^2} = 1 + \frac{1}{Q^2}, \quad (2)$$

is a real, positive quantity > 1.

Hint: At $z = 0$ the parallel components of **E** and **H** are continuous (see (4) of Question 10.6).

(b) Use the definitions of the reflection coefficient R and transmission coefficient T,

$$R = \left| \frac{\langle \mathbf{S}' \cdot \hat{\mathbf{z}} \rangle}{\langle \mathbf{S} \cdot \hat{\mathbf{z}} \rangle} \right| \quad \text{and} \quad T = \left| \frac{\langle \mathbf{S}'' \cdot \hat{\mathbf{z}} \rangle}{\langle \mathbf{S} \cdot \hat{\mathbf{z}} \rangle} \right|, \quad (3)$$

to show that

$$R = \frac{1 + \sqrt{\alpha} - \sqrt{2\sqrt{\alpha} + 2}}{1 + \sqrt{\alpha} + \sqrt{2\sqrt{\alpha} + 2}} \quad \text{and} \quad T = \frac{2\sqrt{2\sqrt{\alpha} + 2}}{1 + \sqrt{\alpha} + \sqrt{2\sqrt{\alpha} + 2}}. \quad (4)$$

(c) Find the limiting forms of (4) for

☞ $Q \ll 1$ (i.e. medium 2 is a good conductor), and

☞ $Q \gg 1$ (i.e. medium 2 is a poor conductor).

(d) Determine the value of α for which $R = T$.

(e) Calculate R and T for copper at 100 GHz taking the order-of-magnitude value of $Q = 10^{-7}$ (see Question 10.11).

(f) Plot $R(\alpha)$ and $T(\alpha)$ for $1 < \alpha \leq 1000$.

(g) To illustrate the effect of the discontinuity in σ at $z = 0$, sketch $E(z)$, $E'(z)$ and $E''(z)$ on the same axes at $t = 0$ for each of the following values of Q: 40.0, 4.0, 0.4 and 0.04. Repeat for $B(z)$, $B'(z)$ and $B''(z)$.

Solution

(a) The fields of the incident, reflected and transmitted waves are as follows:

incident wave

$$\left. \begin{array}{l} \mathbf{E} = E_0 e^{i(kz-\omega t)}\,\hat{\mathbf{x}} \\[2mm] \mathbf{H} = \dfrac{k\hat{\mathbf{k}} \times \mathbf{E}}{\mu_0 \omega} = \sqrt{\dfrac{\epsilon_0}{\mu_0}} E_0 e^{i(kz-\omega t)}\,\hat{\mathbf{y}} \end{array} \right\} \quad \text{since } \dfrac{k}{\omega} = c^{-1} = \sqrt{\epsilon_0 \mu_0}. \tag{5}$$

reflected wave

$$\left. \begin{array}{l} \mathbf{E}' = E_0' e^{-i(kz+\omega t)}\,\hat{\mathbf{x}} \\[2mm] \mathbf{H}' = \dfrac{\hat{\mathbf{k}}' \times \mathbf{E}'}{\mu_0 c} = -\sqrt{\dfrac{\epsilon_0}{\mu_0}} E_0' e^{-i(kz+\omega t)}\,\hat{\mathbf{y}} \end{array} \right\} \quad \text{because } \mathbf{k}' = -\mathbf{k}. \tag{6}$$

transmitted wave

$$\left. \begin{array}{l} \mathbf{E}'' = E_0'' e^{i(k''z-\omega t)}\,\hat{\mathbf{x}} \\[2mm] \mathbf{H}'' = \dfrac{\hat{\mathbf{k}}'' \times \mathbf{E}''}{\mu_0 \omega} = \dfrac{k''}{\mu_0 \omega} E_0'' e^{i(k''z-\omega t)}\,\hat{\mathbf{y}} \end{array} \right\} \quad \text{where } \mathbf{k}'' \text{ is a complex wave vector.} \tag{7}$$

Equations (5)–(7) and the matching conditions $\big($see (4) of Question 10.6$\big)$ yield

$$(E_0 + E_0') = E_0'' \quad \text{and} \quad \sqrt{\dfrac{\epsilon_0}{\mu_0}}\,(E_0 - E_0') = \dfrac{k''}{\mu_0 \omega} E_0''. \tag{8}$$

Substituting‡ $k'' = \sqrt{\dfrac{\epsilon_0 \mu_0 \omega^2}{2}} \left(\sqrt{\sqrt{\alpha}+1} + i\sqrt{\sqrt{\alpha}-1} \right)$ in $(8)_2$ and rearranging give (1).

(b) Solving (1) simultaneously for the relative amplitudes E_0'/E_0 and E_0''/E_0 yields

$$\dfrac{E_0'}{E_0} = \dfrac{1 - \sqrt{\frac{\sqrt{\alpha}+1}{2}} - i\sqrt{\frac{\sqrt{\alpha}-1}{2}}}{1 + \sqrt{\frac{\sqrt{\alpha}+1}{2}} + i\sqrt{\frac{\sqrt{\alpha}-1}{2}}} \quad \text{and} \quad \dfrac{E_0''}{E_0} = \dfrac{2}{1 + \sqrt{\frac{\sqrt{\alpha}+1}{2}} + i\sqrt{\frac{\sqrt{\alpha}-1}{2}}}. \tag{9}$$

‡See (4) and (5) of Question 10.10.

From $(2)_1$ and (10) of Question 7.5 it follows that $R = \dfrac{\left|\frac{1}{2}\sqrt{\frac{\epsilon_0}{\mu_0}}E_0'^{\,2}\right|}{\frac{1}{2}\sqrt{\frac{\epsilon_0}{\mu_0}}E_0^2} = \left|\dfrac{E_0'}{E_0}\right|^2$, so

$$R = \left(\frac{E_0'}{E_0}\right)\left(\frac{E_0'}{E_0}\right)^* = \left(\frac{1 - \sqrt{\frac{\sqrt{\alpha}+1}{2}} - i\sqrt{\frac{\sqrt{\alpha}-1}{2}}}{1 + \sqrt{\frac{\sqrt{\alpha}+1}{2}} + i\sqrt{\frac{\sqrt{\alpha}-1}{2}}}\right)\left(\frac{1 - \sqrt{\frac{\sqrt{\alpha}+1}{2}} + i\sqrt{\frac{\sqrt{\alpha}-1}{2}}}{1 + \sqrt{\frac{\sqrt{\alpha}+1}{2}} - i\sqrt{\frac{\sqrt{\alpha}-1}{2}}}\right), \quad (10)$$

because of $(9)_1$. Expanding (10) gives $(4)_1$. From $(3)_2$ and (24) of Question 10.10

it follows that $T = \dfrac{\left|\frac{1}{2}\sqrt{\frac{\epsilon_0}{\mu_0}}\sqrt{\frac{\sqrt{\alpha}+1}{2}}E_0''^{\,2}\right|}{\frac{1}{2}\sqrt{\frac{\epsilon_0}{\mu_0}}E_0^2} = \sqrt{\frac{\sqrt{\alpha}+1}{2}}\left(\frac{E_0''}{E_0}\right)\left(\frac{E_0''}{E_0}\right)^*$ or

$$T = \sqrt{\frac{\sqrt{\alpha}+1}{2}}\left(\frac{2}{1 + \sqrt{\frac{\sqrt{\alpha}+1}{2}} + i\sqrt{\frac{\sqrt{\alpha}-1}{2}}}\right)\left(\frac{2}{1 + \sqrt{\frac{\sqrt{\alpha}+1}{2}} - i\sqrt{\frac{\sqrt{\alpha}-1}{2}}}\right), \quad (11)$$

because of $(9)_2$. Expanding (11) gives $(4)_2$.

(c) Substituting $\alpha = 1 + 1/Q^2$ in (4) and taking the limit

☞ $Q \to \infty$ gives $R \simeq \dfrac{1}{16Q^2}$ and $T \simeq 1 - \dfrac{1}{16Q^2}$, $\qquad (12)$

☞ $Q \to 0$ gives $R \simeq 1 - 2\sqrt{2Q}$ and $T \simeq 2\sqrt{2Q}$. $\qquad (13)$

(d) Substituting $R = T$ in (4) gives the quadratic equation $(\alpha - 289)(\alpha - 1) = 0$, whose physical root $\alpha = 289$ corresponds to $\omega\tau \approx 0.0589$.

(e) From (13) we have $R = 1 - 2\sqrt{2 \times 10^{-7}} = 0.999$ and $T = 1 - R = 0.001$.

(f)

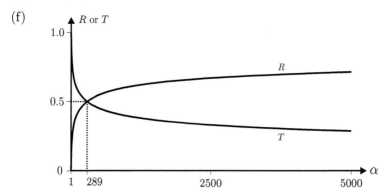

The asymptotic behaviour of these coefficients can be illustrated by increasing the plot range for α by a factor of ~ 1000.

(g) The plots on p. 485 are drawn with normalized amplitudes for the incident **E**- and **B**-fields.

	$Q = 40.0$	E	———
		E′	············
		E″	▪▪▪▪▪

×10
magnification

	$Q = 40.0$	B	———
		B′	············
		B″	▪▪▪▪▪

×10
magnification

	$Q = 4.0$	E	———
		E′	············
		E″	▪▪▪▪▪

	$Q = 4.0$	B	———
		B′	············
		B″	▪▪▪▪▪

	$Q = 0.4$	E	———
		E′	············
		E″	▪▪▪▪▪

	$Q = 0.4$	B	———
		B′	············
		B″	▪▪▪▪▪

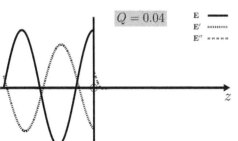

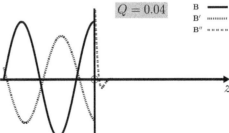

	$Q = 0.04$	E	———
		E′	············
		E″	▪▪▪▪▪

	$Q = 0.04$	B	———
		B′	············
		B″	▪▪▪▪▪

Comments

(i) It is clear from (4) that $R + T = 1$, which we expect, since energy is conserved.

(ii) ☞ As $Q \to \infty$, $\alpha \to 1$ and (9) shows that for normal incidence $E_0' \to 0$, $E_0'' \to E_0$. That is, at the boundary between the two media there is no reflection[#] (see also $(12)_1$): the incident and transmitted waves have the same amplitude and are in phase.

   ☞ As $Q \to 0$, $\alpha \to \infty$ and (9) shows that $E_0' \to -E_0$, $E_0'' \to 0$. That is, at the boundary between the two media there is total internal reflection (see also $(13)_2$). The incident and reflected waves have the same amplitude but are $180°$ out of phase.

(iii) Equation $(13)_1$ confirms a well-known result that metals are good reflectors of electromagnetic waves. We see that for $Q \ll 1$ the reflection coefficient is close to unity. A small fraction of the energy is transmitted which is, of course, ultimately dissipated inside the conductor.

(iv) For an arbitrary angle of incidence θ we find that

$$|k| \sin \theta = |k''| \sin \theta'', \tag{14}$$

where θ'' is the angle of refraction. Equation (14) is a form of Snell's law and is discussed in Question 10.14. Applying it here to a good conductor for which $|k''| = \sqrt{\mu_0 \sigma \omega}$ gives

$$\sin \theta'' = \frac{|k|}{|k''|} \sin \theta = \frac{\omega \sqrt{\epsilon_0 \mu_0}}{\sqrt{\mu_0 \sigma \omega}} \sin \theta = \sqrt{Q} \sin \theta \ll 1, \tag{15}$$

since $Q \ll 1$. Thus $\theta'' \simeq 0$, which shows that the transmitted wave propagates at essentially $90°$ to the interface for all angles of incidence.

Question 10.14

Suppose a plane boundary at $z = 0$ separates a medium $(z < 0)$ having permittivity ϵ_1; permeability μ_1 from a medium $(z > 0)$ having permittivity ϵ_2; permeability μ_2. An electromagnetic wave travelling in medium 1 is partially reflected and partially transmitted at the boundary. The incident, reflected and transmitted fields satisfy matching conditions[‡] at $z = 0$, which lead to the following kinematic properties:

1. The frequencies of the incident, reflected and transmitted waves are equal.

2. The incident, reflected and transmitted wave vectors ($\mathbf{k}$, $\mathbf{k}'$ and $\mathbf{k}''$ respectively) lie in a plane (known as the plane of incidence) normal to the boundary between the two media.

[#] This is not a general result. It is only true here since $\epsilon_1 = \epsilon_2 = \epsilon_0$ and $\mu_1 = \mu_2 = \mu_0$.

[‡] As we have already seen, Maxwell's equations impose certain constraints on the fields at $z = 0$. However, we need not be concerned with their *particular* form in answering this question.

3. The angles of incidence and reflection (θ and θ' respectively) satisfy

$$\theta = \theta'. \tag{1}$$

4. The angle of refraction θ'' is related to θ by the equation

$$n_1 \sin\theta = n_2 \sin\theta'', \tag{2}$$

where $n_1 = c/c_1 = \sqrt{\dfrac{\epsilon_1\mu_1}{\epsilon_0\mu_0}}$ and $n_2 = c/c_2 = \sqrt{\dfrac{\epsilon_2\mu_2}{\epsilon_0\mu_0}}$ are the refractive indices of media 1 and 2.

Prove each of the properties stated above, assuming harmonic plane waves.

Solution

Choose Cartesian axes such that the **k**-vector of the incident wave lies in the yz-plane as shown in the figure below. Then $\mathbf{k} = (0, k_y, k_z)$.

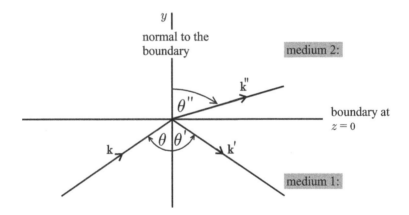

The incident and reflected fields in medium 1 are matched to the corresponding field (**E** or **B**) in medium 2 by an equation of the general form[†]

$$(\cdots)\,e^{i(\mathbf{k}\cdot\mathbf{r}-\omega t)} + (\cdots)\,e^{i(\mathbf{k}'\cdot\mathbf{r}-\omega' t)} = (\cdots)\,e^{i(\mathbf{k}''\cdot\mathbf{r}-\omega'' t)}, \tag{3}$$

where we omit phase changes[b] of π in (3), which might occur for the reflected and transmitted waves at the boundary. Now it is clear that (3) cannot be satisfied for

[†]We use the $(\cdots)$ notation to emphasize, as indicated in the second footnote on p. 486, that we need not be concerned with the particular form which the fields assume in the boundary.

[b]Such phase changes, if they occur at all, have no effect on the ensuing discussion.

arbitrary $\mathbf{r}$ and t, unless the argument of each exponential (the phase term) is the same. So

$$\left(\mathbf{k}\cdot\mathbf{r}-\omega t\right)_{z=0} = \left(\mathbf{k}'\cdot\mathbf{r}-\omega't\right)_{z=0} = \left(\mathbf{k}''\cdot\mathbf{r}-\omega''t\right)_{z=0}. \tag{4}$$

Because $\mathbf{r}$ and t are independent variables, it follows immediately from (4) that

$$\omega = \omega' = \omega''. \tag{5}$$

This proves property 1 . We also obtain from (4) the results $\mathbf{r}\cdot(\mathbf{k}-\mathbf{k}')|_{z=0} = 0$ and $\mathbf{r}\cdot(\mathbf{k}-\mathbf{k}'')|_{z=0} = 0$. Thus

$$\left.\begin{array}{r}-k'_x x + (k_y - k'_y)y = 0 \\ -k''_x x + (k_y - k''_y)y = 0\end{array}\right\}, \tag{6}$$

which (since x and y are independent variables) can only be satisfied if $k'_x = k''_x = 0$ and $k'_y = k''_y = k_y$. The wave vectors $\mathbf{k}'$ and $\mathbf{k}''$ must therefore also lie in the same plane as $\mathbf{k}$ (here the yz-plane), which proves property 2 . Furthermore we see that $(\mathbf{k}-\mathbf{k}') = (0, 0, k_z - k'_z)$ and $(\mathbf{k}-\mathbf{k}'') = (0, 0, k_z - k''_z)$. Hence $(\mathbf{k}-\mathbf{k}') \times \hat{\mathbf{z}} = 0$ and $(\mathbf{k}-\mathbf{k}'') \times \hat{\mathbf{z}} = 0$, implying (respectively) that:

☞ $k\sin\theta = k'\sin\theta'$. Now $\omega/c_1 = \omega'/c_1$ or $k = k'$. Thus $\sin\theta = \sin\theta' \Rightarrow \theta = \theta'$, which proves property 3 .

☞ $k\sin\theta = k''\sin\theta''$. Thus $(\omega/c_1)\sin\theta = (\omega''/c_2)\sin\theta'' = (\omega/c_2)\sin\theta''$. This proves property 4 , since $c_1/c_2 = n_2/n_1$.

Comments

(i) The kinematic properties proved above are well-known results from geometrical optics, namely:

☞ The incident, reflected and refracted rays lie in the same plane,

☞ The angles of incidence and reflection are equal (this is known as the law of reflection) and

☞ Snell's law (also known as the law of refraction).

These results are a consequence only of the plane-wave nature of the fields and the requirement that these fields (or some component thereof) should be continuous at the boundary. As a result, we anticipate that the same results apply elsewhere in physics. They do. For example, with suitable definitions for the refractive index, water and sound waves also obey the laws of reflection and refraction.

(ii) In Question 10.15 we consider some dynamical properties, which explicitly require the *particular* form of the matching conditions.

Question 10.15

Suppose the electric vector of the incident wave described in Question 10.14 is polarized

☞ perpendicular to the plane of incidence (known as the s-polarization state[‡]),

☞ parallel to the plane of incidence (known as the p-polarization state),

as shown in the figures below.

s-polarization   p-polarization

(a) Use the matching conditions given by (8) of Question 10.6 to show that the amplitude reflection and transmission coefficients are

☞ s-polarization

$$r_\mathrm{s} = \frac{E_0'}{E_0} = \frac{n_1 \cos\theta - n_2 \cos\theta''}{n_1 \cos\theta + n_2 \cos\theta''} \; ; \qquad t_\mathrm{s} = \frac{E_0''}{E_0} = \frac{2n_1 \cos\theta}{n_1 \cos\theta + n_2 \cos\theta''}, \qquad (1)$$

☞ p-polarization

$$r_\mathrm{p} = \frac{E_0'}{E_0} = \frac{n_2 \cos\theta - n_1 \cos\theta''}{n_2 \cos\theta + n_1 \cos\theta''} \; ; \qquad t_\mathrm{p} = \frac{E_0''}{E_0} = \frac{2n_1 \cos\theta}{n_2 \cos\theta + n_1 \cos\theta''}. \qquad (2)$$

(b) Hence show that the energy reflection and transmission coefficients are

☞ s-polarization

$$R_\mathrm{s} = \left(\frac{n_1 \cos\theta - n_2 \cos\theta''}{n_1 \cos\theta + n_2 \cos\theta''} \right)^2 \; ; \qquad T_\mathrm{s} = \frac{4 n_1 n_2 \cos\theta \cos\theta''}{(n_1 \cos\theta + n_2 \cos\theta'')^2}. \qquad (3)$$

☞ p-polarization

$$R_\mathrm{p} = \left(\frac{n_2 \cos\theta - n_1 \cos\theta''}{n_2 \cos\theta + n_1 \cos\theta''} \right)^2 \; ; \qquad T_\mathrm{p} = \frac{4 n_1 n_2 \cos\theta \cos\theta''}{(n_2 \cos\theta + n_1 \cos\theta'')^2}. \qquad (4)$$

Hint: Use the definitions of R and T given in (3) of Question 10.13.

[‡]The notation derives from the German word *senkrecht*, meaning perpendicular.

Solution

(a) ☞ s-polarization

The matching conditions $(8)_3$ and $(8)_4$ of Question 10.6 give $(\mathbf{E}+\mathbf{E}'-\mathbf{E}'')\cdot\hat{\mathbf{x}}=0$ and $(\mathbf{H}+\mathbf{H}'-\mathbf{H}'')\cdot\hat{\mathbf{y}}=0$. So

$$E_0 + E_0' = E_0'' \quad \text{and} \quad (H_0 - H_0')\cos\theta = H_0''\cos\theta'', \tag{5}$$

since the exponential terms cancel. (The reader may wish to confirm that the remaining matching conditions do not provide any additional information.) Now $\mathbf{H} = \dfrac{\hat{\mathbf{k}}\times\mathbf{E}}{\mu_0 v} = n\sqrt{\dfrac{\epsilon_0}{\mu_0}}\,\hat{\mathbf{k}}\times\mathbf{E}$, and so $(5)_2$ becomes

$$n_1(E_0 - E_0')\cos\theta = n_2 E_0''\cos\theta''. \tag{6}$$

Solving $(5)_1$ and (6) simultaneously for r_s and t_s gives (1).

☞ p-polarization

Here the matching conditions lead to

$$(E_0 - E_0')\cos\theta = E_0''\cos\theta'' \quad \text{and} \quad \sqrt{\frac{\epsilon_1}{\mu_0}}(E_0 + E_0') = \sqrt{\frac{\epsilon_2}{\mu_0}}E_0'', \tag{7}$$

which, when solved simultaneously, yield (2).

(b) ☞ s-polarization

The definition of R and (10) of Question 7.5 give $R_s = \left|\dfrac{\frac{1}{2}\sqrt{\frac{\epsilon_1}{\mu_0}}E_0'^2}{\frac{1}{2}\sqrt{\frac{\epsilon_1}{\mu_0}}E_0^2}\right| = r_s^2$. Hence

$(3)_1$. Similarly $T_s = \left|\dfrac{\frac{1}{2}\sqrt{\frac{\epsilon_2}{\mu_0}}E_0''^2\cos\theta''}{\frac{1}{2}\sqrt{\frac{\epsilon_1}{\mu_0}}E_0^2\cos\theta}\right| = \dfrac{n_2\cos\theta''}{n_1\cos\theta}\left|\dfrac{E_0''}{E_0}\right|^2$. Hence $(3)_2$.

☞ p-polarization

Similarly for the derivation of (4).

Comments

(i) The results (1)–(4) are collectively known as Fresnel's equations. They are used in optical applications where light travels across a boundary from one medium to another. With $n = n_2/n_1$ (the refractive index of medium 2 relative to medium 1), they can be expressed in the alternative forms:

☞ s-polarization

$$r_s = \frac{\cos\theta - n\cos\theta''}{\cos\theta + n\cos\theta''} \; ; \qquad t_s = \frac{2\cos\theta}{\cos\theta + n\cos\theta''} \tag{8}$$

and

$$R_s = \left(\frac{\cos\theta - n\cos\theta''}{\cos\theta + n\cos\theta''}\right)^2 \; ; \qquad T_s = \frac{4n\cos\theta\cos\theta''}{(\cos\theta + n\cos\theta'')^2} . \tag{9}$$

☞ p-polarization

$$r_p = \frac{n\cos\theta - \cos\theta''}{n\cos\theta + \cos\theta''} \; ; \qquad t_p = \frac{2\cos\theta}{n\cos\theta + \cos\theta''} \tag{10}$$

and

$$R_p = \left(\frac{n\cos\theta - \cos\theta''}{n\cos\theta + \cos\theta''}\right)^2 \; ; \qquad T_p = \frac{4n\cos\theta\cos\theta''}{(n\cos\theta + \cos\theta'')^2} . \tag{11}$$

Notice that (3) and (4)—and (9) and (11)—satisfy energy conservation: $R_s + T_s = 1$ and $R_p + T_p = 1$.

(ii) If the incident wave (having intensity I) is unpolarized then it can be represented as a superposition of equal-amplitude s and p components[#] where $I_s = I_p = \frac{1}{2}I$. The reflection coefficients of the s-component wave, p-component wave and incident wave are $R_s = I_s'/I_s$, $R_p = I_p'/I_p$ and $R = (I_s' + I_p')/I$ respectively. Thus $R_s + R_p = 2I_s'/I + 2I_p'/I = 2R$, or

$$R = \frac{R_s + R_p}{2} . \tag{12}$$

Similarly, the transmission coefficient of the incident wave is

$$T = \frac{T_s + T_p}{2} . \tag{13}$$

(iii) Substituting $\theta = \theta'' = 0$ in $(8)_1$ and $(10)_1$ gives, for normal incidence, $r_s = -r_p = \frac{1-n}{1+n}$. For $n > 1$, $r_s < 0$ ($r_p > 0$) and both $E_{0,s}$ and $E_{0,p}'$ point in opposite directions which corresponds to a phase change of π. For $n < 1$ there is no phase change.

[#]Recall, from Comment (iv) of Question 7.9, that unpolarized radiation can be represented mathematically as a superposition of two arbitrary orthogonal incoherent linearly polarized states of equal amplitude.

(iv) Using Snell's law to eliminate n from $(10)_1$ gives

$$r_p = \frac{\tan(\theta - \theta'')}{\tan(\theta + \theta'')}. \tag{14}$$

The angle θ for which r_p vanishes corresponds to $\theta + \theta'' = \frac{1}{2}\pi$ (or $\tan\theta = n$). This is known as the Brewster angle θ_B and is a well-known result from elementary geometrical optics.

(v) Using Snell's law to eliminate θ'' from (8)–(11) gives

☞ s-polarization

$$r_s = \frac{\cos\theta - \sqrt{n^2 - \sin^2\theta}}{\cos\theta + \sqrt{n^2 - \sin^2\theta}}; \qquad t_s = \frac{2\cos\theta}{\cos\theta + \sqrt{n^2 - \sin^2\theta}} \tag{15}$$

and

$$R_s = \left(\frac{\cos\theta - \sqrt{n^2 - \sin^2\theta}}{\cos\theta + \sqrt{n^2 - \sin^2\theta}}\right)^2; \qquad T_s = 1 - R_s. \tag{16}$$

☞ p-polarization

$$r_p = \frac{n^2\cos\theta - \sqrt{n^2 - \sin^2\theta}}{n^2\cos\theta + \sqrt{n^2 - \sin^2\theta}}; \qquad t_p = \frac{2n\cos\theta}{n^2\cos\theta + \sqrt{n^2 - \sin^2\theta}} \tag{17}$$

and

$$R_p = \left(\frac{n^2\cos\theta - \sqrt{n^2 - \sin^2\theta}}{n^2\cos\theta + \sqrt{n^2 - \sin^2\theta}}\right)^2; \qquad T_p = 1 - R_p. \tag{18}$$

For $n < 1$, $(15)_1$ and $(17)_1$ show that r_s and r_p become complex quantities for $\theta > \sin^{-1}n^{-1}$. The angle $\sin^{-1}n^{-1}$ is known as the critical angle θ_c and is another elementary result. If $\theta \geq \theta_c$, both amplitude reflection coefficients have unit magnitude and can be written as

$$r_s = \frac{\cos\theta - i\sqrt{\sin^2\theta - n^2}}{\cos\theta + i\sqrt{\sin^2\theta - n^2}} = e^{-i\varphi_s} \text{ and } r_p = \frac{n^2\cos\theta - i\sqrt{\sin^2\theta - n^2}}{n^2\cos\theta + i\sqrt{\sin^2\theta - n^2}} = e^{-i\varphi_p},$$

where

$$\varphi_s = \frac{1}{2}\tan^{-1}\left(\frac{\sin^2\theta - n^2}{\cos\theta}\right) \quad\text{and}\quad \varphi_p = \frac{1}{2}\tan^{-1}\left(\frac{\sin^2\theta - n^2}{n^2\cos\theta}\right). \tag{19}$$

Each polarization state undergoes total internal reflection at the boundary $(r_s = r_p = 1)$, with phase changes given by (19).

(vi) The graphs below show plots of the Fresnel equations (15)–(18) for $n = 3/2$ and $n = 2/3$ (typical for light travelling from air to glass and glass to air respectively).

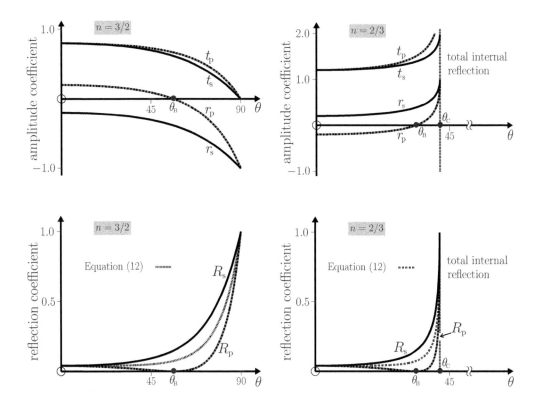

Question 10.16*

Consider a dielectric medium (e.g. a dilute gas) comprising heavy positive ion cores surrounded by electrons which are bound to the cores by a harmonic restoring force $-m\omega_0^2\mathbf{r}$. Here ω_0 is the natural frequency of oscillation of an electron and m its mass. An electromagnetic wave $\mathbf{E} = \mathbf{E}_0 e^{2\pi i(z-ct)/\lambda}$ propagating in the medium causes the electrons to oscillate with amplitude $r_0 \ll \lambda$. Energy loss due to collisions and radiation is incorporated in the above model through a viscous damping force $-m\gamma\dot{\mathbf{r}}$ acting on the electrons (recall, from Question 10.8, that γ is a phenomenological constant).‡

(a) Use Newton's second law to show that the displacement $\mathbf{r}(t)$ of an electron is

$$\mathbf{r}(t) = \frac{-(e/m)}{(\omega_0^2 - \omega^2) - i\gamma\omega} \mathbf{E}_0 e^{-i\omega t}. \tag{1}$$

---

‡The energy loss due to radiation depends on the acceleration $\ddot{\mathbf{r}}$, rather than the velocity $\dot{\mathbf{r}}$ (see Chapter 11). Nevertheless, this form with the damping force $\propto \dot{\mathbf{r}}$ does provide some qualitative insight for the dispersion of electromagnetic waves in the medium.

(b) Hence show that the relative permittivity $\epsilon_r(\omega)$ of the medium is given by

$$\epsilon_r(\omega) = 1 + \frac{Ne^2}{m\epsilon_0} \frac{1}{(\omega_0^2 - \omega^2) - i\gamma\omega}, \tag{2}$$

where N is the number density of participating electrons.

(c) Use (2) to find the real and imaginary parts of the refractive index n of the medium (assume that $n \approx 1$).

Solution

(a) The electronic equation of motion is $m\dfrac{d^2\mathbf{r}}{dt^2} = \mathbf{F}_{\text{restoring}} + \mathbf{F}_{\text{damping}} + \mathbf{F}_{\text{driving}}$. Since the displacement of an electron from its equilibrium position is always very much less than λ, the driving force exerted on the electron by the wave[‡] is independent of position and we put $\mathbf{F}_{\text{driving}} = -e\mathbf{E}_0 e^{-i\omega t}$. Thus,

$$m\frac{d^2\mathbf{r}}{dt^2} = -m\omega_0^2\mathbf{r} - m\gamma\dot{\mathbf{r}} - e\mathbf{E}_0 e^{-i\omega t}. \tag{3}$$

In the steady state the electrons oscillate with the frequency of the incident wave, and so we attempt a trial solution of the form $\mathbf{r}(t) = \mathbf{r}_0 e^{-i\omega t}$. Substituting this in (3) and solving for $\mathbf{r}(t)$ gives (1).

(b) The dipole moment $\mathbf{p}$ of one of these bound electrons is $(-e)\mathbf{r}(t)$. Thus the dipole moment per unit volume follows from (III) of the introduction to Chapter 9, and is $\mathbf{P} = N\mathbf{p} = -Ne\mathbf{r}(t) = \epsilon_0\chi_e\mathbf{E}$. Using (1) gives

$$\chi_e = \frac{Ne^2}{m\epsilon_0} \frac{1}{(\omega_0^2 - \omega^2) - i\gamma\omega}. \tag{4}$$

For an lih dielectric $\epsilon_r = 1 + \chi_e$ (see (4) of Question 9.4) which implies (2).

(c) For an lih dielectric with $\mu_r = 1$, the refractive index $n = \sqrt{\epsilon_r}$. But since $n \approx 1$ the second term of (2) is small, and so

$$n(\omega) \simeq 1 + \frac{Ne^2}{2m\epsilon_0} \frac{1}{(\omega_0^2 - \omega^2) - i\gamma\omega}, \tag{5}$$

which has the real and imaginary parts

$$n_r = 1 + \frac{Ne^2}{2m\epsilon_0} \frac{(\omega_0^2 - \omega^2)}{(\omega_0^2 - \omega^2)^2 + \gamma^2\omega^2}, \qquad n_i = \frac{Ne^2}{2m\epsilon_0} \frac{\gamma\omega}{(\omega_0^2 - \omega^2)^2 + \gamma^2\omega^2}. \tag{6}$$

[‡]For $v \ll c$ the contribution to $\mathbf{F}_{\text{driving}}$ from the magnetic field of the wave is negligible.

Comments

(i) The preceding discussion is for a single-electron resonance. Now suppose, instead, that each ion core has f_j electrons with resonant frequencies ω_{0j} and damping constants γ_j. Then (6) generalizes

$$n_{\mathrm{r}} = 1 + \frac{Ne^2}{2m\epsilon_0} \sum_j \frac{f_j(\omega_{0j}^2 - \omega^2)}{(\omega_{0j}^2 - \omega^2)^2 + \gamma_j^2\omega^2}, \qquad n_{\mathrm{i}} = \frac{Ne^2}{2m\epsilon_0} \sum_j \frac{f_j\gamma_j\omega}{(\omega_{0j}^2 - \omega^2)^2 + \gamma_j^2\omega^2}. \qquad (7)$$

(ii) The classical model presented above is simple to understand, and it offers a qualitatively correct description of the dielectric properties of the medium. With suitable definitions[†] for ω_{0j}, γ_j and f_j, quantum-mechanical calculations yield expressions of a similar form to (7).

(iii) In Question 10.17, we plot n_{r} and n_{i} for a two-electron resonance.

Question 10.17

Suppose the dielectric medium of Question 10.16 has five participating electrons per atom. Three of these electrons have a resonance at ω_0 and a damping constant γ. The remaining two electrons have a resonance at $2\omega_0$ and a damping constant $\frac{1}{2}\gamma$. Plot the graphs of n_{r} and n_{i} vs ω/ω_0 for $\frac{1}{2} \leq \omega/\omega_0 \leq \frac{5}{2}$, taking $\gamma = \omega_0/10^{\ddagger}$ and $\dfrac{Ne^2}{2m\epsilon_0\omega_0^2} = 2 \times 10^{-4}$.

Solution

Substituting the above values in (7) of Question 10.16 and putting $x = \omega/\omega_0$ give

$$n_{\mathrm{r}} = 1 + 0.0002 \left[\frac{3 \times (1 - x^2)}{\left(1 - x^2\right)^2 + \frac{x^2}{100}} + \frac{2 \times \left(1 - \frac{x^2}{4}\right)}{\left(1 - \frac{x^2}{4}\right)^2 + \frac{x^2}{6400}} \right] \qquad (1)$$

and

$$n_{\mathrm{i}} = 0.0002 \left[\frac{3 \times \frac{x}{10}}{\left(1 - x^2\right)^2 + \frac{x^2}{100}} + \frac{2 \times \frac{x}{40}}{\left(1 - \frac{x^2}{4}\right)^2 + \frac{x^2}{6400}} \right]. \qquad (2)$$

The graphs of (1) and (2) are shown below.

[†]For example, we interpret the ω_{0j} as transition frequencies to excited states of the atomic electrons and the f_j represent oscillator strengths.

[‡]This value is chosen for illustrative purposes only. In a dielectric material at optical frequencies the ratio is much smaller than this.

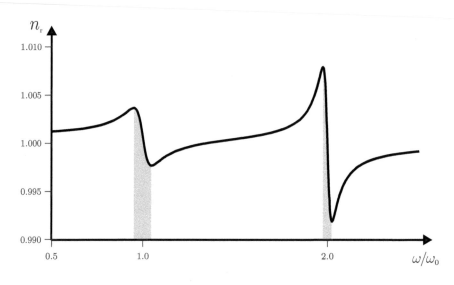

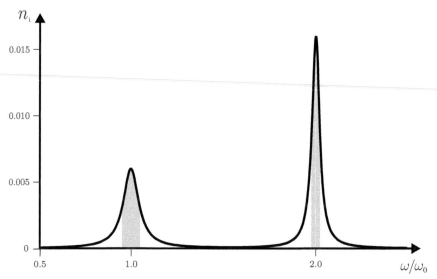

Comments

(i) The graph of n_r vs ω/ω_0 reveals the following:

- ☞ away from resonance the refractive index increases with frequency. This is normal dispersion and is consistent with what we learn in optics. Frequencies in the visible spectrum corresponding to blue light are refracted more than those corresponding to red light.
- ☞ In the vicinity of resonance there are regions (shown shaded in the graph) where $dn_r/d\omega < 0$. That is, the refractive index decreases with increasing

frequency. This is called anomalous dispersion, so named because it is the opposite to what we usually encounter in introductory courses on optics.

(ii) The graph of n_i indicates that, since $n_i \ll 1$ everywhere, the refractive index is essentially real. Often in the vicinity of resonance[#] the validity of this statement does not hold, and n is a complex quantity. In these cases, the material is found to absorb energy from the incident radiation.

(iii) For most atoms and molecules the electronic transition frequencies ω_{0j} correspond to resonances which lie in the visible or ultraviolet region of the electromagnetic spectrum. Vibrational and rotational modes of oscillation of the ions can also be excited. These usually appear at infrared frequencies and extend into the microwave region of the spectrum.

(iv) In the visible region of the spectrum the damping terms γ_j are usually negligible for transparent materials (since here ω is less than the ultraviolet transition frequencies ω_{0j}). The quantity $\dfrac{f_j(\omega_{0j}^2 - \omega^2)}{(\omega_{0j}^2 - \omega^2)^2 + \gamma_j^2\omega^2}$ in $(7)_1$ is therefore approximately $\dfrac{f_j}{\omega_{0j}^2}\left(1 - \dfrac{\omega^2}{\omega_{0j}^2}\right)^{-1}$ which can be written as the series expansion

$$\frac{f_j}{\omega_{0j}^2}\left(1 - \frac{\omega^2}{\omega_{0j}^2}\right)^{-1} = \frac{f_j}{\omega_{0j}^2}\left(1 + \frac{\omega^2}{\omega_{0j}^2} + \frac{\omega^4}{\omega_{0j}^4} + \cdots\right). \tag{3}$$

Substituting (3) in $(7)_1$ of Question 10.16 gives the refractive index

$$n = 1 + \left(\frac{Ne^2}{2m\epsilon_0}\sum_j \frac{f_j}{\omega_{0j}^2}\right) + \omega^2\left(\frac{Ne^2}{2m\epsilon_0}\sum_j \frac{f_j}{\omega_{0j}^4}\right) + \omega^4\left(\frac{Ne^2}{2m\epsilon_0}\sum_j \frac{f_j}{\omega_{0j}^6}\right) + \cdots. \tag{4}$$

Expressed in terms of the vacuum wavelength $\lambda = 2\pi c/\omega$, (4) is

$$n(\lambda) = A + \frac{B}{\lambda^2} + \frac{C}{\lambda^4} + \cdots, \tag{5}$$

where the A, B, C, ... are material-specific coefficients. Equation (5) is known as Cauchy's equation.[†] Because of its simplicity, (5) is useful in certain applications where it is usually sufficient to consider only the first two terms. Cauchy's equation does not apply to regions of anomalous dispersion and it becomes inaccurate in the infrared.

[#]The width of a resonance peak is proportional to the appropriate γ_j. More realistic values of γ_j (see the second footnote on p. 495) produce much sharper resonances than those plotted above.

[†]It was first established empirically in 1836 by the mathematician A. L. Cauchy.

Question 10.18*

Circularly polarized electromagnetic waves of the form

$$\mathbf{E} = E_0(\hat{\mathbf{x}} \pm i\hat{\mathbf{y}}) e^{i(kz - \omega t)} \tag{1}$$

propagate in a tenuous electronic plasma that is permeated by a strong external axial magnetostatic field $B_0 \hat{\mathbf{z}}$. The upper (lower) sign in (1) is for waves having positive (negative) helicity.[‡]

(a) Derive the dispersion relation

$$k = \frac{\omega}{c} \left[1 - \frac{\omega_p^2}{\omega(\omega \mp \omega_B)} \right]^{-\frac{1}{2}}, \tag{2}$$

where $\omega_p = \sqrt{\dfrac{Ne^2}{\epsilon_0 m_e}}$ and $\omega_B = \dfrac{e B_0}{m_e}$ are the plasma and cyclotron frequencies respectively.

Hint: Start with an equation of motion for the electrons. Assume that the amplitude of oscillation r_0 is very much less than the wavelength ($r_0 \ll c/\omega$) and neglect collisions ($\tau_c \gg \omega^{-1}$). Assume also that the **B**-field of the wave is negligible compared to $\mathbf{B}_0$.

(b) Use (2) to derive expressions for the phase and group velocities ($v_\phi = \omega/k$ and $v_g = d\omega/dk$) for these waves.

Solution

(a) We consider a typical electron at which the origin is chosen. Then $\mathbf{r} = 0$, and from Newton's second law with $\mathbf{F} = q(\mathbf{E} + \mathbf{v} \times \mathbf{B}_{\text{total}})$ we obtain

$$m\ddot{\mathbf{r}} = (-e)\big[\mathbf{E} + \dot{\mathbf{r}} \times (\mathbf{B} + \mathbf{B}_0) \big].$$

For a 'strong' magnetostatic field ($\mathbf{B}_0 \gg \mathbf{B}$) the equation of motion becomes

$$m\ddot{\mathbf{r}} \simeq (-e)\big[\mathbf{E} + \dot{\mathbf{r}} \times \mathbf{B}_0 \big]. \tag{3}$$

Now the motion of the electron is obviously confined to the xy-plane, and so we attempt a steady-state solution of (3) of the form

$$\mathbf{r}(t) = r_0(\hat{\mathbf{x}} \pm i\hat{\mathbf{y}}) e^{-i\omega t}. \tag{4}$$

Substituting (4) and (1) in (3) gives $[m\omega^2 r_0 - e E_0 \mp e B_0 \omega r_0](\hat{\mathbf{x}} \pm i\hat{\mathbf{y}}) = 0$. Solving for r_0 then yields

---

[‡]In the terminology of optics these correspond to left (right) circularly polarized waves; see also Question 7.9.

$$\mathbf{r}(t) = \frac{e/m_e}{\omega(\omega \mp \omega_B)} \mathbf{E}. \tag{5}$$

From (4) of Question 4.3, we obtain the current density $\mathbf{J}_f = n(-e)\mathbf{v}$ where $\mathbf{v} = \dot{\mathbf{r}}$ follows from (5). Thus $\mathbf{J}_f = iNe^2/[m_e(\omega \mp \omega_B)]\mathbf{E}$. But $\mathbf{J}_f = \sigma\mathbf{E}$, and so we obtain the conductivity

$$\sigma = \frac{iNe^2}{m_e(\omega \mp \omega_B)}.$$

Substituting this in (3) of Question 10.10 and rearranging yield (2).

(b) The phase velocity follows directly from (2)

$$\left(\frac{\omega}{k}\right)_{\pm} = c\left[1 - \frac{\omega_p^2}{\omega(\omega \mp \omega_B)}\right]^{-\frac{1}{2}}. \tag{6}$$

Differentiating (2) with respect to k yields the group velocity, after a little algebra,

$$\left(\frac{d\omega}{dk}\right)_{\pm} = c\frac{2\omega(\omega \mp \omega_B)^2}{2\omega(\omega \mp \omega_B)^2 \pm \omega_p^2\omega_B}\sqrt{1 - \frac{\omega_p^2}{\omega(\omega \mp \omega_B)}}. \tag{7}$$

Comments

(i) In the presence of an external magnetic field the plasma is circularly birefringent, and the refractive indices for positive-helicity (negative-helicity) waves travelling along the external magnetic field are

$$n_{\pm} = \frac{c}{v_\phi} = \sqrt{1 - \frac{\omega_p^2}{\omega(\omega \mp \omega_B)}}. \tag{8}$$

It is evident from (8) that n_+ and n_- are real only in certain frequency *intervals*. For any given value of ω, one of the following statements hold:

☞ n_+ and n_- are both real,
☞ n_+ and n_- are both pure imaginary,
☞ n_+ is real and n_- is pure imaginary,
☞ n_+ is pure imaginary and n_- is real.

Since travelling waves exist only for real n, it turns out that in some frequency intervals either only one polarization state or neither polarization state can propagate. A numerical example of this is provided in Question 10.19. On the basis of the above results, it is plausible that the propagation of radio waves in the Earth's ionosphere will depend (*inter alia*) on both the frequency and helicity of the radiation. These conclusions are confirmed in practice.

(ii) It is also evident from (8) that as $\omega \to 0$ only positive-helicity waves are present. In this low-frequency limit

$$v_\phi \to c \sqrt{\frac{\omega_B\, \omega}{\omega_p}}$$

and $v_g \to 2v_\phi$. At low frequencies, the plasma exhibits anomalous dispersion because $v_g > v_\phi$.

(iii) In the opposite limit, as $\omega \to \infty$, for waves of either helicity $v_\phi \to c$ from above and $v_g \to c$ from below. The plasma exhibits normal dispersion at high frequencies because $v_g < v_\phi$.

Question 10.19*

Consider the tenuous electronic plasma of Question 10.18. Suppose $\omega_B = 5 \times 10^6$ rad s^{-1} and $\omega_p = 2 \times 10^7$ rad s^{-1} (typical for the Earth's magnetic field and ionosphere respectively).[‡] Plot the following graphs:

☞ the dispersion relation ω vs k,

☞ the phase velocity v_ϕ vs ω and

☞ the group velocity v_g vs ω.

Solution

It is convenient to let $\omega = \alpha\, \omega_p$ where the parameter $0 \leq \alpha < \infty$. Since $\dfrac{\omega_B}{\omega_p} = \dfrac{1}{4}$, we have from (6) of Question 10.18

$$k_\pm = \frac{\alpha\, \omega_p}{c} \sqrt{1 - \frac{1}{\alpha(\alpha \mp \tfrac{1}{4})}}, \tag{1}$$

where k_+ (k_-) is for a wave of positive (negative) helicity.

positive-helicity waves

k_+ is zero if $\alpha_+(\alpha_+ - \tfrac{1}{4}) = 1$ or $4\alpha_+^2 - \alpha_+ - 4 = 0$. The appropriate root of this quadratic is $\alpha_+ \simeq 1.133$.

negative-helicity waves

k_- is zero if $\alpha_-(\alpha_- + \tfrac{1}{4}) = 1$ or $4\alpha_+^2 + \alpha_+ - 4 = 0$. The appropriate root of this quadratic is $\alpha_- \simeq 0.883$.

[‡] Although the values of ω_B and ω_p are 'typical' for the Earth, this does not imply that our simple example should be regarded as a model for the real Earth–ionosphere system (which is, of course, much more complex).

The graphs of ω vs k, v_ϕ vs ω and v_g vs ω are shown below and on p. 502.

Comments

(i) The graphs below reveal that for $\omega < \omega_B$ only positive-helicity waves can propagate (because k is imaginary for negative-helicity waves in this region). In the interval $\omega_B < \omega < \omega_-$ waves of neither helicity can propagate; there is a gap in the dispersion relation and no travelling-wave solutions of the wave equation exist. For $\omega_- < \omega < \omega_+$ only negative-helicity waves can propagate. Above ω_+ both states can be present. Dispersion relations also arise elsewhere in physics, and they sometimes exhibit analogous behaviour; for example, the dispersion relation of a diatomic linear chain has a gap between the acoustic and optical branches.

(ii) The low-frequency behaviour of the dispersion relation (2) of Question 10.18 gives rise to an interesting atmospheric phenomenon called whistler waves. These are produced by lightning: a strike in one hemisphere may generate a broad spectrum of electromagnetic radiation from visible light to very low-frequency radio waves $\sim$(1–100) kHz. Some of this energy travels along geomagnetic-field lines in the positive-helicity state and can echo back and forth between the two hemispheres. Higher frequencies travel faster than lower frequencies since $v_g \propto \sqrt{\omega}$, as discussed in Comment (ii) of Question 10.18. A single lightning strike may produce several whistlers, with each wave being guided along a different magnetic-field line. An audio receiver detects the high-frequency components first, and then the low-frequency components. This is heard as a whistling sound of rapidly decreasing pitch whose duration T depends on the distance d of travel.

☞ dispersion relation

The following graph was obtained using (1) and *Mathematica*'s `ParametricPlot` function.

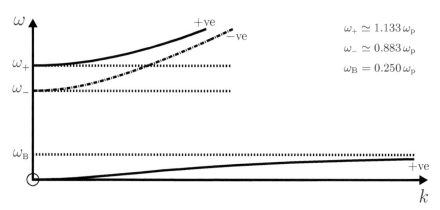

$$\omega_+ \simeq 1.133\,\omega_p$$
$$\omega_- \simeq 0.883\,\omega_p$$
$$\omega_B = 0.250\,\omega_p$$

(iii) An order-of-magnitude estimate for T in the frequency range $(10^3$–$10^5)$Hz follows from the above values of ω_p and ω_B. Taking $d \sim 2 \times 10^7$ m (the distance along the Earth's surface between the North and South poles) gives

$$T = \frac{d\omega_{\mathrm{p}}}{2c\sqrt{\omega_{\mathrm{B}}}}\left(\frac{1}{\sqrt{\omega_{\mathrm{low}}}} - \frac{1}{\sqrt{\omega_{\mathrm{high}}}}\right)$$

$$= \frac{(2\times10^7)\times(2\times10^7)}{2\times(3\times10^8)\times\sqrt{5\times10^6}}\left(\frac{1}{\sqrt{2\pi\times10^3}} - \frac{1}{\sqrt{2\pi\times10^5}}\right) \sim \boxed{2\,\mathrm{s}},$$

which compares favourably with measured values.

☞ phase velocity

The following graph was obtained using (6) of Question 10.18 and *Mathematica*'s Plot function.

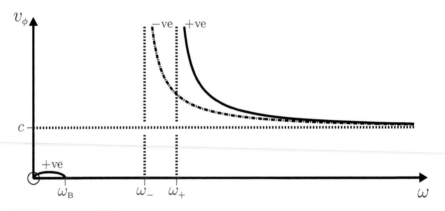

☞ group velocity

The following graph was obtained using (7) of Question 10.18 and *Mathematica*'s Plot function.

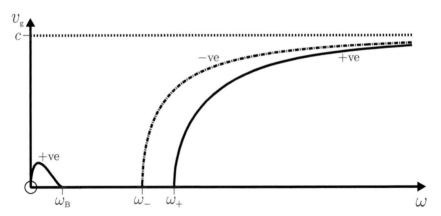

Question 10.20 **

A linearly polarized electromagnetic wave of the form

$$\mathbf{E} = E_0\, e^{i(kz-\omega t)}\, \hat{\mathbf{x}} \tag{1}$$

propagates in a tenuous electronic plasma permeated by a strong axial magnetostatic field $B_0\hat{\mathbf{z}}$.

(a) Show that the plane of polarization is rotated through an angle

$$\theta = \left(\frac{\omega\,\delta n}{2c}\right)z, \tag{2}$$

where z is the distance travelled and δn is the circular birefringence defined as

$$\delta n = n_- - n_+. \tag{3}$$

Hint: Use (1) of Question 10.18 to express (1) as a linear superposition of the positive-helicity and negative-helicity states.

(b) Suppose $\omega = 4\omega_{\mathrm{p}}$. Use the values for ω_{B} and ω_{p} given in Question 10.19 to calculate the angle of rotation per unit distance of travel. Quote your answer in degrees per kilometre.

Solution

(a) At $z = 0$ and time $t = 0$ the wave is polarized along the positive x-axis. From (1) of Question 10.18 we have

$$\mathbf{E} = \tfrac{1}{2}(\mathbf{E}_+ + \mathbf{E}_-)$$

$$= \tfrac{1}{2}E_0\left\{(\hat{\mathbf{x}} + i\hat{\mathbf{y}})e^{i\omega(zn_+/c - t)} + (\hat{\mathbf{x}} - i\hat{\mathbf{y}})e^{i\omega(zn_-/c - t)}\right\}$$

$$= \tfrac{1}{2}E_0\left\{\left[e^{i\omega z n_+/c} + e^{i\omega z n_-/c}\right]\hat{\mathbf{x}} + i\left[e^{i\omega z n_+/c} - e^{i\omega z n_-/c}\right]\hat{\mathbf{y}}\right\}e^{-i\omega t}. \tag{4}$$

Further along the z-axis at $t = 0$, the $\mathbf{E}$ vector makes an angle θ with respect to $\hat{\mathbf{x}}$ as shown below.

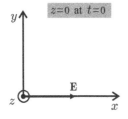

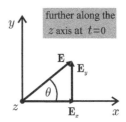

Clearly $\tan \theta = E_y/E_x$, and with the components of $\mathbf{E}$ given by (4) we obtain

$$\tan \theta = \frac{i\left(e^{i\omega z n_+/c} - e^{i\omega z n_-/c}\right)}{\left(e^{i\omega z n_+/c} + e^{i\omega z n_-/c}\right)}. \qquad (5)$$

Using (3) to eliminate n_+ from (5) gives

$$\tan \theta = \frac{i\, e^{-i\omega z\, \delta n/2c}\left(e^{i\omega(n_- - \frac{1}{2}\delta n)\frac{z}{c}} - e^{i\omega(n_- + \frac{1}{2}\delta n)\frac{z}{c}}\right)}{e^{-i\omega z\, \delta n/2c}\left(e^{i\omega(n_R - \frac{1}{2}\delta n)\frac{z}{c}} + e^{i\omega(n_- + \frac{1}{2}\delta n)\frac{z}{c}}\right)}$$

$$= \frac{-i\left(e^{i\omega z\delta n/2c} - e^{-i\omega z\delta n/2c}\right)}{\left(e^{i\omega z\delta n/2c} + e^{-i\omega z\delta n/2c}\right)} = \frac{2\sin\left(\frac{1}{2}\omega z\,\delta n/c\right)}{2\cos\left(\frac{1}{2}\omega z\,\delta n/c\right)}$$

$$= \tan\left(\tfrac{1}{2}\omega z\,\delta n/c\right),$$

which is (2).

(b) From (8) of Question 10.18 and (3) we find that, for $\omega = 4\omega_{\mathrm{p}}$,

$$\delta n = \sqrt{1 - \frac{(\omega_{\mathrm{p}}/\omega)^2}{1 + (\omega_{\mathrm{B}}/\omega)}} - \sqrt{1 - \frac{(\omega_{\mathrm{p}}/\omega)^2}{1 - (\omega_{\mathrm{B}}/\omega)}} \qquad (6)$$

$$= \sqrt{1 - \frac{(1/4)^2}{1 + (1/16)}} - \sqrt{1 - \frac{(1/4)^2}{1 - (1/16)}}$$

$$\simeq 0.00405.$$

The angle of rotation per unit distance is therefore

$$\frac{\theta}{z} = \frac{\omega\,\delta n}{2c}$$

$$\simeq \frac{8 \times 10^7 \times 0.00405}{2 \times 3 \times 10^8} \times \frac{180}{\pi} \times 10^3$$

$$\simeq 31^\circ\,\mathrm{km}^{-1}.$$

Comments

(i) For an arbitrary direction of propagation, the cyclotron frequency $\omega_{\mathrm{B}} = \dfrac{e\mathbf{B}_0 \cdot \hat{\mathbf{k}}}{m_{\mathrm{e}}}$
is a function of angle through the dot product. The medium is both circularly birefringent and anisotropic.

(ii) The rotation of the plane of polarization of an electromagnetic wave propagating through matter in the presence of a strong magnetic field **B** was first discovered experimentally by Faraday in 1845. In these experiments, Faraday found that the plane of linearly polarized light, travelling through a glass rod of length ℓ, was rotated through an angle θ when **B** was applied along the direction of propagation. By varying B and ℓ, Faraday deduced the empirical relationship

$$\theta = \mathcal{V}B\ell, \tag{7}$$

where the factor of proportionality $\mathcal{V}$, called the Verdet constant, is material-dependent. Faraday's experiments were an early indication that optical effects could be explained in terms of electromagnetism: the emerging science of the late nineteenth century.

(iii) This rotation of the plane of vibration of a linearly polarized electromagnetic wave, whether it occurs for a radio wave travelling through a plasma or visible light in a dielectric medium, is known as the Faraday effect (here the angle θ is called the Faraday rotation).

(iv) The quantity

$$\delta n = \sqrt{\frac{\omega(\omega + \omega_B) - \omega_p^2}{\omega + \omega_B}} - \sqrt{\frac{\omega(\omega - \omega_B) - \omega_p^2}{\omega - \omega_B}}$$

(see (6)) can be expanded‡ in powers of ω_B which gives rise to the infinite series

$$\delta n = \frac{\omega_p^2}{\omega^2\sqrt{\omega^2 - \omega_p^2}}\,\omega_B + \frac{\omega_p^2\sqrt{\omega^2 - \omega_p^2}\left(8\omega^4 - 12\omega^2\omega_p^2 + 5\omega_p^4\right)}{8\omega^4(\omega^2 - \omega_p^2)^3}\,\omega_B^3 + \cdots. \tag{8}$$

Substituting (8) in (2) and putting $\omega_B = \dfrac{e B_0}{m_e}$ yield, to first order,

$$\theta \simeq \left[\frac{e}{2m_e c}\frac{\omega_p^2/\omega}{\sqrt{\omega^2 - \omega_p^2}}\right]B_0 z,$$

which agrees with Faraday's empirical result (7) for visible light. The frequency-dependent term in brackets is the Verdet constant $\mathcal{V}$ for the plasma.

---

‡This is most easily done using *Mathematica*'s Series function.

11

Electromagnetic radiation

In earlier chapters of this book, we obtained solutions of Maxwell's equations which describe the propagation of electromagnetic waves in vacuum and in matter. We will now consider questions which relate the electromagnetic field to its sources $\big(\rho(\mathbf{r}',t)$ and $\mathbf{J}(\mathbf{r}',t)\big)$. Two important features which will emerge from our discussion as the chapter develops are:

☞ The electromagnetic field at a field point P reflects the behaviour of the sources at the earlier (retarded) time $t - r/c$, where $\mathbf{r}$ is the vector from an arbitrary origin O to P.

☞ The electromagnetic field has a component that transports energy away from the sources to infinity, in a process which is irreversible. It is this component of the field that is called electromagnetic radiation.

The retarded electromagnetic potentials are an obvious starting point, since they lead directly to the fields from

$$\mathbf{E}(\mathbf{r},t) = -\boldsymbol{\nabla}\Phi - \frac{\partial \mathbf{A}}{\partial t} \qquad \text{and} \qquad \mathbf{B}(\mathbf{r},t) = \boldsymbol{\nabla}\times\mathbf{A}.$$

However, we need not concern ourselves with Φ per se. It suffices to consider the retarded vector potential only, because Φ and $\mathbf{A}$ are related by the Lorenz gauge condition (see Comment (ii) of Question 8.20). With this in mind, we recall the series expansion for the dynamic vector potential

$$A_i(\mathbf{r},t) = \frac{\mu_0}{4\pi}\left\{ \frac{1}{r}\int_v J_i\,dv' + \frac{r_j}{r^3}\left(\int_v r'_j J_i\,dv' + \frac{r}{c}\int_v r'_j \dot{J}_i\,dv'\right) + \cdots \right\}, \qquad (\mathrm{I})$$

derived in Question 8.20. We begin, in Question 11.1, by expressing (I) in terms of electric and magnetic multipole moments. This form of $\mathbf{A}$ will prove to be very useful for discussing the various types of electromagnetic radiation which we will consider in this chapter, namely electric dipole and magnetic dipole–electric quadrupole. We derive the important Larmor formula in Question 11.8, which reinforces an idea that is usually encountered in elementary physics. Namely, that accelerated electric charges are the source of electromagnetic radiation. Some simple applications of Larmor's formula, both in its non-relativistic and relativistic forms, are given. Also considered are examples involving antennas, antenna arrays and the scattering of radiation by a free electron.

Solved Problems in Classical Electromagnetism. J. Pierrus, Oxford University Press (2018).
© J. Pierrus. DOI: 10.1093/oso/9780198821915.001.0001

Question 11.1*

Use the integral transforms derived in Question 1.30,

$$
\left.
\begin{aligned}
\int_v J_i \, dv' &= \int_v r_i' \frac{\partial \rho}{\partial t} \, dv' \\[2mm]
\int_v r_j' J_i \, dv' &= -\tfrac{1}{2} \varepsilon_{ijk} \int_v (\mathbf{r}' \times \mathbf{J})_k \, dv' + \tfrac{1}{2} \int_v r_i' r_j' \frac{\partial \rho}{\partial t} \, dv'
\end{aligned}
\right\},
\tag{1}
$$

to express the series expansion (I) of the introduction in the form

$$
A_i(\mathbf{r}, t) = \frac{\mu_0}{4\pi} \left\{ \frac{\dot{p}_i}{r} + \frac{r_j}{r^3} \left(\left(\tfrac{1}{2} \dot{q}_{ij} - \varepsilon_{ijk} m_k \right) + \frac{r}{c} \left(\tfrac{1}{2} \ddot{q}_{ij} - \varepsilon_{ijk} \dot{m}_k \right) \right) + \cdots \right\}.
\tag{2}
$$

Here $p_i(t)$, $q_{ij}(t)$ and $m_i(t)$ are time-dependent electric dipole, electric quadrupole and magnetic dipole moments respectively.[‡] In (2), the dot notation denotes differentiation with respect to the retarded time at the origin

$$
t' = t - r/c
\tag{3}
$$

(see (4) of Question 8.20).

Solution

From $(1)_1$ and the definition of $\mathbf{p}(t)$ it follows that

$$
\int_v J_i \, dv' = \int_v r_i' \frac{\partial \rho}{\partial t} \, dv' = \frac{d}{dt} \int_v r_i' \rho = \frac{dp_i}{dt}
$$

$$
= \frac{dp_i}{dt'} \frac{dt'}{dt}
$$

$$
= \dot{p}_i,
\tag{4}
$$

because $dt'/dt = 1$. Similarly, the definitions of $\mathbf{m}(t)$ and $q_{ij}(t)$ together with $(1)_2$ give

$$
\int_v r_j' J_i \, dv' = -\varepsilon_{ijk} m_k(t) + \tfrac{1}{2} \dot{q}_{ij}(t).
\tag{5}
$$

Substituting (4) and (5) in the series expansion (I) leads immediately to (2).

[‡] The time-dependent multipole moments are defined as in Questions 2.20 and 4.24, but with $\rho(\mathbf{r}') \to \rho(\mathbf{r}', t)$ and $\mathbf{J}(\mathbf{r}') \to \mathbf{J}(\mathbf{r}', t)$.

Comments

(i) Equation (2) is the multipole expansion of the dynamic vector potential. Unlike the electrostatic and magnetostatic expansions encountered previously, we now see that combinations of electric and magnetic multipole moments appear simultaneously. The leading term of $A_i(\mathbf{r}, t)$ is the electric dipole contribution (which may or may not be zero). This is followed by electric quadrupole–magnetic dipole contributions which are of a similar order. The electric octopole–magnetic quadrupole contributions come next, and so on. In (2), it is understood that these various multipole moments and their derivatives are evaluated at the retarded time t'. It is sometimes useful to express this dependence explicitly as

$$A_i = \frac{\mu_0}{4\pi}\left\{ \frac{[\dot{p}_i]}{r} + \frac{r_j}{2r^3}\left([\dot{q}_{ij}] + \frac{r}{c}[\ddot{q}_{ij}]\right) - \frac{r_j \varepsilon_{ijk}}{r^3}\left([m_k] + \frac{r}{c}[\dot{m}_k]\right) + \cdots \right\}, \quad (6)$$

where the notation $[\ ]_{\mathrm{ret}}$ or simply $[\ ]$ means that the enclosed quantity is to be evaluated at the retarded time t'.

(ii) The multipole expansion of the retarded scalar potential $\Phi(\mathbf{r}, t)$ is easier to derive than (6). By starting with (8) of Question 8.20 and proceeding in a similar way, we obtain the result

$$\Phi(\mathbf{r}, t) = \frac{1}{4\pi\epsilon_0}\left\{ \frac{q}{r} + \frac{r_i}{r^3}\left(p_i + \frac{r}{c}\dot{p}_i\right) + \frac{(3r_ir_j - r^2\delta_{ij})}{2r^5}\left(q_{ij} + \frac{r}{c}\dot{q}_{ij}\right) + \frac{r_ir_j}{2c^2r^3}\ddot{q}_{ij} \right\}. \quad (7)$$

The reader may wish to verify that (6) and (7) satisfy the Lorenz condition

$$\nabla \cdot \mathbf{A} + \frac{1}{c^2}\frac{\partial \Phi}{\partial t} = 0.$$

Because of this, we shall have no further need of (7) as explained in Comment (ii) on p. 408.

(iii) In some of the questions which now follow, the multipole expansion (6) is used to determine the electromagnetic radiation arising from an arbitrary distribution of electric charges and/or currents. This radiation will always carry the signature of the leading non-vanishing multipole. Contributions from the higher multipole moments—when these exist—will be completely insignificant by comparison.

(iv) Low-order multipole expansions may be applied to calculate radiation fields when $\mathbf{J}(\mathbf{r}, t)$ varies sufficiently slowly (on a time scale T, say) compared to the time d/c for the electromagnetic wave to traverse the charge distribution.[‡] Thus $T \gg d/c$ where $T \sim d/u$ and u the average speed of the charges inside v' requires $u \ll c$, which is essentially the non-relativistic limit. In Questions 11.21–11.24 we deal with the radiation emitted from accelerated charges moving relativistically.

[‡]Here $d \sim \sqrt[3]{v'}$ is a typical length scale for the distribution.

Question 11.2

Consider an arbitrary distribution of charges and currents. Suppose the charge and current densities vary harmonically in time with angular frequency ω. Then

$$\left. \begin{array}{ll} \rho(\mathbf{r}',t) & = \rho_0(\mathbf{r}')\,e^{-i\omega t} \\[2mm] \mathbf{J}(\mathbf{r}',t) & = \mathbf{J}_0(\mathbf{r}')\,e^{-i\omega t} \end{array} \right\}. \tag{1}$$

Use (6) of Question 11.1 to show that the electric dipole (ed), magnetic dipole (md) and electric quadrupole (eq) contributions to the vector potential are

$$\left. \begin{array}{ll} \mathbf{A}^{\text{ed}} & = -i\omega \dfrac{\mu_0}{4\pi}\dfrac{[\mathbf{p}]}{r} = \dfrac{\mu_0}{4\pi}\dfrac{[\dot{\mathbf{p}}]}{r} \\[4mm] \mathbf{A}^{\text{md}} & = \dfrac{\mu_0}{4\pi}(1-ikr)\dfrac{[\mathbf{m}]\times\mathbf{r}}{r^3} \\[4mm] A_i^{\text{eq}} & = \dfrac{\mu_0}{4\pi}(1-ikr)\dfrac{r_j[\dot{q}_{ij}]}{2r^3} \end{array} \right\} , \tag{2}$$

where $k = 2\pi \div$ the wavelength λ of the radiation $= \dfrac{\omega}{c}$.

Solution

We begin by recognizing that time-harmonic charge and current densities result in multipole moments which are also time-harmonic. So

$$\left. \begin{array}{ll} \mathbf{p}(t) & = \displaystyle\int_v \mathbf{r}'\rho(\mathbf{r}',t)\,dv' = e^{-i\omega t}\displaystyle\int_v \mathbf{r}'\rho_0(\mathbf{r}')\,dv' = \mathbf{p}_0\,e^{-i\omega t} \\[5mm] \mathbf{m}(t) & = \tfrac{1}{2}\displaystyle\int_v \mathbf{r}'\times\mathbf{J}(\mathbf{r}',t)\,dv' = \tfrac{1}{2}e^{-i\omega t}\displaystyle\int_v \mathbf{r}'\times\mathbf{J}_0(\mathbf{r}')\,dv' = \mathbf{m}_0\,e^{-i\omega t} \\[5mm] q_{ij}(t) & = \displaystyle\int_v r_i'r_j'\rho(\mathbf{r}',t)\,dv' = e^{-i\omega t}\displaystyle\int_v r_i'r_j'\rho_0(\mathbf{r}')\,dv' = q_{0ij}\,e^{-i\omega t} \end{array} \right\} , \tag{3}$$

where $\mathbf{p}_0 = \displaystyle\int_v \mathbf{r}'\rho_0(\mathbf{r}')\,dv'$, $\mathbf{m}_0 = \tfrac{1}{2}\displaystyle\int_v \mathbf{r}'\times\mathbf{J}_0(\mathbf{r}')\,dv'$ and $q_{0ij} = \displaystyle\int_v r_i'r_j'\rho_0(\mathbf{r}')\,dv'$.

Next we calculate time derivatives of the multipole moments. Consider for example the electric dipole moment. From $\mathbf{p} = \mathbf{p}_0\,e^{-i\omega t}$ we obtain

$$\dot{\mathbf{p}} = \frac{d\mathbf{p}}{dt'} = \frac{d\mathbf{p}}{dt}\bigg/\frac{dt'}{dt} = \frac{d\mathbf{p}}{dt} = -i\omega\mathbf{p}, \tag{4}$$

where in the penultimate step we use $dt'/dt = 1$ (see (3) of Question 11.1). It then follows that $\dot{\mathbf{p}}(t') = -i\omega\mathbf{p}(t') = -i\omega[\mathbf{p}]$ or $\dot{p}_i(t') = -i\omega[p_i]$. Similarly, $\dot{m}_i(t') = -i\omega[m_i]$, $\dot{q}_{ij}(t') = -i\omega[q_{ij}]$ and $\ddot{q}_{ij}(t') = -i\omega[\dot{q}_{ij}]$. Substituting these results in (6) of Question 11.1 and putting $1/c = k/\omega$ give

$$
A_i(\mathbf{r},t) = \frac{\mu_0}{4\pi}\left\{ \frac{-i\omega[p_i]}{r} + \frac{r_j}{2r^3}(1-ikr)[\dot{q}_{ij}] - \varepsilon_{ijk}\frac{r_j}{r^3}(1-ikr)[m_k] + \cdots \right\}
$$

$$
= A_i^{ed} + A_i^{eq} + A_i^{md} + \cdots .
$$

Hence (2).

Comments

(i) The contributions to the vector potential, expressed in the form of (2), turn out to be very useful. For example, these equations can be used to calculate the **E**- and **B**- fields due to electric dipole, magnetic dipole and electric quadrupole radiation. See Questions 11.3, 11.4 and 11.9.

(ii) The dimensionless quantity kr, appearing in (2), delineates three distinct regions of space. These are:

   ☞ $kr \ll 1$ (or $r \ll \lambda$), known as the near or static zone,

   ☞ $kr \sim 1$ (or $r \sim \lambda$), known as the intermediate or induction zone, and

   ☞ $kr \gg 1$ (or $r \gg \lambda$), known as the far or radiation zone.

(iii) It is evident from (1) that the time scale T, for the distribution to undergo a change in configuration, is of order ω^{-1}. So $T \sim 2\pi/\omega = \lambda/c$, where λ is a characteristic wavelength of the radiation. Now in Comment (iv) of Question 11.1 we showed that $T \gg d/c$, which implies $\lambda \gg d$. This condition and the far zone requirement can together be expressed in the form

$$
d \ll \lambda \ll r. \tag{5}
$$

We remind the reader that the condition $d \ll r$ (see (9) of Question 8.20) ensures the convergence of our multipole expansions (e.g., (I) of the introduction and (2) of Question 11.1).

(iv) In cases where the condition $d \ll \lambda$ is not satisfied, the multipole approach—based on (2)—fails. See Question 11.14 for an example of how one could determine a radiation field in such circumstances.

(v) The above analysis which treats ρ and $\mathbf{J}$ as time-harmonic quantities involves no loss of generality, because for an arbitrary time variation of the source densities an underlying Fourier analysis can usually be assumed.[1]

---

[1] J. D. Jackson, *Classical electrodynamics*, Chap. 9, p. 407. New York: Wiley, 3rd edn, 1998.

Question 11.3[*]

Consider a harmonically oscillating electric dipole moment, $\mathbf{p} = \mathbf{p}_0 e^{-i\omega t}$. Use the vector potential $(2)_1$ of Question 11.2 to show that the electromagnetic field is

$$\mathbf{E} = \frac{1}{4\pi\epsilon_0} \frac{(1 - ikr)\{3([\mathbf{p}] \cdot \hat{\mathbf{r}})\hat{\mathbf{r}} - [\mathbf{p}]\} - k^2 r^2 \{([\mathbf{p}] \times \hat{\mathbf{r}}) \times \hat{\mathbf{r}}\}}{r^3}, \tag{1}$$

and

$$\mathbf{B} = \frac{\mu_0}{4\pi} (1 - ikr) \frac{[\dot{\mathbf{p}}] \times \hat{\mathbf{r}}}{r^2}. \tag{2}$$

Solution

The magnetic field follows from $\mathbf{B} = \boldsymbol{\nabla} \times \mathbf{A}^{\text{ed}}$ or $B_i = \varepsilon_{ijk} \nabla_j A_k^{\text{ed}} = \frac{\mu_0}{4\pi} \varepsilon_{ijk} \nabla_j \{r^{-1}[\dot{p}_k]\}$. Then

$$\begin{aligned}
B_i &= \frac{\mu_0}{4\pi} \varepsilon_{ijk} \{\nabla_j r^{-1}[\dot{p}_k] + r^{-1} \nabla_j [\dot{p}_k]\} \\
&= \frac{\mu_0}{4\pi} \varepsilon_{ijk} \left\{ \frac{\partial r^{-1}}{\partial r} \frac{\partial r}{\partial r_j} [\dot{p}_k] + \frac{1}{r} \frac{d[\dot{p}_k]}{dt'} \frac{\partial t'}{\partial r_j} \right\}.
\end{aligned} \tag{3}$$

Now $\partial r / \partial r_j = r_j / r$ and $\dfrac{\partial t'}{\partial r_j} = \dfrac{\partial (t - r/c)}{\partial r_j} = -\dfrac{1}{c} \dfrac{\partial r}{\partial r_j} = -\dfrac{r_j}{rc}$. Substituting these results in (3) gives

$$B_i = -\frac{\mu_0}{4\pi} \varepsilon_{ijk} \left\{ \frac{r_j}{r^3} [\dot{p}_k] + \frac{r_j}{r^2 c} [\ddot{p}_k] \right\}. \tag{4}$$

For a harmonically oscillating dipole, $[\ddot{p}_k] = -i\omega[\dot{p}_k]$, and so (4) can be written as

$$\begin{aligned}
B_i &= -\frac{\mu_0}{4\pi} \varepsilon_{ijk} \frac{r_j}{r^3} (1 - ikr)[\dot{p}_k] \\
&= \frac{\mu_0}{4\pi} \varepsilon_{ikj} \frac{[\dot{p}_k] r_j}{r^3} (1 - ikr),
\end{aligned} \tag{5}$$

which is (2). Next we use the Maxwell equation $\boldsymbol{\nabla} \times \mathbf{B} = \dfrac{1}{c^2} \dfrac{\partial \mathbf{E}}{\partial t}$ to calculate the electric field. Since a harmonic time dependence is assumed, it follows that $\mathbf{E} = \dfrac{ic^2}{\omega} (\boldsymbol{\nabla} \times \mathbf{B})$ with $\mathbf{B}$ given by (2). Hence

$$E_i = \frac{ic^2}{\omega} \frac{\mu_0}{4\pi} \varepsilon_{ijk}\varepsilon_{k\ell m}\nabla_j\{(1 - ikr)r^{-3}[\dot{p}_\ell]\,r_m\}$$

$$= \frac{1}{4\pi\epsilon_0}\frac{i}{\omega}(\delta_{i\ell}\delta_{jm} - \delta_{im}\delta_{j\ell})\nabla_j\{(1 - ikr)r^{-3}[\dot{p}_\ell]\,r_m\}. \tag{6}$$

Performing the differentiation in (6) gives

$$\nabla_j\{(1 - ikr)r^{-3}[\dot{p}_\ell]\,r_m\}$$

$$= r^{-3}[\dot{p}_\ell]\,r_m\nabla_j(1 - ikr) + (1 - ikr)\{[\dot{p}_\ell]\,r_m\nabla_j r^{-3} + r^{-3}r_m\nabla_j[\dot{p}_\ell] + r^{-3}[\dot{p}_\ell]\nabla_j r_m\}$$

$$= -ikr^{-4}r_j[\dot{p}_\ell]\,r_m + (1 - ikr)\{-3r^{-5}r_j[\dot{p}_\ell]\,r_m + ikr^{-4}r_j[\dot{p}_\ell]\,r_m + r^{-3}\delta_{jm}[\dot{p}_\ell]\}, \tag{7}$$

where, in the last step, we use

$$\nabla_j[\dot{p}_\ell] = \frac{\partial[\dot{p}_\ell]}{\partial r_j} = \frac{d[\dot{p}_\ell]}{dt'}\frac{\partial t'}{\partial r_j} = -\frac{r_j}{rc}[\ddot{p}_\ell]$$

$$= \frac{ir_j\omega}{rc}[\dot{p}_\ell]$$

$$= \frac{ikr_j}{r}[\dot{p}_\ell].$$

Substituting (7) in (6), contracting tensors and simplifying yield (1).

Comment

An alternative method for obtaining the field, which does not involve Cartesian tensors, is given in the solution to Question 11.4.

Question 11.4*

Consider a harmonically oscillating magnetic dipole moment, $\mathbf{m} = \mathbf{m}_0 e^{-i\omega t}$. Use the vector potential $(2)_2$ of Question 11.2 and the curl operator for spherical polar coordinates $($see $(XI)_3$ of Appendix C$)$ to show that the electromagnetic field is

$$\mathbf{E} = -\frac{1}{4\pi\epsilon_0}(1 - ikr)\frac{[\dot{\mathbf{m}}] \times \hat{\mathbf{r}}}{c^2 r^2} \tag{1}$$

and

$$\mathbf{B} = \frac{\mu_0}{4\pi}\frac{(1 - ikr)\{3([\mathbf{m}]\cdot\hat{\mathbf{r}})\hat{\mathbf{r}} - [\mathbf{m}]\} - k^2 r^2\{([\mathbf{m}] \times \hat{\mathbf{r}}) \times \hat{\mathbf{r}}\}}{r^3}. \tag{2}$$

Solution

Suppose we choose the z-axis of coordinates to coincide with the axis of the dipole. Then $\mathbf{m} = m_0 e^{-i\omega t}\hat{\mathbf{z}} = m_0 e^{-i\omega t}(\hat{\mathbf{r}}\cos\theta - \hat{\boldsymbol{\theta}}\sin\theta)$ and

$$\mathbf{A}^{md} = \frac{\mu_0}{4\pi}(1 - ikr)\frac{[\mathbf{m}]\times\hat{\mathbf{r}}}{r^2} = \frac{\mu_0}{4\pi}\frac{(1 - ikr)}{r^2}m_0 e^{i(kr-\omega t)}\sin\theta\,\hat{\boldsymbol{\phi}}, \tag{3}$$

since $[\mathbf{m}] = m_0 e^{-i\omega(t-r/c)}(\hat{\mathbf{r}}\cos\theta - \hat{\boldsymbol{\theta}}\sin\theta)$. The curl of $\mathbf{A}^{md}$ is

$$\mathbf{B} = \hat{\mathbf{r}}\frac{1}{r\sin\theta}\frac{\partial}{\partial\theta}(\sin\theta A_\phi) - \hat{\boldsymbol{\theta}}\frac{1}{r}\frac{\partial}{\partial r}(rA_\phi)$$

$$= \frac{\mu_0}{4\pi}m_0 e^{-i\omega t}\left\{\hat{\mathbf{r}}\frac{e^{ikr}(1 - ikr)}{r^3\sin\theta}\frac{\partial}{\partial\theta}\sin^2\theta - \hat{\boldsymbol{\theta}}\frac{\sin\theta}{r}\frac{\partial}{\partial r}\frac{e^{ikr}(1 - ikr)}{r}\right\}$$

$$= \frac{\mu_0}{4\pi}\frac{1}{r^3}\left\{(1 - ikr)(2\hat{\mathbf{r}}\cos\theta + \hat{\boldsymbol{\theta}}\sin\theta) - \hat{\boldsymbol{\theta}}k^2 r^2\sin\theta\right\}m_0 e^{i(kr-i\omega t)}, \tag{4}$$

which is (2) since $(2\hat{\mathbf{r}}\cos\theta + \hat{\boldsymbol{\theta}}\sin\theta) = \left(3(\hat{\mathbf{r}}\cos\theta - \hat{\boldsymbol{\theta}}\sin\theta)\cdot\hat{\mathbf{r}}\right)\hat{\mathbf{r}} - (\hat{\mathbf{r}}\cos\theta - \hat{\boldsymbol{\theta}}\sin\theta) = 3(\hat{\mathbf{z}}\cdot\hat{\mathbf{r}})\hat{\mathbf{r}} - \hat{\mathbf{z}}$ and $\hat{\boldsymbol{\theta}}\sin\theta = \left((\hat{\mathbf{r}}\cos\theta - \hat{\boldsymbol{\theta}}\sin\theta)\times\hat{\mathbf{r}}\right)\times\hat{\mathbf{r}} = (\hat{\mathbf{z}}\times\hat{\mathbf{r}})\times\hat{\mathbf{r}}$. We obtain the electric field, as in Question 11.3, from the Maxwell equation $\mathbf{E} = \dfrac{ic^2}{\omega}(\nabla\times\mathbf{B})$, with $\mathbf{B}$ given by (4). Thus

$$\mathbf{E} = \frac{1}{4\pi\epsilon_0}\frac{i}{\omega}\frac{1}{r}\left\{\frac{\partial}{\partial r}(rB_\theta) - \frac{\partial}{\partial\theta}(B_r)\right\}\hat{\boldsymbol{\phi}}$$

$$= \frac{1}{4\pi\epsilon_0}\frac{im_0}{\omega r}\left\{\sin\theta\frac{\partial}{\partial r}\left(\frac{1}{r^2} - \frac{ik}{r} - k^2\right)e^{i(kr-\omega t)} - \frac{2(1 - ikr)e^{i(kr-\omega t)}}{r^3}\frac{\partial}{\partial\theta}(\cos\theta)\right\}\hat{\boldsymbol{\phi}}$$

$$= \frac{1}{4\pi\epsilon_0}\frac{i}{\omega r}\left\{ik\sin\theta\left(\frac{1}{r^2} - \frac{ik}{r} - k^2\right) - \sin\theta\left(\frac{2}{r^3} - \frac{ik}{r^2}\right) + \frac{2(1 - ikr)\sin\theta}{r^3}\right\}[m]\hat{\boldsymbol{\phi}}$$

$$= -\frac{1}{4\pi\epsilon_0}\frac{[\dot{m}]}{c^2 r^2}(1 - ikr)\sin\theta\,\hat{\boldsymbol{\phi}},$$

which is (1) since $\sin\theta\,\hat{\boldsymbol{\phi}} = \hat{\mathbf{z}}\times\hat{\mathbf{r}}$.

Comments

(i) The electromagnetic field can also be derived from the Cartesian tensor approach employed in Question 11.3. The reader must decide, when faced with the circumstances of a particular problem, which method is the more appropriate.

(ii) Comparing (1) and (2) above with (2) and (1) of Question 11.3 reveals that dipole fields have a dual nature:[‡]

$$\left.\begin{array}{c} \mathbf{E}^{md} = -\dfrac{m_0}{p_0}\mathbf{B}^{ed} \\[2mm] \text{and} \\[2mm] \mathbf{B}^{md} = \dfrac{m_0}{p_0 c^2}\mathbf{E}^{ed} \end{array}\right\}. \qquad (5)$$

Question 11.5*

Use (1) and (2) of Questions 11.3 and 11.4 to show that the electromagnetic field in the near zone of a harmonically oscillating dipole is:

☞ electric dipole

$$\mathbf{E} = \frac{1}{4\pi\epsilon_0}\frac{\left(3\mathbf{p}(t)\cdot\hat{\mathbf{r}}\right)\hat{\mathbf{r}} - \mathbf{p}(t)}{r^3} \qquad \text{and} \qquad \mathbf{B} = \frac{\mu_0}{4\pi}\frac{\dot{\mathbf{p}}(t)\times\hat{\mathbf{r}}}{r^2}, \qquad (1)$$

☞ magnetic dipole

$$\mathbf{E} = -\frac{1}{4\pi\epsilon_0}\frac{\dot{\mathbf{m}}(t)\times\hat{\mathbf{r}}}{c^2 r^2} \qquad \text{and} \qquad \mathbf{B} = \frac{\mu_0}{4\pi}\frac{\left(3\mathbf{m}(t)\cdot\hat{\mathbf{r}}\right)\hat{\mathbf{r}} - \mathbf{m}(t)}{r^3}. \qquad (2)$$

The dipole moments in (1) and (2) are unretarded: their explicit dependence on the time t is indicated.

Solution

In the near zone $kr \ll 1$ and so $[\mathbf{p}] = \mathbf{p}(t - r/c) = \mathbf{p}(t - kr/\omega) \simeq \mathbf{p}(t)$. Similarly, $[\mathbf{m}] \simeq \mathbf{m}(t)$. Then (1) and (2) follow immediately from (1) and (2) of Questions 11.3 and 11.4 if we again apply the approximation $kr \ll 1$.

Comment

Because the effects of time retardation are negligible in the near zone, the dipole moments in (1) and (2) are evaluated at time t—this being the unretarded time. We note that $(1)_1$ and $(2)_2$ have the same form as the static dipole fields[♯]; these fields propagate instantaneously from the location of the dipole to the observation point at P. We note also that $(1)_2$ and $(2)_1$ are reminiscent of the Biot–Savart law for a current element $I\,d\boldsymbol{\ell} = \dot{\mathbf{p}}$ or $I\,d\boldsymbol{\ell} = \dot{\mathbf{m}}/c$: hence the alternative name, 'static zone'.

[‡]See also Comment (v) of Question 7.2.

[♯]See (2) of Questions 2.11 and 4.10.

Question 11.6

Use equations (1) and (2) of Questions 11.3 and 11.4 to show that the electromagnetic field in the far zone of a harmonically oscillating dipole is:

☞ electric dipole

$$\mathbf{E} = \frac{1}{4\pi\epsilon_0} \frac{\left(\left[\ddot{\mathbf{p}}\right] \times \hat{\mathbf{r}}\right) \times \hat{\mathbf{r}}}{c^2 r} \qquad \text{and} \qquad \mathbf{B} = \frac{\mu_0}{4\pi} \frac{\left[\ddot{\mathbf{p}}\right] \times \hat{\mathbf{r}}}{cr}, \tag{1}$$

☞ magnetic dipole

$$\mathbf{E} = -\frac{1}{4\pi\epsilon_0} \frac{\left[\ddot{\mathbf{m}}\right] \times \hat{\mathbf{r}}}{c^3 r} \qquad \text{and} \qquad \mathbf{B} = \frac{\mu_0}{4\pi} \frac{\left(\left[\ddot{\mathbf{m}}\right] \times \hat{\mathbf{r}}\right) \times \hat{\mathbf{r}}}{c^2 r}. \tag{2}$$

Note that the dipole moments in (1) and (2) must be evaluated at the retarded time.

Solution

First make the approximation $kr \gg 1$ in (1) and (2) of Question 11.3. Put $k = \omega/c$. The results then follow directly if $-i\omega[\dot{\mathbf{p}}]$ and $-\omega^2[\mathbf{p}]$ are replaced with $[\ddot{\mathbf{p}}]$. Similarly, for the magnetic dipole of Question 11.4 we use $\mathbf{m}$ instead of $\mathbf{p}$.

Comments

(i) Equations (1) and (2) reveal the following features of the field in the far zone:

☞ Radiation fields vary inversely with distance, unlike static fields which vary inversely as distance squared.

☞ $\mathbf{E}$ and $\mathbf{B}$ are mutually orthogonal $\left(\mathbf{E} = c\mathbf{B} \times \hat{\mathbf{r}}\right)$ and the field propagates radially $\left(\mathbf{k} = k\hat{\mathbf{r}}\right)$ as a plane wave with phase velocity $\omega/k = c$.

☞ With the dipole aligned along the z-axis, the field components have a $\sin\theta$ dependence $\left(\text{see (1) of Question 11.8 with } \theta \text{ being the usual spherical polar coordinate}\right)$. Both $\mathbf{E}$ and $\mathbf{B}$ are zero for $\theta = 0$ and assume maximum values in the equatorial plane where $\theta = 90°$.

(ii) The vector potential in the far zone of a point magnetic dipole, $\mathbf{A} = -\dfrac{\mu_0}{4\pi} \dfrac{i\omega[\mathbf{m}] \times \hat{\mathbf{r}}}{cr}$, is obtained from $(2)_2$ of Question 11.2 by taking the limit $kr \to \infty$. Then both dipole potentials in the radiation zone can be expressed as

$$\mathbf{A}^{\text{ed}} = \frac{\mu_0}{4\pi} \frac{[\dot{\mathbf{p}}]}{r} \qquad \text{and} \qquad \mathbf{A}^{\text{md}} = \frac{\mu_0}{4\pi} \frac{[\dot{\mathbf{m}}] \times \hat{\mathbf{r}}}{cr}. \tag{3}$$

It is easily shown that (1) and (2), with $\mathbf{A}$ given by (3), can both be expressed in the alternative (and sometimes useful) forms

$$\mathbf{E} = \left(\left[\frac{d\mathbf{A}}{dt} \right] \times \hat{\mathbf{r}} \right) \times \hat{\mathbf{r}} \quad \text{and} \quad \mathbf{B} = \frac{1}{c} \left[\frac{d\mathbf{A}}{dt} \right] \times \hat{\mathbf{r}}. \tag{4}$$

Question 11.7*

Consider a system of N interacting charged particles moving arbitrarily. Suppose all particles have the same charge-to-mass ratio: $q_1/m_1 = q_2/m_2 = \cdots = q_N/m_N$. Establish the conditions which will guarantee that there is no electric or magnetic dipole radiation.

Solution

It is evident from (1) and (2) of Question 11.6 that there can be no dipole radiation if $\ddot{\mathbf{p}}$ and $\ddot{\mathbf{m}}$ are both zero. We begin by calculating $\mathbf{p}$ and $\mathbf{m}$ relative to an arbitrary origin O.

☞ electric dipole radiation

Suppose $\mathbf{r}_1, \mathbf{r}_2, \ldots$ are the position vectors of $q_1, q_2, \ldots$ relative to O. The electric dipole moment about O (see (8) of Question 2.20) is

$$\mathbf{p} = \sum_{i=1}^{N} q_i \mathbf{r}_i = \sum_{i=1}^{N} \frac{q_i}{m_i} m_i \mathbf{r}_i$$

$$= \frac{q}{m} \sum_{i=1}^{N} m_i \mathbf{r}_i,$$

where q/m is the common charge-to-mass ratio. By definition, the position vector of the centre-of-mass is $\mathbf{r}_{\mathrm{CM}} = \dfrac{\sum_{i=1}^{N} m_i \mathbf{r}_i}{\sum_{i=1}^{N} m_i}$, and so $\mathbf{p} = \dfrac{q}{m}(m_1 + m_2 + \cdots + m_N)\mathbf{r}_{\mathrm{CM}}$. Thus

$$\ddot{\mathbf{p}} = \frac{q}{m}(m_1 + m_2 + \cdots + m_N)\ddot{\mathbf{r}}_{\mathrm{CM}}$$

$$= \frac{q}{m}\mathbf{F}^{(\mathrm{e})}, \tag{1}$$

where $\mathbf{F}^{(\mathrm{e})} = (m_1 + m_2 + \cdots + m_N)\ddot{\mathbf{r}}_{\mathrm{CM}}$ is the net external force acting on the system. Now it is clear from (1) that $\mathbf{F}^{(\mathrm{e})} = 0 \Rightarrow \ddot{\mathbf{p}} = 0$, and so there can be *no electric dipole radiation if the net external force acting on the system is zero.*

☞ magnetic dipole radiation

Calculating the magnetic dipole moment about O (see (1) of Question 4.30) gives

$$\mathbf{m} = \frac{1}{2}\sum_{i=1}^{N} q_i \mathbf{r}_i \times \mathbf{u}_i = \frac{1}{2}\sum_{i=1}^{N} \frac{q_i}{m_i}(\mathbf{r}_i \times m_i \mathbf{u}_i), \tag{2}$$

where $\mathbf{u}_i$ is the velocity of the ith particle. The quantity in brackets in (2) is the angular momentum of the ith charge, and so

$$\mathbf{m} = \frac{1}{2}\frac{q}{m}\sum_{i=1}^{N} \mathbf{L}_i = \frac{1}{2}\frac{q}{m}\mathbf{L}_{\text{net}}, \tag{3}$$

where q/m is the common charge-to-mass ratio and $\mathbf{L}_{\text{net}} = \sum_{i=1}^{N} \mathbf{L}_i$ is the total angular momentum of the system. Differentiating (3) with respect to time gives

$$\dot{\mathbf{m}} = \frac{1}{2}\frac{q}{m}\dot{\mathbf{L}}_{\text{net}}$$

$$= \frac{1}{2}\frac{q}{m}\mathbf{\Gamma}^{(\text{e})}, \tag{4}$$

where $\mathbf{\Gamma}^{(\text{e})}$ is the net external torque acting on the system. It follows from (4) that $\mathbf{\Gamma}^{(\text{e})} = 0 \Rightarrow \dot{\mathbf{m}} = 0$, and so there can be *no magnetic dipole radiation if the net external torque acting on the system is zero.*

Question 11.8

A point charge q performs one-dimensional simple harmonic motion about an origin O with angular frequency ω.

(a) Show that the fields in the far zone due to electric dipole radiation are

$$\mathbf{E} = -\frac{1}{4\pi\epsilon_o}\frac{qa_0}{rc^2}\sin\theta\, e^{i(kr-\omega t)}\hat{\boldsymbol{\theta}} \quad \text{and} \quad \mathbf{B} = -\frac{\mu_0}{4\pi}\frac{qa_0}{rc}\sin\theta\, e^{i(kr-\omega t)}\hat{\boldsymbol{\phi}}, \tag{1}$$

where a_0 is the acceleration amplitude and (r, θ, ϕ) are spherical polar coordinates.

(b) Use (1) to show that the instantaneous Poynting vector $\mathbf{S}$ in the far zone is

$$\mathbf{S} = \frac{\mu_0}{16\pi^2}\frac{q^2|\mathbf{a}(t-r/c)|^2\sin^2\theta}{cr^2}\hat{\mathbf{r}}, \tag{2}$$

where $\mathbf{a}(t-r/c) = \mathbf{a}_0 e^{-i\omega(t-r/c)}$ is the retarded acceleration of the charge.

(c) Hence show that the power radiated by the charge is

$$P = \frac{\mu_0}{6\pi c} q^2 |\mathbf{a}(t - r/c)|^2. \tag{3}$$

(d) Calculate also the time-average power $\langle P \rangle$.

Solution

(a) Suppose the charge oscillates about the origin along the z-axis of Cartesian coordinates. The position vector of q is $\mathbf{z} = \mathbf{z}_0 e^{-i\omega t}$ and its dipole moment about O is

$$\mathbf{p} = q\mathbf{z} = \mathbf{p}_0 e^{-i\omega t}, \tag{4}$$

where $\mathbf{p}_0 = q\mathbf{z}_0$. For the purpose of calculating $[\ddot{\mathbf{p}}]$, we adopt the approach of Question 11.2. This gives $[\ddot{\mathbf{p}}] = -\omega^2[\mathbf{p}] = -\omega^2\mathbf{p}_0 e^{-i\omega[t]} = -\omega^2\mathbf{p}_0 e^{-i\omega(t-r/c)}$. Now $\omega/c = k$, and so

$$[\ddot{\mathbf{p}}] = -\omega^2\mathbf{p}_0 e^{i(kr-\omega t)}, \tag{5}$$

where $\omega^2\mathbf{p}_0 = \omega^2 q\mathbf{z}_0 = \omega^2 q z_0 \hat{\mathbf{z}}$. The acceleration amplitude in simple harmonic motion is $a_0 = \omega^2 z_0$. Thus $\omega^2\mathbf{p}_0 = q a_0 \hat{\mathbf{z}}$, and

$$[\ddot{\mathbf{p}}] = -q a_0 \hat{\mathbf{z}} e^{i(kr-\omega t)} = -q a_0 (\hat{\mathbf{r}} \cos\theta - \hat{\boldsymbol{\theta}} \sin\theta) e^{i(kr-\omega t)}. \tag{6}$$

Substituting (6) in $\mathbf{E} = \dfrac{1}{4\pi\epsilon_0} \dfrac{([\ddot{\mathbf{p}}] \times \hat{\mathbf{r}}) \times \hat{\mathbf{r}}}{c^2 r}$ gives $(1)_1$ since $(\hat{\boldsymbol{\theta}} \times \hat{\mathbf{r}}) \times \hat{\mathbf{r}} = -\hat{\boldsymbol{\theta}}$. In the radiation zone $\mathbf{B} = (\hat{\mathbf{r}} \times \mathbf{E})/c$, and hence $(1)_2$.

(b) By definition $\mathbf{S} = \mu_0^{-1}\mathbf{E} \times \mathbf{B}$ where, in the radiation zone, $E = cB$ and the energy flow is radial. So $\mathbf{S} = (c/\mu_0)|\mathbf{B}|^2\hat{\mathbf{r}}$ with $|\mathbf{B}| = \dfrac{\mu_0}{4\pi} \dfrac{|\ddot{\mathbf{p}}(t - r/c)| \sin\theta}{cr}$ given by $(1)_2$ of Question 11.6. Hence

$$\mathbf{S} = \frac{\mu_0}{16\pi^2} \frac{|\ddot{\mathbf{p}}(t - r/c)|^2}{c^2 r^2} \sin^2\theta \, \hat{\mathbf{r}}. \tag{7}$$

Now we see from (6) that

$$|\ddot{\mathbf{p}}(t - r/c)| = q a_0 e^{i(kr-\omega t)} = q a_0 e^{-i\omega(t-r/c)} = q|\mathbf{a}(t - r/c)|. \tag{8}$$

Substituting (8) in (7) yields (2).

(c) The power radiated into an element of solid angle $d\Omega$ centred on O is $d\mathcal{P} = (\mathbf{S} \cdot \hat{\mathbf{r}})r^2 d\Omega$. Integrating this over the surface of a unit sphere gives

$$\mathcal{P} = \oint_s (\mathbf{S} \cdot \hat{\mathbf{r}})\, r^2 d\Omega = \int_0^{2\pi}\int_0^{\pi} (\mathbf{S} \cdot \hat{\mathbf{r}})\, r^2 \sin\theta\, d\theta\, d\phi. \tag{9}$$

Substituting (2) in (9) yields (3).‡

(d) For time-harmonic fields $\langle \mathbf{S} \rangle = \dfrac{1}{2\mu_0}\mathrm{Re}\,(\mathbf{E} \times \mathbf{B}^*)$. Then using (1) gives

$$\langle \mathbf{S} \rangle = \frac{\mu_0}{32\pi^2}\frac{q^2 a_0^2 \sin^2\theta}{cr^2}\hat{\mathbf{r}}. \tag{10}$$

The average power radiated per unit solid angle is $\left\langle \dfrac{d\mathcal{P}}{d\Omega} \right\rangle = (\langle \mathbf{S} \rangle \cdot \hat{\mathbf{r}})r^2$, and so

$$\langle \mathcal{P} \rangle = \int (\langle \mathbf{S} \rangle \cdot \hat{\mathbf{r}})d\Omega = \int_0^{2\pi}\int_0^{\pi}(\langle \mathbf{S} \rangle \cdot \hat{\mathbf{r}})\, r^2 \sin\theta\, d\theta\, d\phi. \tag{11}$$

Substituting (10) in (11) and integrating yield

$$\langle \mathcal{P} \rangle = \frac{\mu_0}{12\pi c}q^2 a_0^2, \quad \text{or} \quad \langle \mathcal{P} \rangle = \frac{q^2 a_0^2}{12\pi\epsilon_0 c^3}. \tag{12}$$

Comments

(i) Equations (2) and (3) reveal a well-known fact that only charges experiencing an acceleration can produce electromagnetic radiation. Although derived here for a harmonically oscillating electric dipole, these are general results applying to any charge moving non-relativistically. Equation (3) is Larmor's radiation formula and it is sometimes written in the equivalent form

$$\mathcal{P} = \frac{\mu_0 q^2}{6\pi c}|\dot{\mathbf{u}}(t - r/c)|^2, \tag{13}$$

where $\mathbf{u}$ is the velocity of the charge, with $u \ll c$. The generalization of Larmor's formula to include relativistic motion will be discussed in Question 12.25, and is given by

$$\mathcal{P} = \frac{\mu_0 q^2}{6\pi c}\gamma^6(u)\left[a^2 - |\boldsymbol{\beta} \times \mathbf{a}|^2\right], \tag{14}$$

where $\gamma(u) = (1 - u^2/c^2)^{-1/2}$.

‡Let $u = \cos\theta$. Then $\displaystyle\int_0^{\pi}\sin^3\theta\, d\theta = \int_{-1}^{1}(1 - u^2)u\, du = \tfrac{4}{3}$.

(ii) The time-average quantities (10) and (12) do not depend on time, unlike the instantaneous quantities (2) and (3) which obviously do.

(iii) It is evident from (2) and (10) that the angular distribution of the radiated power is proportional to $\sin^2\theta$. It is zero along the direction of oscillation ($\theta = 0$ or $180°$) and a maximum in the equatorial plane ($\theta = \pm 90°$). This is illustrated in the polar diagram below.

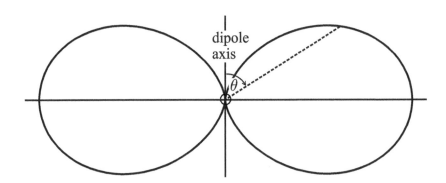

(iv) For time-harmonic fields in the radiation zone ($kr \gg 1$), $\mathbf{B} = \nabla \times \mathbf{A} = i\mathbf{k} \times \mathbf{A}$ then $\mathbf{E} = c\mathbf{B} \times \hat{\mathbf{r}}$, or

$$\left.\begin{aligned}
\mathbf{B} &= -i\frac{\omega}{c}\mathbf{A} \times \hat{\mathbf{r}} \\
\text{and} & \\
\mathbf{E} &= -i\omega\left(\mathbf{A} \times \hat{\mathbf{r}}\right) \times \hat{\mathbf{r}}
\end{aligned}\right\}. \tag{15}$$

In (15), $\mathbf{E}$ and $\mathbf{B}$ are expressed in terms of the vector potential (this is a form which is sometimes useful). Note that (15) agrees with (4) of Question 11.6 for dipole radiation.

(v) We use (15) to express the time-average Poynting vector $\langle \mathbf{S} \rangle = \dfrac{1}{2\mu_0}\mathrm{Re}\left(\mathbf{E} \times \mathbf{B}^*\right)$ for the far zone in terms of the vector potential as follows:

$$\langle \mathbf{S} \rangle = \frac{\omega^2}{2\mu_0 c}\mathrm{Re}\left\{\left((\mathbf{A} \times \hat{\mathbf{r}}) \times \hat{\mathbf{r}}\right) \times (\mathbf{A}^* \times \hat{\mathbf{r}})\right\}$$

$$= \frac{\omega^2}{2\mu_0 c}\left\{\mathbf{A} \cdot \mathbf{A}^* - (\hat{\mathbf{r}} \cdot \mathbf{A})(\hat{\mathbf{r}} \cdot \mathbf{A}^*)\right\}\hat{\mathbf{r}}, \tag{16}$$

where, in arriving at the last step, we make repeated use of the **BAC–CAB** rule.

(Note: it is easily verified that (4) and (16) above, together with $(2)_1$ of Question 11.2, yield (10).)

Question 11.9[*]

Calculate the average power of electric quadrupole radiation $\langle \mathcal{P} \rangle_{eq}$ for the oscillating point charge q described in Question 11.8.

Solution

The only non-zero component of the quadrupole moment tensor q_{ij} is $q_{zz} = qz^2 = qz_0^2 e^{-2i\omega t}$, and $[\dot{q}_{zz}] = -2i\omega q z_0^2 e^{-2i\omega t}$. Clearly only the z-component of $\mathbf{A}$ in $(2)_3$ of Question 11.2 survives, and

$$A_z = \frac{\mu_0}{4\pi}(1 - ikr)\frac{z[\dot{q}_{zz}]}{2r^3} = -\frac{\mu_0}{4\pi}(i + kr)\frac{\omega q z_0^2 z}{r^3}e^{2i(kr - \omega t)}. \tag{1}$$

Now $\mathbf{B} = \nabla \times \mathbf{A} = \hat{\mathbf{x}}\frac{\partial A_z}{\partial y} - \hat{\mathbf{y}}\frac{\partial A_z}{\partial x}$ gives

$$\mathbf{B} = -\frac{\mu_0}{4\pi}\frac{\omega q z_0^2}{r^5}\left\{2ik^2r^2 - (3i + 4kr)\right\}e^{2i(kr - \omega t)}z(y\hat{\mathbf{x}} - x\hat{\mathbf{y}})$$

$$= \frac{\mu_0}{4\pi}\frac{\omega q z_0^2}{2r^3}\left\{2ik^2r^2 - (3i + 4kr)\right\}\sin 2\theta\, e^{2i(kr - \omega t)}\hat{\boldsymbol{\phi}}, \tag{2}$$

since $z = r\cos\theta$ and $y\hat{\mathbf{x}} - x\hat{\mathbf{y}} = -r\sin\theta\hat{\boldsymbol{\phi}}$. In the far zone $kr \gg 1$, and (2) becomes

$$\mathbf{B} = \frac{\mu_0}{4\pi}\frac{i\omega^3 q z_0^2}{c^2 r}\sin 2\theta\, e^{2i(kr - \omega t)}\hat{\boldsymbol{\phi}}, \tag{3}$$

with

$$\mathbf{E} = c\mathbf{B} \times \hat{\mathbf{r}} = \frac{1}{4\pi\epsilon_0}\frac{i\omega^3 q z_0^2}{c^3 r}\sin 2\theta\, e^{2i(kr - \omega t)}\hat{\boldsymbol{\theta}}. \tag{4}$$

Proceeding as before, we obtain

$$\langle \mathbf{S} \rangle = \frac{\mu_0}{32\pi^2}\frac{\omega^6 q^2 z_0^4}{c^3 r^2}\sin^2 2\theta\, \hat{\mathbf{r}}. \tag{5}$$

Substituting (5) in (10) of Question 11.8 yields

$$\langle \mathcal{P} \rangle_{eq} = \frac{\mu_0}{4\pi c^3}\omega^6 q^2 z_0^4 \int_0^\pi \sin^3\theta \cos^2\theta\, d\theta = \frac{\mu_0}{15\pi c^3}\omega^6 q^2 z_0^4, \tag{6}$$

since the definite integral in (6) has the value $\frac{4}{15}$.[‡]

[‡]Let $u = \cos\theta$. Then $\int_0^\pi \sin^3\theta \cos^2\theta\, d\theta = \int_{-1}^1 (1 - u^2)u^2\, du = \frac{4}{15}$.

Comments

(i) It is instructive to compare the electric dipole power with the electric quadrupole power radiated by this charge. Using $(12)_1$ of Question 11.8 and (6) to calculate the ratio of average powers gives:

$$\frac{\langle P\rangle_{eq}}{\langle P\rangle_{ed}} = \frac{16\pi^2}{5}\left(\frac{z_0}{\lambda}\right)^2. \tag{7}$$

Now $z_0/\lambda \ll 1$ (see (5) of Question 11.2 with $z_0 = d$) and we see from (7) that electric dipole radiation is dominant. The quadrupole contribution (and therefore all higher electric multipoles) is completely negligible.

(ii) The time-average Poynting vector of an oscillating electric quadrupole,[‡] also has a $\sin^2 2\theta$ dependence (as in (5)), and the angular distribution of the radiated power is illustrated in the polar diagram below.

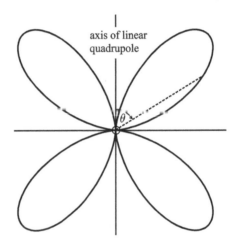

Question 11.10*

A particle having charge q acted upon by a linear restoring force causes a two-dimensional simple-harmonic motion: the particle moves in the xy-plane about the origin O with angular frequency w. Suppose the amplitudes of the motion are x_0 and y_0 and that the particle is at $(x_0,0)$ at time $t = 0$. Calculate the **E**- and **B**-fields in the far zone, and hence show that the average power due to electric dipole radiation is

$$\langle P\rangle_{ed} = \frac{\mu_0}{12\pi}\frac{q^2 w^4}{c}\left(x_0^2 + y_0^2\right). \tag{1}$$

[‡] Consider, for example, the following distribution which has a leading electric quadrupole moment: three point charges $-q$, $2q$ and $-q$ located at $(0,0,-z)$, $(0,0,0)$ and $(0,0,z)$ respectively, with the negative charges oscillating in anti-phase about their equilibrium positions.

Solution

The particle has an elliptical trajectory; its position vector is $\mathbf{r}' = (x_0\cos\omega t, y_0\sin\omega t, 0)$. The electric dipole moment of q about O is $\mathbf{p}(t) = q\mathbf{r}' = q(\hat{\mathbf{x}}x_0\cos\omega t + \hat{\mathbf{y}}y_0\sin\omega t)$ with $[\ddot{\mathbf{p}}] = -\omega^2 q(\hat{\mathbf{x}}x_0\cos\omega[t] + \hat{\mathbf{y}}y_0\sin\omega[t])$. Using the results of Appendix C to introduce the spherical polar coordinates (r, θ, ϕ) gives

$$[\ddot{\mathbf{p}}] = -\omega^2 q\Big\{\hat{\mathbf{r}}\big(x_0\cos\omega[t]\cos\phi + y_0\sin\omega[t]\sin\phi\big)\sin\theta +$$
$$\hat{\boldsymbol{\theta}}\big(x_0\cos\omega[t]\cos\phi + y_0\sin\omega[t]\sin\phi\big)\cos\theta -$$
$$\hat{\boldsymbol{\phi}}\big(x_0\cos\omega[t]\sin\phi - y_0\sin\omega[t]\cos\phi\big)\Big\}, \tag{2}$$

and so

$$\big([\ddot{\mathbf{p}}]\times\hat{\mathbf{r}}\big)\times\hat{\mathbf{r}} = \omega^2 q\Big\{\hat{\boldsymbol{\theta}}\big(x_0\cos\omega[t]\cos\phi + y_0\sin\omega[t]\sin\phi\big)\cos\theta -$$
$$\hat{\boldsymbol{\phi}}\big(x_0\cos\omega[t]\sin\phi - y_0\sin\omega[t]\cos\phi\big)\Big\}. \tag{3}$$

Then (1) of Question 11.6 yields

$$\mathbf{E} = \frac{1}{4\pi\epsilon_0}\frac{q\omega^2}{rc^2}\Big\{\hat{\boldsymbol{\theta}}\big(x_0\cos\omega[t]\cos\phi + y_0\sin\omega[t]\sin\phi\big)\cos\theta -$$
$$\hat{\boldsymbol{\phi}}\big(x_0\cos\omega[t]\sin\phi - y_0\sin\omega[t]\cos\phi\big)\Big\}. \tag{4}$$

Now because $\mathbf{B} = (\hat{\mathbf{r}}\times\mathbf{E})/c$ in the far zone, we obtain

$$\mathbf{B} = \frac{\mu_0}{4\pi}\frac{q\omega^2}{rc}\Big\{\hat{\boldsymbol{\phi}}\big(x_0\cos\omega[t]\cos\phi + y_0\sin\omega[t]\sin\phi\big)\cos\theta +$$
$$\hat{\boldsymbol{\theta}}\big(x_0\cos\omega[t]\sin\phi - y_0\sin\omega[t]\cos\phi\big)\Big\}. \tag{5}$$

Using (4) and (5) to calculate the time-average Poynting vector gives

$$\langle\mathbf{S}\rangle = \frac{1}{\mu_0}\langle\mathbf{E}\times\mathbf{B}^*\rangle = \frac{\mu_0}{16\pi^2}\frac{q^2\omega^4}{cr^2}\Big\{\big\langle\cos^2\theta\big(x_0\cos\omega[t]\cos\phi + y_0\sin\omega[t]\sin\phi\big)^2 +$$
$$\big(x_0\cos\omega[t]\sin\phi - y_0\sin\omega[t]\cos\phi\big)^2\big\rangle\Big\}\hat{\mathbf{r}}. \tag{6}$$

Expanding the term in curly brackets in (6) and using $\langle\cos^2\omega[t]\rangle = \langle\sin^2\omega[t]\rangle = \frac{1}{2}$ yield

$$\langle\mathbf{S}\rangle = \frac{\mu_0}{16\pi^2}\frac{q^2\omega^4}{2cr^2}\Big\{x_0^2(1 - \cos^2\phi\,\sin^2\theta) + y_0^2(1 - \sin^2\phi\,\sin^2\theta)\Big\}\hat{\mathbf{r}}. \tag{7}$$

Substituting (7) in (11) of Question 11.8 gives

$$\langle P \rangle_{ed} = \frac{\mu_0}{32\pi^2} \frac{q^2 \omega^4}{c} \left\{ x_0^2 \int_0^{2\pi} \int_0^{\pi} (1 - \cos^2\phi \, \sin^2\theta) \sin\theta \, d\theta \, d\phi \; + \right.$$

$$\left. y_0^2 \int_0^{2\pi} \int_0^{\pi} (1 - \sin^2\phi \, \sin^2\theta) \sin\theta \, d\theta \, d\phi \right\}. \tag{8}$$

Each double integral in (8) has the value $8\pi/3$. Hence (1).

Comments

(i) For a one-dimensional oscillation (say $y_0 = 0$) the radiated power is $\langle P \rangle_{ed} = \dfrac{\mu_0}{12\pi} \dfrac{q^2 \omega^4}{c} x_0^2 = \dfrac{\mu_0}{12\pi} \dfrac{q^2}{c} a_0^2$, which is $(12)_1$ of Question 11.8.

(ii) If the particle moves in a circle of radius r_0, the dipole moment $p_0 = qr_0$ is constant in magnitude. In this case, the radiation arises because the dipole is continuously changing direction. The average power radiated is $\langle P \rangle_{ed} = \dfrac{\mu_0}{6\pi} \dfrac{q^2 \omega^4}{c} r_0^2 = \dfrac{\mu_0}{6\pi} \dfrac{q^2}{c} a_0^2$, which is double the average power of an electric dipole oscillating in one dimension. This is not a surprising result because a rotating dipole of constant magnitude can be treated as a linear superposition of two orthogonal dipoles $p_0 e^{-i\omega t} \hat{\mathbf{x}}$ and $p_0 e^{-i\omega t} \hat{\mathbf{y}}$ oscillating $90°$ out of phase with each other.

Question 11.11*

Suppose the charged particle of Question 11.10 moves in a circle of radius r_0. Calculate (a) the average power $\langle P \rangle_{md}$ due to magnetic dipole radiation, and (b) the average power $\langle P \rangle_{eq}$ due to electric quadrupole radiation.

Solution

(a) Since $\langle P \rangle$ is proportional to the product of E and B, we have

$$\frac{\langle P \rangle_{md}}{\langle P \rangle_{ed}} = \frac{E_{md} B_{md}}{E_{ed} B_{ed}} = \left(\frac{E_{md}}{B_{ed}} \right) \left(\frac{B_{md}}{E_{ed}} \right) = \left(\frac{m_0}{p_0 c} \right)^2, \tag{1}$$

using the duality transformation (5) of Question 11.4. Clearly

$$p_0 = qr_0, \tag{2}$$

and m_0 is the magnitude of the magnetic dipole moment which we now calculate: let $m_0 = ia$ where $i = q/T = q\omega/2\pi$ and $\mathbf{a} = \pi r_0^2 \hat{\mathbf{z}}$ = the vector area of the circular orbit. Thus

$$m_0 = \tfrac{1}{2}q\omega r_0^2 . \tag{3}$$

Then (1) of Question 11.10 and (1)–(3) give

$$\langle \mathcal{P} \rangle_{\text{md}} = \frac{\mu_0}{24\pi} \frac{q^2\omega^6 r_0^4}{c^3} . \tag{4}$$

(b) All components of q_{ij} are zero except $q_{xx} = qr_0^2\cos^2\omega t$, $q_{yy} = qr_0^2\sin^2\omega t$ and $q_{xy} = q_{yx} = qr_0^2\cos 2\omega t$. The vector potential $\big($see $(2)_3$ of Question 11.2$\big)$ is

$$\mathbf{A} = \frac{\mu_0}{4\pi}\frac{(1-ikr)}{2r^3}\Big\{(x[\dot{q}_{xx}] + y[\dot{q}_{xy}])\,\hat{\mathbf{x}} + (y[\dot{q}_{yy}] + x[\dot{q}_{yx}])\,\hat{\mathbf{y}}\Big\}$$

$$= -\frac{\mu_0}{4\pi}\frac{(1-ikr)}{r^3}q\omega r_0^2\Big\{\tfrac{1}{2}(x\hat{\mathbf{x}} - y\hat{\mathbf{y}}) + (y\hat{\mathbf{x}} + x\hat{\mathbf{y}})\Big\}\sin 2\omega[t]. \tag{5}$$

Converting the Cartesian coordinates in (5) to spherical polar coordinates using

$$x\hat{\mathbf{x}} - y\hat{\mathbf{y}} = r\sin\theta\Big\{\hat{\mathbf{r}}\sin\theta\cos 2\phi + \hat{\boldsymbol{\theta}}\cos\theta\cos 2\phi - \hat{\boldsymbol{\phi}}\sin 2\phi\Big\}$$

and

$$y\hat{\mathbf{x}} + x\hat{\mathbf{y}} = r\sin\theta\Big\{\hat{\mathbf{r}}\sin\theta\sin 2\phi + \hat{\boldsymbol{\theta}}\cos\theta\sin 2\phi + \hat{\boldsymbol{\phi}}\cos 2\phi\Big\}$$

gives

$$\mathbf{A} = -\frac{\mu_0}{4\pi}\frac{(1-ikr)}{r^3}q\omega r_0^2\Big\{\hat{\mathbf{r}}\sin^2\theta\left(\tfrac{1}{2}\cos 2\phi + \sin 2\phi\right) +$$

$$\hat{\boldsymbol{\theta}}\sin\theta\cos\theta\left(\tfrac{1}{2}\cos 2\phi + \sin 2\phi\right) +$$

$$\hat{\boldsymbol{\phi}}\sin\theta\left(\cos 2\phi - \tfrac{1}{2}\sin 2\phi\right)\Big\}\sin 2\omega[t]. \tag{6}$$

This vector potential in (16) of Question 11.8 yields

$$\langle \mathbf{S} \rangle = \frac{\omega^2}{\mu_0 c}\left(\frac{\mu_0}{4\pi}\right)^2\frac{(1-ikr)(1+ikr)}{cr^4}q^2\omega^2 r_0^4\Big\{\sin^2\theta\cos^2\theta\left(\tfrac{1}{2}\cos 2\phi + \sin 2\phi\right)^2 +$$

$$\sin^2\theta\left(\cos 2\phi - \tfrac{1}{2}\sin 2\phi\right)^2\Big\}\langle\sin^2\omega[t]\rangle\,\hat{\mathbf{r}}. \tag{7}$$

Substituting $\langle\sin^2\omega[t]\rangle = \tfrac{1}{2}$ in (7) and making the far zone approximation $kr \gg 1$ give

$$\langle \mathbf{S} \rangle = \frac{\mu_0}{128\pi^2}\frac{q^2\omega^6 r_0^4}{c^3 r^2}\Big\{5\sin^2\theta - (1 + 3\sin^2 2\phi + 2\sin 4\phi)\sin^4\theta\Big\}\,\hat{\mathbf{r}}. \tag{8}$$

It then follows from (11) of Question 11.8 that

$$\langle\mathcal{P}\rangle_{\text{eq}} = \frac{\mu_0}{128\pi^2}\frac{q^2\omega^6 r_0^4}{c^3}\left\{5\int_0^{2\pi}\int_0^{\pi}\sin^3\theta\,d\theta\,d\phi - \right.$$

$$\left. \int_0^{2\pi}\int_0^{\pi}(1+3\sin^2 2\phi + 2\sin 4\phi)\sin^5\theta\,d\theta\,d\phi\right\}. \quad (9)$$

The double integrals in (9) are easy to calculate. Their values are $8\pi/3$ and $16\pi/3$ respectively, and so

$$\langle\mathcal{P}\rangle_{\text{eq}} = \frac{\mu_0}{16\pi}\frac{q^2\omega^6 r_0^4}{c^3}. \quad (10)$$

Comment

From (4) and (10) it is evident that $\langle\mathcal{P}\rangle_{\text{eq}} = \frac{3}{2}\langle\mathcal{P}\rangle_{\text{md}}$. The magnetic dipole–electric quadrupole contributions to the total radiated power are of a similar magnitude. This is not a surprising result, because—as we have already seen—magnetic dipole–electric quadrupole radiation arise from the same order term in the multipole expansion of the vector potential. Furthermore, both of these radiation powers are very much less than $\langle\mathcal{P}\rangle_{\text{ed}}$.[‡]

Question 11.12

A circular loop of wire having diameter d/π[♯] is connected to an oscillator. Suppose that an alternating current $I = I_0 e^{-i\omega t}$, varying harmonically in time with angular frequency $\omega \ll c/d$, is present in the circuit.

(a) Show that the fields in the far zone, due to magnetic dipole radiation, are

$$\mathbf{E} = \frac{1}{4\pi\epsilon_0}\frac{I_0 d^2\omega^2}{4\pi rc^3}\sin\theta\,e^{i(kr-\omega t)}\hat{\boldsymbol\phi} \quad \text{and} \quad \mathbf{B} = -\frac{\mu_0}{4\pi}\frac{I_0 d^2\omega^2}{4\pi rc^2}\sin\theta\,e^{i(kr-\omega t)}\hat{\boldsymbol\theta}. \quad (1)$$

Hint: Instead of commencing the calculation *ab initio*, make use of the dual nature of electric and magnetic dipole fields.

(b) Hence show that the average radiated power is

$$\langle\mathcal{P}\rangle_{\text{md}} = \frac{1}{2}\left(\frac{\mu_0 d^4\omega^4}{96\pi^3 c^3}\right)I_0^2. \quad (2)$$

[‡] To see this, let $x_0 = y_0 = r_0$ in (1) of Question 11.10 and recall that $\lambda \gg r_0$ (because of (5) of Question 11.2).

[♯] We intentionally specify the loop diameter as d/π rather than d, for reasons which will become apparent in Comment (iii) of Question 11.13.

Solution

(a) Substituting (1) of Question 11.8 in the duality transformation (5) of Question 11.4 gives

$$\mathbf{E}^{\mathrm{md}} = -\frac{m_0}{p_0}\mathbf{B}^{\mathrm{ed}} \quad \text{and} \quad \mathbf{B}^{\mathrm{md}} = \frac{m_0}{p_0 c^2}\mathbf{E}^{\mathrm{ed}},$$

where $p_0 = qa_0/\omega^2$ and $m_0 = I_0 d^2/4\pi$. Hence (1).

(b) It follows from (1) and $\langle \mathbf{S} \rangle = \dfrac{1}{2\mu_0}\mathrm{Re}\,(\mathbf{E} \times \mathbf{B}^*)$ that

$$\langle \mathbf{S} \rangle = \frac{\mu_0}{512\pi^4}\frac{I_0^2 d^4 \omega^4 \sin^2\theta}{c^3 r^2}\,\hat{\mathbf{r}}. \tag{3}$$

Integrating (3) as before $\big(\text{see, for example, (d) of Question 11.8}\big)$ yields (2).

Comments

(i) Because $\omega \ll c/d$ it follows that $\lambda \gg d$. The current is spatially constant around the loop and its line-charge density is zero. Consequently, all electric multipole moments are also zero, and there is no electric multipole radiation. The magnetic dipole is thus the leading term in the multipole expansion of the vector potential.

(ii) It is useful to characterize the dipole loop by a radiation resistance R_{rad} which is the coefficient of $\frac{1}{2}I_0^2$ in (2), then

$$R_{\mathrm{rad}} = \frac{\mu_0 d^4 \omega^4}{96\pi^3 c^3}. \tag{4}$$

Substituting $\omega = 2\pi c/\lambda$ in (4) gives

$$R_{\mathrm{rad}} = \frac{\pi}{6}\sqrt{\frac{\mu_0}{\epsilon_0}}\left(\frac{d}{\lambda}\right)^4 \simeq 197\left(\frac{d}{\lambda}\right)^4 \Omega. \tag{5}$$

We noted in (i) above that $\lambda \gg d$, and so it follows from (5) that $R_{\mathrm{rad}} \ll 1\,\Omega$. The effective ohmic resistance of the loop, R_{eff}, is likely to be several orders of magnitude larger than R_{rad}. As a result, only a small fraction of the energy provided by the oscillator appears as radiation; almost all of it is dissipated as heat. Evidently, an oscillating current loop is a highly inefficient way of producing electromagnetic radiation.

(iii) The electric dipole analogue of the loop antenna is the short-dipole antenna which we discuss in Question 11.13.

Question 11.13*

Consider a short-dipole antenna[‡] which comprises two thin conducting wires each of length $\frac{1}{2}d$ and cross-sectional area a, lying along the z-axis of Cartesian coordinates on either side of the origin. Suppose an oscillator excites the antenna at its centre and produces a current,

$$I(z',t) = I_0\left(1 - \frac{2|z'|}{d}\right)e^{-i\omega t}. \tag{1}$$

(a) Show that the vector potential for electric dipole radiation is

$$\mathbf{A}_{\mathrm{ed}} = \frac{\mu_0}{4\pi}\frac{I_0 d}{2r}e^{i(kr-\omega t)}(\hat{\mathbf{r}}\cos\theta - \hat{\boldsymbol{\theta}}\sin\theta), \tag{2}$$

where r and θ are the usual spherical polar coordinates.

(b) Use (2) to determine the angular distribution of radiated power, and hence calculate the radiation resistance of the antenna.

Solution

(a) From (2)$_1$ of Question 11.2 we have $\mathbf{A}_{\mathrm{ed}} = \frac{\mu_0}{4\pi}\frac{[\dot{\mathbf{p}}]}{r}$, where $\mathbf{p}(t)$ is determined from the continuity equation $\nabla'\cdot\mathbf{J} + \frac{\partial\rho}{\partial t} = 0$ as follows. Let $\mathbf{J} = I\hat{\mathbf{z}}/a$ and $\rho = \lambda/a$.[♯]

Then $\nabla'\cdot(I\hat{\mathbf{z}}) + \frac{\partial\lambda}{\partial t} = \frac{\partial I}{\partial z'} + \frac{\partial\lambda}{\partial t} = 0$. Now from (1), $\frac{\partial I}{\partial z'} = \mp\frac{2I_0}{d}e^{-i\omega t}$ where the upper (lower) sign is for positive (negative) values of z. Thus

$$\frac{\partial\lambda}{\partial t} = \pm\frac{2I_0}{d}e^{-i\omega t}, \tag{3}$$

and so

$$\lambda(t) = \pm\frac{2iI_0}{\omega d}e^{-i\omega t}. \tag{4}$$

The definition of $\mathbf{p}$ together with (4) give

$$\mathbf{p}(t) = \int_{-\frac{d}{2}}^{\frac{d}{2}} z'\,\lambda(t)\,dz'\,\hat{\mathbf{z}} = \frac{2iI_0}{\omega d}\left\{-\int_{-\frac{d}{2}}^{0} z'\,dz' + \int_0^{\frac{d}{2}} z'\,dz'\right\}e^{-i\omega t}\,\hat{\mathbf{z}}$$

$$= \frac{iI_0 d}{2\omega}e^{-i\omega t}\,\hat{\mathbf{z}}. \tag{5}$$

[‡]That is, an antenna whose length is very much less than the wavelength of the radiation it produces. Hence the name 'short'.

[♯]Take care to distinguish between the symbols λ (line-charge density) and λ (wavelength).

Then $[\dot{\mathbf{p}}] = \frac{1}{2}I_0 d\, e^{-i\omega[t]}\,\hat{\mathbf{z}} = \frac{1}{2}I_0 d\, e^{i(kr-\omega t)}\,\hat{\mathbf{z}}$ and $\mathbf{A}_{\mathrm{ed}} = \dfrac{\mu_0}{4\pi}\dfrac{I_0 d}{2r}\,e^{i(kr-\omega t)}\,\hat{\mathbf{z}}$, which is (2) since $\hat{\mathbf{z}} = \hat{\mathbf{r}}\cos\theta - \hat{\boldsymbol{\theta}}\sin\theta$.

(b) Substituting (2) in (16) of Question 11.8 gives $\langle\mathbf{S}\rangle$, and hence

$$\left\langle\frac{d\mathcal{P}}{d\Omega}\right\rangle = \frac{\mu_0}{128\pi^2}\,\frac{I_0^2\omega^2 d^2}{c}\,\sin^2\theta. \tag{6}$$

Integrating (6) over all solid angles yields

$$\langle\mathcal{P}\rangle_{\mathrm{ed}} = \frac{\mu_0}{48\pi}\,\frac{I_0^2 d^2\omega^2}{c}, \tag{7}$$

and therefore a radiation resistance of

$$R_{\mathrm{rad}} = \frac{\mu_0}{24\pi}\frac{d^2}{c}\left(\frac{2\pi c}{\lambda}\right)^2 = \frac{\pi}{6}\sqrt{\frac{\mu_0}{\epsilon_0}}\left(\frac{d}{\lambda}\right)^2$$

$$\simeq 197\left(\frac{d}{\lambda}\right)^2\,\Omega. \tag{8}$$

Comments

(i) The current, given by (1), has the profile of a standing wave and is symmetric on the two arms of the antenna. There are nodes at $z = \pm\frac{1}{2}d$ and an anti-node at $z = 0$. In contrast, the charge density λ is spatially constant along the antenna but oscillates harmonically in time.

(ii) The radiation field of this short-dipole antenna is given by (1) of Question 11.8 with $qa_0 \to I_0\omega d$.

(iii) Since the length of the short-dipole antenna and the circumference of the loop antenna (described in Question 11.12) are equal, a direct comparison of their radiation resistances is meaningful. Then from (5) of Question 11.12 and (8) we find that

$$\frac{R_{\mathrm{short}}}{R_{\mathrm{loop}}} = \left(\frac{\lambda}{d}\right)^2 \gg 1, \tag{9}$$

since $d \ll \lambda$. Although the radiation resistance of the short-dipole antenna is several orders of magnitude larger than that of the loop antenna, (8) shows that a short-dipole antenna is still a very inefficient way of producing electromagnetic radiation.

(iv) Obviously, the radiation efficiency can be improved by relaxing the requirement $d \ll \lambda$. But then the multipole approach itself is not valid (see Comment (iv) of Question 11.2) and the retarded vector potential (see $(10)_2$ of Question 8.2) must be integrated explicitly. We consider such a case in Question 11.15.

Question 11.14

Suppose the loop antenna of Question 11.12 and the short-dipole antenna of Question 11.13 are each sealed inside their own 'black box'. Describe an experiment you might devise which would establish the nature of the electromagnetic radiation (i.e. electric dipole vs magnetic dipole) emanating from either box.

Solution

From $(1)_1$ of Questions 11.8 and 11.12 we see that the electric field is a maximum in the equatorial plane ($\theta = 90°$), with $\mathbf{E}_{ed}$ along the direction $\hat{\boldsymbol{\theta}}$ and $\mathbf{E}_{md}$ along the direction $\hat{\boldsymbol{\phi}}$. An experiment which measures the polarization of the E-field in the equatorial plane will establish that if

☞ **E** is normal to the plane the radiation is electric dipole.

☞ **E** lies in the plane the radiation is magnetic dipole.

Question 11.15 **

Consider a thin linear antenna of length d lying along the z-axis with its centre at the origin. Suppose the antenna is centre-fed by an oscillator which produces the current distribution

$$I(z',t) = \begin{cases} I_0 \sin(\frac{1}{2}kd - k|z'|)e^{-i\omega t} & |z'| \leq \frac{1}{2}d \\ 0 & |z'| > \frac{1}{2}d, \end{cases} \tag{1}$$

where $k = 2\pi/\lambda = \omega/c$.

(a) Show that in the radiation zone, the retarded vector potential

$$\mathbf{A}(\mathbf{r},t) = \frac{\mu_0}{4\pi} \int \frac{I(z', t - (|\mathbf{r} - \mathbf{z}'|)/c)}{|\mathbf{r} - \mathbf{z}'|} dz' \tag{2}$$

can be expressed as

$$\mathbf{A}(\mathbf{r},t) = \frac{\mu_0}{2\pi} \frac{e^{i(kr-\omega t)}}{kr} I_0 \left\{ \frac{\cos(\frac{1}{2}kd\cos\theta) - \cos(\frac{1}{2}kd)}{\sin^2\theta} \right\} (\hat{\mathbf{r}}\cos\theta - \hat{\boldsymbol{\theta}}\sin\theta). \tag{3}$$

(b) Hence show that the angular distribution of the radiated power is given by

$$\left\langle \frac{dP}{d\Omega} \right\rangle = \frac{1}{8\pi^2\epsilon_0} \frac{I_0^2}{c} \left\{ \frac{\cos(\frac{1}{2}kd\cos\theta) - \cos(\frac{1}{2}kd)}{\sin\theta} \right\}^2. \tag{4}$$

(c) Suppose $d = m\left(\dfrac{\lambda}{2}\right)$ where $m = 1, 2, 3, \ldots$. Calculate the radiation resistance for:

☞ $m = 1$ (a half-wave antenna), and

☞ $m = 2$ (a full-wave antenna).

Solution

(a) We use (1) to calculate the retarded current $I(z', [t]) = I_0 \sin(\frac{1}{2}kd - k|z'|) e^{-i\omega[t]}$, where $[t] = t - |\mathbf{r} - \mathbf{z}'|/c = t - r/c + \hat{\mathbf{r}} \cdot \mathbf{z}'/c + \cdots$. Applying the following approximation, $[t] \simeq t - r/c + \hat{\mathbf{r}} \cdot \mathbf{z}'/c$, which is valid in the far zone, gives $e^{-i\omega[t]} = e^{i(kr-\omega t)}e^{-ik\hat{\mathbf{r}} \cdot \mathbf{z}'} = e^{i(kr-\omega t)}e^{-ikz'\cos\theta}$, where θ is a spherical polar coordinate. Also in the radiation zone, $|\mathbf{r} - \mathbf{z}'|^{-1} \simeq r^{-1}$. So

$$\mathbf{A}(\mathbf{r}, t) = \frac{\mu_0}{4\pi} \frac{I_0 e^{i(kr-\omega t)}}{r} \int_{-\frac{d}{2}}^{\frac{d}{2}} \sin(\tfrac{1}{2}kd - k|z'|) e^{-ikz'\cos\theta} dz' \,\hat{\mathbf{z}}$$

$$= \frac{\mu_0}{4\pi} \frac{e^{i(kr-\omega t)}}{r} I_0 \left\{ \int_{-\frac{d}{2}}^{0} \sin(\tfrac{1}{2}kd + kz') e^{-ikz'\cos\theta} dz' + \int_{0}^{\frac{d}{2}} \sin(\tfrac{1}{2}kd - kz') e^{-ikz'\cos\theta} dz' \right\} \hat{\mathbf{z}}$$

$$= \frac{\mu_0}{4\pi} \frac{e^{i(kr-\omega t)}}{r} I_0 \left\{ \int_{0}^{\frac{d}{2}} \sin(\tfrac{1}{2}kd - kz') e^{ikz'\cos\theta} dz' + \int_{0}^{\frac{d}{2}} \sin(\tfrac{1}{2}kd - kz') e^{-ikz'\cos\theta} dz' \right\} \hat{\mathbf{z}}$$

$$= \frac{\mu_0}{2\pi} \frac{e^{i(kr-\omega t)}}{r} I_0 \int_{0}^{\frac{d}{2}} \sin(\tfrac{1}{2}kd - k|z'|) \cos(kz'\cos\theta) dz' \,\hat{\mathbf{z}}. \tag{5}$$

Using the trigonometric identity $2\sin\alpha\cos\beta = \sin(\alpha + \beta) + \sin(\alpha - \beta)$ to recast the integrand in (5) gives

$$\mathbf{A}(\mathbf{r}, t) = \frac{\mu_0}{2\pi} \frac{e^{i(kr-\omega t)}}{r} I_0 \left\{ \int_{0}^{\frac{d}{2}} \sin(\tfrac{1}{2}kd - kz' + kz'\cos\theta) dz' + \int_{0}^{\frac{d}{2}} \sin(\tfrac{1}{2}kd - kz' - kz'\cos\theta) dz' \right\} \hat{\mathbf{z}}. \tag{6}$$

Integrating (6) and putting $\hat{\mathbf{z}} = (\hat{\mathbf{r}}\cos\theta - \hat{\boldsymbol{\theta}}\sin\theta)$ yield (3).

(b) The time-average Poynting vector follows immediately from (16) of Question 11.8 and (3), and is

$$\langle \mathbf{S} \rangle = \frac{\mu_0 c\, I_0^2}{8\pi^2\, r^2} \left\{ \frac{\cos(\frac{1}{2}kd\cos\theta) - \cos(\frac{1}{2}kd)}{\sin\theta} \right\}^2 \hat{\mathbf{r}}. \tag{7}$$

Substituting (7) in $\langle dP/d\Omega \rangle = (\langle \mathbf{S} \rangle \cdot \hat{\mathbf{r}})r^2$ gives (4).

(c) Putting $kd = m\pi$ in (2) and integrating over all solid angles yield

$$\mathcal{P} = \begin{cases} \dfrac{\mu_0 c I_0^2}{8\pi^2} \displaystyle\int_0^{2\pi}\!\!\int_0^\pi \dfrac{\cos^2(\frac{\pi}{2}\cos\theta)}{\sin^2\theta}\,\sin\theta\,d\theta\,d\phi & \text{for } m = 1, \\[4mm] \dfrac{\mu_0 c I_0^2}{8\pi^2} \displaystyle\int_0^{2\pi}\!\!\int_0^\pi \dfrac{\cos^4(\frac{\pi}{2}\cos\theta)}{\sin^2\theta}\,\sin\theta\,d\theta\,d\phi & \text{for } m = 2. \end{cases} \tag{8}$$

The integrations over θ in (8) can be performed numerically. The results are

$$\mathcal{P} = \frac{1}{2}\left(\frac{\mu_0 c}{2\pi} \times \begin{cases} 1.21883 \ \text{ for } m = 1 \\ 3.31813 \ \text{ for } m = 2 \end{cases}\right) I_0^2. \tag{9}$$

Taking $\mu_0 c/2\pi = 4\pi \times 10^{-7} \times 3 \times 10^8/2\pi = 60\,\Omega$ gives the radiation resistances

$$R_{\text{rad}} = \frac{\mu_0 c}{2\pi} \times \begin{cases} 1.21883 \ \text{ for } m = 1 \\ 3.31813 \ \text{ for } m = 2 \end{cases}$$

$$\simeq \begin{cases} 73\,\Omega \ \text{ for a half-wave antenna} \\ 199\,\Omega \ \text{ for a full-wave antenna}. \end{cases} \tag{10}$$

Comments

(i) We note from (10) that the radiation resistance of a half- or full-wave antenna is several orders of magnitude greater than that of the short-dipole antenna of Question 11.13. Consequently, half- and full-wave antennas are much more efficient radiators of electromagnetic energy than their shorter cousins.

(ii) It is instructive to calculate the current I_0 required to produce $1\,\text{kW}$ of radiation from these antennas. Using $\sqrt{2\mathcal{P}/R_{\text{rad}}}$ for the current amplitude gives $I_0 = 5.2\,\text{A}$ for a full-wave antenna and $I_0 = 3.2\,\text{A}$ for a half-wave antenna.

(iii) The electromagnetic field in the far zone follows from (15) of Question 11.8 and (3), and is

$$\mathbf{E} = -i\omega \left(\mathbf{A} \times \hat{\mathbf{r}}\right) \times \hat{\mathbf{r}} = -\frac{i}{2\pi\epsilon_0}\frac{I_0}{cr}\left\{\frac{\cos(\frac{1}{2}kd\cos\theta) - \cos(\frac{1}{2}kd)}{\sin\theta}\right\} e^{i(kr-\omega t)}\,\hat{\boldsymbol{\theta}}, \quad (11)$$

$$\mathbf{B} = -i\frac{\omega}{c}\mathbf{A} \times \hat{\mathbf{r}} = -\frac{i\mu_0}{2\pi}\frac{I_0}{r}\left\{\frac{\cos(\frac{1}{2}kd\cos\theta) - \cos(\frac{1}{2}kd)}{\sin\theta}\right\} e^{i(kr-\omega t)}\,\hat{\boldsymbol{\phi}}. \quad (12)$$

It is clear from (11) that the radiation is polarized in the $\hat{\boldsymbol{\theta}}$ direction in the plane containing the antenna and the vector $\hat{\mathbf{r}}$.

(iv) The current distribution (1) is plotted below for both a half- and a full-wave antenna. The solid curves are for time $t = 0$ and the dotted curves are for $t = \pi/\omega$. Positive and negative values of I correspond to opposite directions of the flow.

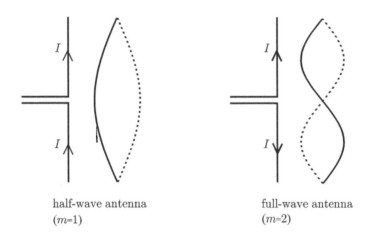

half-wave antenna   full-wave antenna
(m=1)   (m=2)

The simple form represented by equation (1) facilitates a relatively straightforward numerical solution for the radiation field; however, it is an approximation, since it neglects the effects of radiation damping.[2]

(v) The polar diagram below illustrates the angular power distributions for the short-dipole, half-wave and full-wave antennas.

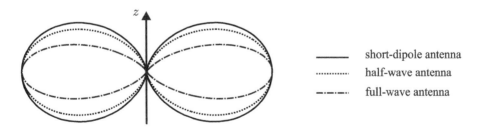

——— short-dipole antenna

············ half-wave antenna

—·—·— full-wave antenna

[2]   J. D. Jackson, *Classical electrodynamics*, Chap. 9, p. 416. New York: Wiley, 3rd edn, 1998.

Notice that $dP/d\Omega$ is similar for the short-dipole and half-wave antennas, whereas for the full-wave antenna the pattern is somewhat more directional. The directivity g (or gain[‡]) of an antenna is defined as follows:

$$g = \frac{\text{maximum power radiated per unit solid angle}}{\text{average power radiated per unit solid angle}}. \tag{13}$$

From (13) we obtain:

short-dipole antenna

$$g = \frac{48 \times 4\pi^2}{128\pi^2} = 1.5 \qquad \left(\text{use (6) and (7) of Question 11.13}\right).$$

half-wave antenna

$$g = \frac{4\pi \times 4\pi}{8\pi^2 \times 1.21883} \simeq 1.6 \qquad \left(\text{use (4) and (9)}_1 \text{ of this question}\right).$$

full-wave antenna

$$g = \frac{4\pi \times 4\pi}{2\pi^2 \times 3.31813} \simeq 2.4 \qquad \left(\text{use (4) and (9)}_2 \text{ of this question}\right).$$

(vi) Further control of the angular power distribution can be achieved using antenna arrays. See Question 11.16.

(vii) Note that (4) reduces, in the case of long wavelengths for which $kd \ll 1$, to (6) of Question 11.13. This follows directly if we replace I_0 in (6) of Question 11.13 with $(\pi d/\lambda) I_0$[♯] and make the approximation

$$\cos(\tfrac{1}{2}kd\cos\theta) - \cos(\tfrac{1}{2}kd) \simeq \left(1 - \tfrac{1}{2}(\tfrac{1}{2}kd\cos\theta)^2\right) - \left(1 - \tfrac{1}{2}(\tfrac{1}{2}kd)^2\right)$$

$$= \frac{\pi^2 d^2}{2\lambda^2}\sin^2\theta.$$

[‡]If losses are included, the directivity and gain of the antenna are not equal.

[♯]The current distributions (1) of Question 11.13 and (1) above have different amplitudes (I_0 and $I_0\sin(\pi d/\lambda)$ respectively). For a valid comparison we require their spatial averages to be equal:

$$\frac{I_{0s}}{d}\int_{-\frac{d}{2}}^{\frac{d}{2}}\left(1 - \frac{2|z'|}{d}\right)dz' = \frac{I_0}{d}\int_{-\frac{d}{2}}^{\frac{d}{2}}\sin(kd - k|z'|)\,dz',$$

where I_{0s} is the current amplitude in the short-dipole antenna. Straightforward integration yields

$$I_{0s} = \frac{4}{\pi}\frac{\lambda}{d}I_0\sin^2\left(\frac{\pi d}{2\lambda}\right).$$

In the long-wavelength limit this becomes $I_{0s} = \dfrac{\pi d}{\lambda}I_0.$

Question 11.16**

Consider two parallel antennas fed with the same signal and separated by a displacement $\mathbf{R}$ ($\ll \mathbf{r}$). Show that the vector potential $\mathbf{A}(\mathbf{r}, t)$ of the second antenna is related to that of the first antenna $\mathbf{A}_0(\mathbf{r}, t)$, located symmetrically about the origin and lying along the z-axis, by

$$\mathbf{A}(\mathbf{r}, t) = \mathbf{A}_0(\mathbf{r}, t) e^{-i\mathbf{k}\cdot\mathbf{R}}, \tag{1}$$

where

$$\mathbf{A}_0 = \frac{\mu_0}{4\pi} \frac{e^{i(kr - \omega t)}}{r} I_0 \int_{-\frac{d}{2}}^{\frac{d}{2}} \sin(\tfrac{1}{2}kd - k|z''|) e^{-ik\,\hat{\mathbf{r}}\cdot\mathbf{z}''} \, d\mathbf{z}''. \tag{2}$$

Solution

For the second antenna, (2) of Question 11.15 gives

$$\mathbf{A}(\mathbf{r}, t) = \frac{\mu_0}{4\pi} \int \frac{I(z'', [t])}{|\mathbf{r} - \mathbf{R} - \mathbf{z}''|} \, d\mathbf{z}''$$

$$= \frac{\mu_0}{4\pi} \int \frac{I_0 \sin(\tfrac{1}{2}kd - k|z''|) e^{-i\omega[t]}}{|\mathbf{r} - \mathbf{R} - \mathbf{z}''|} \, d\mathbf{z}''. \tag{3}$$

In the far zone, $|\mathbf{r} - \mathbf{R} - \mathbf{z}''| \simeq r - \hat{\mathbf{r}} \cdot (\mathbf{R} + \mathbf{z}'')$ and $[t] = t - r/c + k\hat{\mathbf{r}} \cdot (\mathbf{R} + \mathbf{z}'')$. So (3) becomes

$$\mathbf{A}(\mathbf{r}, t) = \left\{ \frac{\mu_0}{4\pi} \frac{e^{i(kr - \omega t)}}{r} I_0 \int_{-\frac{d}{2}}^{\frac{d}{2}} \sin(\tfrac{1}{2}kd - k|z''|) e^{-ik\,\hat{\mathbf{r}}\cdot\mathbf{z}''} \, d\mathbf{z}'' \right\} e^{-i\mathbf{k}\cdot\mathbf{R}}, \tag{4}$$

which proves (1).

Comment

The result (1) can be generalized. For an array of N identical antennas at positions $\mathbf{R}_j$, excited in phase, the resultant vector potential at $\mathbf{r}$ is

$$\mathbf{A}(\mathbf{r}, t) = \mathbf{A}_0(\mathbf{r}, t) \sum_{j=1}^{N} e^{-i\mathbf{k}\cdot\mathbf{R}_j}. \tag{5}$$

If the antennas are driven with different phases φ_j, then

$$\mathbf{A}(\mathbf{r}, t) = \mathbf{A}_0(\mathbf{r}, t) \sum_{j=1}^{N} e^{-i(\mathbf{k}\cdot\mathbf{R}_j + \varphi_j)}. \tag{6}$$

Question 11.17**

Two identical half-wave antennas are aligned parallel to the z-axis of Cartesian coordinates and located $\frac{1}{2}\lambda$ apart along the x-axis as shown in the figure below.

(a) Suppose both antennas are fed with the current (1) of Question 11.15 and excited in phase. Show that the angular power distribution at a point $P(r, \theta, \phi)$ is

$$\left\langle \frac{dP}{d\Omega} \right\rangle = 4\left\langle \frac{dP}{d\Omega} \right\rangle_0 \cos^2\left(\tfrac{1}{2}\pi\sin\theta\,\cos\phi\right), \tag{1}$$

where $\left\langle \dfrac{dP}{d\Omega} \right\rangle_0 = \dfrac{1}{8\pi^2\epsilon_0}\dfrac{I_0^2}{c}\dfrac{\cos^2\left(\tfrac{1}{2}\pi\cos\theta\right)}{\sin^2\theta}$ is the angular power distribution of a single antenna.

(b) Now suppose that the antennas are excited 180° out of phase. Show that

$$\left\langle \frac{dP}{d\Omega} \right\rangle = 4\left\langle \frac{dP}{d\Omega} \right\rangle_0 \cos^2\left(\tfrac{1}{2}\pi(1+\sin\theta\,\cos\phi)\right). \tag{2}$$

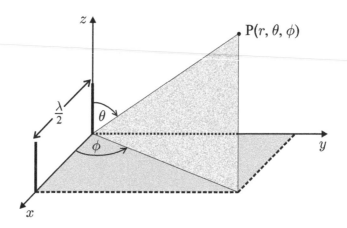

Solution

(a) Choose the coordinate origin such that the vector $\mathbf{R}_j$ (here $j = 1,2$) in (5) of Question 11.16 has the values $\mathbf{R}_1 = 0$ and $\mathbf{R}_2 = \frac{1}{2}\lambda\,\hat{\mathbf{x}}$. Then

$$\mathbf{A} = \mathbf{A}_0\sum_{j=1}^{2} e^{-ik\hat{\mathbf{r}}\cdot\mathbf{R}_j} = \mathbf{A}_0\left(1 + e^{-ik\hat{\mathbf{r}}\cdot\mathbf{R}_2}\right), \tag{3}$$

where $\mathbf{A}_0$ is given by (3) of Question 11.16. Now

$$k\hat{\mathbf{r}}\cdot\mathbf{R}_2 = \pi(\hat{\mathbf{x}}\sin\theta\,\cos\phi + \hat{\mathbf{y}}\sin\theta\,\cos\phi + \hat{\mathbf{z}}\cos\theta)\cdot\hat{\mathbf{x}} = \pi\sin\theta\,\cos\phi,$$

and so (3) is

$$\mathbf{A} = \mathbf{A}_0\left(1 + e^{-i\pi\sin\theta\,\cos\phi}\right) = \mathbf{A}_0 e^{-i\frac{\pi}{2}\sin\theta\,\cos\phi}\left(e^{i\frac{\pi}{2}\sin\theta\,\cos\phi} + e^{-i\frac{\pi}{2}\sin\theta\,\cos\phi}\right)$$

$$= 2\mathbf{A}_0\cos\left(\tfrac{\pi}{2}\sin\theta\,\cos\phi\right)e^{-i\frac{\pi}{2}\sin\theta\,\cos\phi}. \tag{4}$$

In the far zone, (16) of Question 11.8 and (4) give

$$\langle\mathbf{S}\rangle = \frac{1}{2\pi^2\epsilon_0}\frac{I_0^2}{cr^2}\frac{\cos^2\left(\tfrac{1}{2}\pi\cos\theta\right)}{\sin^2\theta}\cos^2\left(\tfrac{1}{2}\pi\sin\theta\,\cos\phi\right)\hat{\mathbf{r}}. \tag{5}$$

Substituting (5) in $\langle d\mathcal{P}/d\Omega\rangle = (\langle\mathbf{S}\rangle\cdot\hat{\mathbf{r}})r^2$ yields (1).

(b) Here the phases φ_j (with $j = 1, 2$) in (6) of Question 11.16 have the values $\phi_1 = 0$ and $\phi_2 = -\pi$, and so

$$\mathbf{A} = \mathbf{A}_0\sum_{j=1}^{2}e^{-i(k\hat{\mathbf{r}}\cdot\mathbf{R}_j + \varphi_j)}$$

$$= \mathbf{A}_0\left(1 + e^{-i(k\hat{\mathbf{r}}\cdot\mathbf{R}_2 + \pi)}\right). \tag{6}$$

Proceeding as above yields (2).

Comments

(i) It follows from L'Hôpital's rule that $\left\langle\dfrac{d\mathcal{P}}{d\Omega}\right\rangle_0 = 0$ along the z-axis ($\theta = 0$), and so (1) and (2) are also zero in this direction.

(ii) The angular power distributions given by (1) and (2) are a maximum as follows:

antennas excited in phase

$\theta = \tfrac{1}{2}\pi,\ \phi = \tfrac{1}{2}\pi$ and $\theta = \tfrac{1}{2}\pi,\ \phi = \tfrac{3}{2}\pi$. This is a 'broadside' array.[3]

antennas excited in anti-phase

$\theta = \tfrac{1}{2}\pi,\ \phi = 0$ and $\theta = \tfrac{1}{2}\pi,\ \phi = \tfrac{3}{2}\pi$. This is an 'end-fire' array.[3]

(iii) The maximum average power radiated per unit solid angle by this two-element array is four ($= 2^2$) times that radiated by either antenna acting alone. This result generalizes: for an N-element array $\left\langle\dfrac{d\mathcal{P}}{d\Omega}\right\rangle_{\text{max}} = N^2\left\langle\dfrac{d\mathcal{P}}{d\Omega}\right\rangle_{0,\,\text{max}}$.

[3]   J. B. Marion and M. A. Heald, *Classical electromagnetic radiation*, Chap. 8, pp. 266–9. New York: Academic Press, 2nd edn, 1980.

(iv) Numerical integration of (1) and (2) over all solid angles gives the total power radiated:

$$P = \frac{\mu_0 c I_0^2}{2\pi} \times \begin{cases} 1.00996 & \text{for the broadside array,} \\ 1.42769 & \text{for the end-fire array.} \end{cases} \tag{7}$$

Taking $\mu_0 c/\pi = 120\,\Omega$ yields the radiation resistances

$$R_{\text{rad}} = \frac{\mu_0 c}{\pi} \times \begin{cases} 1.00996 \\ 1.42769 \end{cases}$$

$$\simeq \begin{cases} 121\,\Omega & \text{for the antennas excited in phase,} \\ 171\,\Omega & \text{for the antennas excited in anti-phase.} \end{cases} \tag{8}$$

(v) Using the directivity of an antenna array defined by (13) of Question 11.15, we obtain

$$g = \frac{8\pi^2}{2\pi^2 \times \begin{cases} 1.00996 \\ 1.42769 \end{cases}}$$

$$\simeq \begin{cases} 4.0 & \text{for the broadside array,} \\ 2.8 & \text{for the end-fire array.} \end{cases} \tag{9}$$

Question 11.18**

Consider a half-wave antenna a_1 lying along the z-axis carrying a current I. It broadcasts a signal which induces an emf $\mathcal{E}$ in an identical antenna a_2, oriented arbitrarily relative to a_1 (see the figure below). Show that if the role of each antenna is reversed,[‡] then a current I in a_2 induces the same emf $\mathcal{E}$ in a_1.

Hint: Use (11) of Question 11.15 with $kd = \pi$ to estimate $\mathcal{E}$. Then express this emf in a symmetrical form.

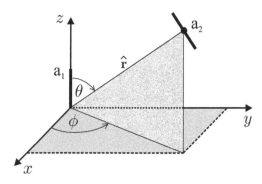

[‡] That is, a_2 broadcasts and a_1 receives.

Solution

The electric field at a_2 due to the radiation emitted by a_1 is

$$\mathbf{E} = -\frac{i}{2\pi\epsilon_0}\frac{I_0}{cr}\frac{\cos(\frac{1}{2}\pi\cos\theta)}{\sin\theta}e^{i(kr-\omega t)}\hat{\boldsymbol{\theta}} = -\frac{i}{2\pi\epsilon_0}\frac{[I]}{cr}\frac{\cos(\frac{1}{2}\pi\cos\theta)}{\sin^2\theta}(\hat{\mathbf{z}}\times\hat{\mathbf{r}})\times\hat{\mathbf{r}}, \quad (1)$$

where $[I]$ is the retarded current. It is convenient to replace $\hat{\mathbf{z}}$ in (1) by the unit vector $\hat{\mathbf{e}}_1$ along a_1, and so

$$\mathbf{E} = -\frac{i}{2\pi\epsilon_0}\frac{[I]}{cr}\frac{\cos(\frac{1}{2}\pi\cos\theta)}{\sin^2\theta}(\hat{\mathbf{e}}_1\times\hat{\mathbf{r}})\times\hat{\mathbf{r}}. \quad (2)$$

Now the tangential component of $\mathbf{E}$ is continuous[#] across the surface of a_2 and the emf induced in this antenna is of order $\mathbf{E}\cdot\mathbf{d}_2$, where $\mathbf{d}_2 = \frac{1}{2}\lambda\hat{\mathbf{e}}_2$ is the (vector) length of a_2. Thus

$$\mathcal{E} \sim \frac{-i}{2\pi\epsilon_0}\frac{[I]\lambda}{2cr}\frac{\cos(\frac{1}{2}\pi\cos\theta)}{\sin^2\theta}\big((\hat{\mathbf{e}}_1\times\hat{\mathbf{r}})\times\hat{\mathbf{r}}\big)\cdot\hat{\mathbf{e}}_2$$

$$= \frac{i}{2\epsilon_0}\frac{[I]}{\omega r}\frac{\cos(\frac{1}{2}\pi\cos\theta)}{\sin^2\theta}(\hat{\mathbf{e}}_1\times\hat{\mathbf{r}})\cdot(\hat{\mathbf{e}}_2\times\hat{\mathbf{r}}), \quad (3)$$

where, in the last step, we use the cyclic nature of the scalar triple product. Equation (3) shows that $\mathcal{E}$ is symmetrical in $\hat{\mathbf{e}}_1$ and $\hat{\mathbf{e}}_2$ and does not change if the direction of $\hat{\mathbf{r}}$ is reversed (i.e. if the roles of a_1 and a_2 are interchanged).

Comments

(i) This question provides an example of a reciprocity relation. Relationships of this sort also arise elsewhere in physics.

(ii) Ref. [4] observes that 'it is intuitively reasonable, especially in the light of the usual reversibility of physical laws, that antennas that broadcast well will also receive signals well under corresponding circumstances'.

Question 11.19[*]

Consider a particle having charge q and mass m moving non-relativistically along the x-axis towards the origin O of Cartesian coordinates. Suppose that at time $t = 0$ the position and velocity of charge q are x_0 (> 0) and $-v_0$ respectively. A second particle having charge Q—which remains at rest at the origin—exerts a repulsive force on q. In analysing the motion of charge q below, we assume Coulomb's law is valid and that the radiation reaction force is negligible.

[#]See (8) of Question 10.6.

[4]   R. H. Good, *Classical electromagnetism*, Chap. 15, p. 382. Philadelphia: Saunders College Publishing, 1999.

(a) Show, using Newton's second law, that the velocity of charge q as a function of position is given by

$$v(u) = \mp v_0\sqrt{1 + \alpha\left(1 - \frac{1}{u}\right)}, \tag{1}$$

where $u = x/x_0$ is a dimensionless position coordinate and $\alpha = \dfrac{1}{2\pi\epsilon_0}\dfrac{Qq}{mv_0^2 x_0}$ is a dimensionless constant $(0 < \alpha < \infty)$. The upper (lower) sign in (1) is for q travelling towards (away from) O.

(b) Hence deduce the distance X_0 of closest approach to the origin and the final velocity v_{final} of charge q. Also, express the acceleration a in terms of u.

(c) Integrate v and show that the position u of q at time t is given (in inverse form) by

$$\left.\begin{array}{l}
t_- = t_0\left[\dfrac{1 - \sqrt{u[(1+\alpha)u - \alpha]}}{1+\alpha} - \dfrac{\alpha}{(1+\alpha)^{3/2}} \times\right.\\[2ex]
\qquad \left.\ln\left\{\sqrt{\dfrac{-\alpha + 2(1+\alpha)u + 2\sqrt{(1+\alpha)u[(1+\alpha)u - \alpha]}}{2 + \alpha + 2\sqrt{1+\alpha}}}\right\}\right] \quad (v < 0)\\[4ex]
t_+ = t_0\left[\dfrac{1 + \sqrt{u[(1+\alpha)u - \alpha]}}{1+\alpha} + \dfrac{\alpha}{(1+\alpha)^{3/2}} \times\right.\\[2ex]
\qquad \left.\ln\left\{\dfrac{\sqrt{(2+\alpha+2\sqrt{1+\alpha})}\left(\sqrt{(1+\alpha)u} + \sqrt{(1+\alpha)u - \alpha}\right)}{\alpha}\right\}\right] \quad (v > 0)
\end{array}\right\}, \tag{2}$$

where $t_0 = x_0/v_0$. Here t_- applies for inward motion (q moves towards O) and t_+ applies for outward motion (q moves away from O).

Hint: Use *Mathematica* to perform the integrations.

(d) Take $\alpha = 0.8$ and plot graphs of x/x_0, v/v_0 and a/a_0 vs t/t_0 for $0 \le t/t_0 \le 3$.

(e) At non-relativistic speeds, the energy E_{rad} radiated by the charge q is a negligible fraction of its initial energy

$$E_0 = \tfrac{1}{2}mv_0^2 + \frac{1}{4\pi\epsilon_0}\frac{Qq}{x_0}. \tag{3}$$

Use Larmor's formula ((13) of Question 11.8) to show that

$$\frac{E_{\text{rad}}}{E_0} = \left[\frac{8(\alpha+1)^{5/2} + 15(\alpha+1)^2 - 10(\alpha+1) + 3}{45(\alpha+1)}\right]\frac{q}{Q}\left(\frac{v_0}{c}\right)^3. \tag{4}$$

Hint: Start with $\dfrac{dE_{\text{rad}}}{dt} = \dfrac{dE_{\text{rad}}}{dv}\dfrac{dv}{dt} = \dfrac{dE_{\text{rad}}}{dv}a$.

Solution

(a) Because of the assumptions, $m\dfrac{dv}{dt} = \dfrac{1}{4\pi\epsilon_o}\dfrac{Qq}{x^2}$ where $\dfrac{dv}{dt} = \dfrac{dx}{dt}\dfrac{dv}{dx} = v\dfrac{dv}{dx}$, which leads to

$$\int_{v_0}^{v} v\,dv = \frac{1}{4\pi\epsilon_o}\frac{Qq}{m}\int_{x_0}^{x}\frac{dx}{x^2},$$

or

$$\tfrac{1}{2}(v^2 - v_0^2) = \frac{1}{4\pi\epsilon_o}\frac{Qq}{m}\left[-\frac{1}{x} + \frac{1}{x_0}\right].$$

Rearranging and using the definitions of u and α give (1).

(b) closest approach

At $x = X_0$ the velocity of q is zero. It is obvious from (1) that this occurs when $1/u = 1 + 1/\alpha$, or equivalently when

$$X_0 = \frac{\alpha}{1+\alpha}x_0. \tag{5}$$

final velocity

Letting $u \to \infty$ in (1) and choosing the lower sign for motion along the positive x-axis give

$$v_{\text{final}} = v_0\sqrt{1+\alpha}. \tag{6}$$

acceleration

From the equation of motion $a = \dfrac{1}{4\pi\epsilon_o}\dfrac{Qq}{mx^2} = \dfrac{1}{2\pi\epsilon_o}\dfrac{Qq}{mv_0^2 x_0}\dfrac{v_0^2}{2x_0}\dfrac{x_0^2}{x^2} = \dfrac{\alpha v_0^2}{2x_0 u^2}$ or

$$a = \frac{\alpha\,a_0}{2u^2}, \tag{7}$$

where $a_0 = v_0^2/x_0$.

(c) From (1) and $v = x_0\dfrac{du}{dt}$ we obtain

$$\int_{t_{\text{init}}}^{t} dt = t - t_{\text{init}} = \mp t_0\int_{u_{\text{init}}}^{u}\frac{du}{\sqrt{1+\alpha\left(1 - 1/u\right)}}, \tag{8}$$

where for q moving towards the origin $t_{\text{init}} = 0$; $u_{\text{init}} = 1$ and for q moving away from the origin $t_{\text{init}} = t_-(\frac{X_0}{x_0}) = t_-(\frac{\alpha}{1+\alpha})$; $u_{\text{init}} = \alpha/(1+\alpha)$. The right-hand side of (8) is conveniently evaluated using *Mathematica's* Integrate function. Hence (2).

(d) The following graphs were plotted parametrically:

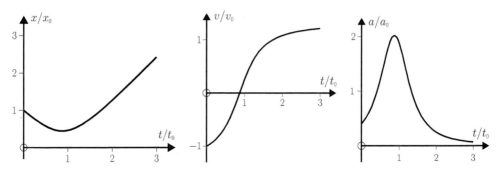

(e) Larmor's formula together with the hint give, $\dfrac{dE_{\mathrm{rad}}}{dv}\,a = \dfrac{\mu_0}{6\pi c}q^2 a^2$ or

$$E_{\mathrm{rad}} = \frac{\mu_0 q^2}{6\pi c}\left[\int_{-v_0}^{0} a\,dv + \int_{0}^{v_{\mathrm{final}}} a\,dv\right]$$

$$= \frac{\mu_0 q^2}{6\pi c}\int_{-v_0}^{v_0\sqrt{1+\alpha}} a\,dv, \qquad (9)$$

where v_{final} is given by (6). In order to evaluate (9), we must express the acceleration in terms of the velocity. This is achieved using (1) to eliminate u from (7). The result is

$$a = \frac{v^4 - 2(1+\alpha)v_0^2 v^2 + (1+\alpha)^2 v_0^4}{2\alpha v_0^2 x_0}$$

$$= \frac{2\pi\epsilon_0 m v_0^2}{2Qq v_0^2}[v^4 - 2(1+\alpha)v_0^2 v^2 + (1+\alpha)^2 v_0^4],$$

where in the last step we use the definition of α. Since the radiated energy is negligible, the total energy of the charge is essentially constant. It is easily shown that E_0 given by (3) can be expressed as $E_0 = (1+\alpha)\frac{1}{2}mv_0^2$, and so

$$a = \frac{2\pi\epsilon_0 E_0}{Qq}\left[\frac{v^4/v_0^2 - 2(1+\alpha)v^2 + (1+\alpha)^2 v_0^2}{(1+\alpha)}\right]. \qquad (10)$$

Substituting (10) in (9) and noting that $\epsilon_0\mu_0 = c^{-2}$ yield

$$\frac{E_{\mathrm{rad}}}{E_0} = \frac{1}{3}\frac{q}{Q}\frac{1}{c^3 v_0^2}\frac{1}{(1+\alpha)}\int_{-v_0}^{v_0\sqrt{1+\alpha}}[v^4/v_0^2 - 2(1+\alpha)v^2 + (1+\alpha)^2 v_0^2]\,dv$$

$$= \frac{1}{3}\frac{q}{Q}\frac{1}{c^3 v_0^2}\frac{1}{(1+\alpha)}\left[\frac{1}{5}v^5/v_0^2 - \frac{2}{3}(1+\alpha)v^3 + (1+\alpha)^2 v_0^2 v\right]_{-v_0}^{v_0\sqrt{1+\alpha}}. \qquad (11)$$

Equation (4) follows from (11) after some simple algebra.

Comment

As $x_0 \to \infty$ the dimensionless parameter $\alpha \to 0$ and $\dfrac{E_{\text{rad}}}{E_0} \to \dfrac{16}{45}\dfrac{q}{Q}\left(\dfrac{v_0}{c}\right)^3$. Clearly, this fraction is very much less than unity for $q \approx Q$ and $v_0 \ll c$. We note that the opposite limit $\alpha \to \infty$ is of no interest here, since the large repulsive force exerted on q as $x_0 \to 0$ would result in $v_{\text{final}} \to c$ (contrary to the non-relativistic assumption underlying the above analysis).

Question 11.20*

A plane unpolarized monochromatic electromagnetic wave propagating along the z-axis of Cartesian coordinates interacts with a free electron (rest mass m_0, charge q) whose equilibrium position is the origin. The electron accelerates and acquires a maximum speed $v_0 \ll c$. Use Larmor's formula (see (13) of Question 11.8) and the differential scattering cross-section $d\sigma/d\Omega$, defined as

$$\frac{d\sigma}{d\Omega} = \frac{\text{average power radiated per unit solid angle}}{\text{incident power per unit area}} = \frac{\left\langle \frac{dP}{d\Omega}\right\rangle}{\frac{1}{2}\sqrt{\frac{\epsilon_0}{\mu_0}}E_0^2}, \tag{1}$$

to derive the scattering cross-section σ_T of the electron, assuming that at all times the displacement of the electron is very much less than λ.

Solution

The force $\mathbf{F}$ exerted on the electron by the electromagnetic wave is $q(\mathbf{E}+\mathbf{v}\times\mathbf{B}) \simeq q\mathbf{E}$ since $B \sim E/c$ and $v \ll c$. Now $\mathbf{E} = E_0 e^{i(kz-\omega t)}\hat{\boldsymbol{\epsilon}}$ where the unit vector $\hat{\boldsymbol{\epsilon}}$, which lies in the xy-plane, is in the (instantaneous) direction of polarization of the wave.

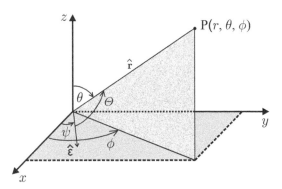

So the acceleration $\mathbf{a} = \mathbf{F}/m = \left(\dfrac{qE_0}{m}\right)e^{i(kz-\omega t)}\hat{\boldsymbol{\epsilon}}$ with $a_0 = \dfrac{qE_0}{m}$. Substituting a_0 in (10) of Question 11.8 gives

$$\langle \mathbf{S} \rangle = \frac{\mu_0}{32\pi^2 c} \frac{q^4 E_0^2 \sin^2 \Theta}{r^2} \hat{\mathbf{r}}, \tag{2}$$

where $\Theta = \cos^{-1}(\hat{\mathbf{e}} \cdot \mathbf{r})$ is the angle between the direction of oscillation of the electron and the direction of observation at the field point P (see the figure shown on p. 543). Substituting (2) in $\langle d\mathcal{P}/d\Omega \rangle = (\langle \mathbf{S} \rangle \cdot \hat{\mathbf{r}}) r^2$ (see Question 11.8) and using (1) yield

$$\frac{d\sigma}{d\Omega} = r_0^2 \sin^2 \Theta, \tag{3}$$

where

$$r_0 = \frac{q^2}{4\pi\epsilon_0 m c^2}. \tag{4}$$

Also shown in the figure on p. 543 is the polarization angle ψ, and because

$$\cos \Theta = \sin\theta \cos(\phi - \psi)$$

it follows that

$$\sin^2 \Theta = 1 - (1 - \cos^2\theta) \cos^2(\phi - \psi). \tag{5}$$

Now the primary wave is unpolarized, and so ψ assumes all values between 0 and 2π with equal probability. Substituting (5) in (3) and averaging over ψ gives

$$\frac{d\sigma}{d\Omega}\bigg|_{\text{unpolarized}} = \frac{\displaystyle\int_0^{2\pi} \frac{d\sigma}{d\Omega} \, d\psi}{\displaystyle\int_0^{2\pi} d\psi}$$

$$= \tfrac{1}{2}(1 + \cos^2\theta) \, r_0^2. \tag{6}$$

Integrating (6) over all solid angles yields

$$\sigma_T = \tfrac{1}{2} r_0^2 \int_0^{2\pi} \int_0^{\pi} (1 + \cos^2\theta) \sin\theta \, d\theta \, d\phi = \frac{8\pi}{3} r_0^2. \tag{7}$$

Comments

(i) The quantity r_0 given by (4) is a physical constant having the dimensions of length. Known as the classical radius of the electron, it has the value $r_0 = 2.82 \times 10^{-15}$ m which is about the size of an atomic nucleus.

(ii) The accelerating electron removes energy from the incident wave and scatters it by radiating in all directions. This scattering of electromagnetic radiation by free, non-relativistic electrons is known as Thomson[‡] scattering and (7) is Thomson's cross-section having the value $\sigma_T = 6.65 \times 10^{-29}$ m^2. The quantity σ_T can be thought of as the area over which the electron intercepts radiation and 'removes' it from the incident wave.

[‡]It was first described by the Nobel laureate J. J. Thomson, who discovered the electron.

(iii) There are several interesting applications of (7). For example, the cosmic microwave background is believed to be linearly polarized because of Thomson scattering and the time required for electromagnetic radiation to diffuse from the core of the Sun to its surface can be estimated to be about 3400 years.[5]

(iv) Thomson scattering is the low-energy limit of Compton scattering. The electron's kinetic energy remains the same before and after the interaction, and the frequency of the scattered radiation is equal to that of the incident wave. This limit is valid provided that $\omega \ll m_0 c^2/\hbar$ where quantum effects are unimportant. At higher frequencies, (6) fails and is replaced by the Klein–Nishina formula which is based on Dirac's relativistic theory of the electron.

Question 11.21**

Consider a point charge q moving relativistically. In the radiation zone, the field is given by

$$\mathbf{E} = \frac{1}{4\pi\epsilon_0} \frac{q}{c} \left[\frac{\hat{\mathbf{R}} \times (\hat{\mathbf{R}} - \boldsymbol{\beta}) \times \dot{\boldsymbol{\beta}}}{R(1 - \boldsymbol{\beta} \cdot \hat{\mathbf{R}})^3} \right], \tag{1}$$

where $\mathbf{R}$ is the vector from the charge to the field point (see (14) of Question 8.14).

(a) Use (1) to show that the instantaneous power per unit solid angle *received* by an observer at time t is

$$\frac{dP(t)}{d\Omega} = \frac{\mu_0 q^2 c}{16\pi^2} \left| \frac{\hat{\mathbf{R}} \times \{(\hat{\mathbf{R}} - \boldsymbol{\beta}) \times \dot{\boldsymbol{\beta}}\}}{(1 - \boldsymbol{\beta} \cdot \hat{\mathbf{R}})^3} \right|^2_{\text{ret}}. \tag{2}$$

(b) Hence show that the instantaneous power per unit solid angle *emitted* by the charge at the retarded time $t' = t - R(t')/c$ is

$$\frac{dP(t')}{d\Omega} = \frac{\mu_0 q^2 c}{16\pi^2} \left| \frac{\hat{\mathbf{R}} \times \{(\hat{\mathbf{R}} - \boldsymbol{\beta}) \times \dot{\boldsymbol{\beta}}\}}{(1 - \boldsymbol{\beta} \cdot \hat{\mathbf{R}})^{5/2}} \right|^2_{\text{ret}}. \tag{3}$$

Hint: The following observation given in Ref. [6] will be helpful when answering this question. The energy lost by the charge at $\mathbf{R}(t')$ radiated into a unit solid angle at θ during the time interval $t' = T_1$ to $t' = T_2$ is measured by an observer during the time interval $t = T_1 + R(T_1)/c$ to $t = T_2 + R(T_2)/c$. Thus we can write

$$\int_{t=T_1 + \frac{R(T_1)}{c}}^{t=T_2 + \frac{R(T_2)}{c}} \left[\mathbf{S} \cdot \hat{\mathbf{R}} \right]_{\text{ret}} dt = \int_{t'=T_1}^{t'=T_2} (\mathbf{S} \cdot \hat{\mathbf{R}}) \frac{\partial t}{\partial t'} dt', \tag{4}$$

where $\mathbf{S} \cdot \hat{\mathbf{R}}$ is the outward component of the Poynting vector evaluated at time t and corresponds to radiation originating from the charge at the retarded time.

[5]   J. Vanderlinde, *Classical electromagnetic theory*, Chap. 10, pp. 288–9. New York: Wiley, 1993.
[6]   J. D. Jackson, *Classical electrodynamics*, Chap. 14, pp. 668–9. New York: Wiley, 3 edn, 1998.

Solution

(a) In calculating the Poynting vector $\mathbf{S}$ it is sufficient to consider the acceleration fields only, because the velocity fields make no contribution to the radiated energy. Now $\mathbf{S} = \dfrac{\mathbf{E}_a \times \mathbf{B}_a}{\mu_0}$ where $c\mathbf{B}_a = \hat{\mathbf{R}} \times \mathbf{E}_a$ (see (17) of Question 8.14). Then

$$\mathbf{S} = \frac{\mathbf{E}_a \times (\hat{\mathbf{R}} \times \mathbf{E}_a)}{\mu_0 c} = \frac{E_a^2}{\mu_0 c} \hat{\mathbf{R}}, \tag{5}$$

since $\mathbf{E}_a \cdot \hat{\mathbf{R}} = 0$. Now with $\mathbf{E}_a$ given by (1) we obtain from (5)

$$\mathbf{S} \cdot \hat{\mathbf{R}} = \frac{1}{\mu_0 c} \frac{q^2}{16\pi^2 \epsilon_0^2 c^2} \left| \frac{\hat{\mathbf{R}} \times (\hat{\mathbf{R}} - \boldsymbol{\beta}) \times \dot{\boldsymbol{\beta}}}{R(1 - \boldsymbol{\beta} \cdot \hat{\mathbf{R}})^3} \right|^2_{\text{ret}}$$

$$= \frac{\mu_0 q^2 c}{16\pi^2} \left| \frac{\hat{\mathbf{R}} \times (\hat{\mathbf{R}} - \boldsymbol{\beta}) \times \dot{\boldsymbol{\beta}}}{R(1 - \boldsymbol{\beta} \cdot \hat{\mathbf{R}})^3} \right|^2_{\text{ret}}.$$

Hence (2) because $d\mathcal{P}/d\Omega = R^2 (\mathbf{S} \cdot \hat{\mathbf{R}})$.

(b) Motivated by the hint, (4) suggests that we define

$$\frac{d\mathcal{P}(t')}{d\Omega} = R^2 (\mathbf{S} \cdot \hat{\mathbf{R}}) \frac{\partial t}{\partial t'} = \frac{d\mathcal{P}(t)}{d\Omega} \frac{\partial t}{\partial t'}, \tag{6}$$

where $\partial t/\partial t'$, given by (7) of Question 8.14, is

$$\frac{\partial t}{\partial t'} = 1 - \boldsymbol{\beta} \cdot \hat{\mathbf{R}}. \tag{7}$$

Substituting (2) and (7) in (6) yields (3).

Comments

(i) The appearance of the factor $\partial t/\partial t'$ in (6) arises because the time interval during which the radiation is emitted is different from the time interval during which it is detected by the observer. Ref. [6] explains that 'the useful and meaningful quantity is $(\mathbf{S} \cdot \hat{\mathbf{R}})(\partial t/\partial t')$, the power radiated per unit area in terms of the charge's own time'.

(ii) The total radiated power, found by integrating (3) over all solid angles, is

$$\mathcal{P} = \frac{\mu_0 q^2}{6\pi c} \gamma^6 \left[a^2 - |\boldsymbol{\beta} \times \mathbf{a}|^2 \right]_{\text{ret}}, \tag{8}$$

which is the relativistic form of Larmor's formula. Instead of attempting this tricky integration here, we will provide an alternative derivation of (8) using the invariance of a radiated power (see Question 12.25).

Question 11.22**

Consider a relativistic charge q whose velocity and acceleration are colinear.

(a) ☞   Use (3) of Question 11.21 to show that the power radiated per unit solid angle

$$\frac{dP}{d\Omega} = \frac{\mu_0 q^2 c}{16\pi^2} \frac{\dot{\beta}^2 \sin^2\theta}{(1 - \beta \cos\theta)^5},$$ (1)

where $\theta = \cos^{-1}(\hat{\beta} \cdot \hat{R}) = \cos^{-1}(\hat{\beta} \cdot \hat{R})$.

☞   Hence show that the total power emitted by the charge is

$$P = \frac{\mu_0 q^2 c}{6\pi} \dot{\beta}^2 \gamma^6,$$ (2)

where $\gamma^2 = (1 - \beta^2)^{-1}$.

(b) Use (1) to show that as $\beta \to 1$ the angular distribution $dP/d\Omega$ is a maximum for

$$\Theta = \cos^{-1}\left(\frac{\sqrt{1 + 15\beta^2} - 1}{3\beta}\right).$$ (3)

Solution

(a) ☞   Since β and $\dot{\beta}$ are parallel, $\beta \times \dot{\beta} = 0$ and (3) of Question 11.21 becomes

$$\frac{dP}{d\Omega} = \frac{q^2}{16\pi^2\epsilon_0 c} \frac{|\hat{R} \times (\hat{R} \times \dot{\beta})|^2}{(1 - \beta \cdot \hat{R})^5}.$$ (4)

Expanding the vector triple product in (4) gives

$$\frac{dP}{d\Omega} = \frac{q^2}{16\pi^2\epsilon_0 c} \frac{\{\hat{R}(\dot{\beta} \cdot \hat{R}) - \dot{\beta}\}^2}{(1 - \beta \cdot \hat{R})^5}$$

$$= \frac{q^2}{16\pi^2\epsilon_0 c} \frac{\dot{\beta}^2 - (\dot{\beta} \cdot \hat{R})^2}{(1 - \beta \cdot \hat{R})^5} = \frac{q^2}{16\pi^2\epsilon_0 c} \frac{\dot{\beta}^2(1 - \cos^2\theta)}{(1 - \beta \cos\theta)^5},$$

which is (1).

☞   Integrating (1) over all solid angles leads to

$$P = \int \frac{dP}{d\Omega} d\Omega$$

$$= \frac{q^2 \dot{\beta}^2}{16\pi^2\epsilon_0 c} \int_0^{2\pi} \int_0^{\pi} \frac{\sin^2\theta}{(1 - \beta \cos\theta)^5} \sin\theta \, d\theta \, d\phi$$

$$= \frac{q^2 \dot{\beta}^2}{8\pi\epsilon_0 c} \int_0^\pi \frac{\sin^3\theta \, d\theta}{(1 - \beta\cos\theta)^5} . \tag{5}$$

Making the substitution $u = \cos\theta$ in (5) gives

$$\mathcal{P} = \frac{q^2\dot{\beta}^2}{8\pi\epsilon_0 c} \int_{-1}^1 \frac{1 - u^2}{(1 - \beta u)^5} \, du. \tag{6}$$

The definite integral in (6) has the value[‡] $\frac{4}{3}(1 - \beta^2)^{-3} = \frac{4}{3}\gamma^6$. Hence (2).

(b) The angle Θ for which $d\mathcal{P}/d\Omega$ is a maximum can be determined by finding the maximum of the function $f(\theta) = \sin^2\theta/(1 - \beta\cos\theta)^5$. Putting $df/d\theta = 0$ yields the quadratic equation $3\beta\cos^2\Theta + 2\cos\Theta - 5\beta = 0$ which has the roots

$$\cos\Theta = \frac{-1 \pm \sqrt{1 + 15\beta^2}}{3\beta}. \tag{7}$$

The positive root[♯] of (7) gives (3).

Comments

(i) In the non-relativistic limit ($\beta \to 0$; $\gamma \to 1$) equations (1) and (2) reproduce the Larmor formula (13) of Question 11.8. But, as $\beta \to 1$, the angular distribution of the radiation elongates in the direction of motion (except for $\theta = 0$; see Ref. [7]), as indicated below in the (normalized) polar plots $(d\mathcal{P}/d\Omega) \div (d\mathcal{P}/d\Omega)_{\theta=\Theta}$.

(ii) By making a suitable binomial expansion of (7), it is easily shown that as $\beta \to 1$ the limiting value of Θ is $\frac{1}{2\gamma}$, and that $\left.\dfrac{d\mathcal{P}}{d\Omega}\right|_{\theta = \frac{1}{2\gamma}} \sim \gamma^8$.

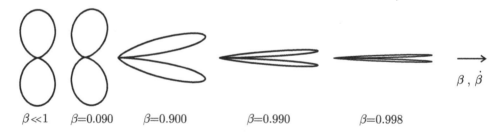

$\beta \ll 1$ $\beta = 0.090$ $\beta = 0.900$ $\beta = 0.990$ $\beta = 0.998$

[‡]The reader may confirm this using *Mathematica*.

[♯]This is the root we require, since $\Theta \to 0$ as $\beta \to 1$. See Comment (i).

[7]   J. B. Marion and M. A. Heald, *Classical electromagnetic radiation*, Chap. 7, p. 217. New York: Academic Press, 2 edn, 1980.

Question 11.23[**]

Consider a relativistic charge q whose velocity $c\boldsymbol{\beta}$ and acceleration $c\dot{\boldsymbol{\beta}}$ are instantaneously perpendicular. Choose Cartesian coordinates such that the charge is moving in the xz-plane and $\boldsymbol{\beta} = \beta\hat{\mathbf{z}}$ and $\dot{\boldsymbol{\beta}} = \dot{\beta}\hat{\mathbf{x}}$.

(a) ☞ Use (3) of Question 11.21 to show that the power radiated per unit solid angle

$$\frac{dP}{d\Omega} = \frac{\mu_0 q^2 \dot{\beta}^2 c}{16\pi^2} \frac{1}{(1 - \beta\cos\theta)^3} \left\{ 1 - \frac{\sin^2\theta\cos^2\phi}{\gamma^2(1 - \beta\cos\theta)^2} \right\}, \tag{1}$$

where θ and ϕ are the usual spherical polar coordinates.

☞ Hence show that the total power emitted by the charge is

$$P = \frac{\mu_0 q^2 c}{6\pi} \dot{\beta}^2 \gamma^4. \tag{2}$$

(b) Use (1) to show that the angle θ_0 at which $dP/d\Omega = 0$ is

$$\theta_0 = \cos^{-1}\beta. \tag{3}$$

Solution

(a) ☞ Since $\boldsymbol{\beta}$ and $\dot{\boldsymbol{\beta}}$ are perpendicular, $\boldsymbol{\beta}\cdot\dot{\boldsymbol{\beta}} = 0$ and (3) of Question 11.21 becomes

$$\frac{dP}{d\Omega} = \frac{q^2}{16\pi^2\epsilon_0 c} \frac{\left\{ \hat{\mathbf{R}} \times \{(\hat{\mathbf{R}} - \boldsymbol{\beta}) \times \dot{\boldsymbol{\beta}}\} \right\}^2}{(1 - \boldsymbol{\beta}\cdot\hat{\mathbf{R}})^5}$$

$$= \frac{q^2}{16\pi^2\epsilon_0 c} \frac{\left\{ (\hat{\mathbf{R}} - \boldsymbol{\beta})(\hat{\mathbf{R}}\cdot\dot{\boldsymbol{\beta}}) - \dot{\boldsymbol{\beta}}(1 - \hat{\mathbf{R}}\cdot\boldsymbol{\beta}) \right\}^2}{(1 - \boldsymbol{\beta}\cdot\hat{\mathbf{R}})^5}$$

$$= \frac{q^2}{16\pi^2\epsilon_0 c} \frac{-(\hat{\mathbf{R}}\cdot\dot{\boldsymbol{\beta}})^2(1 - \beta^2) + \dot{\beta}^2(1 - \hat{\mathbf{R}}\cdot\boldsymbol{\beta})^2}{(1 - \boldsymbol{\beta}\cdot\hat{\mathbf{R}})^5}. \tag{4}$$

Now $\hat{\mathbf{R}} = \sin\theta\cos\phi\,\hat{\mathbf{x}} + \sin\theta\sin\phi\,\hat{\mathbf{y}} + \cos\theta\,\hat{\mathbf{z}}$, and $\hat{\mathbf{R}}\cdot\dot{\boldsymbol{\beta}} = \dot{\beta}\sin\theta\cos\phi$ and $\hat{\mathbf{R}}\cdot\boldsymbol{\beta} = \beta\cos\theta$. Using these results in (4) gives (1).

☞ Proceeding as before (see Question 11.22) yields

$$P = \frac{q^2\dot{\beta}^2}{16\pi^2\epsilon_0 c} \int_0^{2\pi}\int_0^{\pi} \frac{(1 - \beta\cos\theta)^2 - (1 - \beta^2)\sin^2\theta\cos^2\phi}{(1 - \beta\cos\theta)^5} \sin\theta\,d\theta\,d\phi. \tag{5}$$

The definite integral‡ in (5) has the value $\dfrac{8\pi}{3(1-\beta)^2}$, and so (2) follows.

(b) The angle θ_0 at which $dP/d\Omega = 0$ can be determined by finding the zeros of the function

$$f(\theta) = \frac{(1 - \beta \cos\theta)^2 - (1 - \beta^2)\sin^2\theta}{(1 - \beta\cos\theta)^5}. \tag{6}$$

Putting $f(\theta) = 0$ in (6) yields the quadratic equation $(\cos\theta_0 - \beta)^2 = 0$, whose root is given by (3).

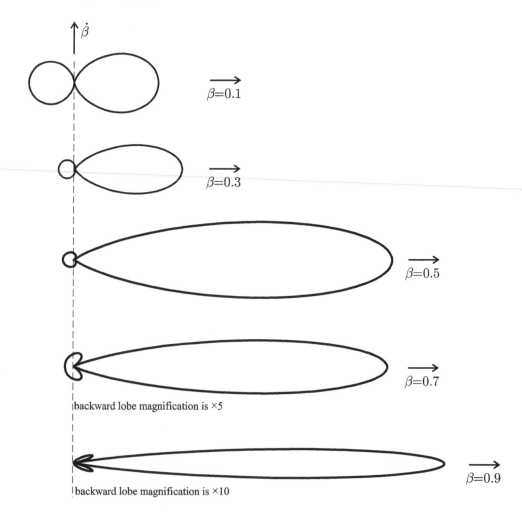

$\beta = 0.1$

$\beta = 0.3$

$\beta = 0.5$

$\beta = 0.7$

backward lobe magnification is ×5

$\beta = 0.9$

backward lobe magnification is ×10

‡For the first integral over θ in (5), put $u = \beta\cos\theta$. For the second integral over θ let $u = 1 - \beta\cos\theta$.

Comments

The following remarks are for a charge moving in a circle in the xz-plane (i.e. $\phi = 0$).

(i) In the non-relativistic limit (1) again reduces to Larmor's result, provided θ is replaced by $\frac{1}{2}\pi - \theta$, which is the angle between $\hat{\mathbf{R}}$ and the acceleration vector.

(ii) The angular distributions of the radiated power are shown in the figure on p. 550 for $\beta = 0.1$, 0.3, 0.5, 0.7 and 0.9. The scale of each diagram is determined by $(dP/d\Omega)_{\theta=0}$. The values corresponding to the given βs are in the (approximate) ratio $1:2:6:27:729$ respectively. Notice the following features:

☞ $dP/d\Omega$ is a maximum in the forward direction ($\theta = 0$).

☞ The angular distributions have forward and backward lobes arising from radiation emitted in the θ-intervals $[-\cos^{-1}\beta, \cos^{-1}\beta]$ and $[\cos^{-1}\beta, 2\pi - \cos^{-1}\beta]$ respectively. For clarity of presentation, the backward lobes of the two largest values of β have been magnified as indicated in the figure.

☞ As $\beta \to 1$ the backward lobe is swept towards forward angles. Almost all the radiation is emitted in the forward lobe whose angular width is of order $2/\gamma$. This narrowing of the forward lobe is sometimes described, for obvious reasons, as the 'lighthouse effect'.

Question 11.24 **

A relativistic particle (rest mass m_0, charge q) is acted upon by a net external force of constant magnitude F.

(a) Suppose the particle moves in a straight line (i.e. $\boldsymbol{\beta}$ and $\dot{\boldsymbol{\beta}}$ are colinear). Use (2) of Question 11.22 to show that the power $\mathcal{P}_{\text{lin}}$ radiated by the particle depends only on F and physical constants.

(b) Suppose the particle moves in a circle (i.e. $\boldsymbol{\beta}$ and $\dot{\boldsymbol{\beta}}$ are perpendicular). Use (2) of Question 11.23 to show that the power $\mathcal{P}_{\text{circ}}$ radiated by the particle is related to $\mathcal{P}_{\text{lin}}$ by

$$\mathcal{P}_{\text{circ}} = \gamma^2\,\mathcal{P}_{\text{lin}}. \tag{1}$$

Solution

(a) The relativistic energy E and momentum $\mathbf{p}$ of the particle are given by $E = \gamma m_0 c^2$ and $\mathbf{p} = \gamma m_0 c \boldsymbol{\beta}$. Now $\dfrac{dE}{dx} = m_0 c^2 \dfrac{d\gamma}{dt}\dfrac{dt}{dx} = m_0 c^2 \dfrac{d\gamma}{dt}\Big/\dfrac{dx}{dt} = m_0 c \gamma^3 \dot{\beta}$, since $\dfrac{dx}{dt} = c\beta$ and $\dfrac{d\gamma}{dt} = \gamma^3 \beta \dot{\beta}$. Also, $\dfrac{d\mathbf{p}}{dt} = m_0 c \boldsymbol{\beta}\dfrac{d\gamma}{dt} + m_0 c \gamma \dfrac{d\boldsymbol{\beta}}{dt} = m_0 c \beta \gamma^3 \beta \dot{\beta} + m_0 c \gamma \dot{\beta} = m_0 c \beta^2 \gamma^3 \dot{\beta} + m_0 c \gamma \dot{\beta} = m_0 c \gamma \dot{\beta}(1 + \gamma^2 \beta^2) = m_0 c \gamma^3 \dot{\beta}$, since $1 + \gamma^2 \beta^2 = \gamma^2$. So

$$\frac{dE}{dx} = \frac{dp}{dt} = m_0 c \gamma^3 \dot{\beta}. \tag{2}$$

Substituting (2) in (2) of Question 11.22 gives

$$\mathcal{P}_{\text{lin}} = \frac{q^2}{6\pi\epsilon_0 m_0^2 c^3} \left(\frac{dp}{dt}\right)^2. \tag{3}$$

This proves the result since $\dfrac{d\mathbf{p}}{dt}$ is **F**.

(b) For relativistic circular motion $F = \gamma m_0 c \dot{\beta}$ and (2) of Question 11.23 gives

$$\mathcal{P}_{\text{circ}} = \frac{q^2}{6\pi\epsilon_0 m_0^2 c^3} \gamma^2 \left(\frac{dp}{dt}\right)^2. \tag{4}$$

Equations (4) and (3) yield (1).

Comments

(i) For a given magnitude of external force, (1) shows that the radiation produced in circular motion is larger, by a factor of γ^2, than that for colinear motion. Evidently, ultra-relativistic particles moving in a circular trajectory can produce significant amounts of synchrotron radiation.

(ii) By contrast, the power radiated by a particle in a linear accelerator is usually insignificant. To see this, we first write (3) as $\mathcal{P}_{\text{lin}} = \dfrac{q^2}{6\pi\epsilon_0 m_0^2 c^3} \left(\dfrac{dE}{dx}\right)^2$ and then calculate the ratio of radiated power to the power $\dot{E}$ provided by external forces. Thus

$$\frac{\mathcal{P}_{\text{lin}}}{\dot{E}} = \frac{q^2}{6\pi\epsilon_0 m_0^2 c^3} \left(\frac{dE}{dx}\right)^2 \Big/ \left(\frac{dE}{dt}\right)$$

$$= \frac{q^2}{6\pi\epsilon_0 m_0^2 c^3} \left(\frac{dE}{dx}\right) \Big/ \left(\frac{dx}{dt}\right)$$

$$= \frac{q^2}{6\pi\epsilon_0 m_0^2 c^4} \frac{1}{\beta} \frac{dE}{dx}$$

$$= \frac{q^2}{6\pi\epsilon_0 m_0^2 c^4} \frac{dE}{dx} \qquad \text{as } \beta \to 1. \tag{5}$$

Typical values[8] of dE/dx are less than $50\,\text{MeV}\,\text{m}^{-1}$, and so $\mathcal{P}_{\text{lin}}/\dot{E} \ll 1$ for electrons, protons and heavier particles.

[8]   J. D. Jackson, *Classical electrodynamics*, Chap. 14, p. 667. New York: Wiley, 3 edn, 1998.

12

Electromagnetism and special relativity

In 1905, when Einstein published his theory of special relativity, Maxwell's work was already about forty years old. Remarkably, after all the earlier debate and controversy surrounding the role of the aether[‡] in wave propagation, classical electrodynamics turned out to be a covariant physical theory. Ironically Newton's second law, known to be form invariant under a Galilean transformation, was exposed to be inadequate at speeds approaching c. Whereas Maxwell's electrodynamics—evolving as it did in the era before Lorentz discovered his transformation and Einstein postulated his relativity principle—survives to this day essentially in its original form. All this in spite of the aether protagonists who insisted that a preferred reference frame, the so-called aether frame, existed solely for the propagation of light waves (this is obviously now known to be wrong). Ref. [1] explains that:

> special relativity has its historic roots in electromagnetism. Lorentz, exploring the electrodynamics of moving charges, was led very close to the final formulation of Einstein. And Einstein's great paper of 1905 was entitled not 'Theory of Relativity', but rather 'On the Electrodynamics of Moving Bodies'. Today we see in the postulates of relativity and their implications a wide framework, one that embraces all physical laws and not solely those of electromagnetism.

See also the second paragraph of Ref. [2] and the remarks of Lévy-Leblond in Ref. [3].

We begin Chapter 12 with a range of questions which establish how certain physical quantities behave under Lorentz transformation. This leads to the important concept of an invariant, and we encounter (and use) a number of these in this chapter. Other topics considered include the transformation of electric and magnetic fields between inertial reference frames, the validity of Gauss's law for an arbitrarily moving point charge (demonstrated numerically), the relativistic Doppler effect, four-vectors and

---

[‡]In the latter part of the nineteenth century, many physicists firmly believed that a medium (called the aether) was necessary for the propagation of electromagnetic waves. This view was based on their experience. They argued that other wave phenomena, like sound waves, could not propagate in vacuum. The aether medium was required to possesses some curious properties. For example, it permeated all space, had negligible density and interacted hardly at all with matter.

---

[1]   E. M. Purcell and D. J. Morin, *Electricity and magnetism*, Chap. 5, p. 236. Cambridge: Cambridge University Press, 3 edn, 2013.
[2]   J. D. Jackson, *Classical electrodynamics*, Chap. 11, p. 514. New York: Wiley, 3 edn, 1998.
[3]   O. L. de Lange and J. Pierrus, *Solved problems in classical mechanics: Analytical and numerical solutions with comments*, Chap. 15, p. 561. Oxford: Oxford University Press, 2010.

Solved Problems in Classical Electromagnetism. J. Pierrus, Oxford University Press (2018).
© J. Pierrus. DOI: 10.1093/oso/9780198821915.001.0001

four-tensors, the electromagnetic-field tensor $F_{\mu\nu}$, Maxwell's equations in covariant form and Larmor's radiation formula for a point charge moving relativistically. Although it is assumed that the reader is familiar with the basic concepts of special relativity (which are usually encountered in introductory courses on the subject), we begin by reviewing some of this material below. Those who find this summary too brief are referred to Chapter 15 of Ref. [4] for a more detailed account.

Einstein's relativity principle is a fundamental result. It asserts that the laws of physics are equally valid in all inertial references frames. So, for example, since the Maxwell–Faraday equation $\nabla \times \mathbf{E} = -\partial\mathbf{B}/\partial t$ is known to be valid in some frame S, it will also be valid in any other frame S′ and will have the same mathematical form: $\nabla' \times \mathbf{E}' = -\partial\mathbf{B}'/\partial t'$. Here the primes denote the space-time coordinates and fields in inertial reference frame S′. Now it is often convenient to choose S and S′ in standard configuration which is illustrated in the figure below. Frame S′ may be regarded as moving relative to S with constant velocity $\mathbf{v}$ along the common xx'-axis (an equivalent interpretation is that S moves relative to S′ with constant velocity $-\mathbf{v}$). For the standard configuration we assume that origins O and O' are coincident at time $t = t' = 0$, and that the planes $y = 0$, $z = 0$ coincide permanently with the planes $y' = 0$, $z' = 0$ respectively.

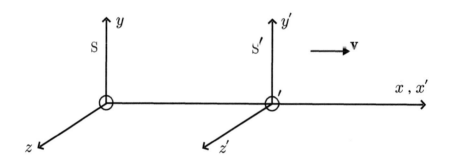

Question 15.1 of Ref. [4] shows that the most general linear homogeneous transformation between the coordinates of S and S′ is given by the Lorentz transformation:

$$
\left.
\begin{array}{ll}
x' = \gamma(x - vt) & \qquad x = \gamma(x' + vt') \\[4pt]
y' = y & \qquad y = y' \\[4pt]
z' = z & \qquad z = z' \\[4pt]
t' = \gamma\!\left(t - vx/c^2\right) & \qquad t = \gamma\!\left(t' + vx'/c^2\right)
\end{array}
\right\} . \qquad (\mathrm{I})
$$

with "or the inverse" between the two sets.

Here the Lorentz factor γ is defined as

[4]   O. L. de Lange and J. Pierrus, *Solved problems in classical mechanics: Analytical and numerical solutions with comments.* Oxford: Oxford University Press, 2010.

$$\gamma(v) = \frac{1}{\sqrt{1 - v^2/c^2}}$$

$$= \frac{1}{\sqrt{1 - \beta^2}}, \tag{II}$$

where $\beta = v/c$ with $0 \le \beta < 1$, and so $1 \le \gamma < \infty$. We obtain from (II) the following useful identities (remember them) which are trivial to prove:

$$\left. \begin{array}{l} \gamma^2(1 - \beta^2) = 1 \\[4pt] \gamma^2(c^2 - v^2) = c^2 \\[4pt] (\gamma^2 - 1) = \gamma^2\beta^2 \end{array} \right\}. \tag{III}$$

Important kinematic consequences arising from the Lorentz transformation include:

☞ The length ℓ_0 of a rod is measured to be greatest when it is at rest relative to an observer (ℓ_0 is said to be the proper length of the rod). If the rod moves relative to the observer with velocity $\mathbf{v}$, its length ℓ in the direction of motion is contracted by the factor γ^{-1}. All dimensions perpendicular to $\mathbf{v}$ are unaffected, and so

$$\ell = \frac{\ell_0}{\gamma}. \tag{IV}$$

☞ A clock ticks at its fastest rate when it is at rest relative to an observer (this being the proper time). If the clock moves with velocity $\mathbf{v}$ relative to the observer, its rate is measured to have slowed down by the factor γ.

☞ Events that are simultaneous in one frame are not necessarily simultaneous in another frame. According to special relativity the simultaneity of spatially separated events is not an absolute property, as it is assumed to be in Newtonian mechanics.

Force is defined as the rate of change of the *relativistic* momentum $\mathbf{F} = d\mathbf{p}/dt$. Consider, for example, a particle having proper mass m_0 moving with velocity $\mathbf{u}(t)$ in an inertial frame S; its relativistic momentum is given by $\mathbf{p} = \gamma(u)\, m_0 \mathbf{u}$, or

$$\mathbf{p} = \frac{m_0 \mathbf{u}}{\sqrt{1 - u^2/c^2}}. \tag{V}$$

Here m_0 (sometimes called the rest mass) is an invariant, meaning that it has the same value in all reference frames. Furthermore, if the particle carries charge q, there is an abundance of convincing experimental evidence to support the invariance of this charge (namely $q = q'$, which is a fundamental result in physics).

Questions 12.1–12.4, which now follow, may also be regarded as summary material. These, together with the introductory review above, should provide the necessary background required to answer the remaining questions of this chapter.

Question 12.1

(a) Consider an event which occurs at $P(\mathbf{r}, t)$ in S and $P'(\mathbf{r}', t')$ in S'. Express the Lorentz transformation (see (I) from the introduction) in the coordinate-free form

$$
\left.
\begin{aligned}
\mathbf{r}' &= \gamma \mathbf{r}_\| + \mathbf{r}_\perp - \gamma \mathbf{v} t \\
t' &= \gamma (t - \mathbf{v} \cdot \mathbf{r}/c^2)
\end{aligned}
\right\}, \tag{1}
$$

where $\mathbf{v}$ is the velocity of S' relative to S, and the subscripts denote components of $\mathbf{r}$ parallel and perpendicular to $\mathbf{v}$.

(b) Hence show that

$$
r'^2 - c^2 t'^2 = r^2 - c^2 t^2. \tag{2}
$$

Solution

(a) Substituting (I) in $\mathbf{r}' = \hat{\mathbf{x}} x' + \hat{\mathbf{y}} y' + \hat{\mathbf{z}} z'$ gives $\mathbf{r}' = \gamma \hat{\mathbf{x}} (x - vt) + \hat{\mathbf{y}} y + \hat{\mathbf{z}} z$. Now $\mathbf{r} = \mathbf{r}_\| + \mathbf{r}_\perp$ by definition, where $\mathbf{r}_\| = \hat{\mathbf{x}} x$ and $\mathbf{r}_\perp = \hat{\mathbf{y}} y + \hat{\mathbf{z}} z$. Since $\mathbf{v} = \hat{\mathbf{x}} v$ we obtain $\mathbf{r}' = \gamma \mathbf{r}_\| - \gamma \mathbf{v} t + \mathbf{r}_\perp$ which is (1)$_1$. The transformation for t' follows immediately from $t' = \gamma(t - vx/c^2)$ because $vx = vr_\| = \mathbf{v} \cdot \mathbf{r}$.

(b) Using (1), and because $\mathbf{r}_\| \cdot \mathbf{r}_\perp = 0$ and $\mathbf{r}_\perp \cdot \mathbf{v} = 0$, it follows that

$$
\begin{aligned}
r'^2 - c^2 t'^2 &= (\gamma \mathbf{r}_\| + \mathbf{r}_\perp - \gamma \mathbf{v} t) \cdot (\gamma \mathbf{r}_\| + \mathbf{r}_\perp - \gamma \mathbf{v} t) - \gamma^2 c^2 (t - \mathbf{r} \cdot \mathbf{v}/c^2)^2 \\
&= \gamma^2 r_\|^2 - 2\gamma^2 r_\| vt + r_\perp^2 + \gamma^2 v^2 t^2 - \gamma^2 c^2 (t^2 - 2 r_\| vt/c^2 + r_\|^2 v^2/c^4) \\
&= \gamma^2 r_\|^2 + (r^2 - r_\|^2) + \gamma^2 v^2 t^2 - \gamma^2 c^2 t^2 - \gamma^2 \beta^2 r_\|^2 \\
&= r^2 + (\gamma^2 - 1) r_\|^2 - \gamma^2 (c^2 - v^2) t^2 - \gamma^2 \beta^2 r_\|^2 \\
&= r^2 + \gamma^2 \beta^2 r_\|^2 - c^2 t^2 - \gamma^2 \beta^2 r_\|^2,
\end{aligned}
$$

where in the last step we use (III) of the introduction. Hence (2).

Comments

(i) The following useful rule relates a direct transformation to its inverse and vice versa:

> interchange primed with unprimed quantities, and replace v with $-v$. (3)

So from (1) and (3),

$$
\left.
\begin{aligned}
\mathbf{r} &= \gamma \mathbf{r}'_\| + \mathbf{r}'_\perp + \gamma \mathbf{v} t' \\
t &= \gamma(t' + \mathbf{v}' \cdot \mathbf{r}'/c^2)
\end{aligned}
\right\}. \tag{4}
$$

(ii) Equivalent forms of (1) and (4) are sometimes required. These are

$$\left.\begin{array}{ll} \mathbf{r}' = \mathbf{r} + (\gamma - 1)\mathbf{r}_{\parallel} - \gamma \mathbf{v}t & \mathbf{r} = \mathbf{r}' + (\gamma - 1)\mathbf{r}'_{\parallel} + \gamma \mathbf{v}t' \\ \text{and inverse} \\ t' = \gamma(t - vr_{\parallel}/c^2) & t = \gamma(t' + vr'_{\parallel}/c^2) \end{array}\right\}. \quad (5)$$

(iii) Equation (2) is known as the invariant interval: it has the same value in all inertial reference frames (in the terminology of relativity, $r^2 - c^2t^2$ is called a *relativistic invariant*). In some of the questions below, we will encounter other quantities that transform in exactly the same way as $\mathbf{r}$ and ct and associated with all of these is a corresponding invariant.

Question 12.2

(a) Consider a particle moving with velocity $\mathbf{u}$ relative to inertial frame S, and $\mathbf{u}'$ relative to S'. Derive the velocity and acceleration transformations:

velocity $$\mathbf{u}' = \frac{\mathbf{u}_{\parallel} + \mathbf{u}_{\perp}/\gamma - \mathbf{v}}{(1 - u_{\parallel}v/c^2)}, \quad (1)$$

acceleration $$\mathbf{a}' = \frac{c^2\mathbf{a}_{\parallel} + \gamma(c^2 - u_{\parallel}v)\mathbf{a}_{\perp} + \gamma a_{\parallel}v\mathbf{u}_{\perp}}{\gamma^3(c^2 - u_{\parallel}v)^3}c^4, \quad (2)$$

where $\mathbf{v}$ is the velocity of S' relative to S.

(b) Hence prove the identity

$$(c^2 - u'^2) = c^2(c^2 - v^2)(c^2 - u^2)(c^2 - u_{\parallel}v)^{-2}. \quad (3)$$

Solution

(a) velocity

By definition $\mathbf{u}' = \dfrac{d\mathbf{r}'}{dt'} = \dfrac{d\mathbf{r}'/dt}{dt'/dt}$, and it follows from (1) of Question 12.1 that

$$\mathbf{u}' = \frac{\gamma(d\mathbf{r}_{\parallel}/dt - \mathbf{v}) + d\mathbf{r}_{\perp}/dt}{\gamma\left(1 - \dfrac{v}{c^2}\dfrac{dr_{\parallel}}{dt}\right)}.$$

Now $\mathbf{u} = \dfrac{d}{dt}(\mathbf{r}_{\parallel} + \mathbf{r}_{\perp}) = \mathbf{u}_{\parallel} + \mathbf{u}_{\perp}$. Hence (1).

acceleration

Again, by definition $\mathbf{a}' = \dfrac{d\mathbf{u}'}{dt'} = \dfrac{d\mathbf{u}'/dt}{dt'/dt} = \dfrac{d\mathbf{u}'/dt}{\gamma(1 - u_{\parallel}v/c^2)}$ where

$$\frac{d\mathbf{u}'}{dt} = \frac{(\mathbf{a}_{\parallel} + \mathbf{a}_{\perp}/\gamma)(1 - u_{\parallel}v/c^2) + (\mathbf{u}_{\parallel} + \mathbf{u}_{\perp}/\gamma - \mathbf{v})a_{\parallel}v/c^2}{(1 - u_{\parallel}v/c^2)^2}. \tag{4}$$

The task of expanding the terms in (4) and completing the steps leading to (2) is tedious, but not difficult. The details are left to the reader.

(b) Start with $u^2 = u_{\parallel}^2 + u_{\perp}^2$. Then from (1) we obtain

$$\gamma^2\left(1 - \frac{u_{\parallel}v}{c^2}\right)^2(c^2 - u'^2) = \gamma^2\left(1 - \frac{u_{\parallel}v}{c^2}\right)^2c^2 - \gamma^2\left(\mathbf{u}_{\parallel} + \frac{\mathbf{u}_{\perp}}{\gamma} - \mathbf{v}\right)\cdot\left(\mathbf{u}_{\parallel} + \frac{\mathbf{u}_{\perp}}{\gamma} - \mathbf{v}\right)$$

$$= \gamma^2\left(c^2 + u_{\parallel}^2\beta^2 - 2u_{\parallel}v\right) - \gamma^2\left(u_{\parallel}^2 - 2u_{\parallel}v + u_{\perp}\gamma^{-2} + v^2\right)$$

$$= \gamma^2\left(c^2 + u_{\parallel}^2\beta^2 - u_{\parallel}^2 - u_{\perp}\gamma^{-2} - v^2\right)$$

$$= \gamma^2(c^2 - v^2) - \gamma^2(1 - \beta^2)u_{\parallel}^2 - u_{\perp}^2$$

$$= c^2 - u_{\parallel}^2 - u_{\perp}^2 \qquad\qquad\text{(because of (III))}$$

$$= c^2 - u^2.$$

Hence (3).

Comments

(i) If the particle is at rest in S (i.e. $\mathbf{u} = 0$), it then follows from (1) that $\mathbf{u}' = -\mathbf{v}$. Substituting this result in the inverse of (2) (see (3) of Question 12.1) yields

$$\mathbf{a} = \frac{c^2\mathbf{a}'_{\parallel} + \gamma(c^2 - v^2)\mathbf{a}'_{\perp}}{\gamma^3(c^2 - v^2)^3}\,c^4 = \gamma^3\left[\mathbf{a}'_{\parallel} + \gamma(1 - \beta^2)\mathbf{a}'_{\perp}\right].$$

Squaring this last equation and noting that $a'_{\parallel}{}^2 = a'^2 - a'_{\perp}{}^2$ give

$$a^2 = \gamma^6\left[a'_{\parallel}{}^2 + \gamma^2(1 - \beta^2)^2 a'_{\perp}{}^2\right] = \gamma^6\left[a'^2 - a'_{\perp}{}^2 + (1 - \beta^2)a'_{\perp}{}^2\right]$$

$$= \gamma^6\left[a'^2 - \beta^2 a'_{\perp}{}^2\right]$$

$$= \gamma^6\left[a'^2 - |\boldsymbol{\beta} \times \mathbf{a}'|^2\right], \tag{5}$$

which is a useful result for calculating the power radiated by an accelerated charge. See Questions 11.22–11.24 and 12.25.

(ii) The identity (3) is another useful result, and we will use it in Question 12.3.

Question 12.3

(a) Consider a rod having proper length ℓ_0 lying along the xx'-axis of frames S and S' (assumed to be in standard configuration). Suppose the velocity of the rod is $\mathbf{u} = u\hat{\mathbf{x}}$ in S and $\mathbf{u}' = u'\hat{\mathbf{x}}$ in S'. Show that the lengths of the rod ℓ, ℓ' in S, S' are related by the transformation

$$\ell' = \frac{\ell\sqrt{1 - v^2/c^2}}{(1 - uv/c^2)}. \tag{1}$$

Hint: Use the identity (3) of Question 12.2.

(b) A particle having proper mass m_0 moves with velocity $\mathbf{u}$, $\mathbf{u}'$ in S, S'. Show that the mass m of the particle in S is related to its mass m' in S' by

$$m' = \frac{m\,(1 - u_{\|}v/c^2)}{\sqrt{1 - v^2/c^2}}. \tag{2}$$

Hint: Start with $m = \gamma(u)\,m_0$.

Solution

(a) From the length-contraction formula (IV) we have $\ell_0 = \gamma(u)\,\ell = \gamma(u')\,\ell'$, where $\gamma(u) = (1 - u^2/c^2)^{-1/2}$. So

$$\ell' = \ell\sqrt{\frac{c^2 - u'^2}{c^2 - u^2}}. \tag{3}$$

Now $(c^2 - u'^2)/(c^2 - u^2) = c^2(c^2 - v^2)(c^2 - u_{\|}v)^{-2} = c^2(c^2 - v^2)(c^2 - uv)^{-2}$. Hence (1).

(b) Clearly $m_0 = m/\gamma(u) = m'/\gamma(u')$, or $m' = m\sqrt{(c^2 - u^2)/(c^2 - u'^2)}$. Again, using the identity (3) of Question 12.2 gives (2).

Question 12.4

(a) Use the definition of relativistic momentum to derive the transformation

$$\mathbf{p}' = \gamma(\mathbf{p}_{\|} - \mathbf{v}\mathcal{E}/c^2) + \mathbf{p}_{\perp}, \tag{1}$$

where $\mathcal{E} = mc^2$ is the relativistic energy.

(b) Also prove that

$$\mathcal{E}' = \gamma(\mathcal{E} - \mathbf{v}\cdot\mathbf{p}). \tag{2}$$

(c) Hence show that the force $\mathbf{F}$ on a particle moving with velocity $\mathbf{u}$ in frame S transforms as follows:

$$\left.\begin{aligned}\mathbf{F}'_\parallel &= \mathbf{F}_\parallel - \frac{(\mathbf{F}_\perp \cdot \mathbf{u}_\perp)\mathbf{v}}{(c^2 - u_\parallel v)}\\[2mm]\mathbf{F}'_\perp &= \frac{\mathbf{F}_\perp}{\gamma(1 - u_\parallel v/c^2)}\end{aligned}\right\}. \tag{3}$$

Hint: The work-energy theorem of classical mechanics also holds relativistically,[‡] so $d\mathcal{E}/dt = \mathbf{F}\cdot\mathbf{u}$.

Solution

(a) Substituting (1) of Question 12.2 and (2) of Question 12.3 in $\mathbf{p}' = m'\mathbf{u}'$ gives

$$\mathbf{p}' = \gamma(1 - u_\parallel v/c^2)m \times \frac{(\mathbf{u}_\parallel + \mathbf{u}_\perp/\gamma - \mathbf{v})}{(1 - u_\parallel v/c^2)}.$$

Now $m = \mathcal{E}/c^2$, and so

$$\mathbf{p}' = \gamma\mathbf{p}_\parallel + \mathbf{p}_\perp - \gamma\mathbf{v}\mathcal{E}/c^2.$$

which is (1).

(b) The transformation for $\mathcal{E}$ is similar:

$$\mathcal{E}' = m'c^2 = \gamma(1 - u_\parallel v/c^2)mc^2 = \gamma(\mathcal{E} - vp_\parallel).$$

Hence (2).

(c) Begin with $\mathbf{F}' = \dfrac{d\mathbf{p}'}{dt'} = \dfrac{d\mathbf{p}'/dt}{dt'/dt} = \dfrac{d\mathbf{p}'/dt}{\gamma(1 - u_\parallel v/c^2)}$. Then using (1), (2) and the hint we obtain

$$\begin{aligned}\mathbf{F}' &= \frac{\dfrac{d}{dt}\left[\gamma(\mathbf{p}_\parallel - \mathbf{v}\mathcal{E}/c^2) + \mathbf{p}_\perp\right]}{\gamma(1 - u_\parallel v/c^2)} = \frac{\gamma\left(\mathbf{F}_\parallel - \dfrac{\mathbf{v}}{c^2}\dfrac{d\mathcal{E}}{dt}\right) + \mathbf{F}_\perp}{\gamma(1 - u_\parallel v/c^2)}\\[4mm]&= \frac{\gamma\left(\mathbf{F}_\parallel - \dfrac{\mathbf{v}}{c^2}\mathbf{F}\cdot\mathbf{u}\right) + \mathbf{F}_\perp}{\gamma(1 - u_\parallel v/c^2)}. \tag{4}\end{aligned}$$

Now $\mathbf{F}'_\parallel = \dfrac{(\mathbf{F}'\cdot\mathbf{v})\mathbf{v}}{v^2}$ and so from (4):

[‡] See (1) of Question 15.10 in Ref. [4].

$$\mathbf{F}'_\| = \frac{\gamma(\mathbf{F}_\| \cdot \mathbf{v} - \beta^2 \mathbf{F} \cdot \mathbf{u})\mathbf{v}}{\gamma v^2(1 - u_\| v/c^2)} \qquad \text{(since } \mathbf{F}_\perp \cdot \mathbf{v} = 0)$$

$$= \frac{\left[F_\| v\mathbf{v} - \beta^2(\mathbf{F}_\| + \mathbf{F}_\perp)\cdot(\mathbf{u}_\| + \mathbf{u}_\perp)\mathbf{v} \right]}{v^2(1 - u_\| v/c^2)}$$

$$= \frac{\left[\mathbf{F}_\| v^2 - \beta^2 \mathbf{F}_\| u_\| v - \beta^2(\mathbf{F}_\perp \cdot \mathbf{u}_\perp)\mathbf{v} \right]}{v^2(1 - u_\| v/c^2)} \qquad (\mathbf{F}_\| \text{ is } \| \text{ to } \mathbf{v} \text{ by definition})$$

$$= \frac{\left[v^2 \mathbf{F}_\|(1 - u_\| v/c^2) - \beta^2(\mathbf{F}_\perp \cdot \mathbf{u}_\perp)\mathbf{v} \right]}{v^2(1 - u_\| v/c^2)},$$

which is $(3)_1$. Equation $(3)_2$ follows immediately from $\mathbf{F}'_\perp = \mathbf{F}' - \mathbf{F}'_\|$, with $\mathbf{F}'$ given by (4) and $\mathbf{F}'_\|$ given by $(3)_1$.

Comments

(i) Notice that $\mathbf{p}$ and $\mathcal{E}/c$ together transform like $\mathbf{r}$ and ct (see (1) of Question 12.1). We therefore expect, and it is easily confirmed, that $p^2 - \mathcal{E}^2/c^2$ is invariant. That is,

$$p'^2 - \mathcal{E}'^2/c^2 = p^2 - \mathcal{E}^2/c^2. \qquad (5)$$

(ii) In the non-relativistic limit, (3) reduces to $\mathbf{F}'_\| = \mathbf{F}_\|$ and $\mathbf{F}'_\perp = \mathbf{F}_\perp$, or

$$\mathbf{F}' = \mathbf{F} \qquad \text{(for } v \ll c) \qquad (6)$$

which is a result we recognize from Galilean relativity.

(iii) An important special case of (3) occurs when the particle is (instantaneously) at rest in S so that $\mathbf{u} = 0$. Then

$$\mathbf{F}'_\| = \mathbf{F}_\| \qquad \text{and} \qquad \mathbf{F}'_\perp = \mathbf{F}_\perp/\gamma. \qquad (7)$$

We note the following:

☞ Equation (7) shows that the force in the instantaneous rest frame of the particle is greater than the corresponding force in any other frame (this is a useful result and is worth remembering).

☞ The transformation (3) does not involve the *coordinates* of the point of application of the force, but rather the *velocity* of the point of application of the force (which is $\mathbf{u}$ in S and $\mathbf{u}'$ in S').

☞ Whenever the force has $\mathbf{F}_\perp \neq 0$, the transformation changes both the magnitude and direction of $\mathbf{F}$. Forces that are colinear in one frame are not necessarily colinear in another frame.

☞ Forces that are equal and opposite in one frame are not necessarily equal and opposite in another frame (i.e. Newton's third law does not, in general, apply relativistically). Ref. [5] explains:

> If the force of A on B at some instant t is $F(t)$, and the force of B on A at the same instant is $-F(t)$; then Newton's third law applies, *in this reference frame*. But a moving observer will report that these equal and opposite forces occurred at *different times*; in his system, therefore, the third law is *violated*. Only in the case of contact interactions, where the two are applied at the *same physical point* (and in the trivial case where the forces are *constant*), can the third law be retained.

Question 12.5

Two identical particles (each carrying a charge q) move alongside each other in the same direction with constant velocity $\mathbf{v} = v\hat{\mathbf{x}}$ as shown in the adjacent figure. Suppose the distance y between the particles is very much greater than the range of nuclear forces. Use (7) of Question 12.4 to determine the force exerted by either particle on the other.

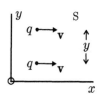

Hint: Start by applying Coulomb's law in the rest frame of the charges.

Solution

Consider the frame S' moving with velocity $\mathbf{v}$ in which the particles are at rest. Suppose S and S' are in the standard configuration. In S' the force between the charges is

$$F'_\| = 0 \quad \text{and} \quad F'_\perp = \frac{1}{4\pi\epsilon_0}\frac{q'^2}{y'^2} = \frac{1}{4\pi\epsilon_0}\frac{q^2}{y^2}, \tag{1}$$

since $q' = q$ (charge is invariant) and $y' = y$ (no Lorentz contraction perpendicular to the velocity). The inverse force transformation of Question 12.4 gives $F_\| = F'_\|$ and $F_\perp = F'_\perp/\gamma$. Hence

$$\left.\begin{aligned} F_\| &= F'_\| = 0 \\[2mm] F_\perp &= \frac{F'_\perp}{\gamma} = \frac{1}{4\pi\epsilon_0}\frac{q^2}{y^2}(1 - v^2/c^2)^{1/2} \end{aligned}\right\}, \tag{2}$$

or

$$\mathbf{F} = \frac{1}{4\pi\epsilon_0}\frac{q^2}{y^2}(1 - v^2/c^2)^{1/2}\hat{\mathbf{y}}. \tag{3}$$

[5] D. J. Griffiths, *Introduction to electrodynamics*, Chap. 12, pp. 517–18. New York: Prentice Hall, 3 edn, 1999.

Comments

(i) Note the following:

- ☞ Strictly speaking, Coulomb's law gives the force between point charges which are *stationary* (as they are here in S'). However, (3) shows that Coulomb's law remains a good approximation for the force at non-relativistic speeds ($v \ll c$).

- ☞ It is also clear from (3) that $F \to 0$ as $v \to c$. The electric force of repulsion is almost (but not entirely) cancelled by a magnetic force of attraction (see Question 12.6). From this we learn an important result: a repulsive force in one inertial frame will be repulsive in *all* inertial reference frames (and similarly for attractive forces).

(ii) In Question 12.6, we consider a more general version of the above problem.

Question 12.6

Consider charges q_1 and q_2, at rest in an inertial frame S', located at $(0,0,0)$ and $(x', y', 0)$ respectively. Use (7) of Question 12.4 to show that the force **F** exerted by q_1 on q_2 in inertial frame S (clearly, this is the frame in which the charges both move with speed v) is

$$\mathbf{F} = q_2 \Big[\mathbf{E}_1 + \mathbf{v} \times \mathbf{B}_1 \Big], \tag{1}$$

where

$$\left. \begin{aligned} \mathbf{E}_1 &= \frac{1}{4\pi\epsilon_0} \frac{\gamma q_1 \mathbf{r}}{(\gamma^2 x^2 + y^2)^{3/2}} \\ \mathbf{B}_1 &= \frac{\mu_0}{4\pi} \frac{\gamma q_1 v y \hat{\mathbf{z}}}{(\gamma^2 x^2 + y^2)^{3/2}} \end{aligned} \right\}. \tag{2}$$

Hint: As for Question 12.5, assume that frames S and S' are in standard configuration and start by applying Coulomb's law.

Solution

It is convenient to determine **F** at *one instant* in time which we take to be $t = t' = 0$. In S' the force **F**' exerted by q_1 on q_2 is given (exactly) by Coulomb's law since the charges are at rest in this frame. Then

$$F'_x = \frac{1}{4\pi\epsilon_0} \frac{q_1 q_2}{r'^3} x', \qquad F'_y = \frac{1}{4\pi\epsilon_0} \frac{q_1 q_2}{r'^3} y', \qquad F'_z = 0. \tag{3}$$

We now use (3) and the inverse force transformation (7) of Question 12.4 $\big(F_\| = F'_\|$; $F_\perp = F'_\perp/\gamma\big)$ to obtain the components of **F** in S, remembering that $\mathbf{u}' = 0$ and electric charge is invariant. This yields

$$F_x = F_x' = \frac{1}{4\pi\epsilon_0}\frac{q_1 q_2}{r'^3}x', \qquad F_y = \frac{F_y'}{\gamma} = \frac{1}{4\pi\epsilon_0}\frac{1}{\gamma}\frac{q_1 q_2}{r'^3}y', \qquad F_z = F_z' = 0. \quad (4)$$

It follows from the Lorentz transformation (I) that: $x' = \gamma x$, $y' = y$, $z' = 0$ with $r' = \sqrt{x'^2 + y'^2}$. Substituting these results in (4) gives (for the non-zero components):

$$F_x = \frac{1}{4\pi\epsilon_0}\frac{\gamma\, q_1 q_2 x}{(\gamma^2 x^2 + y^2)^{3/2}}, \qquad F_y = \frac{1}{4\pi\epsilon_0}\frac{q_1 q_2 y}{(\gamma^2 x^2 + y^2)^{3/2}}\frac{\gamma}{\gamma^2} = \frac{1}{4\pi\epsilon_0}\frac{\gamma q_1 q_2 y(1 - v^2/c^2)}{(\gamma^2 x^2 + y^2)^{3/2}}.$$

Then from $\mathbf{F} = \hat{\mathbf{x}}F_x + \hat{\mathbf{y}}F_y$ we obtain

$$
\begin{aligned}
\mathbf{F} &= \frac{1}{4\pi\epsilon_0}\frac{\gamma q_1 q_2}{(\gamma^2 x^2 + y^2)^{3/2}}\big[\hat{\mathbf{x}}x + \hat{\mathbf{y}}y(1 - v^2/c^2)\big] \\
&= \frac{1}{4\pi\epsilon_0}\frac{\gamma q_1 q_2}{(\gamma^2 x^2 + y^2)^{3/2}}\big[\hat{\mathbf{x}}x + \hat{\mathbf{y}}y + (\hat{\mathbf{x}}\times\hat{\mathbf{z}})yv^2/c^2\big] \\
&= q_2\left[\frac{1}{4\pi\epsilon_0}\frac{\gamma q_1 \mathbf{r}}{(\gamma^2 x^2 + y^2)^{3/2}} + \mathbf{v}\times\frac{1}{4\pi\epsilon_0}\frac{\gamma q_1 vy\hat{\mathbf{z}}}{c^2(\gamma^2 x^2 + y^2)^{3/2}}\right]. \quad (5)
\end{aligned}
$$

But this is (1) with $\mathbf{E}_1$ given by $(2)_1$ and $\mathbf{B}_1$ given by $(2)_2$ since $c^2\epsilon_0 = \mu_0{}^{-1}$.

Comment

That a purely electric field in one frame transforms to an electric *and* a magnetic field in another frame is often surprising and then, upon reflection, the source of some satisfaction for students encountering this result for the first time.[‡] Indeed, even seasoned hands find that 'after almost sixty years, classical electrodynamics still impresses and delights as a beautiful example of the covariance of physical laws under Lorentz transformations'.[6]

Question 12.7

Use the Lorentz transformation for frames in standard configuration to prove that the partial derivatives $\partial/\partial x$, $\partial/\partial y$, $\partial/\partial z$ and $\partial/\partial t$ transform as follows:

$$\frac{\partial}{\partial x} = \gamma\left(\frac{\partial}{\partial x'} - \frac{v}{c^2}\frac{\partial}{\partial t'}\right), \qquad \frac{\partial}{\partial y} = \frac{\partial}{\partial y'}, \qquad \frac{\partial}{\partial z} = \frac{\partial}{\partial z'}, \qquad \frac{\partial}{\partial t} = \gamma\left(\frac{\partial}{\partial t'} - v\frac{\partial}{\partial x'}\right). \quad (1)$$

[‡]It suggests an intimate relationship between electric and magnetic fields—an important matter to which we return in Question 12.22.

[6]   J. D. Jackson, *Classical electrodynamics*, Preface to the 1st edn, p. xii. New York: Wiley, 3 edn, 1998.

Solution

The partial derivatives of a function $F(x_1, x_2, x_3, x_4)$ satisfy

$$\frac{\partial F}{\partial x_i} = \sum_{j=1}^{4} \frac{\partial x'_j}{\partial x_i} \frac{\partial F}{\partial x'_j} \quad \text{or} \quad \frac{\partial}{\partial x_i} = \sum_{j=1}^{4} \frac{\partial x'_j}{\partial x_i} \frac{\partial}{\partial x'_j}.$$

For a given event in S the coordinates (x, y, z, t) are, in general, functions of the coordinates (x', y', z', t') in S'. So, for example,

$$\frac{\partial}{\partial x} = \frac{\partial x'}{\partial x} \frac{\partial}{\partial x'} + \frac{\partial y'}{\partial x} \frac{\partial}{\partial y'} + \frac{\partial z'}{\partial x} \frac{\partial}{\partial z'} + \frac{\partial t'}{\partial x} \frac{\partial}{\partial t'}. \tag{2}$$

Now the Lorentz transformation (I) of the introduction gives

$$\frac{\partial x'}{\partial x} = \gamma, \qquad \frac{\partial y'}{\partial x} = 0, \qquad \frac{\partial z'}{\partial x} = 0, \qquad \frac{\partial t'}{\partial x} = -\frac{\gamma v}{c^2}, \tag{3}$$

because v and γ are constants. Substituting (3) in (2) yields $(1)_1$. The transformations for $\partial/\partial y$, $\partial/\partial z$ and $\partial/\partial t$ follow in a similar way.

Question 12.8

(a) Suppose S and S' are frames in the standard configuration. Use the Maxwell–Faraday equation and the relativity principle to show that the six components of the electromagnetic field (three each for $\mathbf{E}$ and $\mathbf{B}$) transform as follows:

$$\left. \begin{array}{lll} E'_x = E_x, & E'_y = \gamma(E_y - vB_z), & E'_z = \gamma(E_z + vB_y), \\ B'_x = B_x, & B'_y = \gamma(B_y + vE_z/c^2), & B'_z = \gamma(B_z - vE_y/c^2) \end{array} \right\}. \tag{1}$$

Hint: Start with $(\nabla \times \mathbf{E})_y$ and use (1) of Question 12.7. Then consider the z-component. Do the x-component last.

(b) Express (1) in the coordinate-free form:

$$\left. \begin{array}{ll} \mathbf{E}'_\| = \mathbf{E}_\| & \mathbf{E}'_\perp = \gamma[\mathbf{E}_\perp + (\mathbf{v} \times \mathbf{B})] \\ \mathbf{B}'_\| = \mathbf{B}_\| & \mathbf{B}'_\perp = \gamma[\mathbf{B}_\perp - (\mathbf{v} \times \mathbf{E})/c^2] \end{array} \right\}, \tag{2}$$

where, as before, the subscripts $\|$ and $\perp$ refer to components parallel and perpendicular to $\mathbf{v}$ respectively.

Solution

(a) Because of the hint, $\partial E_x/\partial z - \partial E_z/\partial x = -\partial B_y/\partial t$. Then using the transformation equations for $\partial/\partial z$, $\partial/\partial x$, and $\partial/\partial t$ (see (1) of Question 12.7) gives

$$\frac{\partial E_x}{\partial z'} - \gamma\left(\frac{\partial E_z}{\partial x'} - \frac{v}{c^2}\frac{\partial E_z}{\partial t'}\right) = -\gamma\left(\frac{\partial B_y}{\partial t'} - v\frac{\partial B_y}{\partial x'}\right), \tag{3}$$

which after rearrangement yields

$$\frac{\partial E_x}{\partial z'} - \frac{\partial}{\partial x'}\gamma(E_z + vB_y) = -\frac{\partial}{\partial t'}\gamma(B_y + vE_z/c^2). \tag{4}$$

Now the relativity principle (see p. 554) requires that in frame S' the y'-component of the curl of $\mathbf{E}'$ has the form

$$\frac{\partial E'_x}{\partial z'} - \frac{\partial E'_z}{\partial x'} = -\frac{\partial B'_y}{\partial t'}. \tag{5}$$

Comparing (4) and (5) shows that

$$E'_x = E_x, \qquad E'_z = \gamma(E_z + vB_y), \qquad B'_y = \gamma(B_y + vE_z/c^2).$$

Next, we consider $\partial E_y/\partial x - \partial E_x/\partial y = -\partial B_z/\partial t$. Proceeding as before and rearranging yield

$$\frac{\partial}{\partial x'}\gamma(E_y - vB_z) - \frac{\partial E_x}{\partial y'} = -\frac{\partial}{\partial t'}\gamma(B_z - vE_y/c^2). \tag{6}$$

Comparing (6) with $\partial E'_y/\partial x' - \partial E'_x/\partial y' = -\partial B'_z/\partial t'$ shows that

$$E'_x = E_x, \qquad E'_y = \gamma(E_y - vB_z), \qquad B'_z = \gamma(B_z - vE_y/c^2).$$

Finally $\partial E_z/\partial y - \partial E_y/\partial z = -\partial B_x/\partial t$ and (1) of Question 12.7 give

$$\frac{\partial E_z}{\partial y'} - \frac{\partial E_y}{\partial z'} = -\gamma\left(\frac{\partial B_x}{\partial t'} - v\frac{\partial B_x}{\partial x'}\right). \tag{7}$$

In order to progress further, we must make use of $\nabla \cdot \mathbf{B} = 0$. This gives $\partial B_x/\partial x = -\partial B_y/\partial y - \partial B_z/\partial z$. As before, the transformation equations for the derivatives yield

$$v\gamma\left(\frac{\partial B_x}{\partial x'} - \frac{v}{c^2}\frac{\partial B_x}{\partial t'}\right) = -v\frac{\partial B_y}{\partial y'} - v\frac{\partial B_z}{\partial z'}. \tag{8}$$

Adding (7) and (8) and rearranging give

$$\frac{\partial}{\partial y'}(E_z + vB_y) - \frac{\partial}{\partial z'}(E_y - vB_z) = -\gamma\frac{\partial}{\partial t'}\left(B_x - \frac{v^2}{c^2}B_x\right) = -\frac{1}{\gamma}\frac{\partial B_x}{\partial t'}.$$

So

$$\frac{\partial}{\partial y'}\gamma(E_z + vB_y) - \frac{\partial}{\partial z'}\gamma(E_y - vB_z) = -\frac{\partial B_x}{\partial t'}. \tag{9}$$

Comparing (9) with $\partial E_z'/\partial y' - \partial E_y'/\partial z' = -\partial B_x'/\partial t'$ shows that

$$E_y' = \gamma(E_y - vB_z), \qquad E_z' = \gamma(E_z + vB_y), \qquad \boxed{B_x' = B_x}.$$

The boxed equations above are the results we seek.

(b) Substituting (1) in $\mathbf{E}' = \hat{\mathbf{x}}E_x' + \hat{\mathbf{y}}E_y' + \hat{\mathbf{z}}E_z'$ yields

$$\mathbf{E}' = \hat{\mathbf{x}}E_x + \hat{\mathbf{y}}\gamma E_y + \hat{\mathbf{z}}\gamma E_z - \gamma v(\hat{\mathbf{y}}B_z - \hat{\mathbf{z}}B_y).$$

Now $\mathbf{v} = (v, 0, 0)$, and so $(\mathbf{v} \times \mathbf{B}) = -v(\hat{\mathbf{y}}B_z - \hat{\mathbf{z}}B_y)$. Thus

$$\mathbf{E}' = \hat{\mathbf{x}}E_x + \hat{\mathbf{y}}\gamma E_y + \hat{\mathbf{z}}\gamma E_z + \gamma(\mathbf{v} \times \mathbf{B}).$$

By definition, $\hat{\mathbf{x}}E_x = \mathbf{E}_\parallel$ and $(\hat{\mathbf{y}}E_y + \hat{\mathbf{z}}E_z) = \mathbf{E}_\perp$. Furthermore, the vector $(\mathbf{v} \times \mathbf{B})$ is normal to $\mathbf{v}$ and so transforms as part of the perpendicular component. Hence

$$\mathbf{E}' = \mathbf{E}_\parallel + \gamma[\mathbf{E}_\perp + (\mathbf{v} \times \mathbf{B})]. \tag{10}$$

Comparing (10) with $\mathbf{E}' = \mathbf{E}_\parallel' + \mathbf{E}_\perp'$ shows that $\mathbf{E}_\parallel' = \mathbf{E}_\parallel$ and $\mathbf{E}_\perp' = \gamma[\mathbf{E}_\perp + (\mathbf{v} \times \mathbf{B})]$. The proof for the **B**-field transformation is similar.

Comments

(i) The transformation (1) relates the electric and magnetic fields at a space-time point $P(x, y, z, t)$ in S with the electric and magnetic fields at a space-time point $P'(x', y', z', t')$ in S', where P and P' are related by the Lorentz transformation.

(ii) Notice the redundancy in this method, where each of the electric-field equations is duplicated. A more elegant derivation, which avoids this clumsy feature, is presented in Question 12.22.

(iii) Equation (3) can be written in the equivalent form:

$$\left.\begin{aligned} \mathbf{E}' &= \mathbf{E} + (\gamma - 1)\mathbf{E}_\perp + \gamma(\mathbf{v} \times \mathbf{B}) \\ \mathbf{B}' &= \mathbf{B} + (\gamma - 1)\mathbf{B}_\perp - \gamma(\mathbf{v} \times \mathbf{E})/c^2 \end{aligned}\right\}. \tag{11}$$

The non-relativistic limit of (11), obtained from $\beta \to 0$ and $\gamma \to 1$, is

$$\left.\begin{aligned} \mathbf{E}' &= \mathbf{E} + (\mathbf{v} \times \mathbf{B}) \\ \mathbf{B}' &= \mathbf{B} \end{aligned}\right\}, \tag{12}$$

which is the Galilean result (1) of Question 5.5.

(iv) Note that if the magnetic field is zero in some frame ($\mathbf{B}' = 0$) it follows from (2) that

$$\mathbf{B} = (\mathbf{v} \times \mathbf{E})/c^2, \tag{13}$$

whereas if the electric field is zero ($\mathbf{E}' = 0$) we have

$$\mathbf{E} = -(\mathbf{v} \times \mathbf{B}). \tag{14}$$

(v) The inverse transformations of (1) and (2) are sometimes required. They are

$$\left. \begin{array}{lll} E_x = E'_x, & E_y = \gamma(E'_y + vB'_z), & E_z = \gamma(E'_z - vB'_y), \\ B_x = B'_x, & B_y = \gamma(B'_y - vE'_z/c^2), & B_z = \gamma(B'_z + vE'_y/c^2) \end{array} \right\} \tag{15}$$

and

$$\left. \begin{array}{ll} \mathbf{E}_\| = \mathbf{E}'_\| & \mathbf{E}_\perp = \gamma\big[\mathbf{E}'_\perp - (\mathbf{v} \times \mathbf{B}')\big] \\ \mathbf{B}_\| = \mathbf{B}'_\| & \mathbf{B}_\perp = \gamma\big[\mathbf{B}'_\perp + (\mathbf{v} \times \mathbf{E}')/c^2\big] \end{array} \right\}. \tag{16}$$

Question 12.9

Consider an infinite uniform line charge of magnitude λ' at rest along the x'-axis of inertial frame S'. Use the electrostatic field in this frame (see (1) of Question 2.8) to derive the magnetic field in inertial frame S due to a linear current I.

Hint: The charges move with velocity $v\hat{\mathbf{x}}$ in frame S (as usual, we assume that frames S and S' are in standard configuration).

Solution

In frame S' the field is given by

$$\left. \begin{array}{ll} \mathbf{E}'_\| = 0 & \mathbf{E}'_\perp = \dfrac{1}{2\pi\epsilon_0} \dfrac{\lambda'\mathbf{r}'}{r'^2} \\ \\ \mathbf{B}'_\| = 0 & \mathbf{B}'_\perp = 0 \end{array} \right\},$$

where $\mathbf{r}'$ is a vector perpendicular to the line charge. Then, using transformation (16) of Question 12.8, it follows that

$$\left. \begin{array}{ll} \mathbf{E}_\| = \mathbf{E}'_\| = 0 & \mathbf{E}_\perp = \gamma\mathbf{E}'_\perp = \dfrac{\gamma}{2\pi\epsilon_0} \dfrac{\lambda'\mathbf{r}'}{r'^2} \\ \\ \mathbf{B}_\| = \mathbf{B}'_\| = 0 & \mathbf{B}_\perp = \gamma\dfrac{\mathbf{v} \times \mathbf{E}'_\perp}{c^2} = \dfrac{\gamma\lambda'}{2\pi\epsilon_0} \dfrac{(\mathbf{v} \times \mathbf{r}')}{r'^2 c^2} \end{array} \right\},$$

which shows that the magnetic field in S is given by

$$\mathbf{B} = \frac{\gamma \lambda'}{2\pi\epsilon_0} \frac{(\mathbf{v} \times \mathbf{r}')}{r'^2 c^2} = \frac{\gamma \lambda'}{2\pi\epsilon_0} \frac{(\mathbf{v} \times \mathbf{r})}{r^2 c^2}. \tag{1}$$

In the last step we use $\mathbf{r}' = \mathbf{r} - \gamma\mathbf{v}t$ (see (5) of Question 12.1) and $r' = r$ (no Lorentz contraction in the direction perpendicular to $\mathbf{v}$). Because charge is an invariant quantity, $\lambda\ell = \lambda'\ell'$ where $\ell = \ell'/\gamma$ (the mean distance between the charges along the x-axis is Lorentz contracted in frame S). Therefore $\lambda = \lambda'\ell'/\ell = \gamma\lambda'$. Substituting this result in (1) yields

$$\mathbf{B} = \frac{1}{2\pi\epsilon_0} \frac{(\lambda v \hat{\mathbf{x}} \times \hat{\mathbf{r}})}{rc^2} = \frac{\mu_0}{2\pi} \frac{(\lambda v \hat{\mathbf{x}} \times \hat{\mathbf{r}})}{r}.$$

Letting $I = \lambda v$ and defining $\hat{\boldsymbol{\theta}} = \hat{\mathbf{x}} \times \hat{\mathbf{r}}$ give $\mathbf{B} = \dfrac{\mu_0}{2\pi} \dfrac{I}{r} \hat{\boldsymbol{\theta}}$, which is the usual magnetic field of an infinite current (see $(3)_2$ of Question 4.7).

Question 12.10

Prove that the following quantities are relativistically invariant:

(a) $\mathbf{E} \cdot \mathbf{B}$, $\qquad\qquad\qquad\qquad\qquad\qquad\qquad\qquad\qquad\qquad\qquad$ (1)

(b) $\mathbf{E}^2 - c^2\mathbf{B}^2$, $\qquad\qquad\qquad\qquad\qquad\qquad\qquad\qquad\qquad\qquad\quad$ (2)

(c) $\frac{1}{2}\epsilon_0\mathbf{E}^2 - \frac{1}{2}\mu_0^{-1}\mathbf{B}^2$, $\qquad\qquad\qquad\qquad\qquad\qquad\qquad\qquad\qquad\quad$ (3)

(d) $\mathbf{S}^2 - c^2u^2$, $\qquad\qquad\qquad\qquad\qquad\qquad\qquad\qquad\qquad\qquad\qquad$ (4)

where $\mathbf{S}$ is the Poynting vector and u the energy density of the electromagnetic field.

Hints: For (a) and (b) use the transformations of Question 12.8; for (c) use (2); for (d) use (1) and (3).

Solution

(a) Clearly $\mathbf{E}' \cdot \mathbf{B}' = E_x'B_x' + E_y'B_y' + E_z'B_z'$. Then using (1) of Question 12.8 gives

$$\begin{aligned}
\mathbf{E}' \cdot \mathbf{B}' &= E_x'B_x' + E_y'B_y' + E_z'B_z' \\
&= E_xB_x + \gamma^2(E_y - vB_z)(B_y + vE_z/c^2) + \gamma^2(E_z + vB_y)(B_z - vE_y/c^2) \\
&= E_xB_x + \gamma^2[E_yB_y(1 - v^2/c^2) + E_zB_z(1 - v^2/c^2)] \\
&= E_xB_x + E_yB_y + E_zB_z \qquad (\text{since } \gamma^2(1 - v^2/c^2) = 1), \\
&= \mathbf{E} \cdot \mathbf{B},
\end{aligned}$$

as required.

(b) Proceeding as above:

$$
\begin{aligned}
\mathbf{E}'^2 - c^2\mathbf{B}'^2 &= (E_x'^2 + E_y'^2 + E_z'^2) - c^2(B_x'^2 + B_y'^2 + B_z'^2) \\
&= E_x^2 + \gamma^2(E_y - vB_z)^2 + \gamma^2(E_z + vB_y)^2 - \\
&\quad c^2 B_x^2 - c^2\gamma^2(B_y + vE_z/c^2)^2 - c^2\gamma^2(B_z - vE_y/c^2)^2 \\
&= E_x^2 + \gamma^2(1 - \beta^2)E_y^2 + \gamma^2(1 - \beta^2)E_z^2 - \\
&\quad c^2 B_x^2 - \gamma^2(c^2 - v^2)B_y^2 - \gamma^2(c^2 - v^2)B_z^2 \\
&= (E_x^2 + E_y^2 + E_z^2) - c^2(B_x^2 + B_y^2 + B_z^2),
\end{aligned}
$$

which is (2).

(c) Multiplying (2) by $\frac{1}{2}\epsilon_0$ gives $\frac{1}{2}\epsilon_0\mathbf{E}^2 - \frac{1}{2}\epsilon_0 c^2\mathbf{B}^2 = \frac{1}{2}\epsilon_0\mathbf{E}'^2 - \frac{1}{2}\epsilon_0 c^2\mathbf{B}'^2$. Now $\epsilon_0 c^2 = \mu_0^{-1}$. Hence (3).

(d) By definition $\mathbf{S}' = \mu_0^{-1}(\mathbf{E}' \times \mathbf{B}')$ and $S'^2 = \mu_0^{-2}(\mathbf{E}' \times \mathbf{B}') \cdot (\mathbf{E}' \times \mathbf{B}')$. Then, using identity (2) of Question 1.8, we obtain $S'^2 = \mu_0^{-2}[E'^2 B'^2 - (\mathbf{E}' \cdot \mathbf{B}')^2]$. Now $u' = (\frac{1}{2}\epsilon_0 E'^2 + \frac{1}{2}B'^2/\mu_0)$, and so

$$
\begin{aligned}
S'^2 - c^2 u'^2 &= \left[\frac{E'^2 B'^2 - (\mathbf{E}' \cdot \mathbf{B}')^2}{\mu_0^2}\right] - c^2\left[\left(\frac{\epsilon_0 E'^2}{2}\right)^2 + \left(\frac{B'^2}{2\mu_0}\right)^2 + \left(\frac{E'^2 B'^2}{2c^2\mu_0^2}\right)\right] \\
&= -\frac{(\mathbf{E}' \cdot \mathbf{B}')^2}{\mu_0^2} - c^2\left[\left(\frac{\epsilon_0 E'^2}{2}\right) - \left(\frac{B'^2}{2\mu_0}\right)\right]^2 \\
&= -\frac{(\mathbf{E} \cdot \mathbf{B})^2}{\mu_0^2} - c^2\left[\left(\frac{\epsilon_0 E^2}{2}\right) - \left(\frac{B^2}{2\mu_0}\right)\right]^2 \\
&= \left(\frac{\mathbf{E} \times \mathbf{B}}{\mu_0}\right)^2 - c^2\left(\frac{1}{2}\epsilon_0 E^2 + \frac{1}{2}B^2/\mu_0\right),
\end{aligned}
$$

which is (4).

Comments

(i) Some obvious consequences of (1) and (2) are:

   ☞ If $\mathbf{E}$ and $\mathbf{B}$ are perpendicular in an inertial frame S, then they are perpendicular in all other inertial frames (unless either $\mathbf{E}'$ or $c\mathbf{B}'$ is zero; see below).

   ☞ If $\mathbf{E}$ and $\mathbf{B}$ are perpendicular in inertial frame S, then it is possible to find a frame S' in which either $\mathbf{E}'$ or $c\mathbf{B}'$ can be eliminated, depending on whichever is the smaller. See Question 12.11.

   ☞ If the angle between $\mathbf{E}$ and $\mathbf{B}$ is acute (obtuse) in some inertial frame, it will be acute (obtuse) in all other inertial frames.

☞ If $E > cB$ in inertial frame S then $E' > cB'$ in any other frame S', and vice versa. So, for example, if $\mathbf{B}'$ is zero in some frame, then it is impossible to find a frame where $\mathbf{E}$ is zero.

(ii) Notice from (3) that it is the *difference* of the **E**- and **B**-field energy densities (rather than their sum) which is invariant.[‡]

(iii) We discovered in Question 7.5 that for a plane electromagnetic wave in vacuum, **E** and **B** are orthogonal and $E = cB$. So $\mathbf{E} \cdot \mathbf{B} = 0$ and $\mathbf{E}^2 - c^2\mathbf{B}^2 = 0$. Ref. [7] explains that 'if an invariant is zero in any frame, it must be zero in all frames. We see that *any* Lorentz transformation of the wave will leave **E** and $c\mathbf{B}$ perpendicular and equal in magnitude. *A light wave looks like a light wave in any inertial frame of reference*'.

Question 12.11

(a) Consider **E**- and **B**-fields that are perpendicular at some point P in frame S. Show that there exists a frame S' in which the field at P' is either pure electric ($\mathbf{B}' = 0$) or pure magnetic ($\mathbf{E}' = 0$).[♯] In each case, find the velocity **v** of S' relative to S.

(b) Suppose $\mathbf{E} = 10^6\,\hat{\mathbf{y}}\,\mathrm{V\,m^{-1}}$ and $\mathbf{B} = 10^{-3}\,\hat{\mathbf{z}}\,\mathrm{T}$ in frame S. Is it possible to find a primed frame in which either $\mathbf{E}' = 0$ or $\mathbf{B}' = 0$? If so, calculate the velocity **v** of this frame and also the magnitude of the non-zero field.

(c) Suppose a frame S' moves with velocity $\frac{1}{2}\mathbf{v}$ relative to S $\left(\text{using } \mathbf{v} \text{ from (b)}\right)$. Calculate the fields $\mathbf{E}'$ and $\mathbf{B}'$ in this frame.

(d) Repeat (c) for a relative speed of $2\mathbf{v}$ between the frames.

Solution

(a) <u>pure electric</u>

From (13) of Question 12.8 it follows that $v = c^2 B/E$, or

$$v = c\left(\frac{cB}{E}\right) \quad \text{where} \quad E > cB.$$

Because of the cyclic order $\hat{\mathbf{v}} = \dfrac{\mathbf{E} \times \mathbf{B}}{EB}$. Now $\mathbf{v} = v\hat{\mathbf{v}}$, and so

$$\text{for pure electric: } \mathbf{v} = c^2\frac{\mathbf{E} \times \mathbf{B}}{E^2} \quad \text{if} \quad E > cB\,. \tag{1}$$

[‡]The words 'invariant' and 'conserved' have separate meanings and should not be confused.
[♯]Here the coordinates of P and P' are related by the Lorentz transformation.

[7]   E. M. Purcell and D. J. Morin, *Electricity and magnetism*, Chap. 9, p. 453. Cambridge: Cambridge University Press, 3 edn, 2013.

pure magnetic

Now from (14) of Question 12.8 we have $v = E/B$, or

$$v = c\left(\frac{E}{cB}\right) \quad \text{where} \quad E < cB.$$

Again $\hat{\mathbf{v}} = \dfrac{\mathbf{E} \times \mathbf{B}}{EB}$, and so

$$\text{for pure magnetic:} \quad \mathbf{v} = \frac{\mathbf{E} \times \mathbf{B}}{B^2} \quad \text{if} \quad E < cB. \tag{2}$$

(b) Calculating the value of cB yields $3 \times 10^5\,\mathrm{V\,m^{-1}}$. Because $E > cB$ it is possible to find a frame in which the field is pure electric only. So $\beta = 3 \times 10^5/10^6 = 0.3$, or $\mathbf{v} = 9 \times 10^7\,\hat{\mathbf{x}}\,\mathrm{m\,s^{-1}}$. From (2) of Question 12.10: $E'^2 = E^2 - c^2 B^2$ or $E' = \sqrt{10^{12} - 9 \times 10^{10}} = 0.954 \times 10^6\,\mathrm{V\,m^{-1}}$.

(c) With $v = 4.5 \times 10^7\,\mathrm{m\,s^{-1}}$ we obtain $\gamma = 1.011$. Then the transformation equations (2) of Question 12.8 give

$$\left. \begin{aligned}
\mathbf{E}' &= \gamma[\mathbf{E} + (\mathbf{v} \times \mathbf{B})] = 1.011 \times [10^6 \hat{\mathbf{y}} + 4.5 \times 10^7 \times 10^{-3}(-\hat{\mathbf{y}})] \\
&= 0.966 \times 10^6\,\hat{\mathbf{y}}\,\mathrm{V\,m^{-1}} \\[2mm]
\mathbf{B}' &= \gamma[\mathbf{B} - (\mathbf{v} \times \mathbf{E})/c^2] = 1.011 \times [10^{-3}\hat{\mathbf{z}} - 4.5 \times 10^7 \times 10^6/(3 \times 10^8)^2\,\hat{\mathbf{z}}] \\
&= 0.506 \times 10^{-3}\,\hat{\mathbf{z}}\,\mathrm{T}
\end{aligned} \right\}.$$

(d) Now $v = 1.8 \times 10^8\,\mathrm{m\,s^{-1}}$, which gives $\gamma = 1.25$, and we proceed as above. Then

$$\left. \begin{aligned}
\mathbf{E}' &= \gamma[\mathbf{E} + (\mathbf{v} \times \mathbf{B})] = 1.25 \times [10^6 \hat{\mathbf{y}} + 1.8 \times 10^8 \times 10^{-3}(-\hat{\mathbf{y}})] \\
&= 1.025 \times 10^6\,\hat{\mathbf{y}}\,\mathrm{V\,m^{-1}} \\[2mm]
\mathbf{B}' &= \gamma[\mathbf{B} - (\mathbf{v} \times \mathbf{E})/c^2] = 1.25 \times [10^{-3}\hat{\mathbf{z}} - 1.8 \times 10^8 \times 10^6/(3 \times 10^8)^2\,\hat{\mathbf{z}}] \\
&= -1.25 \times 10^{-3}\,\hat{\mathbf{z}}\,\mathrm{T}
\end{aligned} \right\}.$$

Comments

(i) For the fields in (b), we obtain the graph on p. 573. It shows $E'(\beta)$ and $B'(\beta)$ as the speed of frame S$'$ varies in the interval $-0.99 < \beta < 0.99$. We note, in particular, the following:

  ☞ The magnitude of $\mathbf{E}'$ is a minimum in the frame where the field is pure electric ($\beta = 0.3$).

  ☞ The magnitude of $\mathbf{B}'$ decreases as the speed of frame S$'$ increases and reverses direction when β exceeds $cE/B = 0.3$.

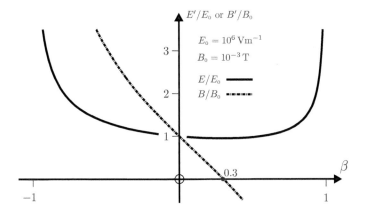

(ii) If **E** and **B** are orthogonal in some frame S and if $E \neq cB$, then it is always possible to find a primed frame where the field is either pure electric or pure magnetic. See Question 12.18 for a case where $E = cB$.

Question 12.12

(a) Suppose that at a point $P(\mathbf{r}, t)$ in frame S there is an electric field **E** and a magnetic field **B**. Show that the velocity **v** of the frame S′ in which **E**′ and **B**′ are parallel to each other, and perpendicular to **v** at the point $P'(\mathbf{r}', t')$,[‡] is given by

$$\frac{\mathbf{v}}{c^2 + v^2} = \frac{\mathbf{E} \times \mathbf{B}}{E^2 + c^2 B^2}. \tag{1}$$

(b) Suppose $\mathbf{E} = E_0 \hat{\mathbf{y}}$ and $\mathbf{B} = E_0(\sqrt{3}\hat{\mathbf{y}} + \hat{\mathbf{z}})/c$. Use (1) to calculate **v**, and hence determine **E**′ and **B**′.

Solution

(a) Since $\mathbf{E}'_{\parallel} = 0$; $\mathbf{B}'_{\parallel} = 0$ it follows immediately from (2) of Question 12.8 that $E_{\parallel} = 0$; $B_{\parallel} = 0$. Then $\mathbf{E}' = \gamma[\mathbf{E} + (\mathbf{v} \times \mathbf{B})]$ and $\mathbf{B}' = \gamma[\mathbf{B} - (\mathbf{v} \times \mathbf{E})/c^2]$, and so $c^2 \mathbf{E}' \times \mathbf{B}' = \gamma^2[c^2 \mathbf{E} \times \mathbf{B} - \mathbf{E} \times (\mathbf{v} \times \mathbf{E}) - c^2 \mathbf{B} \times (\mathbf{v} \times \mathbf{B}) - (\mathbf{v} \times \mathbf{B}) \times (\mathbf{v} \times \mathbf{E})]$. Now **E**′ and **B**′ are parallel vectors and their cross-product is therefore zero. Hence

$$c^2 \mathbf{E} \times \mathbf{B} = \mathbf{E} \times (\mathbf{v} \times \mathbf{E}) + c^2 \mathbf{B} \times (\mathbf{v} \times \mathbf{B}) + (\mathbf{v} \times \mathbf{B}) \times (\mathbf{v} \times \mathbf{E}). \tag{2}$$

Simplifying (2) using the **BAC–CAB** rule $\big($see (3) of Question 1.8$\big)$ yields

[‡]The space-time points $P(\mathbf{r}, t)$ and $P'(\mathbf{r}', t')$ are related by the Lorentz transformation.

$$c^2 \mathbf{E} \times \mathbf{B} = \mathbf{v}E^2 - \mathbf{E}(vE_\parallel) + \mathbf{v}c^2B^2 - c^2\mathbf{B}(vB_\parallel) + \mathbf{v}[\mathbf{E} \cdot (\mathbf{v} \times \mathbf{B})] - \mathbf{E}[\mathbf{v} \cdot (\mathbf{v} \times \mathbf{B})]$$

$$= \mathbf{v}E^2 + \mathbf{v}c^2B^2 + \mathbf{v}[\mathbf{E} \cdot (\mathbf{v} \times \mathbf{B})],$$

where in the last step we use $E_\parallel = 0$ and recognize that $\mathbf{v}$ is perpendicular to $\mathbf{v} \times \mathbf{B}$. Now $\mathbf{E} \cdot (\mathbf{v} \times \mathbf{B}) = -\mathbf{v} \cdot (\mathbf{E} \times \mathbf{B})$ because of the properties of the scalar triple product, and so

$$c^2 \mathbf{E} \times \mathbf{B} = \mathbf{v}E^2 + \mathbf{v}c^2B^2 - \mathbf{v}[\mathbf{v} \cdot (\mathbf{E} \times \mathbf{B})]. \tag{3}$$

Multiplying both sides of (3) with $\mathbf{v} \cdot$ gives

$$c^2 \mathbf{v} \cdot (\mathbf{E} \times \mathbf{B}) = v^2 E^2 + c^2 v^2 B^2 - v^2 [\mathbf{v} \cdot (\mathbf{E} \times \mathbf{B})],$$

and then

$$\mathbf{v} \cdot (\mathbf{E} \times \mathbf{B}) = \frac{v^2(E^2 + c^2 B^2)}{c^2 + v^2}. \tag{4}$$

Using (4) to eliminate $\mathbf{v} \cdot (\mathbf{E} \times \mathbf{B})$ from (3) yields (1).

(b) The velocity is in the direction of $\mathbf{E} \times \mathbf{B}$, which is along $\hat{\mathbf{x}}$. So

$$\frac{v\hat{\mathbf{x}}}{c^2 + v^2} = \frac{E_0^2[\hat{\mathbf{y}} \times (\sqrt{3}\hat{\mathbf{y}} + \hat{\mathbf{z}})]}{c(E_0^2 + 4E_0^2)} = \frac{\hat{\mathbf{x}}}{5c} \qquad \text{or} \qquad \beta^2 - 5\beta + 1 = 0. \tag{5}$$

Solving this quadratic equation gives $\beta = 0.2087$ (obviously we reject the root which gives $\beta > 1$), with a corresponding value of $\gamma = 1.0225$. Then

$$
\left.
\begin{aligned}
\mathbf{E}' &= \gamma[\mathbf{E} + (\mathbf{v} \times \mathbf{B})] = 1.0225 \times [\hat{\mathbf{y}} + 0.2087(\sqrt{3}\hat{\mathbf{z}} - \hat{\mathbf{y}})]E_0 \\
&= \boxed{0.370[2.189\hat{\mathbf{y}} + \hat{\mathbf{z}}]E_0} \\
\mathbf{B}' &= \gamma[\mathbf{B} - (\mathbf{v} \times \mathbf{E})/c^2] = 1.0225 \times [(\sqrt{3}\hat{\mathbf{y}} + \hat{\mathbf{z}}) - 0.2087\hat{\mathbf{z}}]E_0/c \\
&= \boxed{0.809[2.189\hat{\mathbf{y}} + \hat{\mathbf{z}}]E_0/c}
\end{aligned}
\right\}.
$$

Comment

As β increases from zero, the angle between $\mathbf{E}'$ and $\mathbf{B}'$ decreases from 30° (this being the angle between $\mathbf{E}$ and $\mathbf{B}$ in frame S) and becomes zero when $\beta = 0.209$. Then as $\beta \to 1$, the angle θ' between $\mathbf{E}'$ and $\mathbf{B}'$ increases and approaches 90° asymptotically (see Comment (i)$_3$ of Question 12.10).

Question 12.13

Consider a particle having constant proper mass m_0, charge q and moving with velocity $\mathbf{v}$ in an electromagnetic field.

(a) Show that the equation of motion of the particle is

$$\gamma m_0 \frac{d\mathbf{v}}{dt} = q\left[\mathbf{E} + \mathbf{v} \times \mathbf{B} - \mathbf{v}(\mathbf{E} \cdot \mathbf{v})/c^2\right]. \tag{1}$$

(b) Hence show that the component of the acceleration parallel to the electric field tends to zero in the limit $\beta \to 1$.

Solution

(a) From $\mathbf{F} = d\mathbf{p}/dt = q\left[\mathbf{E} + \mathbf{v} \times \mathbf{B}\right]$ with $\mathbf{p} = m_0(1 - v^2/c^2)^{-1/2}\mathbf{v}$ we obtain

$$\gamma m_0 \frac{d\mathbf{v}}{dt} + (\gamma^3 m_0 \mathbf{v}/c^2)\mathbf{v} \cdot \frac{d\mathbf{v}}{dt} = q\left[\mathbf{E} + \mathbf{v} \times \mathbf{B}\right]. \tag{2}$$

Multiplying both sides of (2) by $\mathbf{v}\cdot$ gives

$$q\mathbf{E} \cdot \mathbf{v} = \gamma m_0 \mathbf{v} \cdot \frac{d\mathbf{v}}{dt} + \gamma^3 m_0 \beta^2 \left(\mathbf{v} \cdot \frac{d\mathbf{v}}{dt}\right) = \gamma m_0 \mathbf{v} \cdot \frac{d\mathbf{v}}{dt} + \gamma m_0(\gamma^2 - 1)\mathbf{v} \cdot \frac{d\mathbf{v}}{dt}$$

$$= \gamma^3 m_0 \mathbf{v} \cdot \frac{d\mathbf{v}}{dt}. \tag{3}$$

Using (3) to eliminate the term $\gamma^3 m_0 \mathbf{v} \cdot d\mathbf{v}/dt$ in (2) yields (1).

(b) From (1):

$$\gamma m_0 \frac{d\mathbf{v}_\|}{dt} = q\left[\mathbf{E}_\| - \mathbf{v}(E_\| v)/c^2\right] \qquad \text{(since } \mathbf{v} \times \mathbf{B} \text{ has no component } \| \text{ to } \mathbf{v}\text{)}$$

$$= q\left[\mathbf{E}_\| - \beta^2 \mathbf{E}_\|\right] \qquad \text{($\mathbf{E}_\|$ is parallel to $\mathbf{v}$ by definition)}$$

$$= q(1 - \beta^2)\mathbf{E}_\| . \tag{4}$$

Now it follows from (4) that $\mathbf{a}_\| = \dfrac{q\mathbf{E}_\|}{\gamma^3 m_0} \to 0$ as $\gamma \to \infty$.

Comments

(i) The component of motion perpendicular to $\mathbf{v}$ also follows from (1) and is

$$\gamma m_0 \frac{d\mathbf{v}_\perp}{dt} = q\left[\mathbf{E}_\perp + \mathbf{v} \times \mathbf{B}_\perp\right]. \tag{5}$$

(ii) Putting $\mathbf{E} = 0$ in (1) gives the equation of motion in a uniform magnetic field:

$$\frac{d\mathbf{v}}{dt} = \mathbf{v} \times \boldsymbol{\omega}_{\mathrm{B}}, \tag{6}$$

where

$$\boldsymbol{\omega}_{\mathrm{B}} = \frac{q\mathbf{B}}{\gamma m_0} \tag{7}$$

is the cyclotron frequency. The trajectory of the particle is a circle of radius $v_\perp / \omega_{\mathrm{B}}$ and period $2\pi / \omega_{\mathrm{B}}$ if $\mathbf{v}_\| = 0$, and a helix otherwise. The relativistic motion is described by the same equations as the non-relativistic motion, provided the mass in the formula for the cyclotron frequency is taken to be γm_0.

Comment

See also Question 15.15 in Ref. [4].

Question 12.14

Consider a point charge q moving uniformly with velocity $\mathbf{v}$ in an inertial frame S. Use the transformation equations of Question 12.8 to show that the electromagnetic field at a point $P(\mathbf{r}, t)$ in S is given by

$$
\left.
\begin{aligned}
\mathbf{E}(\mathbf{r}, t) &= \frac{q}{4\pi\epsilon_0} \frac{(1 - \beta^2)\mathbf{R}}{R^3 (1 - \beta^2 \sin^2\theta)^{3/2}} \\[2mm]
\text{and} & \\[2mm]
\mathbf{B}(\mathbf{r}, t) &= \frac{\mu_0}{4\pi} \frac{q(1 - \beta^2)(\mathbf{v} \times \mathbf{R})}{R^3 (1 - \beta^2 \sin^2\theta)^{3/2}}
\end{aligned}
\right\}, \tag{1}
$$

where $\mathbf{R}$ is the vector from the instantaneous position of the charge to P, and θ is the angle between $\mathbf{R}$ and $\mathbf{v}$.

Solution

Let the particle be at the origin of S at time $t = 0$. Choose an inertial frame S' that is coincident with S at $t = t' = 0$ (frames S and S' are in standard configuration with S' moving uniformly with velocity $\mathbf{v}$ relative to S). In S' the particle also has charge q (charge is an invariant) and it is at rest at the origin. So in this frame the electromagnetic field is given by Coulomb's law

$$
\left.
\begin{aligned}
\mathbf{E}' &= \frac{q}{4\pi\epsilon_0} \frac{\mathbf{r}'}{r'^3} \\[2mm]
\text{and} & \\[2mm]
\mathbf{B}' &= 0
\end{aligned}
\right\}. \tag{2}
$$

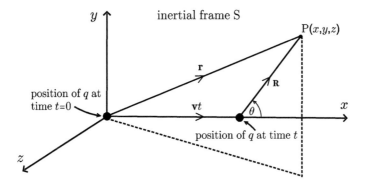

The electric field in S follows from (2) and (16) of Question 12.8, and is

$$\mathbf{E}_\| = \mathbf{E}'_\| = \frac{q}{4\pi\epsilon_0} \frac{\mathbf{r}'_\|}{r'^3} \qquad\qquad \mathbf{E}_\perp = \gamma\mathbf{E}'_\perp = \frac{q}{4\pi\epsilon_0} \frac{\gamma\mathbf{r}'_\perp}{r'^3} \Bigg\}$$

$$= \frac{q}{4\pi\epsilon_0} \frac{\gamma(\mathbf{r}_\| - \mathbf{v}t)}{r'^3} \qquad\qquad = \frac{q}{4\pi\epsilon_0} \frac{\gamma\mathbf{r}_\perp}{r'^3}$$

(here we make use of the Lorentz transformation $\mathbf{r}'_\| = \gamma(\mathbf{r}_\| - \mathbf{v}t)$ and $\mathbf{r}'_\perp = \mathbf{r}_\perp$). Now it is clear from the figure that $\mathbf{r} = \mathbf{R} + \mathbf{v}t$, and so $\mathbf{r}_\| = \mathbf{R}_\| + \mathbf{v}t$ and $\mathbf{r}_\perp = \mathbf{R}_\perp$. Thus

$$\mathbf{E}_\| = \frac{q}{4\pi\epsilon_0} \frac{\gamma\mathbf{R}_\|}{r'^3} \qquad \text{and} \qquad \mathbf{E}_\perp = \frac{q}{4\pi\epsilon_0} \frac{\gamma\mathbf{R}_\perp}{r'^3},$$

which gives

$$\mathbf{E} = \mathbf{E}_\| + \mathbf{E}_\perp = \frac{q}{4\pi\epsilon_0} \frac{\gamma(\mathbf{R}_\| + \mathbf{R}_\perp)}{r'^3} = \frac{q}{4\pi\epsilon_0} \frac{\gamma\mathbf{R}}{r'^3}. \qquad (3)$$

Also, since $\mathbf{r}' = \mathbf{r}'_\| + \mathbf{r}'_\perp = \gamma(\mathbf{r}_\| - \mathbf{v}t) + \mathbf{r}_\perp = \gamma\mathbf{R}_\| + \mathbf{R}_\perp$ we have

$$\begin{aligned}
r'^2 &= (\gamma\mathbf{R}_\| + \mathbf{R}_\perp)\cdot(\gamma\mathbf{R}_\| + \mathbf{R}_\perp) = \gamma^2 R_\|^2 + R_\perp^2 = \gamma^2 R_\|^2 + (R^2 - R_\|^2) \\
&= R^2 + (\gamma^2 - 1)R_\|^2 \\
&= R^2 + (\gamma^2 - 1)R^2 \cos^2\theta \\
&= R^2 + (\gamma^2 - 1)(1 - \sin^2\theta)R^2 \\
&= R^2[\gamma^2 - (\gamma^2 - 1)\sin^2\theta] \\
&= R^2(\gamma^2 - \gamma^2\beta^2 \sin^2\theta) \\
&= \gamma^2(1 - \beta^2 \sin^2\theta)R^2. \qquad (4)
\end{aligned}$$

Substituting (4) in (3) yields and using $\gamma^{-2} = (1 - \beta^2)$ yields $(1)_1$. Next we recognize that because $\mathbf{B}' = 0$, the magnetic field in S is given by (2) of Question 12.8 and is $\mathbf{B} = (\mathbf{v} \times \mathbf{E})/c^2$. Then $(1)_2$ follows from $(1)_1$ and $c^{-2} = \epsilon_0\mu_0$.

Comments

(i) It is evident from $(1)_1$ that the electric field of a moving charge is not spherically symmetric, although symmetry is present in planes perpendicular to **v**. The magnitude of E is a minimum in the forward ($\theta = 0$) and backward ($\theta = \pi$) directions, and is a maximum when $\theta = \pm\frac{1}{2}\pi$. This effect, where the field is reduced along the direction of motion, is referred to as 'contraction of the field', and is illustrated for several values of γ in the figure below (which should be visualized in three dimensions).

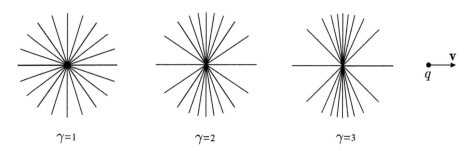

$\gamma=1$ $\gamma=2$ $\gamma=3$

The magnetic-field lines, on the other hand, are circles centred on the trajectory and perpendicular to it (as shown below), and at any instant **B** is symmetrical about the instantaneous position of q.

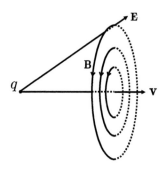

(ii) The direction of **E** is always *radial* and along $\hat{\mathbf{R}}$—the unit vector connecting the current position of the charge to the field point. Further to Comment (ii) of Question 8.15, it is worth noting that:

> The present derivation is far more efficient and sheds some light on the remarkable fact that the field points away from the instantaneous (as opposed to the retarded) position of the charge: $\mathbf{E}_\parallel$ gets a factor of γ from the Lorentz transformation of the *coordinates*; $\mathbf{E}_\perp$ pick up theirs from the transformation of the *field*. It is the balancing of these two γs that leaves **E** parallel to **R**.[8]

[8]   D. J. Griffiths, *Introduction to electrodynamics*, Chap. 12, p. 528. New York: Prentice Hall, 3 edn, 1999.

(iii) As $v \to c$ the electric field is increasingly contracted along $\mathbf{v}$, becoming bunched in the perpendicular direction, whilst the magnetic field continues to remain normal to both $\mathbf{E}$ and $\mathbf{v}$. For an ultra-relativistic charge ($\beta \approx 1$), these three vectors $\mathbf{E}$, $\mathbf{B}$ and $\mathbf{v}$ are mutually orthogonal and satisfy $E = cB$. This is reminiscent of the field of a plane electromagnetic wave in vacuum but without the oscillation. The lateral field of a very fast moving charge 'is almost a shock wave: it is not only very large, but also essentially "radiative" ($cB = E$, $\mathbf{B} \perp \mathbf{E}$)'.[9]

(iv) The field of a point charge moving with constant velocity has now been studied in different ways, not only in this question, but also in Questions 8.13 and 8.15. It is left as an exercise for the reader to establish the equivalence between these different formulations.

Question 12.15 **

A point charge q moving with uniform velocity $\mathbf{v}$ has coordinates (r', θ', ϕ') at time t. Consider an imaginary sphere of radius r_0 and surface s centred on the origin O with $r_0 > r'$ (obviously the charge is inside this sphere at time t).

(a) Show that the electric flux ψ through s is given by

$$\psi = \frac{q}{4\pi\epsilon_0}(1 - \beta^2) \int_0^{2\pi} \int_0^{\pi} \frac{(1 - \alpha \cos \Omega) \sin \theta \, d\theta \, d\phi}{[1 + \alpha^2 - 2\alpha \cos \Omega - \beta^2 \sin^2 \Omega]^{3/2}}, \tag{1}$$

where $\beta = v/c$, $\alpha = r'/r_0$ and $\Omega = \Omega(\theta, \phi)$ is the angle between the vectors $\mathbf{r}'$ and $\mathbf{r} = r_0 \hat{\mathbf{r}}$. Include in your answer a labelled diagram.

(b) Write a *Mathematica* notebook to evaluate (1) numerically by dividing the sphere into a grid of discrete points. Take $0 \leq \theta' \leq \frac{1}{2}\pi$, $\phi' = 0$ and $0 \leq \alpha < 1$ using step sizes $\Delta\theta' = \pi/10$ and $\Delta\alpha = 0.09$. Run your notebook for various values of β in the interval $0 \leq \beta \lesssim 0.999$.

Solution

(a) Because of the symmetry, it is convenient to choose Cartesian axes such that q moves in the xz-plane. Then $\phi' = 0$ and the angle Ω shown in the figure on p. 580 is given by $\cos \Omega = \cos \theta \cos \theta' + \sin \theta \sin \theta' \cos \phi$ because of (VI) of Appendix C. Now, in terms of the present notation, the electric field anywhere on s is

$$\mathbf{E} = \frac{q}{4\pi\epsilon_0} \frac{(1 - \beta^2)}{(1 - \beta^2 \sin^2 \Theta)^{3/2}} \frac{\mathbf{R}}{R^3} \tag{2}$$

[9] W. Rindler, *Introduction to special relativity*, Chap. VI, p. 131. Oxford: Clarendon Press, 1982.

(see $(1)_1$ of Question 12.14). Here $\mathbf{R} = \mathbf{r} - \mathbf{r}' = r_0\hat{\mathbf{r}} - \mathbf{r}'$ and Θ is the angle between $\mathbf{v}$ and $\mathbf{R}$. Then from the definition of flux, $\psi = \oint_s \mathbf{E} \cdot d\mathbf{a}$, where $\mathbf{E}$ is given by (2) and $d\mathbf{a} = r_0^2 \sin\theta \, d\theta \, d\phi \, \hat{\mathbf{r}}$, we obtain

$$\psi = \frac{q}{4\pi\epsilon_0}(1 - \beta^2) r_0^2 \int_0^{2\pi} \int_0^{\pi} \frac{\mathbf{R} \cdot \hat{\mathbf{r}} \sin\theta \, d\theta \, d\phi}{R^3(1 - \beta^2 \sin^2\Theta)^{3/2}}$$

$$= \frac{q}{4\pi\epsilon_0}(1 - \beta^2) r_0^2 \int_0^{2\pi} \int_0^{\pi} \frac{(r_0 - r' \cos\Omega) \sin\theta \, d\theta \, d\phi}{(R^2 - \beta^2 R^2 \sin^2\Theta)^{3/2}}.$$

Now $\sin\Theta = \dfrac{r_0 \sin\Omega}{R}$ where $R = (r_0^2 + r'^2 - 2r_0 r' \cos\Omega)^{1/2}$, and so

$$\psi = \frac{q}{4\pi\epsilon_0}(1 - \beta^2) r_0^2 \int_0^{2\pi} \int_0^{\pi} \frac{(r_0 - r' \cos\Omega) \sin\theta \, d\theta \, d\phi}{(r_0^2 + r'^2 - 2r_0 r' \cos\Omega - \beta^2 r_0^2 \sin^2\Omega)^{3/2}},$$

which is (1).

(b) We use the notebook on p. 581. For β in the interval $0 \leq \beta \lesssim 0.999$ the flux through s evaluates to q/ϵ_0 regardless of where the charge is located inside the sphere (apart from small round-off errors in the calculation). This is not an unexpected result, being a numerical illustration of Gauss's law. (Occasional difficulties with the numerical integration (e.g. slow convergence) will be encountered for certain combinations of α, β and θ'; changing the `PrecisionGoal` and `MaxRecursions` sometimes helps.)

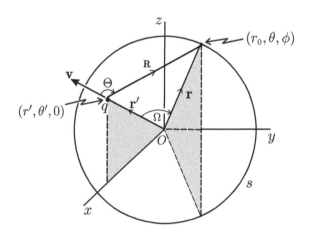

Comment

Of course when $r' > r_0$, with the charge now outside s, we expect $\psi = 0$. The notebook below can easily be modified for this calculation. (Note, however, that in order to avoid computational difficulties, it is advisable to 'trick' *Mathematica* by adding a constant term to the integrand and then subtracting it afterwards.)

```
In[1]:= β = 0.9; φp = 0;

      Ω[θ_, φ_] := ArcCos[Sin[θ] Sin[θp] Cos[φp - φ] + Cos[θ] Cos[θp]]

                        Sin[Ω[θ, φ]]
      θ[α_, Ω_] := ArcSin[ ──────────────────────────── ]
                       √(1 + α² - 2 α Cos[Ω[θ, φ]])

                          (1 - β²)                    (1 - α Cos[Ω[θ, φ]]) Sin[θ]
      Elec[β_, α_, Ω_] := ────────────────────────── ──────────────────────────────
                       (1 - β² Sin[θ[α, Ω]]²)^(3/2) (1 + α² - 2 α Cos[Ω[θ, φ]])^(3/2)

              1
      Do[Int = ─── NIntegrate[Elec[β, α, Ω], {θ, 0, π}, {φ, 0, 2 π},
             4 π
                                                                 180
        PrecisionGoal → 8, MaxRecursion → 10]; Print[α, "      ", N[─── θp], "    ",
                                                                  π

                         q        π    π
      N[φp], "      ", Int x ───], {θp, 0, ─, ──}, {α, 0, 0.9, 0.1}]
                        ε0        2   10
```

Question 12.16*

(This question and (b) and (c) of the solution are based on the approach of Ref. [10].)

Consider a plane electromagnetic wave propagating in vacuum and let S, S' be frames in standard configuration. Suppose that at time $t = t' = 0$ a wavefront WF passing the origin is somehow 'marked'. When this wavefront reaches a point P (at rest) in S, a stationary observer starts counting wavefronts as they pass by. Similarly, counting starts when WF passes P' (at rest) in S'. Counting stops at time t for P and at time t' for P' when $P(x, y, z)$ and $P'(x', y', z')$ are coincident.

(a) Draw a space-time diagram‡ to represent the world lines of P and P' (consider only one spatial dimension (x, say)).

(b) What can one conclude about the number of waves counted by each inertial observer?

(c) Hence prove that the phase of a plane wave is an invariant.

‡If necessary, consult a textbook on special relativity. See, for example, Ref. [11].

[10] W. G. V. Rosser, *Introductory relativity*, Chap. 4, pp. 99–100. London: Butterworths, 1967.
[11] R. Resnick, *Introduction to special relativity*, Chap. A, pp. 188–99. New York: Wiley, 1968.

Solution

(a) We obtain the diagram:

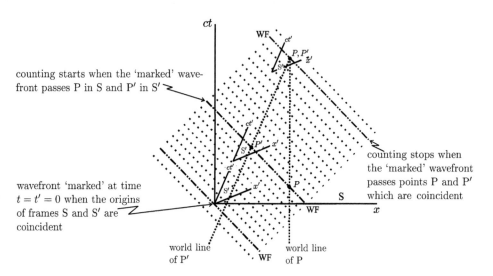

counting starts when the 'marked' wavefront passes P in S and P' in S'

wavefront 'marked' at time
$t = t' = 0$ when the origins
of frames S and S' are
coincident

counting stops when
the 'marked' wavefront
passes points P and P'
which are coincident

world line
of P'

world line
of P

(b) Both observers necessarily count the *same* number of waves, $N = N'$.

(c) In frame S, the phase φ of the wave is $\varphi = \omega t - \mathbf{k} \cdot \mathbf{r} = 2\pi f(t - \mathbf{k} \cdot \mathbf{r}/\omega)$, where f is the frequency and $\omega/k = c$. The figure below shows the wavefront WF when counting begins. Clearly $\mathbf{k} \cdot \mathbf{r} = kr \cos\theta = kR$, and so $\varphi = 2\pi f(t - kR/\omega) = 2\pi f(t - R/c)$. But $t - R/c$ is the counting time and $N = f(t - R/c)$ is the total number of counts. Thus $\varphi = 2\pi N$. Similarly, in frame S' the phase is $\varphi' = 2\pi N'$. But $N = N'$, and therefore $\varphi = \varphi'$, which proves that the phase of a plane wave is invariant.

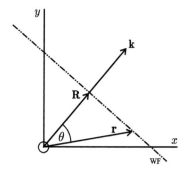

Comment

Because the Lorentz transformation is a linear transformation, a plane wave in frame S transforms to a plane wave in frame S'.

Question 12.17*

A plane electromagnetic wave has angular frequency ω and wave vector $\mathbf{k}$ in frame S. Use the invariance of the phase of a plane wave (see Question 12.16) and the Lorentz transformation to prove that in frame S′

$$
\left.
\begin{aligned}
k'_\| &= \gamma(k_\| - v\omega/c^2) \\
k'_\perp &= k_\perp \\
\omega' &= \gamma(\omega - vk_\|)
\end{aligned}
\quad \text{or} \quad
\begin{aligned}
\mathbf{k}' &= \mathbf{k} + (\gamma - 1)\mathbf{k}_\| - \gamma\mathbf{v}\omega/c^2) \\
\omega' &= \gamma(\omega - \mathbf{v}\cdot\mathbf{k})
\end{aligned}
\right\}.
\tag{1}
$$

Solution

Substituting (4) of Question 12.1 in $(\mathbf{k}\cdot\mathbf{r} - \omega t)$ gives $\mathbf{k}\cdot(\gamma\mathbf{r}'_\| + \mathbf{r}'_\perp + \gamma\mathbf{v}t') - \gamma\omega(t' + r'_\| v/c^2)$. Now the phase of a plane wave is invariant, and so

$$
\mathbf{k}'\cdot\mathbf{r}' - \omega't' = \mathbf{k}'\cdot(\mathbf{r}'_\| + \mathbf{r}'_\perp) - \omega't' = \mathbf{k}\cdot(\gamma\mathbf{r}'_\| + \mathbf{r}'_\perp + \gamma\mathbf{v}t') - \gamma\omega(t' + r'_\| v/c^2).
$$

Rearranging terms gives $(\mathbf{k}' - \gamma\mathbf{k} + \gamma\omega\mathbf{v}/c^2)\cdot\mathbf{r}'_\| + (\mathbf{k}' - \mathbf{k})\cdot\mathbf{r}'_\perp - (\gamma\mathbf{k}\cdot\mathbf{v} - \gamma\omega + \omega')t' = 0$, or

$$
(k'_\| - \gamma k_\| + \gamma\omega v/c^2)r'_\| + (k'_\perp - k_\perp)r'_\perp - (\gamma k_\| v - \gamma\omega + \omega')t' = 0.
\tag{2}
$$

Equation (2) is satisfied for arbitrary $\mathbf{r}'$ and t' only if the coefficients of $r'_\|$, $r'_\perp$ and t' are zero. Thus

$$
\left.
\begin{aligned}
(k'_\| - \gamma k_\| + \gamma v\omega/c^2) &= 0 \\
(k'_\perp - k_\perp) &= 0 \\
(\gamma k_\| v - \gamma\omega + \omega') &= 0
\end{aligned}
\right\}.
$$

Hence the set of equations on the left-hand side of (1). Now since

$$
\mathbf{k}' = \mathbf{k}'_\| + \mathbf{k}'_\perp = \gamma(\mathbf{k}_\| - \mathbf{v}\omega/c^2) + \mathbf{k}_\perp = \gamma(\mathbf{k}_\| - \mathbf{v}\omega/c^2) + (\mathbf{k} - \mathbf{k}_\|),
$$

we obtain the alternative set of equations on the right-hand side of (1).

Comments

(i) It is clear from (1) that $\mathbf{k}$ and ω/c together transform like $\mathbf{r}$ and ct (see (1) of Question 12.1). We therefore expect, and it is easily confirmed, that $k^2 - \omega^2/c^2$ is invariant. That is,

$$
k'^2 - \omega'^2/c^2 = k^2 - \omega^2/c^2.
\tag{3}
$$

(ii) The inverse transformation (obtained by applying the rule (3) of Question 12.1) is:

$$\left.\begin{aligned} k_\| &= \gamma(k_\|' + vw'/c^2) \\ k_\perp &= k_\perp' \\ \omega &= \gamma(\omega' + vk_\|') \end{aligned} \quad \text{or} \quad \begin{aligned} \mathbf{k} &= \mathbf{k}' + (\gamma - 1)\mathbf{k}_\|' + \gamma\mathbf{v}w'/c^2) \\ \omega &= \gamma(\omega' + \mathbf{v}\cdot\mathbf{k}') \end{aligned}\right\}. \quad (4)$$

Question 12.18

Consider a plane polarized electromagnetic wave propagating in vacuum. Suppose that in frame S, the electromagnetic field is given by

$$\mathbf{E} = E_0 e^{i(kx - \omega t)}\,\hat{\mathbf{y}} \qquad \text{and} \qquad \mathbf{B} = \frac{E_0}{c} e^{i(kx - \omega t)}\,\hat{\mathbf{z}}. \quad (1)$$

Obtain the field in frame S' for $\beta = 0.6$, assuming that S and S' are in standard configuration.

Solution

A plane wave transforms to a plane wave,[‡] and so

$$\mathbf{E}' = E_0' e^{i(k'x' - \omega' t')}\,\hat{\mathbf{y}} \qquad \text{and} \qquad \mathbf{B}' = \frac{E_0'}{c} e^{i(k'x' - \omega' t')}\,\hat{\mathbf{z}}. \quad (2)$$

From the relevant transformation equations,[♯] and with $\beta = 0.6$, it is readily shown that

$$\left.\begin{aligned} E_0' &= \sqrt{\frac{1-\beta}{1+\beta}}\,E_0 = \frac{E_0}{2} & k' &= \sqrt{\frac{1-\beta}{1+\beta}}\,k = \tfrac{1}{2}k \\ &\text{and} \\ B_0' &= \sqrt{\frac{1-\beta}{1+\beta}}\,\frac{E_0}{c} = \frac{E_0}{2c} & \omega' &= \sqrt{\frac{1-\beta}{1+\beta}}\,\omega = \tfrac{1}{2}\omega \end{aligned}\right\}. \quad (3)$$

Comment

It is clear from (3) that in a frame where $\beta \to 1$, E_0', B_0', k' and ω' all tend to zero. The wave simply fades away, although its speed remains equal to c as it always does (electromagnetic waves have no rest frame). See also Comment (iii) of Question 12.10.

[‡]See the comment at the end of Question 12.16.

[♯]Use (1) of Question 12.8 for the fields, and (1) of Question 12.17 for k and ω.

Question 12.19*

Consider a plane electromagnetic wave propagating in vacuum. Let θ be the angle between $\mathbf{k}$ and $\mathbf{v}$ (in frame S) and θ' the angle between $\mathbf{k}'$ and $\mathbf{v}$ in S'. Show that

$$\nu' = \frac{(1 - \beta \cos \theta)\nu}{\sqrt{1 - \beta^2}} \; ; \qquad \sin \theta' = \frac{\sqrt{1 - \beta^2} \sin \theta}{1 - \beta \cos \theta} \; ; \qquad \tan \theta' = \frac{\sqrt{1 - \beta^2} \sin \theta}{\cos \theta - \beta} \; , \qquad (1)$$

where the frequency $\nu = \omega/2\pi$.

Solution

The dispersion relation $\omega/k = c$ together with the transformation $\omega' = \gamma(\omega - \mathbf{v} \cdot \mathbf{k}) = \gamma(\omega - vk_{\parallel}) = \gamma(\omega - vk \cos \theta)$ gives

$$\omega' = \gamma(\omega - vk \cos \theta) = \gamma(1 - \beta \cos \theta)\omega. \qquad (2)$$

Hence $(1)_1$. Next we consider the transformation $k'_{\perp} = k_{\perp}$, which can be written as $k' \sin \theta' = k \sin \theta$ or $\sin \theta' = (k/k') \sin \theta = (\omega/\omega') \sin \theta$. Eliminating the frequency ratio in this result using (2) yields $(1)_2$. Lastly, the identity $1 + \cot^2 \theta' = \text{cosec}^2\theta'$ together with $(1)_2$ and elementary trigonometry gives $(1)_3$.

Comments

(i) Equation $(1)_1$ is the relativistic Doppler effect in the general case when the velocity of the source is not necessarily along the line joining the source and the observer. We now consider some limiting cases.

☞ non-relativistic limit $(\beta \ll 1)$

It follows from $(1)_1$ that

$$\nu = \nu' \sqrt{1 - \beta^2} (1 - \beta \cos \theta)^{-1} = \nu'(1 - \tfrac{1}{2}\beta^2 - \cdots)(1 + \beta \cos \theta + \cdots),$$

because of the binomial theorem. To first order in β, we obtain a result familiar from elementary physics: $\nu \simeq \nu'(1 + \beta \cos \theta)$. This becomes obvious if we consider the following:

source moves directly towards the observer $(\theta = 0)$	$\nu = \nu'(1 + \beta)$,
source moves directly away from the observer $(\theta = 180°)$	$\nu = \nu'(1 - \beta)$,
source moves perpendicular to observer's line of sight $(\theta = 90°)$	$\nu = \nu'$.

☞ relativistic limit $(\beta \lesssim 1)$

When the condition $\beta \ll 1$ is not satisfied, we retain relativistic (second-order) effects. It is convenient to think of these effects separately as a longitudinal effect and a transverse effect.

longitudinal Doppler effect in relativity

Let $\theta = 0$ (source and observer move towards each other) or $\theta = 180°$ (source and observer move away from each other). Then

$$
\left.
\begin{aligned}
\nu &= \frac{\sqrt{1-\beta^2}}{(1-\beta)}\,\nu' = \sqrt{\frac{1+\beta}{1-\beta}}\,\nu' \qquad \text{for } \theta = 0 \\[2ex]
\nu &= \frac{\sqrt{1-\beta^2}}{(1+\beta)}\,\nu' = \sqrt{\frac{1-\beta}{1+\beta}}\,\nu' \qquad \text{for } \theta = 180°
\end{aligned}
\right\}. \tag{3}
$$

transverse Doppler effect in relativity

More striking than (3), however, is the fact that the relativistic formula $(1)_1$ predicts a transverse Doppler effect—an effect that is purely relativistic—there is no transverse Doppler effect in classical physics (see $\nu = \nu'$ above). Substituting $\theta = 90°$ in $(1)_1$ gives

$$
\nu = (1 - \beta^2)^{1/2}\,\nu', \tag{4}
$$

which shows that if our line of sight is perpendicular to the relative motion, then we will observe a frequency ν that is *lower* than the proper frequency ν' of the source moving past us.

(ii) The trigonometric identity $\tan^2 \frac{1}{2}\theta = (1 - \cos\theta)/(1 + \cos\theta)$ together with $(1)_2$ and $(1)_3$ can be used to derive the result

$$
\tan\left(\tfrac{1}{2}\theta'\right) = \sqrt{\frac{1+\beta}{1-\beta}}\,\tan\left(\frac{\theta}{2}\right),
$$

which is a well-known formula in astronomy relating to the aberration of starlight.

Question 12.20*

Consider a distribution comprising N stationary charges contained within an infinitesimal volume element δV_0. Suppose this element is a cube of side ℓ_0 aligned with its edges parallel to the coordinate axes of a frame S_0 where it is at rest. In inertial frame S the element has volume δV and moves (together with the enclosed charges) with velocity $(u, 0, 0)$.

(a) Show that the volume $\delta V'$ in a frame S' is related to δV by the transformation

$$\delta V' = \frac{\sqrt{1 - v^2/c^2}}{(1 - \mathbf{u} \cdot \mathbf{v}/c^2)} \delta V, \tag{1}$$

where $\mathbf{v}$ is the velocity of S' relative to S (as usual we assume these frames are in standard configuration).

Hint: Consider the Lorentz contraction of the cube along the xx'-axis.

(b) Use (1) to show that current density $\mathbf{J}$ and charge density ρ of this distribution transform as

$$\left. \begin{array}{l} \mathbf{J}' = \gamma \mathbf{J}_{\parallel} + \mathbf{J}_{\perp} - \gamma \mathbf{v} \rho \\ \rho' = \gamma(\rho - \mathbf{v} \cdot \mathbf{J}/c^2) \end{array} \right\}. \tag{2}$$

Solution

(a) In frame S_0 the sides of the cube each have a length ℓ_0 (this being the proper length) with corresponding proper volume $\delta V_0 = \ell_0^3$. Now in S and S', the edge of the cube is Lorentz contracted along the xx'-axis. So $\delta V = \ell_0^2 \ell$ and $\delta V' = \ell_0^2 \ell'$ or $\delta V' = \ell' \delta V/\ell$, where the ratio of lengths ℓ'/ℓ is given by (1) of Question 12.3. Hence (1).

(b) Because the charge of the volume element is invariant, we have $\rho' \delta V' = \rho \delta V$ or $\rho' = \rho \delta V / \delta V'$. Using (1) to eliminate $\delta V / \delta V'$ from this last result yields

$$\rho' = \gamma \rho(1 - \mathbf{u} \cdot \mathbf{v}/c^2). \tag{3}$$

Now in frame S (S') the current density is $\rho \mathbf{u}$ ($\rho' \mathbf{u}'$) where $\mathbf{u}'$ is the velocity of the charges in the primed frame. Equation $(2)_2$ then follows immediately from (3) and $\mathbf{J} = \rho \mathbf{u}$. Substituting the velocity transformation (1) of Question 12.2 in $(2)_2$ gives

$$\mathbf{J}' = \rho' \mathbf{u}' = \gamma \rho(1 - \mathbf{v} \cdot \mathbf{u}/c^2) \mathbf{u}' = \gamma \rho(1 - \mathbf{v} \cdot \mathbf{u}/c^2) \frac{\mathbf{u}_{\parallel} + \mathbf{u}_{\perp}/\gamma - \mathbf{v}}{(1 - \mathbf{v} \cdot \mathbf{u}/c^2)}$$

$$= \gamma \rho(\mathbf{u}_{\parallel} + \mathbf{u}_{\perp}/\gamma - \mathbf{v}).$$

Hence $(2)_1$.

Comments

(i) Although (2) was derived here for a particular charge distribution, these results are, in fact, valid in general.

(ii) Because $\mathbf{J}$ and $c\rho$ together transform like $\mathbf{r}$ and ct (see (1) of Question 12.1), we expect that $J^2 - c^2\rho^2$ is invariant. The reader can easily confirm that

$$J'^2 - c^2\rho'^2 = J^2 - c^2\rho^2. \tag{4}$$

(iii) So far in this chapter we have met several quantities, all of which behave the same way under Lorentz transformation. They are:

☞ **r** and ct,

☞ **p** and $\mathcal{E}/c$,

☞ **k** and ω/c, and

☞ **J** and $c\rho$.

Quantities which transform in this way are called four-vectors. In the above list we speak of the four-position vector, the four-momentum, the four-wave vector and the four-current density. Notice that the appearance of c in the fourth part (or time part) of the above four-vectors ensures that this part has the same dimensions as the corresponding three-vector.

(iv) Suppose we express the four-position vector in terms of the coordinates

$$x_1 = x, \qquad x_2 = y, \qquad x_3 = z, \qquad x_4 = ict, \tag{5}$$

(here $i = \sqrt{-1}$),‡ then the Lorentz transformation for frames in the standard configuration becomes

$$\left.\begin{aligned} x_1' &= \gamma(x_1 + i\beta x_4) \\ x_2' &= x_2 \\ x_3' &= x_3 \\ x_4' &= \gamma(x_4 - i\beta x_1) \end{aligned}\right\}, \tag{6}$$

or

$$\begin{pmatrix} x_1' \\ x_2' \\ x_3' \\ x_4' \end{pmatrix} = \begin{pmatrix} \gamma & 0 & 0 & i\gamma\beta \\ 0 & 1 & 0 & 0 \\ 0 & 0 & 1 & 0 \\ -i\gamma\beta & 0 & 0 & \gamma \end{pmatrix} \begin{pmatrix} x_1 \\ x_2 \\ x_3 \\ x_4 \end{pmatrix}. \tag{7}$$

If we introduce a Greek index♯ and let it range from one to four, then (7) can be expressed compactly as

$$x_\mu' = a_{\mu\nu}x_\nu, \tag{8}$$

where

$$a_{\mu\nu} = \begin{pmatrix} \gamma & 0 & 0 & i\gamma\beta \\ 0 & 1 & 0 & 0 \\ 0 & 0 & 1 & 0 \\ -i\gamma\beta & 0 & 0 & \gamma \end{pmatrix}. \tag{9}$$

‡The inclusion of the imaginary number i in x_4 enables us to work in a complex Cartesian space known as Minkowski space. The alternative is to work in a real space where $x_4 = ct$, and consequently $x_\mu^2 = x_1^2 + x_2^2 + x_3^2 - x_4^2$. Such a space is Riemannian.
♯Note that a Roman index ranges from one to three, and a repeated index of either type implies summation. So $x_j x_j = x_1^2 + x_2^2 + x_3^2$, and $x_\mu x_\mu = x_1^2 + x_2^2 + x_3^2 + x_4^2$.

Motivated by (8), we define an arbitrary four-vector to be a set of any four quantities A_μ which transform as

$$A'_\mu = a_{\mu\nu} A_\nu \quad \text{with inverse} \quad A_\mu = a_{\nu\mu} A'_\nu. \tag{10}$$

(v) Analogous to their three-vector counterparts (see Appendix A), four-vectors represent a particular type of a more general entity called a four-tensor (namely a first-rank four-tensor). Second-rank four-tensors comprise sixteen quantities $A_{\mu\nu}$ which transform as

$$A'_{\mu\nu} = a_{\mu\rho} a_{\nu\sigma} A_{\rho\sigma} \quad \text{with inverse} \quad A_{\mu\nu} = a_{\rho\mu} a_{\sigma\nu} A'_{\rho\sigma}, \tag{11}$$

and so on. A four-scalar ϕ is a tensor of rank zero and it is unchanged by the transformation $\phi' = \phi$.

(vi) The coefficients $a_{\mu\nu}$ in (8) satisfy the orthogonality relations

$$a_{\mu\nu} a_{\mu\lambda} = \delta_{\nu\lambda}, \quad \text{and} \quad a_{\nu\mu} a_{\lambda\mu} = \delta_{\nu\lambda}. \tag{12}$$

A proof of this result and further discussion of the properties of four-vectors and four-tensors can be found in Questions 15.6 and 15.7 of Ref. [4].

Question 12.21[*]

In the Lorenz gauge, the equations for the potentials **A** and Φ are

$$\left.\begin{array}{l} \Box^2 \mathbf{A} = -\mu_0 \mathbf{J} \\ \Box^2 \Phi = -\rho/\epsilon_0 \end{array}\right\}, \tag{1}$$

where

$$\Box^2 = \nabla^2 - \frac{1}{c^2}\frac{\partial^2}{\partial t^2} \tag{2}$$

$\big($see (11) and (12) of Question 8.2$\big)$.

(a) Express (1) in the form

$$\frac{\partial^2 A_\mu}{\partial x_\mu^2} = -\mu_0 J_\mu \tag{3}$$

where

$$A_\mu = (\mathbf{A}, i\Phi/c) \tag{4}$$

and

$$J_\mu = (\mathbf{J}, ic\rho) \tag{5}$$

is the four-current density $\big($see Question 12.20$\big)$.

(b) Prove that $\Box^2$ is an invariant.

(c) Hence prove that A_μ defined by (4) is a four-vector.

Solution

(a) Expressed in terms of the notation introduced in (5) of Question 12.20, the d'Alembertian operator can be written as $\Box^2 = \partial^2/\partial x_\nu^2$. Then multiplying $(1)_2$ by i/c where $c^{-2} = \epsilon_0 \mu_0$ gives

$$\frac{\partial^2}{\partial x_\nu^2}\left(\frac{i\Phi}{c}\right) = -\mu_0(ic\rho). \tag{6}$$

Equation (3) follows immediately from (4)–(6), since $(1)_1$ is $\dfrac{\partial^2 \mathbf{A}}{\partial x_\nu^2} = -\mu_0 \mathbf{J}$.

(b) Begin with:
$$\frac{\partial^2}{\partial x'^2_\mu} = \left(\frac{\partial x_\nu}{\partial x'_\mu}\frac{\partial}{\partial x_\nu}\right)\left(\frac{\partial x_\lambda}{\partial x'_\mu}\frac{\partial}{\partial x_\lambda}\right)$$

$$= a_{\mu\nu} a_{\mu\lambda}\frac{\partial}{\partial x_\nu}\frac{\partial}{\partial x_\lambda} \qquad\qquad (x_\nu = a_{\mu\nu}x'_\mu)$$

$$= \delta_{\nu\lambda}\frac{\partial}{\partial x_\nu}\frac{\partial}{\partial x_\lambda} \qquad\qquad (a_{\mu\nu}a_{\mu\lambda} = \delta_{\nu\lambda}).$$

Then contracting subscripts in the above gives $\dfrac{\partial^2}{\partial x'^2_\mu} = \dfrac{\partial^2}{\partial x_\nu^2}$, which proves that $\Box^2$ is a scalar (invariant) operator.

(c) Since $\Box^2$ is invariant and J_μ is a four-vector it follows that A_μ defined by (4) is also a four-vector.

Comment

The result $A_\mu = (\mathbf{A}, i\Phi/c)$ is known as the four-vector potential, and because of the properties of four-vectors we expect—as we have seen in similar cases elsewhere in this chapter—that

$$\left.\begin{array}{l} \mathbf{A}' = \gamma\mathbf{A}_\| + \mathbf{A}_\perp - \gamma\mathbf{v}\Phi/c^2 \\[6pt] \Phi' = \gamma(\Phi - \mathbf{v}\cdot\mathbf{A}) \end{array}\right\}, \tag{7}$$

together with the corresponding invariant quantity

$$A'^2 - \Phi'^2/c^2 = A^2 - \Phi^2/c^2. \tag{8}$$

Question 12.22*

(a) Let A_μ be an arbitrary four-vector. Prove that $F_{\nu\mu}$ defined by

$$F_{\nu\mu} = \frac{\partial A_\mu}{\partial x_\nu} - \frac{\partial A_\nu}{\partial x_\mu} \tag{1}$$

is an anti-symmetric second-rank four-tensor.

(b) Suppose $A_\mu = (\mathbf{A}, i\Phi/c)$ is the four-vector potential. Use $\mathbf{E} = -\nabla\Phi - \partial\mathbf{A}/\partial t$ to show that

$$E_j = ic\left(\frac{\partial A_4}{\partial x_j} - \frac{\partial A_j}{\partial x_4}\right), \tag{2}$$

where $j = 1, 2, 3$.

(c) Use (2) and the components of $B_i = \varepsilon_{ijk}\nabla_j A_k$ to show that

$$F_{\nu\mu} = \begin{pmatrix} 0 & B_3 & -B_2 & -iE_1/c \\ -B_3 & 0 & B_1 & -iE_2/c \\ B_2 & -B_1 & 0 & -iE_3/c \\ iE_1/c & iE_2/c & iE_3/c & 0 \end{pmatrix}. \tag{3}$$

Solution

(a) It is obvious, by inspection, that this tensor is anti-symmetric: $F_{\nu\mu} = -F_{\mu\nu}$. In order to prove that $F_{\nu\mu}$ is a four-tensor it is sufficient to show that $f_{\nu\mu} = \partial A_\mu/\partial x_\nu$ is a four-tensor because the difference between two four-tensors is also a four-tensor. Consider

$$f'_{\nu\mu} = \frac{\partial A'_\mu}{\partial x'_\nu} = a_{\mu\lambda}\frac{\partial A_\lambda}{\partial x'_\nu} \qquad\qquad (A'_\mu = a_{\mu\lambda}A_\lambda)$$

$$= a_{\mu\lambda}\frac{\partial x_\phi}{\partial x'_\nu}\frac{\partial A_\lambda}{\partial x_\phi}$$

$$= a_{\mu\lambda}a_{\nu\phi}\frac{\partial A_\lambda}{\partial x_\phi} \qquad\qquad (x_\phi = a_{\nu\phi}x'_\nu)$$

$$= a_{\nu\phi}a_{\mu\lambda}f_{\phi\lambda},$$

which is a second-rank four-tensor $\big($see $(11)_1$ of Question 12.20$\big)$.

(b) Clearly, $E_j = -\nabla_j \Phi - \partial A_j / \partial t$. Multiplying both sides by $-i/c$ gives

$$-iE_j/c = \frac{\partial(i\Phi/c)}{\partial x_j} + \frac{i}{c}\frac{\partial A_j}{\partial t} = \frac{\partial A_4}{\partial x_j} - \frac{\partial A_j}{\partial x_4}, \qquad (4)$$

since $A_4 = i\Phi/c$ and $x_4 = ict$. Hence (2).

(c) The components of **E** which follow from (4) are

$$\left.\begin{array}{l} -\dfrac{i}{c}E_1 = \dfrac{\partial A_4}{\partial x_1} - \dfrac{\partial A_1}{\partial x_4} = F_{14} = -F_{41} \\[2mm] -\dfrac{i}{c}E_2 = \dfrac{\partial A_4}{\partial x_2} - \dfrac{\partial A_2}{\partial x_4} = F_{24} = -F_{42} \\[2mm] -\dfrac{i}{c}E_3 = \dfrac{\partial A_4}{\partial x_3} - \dfrac{\partial A_3}{\partial x_4} = F_{34} = -F_{43} \end{array}\right\},$$

and for **B**

$$\left.\begin{array}{l} B_1 = \dfrac{\partial A_3}{\partial x_2} - \dfrac{\partial A_2}{\partial x_3} = F_{23} = -F_{32} \\[2mm] B_2 = \dfrac{\partial A_1}{\partial x_3} - \dfrac{\partial A_3}{\partial x_1} = F_{31} = -F_{13} \\[2mm] B_3 = \dfrac{\partial A_2}{\partial x_1} - \dfrac{\partial A_1}{\partial x_2} = F_{12} = -F_{21} \end{array}\right\}.$$

Now

$$F_{\nu\mu} = \begin{pmatrix} F_{11} & F_{12} & F_{13} & F_{14} \\ F_{21} & F_{22} & F_{23} & F_{24} \\ F_{31} & F_{32} & F_{33} & F_{34} \\ F_{41} & F_{42} & F_{43} & F_{44} \end{pmatrix},$$

and hence (3) because $F_{11} = F_{22} = F_{33} = F_{44} = 0$ by definition.

Comments

(i) The anti-symmetric second-rank four-tensor $F_{\nu\mu}$ is known as the electromagnetic-field tensor. It 'knits the two 3-vector fields **E** and **B** into a single entity, resolvable on the four dimensions of spacetime in ways that convert electric into magnetic fields, and vice versa, merely by viewing the *electromagnetic field* from relatively moving frames'.[12]

[12]   F. Melia, *Electrodynamics*, Chap. 5, p. 136. Chicago: Chicago University Press, 2001.

(ii) Because $F_{\nu\mu}$ is a second-rank four-tensor it transforms as

$$F'_{\mu\nu} = a_{\mu\rho} a_{\nu\sigma} F_{\rho\sigma} \tag{5}$$

$\big($see $(11)_1$ of Question 12.20$\big)$. We can use (9) of Question 12.20 and (5) to confirm the transformation properties of the electromagnetic field. Consider, for example, F_{12}. Then

$$
\begin{aligned}
F'_{12} &= a_{1\nu} a_{2\mu} F_{\nu\mu} \\
&= a_{11} a_{2\mu} F_{1\mu} + a_{14} a_{2\mu} F_{4\mu} && (a_{12} = a_{13} = 0) \\
&= a_{11} a_{22} F_{12} + a_{14} a_{22} F_{42} && (a_{21} = a_{23} = a_{24} = 0) \\
&= \gamma(F_{12} + i\beta F_{42}).
\end{aligned}
$$

Therefore

$$
\begin{aligned}
B'_3 &= \gamma(B_3 + i\beta F_{42}) \\
&= \gamma(B_3 - vE_2/c^2),
\end{aligned}
$$

which is $(1)_6$ of Question 12.8 since $B'_3 = B'_z$; $B_3 = B_z$; and $E_2 = E_y$. The remaining equations $(1)_1$–$(1)_5$ follow in a similar way.

Question 12.23*

Use the electromagnetic-field tensor

$$
F_{\nu\mu} = \begin{pmatrix}
0 & B_3 & -B_2 & -iE_1/c \\
-B_3 & 0 & B_1 & -iE_2/c \\
B_2 & -B_1 & 0 & -iE_3/c \\
iE_1/c & iE_2/c & iE_3/c & 0
\end{pmatrix} \tag{1}
$$

to express Maxwell's equations as

$$
\left.
\begin{aligned}
\frac{\partial F_{\mu\nu}}{\partial x_\lambda} + \frac{\partial F_{\lambda\mu}}{\partial x_\nu} + \frac{\partial F_{\nu\lambda}}{\partial x_\mu} &= 0 \\[2mm]
\frac{\partial F_{\mu\nu}}{\partial x_\nu} &= \mu_0 J_\mu
\end{aligned}
\right\} \tag{2}
$$

Hint: Consider the homogeneous and inhomogeneous equations separately.

Solution

☞ Maxwell's homogeneous equations

$\nabla \times \mathbf{E} + \dfrac{\partial \mathbf{B}}{\partial t} = 0$ and $\nabla \cdot \mathbf{B} = 0$ represent four equations. These together with (1)

yield

$$
\left.
\begin{aligned}
\frac{\partial E_3}{\partial x_2} - \frac{\partial E_2}{\partial x_3} + \frac{\partial B_1}{\partial t} = 0 \quad &\rightarrow \quad \frac{\partial F_{34}}{\partial x_2} + \frac{\partial F_{42}}{\partial x_3} + \frac{\partial F_{23}}{\partial x_4} = 0 \\[2mm]
\frac{\partial E_1}{\partial x_3} - \frac{\partial E_3}{\partial x_1} + \frac{\partial B_2}{\partial t} = 0 \quad &\rightarrow \quad \frac{\partial F_{14}}{\partial x_3} + \frac{\partial F_{43}}{\partial x_1} + \frac{\partial F_{31}}{\partial x_4} = 0 \\[2mm]
\frac{\partial E_2}{\partial x_1} - \frac{\partial E_1}{\partial x_2} + \frac{\partial B_3}{\partial t} = 0 \quad &\rightarrow \quad \frac{\partial F_{24}}{\partial x_1} + \frac{\partial F_{41}}{\partial x_2} + \frac{\partial F_{12}}{\partial x_4} = 0 \\[2mm]
\frac{\partial B_1}{\partial x_1} + \frac{\partial B_2}{\partial x_2} + \frac{\partial B_3}{\partial x_3} = 0 \quad &\rightarrow \quad \frac{\partial F_{23}}{\partial x_1} + \frac{\partial F_{31}}{\partial x_2} + \frac{\partial F_{12}}{\partial x_3} = 0
\end{aligned}
\right\},
$$

which can be written as $\dfrac{\partial F_{\mu\nu}}{\partial x_\lambda} + \dfrac{\partial F_{\lambda\mu}}{\partial x_\nu} + \dfrac{\partial F_{\nu\lambda}}{\partial x_\mu} = 0$. Of these sixty-four ($= 4^3$) equations,

only the four for which $\mu \neq \nu \neq \lambda$ are non-trivial, because if

- all three indices are equal $F_{11} = F_{22} = F_{33} = F_{44} = 0$ and $(2)_1$ is satisfied automatically.

- any two indices are equal $\dfrac{\partial F_{\mu\nu}}{\partial x_\mu} + \dfrac{\partial F_{\mu\mu}}{\partial x_\nu} + \dfrac{\partial F_{\nu\mu}}{\partial x_\mu} = 0$ (reason: $F_{\mu\nu} = -F_{\nu\mu}$ and $F_{\mu\mu} = 0$).

☞ Maxwell's inhomogeneous equations

$\nabla \times \mathbf{B} - \dfrac{1}{c^2}\dfrac{\partial \mathbf{E}}{\partial t} = \mu_0 \mathbf{J}$ and $\nabla \cdot \mathbf{E} = \dfrac{\rho}{\epsilon_0}$ represent four equations. These along with (1)

yield

$$
\left.
\begin{aligned}
\frac{\partial B_3}{\partial x_2} - \frac{\partial B_2}{\partial x_3} - \frac{1}{c^2}\frac{\partial E_1}{\partial t} = \mu_0 J_1 \quad &\rightarrow \quad \frac{\partial F_{11}}{\partial x_1} + \frac{\partial F_{12}}{\partial x_2} + \frac{\partial F_{13}}{\partial x_3} + \frac{\partial F_{14}}{\partial x_4} = \mu_0 J_1 \\[2mm]
\frac{\partial B_1}{\partial x_3} - \frac{\partial B_3}{\partial x_1} - \frac{1}{c^2}\frac{\partial E_2}{\partial t} = \mu_0 J_2 \quad &\rightarrow \quad \frac{\partial F_{21}}{\partial x_1} + \frac{\partial F_{22}}{\partial x_2} + \frac{\partial F_{23}}{\partial x_3} + \frac{\partial F_{24}}{\partial x_4} = \mu_0 J_2 \\[2mm]
\frac{\partial B_2}{\partial x_1} - \frac{\partial B_1}{\partial x_2} - \frac{1}{c^2}\frac{\partial E_3}{\partial t} = \mu_0 J_3 \quad &\rightarrow \quad \frac{\partial F_{31}}{\partial x_1} + \frac{\partial F_{32}}{\partial x_2} + \frac{\partial F_{33}}{\partial x_3} + \frac{\partial F_{34}}{\partial x_4} = \mu_0 J_3 \\[2mm]
\frac{i}{c}\left(\frac{\partial E_1}{\partial x_1} + \frac{\partial E_2}{\partial x_2} + \frac{\partial E_3}{\partial x_3}\right) = \frac{i}{c}\left(\mu_0 c^2 \rho\right) \quad &\rightarrow \quad \frac{\partial F_{41}}{\partial x_1} + \frac{\partial F_{42}}{\partial x_2} + \frac{\partial F_{43}}{\partial x_3} + \frac{\partial F_{44}}{\partial x_4} = \mu_0 J_4
\end{aligned}
\right\},
$$

which can be written as $(2)_2$.

Comments

(i) Equations $(2)_1$ and $(2)_2$ are the covariant forms of Maxwell's equations, with $F_{\nu\mu}$ given by (1).

(ii) We already know from earlier chapters that charge conservation is implicit in Maxwell's equations. To see this in the present context, suppose we differentiate $(2)_2$. Then:

$$\frac{\partial}{\partial x_\mu}\left(\frac{\partial F_{\mu\nu}}{\partial x_\nu}\right) = \frac{\partial}{\partial x_\mu}(\mu_0 J_\mu) = \mu_0 \frac{\partial J_\mu}{\partial x_\mu}. \tag{3}$$

Manipulating the left-hand side of (3) gives

$$\frac{\partial^2 F_{\mu\nu}}{\partial x_\mu \partial x_\nu} = \frac{\partial^2 F_{\nu\mu}}{\partial x_\nu \partial x_\mu} \qquad \text{(subscripts are arbitrary)}$$

$$= \frac{\partial^2 F_{\nu\mu}}{\partial x_\mu \partial x_\nu} \qquad \text{(interchanging the order of differentiation)}$$

$$= -\frac{\partial^2 F_{\mu\nu}}{\partial x_\mu \partial x_\nu} \qquad \text{(the tensor } F_{\mu\nu} \text{ is anti-symmetric)},$$

implying

$$\frac{\partial^2 F_{\mu\nu}}{\partial x_\mu \partial x_\nu} = 0. \tag{4}$$

Comparing (3) and (4) shows that

$$\frac{\partial J_\mu}{\partial x_\mu} = 0 \quad \text{or} \quad \frac{\partial J_1}{\partial x_1} + \frac{\partial J_2}{\partial x_2} + \frac{\partial J_3}{\partial x_3} + \frac{\partial J_4}{\partial x_4} = 0. \tag{5}$$

Now (5) is the covariant form of the continuity equation, because

$$\nabla \cdot \mathbf{J} = \frac{\partial J_1}{\partial x_1} + \frac{\partial J_2}{\partial x_2} + \frac{\partial J_3}{\partial x_3} \quad \text{and} \quad \frac{\partial \rho}{\partial t} = \frac{\partial(ic\rho)}{\partial(ict)} = \frac{\partial J_4}{\partial x_4}.$$

Question 12.24 *

Consider the four-vector

$$f_\mu = F_{\mu\nu} J_\nu, \tag{1}$$

where $F_{\mu\nu}$ is the electromagnetic-field tensor and J_ν the four-current density. Show that

$$f_\mu = (\mathbf{f}, f_4)$$

where

$$\left.\begin{array}{l} \mathbf{f} = \rho(\mathbf{E} + \mathbf{J} \times \mathbf{B}) \\ f_4 = i\mathbf{J} \cdot \mathbf{E}/c \end{array}\right\}. \tag{2}$$

Solution

It follows from (1) that $f_1 = F_{1\nu}J_\nu = F_{11}J_1 + F_{12}J_2 + F_{13}J_3 + F_{14}J_4$. Substituting the components of $F_{\mu\nu}$ from (3) of Question 12.22 gives $f_1 = J_2B_3 - J_3B_2 - (iE_1/c)(ic\rho) = \rho E_1 + (\mathbf{J} \times \mathbf{B})_1$. Similarly $f_2 = \rho E_2 + (\mathbf{J} \times \mathbf{B})_2$ and $f_3 = \rho E_3 + (\mathbf{J} \times \mathbf{B})_3$. Hence $(2)_1$. Also, $f_4 = F_{4\nu}J_\nu = F_{41}J_1 + F_{42}J_2 + F_{43}J_3 + F_{44}J_4 = (i/c)(E_1J_1 + E_2J_2 + E_3J_3)$. Hence $(2)_2$.

Comments

(i) Equation $(2)_1$ is the Lorentz force density and (1) is its covariant form.

(ii) An appropriate volume integral of f_μ gives the four-force K_μ (known as the Minkowski force) whose space part represents the force on the charge distribution. In Ref. [4] it is shown that for a particle having charge q moving with velocity $\mathbf{v}$, the Minkowski force is

$$K_\mu = \gamma(\mathbf{F}, i\mathbf{F} \cdot \mathbf{v}/c), \tag{3}$$

where $\mathbf{F} = q(\mathbf{E} + \mathbf{v} \times \mathbf{B})$.[‡]

Question 12.25[*]

Derive the relativistic form of Larmor's formula (see Question 11.8)

$$\mathcal{P} = \frac{\mu_0 q^2}{6\pi c} \gamma^6 \left[a^2 - |\boldsymbol{\beta} \times \mathbf{a}|^2\right], \tag{1}$$

assuming that the power radiated by a charged particle is a scalar invariant.[♯] In (1) the quantity in square brackets is evaluated at the retarded time.

Hint: Use (5) of Question 12.2.

[‡]See Questions 15.9 and 15.10 in Ref. [4].
[♯]Since energy and time are both the fourth part of a four-vector, they behave the same way under Lorentz transformation, and so it is plausible that $dE/dt = dE'/dt'$. For a formal proof of this result see, for example, Ref. [13].

[13] A. Zangwill, *Modern electrodynamics*, Chap. 23, p. 884. Cambridge: Cambridge University Press, 2013.

Solution

Consider an arbitrarily moving particle having charge q that is instantaneously at rest relative to an inertial frame S. The power radiated by this particle is given (exactly) by Larmor's formula

$$\mathcal{P} = \frac{\mu_0 q^2 [a^2]}{6\pi c}, \tag{2}$$

where $\mathbf{a}$ is the acceleration in S. Now in an inertial frame S' (suppose S and S' are in standard configuration) the velocity and acceleration of the particle are $-\mathbf{v}$ and $\mathbf{a}'$ respectively. So the power radiated in this frame is

$$\mathcal{P}' = \frac{\mu_0 q^2 [a'^2]}{6\pi c}.$$

Because of the hint,

$$a^2 = \gamma^6 \left(a'^2 - |\boldsymbol{\beta} \times \mathbf{a}'|^2 \right). \tag{3}$$

Substituting (3) in (2) and using $\mathcal{P}' = \mathcal{P}$ give (1).

Comment

See Questions 11.22 and 11.23 for some specific applications of (1).

Epilogue

Classical electrodynamics has turned out to be a highly successful physical theory. It is obviously of great importance, not only for theoretical but also technological reasons. Incredibly, the theory is valid over a vast range of length scales from subatomic dimensions upwards, applies to fields in both vacuum and matter and holds for particles moving at any speed ($v < c$). How one explains this amazing success presumably involves the following features of the theory, all of which have been touched on in this book:

☞ linear superposition,

☞ the gauge invariance of $\mathbf{E}(\mathbf{r}, t)$ and $\mathbf{B}(\mathbf{r}, t)$, and

☞ the covariance of Maxwell's equations.

Appendix A
Vectors and Cartesian tensors

A vector **A** has three components on a set of axes. Relative to Cartesian axes these are A_x, A_y and A_z. It is often convenient to introduce the notation that any one of these components is represented by A_i where a Roman subscript denotes x or y or z. (The use of the subscript i is completely arbitrary and other equivalent choices such as $j, k, l, \ldots, w$ could have been made—except obviously x, y and z, which are reserved.) Thus the position vector **r** has component r_i which is one of x or y or z. Similarly, for the gradient operator, ∇_i represents $\partial/\partial x$ or $\partial/\partial y$ or $\partial/\partial z$. In Cartesian components the scalar (or dot) product of two vectors **A** and **B** is $\mathbf{A} \cdot \mathbf{B} = A_x B_x + A_y B_y + A_z B_z$. Einstein introduced a notation for this summation, namely,

$$A_x B_x + A_y B_y + A_z B_z = A_i B_i = A_j B_j = \cdots = A_w B_w. \qquad (I)$$

Thus two repeated subscripts are understood to be a summation over corresponding Cartesian components. Because of the arbitrary nature of the subscript any repeated pair may be used, as indicated in (I) above.

The use of a single Roman subscript on a quantity, as in A_i, does not in itself guarantee that the quantity is a vector. This is defined formally in terms of the behaviour of components under a transformation of Cartesian axes relative to a fixed origin O. Here we encounter two essentially different types of transformation. One leaves the hand of the axes unchanged and is called a proper transformation, an example being a rotation about O. The other, known as an improper transformation, changes the hand of the axes from right to left or vice versa. Examples are inversion through O and reflection in a plane containing O.[‡]

Also essentially different are two types of vector, both of which possess, on physical grounds, magnitude and direction. These two types are distinguished by their different behaviour under an improper transformation of axes. The position vector **r** changes sign under inversion (see the figure on p. 599) and hence so does the velocity vector $\mathbf{u} = d\mathbf{r}/dt$. These are both examples of *polar* vectors. For any polar vector **v**, we have $\mathbf{v} \xrightarrow{\text{p}} -\mathbf{v}$, where the symbol p represents the parity transformation (inversion through O). On the other hand, the angular momentum vector $\mathbf{L} = m\mathbf{r} \times \mathbf{u}$ does not change sign under inversion: $\mathbf{L} \xrightarrow{\text{p}} \mathbf{L}' = m\mathbf{r}' \times \mathbf{u}' = m(-\mathbf{r}) \times (-\mathbf{u}) = \mathbf{L}$. Quantities of this

---

[‡]Combinations of the above transformations can be devised which are either proper or improper.

sort are termed *axial* vectors or *pseudovectors*.[‡] (Question 1.11 provides examples from electromagnetism of polar and axial vectors.)

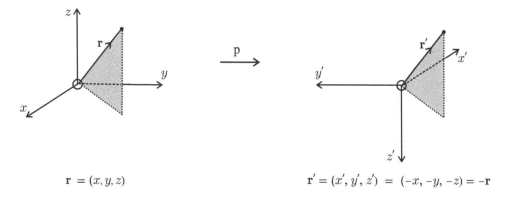

$$\mathbf{r} = (x, y, z) \qquad\qquad \mathbf{r}' = (x', y', z') = (-x, -y, -z) = -\mathbf{r}$$

We are now in a position to state the formal definition of a vector $\mathbf{A}$ that has components A_j relative to Cartesian axes $O(x, y, z)$ and components A_i' relative to Cartesian axes $O(x', y', z')$. For both proper and improper transformations

$$\left.\begin{array}{l}\text{polar vector:}\ \ A_i' = a_{ij}A_j \quad \text{(with inverse transformation } A_j = a_{ij}A_i')\\[2mm] \text{axial vector:}\ \ A_i' = \pm a_{ij}A_j \quad \text{(with inverse transformation } A_j = \pm a_{ij}A_i')\end{array}\right\}, \quad \text{(II)}$$

where in (II)$_2$ the upper (lower) sign is for a proper (improper) transformation of axes. In these equations the a_{ij} are the direction cosines of the i-axis of $O(x', y', z')$ with respect to the j-axis of $O(x, y, z)$, and so $a_{ij} = \cos\theta_{ij}$. An important property of these direction cosines is that they satisfy the orthogonality conditions[1]

$$a_{ik}a_{jk} = \delta_{ij} \qquad \text{and} \qquad a_{ki}a_{kj} = \delta_{ij},$$

where δ_{ij} is the Kronecker delta, which is defined as follows

$$\delta_{ij} = \begin{cases} 1 & \text{if } i = j = x \quad \text{or} \quad \text{if } i = j = y \quad \text{or} \quad \text{if } i = j = z \\ 0 & \text{if } i \neq j. \end{cases} \qquad \text{(III)}$$

The definitions in (II) above can be generalized to Cartesian tensors of any rank.[†] Thus a vector is a tensor of rank one: it has three components A_i that transform in

[‡] In the above, we have made the (reasonable) assumption that time and mass are polar scalars and are unchanged by an improper transformation. This is discussed briefly in the last paragraph before the checklist on p. 600.

[†] The rank of a tensor is the number of its *unrepeated* subscripts. So, for example, T_{ijkl} is a tensor of rank four, T_{ikjk} is a tensor of rank two and T_{ii} is a tensor of rank zero (a scalar).

[1]   O. L. de Lange and J. Pierrus, *Solved problems in classical mechanics: Analytical and numerical solutions with comments*, Chap. 14, p. 537. Oxford: Oxford University Press, 2010.

the manner described by (II). A second-rank polar tensor T_{ij} has nine components that transform as

$$T'_{ij} = a_{ik} a_{jl} T_{kl} \qquad \text{(IV)}$$

for both proper and improper transformation of axes, and so on for higher rank. Similarly, the components of a second-rank axial tensor transform as

$$T'_{ij} = \pm a_{ik} a_{jl} T_{kl}, \qquad \text{(V)}$$

where the upper (lower) sign is for a proper (improper) transformation of axes, and so on for higher rank.

An important special case of (IV) and (V) is for tensors of rank zero. These are more usually known as scalars, and are represented here by the symbol s. If s′ = s for both proper and improper transformation of axes, we call s a polar scalar (some common examples are time, mass and electric charge). An axial scalar, or pseudoscalar, is characterized by s′ = ±s where, as before, the upper (lower) sign is for a proper (improper) transformation of axes (see Comment (ii) of Question 1.8 for an example).

The following checklist may be used to detect and trap errors when using tensors:

☞ Just as the component of a vector does not equal the vector (e.g. $A_x \neq \mathbf{A}$), so is it incorrect to equate a tensor to a vector (e.g. $A_i \neq \mathbf{A}$).

☞ Since $\mathbf{s} \pm \mathbf{A}$ is meaningless, one may not add or subtract tensors of different rank. Only tensors of the same rank may be added or subtracted. Thus the expression $T_{ikjk} - T'_{ij}$ is valid, whereas $T_i + T'_{ij} + T''_{ijk}$ is not.

☞ The particular tensor component on both sides of an equation, or of terms being added or subtracted, must be the same. So $T_{ikkj} = T'_{ji}$ is a valid equation, whereas the expression $T_i - T'_j$ is meaningless.

☞ No subscript may occur more than twice in a single term. Thus terms like T_{ijii} or $\varepsilon_{ijk} T_{jkk}$ or $T_{ij} T'_{il} T''_{im}$ are undefined. However, $T_{ijlm} + T'_{iljm} + T''_{mjli}$ is defined because, although i appears three times, it does not do so more than twice in any one term.

Appendix B
Cartesian coordinates

The figure below shows a *right-handed* system of coordinates (this being the conventional choice in electromagnetism), where for the point $P(x, y, z)$

$$\mathbf{r} = (x, y, z) = \hat{\mathbf{x}}x + \hat{\mathbf{y}}y + \hat{\mathbf{z}}z. \tag{I}$$

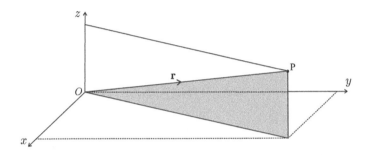

The unit vectors $\hat{\mathbf{x}}$, $\hat{\mathbf{y}}$ and $\hat{\mathbf{z}}$ form an orthogonal set. Clearly, $|\hat{\mathbf{x}}| = |\hat{\mathbf{y}}| = |\hat{\mathbf{z}}| = 1$ and

$$\left.\begin{aligned}
\hat{\mathbf{x}} \cdot \hat{\mathbf{y}} &= 0, & \hat{\mathbf{x}} \times \hat{\mathbf{y}} &= \hat{\mathbf{z}} \\
\hat{\mathbf{y}} \cdot \hat{\mathbf{z}} &= 0, & \hat{\mathbf{y}} \times \hat{\mathbf{z}} &= \hat{\mathbf{x}} \\
\hat{\mathbf{z}} \cdot \hat{\mathbf{x}} &= 0, & \hat{\mathbf{z}} \times \hat{\mathbf{x}} &= \hat{\mathbf{y}}
\end{aligned}\right\}. \tag{II}$$

differential elements

$$\left.\begin{aligned}
&\text{line element } d\boldsymbol{\ell} = \hat{\mathbf{x}}\,dx + \hat{\mathbf{y}}\,dy + \hat{\mathbf{z}}\,dz \\
&\text{area element } d\mathbf{a} = \hat{\mathbf{x}}\,dy\,dz + \hat{\mathbf{y}}\,dz\,dx + \hat{\mathbf{z}}\,dx\,dy \\
&\text{volume element } dv = dx\,dy\,dz
\end{aligned}\right\}. \tag{III}$$

vector operators

$$
\left.
\begin{aligned}
\boldsymbol{\nabla}\psi &= \hat{\mathbf{x}}\frac{\partial\psi}{\partial x} + \hat{\mathbf{y}}\frac{\partial\psi}{\partial y} + \hat{\mathbf{z}}\frac{\partial\psi}{\partial z} \\[2mm]
\boldsymbol{\nabla}\cdot\mathbf{A} &= \frac{\partial A_x}{\partial x} + \frac{\partial A_y}{\partial y} + \frac{\partial A_z}{\partial z} \\[2mm]
\boldsymbol{\nabla}\times\mathbf{A} &= \hat{\mathbf{x}}\left(\frac{\partial A_z}{\partial y} - \frac{\partial A_y}{\partial z}\right) + \hat{\mathbf{y}}\left(\frac{\partial A_x}{\partial z} - \frac{\partial A_z}{\partial x}\right) + \hat{\mathbf{z}}\left(\frac{\partial A_y}{\partial x} - \frac{\partial A_x}{\partial y}\right) \\[2mm]
\nabla^2\psi &= \frac{\partial^2\psi}{\partial x^2} + \frac{\partial^2\psi}{\partial y^2} + \frac{\partial^2\psi}{\partial z^2}
\end{aligned}
\right\} . \quad \text{(IV)}
$$

Appendix C
Spherical polar coordinates

Let $\mathbf{r} = (r, \theta, \phi)$ be the spherical polar coordinates of an arbitrary point P (see the figure below). Then the transformation

$$\left.\begin{array}{l} x = r\sin\theta\cos\phi \\ y = r\sin\theta\sin\phi \\ z = r\cos\theta \end{array}\right\} \tag{I}$$

can be made unique by imposing the restrictions: ☞ $r \geq 0$, ☞ $0 \leq \theta \leq \pi$, ☞ $0 \leq \phi < 2\pi$.

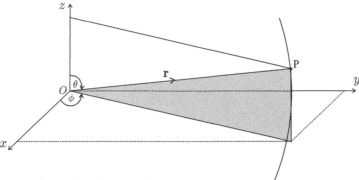

The inverse transformation is given by

$$\left.\begin{array}{l} r = (x^2 + y^2 + z^2)^{1/2} \\ \theta = \tan^{-1}(r^2/z^2 - 1)^{1/2} \\ \phi = \tan^{-1}(y/x) \end{array}\right\}. \tag{II}$$

In these coordinates $\hat{\mathbf{r}}$, $\hat{\boldsymbol{\theta}}$ and $\hat{\boldsymbol{\phi}}$ form an orthogonal system of unit vectors ($|\hat{\mathbf{r}}| = |\hat{\boldsymbol{\theta}}| = |\hat{\boldsymbol{\phi}}| = 1$) that satisfy

$$\left.\begin{array}{ll} \hat{\mathbf{r}} \cdot \hat{\boldsymbol{\theta}} = 0, & \hat{\mathbf{r}} \times \hat{\boldsymbol{\theta}} = \hat{\boldsymbol{\phi}} \\ \hat{\boldsymbol{\theta}} \cdot \hat{\boldsymbol{\phi}} = 0, & \hat{\boldsymbol{\theta}} \times \hat{\boldsymbol{\phi}} = \hat{\mathbf{r}} \\ \hat{\boldsymbol{\phi}} \cdot \hat{\mathbf{r}} = 0, & \hat{\boldsymbol{\phi}} \times \hat{\mathbf{r}} = \hat{\boldsymbol{\theta}} \end{array}\right\}. \tag{III}$$

It is straightforward to show that the spherical polar and Cartesian unit vectors are related by

$$
\left.
\begin{aligned}
\hat{\mathbf{x}} &= \hat{\mathbf{r}}\sin\theta\cos\phi + \hat{\boldsymbol{\theta}}\cos\theta\cos\phi - \hat{\boldsymbol{\phi}}\sin\phi \\
\hat{\mathbf{y}} &= \hat{\mathbf{r}}\sin\theta\sin\phi + \hat{\boldsymbol{\theta}}\cos\theta\sin\phi + \hat{\boldsymbol{\phi}}\cos\phi \\
\hat{\mathbf{z}} &= \hat{\mathbf{r}}\cos\theta - \hat{\boldsymbol{\theta}}\sin\theta
\end{aligned}
\right\}, \qquad \text{(IV)}
$$

and

$$
\left.
\begin{aligned}
\hat{\mathbf{r}} &= \hat{\mathbf{x}}\sin\theta\cos\phi + \hat{\mathbf{y}}\sin\theta\sin\phi + \hat{\mathbf{z}}\cos\theta \\
\hat{\boldsymbol{\theta}} &= \hat{\mathbf{x}}\cos\theta\cos\phi + \hat{\mathbf{y}}\cos\theta\sin\phi - \hat{\mathbf{z}}\sin\theta \\
\hat{\boldsymbol{\phi}} &= -\hat{\mathbf{x}}\sin\phi + \hat{\mathbf{y}}\cos\phi
\end{aligned}
\right\}. \qquad \text{(V)}
$$

angle between two vectors and magnitude of $|\mathbf{r} - \mathbf{r}'|$

Consider the vectors $\mathbf{r} = (r, \theta, \phi)$ and $\mathbf{r}' = (r', \theta', \phi')$. The angle χ between $\mathbf{r}$ and $\mathbf{r}'$ is given by $\cos\chi = \hat{\mathbf{r}} \cdot \hat{\mathbf{r}}' = \sin\theta\sin\theta'\cos\phi\cos\phi' + \sin\theta\sin\theta'\sin\phi\sin\phi' + \cos\theta\cos\theta'$ or

$$
\cos\chi = \cos\theta\cos\theta' + \sin\theta\sin\theta'\cos(\phi - \phi'). \qquad \text{(VI)}
$$

From the cosine rule $|\mathbf{r} - \mathbf{r}'| = \sqrt{r^2 + r'^2 - 2rr'\cos\chi}$, and so

$$
|\mathbf{r} - \mathbf{r}'| = \sqrt{r^2 + r'^2 - 2rr'\big(\cos\theta\cos\theta' + \sin\theta\sin\theta'\cos(\phi - \phi')\big)}. \qquad \text{(VII)}
$$

differential elements

$$
\left.
\begin{aligned}
\text{line element } d\boldsymbol{\ell} &= \hat{\mathbf{r}}\,dr + \hat{\boldsymbol{\theta}}\,r\,d\theta + \hat{\boldsymbol{\phi}}\,r\sin\theta\,d\phi \\
\text{area element } d\mathbf{a} &= \hat{\mathbf{r}}\,r^2\sin\theta\,d\theta\,d\phi \\
\text{volume element } dv &= r^2\sin\theta\,dr\,d\theta\,d\phi \\
&= r^2\,dr\,d(\cos\theta)\,d\phi \quad \text{where } -1 \le \cos\theta \le 1
\end{aligned}
\right\}. \qquad \text{(VIII)}
$$

rates of change

unit vectors

Unlike the Cartesian unit vectors which are constant, these unit vectors have directions that are, in general, dependent on time. So

$$
\left.
\begin{aligned}
\frac{d\hat{\mathbf{r}}}{dt} &= \hat{\boldsymbol{\theta}}\,\dot{\theta} + \hat{\boldsymbol{\phi}}\,\dot{\phi}\sin\theta \\[4pt]
\frac{d\hat{\boldsymbol{\theta}}}{dt} &= -\hat{\mathbf{r}}\,\dot{\theta} + \hat{\boldsymbol{\phi}}\,\dot{\phi}\cos\theta \\[4pt]
\frac{d\hat{\boldsymbol{\phi}}}{dt} &= -(\hat{\mathbf{r}}\,\dot{\phi}\sin\theta + \hat{\boldsymbol{\theta}}\,\dot{\phi}\cos\theta)
\end{aligned}
\right\}. \qquad \text{(IX)}
$$

<u>velocity and acceleration</u>

Using $\mathbf{r} = r\hat{\mathbf{r}}$ and (IX) gives

$$\left.\begin{aligned}
\mathbf{v} &= \dot{\mathbf{r}} = \hat{\mathbf{r}}\,\dot{r} + \hat{\boldsymbol{\theta}}\,r\dot{\theta} + \hat{\boldsymbol{\phi}}\,r\dot{\phi}\sin\theta \\
\mathbf{a} &= \dot{\mathbf{v}} = \hat{\mathbf{r}}\,(\ddot{r} - r\dot{\theta}^2 - r\dot{\phi}^2\sin^2\theta) + \hat{\boldsymbol{\theta}}\,(r\ddot{\theta} + 2\dot{r}\dot{\theta} - r\dot{\phi}^2\sin\theta\cos\theta) + \\
&\qquad\qquad\qquad \hat{\boldsymbol{\phi}}\,(r\ddot{\phi}\sin\theta + 2\dot{r}\dot{\phi}\sin\theta + 2r\dot{\theta}\dot{\phi}\cos\theta)
\end{aligned}\right\}. \qquad \text{(X)}$$

vector operators

$$\left.\begin{aligned}
\boldsymbol{\nabla}\psi &= \hat{\mathbf{r}}\,\frac{\partial\psi}{\partial r} + \hat{\boldsymbol{\theta}}\,\frac{1}{r}\frac{\partial\psi}{\partial\theta} + \hat{\boldsymbol{\phi}}\,\frac{1}{r\sin\theta}\frac{\partial\psi}{\partial\phi} \\[2ex]
\boldsymbol{\nabla}\cdot\mathbf{A} &= \frac{1}{r^2}\frac{\partial}{\partial r}(r^2 A_r) + \frac{1}{r\sin\theta}\frac{\partial}{\partial\theta}(A_\theta\sin\theta) + \frac{1}{r\sin\theta}\frac{\partial A_\phi}{\partial\phi} \\[2ex]
\boldsymbol{\nabla}\times\mathbf{A} &= \hat{\mathbf{r}}\,\frac{1}{r\sin\theta}\left[\frac{\partial}{\partial\theta}(A_\phi\sin\theta) - \frac{\partial A_\theta}{\partial\phi}\right] + \hat{\boldsymbol{\theta}}\left[\frac{1}{r\sin\theta}\frac{\partial A_r}{\partial\phi} - \frac{1}{r}\frac{\partial}{\partial r}(rA_\phi)\right] \\[2ex]
&\qquad\qquad\qquad +\hat{\boldsymbol{\phi}}\,\frac{1}{r}\left[\frac{\partial}{\partial r}(rA_\theta) - \frac{\partial A_r}{\partial\theta}\right] \\[2ex]
\nabla^2\psi &= \frac{1}{r^2}\frac{\partial}{\partial r}\left(r^2\frac{\partial\psi}{\partial r}\right) + \frac{1}{r^2\sin\theta}\frac{\partial}{\partial\theta}\left(\sin\theta\,\frac{\partial\psi}{\partial\theta}\right) + \frac{1}{r^2\sin^2\theta}\frac{\partial^2\psi}{\partial\phi^2}
\end{aligned}\right\}. \qquad \text{(XI)}$$

Appendix D
Cylindrical polar coordinates

We use (r, θ, z) for the cylindrical polar coordinates[‡] of an arbitrary point P (see the figure below), where

$$
\left.
\begin{aligned}
x &= r\cos\theta \\
y &= r\sin\theta \\
z &= z
\end{aligned}
\right.
\qquad
\left.
\begin{aligned}
r &= (x^2 + y^2)^{1/2} \\
\theta &= \tan^{-1}(y/x) \\
z &= z
\end{aligned}
\right\},
\qquad\qquad (I)
$$

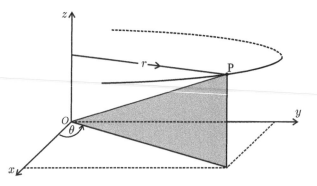

with the unit vectors $\hat{\mathbf{r}}$, $\hat{\boldsymbol{\theta}}$ and $\hat{\mathbf{z}}$ of this (orthogonal) coordinate system satisfying

$$
\left.
\begin{aligned}
\hat{\mathbf{r}} \cdot \hat{\boldsymbol{\theta}} &= 0, \\
\hat{\boldsymbol{\theta}} \cdot \hat{\mathbf{z}} &= 0, \\
\hat{\mathbf{z}} \cdot \hat{\mathbf{r}} &= 0,
\end{aligned}
\right.
\qquad
\left.
\begin{aligned}
\hat{\mathbf{r}} \times \hat{\boldsymbol{\theta}} &= \hat{\mathbf{z}} \\
\hat{\boldsymbol{\theta}} \times \hat{\mathbf{z}} &= \hat{\mathbf{r}} \\
\hat{\mathbf{z}} \times \hat{\mathbf{r}} &= \hat{\boldsymbol{\theta}}
\end{aligned}
\right\}.
\qquad\qquad (II)
$$

It is straightforward to show that the cylindrical polar and Cartesian unit vectors are related as follows:

$$
\left.
\begin{aligned}
\hat{\mathbf{x}} &= \hat{\mathbf{r}}\cos\theta - \hat{\boldsymbol{\theta}}\sin\theta \\
\hat{\mathbf{y}} &= \hat{\mathbf{r}}\sin\theta + \hat{\boldsymbol{\theta}}\cos\theta \\
\hat{\mathbf{z}} &= \hat{\mathbf{z}}
\end{aligned}
\right\}
\quad \text{and} \quad
\left.
\begin{aligned}
\hat{\mathbf{r}} &= \hat{\mathbf{x}}\cos\theta + \hat{\mathbf{y}}\sin\theta \\
\hat{\boldsymbol{\theta}} &= -\hat{\mathbf{x}}\sin\theta + \hat{\mathbf{y}}\cos\theta \\
\hat{\mathbf{z}} &= \hat{\mathbf{z}}
\end{aligned}
\right\}.
\qquad (III)
$$

[‡]The reader should be alert to variations in notation adopted by different textbooks. There are compelling reasons why authors avoid using the symbol r for the radial cylindrical coordinate. Notwithstanding these reasons, and after some equivocation, it was decided to retain the use of r in this book. Note also that the angle θ is sometimes replaced by ϕ.

magnitude of $|\mathbf{r} - \mathbf{r}'|$

Clearly $|\mathbf{r} - \mathbf{r}'| = \sqrt{(r\cos\theta - r'\cos\theta')^2 + (r\sin\theta - r'\sin\theta')^2 + (z - z')^2}$, and so

$$|\mathbf{r} - \mathbf{r}'| = \sqrt{r^2 + r'^2 - 2rr'\cos(\theta - \theta') + (z - z')^2}\,. \qquad \text{(IV)}$$

differential elements

$$\left.\begin{array}{l}
\text{line element } d\boldsymbol{\ell} = \hat{\mathbf{r}}\,dr + \hat{\boldsymbol{\theta}}\,r\,d\theta + \hat{\mathbf{z}}\,dz \\[4pt]
\text{area element } d\mathbf{a} = \hat{\mathbf{r}}\,r\,d\theta\,dz \\[4pt]
\text{volume element } dv = r\,dr\,d\theta\,dz
\end{array}\right\}. \qquad \text{(V)}$$

rates of change

unit vectors

Like the spherical polar unit vectors, these unit vectors also have directions that are, in general, dependent on time. So

$$\left.\begin{array}{l}
\dfrac{d\hat{\mathbf{r}}}{dt} = \hat{\boldsymbol{\theta}}\,\dot{\theta} \\[12pt]
\dfrac{d\hat{\boldsymbol{\theta}}}{dt} = -\hat{\mathbf{r}}\,\dot{\theta} \\[12pt]
\dfrac{d\hat{\mathbf{z}}}{dt} = 0
\end{array}\right\}. \qquad \text{(VI)}$$

velocity and acceleration

Using $\mathbf{r} = r\hat{\mathbf{r}} + z\hat{\mathbf{z}}$ and (VI) gives

$$\left.\begin{array}{l}
\mathbf{v} = \dot{\mathbf{r}} = \hat{\mathbf{r}}\,\dot{r} + \hat{\boldsymbol{\theta}}\,r\dot{\theta} + \hat{\mathbf{z}}\,\dot{z} \\[4pt]
\mathbf{a} = \dot{\mathbf{v}} = \hat{\mathbf{r}}\,(\ddot{r} - r\dot{\theta}^2) + \hat{\boldsymbol{\theta}}\,(r\ddot{\theta} + 2\dot{r}\dot{\theta}) + \hat{\mathbf{z}}\,\ddot{z}
\end{array}\right\}. \qquad \text{(VII)}$$

vector operators

$$\left.\begin{array}{l}
\boldsymbol{\nabla}\psi = \hat{\mathbf{r}}\,\dfrac{\partial\psi}{\partial r} + \hat{\boldsymbol{\theta}}\,\dfrac{1}{r}\dfrac{\partial\psi}{\partial\theta} + \hat{\mathbf{z}}\,\dfrac{\partial\psi}{\partial z} \\[12pt]
\boldsymbol{\nabla}\cdot\mathbf{A} = \dfrac{1}{r}\dfrac{\partial}{\partial r}(rA_r) + \dfrac{1}{r}\dfrac{\partial A_\theta}{\partial\theta} + \dfrac{\partial A_z}{\partial z} \\[12pt]
\boldsymbol{\nabla}\times\mathbf{A} = \hat{\mathbf{r}}\left[\dfrac{1}{r}\dfrac{\partial A_z}{\partial\theta} - \dfrac{\partial A_\theta}{\partial z}\right] + \hat{\boldsymbol{\theta}}\left[\dfrac{\partial A_r}{\partial z} - \dfrac{\partial A_z}{\partial r}\right] + \hat{\mathbf{z}}\,\dfrac{1}{r}\left[\dfrac{\partial}{\partial r}(rA_\theta) - \dfrac{\partial A_r}{\partial\theta}\right] \\[12pt]
\nabla^2\psi = \dfrac{1}{r}\dfrac{\partial}{\partial r}\left(r\dfrac{\partial\psi}{\partial r}\right) + \dfrac{1}{r^2}\dfrac{\partial^2\psi}{\partial\theta^2} + \dfrac{\partial^2\psi}{\partial z^2}
\end{array}\right\}. \qquad \text{(VIII)}$$

Appendix E
The Dirac delta function

The Dirac delta function provides a convenient way of describing certain finite effects that take place over a very short period of time, or which occur at a point in space, along a line or on a surface. In this appendix, we give a brief account of the Dirac delta function (henceforth the 'delta function') and some of its important properties which are often needed in physics. Intuitively, we may regard the one-dimensional delta function $\delta(x)$[‡] to have the following properties: it is infinite at the origin and zero everywhere else. Strictly speaking, $\delta(x)$ is not a function at all; at least not in the usual mathematical sense which requires a function to have a definite value at every point in its domain. Notwithstanding this, a rigorous theory can still be developed by representing $\delta(x)$ as the limit of a sequence of functions (for example, a sequence of Gaussian distributions). For further details, the reader is referred to an appropriate textbook (see, for example, Ref. [1]).

Suppose we let $f(x)$ be any well-behaved function, that is suitably continuous and differentiable. Then, in terms of $f(x)$, we adopt the following definition of the delta function

$$\int_I f(x)\,\delta(x - x_0)\,dx \;=\; \begin{cases} f(x_0) & \text{if } x_0 \in I \\ 0 & \text{if } x_0 \notin I, \end{cases} \tag{I}$$

where I is an interval of integration. Several important properties follow from this definition and are given below. Where necessary, the reader is referred to Ref. [2] for proofs and further information.

☞ If the interval of integration in (I) extends from $-\infty$ to ∞, then

$$\int_{-\infty}^{\infty} f(x)\,\delta(x - x_0)\,dx \;=\; f(x_0). \tag{II}$$

This is a very useful result known as the 'sifting' property of a delta function.

[‡]Take care not to confuse the delta function $\delta(x)$ and the Kronecker delta δ_{ij} which are completely separate entities.

---

[1]   G. B. Arfken and H. J. Weber, *Mathematical methods for physicists*, Chap. 1, pp. 81–5. San Diego: Academic Press, 1995.
[2]   G. Barton, *Elements of Green's functions and propagation*, Chap. 1, pp. 7–40. Oxford: Clarendon Press, 1989.

☞ In (I) let $f(x) = 1$ and $x_0 = 0$. Then

$$\int_{x_1}^{x_2} \delta(x)\, dx \;=\; \begin{cases} 1 & \text{if } x_1 < 0 < x_2 \\ 0 & \text{otherwise,} \end{cases} \tag{III}$$

and so

$$\int_{-\infty}^{\infty} \delta(x)\, dx \;=\; 1. \tag{IV}$$

☞ Assuming that x is a single independent variable, the physical dimensions of $\delta(x)$ are the inverse of those of x (that this is necessary follows from (IV), for example).

☞ The delta function is (weakly) even: $\delta(x - x_0) = \delta(x_0 - x)$.

☞ If α is a real constant then

$$\delta(\alpha x) \;=\; \frac{\delta(x)}{|\alpha|}. \tag{V}$$

Equation (V) may be generalized to include the case when the argument of the delta function is itself a function, $g(x)$, say. Suppose $g(x)$ has real roots x_n (that is, $g(x_n) = 0$) then,

$$\delta\big(g(x)\big) \;=\; \sum_{n} \frac{\delta(x - x_n)}{g'(x_n)} \qquad \text{where } g'(x_n) = \frac{dg}{dx}. \tag{VI}$$

☞ The integral and derivative of $\delta(x)$ are given by

$$\int_{-\infty}^{x} \delta(x')\, dx' = H(x) \qquad \text{and} \qquad \delta(x) = \frac{d}{dx} H(x), \tag{VII}$$

where $H(x)$ is the Heaviside function defined as follows:

$$H(x) \;=\; \begin{cases} 1 & x > 0 \\ 0 & x < 0. \end{cases} \tag{VIII}$$

Delta functions in three dimensions

The above results may be generalized. Suppose $f(\mathbf{r})$ is any continuous function of $\mathbf{r}$. Then

$$\left. \begin{array}{ll} \delta(\mathbf{r}) = 0 & \text{if } \mathbf{r} \neq 0 \\[2mm] \displaystyle\int_{v} f(\mathbf{r})\, \delta(\mathbf{r})\, dv = f(0) & \text{if } 0 \in v \end{array} \right\}. \tag{IX}$$

An important case occurs for $f(\mathbf{r}) = 1$. Then

$$\int_v \delta(\mathbf{r})\, dv = \begin{cases} 1 & \text{if } \mathbf{r} = 0 \in v \\ 0 & \text{if } \mathbf{r} = 0 \notin v. \end{cases} \tag{X}$$

Shifting the singularity from the origin to $\mathbf{r} = \mathbf{r}'$ gives

$$\left. \begin{aligned} \delta(\mathbf{r} - \mathbf{r}') &= 0 & \text{if } \mathbf{r} \neq \mathbf{r}' \\ \int_v f(\mathbf{r}')\,\delta(\mathbf{r} - \mathbf{r}')\, dv' &= f(\mathbf{r}) & \text{if } \mathbf{r} \in v \end{aligned} \right\}. \tag{XI}$$

In Cartesian coordinates, the two- and three-dimensional delta functions are formed from products of their one-dimensional counterparts. So if $\mathbf{r} = (x, y, z)$ and $\mathbf{r}' = (x', y', z')$ then,

$$\delta(\mathbf{r} - \mathbf{r}') = \delta(x - x')\,\delta(y - y')\,\delta(z - z'). \tag{XII}$$

In non-Cartesian coordinate systems the form of the delta function is less obvious. For example, in spherical polar coordinates the volume element is $dv = r^2 dr\, d(\cos\theta)\, d\phi$ (see (VIII) of Appendix C), which suggests[#] that

$$\delta(\mathbf{r} - \mathbf{r}') = r^{-2}\delta(r - r')\,\delta(\cos\theta - \cos\theta')\,\delta(\phi - \phi'). \tag{XIII}$$

Similarly, in cylindrical polar coordinates we have

$$\delta(\mathbf{r} - \mathbf{r}') = r^{-1}\delta(r - r')\,\delta(\theta - \theta')\,\delta(z - z'). \tag{XIV}$$

Because the delta function is even, the terms r^{-2} and r^{-1} in (XII) and (XIII) can be replaced by r'^{-2} and r'^{-1} respectively.

[#]The delta function in an arbitrary system of coordinates can be derived formally by considering the Jacobian of the transformation $J(\xi_i; \eta_i)$ from the generalized coordinates (ξ_1, ξ_2, ξ_3) to (η_1, η_2, η_3).

Appendix F
Legendre polynomials

The Legendre equation

$$(1 - x^2)\frac{d^2y}{dx^2} - 2x\frac{dy}{dx} + n(n+1)y = 0, \tag{I}$$

where n is a constant, is one of the standard differential equations of mathematical physics. It has polynomial solutions called *Legendre polynomials* $P_n(x)$, and one way of finding them is to assume a series solution of (I). The details are readily available elsewhere (see, for example, Ref. [1]). Here we summarize some important results that will be used in several questions in this book.

In Chapter 1 we encounter Legendre's equation in the form

$$\frac{1}{\sin\theta}\frac{d}{d\theta}\left(\sin\theta\frac{dy}{d\theta}\right) + n(n+1)y = 0; \tag{II}$$

the equivalence between (I) and (II) can easily be established by making the change of variable $x = \cos\theta$.

Rodrigues's formula, some explicit forms and a recursion relation

Apart from the series-solution approach mentioned above, the Legendre polynomials for non-negative integer values of n can also be found using Rodrigues's formula:

$$P_n(x) = \frac{1}{2^n n!}\frac{d^n}{dx^n}(x^2 - 1)^n. \tag{III}$$

Some explicit forms are:

$$\left.\begin{array}{ll} P_0(x) = 1 & P_2(x) = \frac{1}{2}(3x^2 - 1) \qquad P_4(x) = \frac{1}{8}(35x^4 - 30x^2 + 3) \\ P_1(x) = x & P_3(x) = \frac{1}{2}(5x^3 - 3x) \qquad P_5(x) = \frac{1}{8}(63x^5 - 70x^3 + 15x) \end{array}\right\}, \tag{IV}$$

and a useful recursion relation for generating them is:

$$nP_n(x) = (2n - 1)xP_{n-1}(x) - (n - 1)P_{n-2}(x). \tag{V}$$

[1]   M. Boas, *Mathematical methods in the physical sciences*, Chap. 12, pp. 566–71. New York: Wiley, 3 edn, 2006.

Completeness and orthogonality[‡]

Two important properties of the Legendre polynomials are:

☞ they form a complete set of functions on the interval $-1 \leq x \leq 1$ (or $0 \leq \theta \leq \pi$).

☞ they satisfy the orthogonality condition

$$\int_{-1}^{1} P_n(x) P_k(x) \, dx \;=\; \frac{2\delta_{nk}}{2n+1}, \tag{VI}$$

where $\delta_{nk} = 1$ if $n = k$ or 0 if $n \neq k$.

Some useful properties

The following general properties can also be proved:

$$\left. \begin{array}{l} |P_n(x)| \leq 1 \\[1.2ex] P_n(1) = 1 \\[1.2ex] P_n(-x) = (-1)^n P_n(x) \\[1.2ex] P_{-(n+1)}(x) = P_n(x) \\[1.2ex] \displaystyle\int_{-1}^{1} P_n(x) \, dx = 2\delta_{0n} \\[2ex] P_n(0) = \begin{cases} 0 \text{ if } n \text{ is odd} \\[1.5ex] \dfrac{(-1)^{n/2} n!}{2^n \left((n/2)!\right)^2} \text{ if } n \text{ is even} \end{cases} \end{array} \right\}. \tag{VII}$$

[‡]If $f_n(x)$ represents a complete set of orthogonal functions, then any other function $f(x)$ can be expressed as the linear combination

$$f(x) = \sum_{n=0}^{\infty} c_n f_n(x), \qquad \text{where the } c_n \text{ are constants.}$$

One of the first examples where this is encountered in physics is for the set of functions $\sin nx$ and $\cos nx$ on the interval $[-\pi, \pi]$; usually within the context of a Fourier series.

Appendix G
Bessel functions

One of the standard forms of Bessel's equation is

$$x^2 \frac{d^2y}{dx^2} + x\frac{dy}{dx} + (x^2 - m^2)y = 0, \tag{I}$$

where the constant m is called the order of the Bessel function y (which is a solution of (I)). Although m is not necessarily an integer, it often is—especially in physical systems which possess azimuthal symmetry. The general solution[1] of (I) is

$$y = a\,J_m(x) + b\,N_m(x), \tag{II}$$

where $J_m(x)$ and $N_m(x)$ are mth-order Bessel functions of the first and second kind respectively, and a and b are arbitrary constants.

It is possible to express $J_m(x)$ and $N_m(x)$ in terms of a power series.[1] The first two Bessel functions of both kinds are drawn in the graph below. At a quick glance, they resemble damped sines and cosines. Now $J_0(0) = 1$ and it oscillates like $\cos x$ but with decreasing amplitude. All the other $J_m(x)$ are zero at $x = 0$ and they oscillate like a damped sine function. The $N_m(x)$, on the other hand, all tend to $-\infty$ as $x \to 0$, and they also oscillate with decreasing amplitude away from the origin.

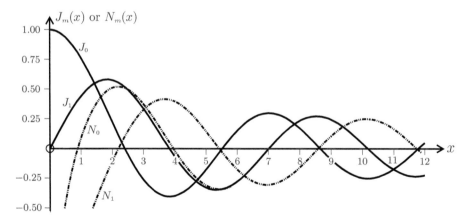

[1]   M. Boas, *Mathematical methods in the physical sciences*, Chap. 12, pp. 587–92. New York: Wiley, 3 edn, 2006.

The zeros of the Bessel functions do not occur at regular intervals like those of $\cos x$ and $\sin x$. Consider for example the roots of $J_m(x)$ and $J'_m(x)$. For $J_m(x)$ these occur at a sequence of points

$$J_m(x) = 0 \qquad \text{where } x = s_{mn}, \quad m = 0, 1, 2, \ldots \quad \text{and} \quad n = 1, 2, 3 \ldots . \qquad \text{(III)}$$

Similarly, for the first derivative

$$J'_m(x) = 0 \qquad \text{where } x = t_{mn}, \quad m = 0, 1, 2, \ldots \quad \text{and} \quad n = 1, 2, 3 \ldots . \qquad \text{(IV)}$$

Appendix H
Parametric representation of a surface

The following provides a brief account of the parametric representation of a smooth (or piecewise smooth) surface s. The reader is referred to an appropriate textbook for a more in-depth discussion (see, for example, Ref. [1]).

Two parameters (u and v, say) are required for a parametric representation of a surface where $u_1 \le u \le u_2$ and $v_1 \le v \le v_2$. Every point in a region R in (u, v) space can be mapped onto the point of s with position vector $\mathbf{r}(u, v)$. In Cartesian coordinates the surface is described by

$$\mathbf{r}(u, v) = \hat{\mathbf{x}}\, x(u, v) + \hat{\mathbf{y}}\, y(u, v) + \hat{\mathbf{z}}\, z(u, v), \tag{I}$$

where $x = x(u, v)$, $y = y(u, v)$ and $z = z(u, v)$ are the parametric equations of the surface.

The flux of a vector field $\mathbf{F}(\mathbf{r})$, which is of particular interest in electromagnetism, can be expressed as

$$\int_s \mathbf{F} \cdot d\mathbf{a} = \int_R \int \mathbf{F}\big(\mathbf{r}(u, v)\big) \cdot \left(\frac{\partial \mathbf{r}}{\partial u} \times \frac{\partial \mathbf{r}}{\partial v} \right) du\, dv, \tag{II}$$

where $\partial \mathbf{r}/\partial u$ and $\partial \mathbf{r}/\partial v$ are tangent vectors (at an arbitrary point P on the surface) along coordinate curves (v constant; u constant respectively). The cross-product of these (linearly independent) vectors defines a vector normal to the surface at P. The magnitude of $\left(\dfrac{\partial \mathbf{r}}{\partial u} \times \dfrac{\partial \mathbf{r}}{\partial v} \right) du\, dv$ is the infinitesimal area of the parallelogram at P spanned by $\partial \mathbf{r}/\partial u$ and $\partial \mathbf{r}/\partial v$.

---

[1] K. F. Riley and M. P. Hobson, *Essential mathematical methods for the physical sciences*, Chap. 2, pp. 94–6. Cambridge: Cambridge University Press, 2011.

Appendix I
The Cauchy–Riemann equations

Here we provide a very brief summary of some essential information. Readers who require a more detailed account should consult a suitable textbook (see, for example, Ref. [1]).

functions of a complex variable

An elementary function f in the real variable x can be applied in the complex plane by replacing x with $z = x + iy$. The resulting complex function $w(z)$ may be regarded as a pair of real functions $u(x,y)$ and $v(x,y)$ of the real variables x and y. That is,

$$w(z) = w(x + iy) = u(x,y) + iv(x,y). \tag{I}$$

analytic functions

The derivative of a complex function is defined as follows:

$$w'(z) = \frac{dw}{dz} = \lim_{\Delta z \to 0} \frac{w(z + \Delta z) - w(z)}{\Delta z}. \tag{II}$$

The function $w(z)$ is *analytic* at a point $z = z_0$ if $w'(z)$ exists, and has a unique value inside some small region in the vicinity of z_0. We note that, in taking the limit above, there are infinitely many ways of approaching z_0. The uniqueness of the derivative at $z = z_0$ implies that (II) must give the *same value* irrespective of how the point z_0 is approached.[‡] We use this important idea next.

[‡] Clearly this is a far more stringent requirement than the equivalent process for differentiating a real function $f(x)$, where x_0 can only be approached from the left or the right.

[1] M. Boas, *Mathematical methods in the physical sciences*, Chap. 14, pp. 666–72. New York: Wiley, 3 edn, 2006.

Cauchy–Riemann equations

Taking Δz along the real axis with $\Delta y = 0$

$$\frac{dw}{dz} = \lim_{\Delta x \to 0} \frac{[u(x + \Delta x, y) - u(x, y)] + i[v(x + \Delta x, y) - v(x, y)]}{\Delta x}$$

$$= \frac{\partial u}{\partial x} + i\frac{\partial v}{\partial x} \ . \tag{III}$$

Taking Δz along the imaginary axis with $\Delta x = 0$

$$\frac{dw}{dz} = \lim_{\Delta y \to 0} \frac{[u(x, y + \Delta y) - u(x, y)] + i[v(x, y + \Delta y) - v(x, y)]}{i\Delta y}$$

$$= -i\frac{\partial u}{\partial y} + \frac{\partial v}{\partial y} \ . \tag{IV}$$

The uniqueness of $w'(z)$ requires that the real (imaginary) part of (III) is equated to the real (imaginary) part of (IV). This gives

$$\frac{\partial u}{\partial x} = \frac{\partial v}{\partial y} \quad \text{and} \quad \frac{\partial v}{\partial x} = -\frac{\partial u}{\partial y}, \tag{V}$$

which are the Cauchy–Riemann equations. They are necessary and sufficient conditions for $w(z)$ to be analytic.

Appendix J
Questions involving computational work

The following questions provide material suitable for project work in computational physics. They require computer algebra/calculations in their solution, or for analysis of the solution, or both.

QUESTION	DESCRIPTION
2.7	Field lines and equipotentials inside a charged non-conducting sphere with a cavity
2.8	Field lines and equipotentials of two infinite parallel line charges
2.9	Field lines and equipotentials of a finite line charge
2.10	Field lines and equipotentials of a circular line charge
2.18–2.19	Mechanical behaviour of three charged particles constrained to move on a circular trajectory
2.24	Field lines and equipotentials of a point electric quadrupole
2.25	Field lines and equipotentials of a physical electric dipole
3.5	Field lines near a plane conductor containing a spherical bulge
3.6	Series expansion of Φ for a thin conducting disc maintained at a potential V_0
3.14	Coefficients of capacitance of three spherical conductors located at the corners of an equilateral triangle
3.15	Capacitance of an anchor ring
3.17	Fourier coefficients of the electrostatic potential inside a conducting box

QUESTION	DESCRIPTION
3.19	Use of the relaxation method to solve Laplace's equation in two dimensions
3.20	Use of the relaxation method to solve Laplace's equation in three dimensions
3.23–3.24	Method of electrostatic images: field of a point charge near a conducting sphere
3.25–3.26	Field of two nearby spherical conductors using an infinite series of image charges
3.27	Capacitance of two parallel conducting cylinders
3.29	Mechanical behaviour of a physical dipole located near a grounded conducting plane
3.32	Fringing field near the edge of a semi-infinite parallel-plate capacitor
3.34	Some examples using a conformal mapping to determine electric-field lines and lines of constant potential
4.11	Magnetic field of a charged spinning non-conducting sphere
4.13	Magnetic field inside a current-carrying cylindrical conductor with a cavity bored parallel to its axis
4.15	Magnetic field of a circular current loop
4.16–4.17	Axial magnetic field in a Helmholtz/anti-Helmholtz coil configuration
4.23	Trajectory of a charged particle moving in a static magnetic dipole field
4.27	Field of a magnetic quadrupole
5.9	Current induced in a coil by a bar magnet falling through it
5.10	Electric field induced by a rotating bar magnet
5.13	Trajectory of a charged particle moving in a time-dependent magnetic field
5.16–5.18	Self inductance calculations for various coil geometries
5.19	Mutual inductance of two coaxial circular coils and force between them

Glossary of symbols

Because there are a large number of physical and mathematical quantities, and only a few hundred Roman and Greek letters (counting different fonts), it is inevitable that many symbols will have more than one meaning. Hopefully, it will always be obvious from the context which meaning is intended. Nevertheless, the reader should be alert to similar-looking symbols which arise from a change in font. For example, P and P represent different quantities (as do λ and λ; ϕ, ϕ and φ). Sometimes these occurrences appear in the same question, although fortunately not very often. The following listing gives the principal meaning of most of the frequently used symbols in this book.

1. Roman letters

$\mathbf{a}$	acceleration, vector area
a	acceleration magnitude, linear dimension (e.g. distance, length, height, radius) or arbitrary constant
a_i	components of $\mathbf{a}$
a_{ij}	anti-symmetric second-rank tensor, direction cosines
$a_{\mu\nu}$	components of Lorentz transformation matrix
$\mathbf{A}$	magnetic vector potential or arbitrary vector field
$A_1, A_2, A_3, \ldots$	constant coefficients
A_i	components of $\mathbf{A}$
A_μ	four-vector potential
$\mathbf{b}$	microscopic magnetic field
b	linear dimension (e.g. distance, length, height, radius) or arbitrary constant
$\mathbf{B}$	magnetic field or arbitrary vector field
$\mathbf{B}_0$	magnetic field amplitude
$B_1, B_2, B_3, \ldots$	constant coefficients
$\mathbf{B}_\mathrm{a}$	radiation magnetic field: acceleration component
B_i	components of $\mathbf{B}$
$\mathbf{B}_\mathrm{v}$	radiation magnetic field: velocity component
c	speed of light in vacuum, contour or arbitrary constant
c_i	speed of light in medium i
$\mathbf{C}$	arbitrary vector field

C	capacitance
C_i	components of $\mathbf{C}$
$\mathbf{d}$	displacement of an origin
d	linear dimension (e.g. distance, length, height, radius)
$\mathbf{D}$	response field (electric displacement)
D_i	components of $\mathbf{D}$
dl or $d\mathbf{l}$	line element
dq	charge element
da or $d\mathbf{a}$	surface element
dv	volume element
$\mathbf{e}$	microscopic electric field
e	base of natural logarithms
$\pm e$	charge of electron or proton
$\mathbf{E}$	electric field
E	energy
$\mathbf{E}_0$	electric-field amplitude
$\mathbf{E}_\mathrm{a}$	radiation electric field: acceleration component
E_i	components of $\mathbf{E}$
E_k	kinetic energy
$\mathbf{E}_\mathrm{v}$	radiation electric field: velocity component
$\mathbf{f}$	force per unit charge
f	cyclic frequency or arbitrary function/scalar field
$\mathbf{F}$	force or arbitrary vector field
F_i	components of $\mathbf{F}$
$F_{\mu\nu}$	electromagnetic-field tensor
$\mathbf{g}$	acceleration due to gravity
g	magnitude of $\mathbf{g}$, arbitrary function/scalar field or gain
G	constant of universal gravitation
$\mathbf{G}$	arbitrary vector field
h	Planck's constant, linear dimension (usually a height)
$\hbar$	reduced Planck's constant
$\mathbf{H}$	response field
H	Heaviside function
H_i	components of $\mathbf{H}$

i	complex current, the imaginary number $\sqrt{-1}$ or an integer
I	direct current, complex current amplitude or moment of inertia
$\mathbf{j}$	microscopic volume density of current
j	usually an integer but sometimes the imaginary number $\sqrt{-1}$, used mainly in Chapter 6 to avoid possible confusion with the complex current i
$\mathbf{J}$	volume density of current
J_i	components of $\mathbf{J}$
$\mathbf{J}_b$	volume density of bound current
$\mathbf{J}_c$ or $\mathbf{J}_f$	volume density of conduction or free current
$\mathbf{J}_d$	volume density of displacement current
J_n	Bessel's function of the first kind
J_μ	four-current density
$\mathbf{k}$	wave vector
k	magnitude of $\mathbf{k}$, Boltzmann's constant, arbitrary constant, and sometimes an integer
k_i	components of $\mathbf{k}$
k_μ	four-wave vector
$\mathbf{K}$	surface density of current
$\mathbf{K}_b$	surface density of bound current
K	elliptic integral
K_i	components of $\mathbf{K}$
K_μ	four-force (Minkowski force)
l	linear dimension (usually a length or distance)
$\mathbf{L}$	angular momentum
L	self-inductance or linear dimension
L_i	components of $\mathbf{L}$
$\mathbf{m}$	magnetic dipole moment
$[\mathbf{m}]$	retarded magnetic dipole moment
m	mass of a particle or an integer
m_e	mass of electron
m_i	components of $\mathbf{m}$
m_{ij}	magnetic quadrupole moment
$\mathbf{M}$	macroscopic magnetic dipole moment density
M	mass of a particle or mutual inductance
M_i	components of $\mathbf{M}$

$\mathbf{n}$	unit vector along outward normal
n	refractive index, number of particles or an integer
n_i, n_i	components of $\mathbf{n}$ or imaginary part of refractive index
n_r	real part of refractive index
N	number of particles or number density
N_n	Bessel's function of the second kind
O	origin
$\bar{O}$	shifted origin
$\mathbf{p}$	linear momentum or electric dipole moment
$[\mathbf{p}]$	retarded electric dipole moment
p	magnitude of $\mathbf{p}$ or sometimes an integer
p_i	components of $\mathbf{p}$
p_μ	four-momentum
$\mathbf{P}$	macroscopic electric dipole moment density
P	field point
P_i	components of $\mathbf{P}$
P_n	Legendre polynomial of order n
P_{ij}	coefficients of potential
q	almost always electric charge but occasionally an integer
$q^{(\alpha)}$	charge of particle α
q_{ij}	electric quadrupole moment
q_{ijk}	electric octopole moment
Q	electric charge or quality factor
$\mathbf{r}$, $\mathbf{r}'$	position vector of field point, source point relative to an arbitrary origin
$\hat{\mathbf{r}}$	unit vector
$\mathbf{r}_q$	trajectory of charge q
$\mathbf{r}'^{(\alpha)}$	position vector of particle α
r	radial coordinate, radius or amplitude reflection coefficient
r_i, r_i'	components of $\mathbf{r}$, $\mathbf{r}'$
$r'^{(\alpha)}_i$	components of $\mathbf{r}'^{(\alpha)}$
$\mathbf{R}$	usually $(\mathbf{r} - \mathbf{r}')$, the vector from source point to field point
R	resistance, energy reflection coefficient or radius
$(\mathbf{r} - \mathbf{r}')_i$	components of $\mathbf{R}$

s	surface
s_{ij}	symmetric second-rank tensor
$\mathbf{S}$	Poynting vector
t	time or amplitude transmission coefficient
t'	retarded time at the origin
$[t]$	retarded time
T	kinetic energy, periodic time, absolute temperature or energy transmission coefficient
T_{ij}	second-rank tensor
$\mathbf{u}$	velocity
$\mathbf{u}^{(\alpha)}$	velocity of particle α
u	field energy density
$u_i{}^{(\alpha)}$	components of $\mathbf{u}^{(\alpha)}$
$u(x, y)$	real part of a complex function
U	electrostatic potential energy
$\mathbf{v}$	velocity
v	volume or complex voltage
v_i	components of $\mathbf{v}$
v_g	group velocity
v_ϕ	phase velocity
$v(x, y)$	imaginary part of a complex function
$\mathbf{V}$	arbitrary vector
V	scalar potential, magnitude of complex voltage, voltage in a dc circuit, potential difference or generalized potential energy
w	linear dimension (e.g. a width)
$w(z)$	analytic function
W	mechanical work
$\hat{\mathbf{x}}$	unit vector
x	Cartesian coordinate
x_μ	four-position vector
X	reactance, dimensionless Cartesian coordinate
$\hat{\mathbf{y}}$	unit vector
y	Cartesian coordinate
Y	admittance, dimensionless Cartesian coordinate

$\hat{\mathbf{z}}$	unit vector
z	Cartesian coordinate, cylindrical polar coordinate, complex impedance or a complex number
Z	magnitude of complex impedance, dimensionless Cartesian coordinate
Z_0	impedance of free space

2. Greek letters

α	polarizability, arbitrary constant, parameter or a coefficient
α_{ij}	polarizability tensor
β	hyperpolarizability, v/c (in relativity), arbitrary constant, parameter or a coefficient
γ	hyperpolarizability, Lorentz factor $(1 - \beta^2)^{-1/2}$, parameter, arbitrary constant or a coefficient
$\boldsymbol{\Gamma}$	torque
δ	skin depth, attenuation distance or Dirac delta function
δ_{ij}	Kronecker delta function
ΔV	macroscopic volume element
$\hat{\boldsymbol{\epsilon}}$	arbitrary unit vector
ϵ	an infinitesimal quantity or permittivity of a dielectric
ϵ_0	permittivity of free space
ϵ_{r}	relative permittivity
ε_{ijk}	Levi-Civita tensor
η	microscopic volume density of charge
$\hat{\boldsymbol{\theta}}$	unit vector
θ	polar coordinate or arbitrary angle
Θ	arbitrary angle
λ	line density of charge, wavelength, arbitrary constant, a parameter or a coefficient
λ	line density of charge (used only if λ is a wavelength)
μ	permeability of a medium
μ_0	permeability of free space

μ_B	Bohr magneton
μ_r	relative permeability
ν	frequency
ξ	generalized coordinate
ρ	volume density of charge or mass
ρ_b	volume density of bound charge
ρ_f	volume density of free charge
σ	surface density of free charge, conductivity or scattering cross-section
σ_b	surface density of bound charge
Σ	summation or (occasionally) conductivity
τ	relaxation time, time constant or proper time
$\hat{\boldsymbol{\phi}}$	unit vector
ϕ	phase term, power factor, angular coordinate or arbitrary scalar field
Φ	magnetic flux
φ	phase term (usually used when ϕ is a polar coordinate)
Φ	electric scalar potential
Φ_m	magnetic scalar potential
χ	gauge function
χ_e	electric susceptibility
χ_m	magnetic susceptibility
ψ	**E**- or **B**-component in a wave equation, wave function or arbitrary scalar field
Ψ	electric (or occasionally, gravitational) flux
Ψ	electric scalar potential (used only with a conformal transformation)
$\boldsymbol{\omega}$	angular velocity
ω	angular frequency
ω_0	resonance frequency
ω_B	Larmor frequency
ω_p	plasma frequency
Ω	solid angle

3. <u>Other letters</u>

$\mathcal{C}$	matrix for coefficients of capacitance
$\mathcal{E}$	emf, relativistic energy
H	Hamiltonian
ℓ	length
ℓ_0	proper length
L	Lagrangian
$\mathcal{N}$	Avogadro's number
$\mathcal{O}(h)$	terms of order h
$\mathcal{P}$	radiated or dissipated power
$\mathcal{R}$	vector displacement
$\mathcal{V}$	Verdet constant

Index